The Fungal Cell Wall Integrity Pathway

The Fungal Cell Wall Integrity Pathway

Editors

María Molina
Humberto Martín

MDPI • Basel • Beijing • Wuhan • Barcelona • Belgrade • Manchester • Tokyo • Cluj • Tianjin

Editors
María Molina
Department of Microbiology
and Parasitology
The Complutense University
of Madrid
Madrid
Spain

Humberto Martín
Department of Microbiology
and Parasitology
The Complutense University
of Madrid
Madrid
Spain

Editorial Office
MDPI
St. Alban-Anlage 66
4052 Basel, Switzerland

This is a reprint of articles from the Special Issue published online in the open access journal *Journal of Fungi* (ISSN 2309-608X) (available at: www.mdpi.com/journal/jof/special_issues/integrity_pathway).

For citation purposes, cite each article independently as indicated on the article page online and as indicated below:

LastName, A.A.; LastName, B.B.; LastName, C.C. Article Title. *Journal Name* **Year**, *Volume Number*, Page Range.

ISBN 978-3-0365-6973-4 (Hbk)
ISBN 978-3-0365-6972-7 (PDF)

Contents

About the Editors

María Molina

María Molina graduated in Pharmacy from the Complutense University of Madrid (UCM) in 1979, obtained her PhD in Pharmacy in 1985, and completed a postdoctoral stay at the University of Nottingham in the field of yeast genetics and molecular biology. She is currently a Full Professor of Microbiology at the UCM and directs a consolidated research group (Signal transduction in *Saccharomyces cerevisiae*) that also belongs to the Ramón y Cajal Institute for Health Research (IRYCIS) since 2009. Her research is focused on the study of cell signaling in yeast, mainly in MAP kinase pathways, as well as on its use as a platform for the expression of heterologous genes involved in human pathology. She has around one hundred articles in JCR (Orcid 0000-0003-0074-3309), three patents, and has supervised 22 doctoral theses.

Humberto Martín

Humberto Martin Brieva was born in 1967; graduated in Pharmacy from the Complutense University of Madrid (UCM) in June 1989. He is a Doctor in Pharmacy as of 1994 with Cum Laude honors. He started a tenure track as a Professor in Microbiology at UCM in February 2002 and as a Full Professor in Microbiology in 2020. He has participated in more than 20 research projects and as a co-principal investigator in two national projects. His research work has focused primarily on the study of cell signaling, particularly in the routes mediated by MAP kinases, using mainly the yeasts *Schizosaccharomyces pombe* and *Saccharomyces cerevisiae* as model organisms. This research has been carried out at the UCM and in three stays in foreign laboratories at Scripps Clinic, USA, National Institute for Medical Research, London, and the University of San Francisco, California, USA. He has published about 50 scientific articles (ORCID 0000-0002-6736-3602) and directed nine doctoral theses.

Journal of *Fungi*

MDPI

Editorial

Special Issue "The Fungal Cell Wall Integrity Pathway"

Humberto Martín * and María Molina *

Departamento de Microbiología y Parasitología, Facultad de Farmacia, Instituto Ramón y Cajal de Investigaciones Sanitarias (IRYCIS), Universidad Complutense de Madrid, Plaza de Ramón y Cajal s/n, 28040 Madrid, Spain
* Correspondence: humberto@ucm.es (H.M.); molmifa@ucm.es (M.M.); Tel.: +34-91-3941888 (H.M. & M.M.)

Citation: Martín, H.; Molina, M. Special Issue "The Fungal Cell Wall Integrity Pathway". *J. Fungi* **2023**, *9*, 293. https://doi.org/10.3390/jof9030293

Received: 14 February 2023
Accepted: 21 February 2023
Published: 24 February 2023

Adaptation to external changes is necessary for all cell types to survive and thrive in diverse environments. Key to these responses are the MAPK-mediated signaling pathways, intracellular communication routes that sense external stimuli at the cell surface and are ubiquitous in the eukaryotic world [1]. Unlike mammalian cells, fungi are surrounded by a rigid structure, the cell wall, whose integrity must be preserved to prevent cell lysis. During the life cycle of fungi, the cell wall must be continuously modified without failing in robustness to allow cells to grow, divide or differentiate. Furthermore, in their natural habitats, the different fungal species are subjected to a wide variety of environmental stresses, including those challenging this outermost cell structure, which they must cope with using appropriate adaptive mechanisms [2]. First, studies in the early 1990s in the budding yeast *Saccharomyces cerevisiae* established that the cell wall integrity (CWI) pathway is the central MAPK cascade required both to monitor the structural and functional status of this essential cell envelope and to trigger an appropriate salvage response under conditions that perturb its integrity [3]. Subsequent studies in other fungal species, including plant and human pathogens, have provided evidence that this MAPK pathway is highly conserved in the fungal kingdom [4]. Therefore, understanding the CWI pathway-mediated compensatory mechanism is key for the development of efficient cell-wall-targeted antifungal therapies. Furthermore, recent efforts to delve deeper into this pathway have revealed that its functional role goes beyond the maintenance of this essential structure, reaching many other physiological aspects that have important implications in fungal growth and virulence [2].

In this Special Issue, expert researchers in this relevant subject have contributed with seven reviews and eleven original articles to advance our understanding of the CWI pathway by covering different structural, regulatory, and functional aspects in distinct yeasts and filamentous fungi. Yoshimi et al. summarize the current knowledge of CWI signaling in several filamentous fungal species, from sensor proteins required for the recognition of environmental changes to MAPK cascades and their targets involved in the regulation of cell wall polysaccharide synthesis genes [5]. To discuss the most important functional aspects, what is known in these fungi is compared to yeast species, such as the widely studied *Saccharomyces cerevisiae* or the pathogens *Candida* and *Cryptococcus*. In addition, they also address the role of the CWI pathway as a target for developing antifungal drugs and for manipulating the surface properties of fungi to improve productivity in industrial processes.

The importance of this signaling pathway in *Candida albicans* and its interconnections with other stress-activated pathways for remodeling the cell envelope under different stresses is reviewed by Ibe & Munro [6]. They analyze the key function of cell wall mannoproteins on the survival, growth, and virulence of this opportunistic pathogen, highlighting the value of understanding their molecular structure, activity, and regulatory mechanisms to identifying the most suitable diagnostic, therapeutic, and vaccine candidates, thus paving the way for better management of candidiasis. De Oliveira et al. focus their review on the CWI signaling in another yeast species, *Cryptococcus neoformans*, and its involvement in some of the virulence traits of this pathogen, such as thermotolerance, melanin synthesis, capsule growth, and Titan cell formation [7]. The ability of *C. neoformans* to enlarge cell size by increasing its protective external structure and undergoing such a characteristic morphological change has been extensively studied, as it allows the fungus to evade and resist the host immune system. These authors also emphasize the role of the CWI pathway as a pharmacological target that can be exploited to improve the effectiveness of existing antifungals as well as to develop new therapies.

The function of the CWI pathway beyond the regulation of cell wall homeostasis is discussed in the reviews on budding and fission yeast models provided by González-Rubio et al. and by Cansado et al., respectively. In the former, the authors summarize the currently available methodologies to identify phosphorylation targets of Slt2, the MAPK of the pathway, and compile all the genuine Slt2 substrates reported to date, whose variety of functions reflects the multiple processes regulated by the CWI pathway in *S. cerevisiae* [8]. They also provide a list of putative Slt2 substrates, ready to be confirmed in future works. The second review also positions the CWI as a multi-faceted pathway that impacts multiple functional aspects of the *Schizosaccharomyces pombe* life cycle both during growth and in response to stress [9]. Among the processes described as regulated by the CWI signaling in these yeasts, besides the maintenance of cell wall integrity, are gene transcription, control of mRNA stability or transport through RNA-binding proteins, regulation of calcium homeostasis, interplay with PKA and TOR pathways, control of cell cycle, and modulation of cytokinesis. These last two aspects are the topics reviewed in detail, respectively, by Quilis et al. and Roncero et al. The CWI pathway is presented by the former authors as a versatile toolbox to arrest cell cycle progression in budding yeast exposed to unfavorable conditions, due to its role in regulating major cell cycle transitions in response to cell surface perturbance or genotoxic stress [10]. They describe the mechanisms by which the CWI pathway impinges on different cell cycle regulators and checkpoints to delay cell cycle progression until the damage is repaired, in order to resume cell division with all the guarantees for cell survival. Roncero et al. first summarize the key steps of yeast cytokinesis, including ring assembly, septum formation, and cell separation, to subsequently analyze the multiple interconnections of the CWI signaling responses with these processes in both *S pombe* and *S. cerevisiae*, highlighting the differences between these two model yeasts with distinct modes of growth [11].

The research papers in this Special Issue present very significant insights into different aspects of various key components of this signaling pathway, ranging from receptors to effectors. Starting with the former, Schöppner et al. address the relationship between the structure and functional properties of the ScWsc1 receptor [12]. By solving a high-resolution crystal structure of the extracellular cysteine-rich domain (CRD) of yeast Wsc1, they show that the protein surface of the CRD contains three aromatic clusters, which play an essential role under cell wall stress conditions. Conservation of these functional hotspots among other fungal Wsc sensors enhances their relevance. In a complementary study, Voskoboynikova et al. succeeded in extracting this mechanosensor from the yeast plasma membrane using a detergent-free procedure into a semi-native lipid environment [13]. The use of the amphipathic styrene-maleic acid (SMA) and SMA-related copolymers allowed for the formation of SMALPS (SMA/lipid particles) that accommodate only single sensor molecules. Analysis of the preparations by dynamic light scattering (DLS), fluorescence correlation spectroscopy (FCS), and single-particle transmission electron microscopy (TEM) allowed for the deduction of the first three-dimensional structural model of Wsc1. This work confirms the tripartite organization of Wsc1, provides novel clues on the structure–function relationship of this sensor, and shows the method for studying other membrane proteins. From a functional perspective, an unresolved issue is the role of the distinct receptors that feed the CWI pathway. The work of Hall et al. provides additional information on this aspect: whereas Mid2 is necessary during shmoo formation, Wsc1 is of critical importance in zygote and diploid survival [14]. Interestingly, activation of Pkc1 can compensate for the absence of the receptor, suggesting that its CWI signaling role is essential for this functionality. Other functions of CWI receptors are described by Montella-Manuel et al., who demonstrate that Mtl1 plays a critical role in sensing reductions in glucose concentration, in order to trigger autophagy during diauxic transition [15]. In addition, it is also involved in the autophagic degradation of mitochondria during the stationary phase. Here, activation of CWI components does not restore the lack of bulk autophagy activation observed in *mtl1* mutants, suggesting a CWI-independent function for this sensor. Unraveling the role of the different receptors of the CWI pathway in cellular physiology promises to be an interesting field of research in the coming years.

This issue also provides new data that refine our understanding of the complex regulation and functional roles of Pkc1, the top kinase of the CWI pathway. Liu et al. demonstrate that this kinase is hyperphosphorylated in response to DNA damage in a Hrr25-dependent manner. Hrr25 (HO and Radiation Repair) is an orthologue of casein kinase 1 (CK1) and is involved in the transcriptional response to DNA damage [16]. Mutation of CK1 consensus sites in Pkc1 prevents Pkc1 hyperphosphorylation and reduces transcriptional induction by genotoxic stress, suggesting that hyperphosphorylation of Pkc1 by Hrr25 contributes to this response. In turn, Sellers-Moya et al. prove that, in contrast to other CWI stimuli, Pkc1 is not essential for the activation of the CWI MAPK module in response to clotrimazole [17]. This imidazole activates both the HOG and CWI pathways and promotes the appearance of a low-mobility form of Slt2 caused by additional phosphorylation to that occurring in the conserved TEY activation motif. The oxidative stress generated by this antifungal drug is linked to Tpk3-mediated PKA activity and is responsible for all these effects. The function of the CWI pathway and in particular the MAPK Slt2 are also expanded in the work of Sánchez-Adriá et al., who show the role of Slt2 in downregulating TORC1 signaling in response to ER (Endoplasmic Reticulum) stress [18]. In this situation, the CWI pathway is involved in the activation of autophagy and the increased synthesis of ATP. Furthermore, these authors show that Slt2 is required for *GFA1* expression, coding the glutamine:fructose-6-phosphate amidotransferase, a key enzyme within the conserved hexosamine biosynthetic pathway (HBP) essential for different metabolic pathways. Taken together, these results suggest the importance of Slt2 in modulating the bioenergetics of the protective response under ER stress, as well as confirm the key role of the CWI pathway in the crosstalk between different signaling pathways.

Novel components participating in the maintenance of cell wall integrity are also uncovered. By performing two high-throughput screenings using the yTHC collection of yeast conditional mutant strains, Sanz et al. provide insights into essential genes required to cope with cell wall damage conditions, as well as those affecting CWI-associated gene expression [19]. This work reveals how the RSC chromatin remodeling complex, unveiled in both screenings, is likely to cooperate with SWI/SNF and SAGA complexes for the chromatin remodeling necessary for the transcriptional activation of CWI-dependent genes upon stress. The paper by Ghanegolmohammadi et al. follows a distinct and novel approach in order to increase our understanding of the function of cell wall mannoproteins [20]. To this end, they investigate 32 mannoprotein mutants in *S. cerevisiae* using high-dimensional morphological phenotyping. Comprehensive analysis of their morphological phenotypes provides strong clues about the functions of the affected proteins belonging to the same cluster of mutants, in particular the mechanistic and functional roles of the distinct glycoproteins in cell morphogenesis.

Finally, two contributions reflect how evolution, using conserved elements, has shaped signaling pathways in fungi. In this line, the work of Navarro et al. shows that the dimorphism-regulating histidine kinase (Drk1) of the pathogenic fungus *Paracoccidioides brasiliensis* plays an important role in cell wall homeostasis, in contrast to the involvement of histidine kinases in the response to high osmolarity in *S. cerevisiae* [21]. In turn, Gomez-Gil et al. investigate the architecture and functionality of the CWI pathway (CIP pathway) of *Schizosaccharomyces japonicus* [22]. They prove that, while certain features of this pathway previously described in *S. pombe* are conserved in *S. japonicus*, others have evolved differently in both fission yeast species.

In sum, this Special Issue covers many, although not all, aspects of the signaling mechanisms and functions of the CWI pathway. It is our sincere hope that the articles presented here serve general readers by providing the most up-to-date information on this important aspect of fungal biology, as well as inspiring new ideas and challenging questions for researchers in the coming years.

Funding: This research was funded by Grant PID2019-105342GB-I00/AEI/10.13039/501100011033 from Ministerio de Ciencia e Innovación (Spain).

Institutional Review Board Statement: Not applicable.

Informed Consent Statement: Not applicable.

Data Availability Statement: Not applicable.

Conflicts of Interest: The authors declare no conflict of interest.

References

1. Chen, R.E.; Thorner, J. Function and regulation in MAPK signaling pathways: Lessons learned from the yeast *Saccharomyces cerevisiae*. *Biochim. Biophys. Acta* **2007**, *1773*, 1311–1340. [CrossRef] [PubMed]
2. Jiménez-Gutiérrez, E.; Alegría-Carrasco, E.; Sellers-Moya, A.; Molina, M.; Martín, H. Not just the wall: The other ways to turn the yeast CWI pathway on. *Int. Microbiol.* **2020**, *23*, 107–119. [CrossRef] [PubMed]
3. Levin, D.E. Regulation of cell wall biogenesis in *Saccharomyces cerevisiae:* The cell wall integrity signaling pathway. *Genetics* **2011**, *189*, 1145–1175. [CrossRef] [PubMed]
4. González-Rubio, G.; Fernández-Acero, T.; Martín, H.; Molina, M. Mitogen-Activated Protein Kinase Phosphatases (MKPs) in Fungal Signaling: Conservation, Function, and Regulation. *Int. J. Mol. Sci.* **2019**, *20*, 1709. [CrossRef]
5. Yoshimi, A.; Miyazawa, K.; Kawauchi, M.; Abe, K. Cell Wall Integrity and Its Industrial Applications in Filamentous Fungi. *J. Fungi* **2022**, *8*, 435. [CrossRef]
6. Ibe, C.; Munro, C.A. Fungal Cell Wall Proteins and Signaling Pathways Form a Cytoprotective Network to Combat Stresses. *J. Fungi* **2021**, *7*, 739. [CrossRef]
7. de Oliveira, H.C.; Rossi, S.A.; García-Barbazán, I.; Zaragoza, Ó.; Trevijano-Contador, N. Cell Wall Integrity Pathway Involved in Morphogenesis, Virulence and Antifungal Susceptibility in Cryptococcus neoformans. *J. Fungi* **2021**, *7*, 831. [CrossRef]
8. González-Rubio, G.; Sastre-Vergara, L.; Molina, M.; Martín, H.; Fernández-Acero, T. Substrates of the MAPK Slt2: Shaping Yeast Cell Integrity. *J. Fungi* **2022**, *8*, 368. [CrossRef]
9. Cansado, J.; Soto, T.; Franco, A.; Vicente-Soler, J.; Madrid, M. The Fission Yeast Cell Integrity Pathway: A Functional Hub for Cell Survival upon Stress and Beyond. *J. Fungi* **2021**, *8*, 32. [CrossRef]
10. Quilis, I.; Gomar-Alba, M.; Igual, J.C. The CWI Pathway: A Versatile Toolbox to Arrest Cell-Cycle Progression. *J. Fungi* **2021**, *7*, 1041. [CrossRef]
11. Roncero, C.; Celador, R.; Sánchez, N.; García, P.; Sánchez, Y. The Role of the Cell Integrity Pathway in Septum Assembly in Yeast. *J. Fungi* **2021**, *7*, 729. [CrossRef] [PubMed]
12. Schöppner, P.; Lutz, A.P.; Lutterbach, B.J.; Brückner, S.; Essen, L.O.; Mösch, H.U. Structure of the Yeast Cell Wall Integrity Sensor Wsc1 Reveals an Essential Role of Surface-Exposed Aromatic Clusters. *J. Fungi* **2022**, *8*, 379. [CrossRef] [PubMed]
13. Voskoboynikova, N.; Karlova, M.; Kurre, R.; Mulkidjanian, A.Y.; Shaitan, K.V.; Sokolova, O.S.; Steinhoff, H.-J.; Heinisch, J.J. A Three-Dimensional Model of the Yeast Transmembrane Sensor Wsc1 Obtained by SMA-Based Detergent-Free Purification and Transmission Electron Microscopy. *J. Fungi* **2021**, *7*, 118. [CrossRef] [PubMed]
14. Hall, A.E.; Lisci, M.; Rose, M.D. Differential Requirement for the Cell Wall Integrity Sensor Wsc1p in Diploids Versus Haploids. *J. Fungi* **2021**, *7*, 1049. [CrossRef]
15. Montella-Manuel, S.; Pujol-Carrion, N.; de la Torre-Ruiz, M.A. The Cell Wall Integrity Receptor Mtl1 Contributes to Articulate Autophagic Responses When Glucose Availability Is Compromised. *J. Fungi* **2021**, *7*, 903. [CrossRef]
16. Liu, L.; Veis, J.; Reiter, W.; Motari, E.; Costello, C.E.; Samuelson, J.C.; Ammerer, G.; Levin, D.E. Regulation of Pkc1 Hyper-Phosphorylation by Genotoxic Stress. *J. Fungi* **2021**, *7*, 874. [CrossRef]
17. Sellers-Moya, Á.; Nuévalos, M.; Molina, M.; Martín, H. Clotrimazole-Induced Oxidative Stress Triggers Novel Yeast Pkc1-Independent Cell Wall Integrity MAPK Pathway Circuitry. *J. Fungi* **2021**, *7*, 647. [CrossRef]
18. Sánchez-Adriá, I.E.; Sanmartín, G.; Prieto, J.A.; Estruch, F.; Randez-Gil, F. Slt2 Is Required to Activate ER-Stress-Protective Mechanisms through TORC1 Inhibition and Hexosamine Pathway Activation. *J. Fungi* **2022**, *8*, 92. [CrossRef]
19. Sanz, A.B.; Díez-Muñiz, S.; Moya, J.; Petryk, Y.; Nombela, C.; Rodríguez-Peña, J.M.; Arroyo, J. Systematic Identification of Essential Genes Required for Yeast Cell Wall Integrity: Involvement of the RSC Remodelling Complex. *J. Fungi* **2022**, *8*, 718. [CrossRef]
20. Ghanegolmohammadi, F.; Okada, H.; Liu, Y.; Itto-Nakama, K.; Ohnuki, S.; Savchenko, A.; Bi, E.; Yoshida, S.; Ohya, Y. Defining Functions of Mannoproteins in *Saccharomyces cerevisiae* by High-Dimensional Morphological Phenotyping. *J. Fungi* **2021**, *7*, 769. [CrossRef]

21. Navarro, M.V.; de Barros, Y.N.; Segura, W.D.; Chaves, A.F.A.; Jannuzzi, G.P.; Ferreira, K.S.; Xander, P.; Batista, W.L. The Role of Dimorphism Regulating Histidine Kinase (Drk1) in the Pathogenic Fungus *Paracoccidioides brasiliensis* Cell Wall. *J. Fungi* **2021**, *7*, 1014. [CrossRef] [PubMed]
22. Gómez-Gil, E.; Franco, A.; Vázquez-Marín, B.; Prieto-Ruiz, F.; Pérez-Díaz, A.; Vicente-Soler, J.; Madrid, M.; Soto, T.; Cansado, J. Specific Functional Features of the Cell Integrity MAP Kinase Pathway in the Dimorphic Fission Yeast *Schizosaccharomyces japonicus*. *J. Fungi* **2021**, *7*, 482. [CrossRef] [PubMed]

Journal of *Fungi*

Review

Cell Wall Integrity and Its Industrial Applications in Filamentous Fungi

Akira Yoshimi [1,2,†], Ken Miyazawa [2,3,†], Moriyuki Kawauchi [1] and Keietsu Abe [2,4,*]

1 Laboratory of Environmental Interface Technology of Filamentous Fungi, Graduate School of Agriculture, Kyoto University, Kyoto 606-8502, Japan; yoshimi.akira.8c@kyoto-u.ac.jp (A.Y.); kawauchi.moriyuki.8c@kyoto-u.ac.jp (M.K.)
2 ABE-Project, New Industry Creation Hatchery Center, Tohoku University, Sendai 980-8579, Japan
3 Laboratory of Filamentous Mycoses, Department of Fungal Infection, National Institute of Infectious Diseases, Tokyo 162-8640, Japan; k-miyazawa@niid.go.jp
4 Laboratory of Applied Microbiology, Graduate School of Agricultural Science, Tohoku University, Sendai 980-8572, Japan
* Correspondence: keietsu.abe.b5@tohoku.ac.jp; Tel.: +81-22-757-4355
† These authors contributed equally to this work.

Abstract: Signal transduction pathways regulating cell wall integrity (CWI) in filamentous fungi have been studied taking into account findings in budding yeast, and much knowledge has been accumulated in recent years. Given that the cell wall is essential for viability in fungi, its architecture has been analyzed in relation to virulence, especially in filamentous fungal pathogens of plants and humans. Although research on CWI signaling in individual fungal species has progressed, an integrated understanding of CWI signaling in diverse fungi has not yet been achieved. For example, the variety of sensor proteins and their functional differences among different fungal species have been described, but the understanding of their general and species-specific biological functions is limited. Our long-term research interest is CWI signaling in filamentous fungi. Here, we outline CWI signaling in these fungi, from sensor proteins required for the recognition of environmental changes to the regulation of cell wall polysaccharide synthesis genes. We discuss the similarities and differences between the functions of CWI signaling factors in filamentous fungi and in budding yeast. We also describe the latest findings on industrial applications, including those derived from studies on CWI signaling: the development of antifungal agents and the development of highly productive strains of filamentous fungi with modified cell surface characteristics by controlling cell wall biogenesis.

Keywords: filamentous fungi; cell wall integrity; signaling pathway; surface sensor; protein kinase C; mitogen-activated protein kinase; plant pathogen; application; fungicide; drug target; culture; productivity

Citation: Yoshimi, A.; Miyazawa, K.; Kawauchi, M.; Abe, K. Cell Wall Integrity and Its Industrial Applications in Filamentous Fungi. *J. Fungi* **2022**, *8*, 435. https://doi.org/10.3390/jof8050435

Academic Editors: María Molina and Humberto Martín

Received: 8 February 2022
Accepted: 20 April 2022
Published: 23 April 2022

1. Introduction

Many microorganisms, especially fungi, have evolved as decomposers of terrestrial plants, which are primary producers. Fungi are considered to be among the most successful taxa in terrestrial ecosystems. The success of fungi is thought to be due to their ability to form filamentous cells, called hyphae, which form a network called mycelium; this ability allows fungi to invade solid substrates and acquire nutrients efficiently from the inside of the substrates that are difficult to penetrate for unicellular microorganisms [1–3]. The invasion of a solid substrate by filamentous fungi begins with contact between substrate surface and the fungi. When filamentous fungi invade solid substrates, their cells are exposed to oxidative stress; changes in osmolality, temperature, and pH; and chemical compounds, including pheromones [4]. Therefore, biochemical reactions at the cell surface affect fungal growth, and fungal cell surface structures, which form the interface between substrates and fungi, play an important role. Understanding the structure of the fungal cell wall and the regulation of its construction may lead to applications in controlling fungal pathogens and the effective utilization of filamentous fungi.

In this review, we focus on the cell wall integrity (CWI) signaling that regulates cell wall construction and remodeling. The cell wall, the outermost layer of the fungal cell, maintains cell morphology, protects the cells, and transmits the external stimuli inside the cell. Fungal CWI signaling has been studied in detail in the budding yeast *Saccharomyces cerevisiae* (reviewed by Levin [5,6], Gustin et al. [7], and Chen and Thomer [8]). In the CWI pathway of *S. cerevisiae*, perturbations of the cell wall are detected by the Wsc-type and Mid-type cell surface sensors. The signal is then consecutively transmitted through the following components: the GDP/GTP exchange factor Rom, the small GTPase Rho1, protein kinase C (PKC), the mitogen-activated protein (MAP) kinase cascade (MAP kinase kinase kinase Bck1; a pair of MAP kinase kinases Mkk1/Mkk2, and the MAP kinase Mpk1/Slt2), and the transcription factors (TFs) Rlm1 and Swi4, a subunit of the Swi4–Swi6 TF complex. The other signaling pathways in *S. cerevisiae* are the high osmolarity glycerol (HOG) pathway (MAP kinase: Hog1 kinase), filamentous and invasive growth (FG) pathway (Kss1 kinase), and pheromone pathway (Fus3 kinase) [9,10]. Extensive crosstalk between these pathways in *S. cerevisiae* has been documented [11,12]. In this review, we refer to the central pathway involved in CWI via PKC–Mpk1/Slt2 or their orthologs as the CWI PKC pathway. When describing the entire system that contributes to the maintenance of CWI, including not only the CWI PKC pathway but also other signaling pathways, we refer to it as CWI signaling.

Our research in this area has resulted in some industrial applications. Here, we discuss the similarities and differences between the functions of CWI signaling factors in filamentous fungi and in yeast, including cell surface sensors in Section 2 and downstream components in Section 3. We describe the development of antifungal agents based on the analysis of CWI signaling in Section 4, and the development of fungal culture technology using strains with modified cell surface structures in Section 5.

2. Cell Surface Sensors of Cell Wall Integrity Signaling Pathway

2.1. Wsc- and Mid-Type Sensors

Filamentous fungi grow by invading and decomposing solid substrates [3], and these features are used for solid-state fermentation in industrial applications [13,14]. These processes are initiated by a contact between the substrate surface and the fungal cell surface. Fungi perceive information at the contact surface and transmit it into the cells. Cell surface sensors embedded in the cell wall are important in this process, and in sensing and responding appropriately to environmental stresses. Perturbation of the cell wall may affect fungal survival, so changes in cell wall structure as such must also be sensed. Cell wall sensors in fungi were first studied in *S. cerevisiae* (for detailed reviews, see [15,16]). We provide an overview of sensor proteins in fungi in Table 1. The membrane-spanning sensors of the *S. cerevisiae* CWI PKC pathway consist of two sub-families: Wsc-type sensors (Wsc1–3) and Mid-type sensors (Mid2 and Mtl1). All of them have a transmembrane region and an extracellular region; the latter is rich in serine and threonine residues and is highly *O*-mannosylated. At the N-terminus, only the Wsc type has the Wsc domain (also referred to as the cysteine-rich domain, CRD), but only the Mid type has an *N*-glycosylated asparagine residue. The glycan chains of the extracellular region of the sensor proteins are interact with cell wall polysaccharides. These proteins function as mechanosensors. Stimuli in the cell wall and the resulting distortion of the plasma membrane are sensed as force that tilts and stretches the serine/threonine-rich region, which acts like a nanospring [15,16]. This structural change results in a conformational change in the cytoplasmic tail, which triggers downstream signal transmission.

The dimorphic fungus *Candida albicans* forms so-called invasive filaments during host invasion. Strains lacking Wsc-type sensors show little change in susceptibility to cell wall stresses, and the formation of invasive filaments does not differ from that of the wild-type strain [17]. These data suggest that the Wsc-type sensors are not crucial for CWI in this fungus.

In filamentous fungi, homologs of the *S. cerevisiae* Wsc1–3 cell wall sensors were identified in silico in a model filamentous fungus *Aspergillus nidulans* [18], and their function was analyzed [19,20]. WscA has a Wsc-domain, a serine- and threonine-rich region, a transmembrane region, and a C-terminal intracellular domain. WscA was considered to be a substrate for *O*-D-mannosyltransferase Pmt because Wsc1 and Mid2 are mannosylated by Pmt; this was confirmed using an HA-tagged WscA-expressing strain [19]. Futagami et al. [20] showed that WscA and WscB, both Wsc1 orthologs in *A. nidulans*, are *N*- and *O*-glycosylated and are localized in the cell wall. Disruption of *wscA* results in abnormal growth and reduced conidiation. The conidial formation is also reduced in the *wscB* deletion strain, but to a lesser extent. The *wscAwscB* double-disruption strain is viable, but its growth retardation is more severe than that caused by *wscA* single deletion [20]. Whereas yeast Wsc1 is involved in stress response under alkaline conditions [21], Wsc-type sensors of *A. nidulans* are thought to sense cell wall changes under acidic conditions [20]. Loss of WscA alters the transcript levels of genes for cell wall α-1,3-glucan synthases (*agsA* and *agsB*), resulting in an increase in the content of alkali-soluble glucan [20]. Loss of Wsc-type sensors also enhances the phosphorylation of a mitogen-activated protein (MAP) kinase, MpkA [20]. These results and the absence of α-1,3-glucan in yeast suggest that *A. nidulans* Wsc-type sensors have a somewhat different sensing spectrum and downstream signaling pathway from those of *S. cerevisiae* [20]. Futagami et al. [22] showed that a Mid-type sensor protein, MtlA, in *A. nidulans* is highly *O*-glycosylated and localized to the cell surface. Loss of MtlA decreases conidial formation, increases sensitivity to cell wall inhibitors, such as calcofluor white (CFW), congo red (CR), and micafungin, an echinocandin antifungal, and decreases cell wall glucan and chitin content [22]. Thus, the CWI sensor MtlA is important for cell wall stress tolerance and cell wall maintenance in this fungus [22].

The function of Wsc1–3 and the Mid-type sensor MidA has been reported in the human pathogenic fungus *Aspergillus fumigatus* [23]. The disruption of *A. fumigatus wsc1*, a gene for a Wsc1 homolog of *S. cerevisiae*, increases sensitivity to caspofungin, an echinocandin antifungal, and additional disruption of *wsc3* reduces colony growth and conidial formation. Disruption of *midA* alone does not affect colony growth, but disruption of *midA* in the *wsc1wsc3* double-disruption strain results in severe growth retardation and severe reduction of conidial formation [23]. Disruption of *wsc2* does not affect colony growth or conidiation. MidA, but not Wsc1–3, is essential for the tolerance to CFW, CR, and high-temperature stress [23]. The functions of Wsc1, Wsc3, and MidA partly overlap, and they are involved in vegetative growth and conidiation [23].

In *Neurospora crassa*, loss of WSC-1, a homolog of *S. cerevisiae* Wsc1, increases sensitivity to caspofungin and CFW and strongly reduces the formation of aerial hyphae and conidia [24]. The *wsc-2* gene encodes another Wsc-type sensor; the *wsc-2* disruption strain has a phenotype similar to that of the wild type, but with a slightly reduced growth rate and conidial formation [24]. Disruption of *wsc-1* also reduces the basal level of phosphorylation and stress-induced activity of MAK-1, a MAP kinase in the CWI PKC pathway in *N. crassa*. Disruption of *wsc-2* has a negligible effect on MAK-1 activation by cell wall stress. The authors of [24] concluded that WSC-1 and WSC-2 are required for MAK-1 activation in *N. crassa* and that both function as cell wall sensors.

The entomopathogen *Beauveria bassiana* has at least nine proteins with a single Wsc-domain [25,26]. Among them, Wsc1A–E are localized in the hyphal cell wall or membrane, and the deletion of each of them increases sensitivity to cell wall perturbation, osmotic stress, oxidative stress, and metal ions, and also delays germination and reduces resistance to UV-B and/or heat stress [25]. None of the deletions have a significant effect on vegetative growth, conidial formation, or virulence [25]. The ninth Wsc sensor, Wsc1I, which contains not only a Wsc domain but also an N-terminal DUF1996 domain (domain of unknown function 1996), is localized to the vacuoles and cell wall/membrane and is involved in sensitivity to osmotic stress, oxidative stress, and cell wall stress compounds [26]. In a *wsc1I* deletion strain, the phosphorylation level of the MAP kinase Hog1 is greatly reduced under osmotic, oxidative, and cell wall stresses, suggesting that Wsc1I senses a variety of cell stresses upstream of the Hog1 pathway [26]. Overall, the data suggest some variations of Wsc- and Mid-type sensors among fungal species.

Table 1. Major fungal surface sensors whose functions have been analyzed.

Phylum	Subphylum	Class	Species	Sensor Name	Type	Typical Phenotype(s) of Deficient Strain in Relation to CWI *	References
Ascomycota	Taphrinomycotina	Schizosaccharomycetes	*Schizosaccharomyces pombe*	Wsc1	Wsc	Slightly sensitive to CFG.	[27]
				Mtl2	Mid	Sensitive to CFG, CAF, vanadate, NaCl, H_2O_2, and SDS. Decreased β-1,3-glucan content in CW.	[27]
	Saccharomycotina	Saccharomycetes	*Saccharomyces cerevisiae*	WSC1	Wsc	Cell lysis defect and thermosensitive growth defect at 37 °C on YPD medium.	[28–30]
				WSC2	Wsc	Deletion of *WSC2* and/or *WSC3* exacerbates the phenotype of the *wsc1*Δ strain.	[29]
				WSC3	Wsc	Deletion of *WSC2* and/or *WSC3* exacerbates the phenotype of the *wsc1*Δ strain.	[29]
				WSC4	Wsc-like	Not generated.	[29,31]
				Mid2	Mid	Resistant to CFW. Changes in growth rate and viability in a number of different cell wall biosynthesis mutants.	[32,33]
				Mtl1	Mid	Not sensitive to thermo-, oxidative, or osmotic stresses or CFW.	[32]
				Msb2	Signaling mucin	Severely osmosensitive in combination with the deficiency in another mucin-like protein, Hkr1.	[34]
			Candida albicans	Wsc1	Wsc	Normal resistance to CR and CFW.	[17]
				Wsc2	Wsc	Normal resistance to CR but lower sensitivity to CFW.	[17]
				Msb2	Signaling mucin	Growth defects at 30 °C and 37 °C and a striking growth defect at 42 °C.	[35,36]
			Pichia pastoris	PpWsc1	Wsc	Sensitive to high temperature and CR.	[37]
				PpWsc2	Wsc	Not sensitive to high temperature or CR.	[37]
				PpWsc3	Wsc	Not sensitive to high temperature or CR.	[37]
			Kluyveromyces lactis	KlWsc1	Wsc	Sensitive to CAF and CR in combination with *KlMid2* disruption.	[38]
				KlWsc2/3	Wsc	Sensitive to CAF and CR in combination with *KlWsc1* and *KlMid2* disruption.	[38]
				KlMid2	Mid	Sensitive to CAF and CR in combination with *KlWsc1* disruption.	[38]
	Pezizomycotina	Eurotiomycetes	*Aspergillus nidulans*	WscA	Wsc	Reduced colony and conidia formation under acidic conditions or not. High frequency of swollen hyphae under hypo-osmotic conditions.	[20]

Table 1. *Cont.*

Phylum	Subphylum	Class	Species	Sensor Name	Type	Typical Phenotype(s) of Deficient Strain in Relation to CWI *	References
				WscB	Wsc	Reduced conidiation and growth inhibition under acidic conditions, but to a lesser extent than those caused by a WscA defect.	[20]
				MtlA	Mid	Reduced conidiation. Growth deficiency in the presence of CW inhibitor. Reduction in the glucan and chitin contents in CW.	[22,39]
				MsbA	Signaling mucin	Sensitive to CR, CFW, and cation stresses ($MnCl_2$).	[40]
			Aspergillus fumigatus	Wsc1	Wsc	Less dense at the colony fringe, but only a mardinal decrease in radial growth. Increased sensitivity to CFG.	[23]
				Wsc2	Wsc	No effect of disruption even in the Δ*wsc1* background.	[23]
				Wsc3	Wsc	Impaired radial growth and reduced conidiation in the Δ*wsc1* background.	[23]
				MidA	Mid	Highly sensitive to CR, CFW, and elevated temperature.	[23]
				MsbA	Signaling mucin	Impaired radial growth. Significant delay in germ tube formation. Sensitive to CR, CFW, nikkomycin Z, and NaCl.	[41]
		Sordariomycetes	*Neurospora crassa*	WSC-1	Wsc	Compact growth. Poor aerial hyphae formation. Almost aconidial. Sensitive to CFG and CFW.	[24]
				WSC-2	Wsc	Slightly reduced growth rate and conidiation.	[24]
				HAM-7	Other	Altered growth and branching pattern. Reduced aerial hyphal formation. No protoperithecia. Defective in cell fusion.	[24]
			Fusarium graminearum	Wsc2B	Wsc	Defects in hyphal growth, virulence, and response to CW stresses (cellulase, lysozyme, and snailase).	[42]
			Fusarium oxysporum	Msb2	Signaling mucin	Significantly slower growth on low-nitrogen medium but not on nutrient-rich medium. Sensitive to CR and CFW.	[43]
			Beauveria bassiana	Wsc1A	Wsc	Increased sensitivity to CW stress, oxidation, high osmolarity. No effect on growth, conidiation, or virulence.	[25]
				Wsc1B	Wsc	Increased sensitivity to CW stress, oxidation, high osmolarity. No effect on growth, conidiation, or virulence.	[25]
				Wsc1C	Wsc	Increased sensitivity to CW stress, oxidation, high osmolarity. No effect on growth, conidiation, or virulence.	[25]

Table 1. *Cont.*

Phylum	Subphylum	Class	Species	Sensor Name	Type	Typical Phenotype(s) of Deficient Strain in Relation to CWI *	References
				Wsc1D	Wsc	Increased sensitivity to CW stress, oxidation, high osmolarity. No effect on growth, conidiation, or virulence.	[25]
				Wsc1E	Wsc	Increased sensitivity to CW stress, oxidation, high osmolarity. No effect on growth, conidiation, or virulence.	[25]
				Wsc1I	Wsc-like?	Increased sensitivity to CW stress, oxidation, high osmolarity. No effect on growth, conidiation, or virulence.	[26]
			Metarhizium rileyi	MrWsc1	Wsc	Targeted knockout has not been successful.	[44]
				MrMid2	Mid	Impaired dimorphic transition, conidiation, and microsclerotium. Sensitive to thermal, CW, and oxidative stresses. Decreased virulence.	[44]
			Pyricularia oryzae	MoMsb2	Signaling mucin	Significantly reduced appressorium formation and virulence. Slightly reduced growth rate.	[45]
		Leotiomycetes	*Botrytis cinerea*	Msb2	Signaling mucin	Normal growth. Almost unable to form appressoria or infection cushions on hard surfaces.	[46]
Basidiomycota	Ustilaginomycotina	Ustilaginomycetes	*Ustilago maydis*	Msb2	Signaling mucin	Impaired host colonization and appressorium formation on plant surface.	[47]
	Agaricomycotina	Tremellomycetes	*Cryptococcus neoformans*	Msb2	Signaling mucin	Resistant to osmotic stress. No thermosensitivity but marginally increased sensitivity to cryostress.	[48]

* Abbreviations: CWI, cell wall integrity; CFG, caspofungin; CAF, caffeine; SDS, sodium dodecyl sulfate; CW, cell wall; CR, congo red; CFW, calcofluor white.

2.2. Other Types of Cell Surface Sensors

In addition to the Wsc- and Mid-type sensors, several other types of sensor proteins function at the cell surface in filamentous fungi (Table 1). In *N. crassa*, HAM-7 was identified as a factor associated with anastomosis and sexual development [49]. It has a typical signal peptide at the N-terminus and a glycosylphosphatidylinositol (GPI) anchor signal at the C-terminus [24,49]; it is GPI-anchored to the plasma membrane, and the N-terminal extracellular domain is thought to be localized in the cell wall space. The loss of HAM-7 affects vegetative growth, hyphal branching pattern, and the formation of protoperithecia, but not the sensitivity to cell wall stress compounds [24]. Similar to WSC-1, HAM-7 is required for the activation of MAK-1 MAP kinase, and a strain deficient in both WSC-1 and HAM-7 shows severe phenotypic alterations such as compact colonies, poor formation of aerial hyphae, almost no conidiation, defective cell fusion, and no formation of protoperithecia [24]. Since these alterations are the same as those caused by the deficiency of the MAK-1 pathway, WSC-1 and HAM-7 are considered to be the major sensors upstream of the MAK-1 pathway, although their functions might differ [24].

Signaling mucins are anchored to the plasma membrane, are localized in the cell wall space, and function upstream of MAP kinases [50]. Signaling mucins have a typical signal peptide, a highly glycosylated extracellular inhibitory mucin domain, a single transmembrane domain, and a short intracellular tail. The signaling mucin Msb2 of *S. cerevisiae* is an upstream sensor of the FG and HOG pathways and is activated by nutrient starvation and by cleavage of the extracellular domain [51]. In addition to Msb2, the mucin-like protein Hkr1 is present in *S. cerevisiae* [34] and is involved in sensing cell wall damage by zymolyase [52], which degrades the β-1,3-glucan network [52,53]. Although Hkr1 orthologs have been found in *Ashbya gossypii*, which is closely related to *S. cerevisiae*, they have not been found in other filamentous fungi examined, such as *A. fumigatus*, *Fusarium graminearum*, *Magnaporthe grisea* (currently *Magnaporthe oryzae*, synonymous to *Pyricularia oryzae*), and *N. crassa* [54]. In *C. albicans*, Msb2 deficiency leads to increased sensitivity to cell wall stresses and loss of invasive phenotypes [17,55]. Together with the data on the Wsc-type sensor-defective strains [17], this finding suggests that the sensing of cell wall changes in this fungus is dependent more on the Msb2 sensor than on the Wsc-type sensors.

In some plant pathogenic fungi, the function of Msb2 homologs has been analyzed in relation to their virulence [43,46,56]. In the soil-borne vascular wilt fungus *Fusarium oxysporum*, loss of Msb2 leads to phenotypic alterations that overlap with those caused by the deficiency of the Fmk1 MAP kinase pathway (ortholog of the FG pathway in *S. cerevisiae*), including defects in penetration of cellophane membranes, adhesion to host plant roots, and virulence to the host plant [43]. Unlike Fmk1 deficiency, *msb2* deletion confers sensitivity to cell wall stress compounds, and this sensitivity is enhanced by a double knock-out of *msb2* and *fmk1* [43]. These observations indicate that Msb2 is involved in invasive growth and infection upstream of Fmk1, and also in the cell wall stress response through a pathway distinct from the CWI PKC pathway [43]. In the rice blast fungus *M. oryzae*, MoMsb2 functions upstream of the Pmk1 MAP kinase pathway as a sensor for hydrophobicity and cutin monomers on the plant surface; MoMsb2 is involved in appressorium formation in cooperation with MoSho1, which is thought to be an osmo-sensor [45]. Appressorium formation and host invasion via Pmk1 activation involve the interaction of MoMsb2 with the small GTPase Ras2, and MoMsb2 function partially overlaps with that of the mucin-like protein Cbp1 [56], which was originally identified as a chitin-binding protein [57] and lacks the mucin and transmembrane domains. As in other fungi, Msb2 of *Botrytis cinerea* functions as a surface sensor upstream of Bmk1 MAP kinase (ortholog of the Kss1 in *S. cerevisiae*) but seems to have little relevance to the CWI PKC pathway [46]. In *A. nidulans*, Msb2 is involved in adhesion and biofilm formation, cell wall stress tolerance, vegetative growth, and conidiation under nutrient deficiency via both the CWI PKC and FG pathways [40]. In *A. fumigatus*, MsbA has a similar function, with a particularly strong effect on the CWI PKC pathway [41]. The deficiency of MsbA in *A. fumigatus* alters host immune responses and increases virulence, which has been attributed to changes in the cell wall structure [41]. Generally, other types of sensors in filamentous fungi are associated with the CWI PKC pathway, but their contribution varies among fungal species.

3. Signal Transduction Downstream of Cell Surface Sensors

3.1. Rom2 and Rho1

In *S. cerevisiae*, the cytoplasmic tails of Wsc- and Mid-type sensors interact with the guanine nucleotide exchange factor (GEF) Rom2 [15,16]. This interaction activates the small GTPase Rho1 by converting it into the GTP-bound state. Rho1-GTP activates PKC [15,16], and is also required for the activity of β-1,3-glucan synthase Fks1 (reviewed by Wagener et al. [58]).

The function of Rom2 has been reported in pathogenic fungi *A. fumigatus* [59] and *Candida* species [60]. Since the deletion of *rom2* was suggested to be lethal in *A. fumigatus*, a conditional strain was used for analysis [59]. Under *rom2*-suppressive conditions, this strain has a severe growth defect, a complete loss of conidiation, and an increased sensitivity to cell wall inhibitors [59]. In *A. fumigatus*, Rom2 is localized to the hyphal tip and septa, and *rom2* suppression increases basal levels of phosphorylation of MpkA MAP kinase [59]. Co-immunoprecipitation of HA-tagged Rom2 with Rho1 confirmed their interaction [59]. These results suggest that Rom2 is involved in the activation of the CWI PKC pathway by acting between Wsc- and Mid-type sensors and Rho1 [59].

In a human pathogen, *Candida glabrata*, a temperature-sensitive (ts) mutation in the *rom2* gene has been identified during the analysis of essential genes in ts mutant strains [60]. In *C. albicans*, a strain carrying the same ts mutation was generated; it had colony defects because of the lysis phenotype at a temperature shift without osmostabilizer, as in the ts mutant of *C. glabrata* [60]. These data and the fact that a heterozygous mutant (*Rom2/rom2*) but not null mutant (*rom2/rom2*) was obtained suggest that the *rom2* gene is essential for viability in these *Candida* species and that *Candida* Rom2 is involved in the CWI PKC pathway, in line with its functional similarities with the yeast Rom2 [60].

The function of Rho1 has been analyzed in *Aspergillus* species [61–64]. In *A. fumigatus*, AfRho1 forms a complex with β-1,3-glucan synthase Fks1 [61] and, together with Rho3, is localized to the hyphal tip under normal growth conditions and seems to control the CWI PKC pathway and the cytoskeleton [62]. Among five subfamilies of small GTPases, the Rho subfamily is most extensively characterized [65]. The industrial fungus *Aspergillus niger* has six Rho GTPases, and RhoA plays a central role in polarity establishment and survival, RhoB and RhoD are important for the CWI PKC pathway, and RhoD is important for septum formation, while RhoC has a minor function [63]. RacA and CftA (Cdc42) also maintain polarity, but RacA seems to contribute more than Cdc42 in *A. niger* [63]. In *A. nidulans*, RhoA is involved in polar growth, branching, and cell wall biogenesis [64]. In *F. oxysporum*, the loss of the *rho1* gene is not lethal but results in severe growth defects with abnormal cell walls; the cell wall alteration is thought to activate immune responses in host plants [66].

The basidiomycete *Ustilago maydis* causes corn smut disease; an Rho1 homolog of *U. maydis* is required for vegetative growth and is associated with cell polarity and cytokinesis, and Rho1 loss results in abnormalities in budding and chitin deposition [67]. In another basidiomycete, the edible mushroom *Grifola frondosa*, loss of Rho1 results in reduced mycelial growth, decreased amount of cell wall polysaccharides, and increased sensitivity to cell wall stress [68].

3.2. Protein Kinase C

At the N-terminus, PKC has C1 and C2 cysteine-rich domains, and Rho1 is thought to interact with the C1 domain to regulate the activity of the CWI PKC pathway [69]. Rho1 also binds the N-terminal HR1A domain of PKC, but this binding seems to be involved in a pathway independent of the CWI PKC pathway [70,71]. At the C-terminus, PKC has a serine/threonine kinase domain and a hydrophobic tail with an NFD (Asn-Phe-Asp) motif, the phenylalanine residue of which coordinates the substrate ATP to activate PKC [69].

Comparative genomics has been applied to factors associated with the CWI PKC pathway in human pathogenic fungi such as *C. albicans* and *A. fumigatus* and in plant pathogens such as *M. grisea* and *U. maydis* vs. those in *S. cerevisiae* [54]. The CWI PKC pathway seems to be conserved in most fungal species [54,72]. In *C. albicans*, PKC deficiency leads to cell lysis in both the budding and hyphal growth forms that can be ameliorated by osmotic stabilization [73].

The function of PKC of filamentous fungi has been extensively studied in aspergilli, in particular in *A. nidulans* (see our detailed review [74]). Loss of *pkcA* in *A. nidulans* is lethal [75,76]. Repression of this gene increases sensitivity to cell wall stress agents such as caspofungin and CFW and leads to abnormal cell wall structure [77,78]. PkcA has pleiotropic effects and regulates mitosis, germination, secondary metabolism, and farnesol-induced cell death [75–78]. PkcA inhibits apoptosis induction via the MpkA [79]. Expression of a constitutively active PkcA mutant in *A. nidulans* increases transcript levels of several chitin synthase genes (*chsB*, *chsC*, *chsD*, *csmA*, and *csmB*) and the α-1,3-glucan synthase gene *agsB* [80]. These findings indicate that PkcA in *A. nidulans* regulates the transcription of cell wall-related genes, and at least in this fungus, PkcA seems to play a central role in the CWI PKC pathway. In *A. fumigatus*, *pkcA* is thought to be an essential gene, and the analysis using a non-essential mutant of *pkcA* suggests that PkcA functions upstream of the MAP kinase MpkA [81]. In *N. crassa*, PKC is associated with the light-response signaling pathway and is essential for growth [82,83].

The loss of PKC is lethal in many filamentous fungi, so there are not many examples of functional analysis of PKC in plant pathogens. In *M. oryzae*, RNAi-based analysis has shown that the repression of *pkc1* causes severe growth retardation and considerably affects the transcription of genes involved in cell wall remodeling, autophagy, signal transduction, and secondary metabolism [84]. Sugahara et al. [85] found that a filamentous fungus-specific PKC inhibitor suppresses hyphal melanization in *M. grisea* by suppressing the expression of melanin synthesis-related genes, which are required for pathogenicity of some plant pathogenic fungi [86,87]. Overall, PKC is important in survival and pathogenesis and the development of drugs targeting fungal PKC may be an effective strategy.

3.3. MAP Kinase Cascades Involved in Cell Wall Integrity and Their Targets

In *S. cerevisiae*, the MAP kinase kinase kinase Bck1 activates a pair of redundant MAP kinase kinases Mkk1/Mkk2, and they activate the MAP kinase Mpk1/Slt2 [8]. Scaffold protein Spa2 mediates the interaction between the MAP kinase kinases and MAP kinase [8]. Mpk1/Slt2 phosphorylates the transcription factors (TFs) Rlm1 and Swi4, a subunit of the Swi4–Swi6 TF complex. At least 25 CW-related genes, including genes for β-1,3-glucan synthase and chitin synthases, are regulated by Mpk1/Slt2 [88]. Cell wall stress compounds such as CFW, CR, and zymolyase lead to Mpk1/Slt2 activation [89–91]. The transcriptional response to CR is almost exclusively dependent on Mpk1/Slt2 and Rlm1 [92], but the response to cell damage caused by zymolyase requires both CWI and HOG pathways [91]. In response to CWI damage, a complex transcriptional response program associated with altering metabolism and remodeling the cell wall is elaborately implemented [92,93]. Activation of the Mpk1/Slt2 pathway also required for stimulation of calcium influx through the plasma membrane Ca^{2+} channels Cch1–Mid1, resulting in calcineurin activation, TF Crz1 dephosphorylation, its nuclear translocation and transcriptional regulation of genes related to adaptation to cell wall and cytoplasmic stresses [8]. Mpk1/Slt2 is also activated by hyperosmotic stress, which is dependent on the activation of the Mid2 sensor and Hog1 MAP kinase [8]. The regulation of cell wall biogenesis in fungi is linked to various aspects of morphological control and stress responses through active crosstalk with other signaling pathways [8,94].

Although not all of the MAPK orthologs of fungal pathogens such as *C. albicans* have been functionally analyzed fully, their position in the pathway seems to reflect the *S. cerevisiae* paradigm [72]. In *C. albicans*, the response of the CaSko1 transcription factor to caspofungin depends on the Psk1 PAK kinase but not on the Hog1 MAP kinase [72]. In contrast to *S. cerevisiae* Ste11, a MAP kinase kinase kinase of the Kss1 pathway, there is no evidence that *C. albicans* Ste11 activates Hog1 [72]. The Cas5 transcription factor also contributes to the transcriptional response to caspofungin and has no ortholog in *S. cerevisiae* [72].

Among MAP kinases of filamentous fungi, the orthologs of the genes of the CWI PKC pathway have been analyzed in *A. nidulans*, and the differences in their functions between *A. nidulans* and *S. cerevisiae* have been discussed [18]. In *A. nidulans*, the loss of the Mpk1/Slt2 ortholog MpkA modulates conidial germination and polar growth and increases sensitivity to cell wall stress compounds such as micafungin and CFW [18]. The most distinctive difference between *A. nidulans* MpkA and *S. cerevisiae* Mpk1/Slt2 is in their target genes. The transcription of most cell wall-related genes is MpkA-independent, whereas transcription of synthase genes for α-1,3-glucan, which is absent in the cell wall of *S. cerevisiae*, depends on the TF RlmA via MpkA [18]. Transcription of *fksA* for β-1,3-glucan synthase and *chsB* for chitin synthase is MpkA-dependent under some cell wall stresses [22,80], but factors involved in the transcriptional regulation of many other cell wall-related genes are largely unknown.

In *A. nidulans*, MpkB, an ortholog of Kss1 and Fus3 MAP kinases of the FG and pheromone pathways of *S. cerevisiae* and MpkA have the same phosphorylation motif, and MpkB deletion increases sensitivity to micafungin [95]. Similarly, an MpkB-deficient strain of *A. fumigatus* has an increased sensitivity to caspofungin [96], but at least in *A. nidulans*, MpkB is not involved in the transcriptional regulation of cell wall-related genes [95], suggesting that MpkB may be involved in CWI in a different way. The Kss1/Fus3 orthologous pathway has been extensively studied as a virulence-related factor (see Jiang et al. [97] for a concise and systematic review on this and Mpk1/Slt2 pathways). This pathway is important for appressorium formation and invasive growth of the rice blast fungus *M. oryzae*, and orthologous pathways function similarly in many other appressorium-forming plant pathogens. An MpkB ortholog Chk1 in the southern corn leaf blight fungus *Cochliobolus heterostrophus* regulates not only sexual-asexual development and pathogenicity, but also adaptation to oxidative and heavy-metal stresses [98].

The function of the orthologous Mpk1/Slt2 pathway, especially its involvement in virulence, varies among species in plant pathogenic fungi [97]. Mps1 MAP kinase is dispensable for appressorium formation in *M. oryzae*, but the Mpk1/Slt2 orthologs play a pivotal role in the early stages of appressorium formation in *Colletotrichum lagenarium* and *Colletotrichum gloeosporioides*. They are also involved in various growth processes and pathogenicity-related functions in plant pathogenic fungi: the loss of the Mpk1/Slt2 ortholog leads to severe defects in aerial hyphal formation and sporulation in *M. oryzae* and increases formation of aerial hyphae and decreases that of sclerotium in *Sclerotinia sclerotiorum*.

Among basidiomycetes, the Mpk1/Slt2 orthologous pathway has been analyzed in the pathogenic yeast *Cryptococcus neoformans*. Most of the components of the CWI PKC pathway are conserved except for the sensor proteins in comparison with ascomycete fungal pathogens, and the Mpk1 orthologs seem to share common functions related to cell wall biogenesis, heat stress response, and virulence [99,100]. In *C. neoformans*, Pkc1 activity is important in dynamic morphological changes during infection, especially in changes to the large (so-called Titan) cells and capsule formation [100]. In the edible mushroom *Ganoderma lucidum*, the target of rapamycin (TOR) pathway, which plays a central role in cell growth, regulates cell wall synthesis via an Mpk1/Slt2 ortholog, indicating the potential relationship between the TOR and CWI PKC pathway [101]. In this fungus, the Swi6 ortholog appears to function downstream of the Mpk1/Slt2 pathway; Swi6 has two splice variants, and the variant Swi6B appears to be associated with regulation of the CWI

PKC pathway [102]. The development of various CWI regulatory mechanisms in different fungi is due to the complex evolution of the CWI pathway to suit their survival strategies. The development of antifungal agents targeting these unique systems may be effective and is described in Section 4. In addition, the unique technology to increase production by controlling the cell surface structure, which has been developed on the basis of the analysis of CWI signaling, is discussed in Section 5.

4. Cell Wall Integrity as a Drug Target for Antifungal Agents

4.1. Compounds That Inhibit the Synthesis of Cell Wall Polysaccharides

The development of antifungal drugs is important in medicine for fungal disease treatment and in agriculture for crop protection. Because the fungal cell wall is essential for survival and its architecture is fungus-specific, factors involved in its construction may be effective targets for antifungal agents. The echinocandin class compounds, such as caspofungin, micafungin, and anidulafungin are semisynthetic lipopeptides; they have been widely used for more than 30 years since their development [103]. Echinocandins have superior antifungal activity against *Candida* spp. and *Aspergillus* spp. and are therapeutic agents in particular for esophageal candidiasis, invasive candidiasis, and invasive aspergillosis [103]. They are also active against some other ascomycetes, including *Alternaria* spp. and *Bipolaris* spp., but not against basidiomycete *C. neoformans* or any zygomycetes [103,104]. Echinocandins inhibit the synthesis of β-1,3-glucan, an essential cell wall component in many fungi; for example, they impair the activity of glucan synthases encoded by the *FKS1* and *FKS2* genes in *S. cerevisiae* [103,104]. The emergence of resistant strains has been reported in some *Candida* spp., and an amino acid substitution in Fks1 seems to contribute to the resistance [104]. Because *Aspergillus* spp. growth is not completely suppressed by echinocandins, it is rather difficult to distinguish whether they are resistant or not, but resistant mutants have been generated in the laboratory [105]. Resistant strains have been also isolated in clinical situations, raising concerns about the increase in the incidence of such strains [105].

The nucleoside antibiotics, blasticidin S and polyoxins, are known as forerunners for antibiotics used for agriculture [106]. Blasticidin S inhibits protein synthesis, whereas polyoxins inhibit cell wall synthesis in target fungi [106–108]. Polyoxin A was isolated from *Streptomyces cacaoi* in 1965 [109,110] as a new nucleoside compound and was marketed in 1967 [106]. Polyoxins are structurally similar to the substrate for biosynthesis of chitin (UDP-*N*-acetylglucosamine) [107,108,111], which is an essential cell wall component of plant pathogenic fungi [106–108]. Nikkomycins are structurally related to polyoxins [111]. Polyoxins and nikkomycins are taken up by the fungi and mimic the substrate of chitin synthase, antagonistically inhibiting cell wall chitin synthesis [106–108,111,112].

Some dyes such as CFW and CR are used for laboratory experiments. CFW and CR bind to the fungal cell wall components chitin and glucans [113] and inhibit cell wall synthesis [114]. Their mechanism of action is well summarized by Ram and Klis [114]. Both compounds have two sulfonic acid groups, which are negatively charged under slightly acidic to basic conditions; this makes the compounds soluble and active against fungi [114]. Under these conditions, they cannot pass through the plasma membrane because they each carry two negative charges and are thought to target compounds on the outside of the cell wall [114]. CFW and CR bind to β-linked-glucans in vitro, but CFW preferentially stains chitin in fungal cell wall [114]. Among the cell wall polysaccharides, β-1,3-glucan interacts strongly with CR but not as strongly with CFW in vitro [113]. In *A. nidulans*, the strain with increased cell surface exposure of β-1,3-glucan due to the loss of α-1,3-glucan from the cell wall has increased CR adsorption [115]. In addition, CR adsorbs more on purified β-1,3-glucan or chitin and less on mutan (bacterial α-1,3-glucan) in vitro [115]. Overall, CFW and CR are thought to act by binding to chitin and β-linked-glucan chains, thereby inhibiting the assembly of chitin and β-linked-glucans and weakening the cell wall [114]. Recently, several transcription factors involved in CR sensitivity and CR dynamics in fungal cells have been analyzed [116]. The amorphous cell surface polysaccharide, galactosaminogalactan (GAG), interferes with the uptake of CR into the fungal cell [116]. CR-resistant strains form larger abnormal swollen ("Quasimodo") cells than the wild-type or CR-sensitive strains [116]. Those cells adsorb more CR, leading to CR removal from the culture media and resulting in the acquisition of CR resistance [116]. CR affects the transcription of the genes related to primary and secondary metabolism and toxin efflux systems, suggesting that damage to the fungal cell wall can cause serious adverse effects for fungal growth [116].

4.2. Compounds That Act on the CWI Signaling

Factors involved in fungal signaling systems are often unique to fungi and so may be effective targets for antifungal drugs. For example, dicarboximide fungicides such as iprodione and procymidone, and phenylpyrrole fungicides such as fludioxonil have been used for many years to control crop diseases [117–120]. These fungicides convert type III histidine kinases to phosphatases, which deactivate the histidine-containing phosphotransfer intermediator Ypd1, resulting in abnormal activation of the downstream HOG pathway, and disturb the fungal osmotic response signaling system [117–119].

Recently, Beattie and Krysan [121] developed a high-throughput screening system for antifungal agents based on the adenylate kinase (AK) assay. AK released during fungal cell lysis phosphorylates an ADP-containing reagent, and the generated ATP is detected by luciferase. Using this method, the authors found that compound PIK-75 inhibits growth of not only *A. fumigatus* but also *C. neoformans*, and that PIK-75 activity at least in part is due to the loss of CWI. This assay allows cell-wall active compounds to be identified even if their action on the cell wall is indirect [121].

As described in *3.1*, PKC is an important signaling factor associated with cell wall biogenesis. Therefore, PKC inhibitors, such as staurosporine, enzastaurin, and ruboxistaurin, can be expected to have excellent antifungal activity. For example, staurosporine strongly inhibits PKC in filamentous fungi [85]. However, these inhibitors cannot be used as antifungal agents because they also inhibit human PKC. To screen for specific inhibitors of fungal cell wall biogenesis, Sugahara et al. [85] conducted an in silico screening to target PKC of *M. oryzae*. The overall concept of the study is depicted in Figure 1. A three-dimensional MgPkc1 structure was modeled to screen for compounds that might inhibit its kinase domain, and the candidate compounds were tested for antifungal activity against *M. grisea*. Among them, Z-705 had the highest inhibitory effect. Chimeric *PKCs* encoding the regulator domain from *S. cerevisiae* and the kinase domain from *S. cerevisiae* (control), *M. grisea* or *A. nidulans* were integrated in the *S. cerevisiae* genome, and Z-705 specifically inhibited chimeric PKCs with the kinase domain from filamentous fungi, but not with that from *S. cerevisiae* [85]. The inhibitory effect was comparable to that of staurosporine, a well-known PKC inhibitor. This compound also inhibited hyphal melanization induced by cell wall stress in *M. grisea*, which is necessary for the infection process [86,87] (Figure 1). In *M. grisea*, Mps1 acts downstream of Pkc1 and regulates gene expression for synthase of α-1,3-glucan that confers the ability to evade the host plant immune system [122,123]. From this perspective, Pkc1 inhibition is also important for reducing pathogenicity. We believe that the efficacy of drugs targeting the cell wall of filamentous fungi is increasing, and the development of such drugs should be further promoted in the future.

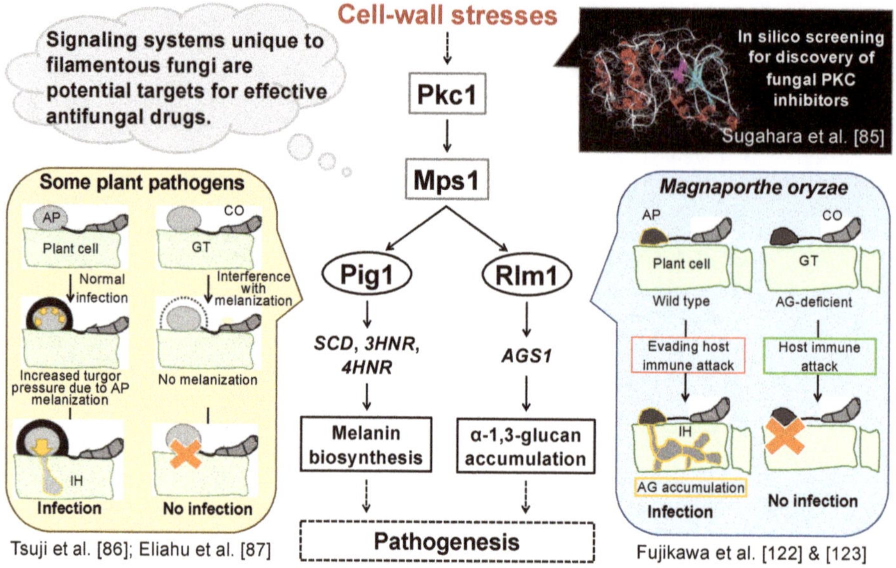

Figure 1. Diagram of drug development to target the CWI signaling pathway in filamentous fungi. In *M. oryzae*, Pkc1 is a protein kinase C, Mps1 is a MAP kinase, and Pig1 and Rlm1 are transcription factors. *SCD* (encoding scytalone dehydratase), *3HNR* (trihydroxy-naphthalene reductase), and *4HNR* (1,3,6,8-tetrahydroxy-naphthalene reductase) are involved in the biosynthesis of 1,8-dihydroxynaphthalene (DHN) melanin in several plant pathogens, including *M. oryzae*. *AGS1* encodes α-1,3-glucan synthase. Abbreviations: AP, appressorium; GT, germ tube; CO, conidium; IH, invasive hyphae; AG, α-1,3-glucan.

5. Improvement of Productivity by Modification of Macromorphology in Filamentous Fungi

Another applied approach derived from studies on cell wall biogenesis in filamentous fungi is the improvement of culture characteristics to increase productivity by controlling the surface properties of fungal cells and fungal morphology. Here, we describe some examples of this approach.

5.1. Phenotypes of α-1,3-Glucan-Deficient Mutants

As described above, expression of α-1,3-glucan synthase genes (*agsA*, *agsB*) is controlled by MAP kinase MpkA in *A. nidulans* [18]. Single or double disruption of the two α-1,3-glucan synthase genes of *A. nidulans* has revealed that the single disruption of *agsB* and the double disruption of *agsA* and *agsB* cause complete loss of cell wall α-1,3-glucan but are not lethal [115]. *Aspergillus fumigatus* has three α-1,3-glucan synthase genes, and disruption of all the three genes (Δ*ags*) is not lethal [124,125]. The germinating conidia of *A. fumigatus* Δ*ags* do not aggregate [124]. The hyphae of α-1,3-glucan-deficient mutants of *A. nidulans* such as Δ*agsB* and Δ*agsA*Δ*agsB* are dispersed in shake-flask cultures, whereas those of the parental strain form tightly aggregated pellets [18]. In *A. nidulans*, the *agsB* gene is clustered with the α-amylase-encoding genes *amyD* and *amyG* [126]. An intracellular α-amylase AmyG hypothetically contributes to synthesis of the primer molecule for α-1,3-glucan polymerization by the α-1,3-glucan synthase [126,127]. Disruption of the *amyG* gene results in a substantial decrease in the content of cell wall α-1,3-glucan and lead to hyphal dispersion or formation of tiny hyphal pellets [126,127]. Disruption or overexpression of *amyD*, which encodes GPI-anchored α-amylase, increases or decreases cell wall α-1,3-glucan, respectively [126]. The hyphae of the *amyD* overexpression strain show a phenotype similar to those of the α-1,3-glucan-deficient mutants [126,128]. These results suggest that AmyD represses α-1,3-glucan biosynthesis. Overexpression of AmyD without the C-terminal GPI-anchor in *A. nidulans* scarcely affects cell wall α-1,3-glucan, suggesting the importance of the GPI anchor for correct cellular localization and function of AmyD [129]. In *A. fumigatus*, treatment with α-1,3-glucanase removes α-1,3-glucan from conidia and leads to their dispersion in medium, indicating the involvement of α-1,3-glucan in conidial aggregation [130]. Dispersion of hyphae in *A. nidulans* in α-1,3-glucan-deficient mutants and that of germinating conidia in *A. fumigatus* α-1,3-glucan-deficient mutants suggests that α-1,3-glucan functions as an aggregation factor for hyphae and conidia. Interestingly, *A. nidulans* Δ*agsB* and Δ*agsA*Δ*agsB* mutants produce considerably more hyphal cells than the wild-type strain does under submerged culture conditions, implying that the dispersed hyphal cells of the α-1,3-glucan deficient mutants can be used for fermentation of valuable products. The α-1,3-glucan deficient mutants of *A. nidulans* show better production of endogenous penicillin and α-amylase than the wild type [131].

In the α-1,3-glucan-deficient mutant of the industrial fungus *A. oryzae*, three α-1,3-glucan synthase genes are disrupted [132]. The *A. oryzae* Δ*agsA*Δ*agsB*Δ*agsC* mutant forms smaller hyphal pellets than the parental wild-type strain, suggesting that α-1,3-glucan is also an aggregation factor in *A. oryzae*. This mutant produces more recombinant protein than the wild-type strain [133]. Jeennor et al. [134] reported that the disruption of the *ags1* gene (probably *agsB* in Miyazawa's work [133]) in *A. oryzae* significantly improves lipid production in a stirred-tank bioreactor. The disruptant of the *Aspergillus luchuensis agsE* gene, an ortholog to *A. nidulans agsB*, shows better protoplast formation than the wild-type strain when treated with the cell wall lytic enzyme Yatalase [135]. The *A. fumigatus* mutants in which the *ags1* gene, an ortholog of *A. nidulans agsB*, is disrupted, form smaller hyphal pellets than the wild type [136]. Taken together, α-1,3-glucan is an aggregation factor for hyphae and conidia in *Aspergillus* fungi.

5.2. Phenotypes of Galactosaminogalactan-Deficient Mutants

In *Aspergillus* species, GAG is one of the components of the extracellular matrix and is essential for biofilm formation [137]. In the background of the defect of α-1,3-glucan biosynthesis (Δ*agsA*Δ*agsB*Δ*agsC*) in *A. oryzae*, disruption of the *sphZ* and *ugeZ* genes (AGΔ-GAGΔ), which are speculative GAG biosynthetic genes of *A. oryzae*, leads to dispersion of hyphae under submerged culture conditions, suggesting that GAG also contributes to aggregation in *A. oryzae* [138]. A simultaneous defect of α-1,3-glucan and GAG biosynthesis also leads to hyphal dispersion in *A. fumigatus* [136]. In *B. cinerea* and *Cochlioborus heterostrophus*, GAG also contributes to hyphal aggregation [139]. Recently, Mei et al. [140] reported that the insect pathogenic fungus *Metarhizium robertsii* has GAG biosynthetic genes, and defects of GAG biosynthesis lead to hyphal dispersion. In ascomycetes, GAG biosynthetic genes are found in some *Pezizomycotina*, and only in *Trichosporon asahii* in basidiomycetes [141]. Expression of GAG biosynthetic genes is thought to be regulated by transcription factors such as StuA, MedA, and SomA in *A. fumigatus* [142–144]. Since disruption of the *agdZ* gene increases GAG secretion in *A. oryzae* and *A. fumigatus* (Miyazawa et al., unpublished results), some mechanisms might sense adhesion of hyphae and subsequently downregulate GAG biosynthesis.

5.3. Improvement of Productivity Using a Mutant Lacking both α-1,3-Glucan and GAG

Regulation of macromorphology such as hyphal pellets and pulp form has been a key issue in fermentation using filamentous fungi [145]. Macromorphology of filamentous fungi is controlled by adjusting culture conditions such as agitation speed, pH, and medium composition [146,147]. Recently, addition of microparticles such as titanate and talc to liquid culture media has been found to promote the formation of micro-pellets that can improve productivity of fermentation of filamentous fungi [148]. We here illustrate our strategy for improving productivity with cell wall mutants of *A. oryzae* (Figure 2). Miyazawa et al. [138] showed that hyphae of the AGΔ-GAGΔ mutant are fully dispersed under submerged culture conditions, and production of recombinant polyesterase CutL1 is significantly higher in AGΔ-GAGΔ than in the parental wild-type strain in shake-flask culture [138]. Ichikawa et al. [149] showed that the production of secreted CutL1 was higher in AGΔ-GAGΔ than in the wild type or mutants lacking α-1,3-glucan (AGΔ) or GAG (GAGΔ) in batch culture in a 5 L lab-scale bioreactor. The apparent viscosity of the AGΔ-GAGΔ culture tended to be lower than that of the wild-type strain culture at each agitation speed examined (200–600 rpm), suggesting that the lack of α-1,3-glucan and GAG in the hyphae improves culture rheology, increasing recombinant protein production [149]. Sakuragawa et al. [150] reported that the AGΔ-GAGΔ strain produces more recombinant cellulase CBHI than the wild-type strain in a 250 mL bioreactor. The AGΔ-GAGΔ strain shows rapid glucose consumption, increased mycelial dry weight, and higher respiration activity in comparison with the wild-type strain. The levels of metabolites of glycolysis and TCA cycle are lower in AGΔ-GAGΔ than in the wild type in liquid culture, suggesting that AGΔ-GAGΔ shows higher metabolic flux than the wild type [150]. Since the production of beneficial compounds from fungal cells is attributable to complex physiological events, the mechanisms underlying the productivity of AGΔ-GAGΔ in the bioreactor are presently being analyzed. Further improvement of the productivity is expected to be achieved by conferring stress susceptibility to the AGΔ-GAGΔ mutant and fine tuning the culture conditions through the screening for stress factors and multi-omics analyses in the cultivation.

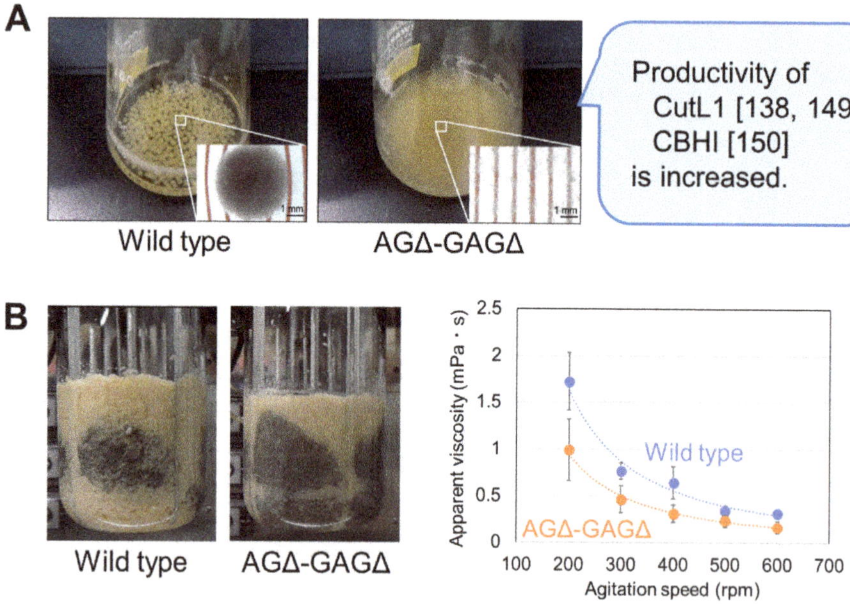

Figure 2. Improvement of productivity with the *Aspergillus oryzae* mutant lacking both α-1,3-glucan and GAG (AGΔ-GAGΔ). (**A**) Growth of the wild-type and AGΔ-GAGΔ strains in liquid culture. Although the wild type forms pellets of several millimeters, the AGΔ-GAGΔ hyphae are fully dispersed. This unique macromorphology of AGΔ-GAGΔ results in increased production of secreted recombinant polyesterase CutL1 and recombinant cellulase CBHI. Conidia (1.0×10^5/mL) of each strain were inoculated into 50 mL of YPD (2% peptone, 1% yeast extract and 2% glucose) medium in a 200 mL Erlenmeyer flask and rotated at 120 rpm at 30 °C. Magnified images (bottom right) were taken under a stereomicroscope. (**B**) AGΔ-GAGΔ culture has improved rheological properties. The wild type and AGΔ-GAGΔ expressing recombinant *cutL1* gene were cultured in YPDS (6% peptone, 1% yeast extract, 6% glucose and 20 mM succinate buffer) in a 5 L lab-scale bioreactor. Left panels: Chinese ink was dropped onto the culture surface at 60 h, and diffusion was imaged at 6 s. Right panel: Apparent viscosity of the culture at 36 h. Torque values were measured with a mixing torquemeter, and apparent viscosity was calculated from the *Np-Re* diagram at the indicated agitation speeds.

5.4. Improvement of Productivity by Mutations in Cell Wall-Related Genes

Both extracellular hydrolytic enzymes such as amylases and proteases and cell wall synthesizing enzymes are packaged in vesicles and delivered from the Golgi to the hyphal tip of filamentous fungi [151]. Delivery of cell wall synthesizing enzymes to the hyphal tip balances necessity to secrete extracellular enzymes for nutrient acquisition [151]. Since secretion of enzymes and cell wall biogenesis are linked, perturbation to cell wall biogenesis seems to considerably affect enzyme secretion [151].

The *A. niger* SH2 strain is widely used in industrial enzyme production [151,152]. In the SH2 genome sequence, Yin et al. [152] found frame-shift mutations and non-synonymous SNPs in genes of CWI signaling, β-1,3-glucan synthesis and chitin synthesis and suggested that they affect hyphal development and hyphal fragmentation during industrial fermentation. Sun et al. [153] constructed *A. niger* mutants with the silenced chitin synthase gene *chsC*. The mutants showed shorter hyphae with lower proportion of dispersed mycelia, decreased viscosity and improved oxygen and mass transfer efficiency, which consequently improved production of citric acid [153]. Yin et al. [154] evaluated citrate production by *A. niger* H915-1 (an industrial producer) and by A1 and L2 ("degenerated" isolates of H915-1) strains. The H915-1 forms bulbous hyphae with short, swollen branches during citrate fermentation, and has the highest citrate titer, whereas A1 forms fewer compact pellets and L2 forms mycelial clumps [154]. Yin et al. [154] indicated that these differences in morphology may influence medium viscosity and hyphal respiration [154]. For citrate generation, the tight pellet form but not the diffuse filamentous form is preferred [154]. Liu et al. [155] reported that silencing of the *chs4* gene encoding class III chitin synthase in *Penicillium chrysogenum* by RNA interference causes formation of a smaller pellet, hyper-branched hyphae, and improves penicillin production. To find *N. crassa* mutants with decreased viscosity in submerged culture, Lin et al. [156] screened 90 morphological mutants and found two such mutants. The causing gene *gul-1* encodes an mRNA-binding protein. Disruption of this gene downregulates GPI-anchored cell wall proteins, upregulates non-GPI cell wall proteins, and alters expression of the hydrophobin gene. Disruption of *gul-1* in the hyper-cellulase–producing strain significantly decreases culture viscosity compared to the parental strain. Fiedler et al. [157] analyzed the transcriptomics of *A. niger* cells treated with inhibitors of synthesis of chitin (CFW), glucan (caspofungin), sphingolipids (aureobasidin A), and ergosterol (fenpropimorph), and of calcium/calcineurin signaling (FK506), which directly or indirectly interfere with CWI. The analysis suggests that (i) the CWI PKC pathway as a main compensatory response is induced by caspofungin via RhoB and by aureobasidin A via RhoD, followed by activation of the MAPKK MkkA and the TF RlmA; (ii) RlmA is the main TF for protection against CFW, but it cooperates with MsnA and CrzA for protection against caspofungin and aureobasidin A; (iii) aureobasidin A, but not fenpropimorph, induces cell wall stress.

Overall, the macromorphology of filamentous fungi closely relates to productivity. Although several components regulated by the CWI PKC pathway in the production strains have been revealed, how to regulate the CWI PKC pathway to improve productivity is scarcely understood. Combining the screening of phenotypic mutants and analysis of the mechanisms underlying cellular physiology as described by Lin et al. [156] could lead to a breakthrough technology to further improve fungal productivity.

6. Conclusions and Perspectives

The cell wall of filamentous fungi is constantly exposed to the environment and is closely involved in interactions with other microorganisms, plants and animals. The fungal cell wall, as well as those of bacteria and plants, is mainly composed of polysaccharides, but these polysaccharides and their structures are quite different from those of bacteria and plants. Although the PKC is conserved in all eukaryotes, CWI PKC pathway has evolved independently in fungi and varies even at the species level. Perturbing CWI signaling is an effective strategy for controlling fungal growth. Chemical compounds that target certain signaling factors of CWI signaling can be used to control pathogens of plants and animals. Effective antifungal drugs targeting the cell wall biosynthesis of filamentous fungi are now on the market, and the screening for and consequent development of such chemicals are underway. Since the genomic information of filamentous fungi is continuously accumulated and artificial intelligence (AI)-based analyses are advancing in various fields, the development of antifungal drugs targeting CWI signaling will be further accelerated by utilizing AI technology in the analysis of genomic information.

The studies of CWI signaling have revealed that polysaccharides such as α-1,3-glucan and GAG function as adhesive factors for hyphae in aspergilli and cause the formation of hyphal pellets. Regulation of the display of these polysaccharides on the cell surface enables filamentous fungi to control their macromorphology such as pellets and pulp forms. Filamentous fungi are extensively used for large-scale industrial cultivation in submerged culture for production of proteins and low-molecular-weight chemicals. However, the capacity of production by filamentous fungi does not reach that by the unicellular fungus *S. cerevisiae* or bacteria *Escherichia coli* and *Bacillus subtilis*, because of the unstable macromorphology of filamentous fungi during liquid cultivation. Several attempts have been made to control the hyphal morphology in filamentous fungi to improve the cultivation characteristics, but the fundamental technology to control hyphal pellet formation has not been established. Modifying polysaccharide contents of the cell surface has led to strains with dispersed hyphae and normal growth, which ensures the efficient acquisition of nutrients and dissolved oxygen. Further analysis of the mechanisms of cell wall biogenesis in filamentous fungi will generate knowledge that will lead to the development of antifungal agents and may also lead to innovative technology for industrial cultivation using filamentous fungi. Therefore, studies on the cell wall biogenesis of filamentous fungi should be continuously promoted, so that the ensuing fruitful achievements can contribute to the improvement of human life.

Author Contributions: A.Y., K.M. and K.A. wrote the manuscript. M.K. checked the manuscript. K.A. supervised this research. All authors have read and agreed to the published version of the manuscript.

Funding: This work was supported by the Institute for Fermentation, Osaka, Japan (Grant numbers K-2019-002 to A.Y. and M.K. and L-2018-2-014 to K.A.). This work was also supported by a project JPNP20011 (K.A.) commissioned by the New Energy and Industrial Technology Development Organization (NEDO), the Japan Society for the Promotion of Science (JSPS) KAKENHI Grant Numbers JP20H02895, JP26292037 (K.A.), JP18K05384 (A.Y.), JP18J11870 and JP20K22773 (K.M.). This work also supported by ISHIZUE 2021 of Kyoto University Research Development Program (A.Y.) and the Kyoto University Foundation (M.K.).

Institutional Review Board Statement: Not applicable.

Informed Consent Statement: Not applicable.

Conflicts of Interest: The authors declare no conflict of interest.

References

1. Cairns, T.C.; Zheng, X.; Zheng, P.; Sun, J.; Meyer, V. Turning inside out: Filamentous fungal secretion and its applications in biotechnology, agriculture, and the clinic. *J. Fungi* **2021**, *7*, 535. [CrossRef] [PubMed]
2. Treseder, K.K.; Lennon, J.T. Fungal traits that drive ecosystem dynamics on land. *Microbiol. Mol. Biol. Rev.* **2015**, *79*, 243–262. [CrossRef] [PubMed]
3. Gadd, G.M. The geomycology of elemental cycling and transformations in the environment. *Microbiol. Spectr.* **2017**, *5*, FUNK-0010-2016. [CrossRef] [PubMed]
4. Hagiwara, D.; Yoshimi, A.; Sakamoto, K.; Gomi, K.; Abe, K. Response and adaptation to cell wall stress and osmotic stress in *Aspergillus* species. In *Stress Biology of Yeasts and Fungi: Applications for Industrial Brewing and Fermentation*; Takagi, H., Kitagaki, H., Eds.; Springer: Tokyo, Japan, 2015; pp. 199–218.
5. Levin, D.E. Cell wall integrity signaling in *Saccharomyces cerevisiae*. *Microbiol. Mol. Biol. Rev.* **2005**, *69*, 262–291. [CrossRef] [PubMed]
6. Levin, D.E. Regulation of cell wall biogenesis in *Saccharomyces cerevisiae*: The cell wall integrity signaling pathway. *Genetics* **2011**, *189*, 1145–1175. [CrossRef] [PubMed]
7. Gustin, M.C.; Albertyn, J.; Alexander, M.; Davenport, K. MAP kinase pathways in the yeast *Saccharomyces cerevisiae*. *Microbiol. Mol. Biol. Rev.* **1998**, *62*, 1264–1300. [CrossRef] [PubMed]

8. Chen, R.E.; Thorner, J. Function and regulation in MAPK signaling pathways: Lessons learned from the yeast *Saccharomyces cerevisiae*. *Biochim. Biophys. Acta* **2007**, *1773*, 1311–1340. [CrossRef]
9. Hohmann, S. Control of high osmolarity signalling in the yeast *Saccharomyces cerevisiae*. *FEBS Lett.* **2009**, *583*, 4025–4029. [CrossRef]
10. Cullen, P.J.; Sprague, G.F., Jr. The regulation of filamentous growth in yeast. *Genetics* **2012**, *190*, 23–49. [CrossRef]
11. McClean, M.N.; Mody, A.; Broach, J.R.; Ramanathan, S. Cross-talk and decision making in MAP kinase pathways. *Nat. Genet.* **2007**, *39*, 409–414. [CrossRef]
12. Saito, H. Regulation of cross-talk in yeast MAPK signaling pathways. *Curr. Opin. Microbiol.* **2010**, *13*, 677–683. [CrossRef] [PubMed]
13. Abe, K.; Gomi, K.; Hasegawa, F.; Machida, M. Impact of *Aspergillus oryzae* genomics on industrial production of metabolites. *Mycopathologia* **2006**, *162*, 143–153. [CrossRef] [PubMed]
14. Machida, M.; Yamada, O.; Gomi, K. Genomics of *Aspergillus oryzae*: Learning from the history of Koji mold and exploration of its future. *DNA Res.* **2008**, *15*, 173–183. [CrossRef] [PubMed]
15. Kock, C.; Dufrene, Y.F.; Heinisch, J.J. Up against the wall: Is yeast cell wall integrity ensured by mechanosensing in plasma membrane microdomains? *Appl. Environ. Microbiol.* **2015**, *81*, 806–811. [CrossRef] [PubMed]
16. Jendretzki, A.; Wittland, J.; Wilk, S.; Straede, A.; Heinisch, J.J. How do I begin? Sensing extracellular stress to maintain yeast cell wall integrity. *Eur. J. Cell Biol.* **2011**, *90*, 740–744. [CrossRef] [PubMed]
17. Zucchi, P.C.; Davis, T.R.; Kumamoto, C.A. A *Candida albicans* cell wall-linked protein promotes invasive filamentation into semi-solid medium. *Mol. Microbiol.* **2010**, *76*, 733–748. [CrossRef]
18. Fujioka, T.; Mizutani, O.; Furukawa, K.; Sato, N.; Yoshimi, A.; Yamagata, Y.; Nakajima, T.; Abe, K. MpkA-dependent and -independent cell wall integrity signaling in *Aspergillus nidulans*. *Eukaryot. Cell* **2007**, *6*, 1497–1510. [CrossRef]
19. Goto, M.; Harada, Y.; Oka, T.; Matsumoto, S.; Takegawa, K.; Furukawa, K. Protein *O*-mannosyltransferases B and C support hyphal development and differentiation in *Aspergillus nidulans*. *Eukaryot. Cell* **2009**, *8*, 1465–1474. [CrossRef]
20. Futagami, T.; Nakao, S.; Kido, Y.; Oka, T.; Kajiwara, Y.; Takashita, H.; Omori, T.; Furukawa, K.; Goto, M. Putative stress sensors WscA and WscB are involved in hypo-osmotic and acidic pH stress tolerance in *Aspergillus nidulans*. *Eukaryot. Cell* **2011**, *10*, 1504–1515. [CrossRef]
21. Serrano, R.; Martin, H.; Casamayor, A.; Arino, J. Signaling alkaline pH stress in the yeast *Saccharomyces cerevisiae* through the Wsc1 cell surface sensor and the Slt2 MAPK pathway. *J. Biol. Chem.* **2006**, *281*, 39785–39795. [CrossRef]
22. Futagami, T.; Seto, K.; Kajiwara, Y.; Takashita, H.; Omori, T.; Takegawa, K.; Goto, M. The putative stress sensor protein MtlA is required for conidia formation, cell wall stress tolerance, and cell wall integrity in *Aspergillus nidulans*. *Biosci. Biotechnol. Biochem.* **2014**, *78*, 326–335. [CrossRef] [PubMed]
23. Dichtl, K.; Helmschrott, C.; Dirr, F.; Wagener, J. Deciphering cell wall integrity signalling in *Aspergillus fumigatus*: Identification and functional characterization of cell wall stress sensors and relevant Rho GTPases. *Mol. Microbiol.* **2012**, *83*, 506–519. [CrossRef] [PubMed]
24. Maddi, A.; Dettman, A.; Fu, C.; Seiler, S.; Free, S.J. WSC-1 and HAM-7 are MAK-1 MAP kinase pathway sensors required for cell wall integrity and hyphal fusion in *Neurospora crassa*. *PLoS ONE* **2012**, *7*, e42374. [CrossRef] [PubMed]
25. Tong, S.M.; Chen, Y.; Zhu, J.; Ying, S.H.; Feng, M.G. Subcellular localization of five singular WSC domain-containing proteins and their roles in *Beauveria bassiana* responses to stress cues and metal ions. *Environ. Microbiol. Rep.* **2016**, *8*, 295–304. [CrossRef] [PubMed]
26. Tong, S.M.; Wang, D.Y.; Gao, B.J.; Ying, S.H.; Feng, M.G. The DUF1996 and WSC domain-containing protein Wsc1I acts as a novel sensor of multiple stress cues in *Beauveria bassiana*. *Cell. Microbiol.* **2019**, *21*, e13100. [CrossRef]
27. Cruz, S.; Munoz, S.; Manjon, E.; Garcia, P.; Sanchez, Y. The fission yeast cell wall stress sensor-like proteins Mtl2 and Wsc1 act by turning on the GTPase Rho1p but act independently of the cell wall integrity pathway. *MicrobiologyOpen* **2013**, *2*, 778–794. [CrossRef]
28. Gray, J.V.; Ogas, J.P.; Kamada, Y.; Stone, M.; Levin, D.E.; Herskowitz, I. A role for the Pkc1 MAP kinase pathway of *Saccharomyces cerevisiae* in bud emergence and identification of a putative upstream regulator. *EMBO J.* **1997**, *16*, 4924–4937. [CrossRef]
29. Verna, J.; Lodder, A.; Lee, K.; Vagts, A.; Ballester, R. A family of genes required for maintenance of cell wall integrity and for the stress response in *Saccharomyces cerevisiae*. *Proc. Natl. Acad. Sci. USA* **1997**, *94*, 13804–13809. [CrossRef]
30. Jacoby, J.J.; Nilius, S.M.; Heinisch, J.J. A screen for upstream components of the yeast protein kinase C signal transduction pathway identifies the product of the *SLG1* gene. *Mol. Gen. Genet.* **1998**, *258*, 148–155. [CrossRef]
31. Wei, Y.F.; Chen, B.J.; Samson, L. Suppression of *Escherichia coli alkB* mutants by *Saccharomyces cerevisiae* genes. *J. Bacteriol.* **1995**, *177*, 5009–5015. [CrossRef]
32. Ketela, T.; Green, R.; Bussey, H. *Saccharomyces cerevisiae* Mid2p is a potential cell wall stress sensor and upstream activator of the PKC1-MPK1 cell integrity pathway. *J. Bacteriol.* **1999**, *181*, 3330–3340. [CrossRef] [PubMed]
33. Rajavel, M.; Philip, B.; Buehrer, B.M.; Errede, B.; Levin, D.E. Mid2 is a putative sensor for cell integrity signaling in *Saccharomyces cerevisiae*. *Mol. Cell. Biol.* **1999**, *19*, 3969–3976. [CrossRef] [PubMed]
34. Tatebayashi, K.; Tanaka, K.; Yang, H.Y.; Yamamoto, K.; Matsushita, Y.; Tomida, T.; Imai, M.; Saito, H. Transmembrane mucins Hkr1 and Msb2 are putative osmosensors in the SHO1 branch of yeast HOG pathway. *EMBO J.* **2007**, *26*, 3521–3533. [CrossRef] [PubMed]

35. Puri, S.; Kumar, R.; Chadha, S.; Tati, S.; Conti, H.R.; Hube, B.; Cullen, P.J.; Edgerton, M. Secreted aspartic protease cleavage of *Candida albicans* Msb2 activates Cek1 MAPK signaling affecting biofilm formation and oropharyngeal candidiasis. *PLoS ONE* **2012**, *7*, e46020. [CrossRef] [PubMed]
36. Saraswat, D.; Kumar, R.; Pande, T.; Edgerton, M.; Cullen, P.J. Signalling mucin Msb2 Regulates adaptation to thermal stress in *Candida albicans*. *Mol. Microbiol.* **2016**, *100*, 425–441. [CrossRef] [PubMed]
37. Ohsawa, S.; Yurimoto, H.; Sakai, Y. Novel function of Wsc proteins as a methanol-sensing machinery in the yeast *Pichia pastoris*. *Mol. Microbiol.* **2017**, *104*, 349–363. [CrossRef]
38. Rodicio, R.; Buchwald, U.; Schmitz, H.P.; Heinisch, J.J. Dissecting sensor functions in cell wall integrity signaling in *Kluyveromyces lactis*. *Fungal Genet. Biol.* **2008**, *45*, 422–435. [CrossRef]
39. Futagami, T.; Goto, M. Putative cell wall integrity sensor proteins in *Aspergillus nidulans*. *Commun. Integr. Biol.* **2012**, *5*, 206–208. [CrossRef]
40. Brown, N.A.; Dos Reis, T.F.; Goinski, A.B.; Savoldi, M.; Menino, J.; Almeida, M.T.; Rodrigues, F.; Goldman, G.H. The *Aspergillus nidulans* signalling mucin MsbA regulates starvation responses, adhesion and affects cellulase secretion in response to environmental cues. *Mol. Microbiol.* **2014**, *94*, 1103–1120. [CrossRef]
41. Gurgel, I.L.d.S.; Jorge, K.T.d.O.S.; Malacco, N.L.S.d.O.; Souza, J.A.M.; Rocha, M.C.; Fernandes, M.F.; Martins, F.R.B.; Malavazi, I.; Teixeira, M.M.; Soriani, F.M.; et al. The *Aspergillus fumigatus* mucin MsbA regulates the cell wall integrity pathway and controls recognition of the fungus by the immune system. *mSphere* **2019**, *4*, e00350-19. [CrossRef]
42. Xu, L.; Wang, M.; Tang, G.; Ma, Z.; Shao, W. The endocytic cargo adaptor complex is required for cell-wall integrity via interacting with the sensor FgWsc2B in *Fusarium graminearum*. *Curr. Genet.* **2019**, *65*, 1071–1080. [CrossRef] [PubMed]
43. Perez-Nadales, E.; Di Pietro, A. The membrane mucin Msb2 regulates invasive growth and plant infection in *Fusarium oxysporum*. *Plant Cell* **2011**, *23*, 1171–1185. [CrossRef] [PubMed]
44. Xin, C.; Xing, X.; Wang, F.; Liu, J.; Ran, Z.; Chen, W.; Wang, G.; Song, Z. MrMid2, encoding a cell wall stress sensor protein, is required for conidium production, stress tolerance, microsclerotium formation and virulence in the entomopathogenic fungus *Metarhizium rileyi*. *Fungal Genet. Biol.* **2020**, *134*, 103278. [CrossRef] [PubMed]
45. Liu, W.; Zhou, X.; Li, G.; Li, L.; Kong, L.; Wang, C.; Zhang, H.; Xu, J.R. Multiple plant surface signals are sensed by different mechanisms in the rice blast fungus for appressorium formation. *PLoS Pathog.* **2011**, *7*, e1001261. [CrossRef]
46. Leroch, M.; Mueller, N.; Hinsenkamp, I.; Hahn, M. The signalling mucin Msb2 regulates surface sensing and host penetration via BMP1 MAP kinase signalling in *Botrytis cinerea*. *Mol. Plant Pathol.* **2015**, *16*, 787–798. [CrossRef]
47. Lanver, D.; Mendoza-Mendoza, A.; Brachmann, A.; Kahmann, R. Sho1 and Msb2-related proteins regulate appressorium development in the smut fungus *Ustilago maydis*. *Plant Cell* **2010**, *22*, 2085–2101. [CrossRef]
48. So, Y.S.; Jang, J.; Park, G.; Xu, J.; Olszewski, M.A.; Bahn, Y.S. Sho1 and Msb2 play complementary but distinct roles in stress responses, sexual differentiation, and pathogenicity of *Cryptococcus neoformans*. *Front. Microbiol.* **2018**, *9*, 2958. [CrossRef]
49. Fu, C.; Iyer, P.; Herkal, A.; Abdullah, J.; Stout, A.; Free, S.J. Identification and characterization of genes required for cell-to-cell fusion in *Neurospora crassa*. *Eukaryot. Cell* **2011**, *10*, 1100–1109. [CrossRef]
50. Cullen, P.J. Signaling mucins: The new kids on the MAPK block. *Crit. Rev. Eukaryot. Gene Expr.* **2007**, *17*, 241–257. [CrossRef]
51. Vadaie, N.; Dionne, H.; Akajagbor, D.S.; Nickerson, S.R.; Krysan, D.J.; Cullen, P.J. Cleavage of the signaling mucin Msb2 by the aspartyl protease Yps1 is required for MAPK activation in yeast. *J. Cell Biol.* **2008**, *181*, 1073–1081. [CrossRef]
52. Rodriguez-Pena, J.M.; Diez-Muniz, S.; Bermejo, C.; Nombela, C.; Arroyo, J. Activation of the yeast cell wall integrity MAPK pathway by zymolyase depends on protease and glucanase activities and requires the mucin-like protein Hkr1 but not Msb2. *FEBS Lett.* **2013**, *587*, 3675–3680. [CrossRef] [PubMed]
53. Kitamura, K.; Kaneko, T.; Yamamoto, Y. Lysis of viable yeast cells by enzymes of *Arthrobacter luteus*: II. Purification and properties of an enzyme, zymolyase, which lyses viable yeast cells. *J. Gen. Appl. Microbiol.* **1974**, *20*, 323–344. [CrossRef]
54. Rispail, N.; Soanes, D.M.; Ant, C.; Czajkowski, R.; Grunler, A.; Huguet, R.; Perez-Nadales, E.; Poli, A.; Sartorel, E.; Valiante, V.; et al. Comparative genomics of MAP kinase and calcium-calcineurin signalling components in plant and human pathogenic fungi. *Fungal Genet. Biol.* **2009**, *46*, 287–298. [CrossRef] [PubMed]
55. Roman, E.; Cottier, F.; Ernst, J.F.; Pla, J. Msb2 signaling mucin controls activation of Cek1 mitogen-activated protein kinase in *Candida albicans*. *Eukaryot. Cell* **2009**, *8*, 1235–1249. [CrossRef] [PubMed]
56. Wang, G.; Li, G.; Zhang, S.; Jiang, C.; Qin, J.; Xu, J.R. Activation of the signalling mucin MoMsb2 and its functional relationship with Cbp1 in *Magnaporthe oryzae*. *Environ. Microbiol.* **2015**, *17*, 2969–2981. [CrossRef] [PubMed]
57. Kamakura, T.; Yamaguchi, S.; Saitoh, K.-i.; Teraoka, T.; Yamaguchi, I. A novel gene, CBP1, encoding a putative extracellular chitin-binding protein, may play an important role in the hydrophobic surface sensing of *Magnaporthe grisea* during appressorium differentiation. *Mol. Plant Microbe Interact.* **2002**, *15*, 437–444. [CrossRef]
58. Wagener, J.; Striegler, K.; Wagener, N. α- and β-1,3-Glucan synthesis and remodeling. In *The Fungal Cell Wall: An Armour and a Weapon for Human Fungal Pathogens*; Latgé, J.-P., Ed.; Springer International Publishing: Cham, Switzerland, 2020; pp. 53–82.
59. Samantaray, S.; Neubauer, M.; Helmschrott, C.; Wagener, J. Role of the guanine nucleotide exchange factor Rom2 in cell wall integrity maintenance of *Aspergillus fumigatus*. *Eukaryot. Cell* **2013**, *12*, 288–298. [CrossRef]
60. Kanno, T.; Takekawa, D.; Miyakawa, Y. Analysis of the essentiality of ROM2 genes in the pathogenic yeasts *Candida glabrata* and *Candida albicans* using temperature-sensitive mutants. *J. Appl. Microbiol.* **2015**, *118*, 851–863. [CrossRef]

61. Beauvais, A.; Bruneau, J.M.; Mol, P.C.; Buitrago, M.J.; Legrand, R.; Latgè, J.P. Glucan synthase complex of *Aspergillus fumigatus*. *J. Bacteriol.* **2001**, *183*, 2273–2279. [CrossRef]
62. Dichtl, K.; Ebel, F.; Dirr, F.; Routier, F.H.; Heesemann, J.; Wagener, J. Farnesol misplaces tip-localized Rho proteins and inhibits cell wall integrity signalling in *Aspergillus fumigatus*. *Mol. Microbiol.* **2010**, *76*, 1191–1204. [CrossRef]
63. Kwon, M.J.; Arentshorst, M.; Roos, E.D.; van den Hondel, C.A.; Meyer, V.; Ram, A.F. Functional characterization of Rho GTPases in *Aspergillus niger* uncovers conserved and diverged roles of Rho proteins within filamentous fungi. *Mol. Microbiol.* **2011**, *79*, 1151–1167. [CrossRef] [PubMed]
64. Guest, G.M.; Lin, X.; Momany, M. *Aspergillus nidulans* RhoA is involved in polar growth, branching, and cell wall synthesis. *Fungal Genet. Biol.* **2004**, *41*, 13–22. [CrossRef] [PubMed]
65. Ridley, A.J. Rho GTPases and actin dynamics in membrane protrusions and vesicle trafficking. *Trends Cell Biol.* **2006**, *16*, 522–529. [CrossRef] [PubMed]
66. Martinez-Rocha, A.L.; Roncero, M.I.; Lopez-Ramirez, A.; Marine, M.; Guarro, J.; Martinez-Cadena, G.; Di Pietro, A. Rho1 has distinct functions in morphogenesis, cell wall biosynthesis and virulence of *Fusarium oxysporum*. *Cell. Microbiol.* **2008**, *10*, 1339–1351. [CrossRef]
67. Pham, C.D.; Yu, Z.; Sandrock, B.; Bolker, M.; Gold, S.E.; Perlin, M.H. *Ustilago maydis* Rho1 and 14-3-3 homologues participate in pathways controlling cell separation and cell polarity. *Eukaryot. Cell* **2009**, *8*, 977–989. [CrossRef]
68. Zan, X.Y.; Zhu, H.A.; Jiang, L.H.; Liang, Y.Y.; Sun, W.J.; Tao, T.L.; Cui, F.J. The role of Rho1 gene in the cell wall integrity and polysaccharides biosynthesis of the edible mushroom *Grifola frondosa*. *Int. J. Biol. Macromol.* **2020**, *165*, 1593–1603. [CrossRef]
69. Heinisch, J.J.; Rodicio, R. Protein kinase C in fungi-more than just cell wall integrity. *FEMS Microbiol. Rev.* **2018**, *42*, 22–39. [CrossRef]
70. Schmitz, H.P.; Jockel, J.; Block, C.; Heinisch, J.J. Domain shuffling as a tool for investigation of protein function: Substitution of the cysteine-rich region of Raf kinase and PKC η for that of yeast Pkc1p. *J. Mol. Biol.* **2001**, *311*, 1–7. [CrossRef]
71. Schmitz, H.-P.; Lorberg, A.; Heinisch, J.J. Regulation of yeast protein kinase C activity by interaction with the small GTPase Rho1p through its amino-terminal HR1 domain. *Mol. Microbiol.* **2002**, *44*, 829–840. [CrossRef]
72. Gow, N.A.R.; Latge, J.P.; Munro, C.A. The fungal cell wall: Structure, biosynthesis, and function. *Microbiol. Spectr.* **2017**, *5*, FUNK-0035-2016. [CrossRef]
73. Paravicini, G.; Mendoza, A.; Antonsson, B.; Cooper, M.; Losberger, C.; Payton, M.A. The *Candida albicans* PKC1 gene encodes a protein kinase C homolog necessary for cellular integrity but not dimorphism. *Yeast* **1996**, *12*, 741–756. [CrossRef]
74. Yoshimi, A.; Miyazawa, K.; Abe, K. Cell wall structure and biogenesis in *Aspergillus* species. *Biosci. Biotechnol. Biochem.* **2016**, *80*, 1700–1711. [CrossRef] [PubMed]
75. Herrmann, M.; Sprote, P.; Brakhage, A.A. Protein kinase C (PkcA) of *Aspergillus nidulans* is involved in penicillin production. *Appl. Environ. Microbiol.* **2006**, *72*, 2957–2970. [CrossRef] [PubMed]
76. Ichinomiya, M.; Uchida, H.; Koshi, Y.; Ohta, A.; Horiuchi, H. A protein kinase C-encoding gene, *pkcA*, is essential to the viability of the filamentous fungus *Aspergillus nidulans*. *Biosci. Biotechnol. Biochem.* **2007**, *71*, 2787–2799. [CrossRef] [PubMed]
77. Ronen, R.; Sharon, H.; Levdansky, E.; Romano, J.; Shadkchan, Y.; Osherov, N. The *Aspergillus nidulans pkcA* gene is involved in polarized growth, morphogenesis and maintenance of cell wall integrity. *Curr. Genet.* **2007**, *51*, 321–329. [CrossRef] [PubMed]
78. Teepe, A.G.; Loprete, D.M.; He, Z.; Hoggard, T.A.; Hill, T.W. The protein kinase C orthologue PkcA plays a role in cell wall integrity and polarized growth in *Aspergillus nidulans*. *Fungal Genet. Biol.* **2007**, *44*, 554–562. [CrossRef]
79. Katayama, T.; Uchida, H.; Ohta, A.; Horiuchi, H. Involvement of protein kinase C in the suppression of apoptosis and in polarity establishment in *Aspergillus nidulans* under conditions of heat stress. *PLoS ONE* **2012**, *7*, e50503. [CrossRef]
80. Katayama, T.; Ohta, A.; Horiuchi, H. Protein kinase C regulates the expression of cell wall-related genes in RlmA-dependent and independent manners in *Aspergillus nidulans*. *Biosci. Biotechnol. Biochem.* **2015**, *79*, 321–330. [CrossRef]
81. Rocha, M.C.; Godoy, K.F.; de Castro, P.A.; Hori, J.I.; Bom, V.L.; Brown, N.A.; Cunha, A.F.; Goldman, G.H.; Malavazi, I. The *Aspergillus fumigatus* $pkcA^{G579R}$ mutant is defective in the activation of the cell wall integrity pathway but is dispensable for virulence in a neutropenic mouse infection model. *PLoS ONE* **2015**, *10*, e0135195. [CrossRef]
82. Arpaia, G.; Cerri, F.; Baima, S.; Macino, G. Involvement of protein kinase C in the response of *Neurospora crassa* to blue light. *Mol. Gen. Genet.* **1999**, *262*, 314–322. [CrossRef]
83. Franchi, L.; Fulci, V.; Macino, G. Protein kinase C modulates light responses in *Neurospora* by regulating the blue light photoreceptor WC-1. *Mol. Microbiol.* **2005**, *56*, 334–345. [CrossRef] [PubMed]
84. Penn, T.J.; Wood, M.E.; Soanes, D.M.; Csukai, M.; Corran, A.J.; Talbot, N.J. Protein kinase C is essential for viability of the rice blast fungus *Magnaporthe oryzae*. *Mol. Microbiol.* **2015**, *98*, 403–419. [CrossRef] [PubMed]
85. Sugahara, A.; Yoshimi, A.; Shoji, F.; Fujioka, T.; Kawai, K.; Umeyama, H.; Komatsu, K.; Enomoto, M.; Kuwahara, S.; Hagiwara, D.; et al. Novel antifungal compound Z-705 specifically inhibits protein kinase C of filamentous fungi. *Appl. Environ. Microbiol.* **2019**, *85*, e02923-18. [CrossRef] [PubMed]
86. Tsuji, G.; Kenmochi, Y.; Takano, Y.; Sweigard, J.; Farrall, L.; Furusawa, I.; Horino, O.; Kubo, Y. Novel fungal transcriptional activators, Cmr1p of *Colletotrichum lagenarium* and pig1p of *Magnaporthe grisea*, contain Cys2His2 zinc finger and Zn(II)2Cys6 binuclear cluster DNA-binding motifs and regulate transcription of melanin biosynthesis genes in a developmentally specific manner. *Mol. Microbiol.* **2000**, *38*, 940–954. [CrossRef]

87. Eliahu, N.; Igbaria, A.; Rose, M.S.; Horwitz, B.A.; Lev, S. Melanin biosynthesis in the maize pathogen *Cochliobolus heterostrophus* depends on two mitogen-activated protein kinases, Chk1 and Mps1, and the transcription factor Cmr1. *Eukaryot. Cell* **2007**, *6*, 421–429. [CrossRef]
88. Jung, U.S.; Levin, D.E. Genome-wide analysis of gene expression regulated by the yeast cell wall integrity signalling pathway. *Mol. Microbiol.* **1999**, *34*, 1049–1057. [CrossRef]
89. Martin, H.; Rodriguez-Pachon, J.M.; Ruiz, C.; Nombela, C.; Molina, M. Regulatory mechanisms for modulation of signaling through the cell integrity Slt2-mediated pathway in *Saccharomyces cerevisiae*. *J. Biol. Chem.* **2000**, *275*, 1511–1519. [CrossRef]
90. de Nobel, H.; Ruiz, C.; Martin, H.; Morris, W.; Brul, S.; Molina, M.; Klis, F.M. Cell wall perturbation in yeast results in dual phosphorylation of the Slt2/Mpk1 MAP kinase and in an Slt2-mediated increase in FKS2-lacZ expression, glucanase resistance and thermotolerance. *Microbiology* **2000**, *146 Pt 9*, 2121–2132. [CrossRef]
91. Bermejo, C.; Rodriguez, E.; Garcia, R.; Rodriguez-Pena, J.M.; Rodriguez de la Concepcion, M.L.; Rivas, C.; Arias, P.; Nombela, C.; Posas, F.; Arroyo, J. The sequential activation of the yeast HOG and SLT2 pathways is required for cell survival to cell wall stress. *Mol. Biol. Cell* **2008**, *19*, 1113–1124. [CrossRef]
92. Garcia, R.; Bermejo, C.; Grau, C.; Perez, R.; Rodriguez-Pena, J.M.; Francois, J.; Nombela, C.; Arroyo, J. The global transcriptional response to transient cell wall damage in *Saccharomyces cerevisiae* and its regulation by the cell integrity signaling pathway. *J. Biol. Chem.* **2004**, *279*, 15183–15195. [CrossRef]
93. Garcia, R.; Rodriguez-Pena, J.M.; Bermejo, C.; Nombela, C.; Arroyo, J. The high osmotic response and cell wall integrity pathways cooperate to regulate transcriptional responses to zymolyase-induced cell wall stress in *Saccharomyces cerevisiae*. *J. Biol. Chem.* **2009**, *284*, 10901–10911. [CrossRef] [PubMed]
94. Hayes, B.M.; Anderson, M.A.; Traven, A.; van der Weerden, N.L.; Bleackley, M.R. Activation of stress signalling pathways enhances tolerance of fungi to chemical fungicides and antifungal proteins. *Cell. Mol. Life Sci.* **2014**, *71*, 2651–2666. [CrossRef] [PubMed]
95. Yoshimi, A.; Fujioka, T.; Mizutani, O.; Marui, J.; Hagiwara, D.; Abe, K. Mitogen-activated protein kinases MpkA and MpkB independently affect micafungin sensitivity in *Aspergillus nidulans*. *Biosci. Biotechnol. Biochem.* **2015**, *79*, 836–844. [CrossRef] [PubMed]
96. Manfiolli, A.O.; Siqueira, F.S.; Dos Reis, T.F.; Van Dijck, P.; Schrevens, S.; Hoefgen, S.; Foge, M.; Strassburger, M.; de Assis, L.J.; Heinekamp, T.; et al. Mitogen-activated protein kinase cross-talk interaction modulates the production of melanins in *Aspergillus fumigatus*. *mBio* **2019**, *10*, e00215-19. [CrossRef] [PubMed]
97. Jiang, C.; Zhang, X.; Liu, H.; Xu, J.R. Mitogen-activated protein kinase signaling in plant pathogenic fungi. *PLoS Pathog.* **2018**, *14*, e1006875. [CrossRef]
98. Izumitsu, K.; Yoshimi, A.; Kubo, D.; Morita, A.; Saitoh, Y.; Tanaka, C. The MAPKK kinase ChSte11 regulates sexual/asexual development, melanization, pathogenicity, and adaptation to oxidative stress in *Cochliobolus heterostrophus*. *Curr. Genet.* **2009**, *55*, 439–448. [CrossRef]
99. Dichtl, K.; Samantaray, S.; Wagener, J. Cell wall integrity signalling in human pathogenic fungi. *Cell. Microbiol.* **2016**, *18*, 1228–1238. [CrossRef]
100. de Oliveira, H.C.; Rossi, S.A.; García-Barbazán, I.; Zaragoza, Ó.; Trevijano-Contador, N. Cell wall integrity pathway involved in morphogenesis, virulence and antifungal susceptibility in *Cryptococcus neoformans*. *J. Fungi* **2021**, *7*, 831. [CrossRef]
101. Chen, D.D.; Shi, L.; Yue, S.N.; Zhang, T.J.; Wang, S.L.; Liu, Y.N.; Ren, A.; Zhu, J.; Yu, H.S.; Zhao, M.W. The Slt2-MAPK pathway is involved in the mechanism by which target of rapamycin regulates cell wall components in *Ganoderma lucidum*. *Fungal Genet. Biol.* **2019**, *123*, 70–77. [CrossRef]
102. Lian, L.; Zhang, G.; Zhu, J.; Wang, Y.; Wang, L.; Liu, R.; Shi, L.; Ren, A.; Zhao, M. Swi6B, an alternative splicing isoform of Swi6, mediates the cell wall integrity of *Ganoderma lucidum*. *Environ. Microbiol.* **2021**, *23*, 4405–4417. [CrossRef]
103. Denning, D.W. Echinocandin antifungal drugs. *Lancet* **2003**, *362*, 1142–1151. [CrossRef]
104. Kauffman, C.A.; Carver, P.L. Update on echinocandin antifungals. *Semin. Respir. Crit. Care Med.* **2008**, *29*, 211–219. [CrossRef] [PubMed]
105. Walker, L.A.; Gow, N.A.R.; Munro, C.A. Fungal echinocandin resistance. *Fungal Genet. Biol.* **2010**, *47*, 117–126. [CrossRef] [PubMed]
106. Osada, H. Special issue: Nucleoside antibiotics, polyoxin and beyond. *J. Antibiot* **2019**, *72*, 853–854. [CrossRef] [PubMed]
107. Endo, A.; Misato, T. Polyoxin D, a competitive inhibitor of UDP-N-acetylglucosamine: Chitin N-acetylglycosaminyltransferase in *Neurospora crassa*. *Biochem. Biophys. Res. Commun.* **1969**, *37*, 718–722. [CrossRef]
108. Endo, A.; Kakiki, K.; Misato, T. Mechanism of action of the antifugal agent polyoxin D. *J. Bacteriol.* **1970**, *104*, 189–197. [CrossRef]
109. Isono, K.; Nagatsu, J.; Kawashima, Y.; Suzuki, S. Studies on polyoxins, antifungal antibiotics: Part I. Isolation and characterization of polyoxins A and B. *Agric. Biol. Chem.* **1965**, *29*, 848–854. [CrossRef]
110. Isono, K.; Asahi, K.; Suzuki, S. Polyoxins, antifungal antibiotics. XIII. Structure of polyoxins. *J. Am. Chem. Soc.* **1969**, *91*, 7490–7505. [CrossRef]
111. Serpi, M.; Ferrari, V.; Pertusati, F. Nucleoside Derived Antibiotics to Fight Microbial Drug Resistance: New Utilities for an Established Class of Drugs? *J. Med. Chem.* **2016**, *59*, 10343–10382. [CrossRef]
112. Fiedler, H.-P.; Kurth, R.; Langhärig, J.; Delzer, J.; Zähner, H. Nikkomycins: Microbial inhibitors of chitin synthase. *J. Chem. Technol. Biotechnol.* **1982**, *32*, 271–280. [CrossRef]
113. Wood, P.J. Specificity in the interaction of direct dyes with polysaccharides. *Carbohydr. Res.* **1980**, *85*, 271–287. [CrossRef]

114. Ram, A.F.J.; Klis, F.M. Identification of fungal cell wall mutants using susceptibility assays based on Calcofluor white and Congo red. *Nat. Protoc.* **2006**, *1*, 2253–2256. [CrossRef] [PubMed]
115. Yoshimi, A.; Sano, M.; Inaba, A.; Kokubun, Y.; Fujioka, T.; Mizutani, O.; Hagiwara, D.; Fujikawa, T.; Nishimura, M.; Yano, S.; et al. Functional analysis of the α-1,3-glucan synthase genes *agsA* and *agsB* in *Aspergillus nidulans*: AgsB is the major α-1,3-glucan synthase in this fungus. *PLoS ONE* **2013**, *8*, e54893. [CrossRef] [PubMed]
116. Liu, Z.; Raj, S.; van Rhijn, N.; Fraczek, M.; Michel, J.P.; Sismeiro, O.; Legendre, R.; Varet, H.; Fontaine, T.; Bromley, M.; et al. Functional genomic and biochemical analysis reveals pleiotropic effect of congo red on *Aspergillus fumigatus*. *mBio* **2021**, *12*, e00863-21. [CrossRef] [PubMed]
117. Lawry, S.M.; Tebbets, B.; Kean, I.; Stewart, D.; Hetelle, J.; Klein, B.S. Fludioxonil induces Drk1, a fungal group III hybrid histidine kinase, to dephosphorylate its downstream target, Ypd1. *Antimicrob. Agents Chemother.* **2017**, *61*, e01414–e01416. [CrossRef]
118. Brandhorst, T.T.; Kean, I.R.L.; Lawry, S.M.; Wiesner, D.L.; Klein, B.S. Phenylpyrrole fungicides act on triosephosphate isomerase to induce methylglyoxal stress and alter hybrid histidine kinase activity. *Sci. Rep.* **2019**, *9*, 5047. [CrossRef]
119. Yoshimi, A.; Hagiwara, D.; Ono, M.; Fukuma, Y.; Midorikawa, Y.; Furukawa, K.; Fujioka, T.; Mizutani, O.; Sato, N.; Miyazawa, K.; et al. Downregulation of the *ypdA* gene encoding an intermediate of His-Asp phosphorelay signaling in *Aspergillus nidulans* induces the same cellular effects as the phenylpyrrole fungicide fludioxonil. *Front. Fungal Biol.* **2021**, *2*, 675459. [CrossRef]
120. Kojima, K.; Takano, Y.; Yoshimi, A.; Tanaka, C.; Kikuchi, T.; Okuno, T. Fungicide activity through activation of a fungal signalling pathway. *Mol. Microbiol.* **2004**, *53*, 1785–1796. [CrossRef]
121. Beattie, S.R.; Krysan, D.J. A unique dual-readout high-throughput screening assay to identify antifungal compounds with *Aspergillus fumigatus*. *mSphere* **2021**, *6*, e00539-21. [CrossRef]
122. Fujikawa, T.; Kuga, Y.; Yano, S.; Yoshimi, A.; Tachiki, T.; Abe, K.; Nishimura, M. Dynamics of cell wall components of *Magnaporthe grisea* during infectious structure development. *Mol. Microbiol.* **2009**, *73*, 553–570. [CrossRef]
123. Fujikawa, T.; Sakaguchi, A.; Nishizawa, Y.; Kouzai, Y.; Minami, E.; Yano, S.; Koga, H.; Meshi, T.; Nishimura, M. Surface α-1,3-glucan facilitates fungal stealth infection by interfering with innate immunity in plants. *PLoS Pathog.* **2012**, *8*, e1002882. [CrossRef] [PubMed]
124. Henry, C.; Latgè, J.P.; Beauvais, A. α1,3 Glucans are dispensable in *Aspergillus fumigatus*. *Eukaryot. Cell* **2012**, *11*, 26–29. [CrossRef] [PubMed]
125. Beauvais, A.; Bozza, S.; Kniemeyer, O.; Formosa, C.; Balloy, V.; Henry, C.; Roberson, R.W.; Dague, E.; Chignard, M.; Brakhage, A.A.; et al. Deletion of the α-(1,3)-glucan synthase genes induces a restructuring of the conidial cell wall responsible for the avirulence of *Aspergillus fumigatus*. *PLoS Pathog.* **2013**, *9*, e1003716. [CrossRef]
126. He, X.X.; Li, S.N.; Kaminskyj, S.G.W. Characterization of *Aspergillus nidulans* α-glucan synthesis: Roles for two synthases and two amylases. *Mol. Microbiol.* **2014**, *91*, 579–595. [CrossRef]
127. Miyazawa, K.; Yoshimi, A.; Kasahara, S.; Sugahara, A.; Koizumi, A.; Yano, S.; Kimura, S.; Iwata, T.; Sano, M.; Abe, K. Molecular mass and localization of α-1,3-glucan in cell wall control the degree of hyphal aggregation in liquid culture of *Aspergillus nidulans*. *Front. Microbiol.* **2018**, *9*, 2623. [CrossRef]
128. He, X.; Li, S.; Kaminskyj, S. An amylase-like protein, AmyD, is the major negative regulator for α-glucan synthesis in *Aspergillus nidulans* during the asexual life cycle. *Int. J. Mol. Sci.* **2017**, *18*, 695. [CrossRef]
129. Miyazawa, K.; Yamashita, T.; Takeuchi, A.; Kamachi, Y.; Yoshimi, A.; Tashiro, Y.; Koizumi, A.; Ogata, M.; Yano, S.; Kasahara, S.; et al. A glycosylphosphatidylinositol-anchored α-amylase encoded by *amyD* contributes to a decrease in the molecular mass of cell wall α-1,3-glucan in *Aspergillus nidulans*. *Front. Fungal Biol.* **2022**, *2*, 821946. [CrossRef]
130. Fontaine, T.; Beauvais, A.; Loussert, C.; Thevenard, B.; Fulgsang, C.C.; Ohno, N.; Clavaud, C.; Prevost, M.C.; Latgè, J.P. Cell wall α1-3glucans induce the aggregation of germinating conidia of *Aspergillus fumigatus*. *Fungal Genet. Biol.* **2010**, *47*, 707–712. [CrossRef]
131. Abe, K.; Gomi, K.; Yoshimi, A. Method for Manuifacturing Useful Substance in Which High-Density Cultured Strain of Filamentous Fungi Is Used. U.S. Patent 11,015,175, 25 May 2021.
132. Zhang, S.; Sato, H.; Ichinose, S.; Tanaka, M.; Miyazawa, K.; Yoshimi, A.; Abe, K.; Shintani, T.; Gomi, K. Cell wall α-1,3-glucan prevents α-amylase adsorption onto fungal cell in submerged culture of *Aspergillus oryzae*. *J. Biosci. Bioeng.* **2017**, *124*, 47–53. [CrossRef]
133. Miyazawa, K.; Yoshimi, A.; Zhang, S.; Sano, M.; Nakayama, M.; Gomi, K.; Abe, K. Increased enzyme production under liquid culture conditions in the industrial fungus *Aspergillus oryzae* by disruption of the genes encoding cell wall α-1,3-glucan synthase. *Biosci. Biotechnol. Biochem.* **2016**, *80*, 1853–1863. [CrossRef]
134. Jeennor, S.; Anantayanon, J.; Panchanawaporn, S.; Chutrakul, C.; Laoteng, K. Morphologically engineered strain of *Aspergillus oryzae* as a cell chassis for production development of functional lipids. *Gene* **2019**, *718*, 144073. [CrossRef] [PubMed]
135. Tokashiki, J.; Hayashi, R.; Yano, S.; Watanabe, T.; Yamada, O.; Toyama, H.; Mizutani, O. Influence of α-1,3-glucan synthase gene *agsE* on protoplast formation for transformation of *Aspergillus luchuensis*. *J. Biosci. Bioeng.* **2019**, *128*, 129–134. [CrossRef]
136. Miyazawa, K.; Umeyama, T.; Hoshino, Y.; Abe, K.; Miyazaki, Y. Quantitative monitoring of mycelial growth of *Aspergillus fumigatus* in liquid culture by optical density. *Microbiol. Spectr.* **2022**, *10*, e00063-21. [CrossRef] [PubMed]
137. Speth, C.; Rambach, G.; Lass-Flörl, C.; Howell, P.L.; Sheppard, D.C. Galactosaminogalactan (GAG) and its multiple roles in *Aspergillus* pathogenesis. *Virulence* **2019**, *10*, 976–983. [CrossRef] [PubMed]

138. Miyazawa, K.; Yoshimi, A.; Sano, M.; Tabata, F.; Sugahara, A.; Kasahara, S.; Koizumi, A.; Yano, S.; Nakajima, T.; Abe, K. Both galactosaminogalactan and α-1,3-glucan contribute to aggregation of *Aspergillus oryzae* hyphae in liquid culture. *Front. Microbiol.* **2019**, *10*, 2090. [CrossRef] [PubMed]
139. Abe, K.; Yoshimi, A.; Miyazawa, K.; Tabata, F.; Gomi, K.; Sano, M. Mutant Filamentous Fungus and Substance Production Method in Which Said Mutant Filamentous Fungus Is Used. U.S. Patent 11,021,725, 1 June 2021.
140. Mei, L.; Wang, X.; Yin, Y.; Tang, G.; Wang, C. Conservative production of galactosaminogalactan in *Metarhizium* is responsible for appressorium mucilage production and topical infection of insect hosts. *PLoS Pathog.* **2021**, *17*, e1009656. [CrossRef] [PubMed]
141. Miyazawa, K.; Yoshimi, A.; Abe, K. The mechanisms of hyphal pellet formation mediated by polysaccharides, α-1,3-glucan and galactosaminogalactan, in *Aspergillus* species. *Fungal Biol. Biotechnol.* **2020**, *7*, 10. [CrossRef]
142. Gravelat, F.N.; Ejzykowicz, D.E.; Chiang, L.Y.; Chabot, J.C.; Urb, M.; Macdonald, K.D.; al-Bader, N.; Filler, S.G.; Sheppard, D.C. *Aspergillus fumigatus* MedA governs adherence, host cell interactions and virulence. *Cell. Microbiol.* **2010**, *12*, 473–488. [CrossRef]
143. Chen, Y.; Le Mauff, F.; Wang, Y.; Lu, R.; Sheppard, D.C.; Lu, L.; Zhang, S. The transcription factor SomA synchronously regulates biofilm formation and cell wall homeostasis in *Aspergillus fumigatus*. *mBio* **2020**, *11*, e02329-20. [CrossRef]
144. Lee, M.J.; Geller, A.M.; Bamford, N.C.; Liu, H.; Gravelat, F.N.; Snarr, B.D.; Le Mauff, F.; Chabot, J.; Ralph, B.; Ostapska, H.; et al. Deacetylation of fungal exopolysaccharide mediates adhesion and biofilm formation. *mBio* **2016**, *7*, e00252-16. [CrossRef]
145. Antecka, A.; Blatkiewicz, M.; Bizukojc, M.; Ledakowicz, S. Morphology engineering of basidiomycetes for improved laccase biosynthesis. *Biotechnol. Lett.* **2016**, *38*, 667–672. [CrossRef] [PubMed]
146. Zhang, J.; Zhang, J. The filamentous fungal pellet and forces driving its formation. *Crit. Rev. Biotechnol.* **2016**, *36*, 1066–1077. [CrossRef] [PubMed]
147. Antecka, A.; Bizukojc, M.; Ledakowicz, S. Modern morphological engineering techniques for improving productivity of filamentous fungi in submerged cultures. *World J. Microbiol. Biotechnol.* **2016**, *32*, 193. [CrossRef] [PubMed]
148. Driouch, H.; Hansch, R.; Wucherpfennig, T.; Krull, R.; Wittmann, C. Improved enzyme production by bio-pellets of *Aspergillus niger*: Targeted morphology engineering using titanate microparticles. *Biotechnol. Bioeng.* **2012**, *109*, 462–471. [CrossRef]
149. Ichikawa, H.; Miyazawa, K.; Komeiji, K.; Susukida, S.; Zhang, S.; Muto, K.; Orita, R.; Takeuchi, A.; Kamachi, Y.; Hitosugi, M.; et al. Improved recombinant protein production in *Aspergillus oryzae* lacking both α-1,3-glucan and galactosaminogalactan in batch culture with a lab-scale bioreactor. *J. Biosci. Bioeng.* **2022**, *133*, 39–45. [CrossRef]
150. Sakuragawa, T.; Wakai, S.; Zhang, S.; Kawaguchi, H.; Ogino, C.; Kondo, A. Accelerated glucose metabolism in hyphae-dispersed *Aspergillus oryzae* is suitable for biological production. *J. Biosci. Bioeng.* **2021**, *132*, 140–147. [CrossRef]
151. Cairns, T.C.; Zheng, X.M.; Zheng, P.; Sun, J.B.; Meyer, V. Moulding the mould: Understanding and reprogramming filamentous fungal growth and morphogenesis for next generation cell factories. *Biotechnol. Biofuels* **2019**, *12*, 77. [CrossRef]
152. Yin, C.; Wang, B.; He, P.; Lin, Y.; Pan, L. Genomic analysis of the aconidial and high-performance protein producer, industrially relevant *Aspergillus niger* SH2 strain. *Gene* **2014**, *541*, 107–114. [CrossRef]
153. Sun, X.W.; Wu, H.F.; Zhao, G.H.; Li, Z.M.; Wu, X.H.; Liu, H.; Zheng, Z.M. Morphological regulation of *Aspergillus niger* to improve citric acid production by *chsC* gene silencing. *Bioprocess Biosyst. Eng.* **2018**, *41*, 1029–1038. [CrossRef]
154. Yin, X.; Shin, H.D.; Li, J.; Du, G.; Liu, L.; Chen, J. Comparative genomics and transcriptome analysis of *Aspergillus niger* and metabolic engineering for citrate production. *Sci. Rep.* **2017**, *7*, 41040. [CrossRef]
155. Liu, H.; Zheng, Z.; Wang, P.; Gong, G.; Wang, L.; Zhao, G. Morphological changes induced by class III chitin synthase gene silencing could enhance penicillin production of *Penicillium chrysogenum*. *Appl. Microbiol. Biotechnol.* **2013**, *97*, 3363–3372. [CrossRef] [PubMed]
156. Lin, L.; Sun, Z.; Li, J.; Chen, Y.; Liu, Q.; Sun, W.; Tian, C. Disruption of *gul-1* decreased the culture viscosity and improved protein secretion in the filamentous fungus *Neurospora crassa*. *Microb. Cell Fact.* **2018**, *17*, 96. [CrossRef] [PubMed]
157. Fiedler, M.R.; Lorenz, A.; Nitsche, B.M.; van den Hondel, C.A.; Ram, A.F.; Meyer, V. The capacity of *Aspergillus niger* to sense and respond to cell wall stress requires at least three transcription factors: RlmA, MsnA and CrzA. *Fungal Biol. Biotechnol.* **2014**, *1*, 5. [CrossRef] [PubMed]

Journal of *Fungi*

MDPI

Review

Fungal Cell Wall Proteins and Signaling Pathways Form a Cytoprotective Network to Combat Stresses

Chibuike Ibe [1,*] and Carol A. Munro [2]

1 Department of Microbiology, Faculty of Biological Sciences, Abia State University, Uturu 441107, Nigeria
2 Aberdeen Fungal Group, Institute of Medical Sciences, School of Medicine, Medical Sciences and Nutrition, University of Aberdeen, Aberdeen AB24 3FX, UK; c.a.munro@abdn.ac.uk
* Correspondence: c.ibe@abiastateuniversity.edu.ng

Abstract: *Candida* species are part of the normal flora of humans, but once the immune system of the host is impaired and they escape from commensal niches, they shift from commensal to pathogen causing candidiasis. *Candida albicans* remains the primary cause of candidiasis, accounting for about 60% of the global candidiasis burden. The cell wall of *C. albicans* and related fungal pathogens forms the interface with the host, gives fungal cells their shape, and also provides protection against stresses. The cell wall is a dynamic organelle with great adaptive flexibility that allows remodeling, morphogenesis, and changes in its components in response to the environment. It is mainly composed of the inner polysaccharide rich layer (chitin, and β-glucan) and the outer protein coat (mannoproteins). The highly glycosylated protein coat mediates interactions between *C. albicans* cells and their environment, including reprograming of wall architecture in response to several conditions, such as carbon source, pH, high temperature, and morphogenesis. The mannoproteins are also associated with *C. albicans* adherence, drug resistance, and virulence. Vitally, the mannoproteins contribute to cell wall construction and especially cell wall remodeling when cells encounter physical and chemical stresses. This review describes the interconnected cell wall integrity (CWI) and stress-activated pathways (e.g., Hog1, Cek1, and Mkc1 mediated pathways) that regulates cell wall remodeling and the expression of some of the mannoproteins in *C. albicans* and other species. The mannoproteins of the surface coat is of great importance to pathogen survival, growth, and virulence, thus understanding their structure and function as well as regulatory mechanisms can pave the way for better management of candidiasis.

Keywords: fungi; cell wall; cell wall proteins; signaling pathways; stress tolerance

Citation: Ibe, C.; Munro, C.A. Fungal Cell Wall Proteins and Signaling Pathways Form a Cytoprotective Network to Combat Stresses. *J. Fungi* **2021**, *7*, 739. https://doi.org/10.3390/jof7090739

Academic Editors: María Molina and Humberto Martín

Received: 29 July 2021
Accepted: 4 September 2021
Published: 8 September 2021

1. Introduction

Candida albicans is abundantly found in mammals. It often resides on the skin and mucosal layers of individuals as part of their normal flora. *C. albicans* causes a range of infections from superficial to life-threatening and systemic, dependent upon the host's immune system [1] *C. albicans* uses an arsenal of pathogenic mechanisms to subdue or evade host immune responses [2,3]. The mannosylated surface protein coat is covalently linked to the skeletal cell wall polysaccharides and plays a vital role in mediating *C. albicans* interaction with the host. The proteins are not only important in maintaining cell wall integrity, masking the polysaccharide rich layer, therefore preventing recognition by dectin-1, but also contribute to virulence of this pathogen in many ways. They mediate adherence to host cells and indwelling medical devices, enable invasion of epithelial cells, facilitate biofilm formation, protect *C. albicans* against immune attack, coordinate communication between host cells and *C. albicans*, and are important in nutrient scavenging including zinc and iron [3]. Given the important roles of the cell surface proteins at every stage of *C. albicans* infection process, research has been focused on expanding our understanding of their biology and structure as well as their function in the cell wall [4]. This area is, however, rapidly expanding as the cell surface proteins have the potential to be a unique drug/vaccine target [5–7]. Proteomics analysis of purified cell wall material has shown that the walls contain about 20 different types of covalently bound cell wall proteins (CWPs) at any time and the protein profiles can change dramatically depending on the growth conditions [8]. In addition, the presence of particular cell surface proteins morphologically depends and correlates with either *C. albicans* yeast or hyphal form [8]. The aim of this review is to discuss the characteristics and functions of covalently bound CWPs, and how they are important for fitness and virulence, and enable the fungus to cope with host infection-induced stress conditions. The review will also discuss the regulatory mechanisms that control expression of cell wall-related genes and relate what is known in *C. albicans* and other *Candida* species.

2. Function of Cell Wall Proteins

Based on the existing model of the cell wall, it is made up of an inner polysaccharide rich layer and the outer protein coat [9–11]. A 3-D nanoscalar model of the *C. albicans* cell wall has been developed to probe accurate thickness and structure of the cell wall [4,10]. The investigators used an optimized 3-D electron tomogram and computer vision technique to make accurate measurements of cell wall thickness [4]. The scalar model developed gave a more refined prediction of the thickness of each cell wall layer and the precise structure of some of the wall components [4]. The inner layer of the cell wall is composed mainly of β-glucans (β-1,3-glucan and β-1,6-glucans), chitin microfibrils, and a small amount of mannosylated proteins is distributed throughout the inner layer [4]. Chitin (β-1,4-*N*-acetyl glucosamine) and β-glucans (β-1,3-glucan) are the main structural polysaccharides of the cell wall [12]. β-1,3-glucan forms a three-dimensional network comparable to a flexible wire spring, which explains the elastic nature of the cell wall and provides the platform for the attachment of β-1,6-glucan, CWPs, and chitin [13]. Chitin is covalently cross-linked to the β-1,3-glucan network and contributes to the rigidity and physical strength of the fungal cell wall [12]. The outer coat is made up of glycan fibrils post-translationally attached to CWPs that are vertically arranged perpendicular to the inner layer [4].

The outer coat of mannoproteins determines cell wall permeability and surface charge [9]. Restriction of cell wall permeability is due to the densely packed CWPs, the presence of bulky protein sidechains, and the formation of disulfide bridges between CWPs [12,14,15]. This feature protects the structural polysaccharides against enzyme degradation and dectin-1 receptor recognition [15,16]. The use of genomic and proteomic techniques has advanced our knowledge of the nature and abundance of these surface proteins. CWPs have a unique structure, they generally contain: an N terminus with a secretory motif and a C terminus [17]. They bear serine/threonine-*O*-manno-oligosaccharide and/or asparagine-*N*-glycan and may contain internal repeats and/or a glycosylphosphatidylinositol, GPI anchor attachment sequences [18,19]. The most abundant cell proteins are the GPI-modified proteins, which receive a GPI anchor during their passage through the secretory pathway [20–24] and constitute about 88% of the total wall mannosylated protein classes [25] (Table 1). The second class of CWPs are those with internal repeats, PIR-CWPs [18,26] (Table 1).

Table 1. Characteristics of specific *Candida albicans* surface proteins.

Protein/Family	Features and Functions	Regulation
GPI modified CWPs		
Adhesins, invasins **Als family** (Als1-7, 9)	N-terminal c. 300-residue Ig domain, bind variety of substrates [27,28]; high (Als1, Als2), intermediate (Als4, Als9), and low (Als5-7) levels of gene expression [27,29–31]. Als3 is expressed uniformly all over hyphae [32]. Als1 and Als3 N terminal sequences are used as vaccine antigen [7]. Als1 and Als3 contribute to biofilm formation, and Als3 functions as an invasin, and as a ferritin receptor [27,30,31,33].	Als proteins are differentially expressed, *ALS1*, *ALS3*, and *HWP1* are under the positive regulatory control of Bcr1 [34–36]. Tup1 (repressor of filamentation] and Ahr1 are required for full expression of *ALS3* [37].
Hwp1, Hwp2, Eap1, Ihd1, and Hyr1	Hwp1 level is induced by oxygen and iron restriction [38]. N terminal is recognised as substrate for epithelial transglutaminases [39]. Hwp1 facilitates cell to cell interaction important in biofilm development [33]. N terminal 14-mer peptide and recombinant N terminal fragment are used in vaccine and diagnostic development, respectively [40,41]. Hwp2 has sequence identity with Hwp1 and can function in adhesion and invasion; it is also involved in oxidative stress tolerance and protein aggregation [42,43]. Hwp1, Hwp2, Eap1, and Ihd1 contribute to initial cell attachment and adhesion maintenance during biofilm formation [44]. N terminal of Hyr1 has been used in vaccines and diagnostics development [5,45].	
Carbohydrate active enzymes		
1,3-β-Glucan processing Phr1-3, Pga4, and Pga5	N terminal glycoside hydrolase (GH) 72 domain; play a role in cell wall construction (β-1,3-glucan modification); incorporated at acidic pH (Phr2) and neutral/alkaline pH (Phr1) [45,46]. Pga4 is transcribed independent of pH, and Phr3 and Pga5 have low expression levels [47]. Pga4 is serum- and host infection-inducible [48].	*PHR1* and *PHR2* are differentially regulated by extracellular pH [49].
Chitin-glucan cross-linkersChr family	N terminal GH16 domain; involved in cell wall organization and integrity; cross-linking β-1,3-glucan and chitin; involved in protoplast regeneration [50,51]. Control cell wall elasticity in osmotic resistance [52].	*UTR2* expression is regulated by calcineurin and Crz1 [53]. *CRH11* is subject to caspofungin-induced Cas5 regulation [54]
Others **Dfg5 and Dcw1**	Putative glycosyltransferase enzyme activity; involved in the incorporation of GPI anchored proteins into the cell wall [55,56]. Dfg5 and Dcw1 are involved in hyphal morphogenesis and biofilm formation; Dfg5 is required for growth; Dcw1 is required for cell wall integrity response; Dfg5 has synthetic lethality with Dcw1 [56].	*DFG5* has been shown to be regulated by Rlm1 in *S. cerevisaie*, but not in *C. albicans* [54].
Pga31-like (Pga29-31)	Enriched in pathogenic fungi [57]. Pga31 has predicted transmembrane domain and with Pga30 they have three conserved cysteine residues (http://www.candidagenome.org/ (accessed on 12 July 2021)). Pga29 and Pga31 are echinocandin induced; Pga29 is required for normal cell surface property [58]. Pga31 is induced during protoplast regeneration [59] and may be involved in cell wall chitin synthesis during remodelling in response to stress [60].	*PGA31* is upregulated by the Pkc pathway [60].
Sod4 and Sod5	Superoxide dismutase; contribute to combating oxidative stress by clearing reactive oxygen species [61].	Rim101 is required for induction of *SOD5* under certain conditions, and Efg1 is required specifically for serum-modulated expression [62].
Sap9 and Sap10	Yapsin-like proteins are mainly found in the cell membrane (Sap9) and cell wall (Sap10); required for full cell wall integrity [63].	
Pga59	Cell wall localised [64]; abundant in the cell wall protein coat; mature protein consists of three cysteine residues and cross-links cell wall proteins through disulphide bridges [64].	
Rbt5, Pga10, and Pga7	N terminal CFEM domain; cell membrane (Pga7 and Rbt5) localised; loss of function results in fragile biofilms (Pga10 and Rbt5) [65,66]; function as haeme receptors and involved in haeme-iron utilization [67,68]. Rbt5 levels increase following iron and oxygen restriction [38].	The proteins have been shown to be expressed during yeast to hyphae switch and thus are regulated by Tup1 [66].
Non-GPI modified CWPs		
Pir1	C terminal conserved four cysteine pattern and seven repeats; predicted to cross-link β-1,3-glucan chains [69]; protein levels increase in hypoxic conditions [38].	
Mp65	C terminal GH17z domain; present in fibrillar material with putative transglycosylase activity; potential vaccine candidate [70].	

A 3-D electron tomogram was used to determine the structure of the outer coat of mannosylated proteins. The scalar architectural model of mannosylated proteins gave a more precise detail of their structure, location and molecular size including measurements of their length and branching [4,71]. The cell wall through the outer proteins mediates host pathogen-interaction. The scalar architectural model may be useful in investigating the structure–function relationships that support the fungal infection strategy [4,10].

CWPs have both enzymatic and structural functions and their population may differ in their abundances depending on environmental conditions, developmental stage and phase of the cell cycle [9]. During cell wall synthesis, the cell wall polysaccharides, chitin and β-1,3-glucan are synthesized by enzymes localized in the plasma membrane and are extruded out to the cell exterior and are then acted upon by wall-localized cell wall remodeling enzymes [17]. CWPs modify these cell wall polysaccharides and cross-link them, thus maintaining cell wall integrity [9]. The cross-linking between cell wall macromolecules extruded into the wall space is catalyzed by carbohydrate active cell wall remodeling enzymes, hydrolases, transferases, and transglycosidases that are located in the cell wall space [17,72]. Some of these enzymes include β-1,3-glucanosyltransferases, e.g., Phr family (see Table 1), which are a Gas-like family, Bgl2 (GH17), and Crh family representing chitin-glucanosyltransferases, these are cell wall-localized GPI anchored proteins [17,50,51,73–76]. *C. albicans phr1Δ/Δ* and *phr2Δ/Δ* mutants showed hypersensitivity to cell wall stressors such as Calcofluor white, CFW [45]. In *C. albicans*, synthetically lethal GPI-anchored proteins such as Dfg5 and Dcw1 (glycoside hydrolase (GH) family 76) are required for the incorporation of mannosylated proteins into the cell wall [55].

Structural surface proteins with no enzymatic activities such as flocculins (e.g., Flo1, Pga24), agglutinins (e.g., Als1, Rbt1, Hwp1), or β-1,3-glucan cross connectors (e.g., Pir1) that can form a scaffold for the attachment of other wall components, are important for cell:cell interactions and wall integrity [27,30,31,33,77–81]. Ssr1, a structural protein has been shown to contribute to normal cell wall architecture [82]. Pga59 is thought to be associated with the formation of a coat around the cell wall that can restrict cell wall permeability [64]. CWPs are also associated with virulence, biofilm formation, and coping with stress in fungi [33,36,61,67,68,83–85]. The following are some examples. The *ALS* gene family encodes eight GPI modified cell surface glycoproteins with peptide binding ability Ig-fold domain at the N terminus [86]. The Ig-fold mediates adhesion to fibronectin and other specific host proteins [87], and cell to cell aggregation through Als to Als interaction (Nobile et al., 2008). Heterologous expression of Als proteins in a nonadherent *S. cerevisiae* strain demonstrated that the Als proteins promote attachment to different surfaces (Nobile et al., 2008). The *als3Δ/Δ* mutant has reduced virulence in a murine model of oropharyngeal candidiasis [31]. Hwp1 N terminus contains a secretory signal sequence rich in proline and glutamine that is cross-linked by host transglutaminase to epithelial cells enabling the attachment of *C. albicans* to human buccal epithelial cells [39,88,89]. *C. albicans hwp1Δ* mutant has reduced ability to bind to human buccal epithelial cells and has poor translocation from the mouse intestine into the bloodstream, demonstrating a role for Hwp1 in disseminated candidiasis [31]. Attachment to host cells by *C. albicans* can also be due to morphology-independent covalently bound wall proteins, Hyr1, Ecm33, Iff4, and Eap1, covalently bound wall proteins, Phr1, and cell-surface associated proteases, Sap9 and Sap10 [90,91].

C. albicans can use endocytosis (through interaction of Als3 with host cadherins) or active penetration to invade the host cell [31]. After *C. albicans* adhesion to the host cell surface and hyphal germination and growth, there are hyphal-induced hydrolytic enzymes that facilitate host cell degradation. They particularly aid active penetration into host cells and damage tissues [92].

C. albicans expresses ten secreted aspartyl proteinase (Sap) isoenzymes. Each mature Sap protein contains two aspartic acid residues conserved within the active site and a conserved cysteine residue that plays a structural role. Sap1–8 are secreted and released to the environment, whereas Sap9 and Sap10 are cell surface bound [63,93,94]. Sap proteins have been linked to the ability of *C. albicans* to adhere to and damage host tissue as well as the ability to evade the host immune response [95]. Sap9 and Sap10 have proteolytic activity on non-basic, basic, and dibasic peptides and have targeted Cht2, Ecm33, Pga4, Ywp1, Als2, Rhd3, Rbt5, and glucan cross-linked protein, Pir1 as substrates. *C. albicans sap9*Δ/Δ and *sap10*Δ/Δ mutants demonstrated reduction in cell wall-associated Cht2 activity suggesting a direct influence of Sap9 and Sap10 activity on Cht2 function and a role in maintaining cell wall integrity [96].

During *C. albicans* infection, Als family, and Eap1 adhesin, are involved in the development of biofilms, an important virulent attribute. The fungus forms biofilms when it encounters solid surfaces such as indwelling medical devices, where fungal cells are encapsulated in a dense extracellular matrix, which sequesters antifungal drugs promoting drug resistance and persistence in the host [97,98]. Biofilm formation by *C. albicans* has been shown to be under the positive regulatory control of the transcription factor, Bcr1. Bcr1 regulates the expression of Als1, Als3, and Hwp1 [34–36]. These proteins in addition to Als2 are associated with various stages of biofilm formation in *C. albicans* [28,99].

The cell wall during growth requires continuous remodeling of its macromolecular network [17]. During cell wall stress, a fungus can also rapidly remodel its wall and adapt the composition of the new cell wall [52,73,100]. For example, in exposure to cell damaging antifungal drugs, *C. albicans* triggers cell wall rescue mechanisms that influence the expression of wall biosynthetic genes and CWPs [4,60,101]. Rescuing the cell wall requires stress signals that activate the cell wall integrity (CWI) pathways. Cell surface proteins that function as mechanosensors primarily are responsible for activating these CWI pathways. These proteins (Wsc1-3 and Mtl1) act like linear nanosprings that detect and transmit cell wall damage or stress [102–105] to the downstream receptors in the signaling pathways. The sensors have an overall similar structure in that they contain in their sequences: short C terminal cytoplasmic domains, a single transmembrane domain, and a periplasmic ectodomain that is rich in Ser/Thr residues [106]. The Ser/Thr-rich regions are highly *O*-mannosylated, accounting for extension and stiffening of the proteins. Thus, these polypeptides have been proposed as mechanosensors that act as rigid probes of the extracellular matrix [106,107]. Functionally, signals are received and transmitted through the highly *O*-mannosylated extracellular domains and phosphatidylinositol (PI)-4,5-bisphosphate, which recruits the N terminal domains of the Rom1/2-guanine nucleotide exchange factors through the plasma membrane, the sensors stimulate nucleotide exchange on Rho1 [102,105]. The various effectors of Rho1 include β-1,3-glucan synthase, β-1,3-glucan synthase activity, and Pkc1-activated MAPK cascade [104].

In summary, CWPs have a wide range of diverse functions that contribute to virulence, to maintenance of wall structure to ensure cellular integrity remains intact, and to sensing and transmitting signals from the environment. Many CWPs have been functionally characterized and their amino acid sequences are known, but only a handful have had their structures fully elucidated. Structure has a functional implication and understanding CWP structure can increase our knowledge of their functions, including roles in cell wall biogenesis.

3. Fungal Cell Wall Remodeling and Signaling Pathways That Are Activated in Response to Stress

C. albicans has been shown to grow at a high concentration of caspofungin a phenomenon called paradoxical growth. Paradoxical growth, in *C. albicans* is associated with induced cell aggregation and an increase in cell volume and cell wall chitin content [108]. In *C. auris*, however, it only induced an increase in cell wall chitin content [108]. Genes encoding Fks1 and Fks2 harboring the single nucleotide polymorphisms hot spot regions have been identified in *C. auris* [108]. The Fks2 carries the F635Y mutation that confers intrinsic echinocandin resistance on *Candida glabrata* [108]. Interestingly, *C. auris* RNA-seq data showed that paradoxical growth activates genes encoding cell membrane proteins and GPI-modified proteins required for cell wall damage response, chitin synthase, and

MAPKs such as Mkc1, and Hog1 involved in maintaining cell wall integrity [108]. Fungal pathogens activate a lot of pathways to successfully adapt to caspofugin stress.

Deletion of cell wall biosynthetic pathway genes in fungi often results in increased susceptibility of the cell wall to wall perturbing agents as well as alterations in chitin and β-1,3-glucan contents and linkages in the cell wall, synthesis of new wall proteins, and changes in the crosslinking to alternative wall polysaccharides [109,110]. Inhibition of β-1,3-glucan synthesis has been associated with altered crosslinking of chitin to β-1,6-glucan-GPI-modified proteins in the cell wall [109]. The amount of chitin→β-glucan←GPI-CWP complexes in the cell wall increased to 40% in wall defective mutants, indicating this is a repair mechanism protecting the cell wall from degrading enzymes and other stresses [109]. Most cell wall restructuring processes do not involve activation of the signaling pathways. For example, the carbon-source-induced alteration in osmotic tolerance in *C. albicans* was shown to be independent of the CWI pathways, but rather mediated by alterations in the architecture and biophysical properties of the cell wall [111]. However, during the cell wall response to most stressors, signals that indicate weaknesses in the wall are received by the surface sensors and transmitted leading to activation of the corresponding CWI pathways. In *Saccharomyces cerevisiae* and *C. albicans*, signaling pathways are activated in response to a wide range of stresses such as CFW, harsh temperatures, oxygen starvation, host immune response during infection and antifungal such as echinocandins, altered nutrient levels, and carbon source [109,112–114]. Cell wall stress response is mediated through the protein kinase C, PKC cell integrity mitogen-activated protein (MAP) kinase cascade, and its downstream transcription factors [112,114,115] (Figure 1). Other MAP kinase cascades, the high osmolarity glycerol response, HOG, and Candida ERK-like kinase, Cek1 mediated pathways, have also been shown to play a role in the cell wall reconstruction process [112,116,117] (Figure 1). MAP kinase defective *C. albicans* mutants display attenuated virulence in infection models showing that MAP kinase pathways are also important for virulence [118–120].

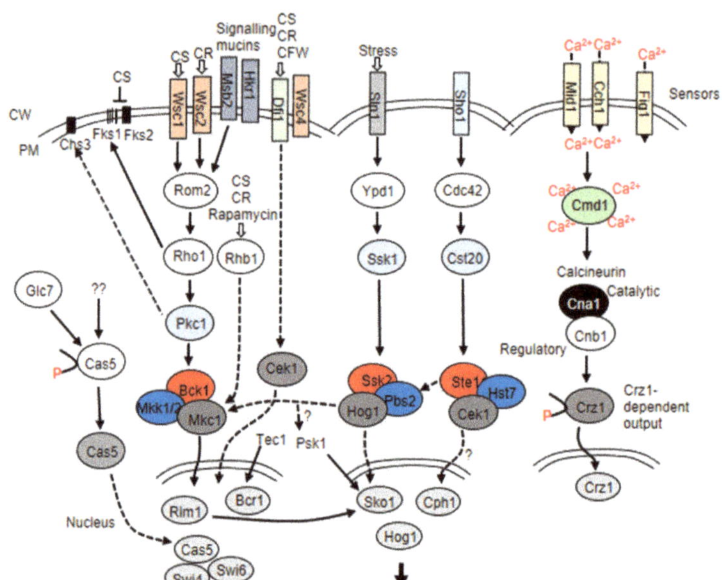

Figure 1. Signaling pathways that regulate cell wall remodeling of *S. cerevisiae* and *C. albicans*. The Hog1, Cek1, and Pkc MAP kinase cascades and the Ca^{2+}/calcineurin signaling pathway control a number of

cellular processes including cell wall synthesis and maintenance. Upstream membrane sensors of the MAP kinase cascades include Wsc family, Dfi1, Sho1, and Sln1, detect signals reporting weakened cell wall or alterations in the wall, and convey the signal to the downstream components of the pathway. The PKC pathway plays an important role in response to caspofungin and activates Rho1, a regulatory sub-unit of β-1,3-glucan synthase. Rhb1, an Rheb-related GTPase, activate the CWI MAP kinase Mkc1 in response to cell wall stress. An Rhb1 deletion mutant is hypersensitive to cell wall stress and to rapamycin [121]. Rho1 activates protein kinase C, which phosphorylates and activates Bck1 in the MAP kinase cascade. Bck1 in turn activates the MAP kinase kinases Mkk1/2, which then phosphorylate Mkc1, which may hypothetically target Rlm1 in *C. albicans*. Although the Mkc1–Rlm1 relationship has been shown in *S. cerevisiae*, there is no evidence in *C. albicans* that Rlm1 is downstream of the Pkc pathway. A number of transcription factors contribute to the echinocandin stress response including Cas5 and Bcr1 [54]. In *C. albicans*, Cas5 is activated through an unknown mechanism involving dephosphorylation by Glc7 phosphatase [122]. Cas5 interacts with Swi4 and Swi6 to activate Cas5-dependent gene transcription leading to the upregulation of genes involved in cell biogenesis/integrity and cellular metabolism [122]. Cas5 and Efg1 have been shown to interact in response to caspofungin stress. Efg1 regulates the transcriptional response to cell wall damage by caspofungin [123]. *C. albicans efg1*Δ/Δ mutant is hypersensitive to caspofungin [123,124]. Cas5 and Efg1 coregulate the expression of caspofungin-inducible genes. Cek1 pathway impinges on cell wall regulation and has also been implicated in systemic candidiasis [119,125]. *C. albicans* Dfi1, a homologue of *S. cerevisiae* Mid2/Mtl1 is known to partly activate the MAP kinase Cek1 and confer tolerance to caspofungin, CR, and CFW [126]. A Dfi1 deletion mutant is severely affected in invasive filamentation and virulence in a murine infection model. Msb2 in cooperation with Sho1 is also thought to play a role in Cek1 activation [127]. It is predicted that the transcription factor, Cph1 a homologue of *Sc*Stel2, is downstream of the Cek1 mediated pathway [112,119,127]. Cph1 is associated with regulation of filamentation [127]. The Rlm1 and Bcr1 transcription factors control the expression of a number of cell wall-related genes [34,128] with Bcr1 playing a dominant role in the regulation of biofilm formation by controlling expression of several important adhesins. In *C. albicans*, the Rlm1 activation mechanism is unknown, but once localized in the nucleus, activated Rlm1 leads to the upregulation of genes involves in cell wall biogenesis/integrity, macromolecular localization, and organelle localization [129]. Putative Rlm1 binding motifs in the promoters of *CHS2* and *CHS8* contribute to their cell wall stress-activated regulation [10,130,131]. In *S. cerevisiae*, Pkc1 is involved in targeting Chs3 to the plasma membrane in response to heat shock [129,132]. Significant re-wiring of signaling pathways is evident in *C. albicans*, compared to the *S. cerevisiae* paradigm, for example, the role of the Sko1 transcription factor in response to caspofungin is independent of Hog1 MAP kinase, but involves the Psk1 PAK kinase [133] and Rlm1. In *C. albicans*, Sko1 regulates the expression of some genes involved in cell wall biogenesis and remodeling, and osmoadaptation [133]. Sko1 binding motif has been identified for regulating Sko1-dependent genes. Sko1 also binds to its motif to promote self-activation. The calcineurin pathway is activated by calcium that may enter the cells through membrane-localized channels Cch1 and Mid1 or a third minor channel Fig1. Alternatively, the pathway may be activated by calcium released from intracellular stores. Ca^{2+} binds to and activates calmodulin (Cmd1) that in turn activates the phosphatase calcineurin. The calcineurin is made up of two sub-units, Cna1 and Cnb1. Calcineurin dephosphorylates the transcription factor Crz1, which moves into the nucleus and induces expression of genes through binding to CDREs (calcium dependent response elements) within their promoter sequences. Two Crz1 DNA binding motifs have been identified in some genes regulated by Crz1. Adapted from [112,114,126]. CR = Congo red, CS = caspofungin, CFW calcofluor white, CWM = cell wall matrix, PM = plasma membrane.

There is some redundancy in the regulatory networks responding to echinocandin-induced cell wall damage where more than one transcription factor controls overlapping sets of downstream target genes to control changes in the cell wall [54,122,133,134]. Three transcription factors, Cas5, Sko1, and Rlm1 have been implicated in echinocandin-induced cell wall damage signaling [54,133].

Cas5 has been shown to be involved in cell wall remodeling in *C. albicans* during cell growth, morphology, and virulence [54,122,135,136]. *C. albicans cas5*Δ/Δ mutants and including mutants with a missense mutation in Cas5 DNA-binding domain is hypersensitive to caspofungin and other cell wall stressors such as CFW [54,122]. A *cas5Δ/Δ* deletion mutant has also been shown to have attenuated virulence in both murine and invertebrate models of systemic candidiasis [135]. Genome-wide microarray studies showed that Cas5 regulates about 50% of the highly expressed caspofungin-inducible genes, including some cell wall integrity genes [54]. Studies using RNA polymerase II chromatin immunoprecipitation and sequencing analyses showed that the number of caspofungin-inducible genes is markedly higher and genes with cell wall-associated functions were markedly overrepresented [122]. Furthermore, Cas5 was found to regulate over 60% of caspofungin-inducible genes, including those involved in cell wall integrity [122].

Information on the upstream regulation of Cas5 is limited in *C. albicans*, but available data suggest that Cas5 is dephosphorylated by phosphatase Glc7 following caspofungin-induced cell wall damage [122]. The study further showed that upon dephosphorylation of Cas5, it is activated and interacts with Swi4 and Swi6 to activate the transcription of Cas5-dependent genes [122]. This leads to the upregulation of genes involved in cell wall synthesis/integrity and cell metabolism [122].

Cas5 together with Efg1 regulate the transcriptional response to cell wall damage by caspofungin [123]. Efg1 is a member of the APSES family of basic helix-loop-helix transcriptional regulators that is proposed to function downstream of the cAMP/protein kinase A (PKA) pathway to induce a hyphal transcription program [137,138]. Likewise, Efg1 is important for transcriptional responses to echinocandins and *C. albicans efg1Δ/Δ* mutant is hypersensitive to caspofungin [124]. Efg1 also required for the induction of *CAS5* in response to cell wall damage by caspofungin [125]. Deletion of *EFG1* in a *cas5Δ/Δ* mutant exacerbates caspofungin hypersensitivity and make caspofugin-resistant *C. albicans* sensitive again. The ectopical expression of *CAS5* could not salvage the growth defect of *C. albicans efg1Δ/Δ* mutant treated with caspofungin [123]. Genome wide transcription profiling of *C. albicans cas5Δ/Δ* and *efg1Δ/Δ* mutants using RNA-Seq showed that Cas5 and Efg1 can coregulate the expression of caspofungin-inducible genes and can also independently regulate some genes [123]. Using yeast two-hybrid and in vivo immunoprecipitation, Cas5 and Efg1 were shown to interact and bind to the promoter of some caspofungin-inducible genes to coordinately activate their expression [123].

Efg1 has also been shown to regulate Czf1 expression [139,140]. Czf1, a *C. albicans* zinc finger cluster transcription factor, is required for white-opaque switching and filamentation [141]. Efg1 and Czf1 interact in a yeast two-hybrid experiment [140] and coordinate responses to farnesol during quorum sensing and white-opaque thermal dimorphism [142]. In the screen of a library of genetically activated forms of zinc cluster transcription factors, hyperactive Czf1 was found to have a cell wall associated function in *C. albicans* [143]. Hyperactive Czf1 drives the expression of many CWPs with cell wall associated functions that can induce a physical change in the cell wall architecture and rescue the hypersensitivity of different CWI partway deletion mutants to cell wall perturbing agents [143]. In addition, *C. albicans czf1Δ/Δ* mutant is hypersensitive to caspofungin [143].

Downstream of the Pkc pathway is the transcription factor, Rlm1. Rlm1 has been extensively studied in *S. cerevisiae* where it is the main transcriptional regulator of the Pkc CWI pathway [144,145]. However, our understanding of the function of the protein is limited in *C. albicans*. *C. albicans rlm1Δ/Δ* mutant is hypersensitive to CFW and Congo red [54] and analysis of mutant cell wall composition compared to wild type showed marked reduction in mannan composition and an increase in chitin levels [128]. This suggested that Rlm1 is involved in caspofungin induced CWI signaling. These characteristics of *rlm1Δ/Δ* in *C. albicans* have not been observed in *S. cerevisiae*, showing divergence of these orthologues [128]. In *C. glabrata*, which is more closely related to *S. cerevisiae* than *C. albicans*, *rlm1Δ/Δ*, *mkk1Δ/Δ*, and *bck1Δ/Δ* mutants are sensitive to caspofungin, but not to CFW or Congo red [146] and the full influence of this pathway on cell wall regulation is yet to be studied. Genome-wide microarray studies in *C. albicans* showed that Rlm1 only induced the expression of five genes under basal condition and only two of these genes were caspofungin-inducible [54]. Another genome-wide study demonstrated that Rlm1 regulated the expression of 773 genes under basal conditions [128] and some of the highly upregulated genes have cell wall associated function. These data suggest that Rlm1 may have a more general regulatory role in controlling cell wall associated gene during non-stressed physiological activities. Genome-wide ChIP Seq data revealed that Rlm1-target genes encode proteins that have cell wall-associated function [134]. Rlm1 bound to the upstream intergenic regions of 25 genes and 18 of the genes were highly caspofungin-inducible [134]. Furthermore, a *rlm1*Δ/Δ mutant attenuated virulence in a murine model of systemic candidiasis [128].

Orthologues of Pkc pathway are conserved in *C. albicans*; however, it is not known if the Mkc1 directly or indirectly activates Rlm1. Genomic, biochemical, and cellular data suggest circuit rewiring in Rlm1 and Sko1 CWI signaling [134]. Sko1 function has been extensively studied in *S. cerevisiae* and shown to be part of the MAP kinase high osmolarity glycerol, Hog, signaling pathway with a role in osmotic and oxidative stress responses [147]. The Hog pathway in *C. albicans* is associated with pathogenicity traits and it is involved in the control of both pathogenic and commensal state programs [148]. Sko1 function as the regulator of osmotic stress is conserved in *C. albicans* and it is phophorylated by the MAP kinase Hog1 following osmotic shock [133]. However, Sko1 regulates genes in *C. albicans* whose orthologues in *S. cerevisiae* are not involved in osmotic stress response, therefore showing circuitry rewiring [149]. Sko1 function in regulating the oxidative stress response is also conserved in *C. albicans* [150].

A *sko1*Δ/Δ mutant is hypersensitive to caspofungin, but not to Congo red, CFW, or SDS [148,150], suggesting Sko1 may not have such a global role in cell wall architecture as Rlm1 or Cas5. Microarray and RT-qPCR data demonstrated that Sko1 regulates 81 caspofungin-inducible genes and 26 of these genes are upregulated by Sko1 [133]. Several of the genes regulated by Sko1 have cell wall-associated function (*PGA13* and *CRH11*), and cell metabolism functions [133,134].

The upstream regulatory mechanisms controlling Sko1 expression in *C. albicans* are more complex than in *S. cerevisiae*, where the Hog pathway principally regulates Sko1 transcriptional activity. Caspofungin-induced Sko1 activity is independent of Hog pathway function [133]. RT-qPCR data demonstrated that caspofungin markedly induces *SKO1* transcription and this requires the glucose-partitioning PAS kinase, Psk1 [133], but Psk1 does not regulate transcription directly. Microarray data indicate that Rlm1 regulates *SKO1* expression under basal conditions [128]. Furthermore, caspofungin-induced *SKO1* expression is markedly reduced in *rlm1*Δ/Δ mutant but not in *cas5*Δ/Δ mutant. It is unknown if Psk1 binds directly to Rlm1 to regulate its activation of *SKO1* expression. However, a DNA-binding consensus has been identified in the Sko1 promoter sequence for regulating Sko1 inducible genes and also for autoactivation of *SKO1* [134].

The Hog pathway has not been well studied in other *Candida* species. However, in clinical strains of *C. auris*, Hog1 and Ssk1 have been shown to have variable activities, which suggested some sort of genetic flexibility with effects on cell wall function and stress adaptation [151]. A *C. auris ssk1*Δ/Δ *hog1*Δ/Δ mutant had altered tolerance to caspofungin and amphotericin B, with increased echinocandin susceptibility [151]. The mutant also had altered cell wall mannan content and altered hyper-resistance to cell wall stressors [151]. Targeting these two signaling components of the Hog pathway may provide options for an effective combination therapy or enhancement of echinocandin susceptibility.

Phosphotransferase regulator Ypd1 and phosphatase Ptp2 have been identified as the Sko1 targets following caspofungin treatment of *C. albicans* [134]. Both Ypd1 and Ptp2 are known to inhibit the Hog pathway, indicating that Sko1 blocks the Hog pathway following caspofungin treatment [152,153]. There is also cross communication between the Hog1 and Cek1 pathways under basal condition [154] and *C. albicans hog1*Δ/Δ mutants have constitutively higher levels of Cek1 phosphorylation [117].

The Cek1 pathway is involved in cell pathogenesis and participate in cell wall construction [155,156]. Cell surface signals that activate the Cek1 pathway are transmitted by membrane bound sensor: Sho1, Msb2, and Opy2 [154–156], and mediated through Cph1 and Tec1 [157,158]. Signals through the sensors trigger stimulus through Cst20 to the Ste11-Hst7-Cek1 MAPK cascade [119]. Deletion of any of these downstream elements as well as Cph1 does not affect filamentation [159]. Cek1 has also been shown to target another transcription factor, Ace2 to upregulate genes encoding protein *O*-mannosyltransferases in response to defective protein *N*- or *O*-glycosylation activities [160]. Cell surface proteins are post-translationally modified to maintain cell wall structure. Genes encoding components in the Cek1 pathway, *MSB2*, *CST20*, *HST7*, *CEK1*, and *ACE2* are Ace2 targets, indicating Ace2-mediated transcriptional upregulation of pathway genes under *N*-glycosylation stress [160].

In *C. albicans* and most other fungi, cell damage through the inhibition of β-1,3-glucans synthesis triggers compensatory chitin synthesis [101,114,161–163]. We have shown that Pkc, Hog, and Ca^{2+} signaling pathways co-ordinately regulate chitin synthesis in response to cell wall stress [72,110,163]. These pathways regulate *CHS* gene expression and chitin synthesis individually and in concert, leading to rearrangement of wall macromolecules in response to cell wall stresses [110]. A *lacZ* reporter gene was fused to the putative promoters of each of the *CHS* genes of *C. albicans* to monitor the expression of *CHS* genes when treated with cell wall perturbing agents such as CFW and showed that exogenous Ca^{2+}, which induces the calcineurin pathway, activated all the *CHS* genes in a Crz1-dependent manner [110]. Crz1 is the downstream transcription factor in the Ca^{2+}/calcineurin signaling pathway. Treating *C. albicans* cells with CFW, which activates the Pkc pathway, results in a three-fold increase in chitin content [101]. However, hyper-stimulation of *CHS* gene expression was observed when Pkc and Ca^{2+} pathways were simultaneously activated, and this resulted in increased chitin in the cell wall [110]. In *S. cerevisiae* and *C. albicans*, the Pkc and Hog MAP kinase cascades and the Ca^{2+}/calcineurin pathway have been shown to regulate CWPs, such as Sed1, Pst1, Crh1, Cwp1, Ssr1, Yps1, Pir1, and Pir3, involved in cell wall remodeling activities [51,134,145,164–167].

The Ca^{2+}/calcineurin signaling pathway is implicated in the activation of cell wall remodeling processes in response to damage to the cell wall [52,101] (Figure 1). The proposed model for Crz1 regulation in *C. albicans* is that the influx of Ca^{2+} activates calcineurin that then dephosphorylates and activates Crz1. The activated Crz1 enters the nucleus and binds to one or both Crz1 binding motifs in the promoter of target genes leading to their expression [168]. Crz1 has been shown to regulate the expression of 34 genes involved in cell wall biosynthesis in response to calcium stress and 12 of these genes encode proteins that are covalently bound to the cell wall: *CRH11*, *UTR2*, *PGA1*, *PGA6*, *PGA13*, *PGA23*, *PGA39*, *PGA52*, *PGA20*, *ECM331*, *PHR2*, *DFG5* [168]. Microarray and RNA sequencing data have reveal that Crz1 binds in vitro and in vivo to two identified motifs (calcineurin dependent response element, CDRE) in the promoter of some of the target genes [53,168] to induce their expression. The promoter of 79 genes regulated by Crz1 have two binding motifs for Crz1, while 104 Crz1-regulated genes have only one motif. Meanwhile, 36 Crz1 regulated genes have no discernible Crz1 binding motifs [168]. This suggests that the expression of Crz1 target genes is differentially regulated. It has been shown that Crz1 binds to two motifs in the promoter region of *UTR2* to induce expression in response to calcium stress [168].

C. albicans lacking calcineurin is markedly attenuated in virulence in a murine model of systemic candidiasis and cannot survive in the presence of cell membrane stressors [169–171]. *C. albicans* lacking Crz1, the major target of calcineurin is partially virulent in a murine model of systemic candidiasis, indicating the existence of other calcineurin targets that are important for virulence [168,172,173].

Another determinant of caspofungin sensitivity is the transcription factor, Cup9, which is required for normal caspofungin tolerance in hyphae alone and activates the expression of CWPs with cell wall function [174]. *C. albicans cup9*Δ/Δ mutant is hypersensitensive to caspofungin stress. RNA-seq data from *C. albicans cup9*Δ/Δ mutant with or without caspofungin demonstrated that Cup9 has a narrow rather than global effect in the cell wall damage response and activates proteins such as *PGA31* and *IFF11* with a known role in cell wall integrity [174].

Generally, these signaling pathways have not been studied in detail in other *Candida* species; however, in *C. glabrata*, 3 genes: *SLT2*, *YPK2*, and *YPK1*, whose protein products are involved in cell wall maintenance are associated with in vivo and in vitro echinocandins tolerance [175]. A *C. glabrata* strain lacking these three genes was susceptible to caspofungin treatment in a murine model of gastrointestinal candidiasis [175]. Furthermore, genes encoding ortholoques of kinases in the cell wall signaling pathway *SLT2*, *MKK1*, *BEM2*, and *SW14* were identified in *C. glabrata* as well as genes encoding ortholoques of calcineurin pathway membrane components: *CCH1* and *MID1* [146]. Mutants lacking any of these genes were hypersensitive to caspofungin [146]. Deletion of genes representing all stages of CWI pathway from surface sensing to transcription regulation resulted in various degrees of susceptibility to caspofungin and cell wall degrading enzymes [146]. Although more studies are required to understand caspofungin-induced cell wall stress responses in other *Candida* species, available data suggest similarity in signaling pathways, the response strategies deployed, and the wall proteins involved in maintaining cell wall integrity.

4. Cell Wall Remodeling in Response to Thermal Stress

The fungal response to heat shock has been well characterized [176–178]. Temperature stress signals are thought to be sensed by signaling mucins. Signaling mucins are transmembrane glycoproteins that receive and transmit surface signals to signaling pathways (Figure 2). Signaling mucin, Msb2 is known to regulate environmental stress, cell wall biogenesis, and the Cek1 and Pkc pathways in most fungi [178,179]. Msb2 is a global regulator of temperature stress in *C. albicans* [113]. Msb2 is required for fungal survival and hyphae formation at 42 °C. Msb2 also regulates temperature-dependent activation of genes involved in MAP kinase and unfolded protein response pathways (Figure 2) [113].

Generally, the temperature stress response is controlled by an essential protein, the heat shock transcription factor, Hsf1, which is phosphorylated upon sudden temperature rise [180]. Following temperature rise from 30 to 42 °C, Hsf1 is phosphorylated rapidly within 60 s and upon adaptation, downregulated [181]. Under normal growth conditions, Hsf1 binds as a trimer to heat shock elements (HSEs) in the promoters of target heat shock protein (*HSP*) genes [182]. When *S. cerevisiae* or *C. albicans* cells experience an acute heat shock, Hsf1 is hyper-phosphorylated and activated, resulting in the transcriptional induction of the target *HSP* genes, thus stimulating cellular adaptation to the thermal insult [183]. Most heat shock proteins, Hsp, are molecular chaperones that promote client proteins folding, assembly, or cellular localization. They also often target unfolded or damaged proteins for degradation [184]. In *C. albicans*, Hsf1 interacts with Hsp such as Hsp90 under steady-state conditions, and upon thermal shocks, this interaction is strengthened, suggesting existence of a Hsf1-Hsp90 autoregulatory circuit [177]. Hsp90 is localized to the nucleus during elevated temperatures. It is possible that the Hsf1-Hsp90 regulon is critical for the maintenance of thermal homeostasis, not merely for adaptation to acute heat shocks. This suggests that the Hsf1-Hsp90 interaction is important for regulation of short-term responses to heat shock (Figure 2).

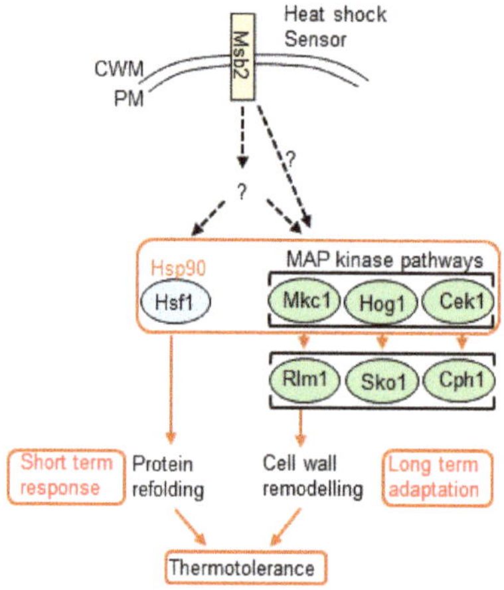

Figure 2. Hsp90 acts as a biological transistor, modulating Hsf1 and the MAPKs transcription factors in response to thermal fluctuations. Msb2 plays a vital role in thermotolerance in *C. albicans*. The protein transmits heat shock signals through unknown mechanisms that induce downstream targets such as the Pkc pathways in response to high temperatures. Hsf1 activation is required for thermotolerance. The MAP kinase signaling pathways are also required to promote thermotolerance through remodeling the cell wall [117,185]. Because Hsp90 coordinates much of this activity, Hsf1, Hog1, Mkc1, and Cek1 are all thought to be Hsp90 client proteins [177,181]. Fluctuations in ambient temperature affect interactions between Hsp90 and Hsf1, and probably affect Hsp90 interactions with the MAP kinase transcription factors [181], thus modulating the role of the signaling pathways and thermal adaptation outcome. Temperature upshifts activate Hsf1, which induces the expression of protein chaperones (HSPs), including Hsp90, which promotes shorter term thermal adaptation. It is thought that Hsp90 then down-regulates Hsf1 and modulates MAP kinase signaling, to alter cell wall architecture, which leads to long term thermotolerance in *C. albicans*. Adapted from [177]. Broken lines indicate unconfirmed regulatory mechanisms.

Cell wall integrity is compromised at elevated temperatures. Temperature affects cell wall polysaccharide composition and the incorporation levels of covalently anchored proteins [186]. Yeasts cells are thought to adapt to heat stress in the longer term by activating the Hog1, Mkc1, and Cek1 MAP kinase pathways, which contribute to thermotolerance [177,186] (Figure 2). These MAP kinase pathways, even though they contribute to thermal adaptation in the longer term through cell wall remodeling, are not essential for Hsf1 activation. Genetic depletion of Hsp90 affects cell wall remodeling activities, suggesting that Hog1, Mkc1, and Cek1 may be client proteins of Hsp90. Hsp90 is thought to be able to integrate both the short term and longer-term molecular responses that underpin thermotolerance [177] (Figure 2).

In *S. cerevisiae*, MAP kinase pathways have been shown to contribute to thermotolerance [132,187], through localization of Chs3 to the plasma membrane in response to heat shock [129]. Each of these MAP kinase pathways is known to contribute to cell wall remodeling and mutations that interfere with cell wall synthesis increase sensitivity of *C. albicans* to elevated temperatures. For example, the deletion of certain protein mannosyltransferases of the PMT family, or the inactivation of *OCH1* can increase susceptibility to temperature [188,189]. Furthermore, deletion of *SSR1* causes elevated susceptibility to temperatures [60].

In a study, thermal upshift was shown to cause reduced secretion of chitinases and have a huge impact on cell wall *N* mannan composition [190]. Analysis of the cell wall phospholipomannan moiety revealed reduction in *N* mannan composition of β-1,2-mannose [190]. *C. albicans* is more susceptible to cell wall stressors when grown at

42 °C [186]. Coping with this thermal stress leads to increased phosphorylation of Mkc1, which mediates activation of the CWI pathways. Consequently, the levels of Sap9, the chitin transglycosylases Crh11 and Utr2, and the cell wall maintenance protein, Ecm33, increased, and cells reinforce their walls with chitin through increased chitin synthesis and reduced chitin degradation [186]. Ecm33 is required for growth at high temperatures and *S. cerevisiae* and *C. albicans ecm33*Δ/Δ disruptant strains exhibit a temperature sensitive growth defect [191,192].

The Mkc1, Hog1, and Cek1 signaling pathways and associated cell wall remodeling mannoproteins have been proposed to promote longer term thermotolerance through the maintenance of a robust cell wall (Figure 2).

5. Echinocandin-Induced Cell Wall Remodeling in Yeast

β-1,3-glucan is a hallmark component of most yeast cell walls and is synthesized by β-1,3-glucan synthase. The protein has an integral membrane catalytic subunit, Fks [193]. *C. albicans* has three *FKS* genes, but the main activity is from the *FKS1* gene product, Fks1. Fks1 is essential and found in association with the regulatory subunit, Rho1 GTPase [194]. Rho1 is required to activate Fks1 for β-1,3-glucan synthesis (Figure 1). Echinocandins non-competitively inhibit β-1,3-glucan synthesis by inhibiting the catalytic function of Fks1, leading to a weak cell wall [195]. Echinocandins are fungicidal against *Candida* species and resistance to the drug has been predominantly associated with point mutations in the *FKS1* gene. However, most yeast have been shown to withstand caspofungin treatment, becoming more tolerant to the drug both in vivo and in vitro by inducing the upregulation of chitin synthesis, the second wall structural polysaccharide [162,163].

Chitin is synthesized by chitin synthase enzymes and *C. albicans* has four chitin synthase proteins comprising of *Chs1*, *Chs2*, *Chs3*, and *Chs8*. Elevated cell wall chitin is a cell wall rescue mechanism shown to be orchestrated by the CWI pathways [101,110,196]. Pkc, Hog, and Ca^{2+} signaling pathways have been shown to control the expression of *CHS2* and *CHS8* through binding motifs in their promoter sequences [131]. Hyper-stimulation of *CHS* gene expression was observed when the three signaling pathways were activated at the same time and this leads to elevated cell wall chitin content [110]. Cell wall mutants with higher basal chitin contents are also less susceptible to caspofungin [60,197]. Chitin synthase proteins can also synthesize alternative septa that restore *C. albicans* capacity to bud during cell wall stress [198].

Genome wide studies have been carried out to study the response of fungal cells to echinocandin drugs treatment and to identify genes whose upregulation is required for adaptive growth in the presence of sub-MIC concentrations of echinocandins. DNA microarrays studies identified genes that are activated in *S. cerevisiae* and *C. albicans* when they are challenged with sub-MIC concentrations of caspofungin [54]. The induced genes include those genes that are typically upregulated following the activation of the Pkc pathway. In *C. albicans* and *S. cerevisiae*, some of the Pkc pathway signature genes: *CRH11*/*CRH1*, *ECM331*/*PST1*, *DFG5*, encode GPI anchored cell surface proteins that have been implicated in cell wall biogenesis or repair [51,55,199–202]. Pga31, a predicted GPI anchored wall protein, is upregulated during caspofungin stress, and *pga31*Δ/Δ mutants have thinner cell walls, reduced chitin content, and are hypersensitive to caspofungin [60]. Cas5 regulates the expression of some CWPs in response to caspofungin, including Crh11, Ecm331, Pga13, and Pga23 [54]. Pga13 plays a role in cell wall architecture [203] and may be required for cell wall repair.

The phosphorylated form of the Pkc pathway component Mkc1/Slt2 and phosphorylated form of Cek1 have been detected in *S. cerevisiae* and *C. albicans* [155] following caspofungin challenge. Furthermore, *C. albicans mkc1*Δ/Δ mutant is hypersensitive to caspofungin [101]. This suggests that the Pkc pathway is a major signaling pathway for triggering cell wall macromolecule rearrangement in response to caspofungin stress in *S. cerevisiae* and *C. albicans* [101,196].

6. Cell Wall Remodeling and Protein Abundance

In an analysis of the cell wall proteome of *C. albicans* growing on minimal medium without stress using liquid chromatography-mass spectrometry, LC-MS revealed 21 covalently bound CWPs. Out of the 21 CWPs identified, 19 had predicted GPI anchor sequence with cell wall associated function [204]. In other studies, the proteomics technique was used to study the impact of carbon source on the *C. albicans* cell wall proteome and secretome when cells were grown in minimal medium containing 2% glucose, lactate, or glucose plus lactate [73,205]. The results revealed higher amounts of predicted GPI anchored CWPs with functions in cell wall biogenesis/integrity in the secretomes and proteomes. Major differences were seen in the profiles of secreted and CWPs in lactate and glucose-grown *C. albicans* cells. Many of the differences suggested that specific cellular processes associated with the cell surface such as cell wall remodeling, adherence, and biofilm formation, may be affected by the change in carbon source [73]. The secretome and proteome of lactate grown cells had increased levels of proteins involved in the remodeling of β-glucan [73]. Lactate grown cells were more adherent, and consequently, more virulent in in vivo models of systemic candidiasis and vaginitis, and display increased resistance to caspofungin as well as other stressors [111]. Lactate signaling regulates glucan masking and modulates the immune response [206]. Furthermore, elevated stress resistance did not correlate with increased activation of the CWI pathways, thus the observed phenotypes may be due to the alteration in the architecture as well as the biochemical and biophysical properties of the cell wall [111]. However, Hog1 or Mkc1 signaling pathways mediate expression of CWPs that promote cell wall elasticity required for adaptation to hyperosmotic stress [52]. Interestingly, alterations in the cell wall in response to different media or carbon sources have been shown to involve changes in the molecular weight of mannoproteins [207]. Mannoproteins from *C. albicans* cultivated on blood or serum have increased molecular weight, when compared with mannoproteins from cells grown on YPD at 30 and 37 °C [207].

A microarray study using DAY185 *C. albicans* strain with or without caspofungin treatment identified 216 caspofungin-inducible genes with an expression change of at least two-fold following 1-h caspofungin treatment [54]. A core set of 34 caspofungin stress inducible genes included genes that are known to be involved in cell wall remodeling such as *PGA13*, *CRH11*, and *PHR1* [54]. In addition, *C. albicans* grown in vagina-simulative medium, aerated with a gas mixture reflecting the gas composition in the vaginal environment had five CWPs [Als3, Hwp1, Sim1, Tos1, Utr2) in the wall that were absent in the YPD grown control [38]. However, O_2 restriction led to higher levels of the non-GPI protein Pir1, β-1,3-glucan cross-linking protein, and of the GPI anchor protein, Hwp1, an adhesion protein [38].

Environmental pH has also been shown to greatly alter the fungal cell wall proteome. Klis lab used a system that mimics mucosal surfaces to investigate the influence of host pH on *C. albicans* cell wall proteome [208]. At pH 4.0, yeast cells and pseudohyphae were predominantly seen while at pH 7.0, hyphal growth was mainly seen. Relative quantitation of ^{15}N-labelled CWPs using ESI-FT-MS revealed the identity of 21 covalently linked CWPs, most of which are GPI anchored, excluding Tos1, Mp65, and Pir1. At pH 7.0, Als1, Als3, Hyr1, Phr1, Rbt1, Sod5, and Tos1 were identified, while only the transglycosidase, Phr2 was found at pH 4.0. Furthermore, at pH 4.0, 12 out of the 21 CWPs were overexpressed, whereas at pH 7.0, 9 proteins were overexpressed. The proteome of the *C. albicans* cell wall is constantly reshuffled to enable cells to adapt to prevailing environmental conditions. The consequences of not adapting to that changing environment is cell death. This is why the cell wall, and its components, are attractive targets for developing more effective diagnostics and therapeutics.

7. Perspective

The covalently bound CWPs in the protein coat are indispensable for the survival of *C. albicans* in the environment and during infection. They also play a major role in the development of biofilms and are regulated by signaling pathways that help remodel the cell wall during stress. However, our knowledge of their structure, which may influence their function regarding structure-function relation is limited and our understanding of their exact function in many cases is still poor. This calls for a continued functional analysis of fungal CWPs. The regulatory mechanisms associated with the construction of the cell wall protein coat are not well understood. The precise mechanism of coupling these proteins to cell wall and their method of interaction with wall polysaccharide and other proteins in the cell wall, which may affect their localization and hence their function, are still not clear. Understanding the function and regulatory mechanisms of these CWPs will ultimately inform our knowledge of fungal pathogenesis and host-pathogen interactions.

CWPs have carbohydrate-binding motifs and may thus be involved in cell wall synthesis and remodeling, in biofilm formation, or even in the interaction with host cell receptors or other environmental signals. Most importantly, our knowledge of the exact roles CWPs play in CWI pathways, their downstream signaling activities, and the extent of their involvement in the cross interactions between the pathways during cell wall stress is relatively unexploited. The cell wall proteome can change significantly in response to specific environmental stress, including during infection. The fungal cell wall proteome changes associated with infection conditions need more extensive studies, as the cell wall in vivo is likely to be very different to the wall generated under laboratory growth conditions. Finally, the relative and absolute quantitation of CWPs under host-related conditions and an extensive understanding of their exact structure and functions will be vital in identifying the most suitable diagnostic, therapeutic, and vaccine candidates.

Author Contributions: Conceptualization, C.I. and C.A.M.; original draft preparation, C.I.; review and editing, C.I. and C.A.M. All authors have read and agreed to the published version of the manuscript.

Funding: This research was funded by Abia State University and the University of Aberdeen.

Institutional Review Board Statement: Not applicable.

Informed Consent Statement: Not applicable.

Conflicts of Interest: The authors declare no conflict of interest.

References

1. Brown, G.D.; Denning, D.W.; Gow, N.A.R.; Levitz, S.M.; Netea, M.G.; White, T.C. Hidden killers: Human fungal infections. *Sci. Transl. Med.* **2012**, *4*, 165rv13. [CrossRef]
2. Mayer, F.L.; Wilson, D.; Hube, B. *Candida albicans* pathogenicity mechanisms. *Virulence* **2013**, *4*, 119–128. [CrossRef]
3. D'Enfert, C.; Kaune, A.K.; Alaban, L.R.; Chakraborty, S.; Cole, N.; Delavy, M.; Kosmala, D.; Marsaux, B.; Frois-Martins, R.; Morelli, M.; et al. The impact of the Fungus-Host-Microbiota interplay upon *Candida albicans* infections: Current knowledge and new perspectives. *FEMS Microbiol. Rev.* **2021**, *45*, 1–55.
4. Lenardon, M.D.; Sood, P.; Dorfmueller, H.C.; Brown, A.J.; Gow, N.A. Scalar nanostructure of the *Candida albicans* cell wall; A molecular, cellular and ultrastructural analysis and interpretation. *Cell Surf.* **2020**, *6*, 100047. [CrossRef]
5. Luo, G.; Ibrahim, A.S.; Spellberg, B.; Nobile, C.; Mitchell, A.P.; Fu, Y. *Candida albicans* Hyr1p confers resistance to neutrophil killing and is a potential vaccine target. *J. Infect. Dis.* **2010**, *201*, 1718–1728. [CrossRef]
6. Edwards, J.E., Jr.; Schwartz, M.M.; Schmidt, C.S.; Sobel, J.D.; Nyirjesy, P.; Schodel, F.; Marchus, E.; Lizakowski, M.; DeMontigny, E.A.; Hoeg, J.; et al. A fungal immunotherapeutic vaccine (NDV-3A) for treatment of recurrent vulvovaginal candidiasis-A Phase 2 randomized, double-blind, placebo-controlled trial. *Clin. Infect. Dis.* **2018**, *66*, 1928–1936. [CrossRef]
7. Spellberg, B.J.; Ibrahim, A.S.; Avanesian, V.; Fu, Y.; Myers, C.; Phan, Q.T.; Filler, S.G.; Yeaman, M.R.; Edwards, J.E., Jr. Efficacy of the anti-Candida rAls3p-N or rAls1p-N vaccines against disseminated and mucosal candidiasis. *J. Infect. Dis.* **2006**, *194*, 256–260. [CrossRef]
8. Klis, F.M.; Sosinska, G.J.; De Groot, P.; Brul, S. Covalently linked cell wall proteins of *Candida albicans* and their role in fitness and virulence. *FEMS Yeast Res.* **2009**, *9*, 1013–1028. [CrossRef]
9. Klis, F.M.; de Groot, P.; Hellingwerf, K. Molecular organization of the cell wall of *Candida albicans*. *Med. Mycol.* **2001**, *39* (Suppl. S1), 1–8. [CrossRef]
10. Sherrington, S.L.; Sorsby, E.; Mahtey, N.; Kumwenda, P.; Lenardon, M.D.; Brown, I.; Ballou, E.R.; Maccallum, D.M.; Hall, R.A. Adaptation of *Candida albicans* to environmental pH induces cell wall remodeling and enhances innate immune recognition. *PLoS Pathog.* **2017**, *13*, e1006403. [CrossRef]

11. Cheng, S.-C.; van de Veerdonk, F.L.; Lenardon, M.; Stoffels, M.; Plantinga, T.; Smeekens, S.; Rizzetto, L.; Mukaremera, L.; Preechasuth, K.; Cavalieri, D.; et al. The dectin-1/inflammasome pathway is responsible for the induction of protective T-helper 17 responses that discriminate between yeasts and hyphae of *Candida albicans*. *J. Leukoc. Biol.* **2011**, *90*, 357–366. [CrossRef]
12. Gow, N.A.R.; Latge, J.-P.; Munro, C.A. The fungal cell wall: Structure, biosynthesis, and function. *Microbiol. Spectr.* **2017**, *5*, 267–292. [CrossRef]
13. Kollar, R.; Petrakova, E.; Ashwell, G.; Robbins, P.W.; Cabib, E. Architecture of the yeast cell wall. The linkage between chitin and beta(1–>3)-glucan. *J. Biol. Chem.* **1995**, *270*, 1170–1178. [CrossRef]
14. Gow, N.A.; Hube, B. Importance of the *Candida albicans* cell wall during commensalism and infection. *Curr. Opin. Microbiol.* **2012**, *15*, 406–412. [CrossRef]
15. Zlotnik, H.; Fernandez, M.P.; Bowers, B.; Cabib, E. *Saccharomyces cerevisiae* mannoproteins form an external cell wall layer that determines wall porosity. *J. Bacteriol.* **1984**, *159*, 1018–1026. [CrossRef]
16. Gow, N.A.; Netea, M.G.; Munro, C.A.; Ferwerda, G.; Bates, S.; Mora-Montes, H.M.; Walker, L.; Jansen, T.; Jacobs, L.; Tsoni, V.; et al. Immune recognition of *Candida albicans* beta-glucan by dectin-1. *J. Infect. Dis.* **2007**, *196*, 1565–1571. [CrossRef]
17. Munro, C.; Richard, M.L. The cell wall: Glycoproteins, remodeling, and regulation. In *Candida and Candidiasis*, 2nd ed.; Calderone, R.A., Clancy, C.J., Eds.; ASM Press: Washington, DC, USA, 2012; pp. 197–223.
18. De Groot, P.W.; Ram, A.F.; Klis, F.M. Features and functions of covalently linked proteins in fungal cell walls. *Fungal Genet. Biol.* **2005**, *42*, 657–675. [CrossRef]
19. Ibe, C.; Walker, L.A.; Gow, N.; Munro, C.A. Unlocking the therapeutic potential of the fungal cell wall: Clinical implications and drug resistance. In *Candida albicans: Cellular and Molecular Biology*; Prasad, R., Ed.; Springer International Publishing: London, UK, 2017; pp. 313–346. [CrossRef]
20. Orlean, P. Architecture and biosynthesis of the *Saccharomyces cerevisiae* cell wall. *Genetics* **2012**, *192*, 775–818. [CrossRef]
21. Bowman, S.M.; Free, S.J. The structure and synthesis of the fungal cell wall. *BioEssays* **2006**, *28*, 799–808. [CrossRef]
22. Ferguson, M.A. The structure, biosynthesis and functions of glycosylphosphatidylinositol anchors, and the contributions of trypanosome research. *J. Cell Sci.* **1999**, *112 Pt 17*, 2799–2809. [CrossRef]
23. Kinoshita, T.; Inoue, N. Dissecting and manipulating the pathway for glycosylphosphatidylinositol-anchor biosynthesis. *Curr. Opin. Chem. Biol.* **2000**, *4*, 632–638. [CrossRef]
24. Richard, M.L.; Plaine, A. Comprehensive analysis of glycosylphosphatidylinositol-anchored proteins in *Candida albicans*. *Eukaryot. Cell* **2007**, *6*, 119–133. [CrossRef]
25. Ruiz-Herrera, J.; Elorza, M.V.; Valentin, E.; Sentandreu, R. Molecular organization of the cell wall of *Candida albicans* and its relation to pathogenicity. *FEMS Yeast Res.* **2006**, *6*, 14–29. [CrossRef]
26. Kapteyn, J.C.; Van Egmond, P.; Sievi, E.; Ende, H.V.D.; Makarow, M.; Klis, F.M. The contribution of the O-glycosylated protein Pir2p/Hsp150 to the construction of the yeast cell wall in wild-type cells and beta1,6-glucan-deficient mutants. *Mol. Microbiol.* **1999**, *31*, 1835–1844. [CrossRef]
27. Zhao, X.; Daniels, K.J.; Oh, S.-H.; Green, C.B.; Yeater, K.M.; Soll, D.R.; Hoyer, L.L. *Candida albicans* Als3p is required for wild-type biofilm formation on silicone elastomer surfaces. *Microbiology* **2006**, *152*, 2287–2299. [CrossRef]
28. Zhao, X.; Oh, S.-H.; Yeater, K.M.; Hoyer, L.L. Analysis of the *Candida albicans* Als2p and Als4p adhesins suggests the potential for compensatory function within the Als family. *Microbiology* **2005**, *151*, 1619–1630. [CrossRef]
29. Hoyer, L.L.; Green, C.B.; Oh, S.H.; Zhao, X. Discovering the secrets of the *Candida albicans* agglutinin-like sequence (ALS) gene family—A sticky pursuit. *Med. Mycol.* **2008**, *46*, 1–15. [CrossRef]
30. Almeida, R.S.; Brunke, S.; Albrecht, A.; Thewes, S.; Laue, M.; Edwards, J.E., Jr.; Filler, S.G.; Hube, B. The hyphal-associated adhesin and invasin Als3 of *Candida albicans* mediates iron acquisition from host ferritin. *PLoS Pathog.* **2008**, *4*, e1000217. [CrossRef]
31. Phan, Q.T.; Myers, C.L.; Fu, Y.; Sheppard, D.C.; Yeaman, M.R.; Welch, W.H.; Ibrahim, A.S.; Edwards, J.E., Jr.; Filler, S.G. Als3 Is a *Candida albicans* invasin that binds to cadherins and induces endocytosis by host cells. *PLoS Biol.* **2007**, *5*, e64. [CrossRef]
32. Laforce-Nesbitt, S.S.; Sullivan, M.A.; Hoyer, L.L.; Bliss, J.M. Inhibition of *Candida albicans* adhesion by recombinant human antibody single-chain variable fragment specific for Als3p. *FEMS Immunol. Med. Microbiol.* **2008**, *54*, 195–202. [CrossRef]
33. Nobile, C.; Schneider, H.A.; Nett, J.E.; Sheppard, D.C.; Filler, S.G.; Andes, D.; Mitchell, A.P. Complementary adhesin function in *C. albicans* biofilm formation. *Curr. Biol.* **2008**, *18*, 1017–1024. [CrossRef] [PubMed]
34. Nobile, C.; Andes, D.; Nett, J.E.; Smith, F.J.; Yue, F.; Phan, Q.-T.; Edwards, J.E., Jr.; Filler, S.G.; Mitchell, A.P. Critical role of Bcr1-dependent adhesins in *C. albicans* biofilm formation In Vitro and In Vivo. *PLoS Pathog.* **2006**, *2*, e63. [CrossRef] [PubMed]
35. Nobile, C.; Mitchell, A.P. Regulation of cell-surface genes and biofilm formation by the *C. albicans* transcription factor Bcr1p. *Curr. Biol.* **2005**, *15*, 1150–1155. [CrossRef]
36. Nobile, C.; Nett, J.E.; Andes, D.; Mitchell, A.P. Function of *Candida albicans* adhesin Hwp1 in biofilm formation. *Eukaryot. Cell* **2006**, *5*, 1604–1610. [CrossRef]
37. Ruben, S.; Garbe, E.; Mogavero, S.; Albrecht-Eckardt, D.; Hellwig, D.; Häder, A.; Krüger, T.; Gerth, K.; Jacobsen, I.D.; Elshafee, O.; et al. Ahr1 and Tup1 contribute to the transcriptional control of virulence-associated genes in *Candida albicans*. *mBio* **2020**, *11*, e00206-20. [CrossRef] [PubMed]
38. Sosinska, G.J.; De Groot, P.; De Mattos, M.J.T.; Dekker, H.L.; de Koster, C.; Hellingwerf, K.J.; Klis, F.M. Hypoxic conditions and iron restriction affect the cell-wall proteome of *Candida albicans* grown under vagina-simulative conditions. *Microbiology* **2008**, *154 Pt 2*, 510–520. [CrossRef]

39. Staab, J.F.; Bradway, S.D.; Fidel, P.L.; Sundstrom, P. Adhesive and mammalian transglutaminase substrate properties of *Candida albicans*. *HwpScience* **1999**, *283*, 1535–1538. [CrossRef]
40. Xin, H.; Dziadek, S.; Bundle, D.R.; Cutler, J.E. Synthetic glycopeptide vaccines combining beta-mannan and peptide epitopes induce protection against candidiasis. *Proc. Natl. Acad. Sci. USA* **2008**, *105*, 13526–13531. [CrossRef]
41. Lain, A.; Elguezabal, N.; Brena, S.; Garcia-Ruiz, J.C.; Del Palacio, A.; Moragues, M.D.; Pontón, J. Diagnosis of invasive candidiasis by enzyme-linked immunosorbent assay using the N-terminal fragment of *Candida albicans* hyphal wall protein-1. *BMC Microbiol.* **2007**, *7*, 35. [CrossRef]
42. Hayek, P.; Dib, L.; Yazbeck, P.; Beyrouthy, B.; Khalaf, R.A. Characterization of Hwp2, a *Candida albicans* putative GPI-anchored cell wall protein necessary for invasive growth. *Microbiol. Res.* **2010**, *165*, 250–258. [CrossRef]
43. Younes, S.; Bahnan, W.; Dimassi, H.; Khalaf, R.A. The *Candida albicans* Hwp2 is necessary for proper adhesion, biofilm formation and oxidative stress tolerance. *Microbiol. Res.* **2011**, *166*, 430–436. [CrossRef] [PubMed]
44. McCall, A.D.; Pathirana, R.; Prabhakar, A.; Cullen, P.J.; Edgerton, M. *Candida albicans* biofilm development is governed by cooperative attachment and adhesion maintenance proteins. *NPJ Biofilms Microbiomes* **2019**, *5*, 1–12. [CrossRef] [PubMed]
45. Fonzi, W.A. PHR1 and PHR2 of *Candida albicans* encode putative glycosidases required for proper cross-linking of beta-1,3- and beta-1,6-glucans. *J. Bacteriol.* **1999**, *181*, 7070–7079. [CrossRef] [PubMed]
46. Mouyna, I.; Fontaine, T.; Vai, M.; Monod, M.; Fonzi, W.A.; Diaquin, M.; Popolo, L.; Hartland, R.P.; Latgé, J.-P. Glycosylphosphatidylinositol-anchored glucanosyltransferases play an active role in the biosynthesis of the fungal cell wall. *J. Biol. Chem.* **2000**, *275*, 14882–14889. [CrossRef] [PubMed]
47. Kovacova, K.; Degani, G.; Stratilová, E.; Farkaš, V.; Popolo, L. Catalytic properties of Phr family members of cell wall glucan remodeling enzymes: Implications for the adaptation of *Candida albicans* to ambient pH. *FEMS Yeast Res.* **2015**, *15*, fou011. [CrossRef]
48. Eckert, S.E.; Heinz, W.J.; Zakikhany, K.; Thewes, S.; Haynes, K.; Hube, B.; Mühlschlegel, F.A. PGA4, a GAS homologue from *Candida albicans*, is up-regulated early in infection processes. *Fungal Genet. Biol.* **2007**, *44*, 368–377. [CrossRef]
49. Calderon, J.; Zavrel, M.; Ragni, E.; Fonzi, W.A.; Rupp, S.; Popolo, L. PHR1, a pH-regulated gene of *Candida albicans* encoding a glucan-remodeling enzyme, is required for adhesion and invasion. *Microbiology* **2010**, *156*, 2484–2494. [CrossRef]
50. Cabib, E.; Farkas, V.; Kosík, O.; Blanco, N.; Arroyo, J.; McPhie, P. Assembly of the yeast cell wall: Crh1p AND Crh2p act as transglycosylases in vivo and in vitro. *J. Biol. Chem.* **2008**, *283*, 29859–29872. [CrossRef]
51. Pardini, G.; De Groot, P.; Coste, A.; Karababa, M.; Klis, F.M.; de Koster, C.; Sanglard, D. The CRH family coding for cell wall Glycosylphosphatidylinositol proteins with a predicted transglycosidase domain affects cell wall organization and virulence of *Candida albicans*. *J. Biol. Chem.* **2006**, *281*, 40399–40411. [CrossRef]
52. Ene, I.V.; Walker, L.A.; Schiavone, M.; Lee, K.K.; Martin-Yken, H.; Dague, E.; Gow, N.A.R.; Munro, C.A.; Brown, A.J.P. Cell wall remodeling enzymes modulate fungal cell wall elasticity and osmotic stress resistance. *mBio* **2015**, *6*, e00986-15. [CrossRef]
53. Karababa, M.; Valentino, E.; Pardini, G.; Coste, A.T.; Bille, J.; Sanglard, D. CRZ1, a target of the calcineurin pathway in *Candida albicans*. *Mol. Microbiol.* **2006**, *59*, 1429–1451. [CrossRef] [PubMed]
54. Bruno, V.M.; Kalachikov, S.; Subaran, R.; Nobile, C.J.; Kyratsous, C.; Mitchell, A.P. Control of the *C. albicans* cell wall damage response by transcriptional regulator Cas5. *PLoS Pathog.* **2006**, *2*, e21. [CrossRef] [PubMed]
55. Ao, J.; Chinnici, J.L.; Maddi, A.; Free, S.J. The N-linked outer chain mannans and the Dfg5p and Dcw1p Endo-α-1,6-Mannanases are needed for incorporation of *Candida albicans* glycoproteins into the cell wall. *Eukaryot. Cell* **2015**, *14*, 792–803. [CrossRef] [PubMed]
56. Mancuso, R.; Chinnici, J.; Tsou, C.; Busarajan, S.; Munnangi, R.; Maddi, A. Functions of *Candida albicans* cell wall glycosidases Dfg5p and Dcw1p in biofilm formation and HOG MAPK pathway. *PeerJ* **2018**, *6*, e5685. [CrossRef] [PubMed]
57. Butler, G.; Rasmussen, M.D.; Lin, M.F.; Santos, M.; Sakthikumar, S.; Munro, C.; Rheinbay, E.; Grabherr, M.; Forche, A.; Reedy, J.L.; et al. Evolution of pathogenicity and sexual reproduction in eight *Candida* genomes. *Nature* **2009**, *459*, 657–662. [CrossRef]
58. De Boer, A.D.; de Groot, P.W.; Weindl, G.; Schaller, M.; Riedel, D.; Diez-Orejas, R.; Klis, F.M.; de Koster, C.G.; Dekker, H.L.; Gross, U.; et al. The *Candida albicans* cell wall protein Rhd3/Pga29 is abundant in the yeast form and contributes to virulence. *Yeast* **2010**, *27*, 611–624. [CrossRef] [PubMed]
59. Castillo, L.; Martínez, A.I.; Garcerá, A.; Garcia-Martinez, J.; Ruiz-Herrera, J.; Valentín, E.; Sentandreu, R. Genomic response programs of *Candida albicans* following protoplasting and regeneration. *Fungal Genet. Biol.* **2006**, *43*, 124–134. [CrossRef]
60. Plaine, A.; Walker, L.; Da Costa, G.; Mora-Montes, H.M.; McKinnon, A.; Gow, N.; Gaillardin, C.; Munro, C.; Richard, M.L. Functional analysis of *Candida albicans* GPI-anchored proteins: Roles in cell wall integrity and caspofungin sensitivity. *Fungal Genet. Biol.* **2008**, *45*, 1404–1414. [CrossRef]
61. Frohner, I.E.; Bourgeois, C.; Yatsyk, K.; Majer, O.; Kuchler, K. *Candida albicans* cell surface superoxide dismutases degrade host-derived reactive oxygen species to escape innate immune surveillance. *Mol. Microbiol.* **2009**, *71*, 240–252. [CrossRef]
62. Martchenko, M.; Alarco, A.-M.; Harcus, D.; Whiteway, M. Superoxide dismutases in *Candida albicans*: Transcriptional regulation and functional characterization of the hyphal-induced *SOD5* gene. *Mol. Biol. Cell* **2004**, *15*, 456–467. [CrossRef]
63. Albrecht, A.; Felk, A.; Pichova, I.; Naglik, J.; Schaller, M.; De Groot, P.; MacCallum, D.; Odds, F.C.; Schäfer, W.; Klis, F.; et al. Glycosylphosphatidylinositol-anchored proteases of *Candida albicans* target proteins necessary for both cellular processes and host-pathogen interactions. *J. Biol. Chem.* **2006**, *281*, 688–694. [CrossRef] [PubMed]

64. Moreno-Ruiz, E.; Ortu, G.; De Groot, P.; Cottier, F.; Loussert, C.; Prévost, M.-C.; De Koster, C.; Klis, F.M.; Goyard, S.; D'Enfert, C. The GPI-modified proteins Pga59 and Pga62 of *Candida albicans* are required for cell wall integrity. *Microbiology* **2009**, *155*, 2004–2020. [CrossRef] [PubMed]
65. Perez, A.; Pedros, B.; Murgui, A.; Casanova, M.; Lopez-Ribot, J.L.; Martinez, J.P. Biofilm formation by *Candida albicans* mutants for genes coding fungal proteins exhibiting the eight-cysteine-containing CFEM domain. *FEMS Yeast Res.* **2006**, *6*, 1074–1084. [CrossRef] [PubMed]
66. Braun, B.R.; Head, W.S.; Wang, M.X.; Johnson, A.D. Identification and characterization of tup1-regulated genes in *Candida albicans*. *Genetics* **2000**, *156*, 31–44. [CrossRef] [PubMed]
67. Weissman, Z.; Kornitzer, D. A family of *Candida* cell surface haem-binding proteins involved in haemin and haemoglobin-iron utilization. *Mol. Microbiol.* **2004**, *53*, 1209–1220. [CrossRef] [PubMed]
68. Kuznets, G.; Vigonsky, E.; Weissman, Z.; Lalli, D.; Gildor, T.; Kauffman, S.J.; Turano, P.; Becker, J.; Lewinson, O.; Kornitzer, D. A relay network of extracellular heme-binding proteins drives, *C. albicans* iron acquisition from hemoglobin. *PLOS Pathog.* **2014**, *10*, e1004407. [CrossRef] [PubMed]
69. Martínez, A.I.; Castillo, L.; Garcerá, A.; Elorza, M.V.; Valentín, E.; Sentandreu, R. Role of Pir1 in the construction of the *Candida albicans* cell wall. *Microbiology* **2004**, *150*, 3151–3161. [CrossRef] [PubMed]
70. Pietrella, D.; Lupo, P.; Rachini, A.; Sandini, S.; Ciervo, A.; Perito, S.; Bistoni, F.; Vecchiarelli, A. A *Candida albicans* mannoprotein deprived of its mannan moiety is efficiently taken up and processed by human dendritic cells and induces T-cell activation without stimulating proinflammatory cytokine production. *Infect. Immun.* **2008**, *76*, 4359–4367. [CrossRef]
71. Willaert, R.G. Adhesins of Yeasts: Protein Structure and Interactions. *J. Fungi* **2018**, *4*, 119. [CrossRef]
72. Munro, C.A. Chitin and glucan, the yin and yang of the fungal cell wall, implications for antifungal drug discovery and therapy. *Adv. Appl. Microbiol.* **2013**, *83*, 145–172. [CrossRef]
73. Ene, I.V.; Heilmann, C.J.; Sorgo, A.G.; Walker, L.A.; de Koster, C.G.; Munro, C.A.; Klis, F.M.; Brown, A.J.P. Carbon source-induced reprogramming of the cell wall proteome and secretome modulates the adherence and drug resistance of the fungal pathogen *Candida albicans*. *Proteomics* **2012**, *12*, 3164–3179. [CrossRef] [PubMed]
74. Rolli, E.; Ragni, E.; Calderon, J.; Porello, S.; Fascio, U.; Popolo, L. Immobilization of the glycosylphosphatidylinositol-anchored gas1 protein into the chitin ring and septum is required for proper morphogenesis in yeast. *Mol. Biol. Cell* **2009**, *20*, 4856–4870. [CrossRef]
75. Sarthy, A.V.; McGonigal, T.; Coen, M.; Frost, D.J.; Meulbroek, J.A.; Goldman, R.C. Phenotype in *Candida albicans* of a disruption of the BGL2 gene encoding a 1,3-β-glucosyltransferase. *Microbiology* **1997**, *143*, 367–376. [CrossRef] [PubMed]
76. Cantarel, B.L.; Coutinho, P.M.; Rancurel, C.; Bernard, T.; Lombard, V.; Henrissat, B. The Carbohydrate-Active EnZymes database (CAZy): An expert resource for Glycogenomics. *Nucleic Acids Res.* **2009**, *37*, D233–D238. [CrossRef] [PubMed]
77. Klis, F.M.; Boorsma, A.; De Groot, P. Cell wall construction in *Saccharomyces cerevisiae*. *Yeast* **2006**, *23*, 185–202. [CrossRef] [PubMed]
78. Klis, F.M.; Brul, S.; De Groot, P.W.J. Covalently linked wall proteins in ascomycetous fungi. *Yeast* **2010**, *27*, 489–493. [CrossRef]
79. Dranginis, A.M.; Rauceo, J.M.; Coronado, J.E.; Lipke, P.N. A Biochemical guide to yeast adhesins: Glycoproteins for social and antisocial occasions. *Microbiol. Mol. Biol. Rev.* **2007**, *71*, 282–294. [CrossRef]
80. Kapteyn, J.C.; Hoyer, L.; Hecht, J.E.; Muller, W.H.; Andel, A.; Verkleij, A.J.; Makarow, M.; Ende, H.V.D.; Klis, F.M. The cell wall architecture of *Candida albicans* wild-type cells and cell wall-defective mutants. *Mol. Microbiol.* **2000**, *35*, 601–611. [CrossRef]
81. García-Sánchez, S.; Aubert, S.; Iraqui, I.; Janbon, G.; Ghigo, J.-M.; D'Enfert, C. *Candida albicans* Biofilms: A Developmental State Associated With Specific and Stable Gene Expression Patterns. *Eukaryot. Cell* **2004**, *3*, 536–545. [CrossRef]
82. Garcerá, A.; Martínez, A.I.; Castillo, L.; Elorza, M.V.; Sentandreu, R.; Valentín, E. Identification and study of a *Candida albicans* protein homologous to *Saccharomyces cerevisiae* Ssr1p, an internal cell-wall protein. *Microbiology* **2003**, *149 Pt 8*, 2137–2145. [CrossRef]
83. Li, F.; Palecek, S.P. *EAP1*, a *Candida albicans* gene involved in binding human epithelial cells. *Eukaryot. Cell* **2003**, *2*, 1266–1273. [CrossRef] [PubMed]
84. Weissman, Z.; Shemer, R.; Conibear, E.; Kornitzer, D. An endocytic mechanism for haemoglobin-iron acquisition in *Candida albicans*. *Mol. Microbiol.* **2008**, *69*, 201–217. [CrossRef] [PubMed]
85. Munro, C. *Candida albican* cell wall mediated virulence. The Yeast Handbook Series; In *Pathogenic Yeast*; Ashbee, R.H., Bignell, E.M., Eds.; Springer: Berlin/Heidelberg, Germany, 2010; pp. 69–95.
86. Gaur, N.K.; Klotz, S.A. Accessibility of the peptide backbone of protein ligands is a key specificity determinant in *Candida albicans* SRS adherence. *Microbiology* **2004**, *150*, 277–284. [CrossRef] [PubMed]
87. Sheppard, D.C.; Yeaman, M.R.; Welch, W.H.; Phan, Q.T.; Fu, Y.; Ibrahim, A.S.; Filler, S.G.; Zhang, M.; Waring, A.J.; Edwards, J.E. Functional and structural diversity in the Als protein family of *Candida albicans*. *J. Biol. Chem.* **2004**, *279*, 30480–30489. [CrossRef]
88. Calderone, R.A.; Fonzi, W.A. Virulence factors of *Candida albicans*. *Trends Microbiol.* **2001**, *9*, 327–335. [CrossRef]
89. Staab, J.F.; Datta, K.; Rhee, P. Niche-specific requirement for hyphal wall protein 1 in virulence of *Candida albicans*. *PLoS ONE* **2013**, *8*, e80842. [CrossRef]
90. Naglik, J.R.; Moyes, D.L.; Wachtler, B.; Hube, B. *Candida albicans* interactions with epithelial cells and mucosal immunity. *Microbes Infect.* **2011**, *13*, 963–976. [CrossRef]
91. Zhu, W.; Filler, S.G. Interactions of *Candida albicans* with epithelial cells. *Cell Microbiol.* **2010**, *12*, 273–282. [CrossRef]

92. Wächtler, B.; Citiulo, F.; Jablonowski, N.; Förster, S.; Dalle, F.; Schaller, M.; Wilson, D.; Hube, B. *Candida albicans*-epithelial interactions: Dissecting the roles of active penetration, induced endocytosis and host factors on the infection process. *PLoS ONE* **2012**, *7*, e36952. [CrossRef]
93. Naglik, J.R.; Challacombe, S.J.; Hube, B. *Candida albicans* secreted aspartyl proteinases in virulence and pathogenesis. *Microbiol. Mol. Biol. Rev.* **2003**, *67*, 400–428. [CrossRef]
94. Taylor, B.N.; Hannemann, H.; Sehnal, M.; Biesemeier, A.; Schweizer, A.; Röllinghoff, M.; Schröppel, K. Induction of *SAP7* correlates with virulence in an intravenous infection model of candidiasis but not in a vaginal infection model in mice. *Infect. Immun.* **2005**, *73*, 7061–7063. [CrossRef] [PubMed]
95. Hube, B.; Naglik, J. *Candida albicans* proteinases: Resolving the mystery of a gene family. *Microbiology* **2001**, *147 Pt 8*, 1997–2005. [CrossRef]
96. Schild, L.; Heyken, A.; de Groot, P.W.; Hiller, E.; Mock, M.; de Koster, C.; Horn, U.; Rupp, S.; Hube, B. Proteolytic cleavage of covalently linked cell wall proteins by *Candida albicans* Sap9 and Sap10. *Eukaryot. Cell.* **2011**, *10*, 98–109. [CrossRef] [PubMed]
97. Ramage, G.; Saville, S.P.; Thomas, D.P.; López-Ribot, J.L. *Candida* Biofilms: An Update. *Eukaryot. Cell* **2005**, *4*, 633–638. [CrossRef] [PubMed]
98. Nobile, C.J.; Nett, J.E.; Hernday, A.D.; Homann, O.R.; Deneault, J.S.; Nantel, A.; Andes, D.R.; Johnson, A.D.; Mitchell, A.P. Biofilm matrix regulation by *Candida albicans* Zap1. *PLoS Biol.* **2009**, *7*, e1000133. [CrossRef]
99. Green, C.B.; Cheng, G.; Chandra, J.; Mukherjee, P.; Ghannoum, M.A.; Hoyer, L.L. RT-PCR detection of *Candida albicans* ALS gene expression in the reconstituted human epithelium (RHE) model of oral candidiasis and in model biofilms. *Microbiology* **2004**, *150 Pt 2*, 267–275. [CrossRef]
100. Brown, A.J.P.; Budge, S.; Kaloriti, D.; Tillmann, A.; Jacobsen, M.D.; Yin, Z.; Ene, I.V.; Bohovych, I.; Sandai, D.; Kastora, S.; et al. Stress adaptation in a pathogenic fungus. *J. Exp. Biol.* **2014**, *217*, 144–155. [CrossRef] [PubMed]
101. Walker, L.A.; Munro, C.; de Bruijn, I.; Lenardon, M.D.; McKinnon, A.D.; Gow, N.A.R. Stimulation of chitin synthesis rescues *Candida albicans* from echinocandins. *PLoS Pathog.* **2008**, *4*, e1000040. [CrossRef]
102. Heinisch, J.J.; Dupres, V.; Wilk, S.; Jendretzki, A.; Dufrene, Y.F. Single-molecule atomic force microscopy reveals clustering of the yeast plasma-membrane sensor Wsc1. *PLoS ONE* **2010**, *5*, e11104. [CrossRef]
103. Rodicio, R.R.; Heinisch, J.J. Together we are strong-cell wall integrity sensors in yeasts. *Yeast* **2010**, *27*, 531–540. [CrossRef] [PubMed]
104. Levin, D.E. Regulation of cell wall biogenesis in *Saccharomyces cerevisiae*: The cell wall integrity signaling pathway. *Genetics* **2011**, *189*, 1145–1175. [CrossRef] [PubMed]
105. Dupres, V.; Alsteens, D.; Wilk, S.; Hansen, B.; Heinisch, J.J.; Dufrêne, Y. The yeast Wsc1 cell surface sensor behaves like a nanospring In Vivo. *Nat. Chem. Biol.* **2009**, *5*, 857–862. [CrossRef]
106. Philip, B.; Levin, D.E. Wsc1 and Mid2 are cell surface sensors for cell wall integrity signaling that act through Rom2, a Guanine Nucleotide Exchange Factor for Rho1. *Mol. Cell. Biol.* **2001**, *21*, 271–280. [CrossRef] [PubMed]
107. Rajavel, M.; Philip, B.; Buehrer, B.M.; Errede, B.; Levin, D.E. Mid2 is a putative sensor for cell integrity signaling in *Saccharomyces cerevisiae*. *Mol. Cell. Biol.* **1999**, *19*, 3969–3976. [CrossRef]
108. Lara-Aguilar, V.; Rueda, C.; García-Barbazán, I.; Varona, S.; Monzón, S.; Jiménez, P.; Cuesta, I.; Zaballos, Á.; Zaragoza, Ó. Adaptation of the emerging pathogenic yeast *Candida auris* to high caspofungin concentrations correlates with cell wall changes. *Virulence* **2021**, *12*, 1400–1417. [CrossRef] [PubMed]
109. Kapteyn, J.C.; Ram, A.F.; Groos, E.M.; Kollar, R.; Montijn, R.C.; Ende, H.V.D.; Llobell, A.; Cabib, E.; Klis, F.M. Altered extent of cross-linking of beta1,6-glucosylated mannoproteins to chitin in *Saccharomyces cerevisiae* mutants with reduced cell wall beta1,3-glucan content. *J. Bacteriol.* **1997**, *179*, 6279–6284. [CrossRef]
110. Munro, C.A.; Selvaggini, S.; de Bruijn, I.; Walker, L.; Lenardon, M.D.; Gerssen, B.; Milne, S.; Brown, A.J.; Gow, N. The PKC, HOG and Ca^{2+} signalling pathways co-ordinately regulate chitin synthesis in *Candida albicans*. *Mol. Microbiol.* **2007**, *63*, 1399–1413. [CrossRef]
111. Ene, I.V.; Adya, A.K.; Wehmeier, S.; Brand, A.C.; MacCallum, D.M.; Gow, N.A.; Brown, A.J.P. Host carbon sources modulate cell wall architecture, drug resistance and virulence in a fungal pathogen. *Cell Microbiol.* **2012**, *14*, 1319–1335. [CrossRef]
112. Monge, R.A.; Román, E.; Nombela, C.; Pla, J. The MAP kinase signal transduction network in *Candida albicans*. *Microbiology* **2006**, *152 Pt 4*, 905–912. [CrossRef]
113. Saraswat, D.; Kumar, R.; Pande, T.; Edgerton, M.; Cullen, P.J. Signalling mucin Msb2 Regulates adaptation to thermal stress in *Candida albicans*. *Mol. Microbiol.* **2016**, *100*, 425–441. [CrossRef]
114. Walker, L.A.; Gow, N.; Munro, C.A. Fungal echinocandin resistance. *Fungal Genet. Biol.* **2010**, *47*, 117–126. [CrossRef] [PubMed]
115. Navarro-García, F.; Eisman, B.; Fiuza, S.; Nombela, C.; Pla, J. The MAP kinase Mkc1p is activated under different stress conditions in *Candida albicans*. *Microbiology* **2005**, *151 Pt 8*, 2737–2749. [CrossRef]
116. Bermejo, C.; Rodríguez, E.; Garcia, R.; Peña, J.M.R.; de la Concepción, M.L.R.; Rivas, C.; Arias, P.; Nombela, C.; Posas, F.; Arroyo, J. The Sequential Activation of the Yeast HOG and SLT2 Pathways is required for cell survival to cell wall stress. *Mol. Biol. Cell* **2008**, *19*, 1113–1124. [CrossRef] [PubMed]
117. Eisman, B.; Alonso-Monge, R.; Román, E.; Arana, D.; Nombela, C.; Pla, J. The Cek1 and Hog1 Mitogen-Activated Protein Kinases Play Complementary Roles in Cell wall biogenesis and chlamydospore formation in the fungal pathogen *Candida albicans*. *Eukaryot. Cell* **2006**, *5*, 347–358. [CrossRef] [PubMed]

118. Diez-Orejas, R.; Molero, G.; Navarro-Garcia, F.; Pla, J.; Nombela, C.; Sanchez-Perez, M. Reduced virulence of *Candida albicans* MKC1 mutants: A role for mitogen-activated protein kinase in pathogenesis. *Infect Immun.* **1997**, *65*, 833–837. [CrossRef] [PubMed]
119. Csank, C.; Schröppel, K.; Leberer, E.; Harcus, D.; Mohamed, O.; Meloche, S.; Thomas, D.Y.; Whiteway, M. Roles of the *Candida albicans* mitogen-activated protein kinase homolog, Cek1p, in hyphal development and systemic candidiasis. *Infect. Immun.* **1998**, *66*, 2713–2721. [CrossRef] [PubMed]
120. Alonso-Monge, R.; Navarro-Garcia, F.; Molero, G.; Diez-Orejas, R.; Gustin, M.; Pla, J.; Sanchez, M.; Nombela, C. Role of the mitogen-activated protein kinase Hog1p in morphogenesis and virulence of *Candida albicans*. *J. Bacteriol.* **1999**, *181*, 3058–3068. [CrossRef] [PubMed]
121. Tsao, C.-C.; Chen, Y.-T.; Lan, C.-Y. A small G protein Rhb1 and a GTPase-activating protein Tsc2 involved in nitrogen starvation-induced morphogenesis and cell wall integrity of *Candida albicans*. *Fungal Genet. Biol.* **2009**, *46*, 126–136. [CrossRef] [PubMed]
122. Xie, J.L.; Qin, L.; Miao, Z.; Grys, B.T.; Diaz, J.D.L.C.; Ting, K.; Krieger, J.R.; Tong, J.; Tan, K.; Leach, M.D.; et al. The *Candida albicans* transcription factor Cas5 couples stress responses, drug resistance and cell cycle regulation. *Nat. Commun.* **2017**, *8*, 1–18. [CrossRef]
123. Xiong, K.; Su, C.; Sun, Q.; Lu, Y. Efg1 and Cas5 orchestrate cell wall damage response to caspofungin in *Candida albicans*. *Antimicrob. Agents Chemother.* **2021**, *65*, e01584-20. [CrossRef] [PubMed]
124. Gregori, C.; Glaser, W.; Frohner, I.E.; Reinoso-Martin, C.; Rupp, S.; Schuller, C.; Kuchler, K. Efg1 Controls caspofungin-induced cell aggregation of *Candida albicans* through the adhesin Als1. *Eukaryot Cell.* **2011**, *10*, 1694–1704. [CrossRef] [PubMed]
125. Zucchi, P.C.; Davis, T.R.; Kumamoto, C.A. A *Candida albicans* cell wall-linked protein promotes invasive filamentation into semi-solid medium. *Mol. Microbiol.* **2010**, *76*, 733–748. [CrossRef] [PubMed]
126. Dichtl, K.; Samantaray, S.; Wagener, J. Cell wall integrity signalling in human pathogenic fungi. *Cell. Microbiol.* **2016**, *18*, 1228–1238. [CrossRef] [PubMed]
127. Liu, H.; Kohler, J.; Fink, G.R. Suppression of hyphal formation in *Candida albicans* by mutation of a STE12 homolog. *Science* **1994**, *266*, 1723–1726. [CrossRef]
128. Silva, Y.B.D.; Vaz, C.; Pereira, J.; Carneiro, C.; Nogueira, E.; Correia, A.; Carreto, L.; Silva, S.; Faustino, A.; Pais, C.; et al. Participation of *Candida albicans* transcription factor RLM1 in cell wall biogenesis and virulence. *PLoS ONE* **2014**, *9*, e86270. [CrossRef]
129. Valdivia, R.H.; Schekman, R. The yeasts Rho1p and Pkc1p regulate the transport of chitin synthase III (Chs3p) from internal stores to the plasma membrane. *Proc. Natl. Acad. Sci. USA* **2003**, *100*, 10287–10292. [CrossRef]
130. Lenardon, M.D.; Munro, C.; Gow, N.A. Chitin synthesis and fungal pathogenesis. *Curr. Opin. Microbiol.* **2010**, *13*, 416–423. [CrossRef]
131. Lenardon, M.D.; Lesiak, I.; Munro, C.; Gow, N.A.R. Dissection of the *Candida albicans* class I chitin synthase promoters. *Mol. Genet. Genom.* **2009**, *281*, 459–471. [CrossRef] [PubMed]
132. Kamada, Y.; Jung, U.S.; Piotrowski, J.; Levin, D. The protein kinase C-activated MAP kinase pathway of Saccharomyces cerevisiae mediates a novel aspect of the heat shock response. *Genes Dev.* **1995**, *9*, 1559–1571. [CrossRef]
133. Rauceo, J.M.; Blankenship, J.R.; Fanning, S.; Hamaker, J.J.; Deneault, J.S.; Smith, F.J.; Nantel, A.; Mitchell, A.P.; Bloom, K. Regulation of the *Candida albicans* cell wall damage response by transcription factor Sko1 and PAS kinase Psk1. *Mol. Biol. Cell.* **2008**, *19*, 2741–2751.
134. Heredia, M.Y.; Ikeh, M.A.C.; Gunasekaran, D.; Conrad, K.A.; Filimonava, S.; Marotta, D.H.; Nobile, C.J.; Rauceo, J.M. An expanded cell wall damage signaling network is comprised of the transcription factors Rlm1 and Sko1 in *Candida albicans*. *PLoS Genet.* **2020**, *16*, e1008908. [CrossRef]
135. Chamilos, G.; Nobile, C.J.; Bruno, V.M.; Lewis, R.E.; Mitchell, A.P.; Kontoyiannis, D.P. *Candida albicans* Cas5, a regulator of cell wall integrity, is required for virulence in murine and toll mutant fly models. *J. Infect. Dis.* **2009**, *200*, 152–157. [CrossRef] [PubMed]
136. Vasicek, E.M.; Berkow, E.L.; Bruno, V.M.; Mitchell, A.P.; Wiederhold, N.; Barker, K.S.; Rogers, P.D. Disruption of the transcriptional regulator Cas5 results in enhanced killing of *Candida albicans* by fluconazole. *Antimicrob. Agents Chemother.* **2014**, *58*, 6807–6818. [CrossRef] [PubMed]
137. Bockmuhl, D.P.; Ernst, J.F. A potential phosphorylation site for an A-type kinase in the Efg1 regulator protein contributes to hyphal morphogenesis of *Candida albicans*. *Genetics* **2001**, *157*, 1523–1530. [CrossRef]
138. Nantel, A.; Dignard, D.; Bachewich, C.; Harcus, D.; Marcil, A.; Bouin, A.-P.; Sensen, C.W.; Hogues, H.; Hoog, M.V.H.; Gordon, P.; et al. Transcription profiling of *Candida albicans* cells undergoing the yeast-to-hyphal transition. *Mol. Biol. Cell* **2002**, *13*, 3452–3465. [CrossRef]
139. Zordan, R.E.; Galgoczy, D.J.; Johnson, A.D. Epigenetic properties of white-opaque switching in *Candida albicans* are based on a self-sustaining transcriptional feedback loop. *Proc. Natl. Acad. Sci. USA* **2006**, *103*, 12807–12812. [CrossRef]
140. Noffz, C.S.; Liedschulte, V.; Lengeler, K.; Ernst, J.F. Functional mapping of the *Candida albicans* Efg1 regulator. *Eukaryot. Cell* **2008**, *7*, 881–893. [CrossRef] [PubMed]
141. Stichternoth, C.; Ernst, J.F. Hypoxic Adaptation by Efg1 Regulates biofilm formation by *Candida albicans*. *Appl. Environ. Microbiol.* **2009**, *75*, 3663–3672. [CrossRef]

142. Langford, M.L.; Hargarten, J.C.; Patefield, K.D.; Marta, E.; Blankenship, J.; Fanning, S.; Nickerson, K.W.; Atkin, A.L. *Candida albicans* Czf1 and Efg1 coordinate the response to farnesol during quorum sensing, white-opaque thermal dimorphism, and cell death. *Eukaryot. Cell* **2013**, *12*, 1281–1292. [CrossRef]
143. Mottola, A.; Ramírez-Zavala, B.; Hünniger, K.; Kurzai, O.; Morschhäuser, J. The zinc cluster transcription factor Czf1 regulates cell wall architecture and integrity in *Candida albicans*. *Mol. Microbiol.* **2021**, *116*, 483–497. [CrossRef]
144. Dodou, E.; Treisman, R. The *Saccharomyces cerevisiae* MADS-box transcription factor Rlm1 is a target for the Mpk1 mitogen-activated protein kinase pathway. *Mol. Cell. Biol.* **1997**, *17*, 1848–1859. [CrossRef]
145. Jung, U.S.; Levin, D. Genome-wide analysis of gene expression regulated by the yeast cell wall integrity signalling pathway. *Mol. Microbiol.* **1999**, *34*, 1049–1057. [CrossRef]
146. Rosenwald, A.G.; Arora, G.; Ferrandino, R.; Gerace, E.L.; Mohammednetej, M.; Nosair, W.; Rattila, S.; Subic, A.Z.; Rolfes, R. Identification of genes in *Candida glabrata* conferring altered responses to caspofungin, a cell wall synthesis inhibitor. *G3 Genes Genomes Genet.* **2016**, *6*, 2893–2907. [CrossRef]
147. Saito, H.; Posas, F. Response to Hyperosmotic Stress. *Genetics* **2012**, *192*, 289–318. [CrossRef] [PubMed]
148. Román, E.; Correia, I.; Prieto, A.D.; Alonso, R.; Pla, J. The HOG MAPK pathway in *Candida albicans*: More than an osmosensing pathway. *Int. Microbiol.* **2019**, *23*, 23–29. [CrossRef] [PubMed]
149. Marotta, D.H.; Nantel, A.; Sukala, L.; Teubl, J.R.; Rauceo, J.M. Genome-wide transcriptional profiling and enrichment mapping reveal divergent and conserved roles of Sko1 in the *Candida albicans* osmotic stress response. *Genomics* **2013**, *102*, 363–371. [CrossRef] [PubMed]
150. Alonso-Monge, R.; Román, E.; Arana, D.M.; Prieto, A.D.; Urrialde, V.; Nombela, C.; Pla, J. The Sko1 protein represses the yeast-to-hypha transition and regulates the oxidative stress response in *Candida albicans*. *Fungal Genet. Biol.* **2010**, *47*, 587–601. [CrossRef] [PubMed]
151. Shivarathri, R.; Jenull, S.; Stoiber, A.; Chauhan, M.; Mazumdar, R.; Singh, A.; Nogueira, F.; Kuchler, K.; Chowdhary, A.; Chauhan, N. The two-component response regulator Ssk1 and the mitogen-activated protein Kinase Hog1 control antifungal drug resistance and cell wall architecture of *Candida auris*. *mSphere* **2020**, *5*, e00973-20. [CrossRef]
152. Day, A.M.; Smith, D.A.; Ikeh, M.A.C.; Haider, M.; Herrero-De-Dios, C.M.; Brown, A.J.; Morgan, B.A.; Erwig, L.P.; Maccallum, D.M.; Quinn, J. Blocking two-component signalling enhances *Candida albicans* virulence and reveals adaptive mechanisms that counteract sustained SAPK activation. *PLoS Pathog.* **2017**, *13*, e1006131. [CrossRef] [PubMed]
153. Mavrianos, J.; Desai, C.; Chauhan, N. Two-component histidine phosphotransfer protein Ypd1 Is not essential for viability in *Candida albicans*. *Eukaryot. Cell* **2014**, *13*, 452–460. [CrossRef]
154. Herrero, C.; Alonso-Monge, R.; Pla, J. The lack of upstream elements of the Cek1 and Hog1 mediated pathways leads to a synthetic lethal phenotype upon osmotic stress in *Candida albicans*. *Fungal Genet. Biol.* **2014**, *69*, 31–42. [CrossRef] [PubMed]
155. Román, E.; Cottier, F.; Ernst, J.F.; Pla, J. Msb2 signaling mucin controls activation of Cek1 mitogen-activated protein kinase in *Candida albicans*. *Eukaryot. Cell* **2009**, *8*, 1235–1249. [CrossRef]
156. Román, E.; Nombela, C.; Pla, J. The Sho1 adaptor protein links oxidative stress to morphogenesis and cell wall biosynthesis in the fungal pathogen *Candida albicans*. *Mol. Cell. Biol.* **2005**, *25*, 10611–10627. [CrossRef]
157. Ramírez-Zavala, B.; Weyler, M.; Gildor, T.; Schmauch, C.; Kornitzer, D.; Arkowitz, R.; Morschhäuser, J. Activation of the Cph1-dependent MAP kinase signaling pathway induces white-opaque switching in *Candida albicans*. *PLoS Pathog.* **2013**, *9*, e1003696. [CrossRef]
158. Sahni, N.; Yi, S.; Daniels, K.J.; Huang, G.; Srikantha, T.; Soll, D.R. Tec1 mediates the pheromone response of the white phenotype of *Candida albicans*: Insights into the evolution of new signal transduction pathways. *PLoS Biol.* **2010**, *8*, e1000363. [CrossRef]
159. Huang, H.; Harcus, D.; Whiteway, M. Transcript profiling of a MAP kinase pathway in *C. albicans*. *Microbiol. Res.* **2008**, *163*, 380–393. [CrossRef] [PubMed]
160. Van Wijlick, L.; Swidergall, M.; Brandt, P.; Ernst, J.F. *Candida albicans* responds to glycostructure damage by Ace2-mediated feedback regulation of Cek1 signaling. *Mol. Microbiol.* **2016**, *102*, 827–849. [CrossRef] [PubMed]
161. Walker, L.A.; Lee, K.K.; Munro, C.A.; Gow, N.A. Caspofungin treatment of *Aspergillus fumigatus* results in ChsG-dependent upregulation of chitin synthesis and the formation of chitin-rich microcolonies. *Antimicrob. Agents Chemother.* **2015**, *59*, 5932–5941. [CrossRef]
162. Lee, K.K.; MacCallum, D.M.; Jacobsen, M.D.; Walker, L.A.; Odds, F.C.; Gow, N.A.R.; Munro, C.A. Elevated cell wall chitin in *Candida albicans* confers echinocandin resistance In Vivo. *Antimicrob. Agents Chemother.* **2012**, *56*, 208–217. [CrossRef]
163. Walker, L.A.; Gow, N.A.R.; Munro, C.A. Elevated chitin content reduces the susceptibility of *Candida* species to caspofungin. *Antimicrob. Agents Chemother.* **2013**, *57*, 146–154. [CrossRef]
164. Cabib, E.; Blanco, N.; Grau, C.; Rodriguez-Pena, J.M.; Arroyo, J. Crh1p and Crh2p are required for the cross-linking of chitin to beta(1–6)glucan in the *Saccharomyces cerevisiae* cell wall. *Mol. Microbiol.* **2007**, *63*, 921–935. [CrossRef]
165. Cabib, E. Two novel techniques for determination of polysaccharide cross-links show that Crh1p and Crh2p attach chitin to both beta(1–6)- and beta(1–3)glucan in the *Saccharomyces cerevisiae* cell wall. *Eukaryot. Cell.* **2009**, *8*, 1626–1636. [CrossRef]
166. Terashima, H.; Yabuki, N.; Arisawa, M.; Hamada, K.; Kitada, K. Up-regulation of genes encoding glycosylphosphatidylinositol (GPI)-attached proteins in response to cell wall damage caused by disruption of *FKS1* in *Saccharomyces cerevisiae*. *Mol. Genet. Genom.* **2000**, *264*, 64–74. [CrossRef] [PubMed]

167. Yoshimoto, H.; Saltsman, K.; Gasch, A.P.; Li, H.X.; Ogawa, N.; Botstein, D.; Brown, P.O.; Cyert, M.S. Genome-wide analysis of gene expression regulated by the calcineurin/Crz1p signaling pathway in *Saccharomyces cerevisiae*. *J. Biol. Chem.* **2002**, *277*, 31079–31088. [CrossRef] [PubMed]
168. Xu, H.; Fang, T.; Omran, R.P.; Whiteway, M.; Jiang, L. RNA sequencing reveals an additional Crz1-binding motif in promoters of its target genes in the human fungal pathogen *Candida albicans*. *Cell Commun. Signal.* **2020**, *18*, 1–14. [CrossRef] [PubMed]
169. Sanglard, D.; Ischer, F.; Marchetti, O.; Entenza, J.; Bille, J. Calcineurin A of *Candida albicans*: Involvement in antifungal tolerance, cell morphogenesis and virulence. *Mol. Microbiol.* **2003**, *48*, 959–976. [CrossRef]
170. Blankenship, J.R.; Wormley, F.L.; Boyce, M.K.; Schell, W.A.; Filler, S.G.; Perfect, J.R.; Heitman, J. Calcineurin is essential for *Candida albicans* survival in serum and virulence. *Eukaryot. Cell.* **2003**, *2*, 422–430. [CrossRef] [PubMed]
171. Cruz, M.C.; Goldstein, A.L.; Blankenship, J.R.; Del Poeta, M.; Davis, D.; Cardenas, M.E.; Perfect, J.R.; McCusker, J.H.; Heitman, J. Calcineurin is essential for survival during membrane stress in *Candida albicans*. *EMBO J.* **2002**, *21*, 546–559. [CrossRef]
172. Jiang, L.; Alber, J.; Wang, J.; Du, W.; Yang, X.; Li, X.; Sanglard, D.; Geyer, J. The *Candida albicans* plasma membrane protein Rch1p, a member of the vertebrate SLC10 carrier family, is a novel regulator of cytosolic Ca^{2+} homoeostasis. *Biochem. J.* **2012**, *444*, 497–502. [CrossRef] [PubMed]
173. Mille, C.; Janbon, G.; Delplace, F.; Ibata-Ombetta, S.; Gaillardin, C.; Strecker, G.; Jouault, T.; Trinel, P.-A.; Poulain, D. Inactivation of CaMIT1 inhibits *Candida albicans* phospholipomannan beta-mannosylation, reduces virulence, and alters cell wall protein beta-mannosylation. *J. Biol. Chem.* **2004**, *279*, 47952–47960. [CrossRef]
174. Ichikawa, Y.; Bruno, V.M.; Woolford, C.A.; Kim, H.; Do, E.; Brewer, G.; Mitchell, A.P. Environmentally contingent control of *Candida albicans* cell wall integrity by transcriptional regulator Cup9. *Genetics* **2021**, *218*, iyab075. [CrossRef]
175. Garcia-Rubio, R.; Hernandez, R.Y.; Clear, A.; Healey, K.R.; Shor, E.; Perlin, D.S. Critical assessment of cell wall integrity factors contributing to in vivo echinocandin tolerance and resistance in *Candida glabrata*. *Front. Microbiol.* **2021**, *12*, 702779. [CrossRef]
176. Brown, D.H., Jr.; Giusani, A.D.; Chen, X.; Kumamoto, C.A. Filamentous growth of *Candida albicans* in response to physical environmental cues and its regulation by the unique CZF1 gene. *Mol. Microbiol.* **1999**, *34*, 651–662. [CrossRef]
177. Leach, M.D.; Budge, S.; Walker, L.; Munro, C.; Cowen, L.; Brown, A.J. Hsp90 orchestrates transcriptional regulation by Hsf1 and cell wall remodelling by MAPK signalling during thermal adaptation in a pathogenic yeast. *PLoS Pathog.* **2012**, *8*, e1003069. [CrossRef] [PubMed]
178. Puri, S.; Kumar, R.; Chadha, S.; Tati, S.; Conti, H.R.; Hube, B.; Cullen, P.J.; Edgerton, M. Secreted aspartic protease cleavage of *Candida albicans* Msb2 activates Cek1 MAPK signaling affecting biofilm formation and oropharyngeal candidiasis. *PLoS ONE* **2012**, *7*, e46020. [CrossRef]
179. Cullen, P.J.; Sabbagh, W.; Graham, E.; Irick, M.M.; Van Olden, E.K.; Neal, C.; Delrow, J.; Bardwell, L.; Sprague, G.F., Jr. A signaling mucin at the head of the Cdc42- and MAPK-dependent filamentous growth pathway in yeast. *Genes Dev.* **2004**, *18*, 1695–1708. [CrossRef] [PubMed]
180. Sorger, P.; Pelham, H.R. Yeast heat shock factor is an essential DNA-binding protein that exhibits temperature-dependent phosphorylation. *Cell* **1988**, *54*, 855–864. [CrossRef]
181. Leach, M.; Tyc, K.M.; Brown, A.J.P.; Klipp, E. Modelling the regulation of thermal adaptation in *Candida albicans*, a major fungal pathogen of humans. *PLoS ONE* **2012**, *7*, e32467. [CrossRef] [PubMed]
182. Jakobsen, B.K.; Pelham, H.R. Constitutive binding of yeast heat shock factor to DNA in vivo. *Mol. Cell Biol.* **1988**, *8*, 5040–5042. [CrossRef]
183. Sorger, P.; Lewis, M.J.; Pelham, H.R.B. Heat shock factor is regulated differently in yeast and HeLa cells. *Nature* **1987**, *329*, 81–84. [CrossRef] [PubMed]
184. Feder, M.E.; Hofmann, G.E. Heat-shock proteins, molecular chaperones, and the stress response: Evolutionary and ecological physiology. *Annu. Rev. Physiol.* **1999**, *61*, 243–282. [CrossRef] [PubMed]
185. Navarro-García, F.; Monge, R.A.; Rico, H.; Pla, J.; Sentandreu, R.; Nombela, C. A role for the MAP kinase gene MKC1 in cell wall construction and morphological transitions in *Candida albicans*. *Microbiology* **1998**, *144 Pt 2*, 411–424. [CrossRef]
186. Heilmann, C.J.; Sorgo, A.G.; Mohammadi, S.; Sosinska, G.J.; de Koster, C.; Brul, S.; De Koning, L.J.; Klis, F.M. Surface Stress Induces a Conserved Cell Wall Stress response in the pathogenic fungus *Candida albicans*. *Eukaryot. Cell* **2013**, *12*, 254–264. [CrossRef]
187. Winkler, A.; Arkind, C.; Mattison, C.P.; Burkholder, A.; Knoche, K.; Ota, I. Heat stress activates the yeast high-osmolarity glycerol mitogen-activated protein kinase pathway, and protein tyrosine phosphatases are essential under heat stress. *Eukaryot. Cell* **2002**, *1*, 163–173. [CrossRef] [PubMed]
188. Prill, S.K.-H.; Klinkert, B.; Timpel, C.; Gale, C.A.; Schröppel, K.; Ernst, J.F. PMT family of *Candida albicans*: Five protein mannosyltransferase isoforms affect growth, morphogenesis and antifungal resistance. *Mol. Microbiol.* **2005**, *55*, 546–560. [CrossRef] [PubMed]
189. Bates, S.; Hughes, H.B.; Munro, C.; Thomas, W.P.; MacCallum, D.; Bertram, G.; Atrih, A.; Ferguson, M.; Brown, A.J.; Odds, F.C.; et al. Outer chain n-glycans are required for cell wall integrity and virulence of *Candida albicans*. *J. Biol. Chem.* **2006**, *281*, 90–98. [CrossRef]
190. Okawa, Y.; Goto, K. Antigenicity of cell wall mannans of *Candida albicans* and *Candida stellatoidea* cultured at high temperatures in BACTEC medium. *Biol. Pharm. Bull.* **2006**, *29*, 1723–1727. [CrossRef] [PubMed]

191. Terashima, H.; Hamada, K.; Kitada, K. The localization change of Ybr078w/Ecm33, a yeast GPI-associated protein, from the plasma membrane to the cell wall, affecting the cellular function. *FEMS Microbiol. Lett.* **2003**, *218*, 175–180. [CrossRef] [PubMed]
192. Gil-Bona, A.; Reales-Calderon, J.A.; Giraldo, C.M.P.; Martinez, R.; Monteoliva, L.; Gil, C. The cell wall protein Ecm33 of *Candida albicans* is involved in chronological life span, morphogenesis, cell wall regeneration, stress tolerance, and host–cell interaction. *Front. Microbiol.* **2016**, *7*, 64. [CrossRef] [PubMed]
193. Douglas, C.M.; Foor, F.; Marrinan, J.A.; Morin, N.; Nielsen, J.B.; Dahl, A.M.; Mazur, P.; Baginsky, W.; Li, W.; El-Sherbeini, M. The *Saccharomyces cerevisiae* FKS1 (ETG1) gene encodes an integral membrane protein which is a subunit of 1,3-beta-D-glucan synthase. *Proc. Natl. Acad. Sci. USA* **1994**, *91*, 12907–12911. [CrossRef]
194. Qadota, H.; Python, C.P.; Inoue, S.B.; Arisawa, M.; Anraku, Y.; Zheng, Y.; Watanabe, T.; Levin, D.E.; Ohya, Y. Identification of yeast Rho1p GTPase as a regulatory subunit of 1,3-beta-glucan Synthase. *Science* **1996**, *272*, 279–281. [CrossRef]
195. Denning, D.W. Echinocandin antifungal drugs. *Lancet* **2003**, *362*, 1142–1151. [CrossRef]
196. Reinoso-Martín, C.; Schüller, C.; Schuetzer-Muehlbauer, M.; Kuchler, K. The yeast protein kinase C cell integrity pathway mediates tolerance to the antifungal drug caspofungin through activation of Slt2p mitogen-activated protein kinase signaling. *Eukaryot. Cell* **2003**, *2*, 1200–1210. [CrossRef] [PubMed]
197. Tams, R.N.; Wagner, A.S.; Jackson, J.W.; Gann, E.R.; Sparer, T.E.; Reynolds, T.B. Pathways that synthesize phosphatidylethanolamine impact *Candida albicans* hyphal length and cell wall composition through transcriptional and posttranscriptional mechanisms. *Infect. Immun.* **2020**, *88*. [CrossRef]
198. Walker, L.A.; Lenardon, M.D.; Preechasuth, K.; Munro, C.; Gow, N.A.R. Cell wall stress induces alternative fungal cytokinesis and septation strategies. *J. Cell Sci.* **2013**, *126 Pt 12*, 2668–2677. [CrossRef]
199. Blanco, N.; Sanz, A.B.; Peña, J.M.R.; Nombela, C.; Farkas, V.; Hurtado-Guerrero, R.; Arroyo, J. Structural and functional analysis of yeast Crh1 and Crh2 transglycosylases. *FEBS J.* **2014**, *282*, 715–731. [CrossRef]
200. Pardo, M.; Monteoliva, L.; Vázquez, P.; Martinez, R.; Molero, G.; Nombela, C.; Gil, C. PST1 and ECM33 encode two yeast cell surface GPI proteins important for cell wall integrity. *Microbiology* **2004**, *150 Pt 12*, 4157–4170. [CrossRef]
201. Peña, J.M.R.; Cid, V.J.; Arroyo, J.; Nombela, C. A novel family of cell wall-related proteins regulated differently during the yeast life cycle. *Mol. Cell. Biol.* **2000**, *20*, 3245–3255. [CrossRef]
202. Spreghini, E.; Davis, D.A.; Subaran, R.; Kim, M.; Mitchell, A.P. Roles of *Candida albicans* Dfg5p and Dcw1p Cell Surface Proteins in Growth and Hypha Formation. *Eukaryot. Cell* **2003**, *2*, 746–755. [CrossRef] [PubMed]
203. Gelis, S.; de Groot, P.W.; Castillo, L.; Moragues, M.-D.; Sentandreu, R.; Gómez, M.-M.; Valentín, E. Pga13 in *Candida albicans* is localized in the cell wall and influences cell surface properties, morphogenesis and virulence. *Fungal Genet. Biol.* **2012**, *49*, 322–331. [CrossRef] [PubMed]
204. Castillo, L.; Calvo, E.; Martínez, A.I.; Ruiz-Herrera, J.; Valentín, E.; Lopez, J.A.; Sentandreu, R. A study of the *Candida albicans* cell wall proteome. *Proteomics* **2008**, *8*, 3871–3881. [CrossRef] [PubMed]
205. Childers, D.S.; Avelar, G.M.; Bain, J.M.; Pradhan, A.; Larcombe, D.E.; Netea, M.G.; Erwig, L.P.; Gow, N.A.R.; Brown, A.J.P. Epitope shaving promotes fungal immune evasion. *mBio* **2020**, *11*, e00984-20. [CrossRef] [PubMed]
206. Ballou, E.R.; Avelar, G.M.; Childers, D.S.; Mackie, J.; Bain, J.M.; Wagener, J.; Kastora, S.L.; Panea, M.D.; Hardison, S.E.; Walker, L.A.; et al. Lactate signalling regulates fungal beta-glucan masking and immune evasion. *Nat. Microbiol.* **2016**, *2*, 16238. [CrossRef] [PubMed]
207. Kruppa, M.; Greene, R.R.; Noss, I.; Lowman, D.W.; Williams, D.L. *C. albicans* increases cell wall mannoprotein, but not mannan, in response to blood, serum and cultivation at physiological temperature. *Glycobiology* **2011**, *21*, 1173–1180. [CrossRef] [PubMed]
208. Sosinska, G.J.; De Koning, L.J.; De Groot, P.; Manders, E.M.M.; Dekker, H.L.; Hellingwerf, K.J.; De Koster, C.G.; Klis, F.M. Mass spectrometric quantification of the adaptations in the wall proteome of *Candida albicans* in response to ambient pH. *Microbiology* **2011**, *157 Pt 1*, 136–146. [CrossRef]

Journal of *Fungi*

Review

Cell Wall Integrity Pathway Involved in Morphogenesis, Virulence and Antifungal Susceptibility in *Cryptococcus neoformans*

Haroldo Cesar de Oliveira [1], Suelen Andreia Rossi [2,3], Irene García-Barbazán [2], Óscar Zaragoza [2] and Nuria Trevijano-Contador [2,*,†]

1 Instituto Carlos Chagas, Fundação Oswaldo Cruz (Fiocruz), Curitiba 81350-010, Brazil; haroldo.oliveira@fiocruz.br

2 Mycology Reference Laboratory, National Centre for Microbiology, Instituto de Salud Carlos III, Majadahonda, 28222 Madrid, Spain; su.arossi@gmail.com (S.A.R.); igarcia@isciii.es (I.G.-B.); ozaragoza@isciii.es (Ó.Z.)

3 Department of Microbiology, Biomedical Sciences Institute, University of São Paulo, São Paulo 05508-000, Brazil

* Correspondence: ntrevijano@isciii.es

† Nuria Trevijano-Contador would like to dedicate this review to her daughters Carla and Lucía for being the best of her life.

Abstract: Due to its location, the fungal cell wall is the compartment that allows the interaction with the environment and/or the host, playing an important role during infection as well as in different biological functions such as cell morphology, cell permeability and protection against stress. All these processes involve the activation of signaling pathways within the cell. The cell wall integrity (CWI) pathway is the main route responsible for maintaining the functionality and proper structure of the cell wall. This pathway is highly conserved in the fungal kingdom and has been extensively characterized in *Saccharomyces cerevisiae*. However, there are still many unknown aspects of this pathway in the pathogenic fungi, such as *Cryptococcus neoformans*. This yeast is of particular interest because it is found in the environment, but can also behave as pathogen in multiple organisms, including vertebrates and invertebrates, so it has to adapt to multiple factors to survive in multiple niches. In this review, we summarize the components of the CWI pathway in *C. neoformans* as well as its involvement in different aspects such as virulence factors, morphological changes, and its role as target for antifungal therapies among others.

Keywords: cell wall; cell wall integrity (CWI) pathway; PKC; GTPases; MAP kinase; morphogenesis; virulence; pathogenesis

Citation: de Oliveira, H.C.; Rossi, S.A.; García-Barbazán, I.; Zaragoza, Ó.; Trevijano-Contador, N. Cell Wall Integrity Pathway Involved in Morphogenesis, Virulence and Antifungal Susceptibility in *Cryptococcus neoformans*. *J. Fungi* **2021**, 7, 831. https://doi.org/10.3390/jof7100831

Academic Editors: María Martín and Molina Humberto

Received: 1 September 2021
Accepted: 30 September 2021
Published: 5 October 2021

Publisher's Note: MDPI stays neutral with regard to jurisdictional claims in published maps and institutional affiliations.

1. Introduction

Cryptococcus neoformans is a basidiomycetes yeast widely distributed in the environment that can behave as a pathogen in susceptible patients [1,2]. *C neoformans* infection is very common, healthy people with intact immunity are resistant to the infection. However, in those with altered immunity, this pathogen causes disease, principally cryptococcal meningoencephalitis. The most common predisposing condition for this disease is HIV/AIDS, although it also occurs in people with other types of immune impairment [3]. Based on a 2017 estimation, globally, there are around 215,000 cases and 180,000 deaths due to cryptococcal meningoencephalitis each year, most of which occur in Sub-Saharan Africa [4].

Cryptococcus neoformans has a cosmopolitan distribution [1] and has been isolated from multiple niches, mainly from pigeon guano and in swamps [5], so it can cause disease in a wide variety of organisms [6–8]. This implies that *Cryptococcus* has to adapt to many changes and different types of hosts. This is different from other fungi such as *Candida albicans* which practically only infects humans. *C. neoformans* is acquired by inhalation of spores that initially establish infection in the lungs [2,9,10], but under immunosuppressive conditions this pathogen can disseminate to the central nervous system (CNS). The capsule is the main phenotypic characteristic of *C. neoformans* and its principal virulence factor since it has a large number of effects on the host and the immune system [11–13]. Among others, the capsule inhibits phagocytosis, blocks the migration of lymphocytes and the production of antibodies [14–17]. The polysaccharide capsule is anchored to the outer layer of the cell wall [18–20] and this union needs to happen correctly since the cell wall in *C. neoformans* has an additional function that is to maintain the capsule [18,19]. There are different factors that affect the cell wall such as temperature or interaction with the host; therefore, the routes that regulate the structure of the wall are of great importance to understand virulence.

The cryptococcal cell wall is an essential structure composed mainly of glucans, chitin, chitosan and glycoproteins. All these components contribute to the cell wall rigidity and integrity [20–23]. The cell wall plays an important role in protecting the cell against different types of stress [21–24]. In addition, the cell wall mediates the interaction with the external environment through different receptors whose activation will trigger a different signaling cascade in the cell [20,21].

The cell wall integrity (CWI) pathway is the main route responsible for cell wall biogenesis and the maintenance of integrity in fungi [25,26]. This pathway was initially described as activated in case of *"cell wall problems"* [27,28]. This route has been well studied and characterized in the yeast *Saccharomyces cerevisiae* and is conserved among different fungal species [28,29]. Furthermore, this via is very similar to the PKC pathway in humans. The cellular integrity pathway is essential for the aging process, oxidative stress responses and cellular morphogenesis among others. This pathway is integrated by different receptors and factors that in a last step activate the MAPK (mitogen-activated protein kinase) module [27,28].

In this review, we will summarize the knowledge about the cell wall integrity pathway and its role in *C. neoformans*.

2. How Does the CWI Signaling Pathway Work in *C. neoformans*?

When the integrity of the cell wall is altered, there are specific surface receptors, although it is not known how these receptors detect stress. Once activated, they interact with the guanine nucleotide exchange factor (GEF) Rom2 that regulate the Rho1 GTPase, which is highly regulated by many subunits. In the case of *C. neoformans*, little is known about the cell surface sensors. Only a putative Mtl1/mid2 homologue encoded by the CNAG_03308 gene (Figure 1) has been identified but its role in the CWI signaling pathway has not been established [28]. However, three homologues of *S. cerevisiae* Rom2 have been found in *C. neoformans*. They are called Rom2, Rom20 and Rom21 [25,28]. The interaction of the membrane sensors with Rom2 activates the small GTPase Rho1, which then activates the protein kinase C (Pkc1). Three Rho GTPases have been identified in *C. neoformans*, Rho1, Rho10 and Rho11 [28,30]. Pkc1, the major kinase involved in cell integrity triggers the activation of a MAPK cascade. In *C. neoformans*, the MAPK module consists of three members; MAPKKK Bck1, MAPKK Mkk2 and MAPK Mpk1 [28]. Absence of these kinases results in alterations at cell wall, such as the content of chitin or chitosan [30–32], which highlights their importance in the maintenance of a proper cell wall structure. The differences and similarities of the CWI pathway in *S. cerevisiae* and *C. neoformans* are summarized in Table S1 of Supplementary Material.

The CWI pathway activation by several stress stimuli will result in the regulation of different transcription factors that will trigger an adaptive response. Classical studies on *S. cerevisiae* transcription factors activation through the CWI pathway revealed that the transcription factors ScRlm1 and the ScSwi4/ScSwi6 complex induce the expression of several genes related to cell wall biogenesis, [27,33]. Although not fully characterized, *C. neoformans* has homologues to ScRlm1, ScSwi4 and ScSwi6, named CNAG_03998, CNAG_07464 and CNAG_01438 respectively, that may play a role in the CWI pathway response and activation [27].

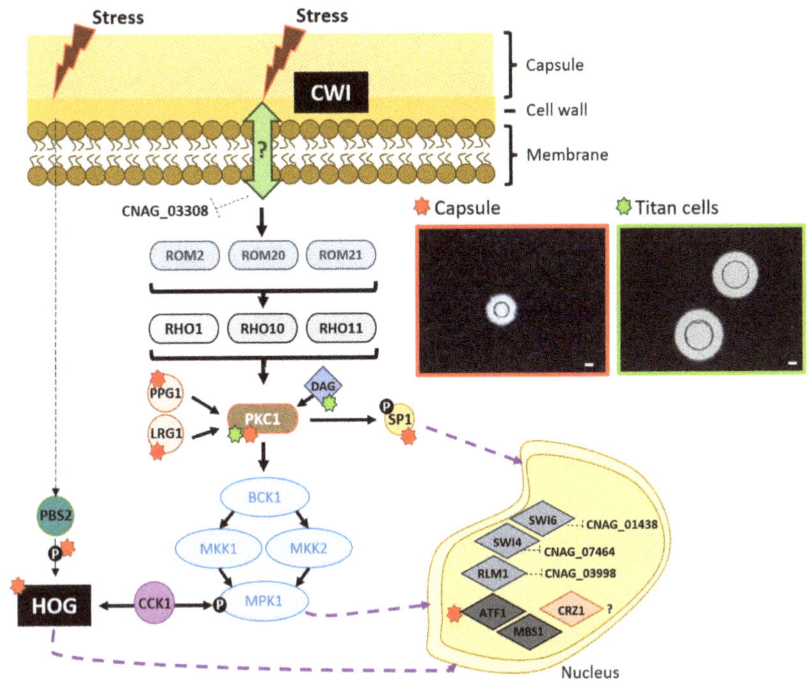

Figure 1. Cryptococcal CWI pathway and its impact on *Cryptococcus* morphological changes. The complete map of the cryptococcal CWI pathway is still under construction. In *Cryptococcus*, the plasma membrane are sensors that capture the environmental stress stimuli is not known, being just identified a homologue to Mid2, that is a transmembrane CWI pathway receptor in *S. cerevisiae*, named CNAG_03308, however, its role in CWI pathway is still unknown. The CWI pathway follows through the interaction with guanine nucleotide exchange factor (GEF), that *Cryptococcus* presents three homologues Rom2, Rom20 and Rom21, that may activate small GTPases Rho1, Rho10 and Rho11, which activates Pkc1, that is regulated by Ppg1 and Lgr1. Besides the small GTPases, DAG is also able to activate Pkc1. Pkc1 will be responsible to continue the pathway through regulation of the MAPK (mitogen-activated protein kinase) cascade (Bck1, Mkk1, Mkk2 and Mpk1) essential to the cell wall maintenance. Besides this, Pkc1 is also important to the phosphorylation of the cryptococcal transcriptional factor Sp1, important to regulate virulence factors expression in *Cryptococcus*. At the end of the pathway, several transcription factors (represented in the nucleus) will control the expression of factors that will work on the maintenance of the cell wall integrity. In *Cryptococcus*, some integrants of the CWI pathway are important for morphological changes that cryptococcal cells suffer during the interaction with host: capsule (red signal) and Titan cells (green signal) formation. Some factors such as Cck1 connect CWI pathway to other cell wall integrity pathway HOG, whose factors also contribute to morphological changes in *Cryptococcus*. In the figure, the bars represent 20 μm. This figure was elaborated based on [18,27,28].

In addition, disruption of the transcription factor Crz1 in *C. neoformans* (homologue to calcineurin-responsive zinc finger ScCrz1 in *S. cerevisiae*), causes a phenotype similar to a CWI mutant (*cna1*), such as with increased susceptibility to cell wall stressors. This suggests that in *C. neoformans* Crz1 could be linked to CWI pathway [34,35], revealing CWI pathway as an open field of study in the cryptococcal research.

3. Cell Wall Integrity and Virulence Factors in *Cryptococcus neoformans*

Cryptococcus neoformans has developed and acquired several features that allow the infection and virulence in a wide of hosts, such as the polysaccharide capsule, growth at 37 °C, melanin production and morphological changes (capsule increase and Titan cells formation). Cell wall integrity is crucial to all of these processes. Responses to environmental stresses are mediated by MAPK phosphorylation cascade [31] and the components of the PKC1-MAPK signaling pathway are vital to maintain the integrity of the cell.

3.1. Adaptation at High Temperature

The ability of *C. neoformans* to grow at 37 °C turns it into an invasive pathogenic fungus and excellent model to study virulence, pathogenesis and host interaction. Most fungal species have the ability to grow optimally between 25–35 °C, however, only a few species of fungi are considered thermotolerant (growth > 35 °C and until 40 °C) [36].

The Pkc1 signaling pathway plays an important role in the response to thermal stress regulated by Rho GTPases [30]. In *C. neoformans*, three Rho GTPases (Rho 1, Rho10 and Rho11) have been identified that are important for cell growth and in response to temperature stress [25,30,31]. Ballou et al. suggested that CdC42 proteins perform complementary functions with other Rho-like GTPases in response to host and environmental temperatures [37]. Furthermore, recent studies have shown that Crz1 and Had1 transcription factors play an important role in cell wall integrity, thermotolerance and virulence in *C. neoformans* [38].

Besides, all mutants of MAPK module (*Bck1*, *Mkk2* and *Mpk1*) have a growth defect at host temperature and are consequently less virulent in mouse models [30,32].

3.2. Capsule and CWI

The capsule is the main phenotypic feature of *C. neoformans*. It is mainly composed of two polysaccharides, glucuronoxymannan (GXM), which constitutes 90–95%, and glucuronoxylomanogalactan (GXMGal) which is 5–8% [13,39]. It also contains a small proportion of mannoproteins (<1%) [39,40]. Despite its importance, the mechanisms involved in capsule synthesis are still not fully understood. The main model suggests that the main components are synthesized intracellularly in the endoplasmic reticulum (ER), and exported to the extracellular space in vesicles. Then, the content of these vesicles is released, and the polysaccharide fibbers attach to the cell wall, in particular, to the α-1,3-glucan [41,42]. At the moment, many genes have been involved in capsule synthesis. Furthermore, several transduction pathways are also required for this process. The main ones are the Pka1-AMPc pathway and other MAPKs such as Hog1. Relatively recent studies speak about the possibility that the MAPK *Bck1* module may have a role in the regulation of capsule synthesis [43].

3.3. CWI and Melanin

Cryptococcus neoformans has also the capacity to accumulate melanin, which is an insoluble dark pigment widely found in nature. Melanin synthesis in this fungus occurs only in the presence of dipholic compounds, such as L-DOPA and is dependent on the enzyme diphenol oxidase, which is encoded by two genes, *LAC1* and *LAC2*, with *LAC1* being the main producer of melanin [44]. These enzymes localize the melanin that accumulates in the cell wall and contributes to the survival of *C. neoformans*, maintaining the integrity of this structure [45]. Melanin protects fungal cells against stress factors, which facilitates the survival of this pathogen in the host [46–50]. Melanin also contributes to virulence [51] and dissemination from the lungs to other organs [48]. The synthesis of this pigment is dependent on activation of the Pkc1 pathway by diacylglycerol (DAG) [52]. Furthermore, the DAG C1 binding domain from Pkc1 is necessary for proper laccase localization at the cryptococcal cell wall and for melanin synthesis [31,53].

4. CWI Pathway and Morphological Changes in *Cryptococcus neoformans*

During infection, *C. neoformans* undergoes morphological changes that allow the fungus to evade and resist the host's immune system [54–56]. *Cryptococcus* can increase its size in two different ways: by enlarging the size of the capsule without changing the size of the cell body, or by increasing both structures [57–59]. This last process results in the appearance of the so-called Titan cells [57,60], which can reach a size of up to 70 μm [58,60,61]. These changes have a great impact on the *C. neoformans* infectious process, as Titan cells are resistant to stress factors [60,62], they are not phagocytosed by host cells [60,63], and polarize the host immune response to a Th2 type response, a non-protective response against *Cryptococcus* infection [64].

Recent studies showed that some enzymes in the CWI pathway and in particular the Pkc1 are important in cryptococcal morphology [18,29,31,65]. In *Cryptococcus*, PKC can be activated by diacylglycerol (DAG) which is produced by the degradation of phospholipids [52] and by the Rho1 GTPase, activated by the Rom2 protein [18,25,31,35]. Cryptococcal PKC besides essential to CWI pathway activation and maintenance of cell integrity [31]. Pkc1 activation also plays an important role to the capsule and Titan cells formation, among other important virulence factors to *Cryptococcus*.

5. CWI Pathway and *Cryptococcus* Capsule Growth

Disruption of the *PKC1* gene in *C. neoformans* results in an aberrant capsule, and this phenotype is restored after complementation of the mutant strain with the wild type gene [31]. The effect on capsule size is not related to the production of the polysaccharide, which surprisingly is overproduced by *C. neoformans Pkc1* mutants, but to the polysaccharide attachment to the cell wall [31]. Heung et al. [52] described that deletion of the C1 domain of Pkc1, which prevents the activation of Pkc1 by DAG, leads to a significant decrease in capsule size with a significant decrease in the density of the polysaccharide density attached to the cryptococcal cells [52].

Cryptococcus neoformans mutants of different proteins from the Pkc1 cascade also revealed a strong role of this pathway in the integrity and maintenance of the cell wall, influencing cryptococcal morphology [25]. Ppg1 is a phosphatase, homolog to *S. cerevisiae* Sit4p that participates in the regulation of the Pkc1 pathway, and its disruption in *C. neoformans* showed it is important for cell integrity and results in a significant decrease in the capsule size, with a genotype resembling the *C. neoformans* capsule-deficient *cap59* strain [25]. A decrease in capsule size is also observed in a *C. neoformans* mutant to Lrg1, a GTPase-activating protein that facilitates the conversion of GTP to GDP on the Rho1p regulating the Pkc1 pathway [25].

Interestingly, a cryptococcal Pkc1-dependent phosphorylation transcription factor Sp1 appears to be important to regulate the expression of some cryptococcal virulence factors, including capsule. *C. neoformans sp1* mutant strain displays a significant increase in capsule size, even in non-inducing conditions, indicating that this transcription factor negatively regulates capsule formation in *C. neoformans* [35]. In contrast, a *sp1 pkc1* double mutant displays a hypocapsular phenotype, which correlated with a reduction of other enzymes involved in cell wall synthesis, such as the Fks1 β-glucan synthase. For this reason, the authors hypothesized that the decreased capsule size in this strain may be the result of a decrease in attachment sites to the capsular polysaccharide, impairing the correct capsule assembling [35].

Pkc1 plays an important role in connecting different cell wall integrity pathways. Pkc1-dependent Cck1 casein kinase I protein is a regulator of the dephosphorylation of Hog1 under stress conditions [66], being the Pkc1 pathway constitutively activated when *HOG1* is disrupted in *C. neoformans* [67] and this intimately connects the high-osmolarity glycerol response (HOG) and PKC pathways, essential for cell integrity. Although disruption of *CCK1* does not affect *C. neoformans* capsule, *CCK1* is also responsible for regulating phosphorylation of Mpk1, a *C. neoformans* MAPK essential to cell wall integrity [32]. However, the HOG pathway plays an important role in the encapsulation of cryptococcal cells [18], revealing the importance of cell wall pathways in capsule assembling.

The HOG pathway, one of the most important components of the MAPK cascade, regulates stress, sexual differentiation and virulence in *C. neoformans* through sensing external challenges, leading to protective responses against different stresses such as high osmolarity, ion concentration and temperature among others, ensuring cell integrity [32,67–69]. HOG pathway seems to play a role in the negative regulation of the cryptococcal capsule. Bahn et al. [67] showed that a *C. neoformans hog1* mutant strain presents an excessive growth in the capsule when compared to the wild type strain. Under normal conditions, Hog1 is phosphorylated by Pbs2 [32,67,70] and under stress, Hog1 is rapidly dephosphorylated. The phosphorylated status is important in the induction or repression of the capsule in *C. neoformans*, being the capsule repressed when Hog1 is phosphorylated [32], showing that the capsule is probably induced when cells are under stress when Hog1 is dephosphorylated. In contrast, Huang et al. [68] showed that deletion of Hog1 in the closely-related species *C. gattii* leads to a decrease in capsules size, which suggests that Hog1 function in virulence is strongly dependent on the genetic background or species from the *Cryptococcus* genus [68].

In *C. neoformans*, Hog1 regulates two transcriptional factors, Atf1 e Mbs1 [71,72]. Disruption of *ATF1* in *C. neoformans* leads a significant increase in capsule size, which resembles the phenotype of *hog1* mutants [72]. However, disruption of *MBS1* does not have any effect on capsule size, indicating that Hog1 under normal conditions represses capsule synthesis through Mbs1 [71]. Analysis of global gene transcription in a *C. neoformans hog1* mutant showed a modest increase of expression of capsule-associated (*CAP*) genes that does not explain this increase in capsule size, so the authors discussed that Hog1 regulates capsule size in *C. neoformans* through other non-characterized factors [73].

6. CWI Pathway and Titan Cells Formation

Titan cells formation is characterized by the enlargement not only of the capsule, but also of the cell body. The molecular characterization of Titan cells has been limited by the lack conditions to obtain them in vitro [74]. However, three independent groups have described several in vitro conditions in which *C. neoformans* significantly increases its cell size [65,75,76], which has facilitated to gain knowledge on the molecular mechanisms involved in Titan cells formation.

In vitro pharmacological inhibition of Pkc1 in *C. neoformans* impairs Titan cells formation [65], indicating that this pathway plays a role in cell growth in this fungus. Interestingly, induction of Titan cells in vitro occurs in the presence of serum, and it has been shown that the polar lipidic fraction of this component induces Titan cells formation in vitro. For this reason, it has been hypothesized that DAG present in the serum induces Pkc1 activation and Titan cells formation in vitro [65].

All these studies show that the morphological changes in *C. neoformans* are responses to extracellular environmental factors and that the proteins of the cell wall integrity pathways play a very important role in this process. Figure 1, brings an overview on cryptococcal CWI pathway and its impact in the *C. neoformans* morphological changes.

7. Cell Wall Integrity Signaling and Antifungal Therapies

The fungal cell wall is composed of molecules that are not present in the human body and since it is an essential structure for the viability, it constitutes an ideal target for antifungal compounds and immunotherapies [20]. The antifungal drugs currently used in the clinic are insufficient and limited and furthermore, some developing countries do not have access to all antifungals due to their high cost [77–79]. Some of the limitations of the study and approval of new antifungals are that it is a slow and expensive process. Moreover, selection of intrinsic and acquired resistance by some species of fungi is becoming a clinical problem, which raises the challenge of finding effective new therapeutic options [80,81].

Echinocandins (caspofungin, micafungin and anidulafungin) are a class of antifungals that inhibit the synthesis of β-glucan in the fungal cell wall through non-competitive inhibition of the enzyme β-1, 3-glucan synthase. This antifungal treatment is ineffective against *C. neoformans* because basiodiomycetes contain low amounts of β-glucan at the cell wall [20]. According to the literature, there are different studies on the effects of echinocandins on some components of the CWI pathway in *C. neoformans*. For example, mutants of the Pkc1 pathway of *S. cerevisiae* are more sensitive to caspofungin [82,83]. However, Gerik et al. described that there were no significant differences between mutants of the components of the Pkc1 pathway and wild strains exposed to caspofungin in *C. neoformans* [25]. The authors refer to the fact that there are some differences between the species, since *C. neoformans* has a single MAP kinase kinase (MKK2) involved in the phosphorylation cascade of Pkc1, in contrast to *S. cerevisiae*, where there are two functional ones, Mkk1 and Mkk2 [25]. The difference in resistance to caspofungin between these two fungi may be due to the fact that in *S. cerevisiae* there are many genes of the Pkc1 pathway that are missing, while this has not been observed in any mutant of *C. neoformans* in susceptibility to caspofungin [25,82].

Exposure to caspofungin in *S. cerevisiae* leads to rapid activation of the Pkc1 pathway through *SLG1*, which is a surface sensor that has no homologue in *C. neoformans* [82].

It has also been described that *MPK1* is essential for the survival of *C. neoformans* in the presence of caspofungin. There is a relationship between calcium/calmodulin-dependent protein phosphatase calcineurin and *MPK1* in regulating cell integrity, since in the absence of functional calcineurin, *MPK1* helps to protect against caspofungin. Furthermore, mutants lacking *MPK11* are more sensitive to caspofungin compared with the wild type strains [32].

In *C. neoformans*, there is a relationship between calcineurin and the CWI signaling pathway at the transcriptional level. The *crz1* mutant has other phenotypes associated with CWI, as sensitivity to caspofungin. *CRZ1* in *S. cerevisiae* is associated with calcineurin and may play an important role in CWI signaling pathway [28,35].

Nowadays the treatment used for cryptococcosis is Amphotericin B (AmB), Fluconazole (FLZ) and 5-flucytosine (5FC). In severe cases, a combination of AmB and 5FC is applied and FLZ is used as maintenance treatment [84]. However, the treatment used with AmB and 5FC has shown some difficulties since in developing countries, 5FC is not available and the toxicity of the treatment persists. In regions with limited resources, the treatment available is often only with FLZ [85].

Fluconazole is a triazole antifungal and acts on the enzyme P450 lanosterol 14α-demethylase (ERG11), inhibiting the synthesis of ergosterol. Prolonged administration of FLZ can result in failure of the treatment of cryptococcosis and induce relapses in the patient. Besides, Sionor et al. in 2009 identified in clinical isolates of *C. neoformans* a pattern of intrinsic resistance to FLZ, called heteroresistance [86]. PKC-MAPK cascade has been demonstrated to affect the susceptibility to FLZ in *C. albicans* and *S. cerevisiae* [87,88]. Lee et al. suggested that in *C. neoformans* PKC-MAPK pathway is also involved in susceptibility to FLZ. The study analyzed the effects of deletions of the components of the Pkc1 pathway, Mkk2, Bck1 and Mpk1 pathway in *C. neoformans* mutants and they observed that these mutants showed hypersensitivity to FLZ compared to wild type strains [89].

In a recent work published by Sellers-Moya et al. the authors showed how imidazoles, and particularly clotrimazole, trigger more changes in MAPK phosphorylation than triazoles, which implies an activation of MAPKs Hog1 and Slt2 high HOG and CWI pathways, respectively in *S. cerevisiae* [90]. They showed that clotrimazole induces a high level of ROS in this yeast compared with fluconazole. In addition, resistance to this drug depends on both routes, HOG and CWI pathways. Regarding CWI they observed the activation by clotrimazole of Slt2 does not depend on pkc1 [90]. Amphotericin B alters the permeability of the fungal membrane by forming ion channels. It has been described that mutants with defects in cell integrity demonstrated increased sensitivity to AmB [91]. In particular, in a work carried out with different strains of *C. neoformans*, the *bck1* mutant (absence of

MAPKKK in the cell integrity MAP kinase cascade) was more sensitive to AmB compared to wild type strains [91].

As mentioned above, the components of the fungal cell wall are not present in humans, so this structure is an excellent target for antifungal therapy. The CWI pathway is required for the adaptation of all types of stresses that disturb the cell wall, among which are included antifungals [28]. Although humans have homologous targets to the core components of the CWI pathway (Pkc1, MAP kinases, or Rho GTPases), studies in this pathway can optimize the specificity of the inhibitory components as will be described below.

Compounds that have a mechanism of action targeting the inhibition of CWI signaling pathway appear as promising options for the development of new antifungals, since it has been shown that *mpk1* mutants, which do not have functional CWI, are avirulent in murine models of cryptococcosis and do not survive human body temperature [32,92].

To find molecules with antifungal action against *C. neoformans*, Hartland et al. performed a screening with more than 360,000 molecules. As a result, the authors highlight molecules that can interfere with the fungal cell wall integrity, impacting the signaling of the cell wall integrity MAP kinase cascade [92].

8. Perspectives

The fungal cell wall is an essential structure for the cell because it maintains the cellular integrity and viability but also mediates the connections of the cell with the environment. The cell wall protects cells against stress conditions from the environment or the host. Cells respond and adapt to stress through a complex transcriptional program regulated by signaling pathways. The cell wall integrity pathway is the key for regulating this adaptive response. The structure of the cell wall and CWI is conserved among fungi, however, in the case of *C. neoformans* there are many aspects of this pathway that are still unknown. In this review, we have summarized how this pathway is involved in different aspects of the biology, including those involved in virulence (polysaccharide capsule and melanin). Several MAPK kinase pathways (such as Hog1 or Pkc1) are important in these processes. In addition, one of the most striking characteristics of *C. neoformans* is the ability to significantly increase its size (appearance of Titan cells), and some enzymes from the CWI pathway are important in these morphological changes. Finally, the use of the CWI pathway as a pharmacological target can help to improve existing antifungals as well as to develop new and more effective treatments. Therefore, the CWI pathway is primordial for the intrinsic adaptation to stress by temperature or antifungals and is crucial for virulence factors of *C. neoformans*.

Supplementary Materials: The following are available online at https://www.mdpi.com/article/10.3390/jof7100831/s1. Table S1: Components of the CWI pathway in Saccharomyces cerevisiae and Cryptococcus neoformans.

Author Contributions: H.C.d.O., S.A.R., Ó.Z. and N.T.-C. have prepared (data curation and writing-review) of this manuscript. I.G.-B. and N.T.-C. participated in the manuscript editing. N.T.-C. have participated in experimental planning of this manuscript. All authors have read and agreed to the published version of the manuscript.

Funding: This work was supported by Grant PID2020-114546RB-100 from the Spanish Ministry for Science and Innovation. De Oliveira H.C. was supported by Inova Fiocruz/Fundação Oswaldo Cruz. Rossi SA is funded by postdoctoral fellowship from Fundacão de Amparo á pesquisa do Estado de São Paulo (reference FAPESP-BEPE 2020/09919-0). García-Barbazán Iwas supported by a FPI fellowship (reference PRE2018-083436). Trevijano-Contador N is funded by a "Ayudas de Atracción de Talento Investigador" Contract of Community of Madrid (reference 2019-T2/BMD-14926).

Institutional Review Board Statement: Not applicable.

Informed Consent Statement: Not applicable.

Data Availability Statement: Not applicable.

Conflicts of Interest: The authors declare no conflict of interest.

References

1. Lazera, M.S.; Salmito Cavalcanti, M.A.; Londero, A.T.; Trilles, L.; Nishikawa, M.M.; Wanke, B. Possible primary ecological niche of *Cryptococcus neoformans*. *Med. Mycol.* **2000**, *38*, 379–383. [CrossRef]
2. Casadevall, A.; Perfect, J. *Cryptococcus Neoformans*; ASM: Washington, DC, USA, 1998; p. 541.
3. Rohatgi, S.; Pirofski, L.A. Host immunity to *Cryptococcus neoformans*. *Future Microbiol.* **2015**, *10*, 565–581. [CrossRef]
4. Rajasingham, R.; Smith, R.M.; Park, B.J.; Jarvis, J.N.; Govender, N.P.; Chiller, T.M.; Denning, D.W.; Loyse, A.; Boulware, D.R. Global burden of disease of HIV-associated cryptococcal meningitis: An updated analysis. *Lancet Infect. Dis.* **2017**, *17*, 873–881. [CrossRef]
5. Casadevall, A. Antibody-mediated protection against intracellular pathogens. *Trends Microbiol.* **1998**, *6*, 102–107. [CrossRef]
6. Feldmesser, M.; Kress, Y.; Novikoff, P.; Casadevall, A. *Cryptococcus neoformans* is a facultative intracellular pathogen in murine pulmonary infection. *Infect. Immun.* **2000**, *68*, 4225–4237. [CrossRef] [PubMed]
7. Mylonakis, E.; Ausubel, F.M.; Perfect, J.R.; Heitman, J.; Calderwood, S.B. Killing of *Caenorhabditis elegans* by *Cryptococcus neoformans* as a model of yeast pathogenesis. *Proc. Natl. Acad. Sci. USA* **2002**, *99*, 15675–15680. [CrossRef]
8. Warpeha, K.M.; Park, Y.D.; Williamson, P.R. Susceptibility of intact germinating *Arabidopsis thaliana* to human fungal pathogens *Cryptococcus neoformans* and *C. gattii*. *Appl. Environ. Microbiol.* **2013**, *79*, 2979–2988. [CrossRef] [PubMed]
9. Heitman, J.; Kozel, T.R.; Kwon-Chung, K.J.; Perferct, J.R.; Casadevall, A. *Cryptococcus. From Human Pathogen to Model Yeast*; ASM Press: Washington, DC, USA, 2011; p. 620.
10. Velagapudi, R.; Hsueh, Y.P.; Geunes-Boyer, S.; Wright, J.R.; Heitman, J. Spores as infectious propagules of *Cryptococcus neoformans*. *Infect. Immun.* **2009**, *77*, 4345–4355. [CrossRef]
11. McFadden, D.C.; Casadevall, A. Capsule and melanin synthesis in *Cryptococcus neoformans*. *Med. Mycol.* **2001**, *39* (Suppl. 1), 19–30. [CrossRef]
12. Vecchiarelli, A. Immunoregulation by capsular components of *Cryptococcus neoformans*. *Med. Mycol.* **2000**, *38*, 407–417. [CrossRef] [PubMed]
13. Zaragoza, O.; Rodrigues, M.L.; De Jesus, M.; Frases, S.; Dadachova, E.; Casadevall, A. The capsule of the fungal pathogen *Cryptococcus neoformans*. *Adv. Appl. Microbiol.* **2009**, *68*, 133–216. [CrossRef] [PubMed]
14. Ellerbroek, P.M.; Hoepelman, A.I.; Wolbers, F.; Zwaginga, J.J.; Coenjaerts, F.E. Cryptococcal glucuronoxylomannan inhibits adhesion of neutrophils to stimulated endothelium in vitro by affecting both neutrophils and endothelial cells. *Infect. Immun.* **2002**, *70*, 4762–4771. [CrossRef]
15. Lipovsky, M.M.; Tsenova, L.; Coenjaerts, F.E.; Kaplan, G.; Cherniak, R.; Hoepelman, A.I. Cryptococcal glucuronoxylomannan delays translocation of leukocytes across the blood-brain barrier in an animal model of acute bacterial meningitis. *J. Neuroimmunol.* **2000**, *111*, 10–14. [CrossRef]
16. Mitchell, T.G.; Friedman, L. In vitro phagocytosis and intracellular fate of variously encapsulated strains of *Cryptococcus neoformans*. *Infect. Immun.* **1972**, *5*, 491–498. [CrossRef]
17. Murphy, J.W.; Cozad, G.C. Immunological unresponsiveness induced by cryptococcal capsular polysaccharide assayed by the hemolytic plaque technique. *Infect. Immun.* **1972**, *5*, 896–901. [CrossRef]
18. O'Meara, T.R.; Alspaugh, J.A. The *Cryptococcus neoformans* capsule: A sword and a shield. *Clin. Microbiol. Rev.* **2012**, *25*, 387–408. [CrossRef] [PubMed]
19. Wang, Z.A.; Li, L.X.; Doering, T.L. Unraveling synthesis of the cryptococcal cell wall and capsule. *Glycobiology* **2018**, *28*, 719–730. [CrossRef]
20. Garcia-Rubio, R.; de Oliveira, H.C.; Rivera, J.; Trevijano-Contador, N. The Fungal Cell Wall: Candida, Cryptococcus, and Aspergillus Species. *Front. Microbiol.* **2019**, *10*, 2993. [CrossRef] [PubMed]
21. Ponton, J. The fungal cell wall and the mechanism of action of anidulafungin. *Rev. Iberoam. Micol.* **2008**, *25*, 78–82. [CrossRef]
22. Gow, N.A.R.; Latge, J.P.; Munro, C.A. The Fungal Cell Wall: Structure, Biosynthesis, and Function. *Microbiol. Spectr.* **2017**, *5*. [CrossRef] [PubMed]
23. Agustinho, D.P.; Miller, L.C.; Li, L.X.; Doering, T.L. Peeling the onion: The outer layers of Cryptococcus neoformans. *Mem. Inst. Oswaldo. Cruz.* **2018**, *113*, e180040. [CrossRef]
24. Free, S.J. Fungal cell wall organization and biosynthesis. *Adv. Genet.* **2013**, *81*, 33–82. [CrossRef] [PubMed]
25. Gerik, K.J.; Donlin, M.J.; Soto, C.E.; Banks, A.M.; Banks, I.R.; Maligie, M.A.; Selitrennikoff, C.P.; Lodge, J.K. Cell wall integrity is dependent on the PKC1 signal transduction pathway in Cryptococcus neoformans. *Mol. Microbiol.* **2005**, *58*, 393–408. [CrossRef]
26. Garcia, R.; Bravo, E.; Diez-Muniz, S.; Nombela, C.; Rodriguez-Pena, J.M.; Arroyo, J. A novel connection between the Cell Wall Integrity and the PKA pathways regulates cell wall stress response in yeast. *Sci. Rep.* **2017**, *7*, 5703. [CrossRef] [PubMed]
27. Sanz, A.B.; Garcia, R.; Rodriguez-Pena, J.M.; Arroyo, J. The CWI Pathway: Regulation of the Transcriptional Adaptive Response to Cell Wall Stress in Yeast. *J. Fungi* **2017**, *4*. [CrossRef]
28. Dichtl, K.; Samantaray, S.; Wagener, J. Cell wall integrity signalling in human pathogenic fungi. *Cell. Microbiol.* **2016**, *18*, 1228–1238. [CrossRef]
29. Levin, D.E. Regulation of cell wall biogenesis in Saccharomyces cerevisiae: The cell wall integrity signaling pathway. *Genetics* **2011**, *189*, 1145–1175. [CrossRef]

30. Lam, W.C.; Gerik, K.J.; Lodge, J.K. Role of Cryptococcus neoformans Rho1 GTPases in the PKC1 signaling pathway in response to thermal stress. *Eukaryot. Cell* **2013**, *12*, 118–131. [CrossRef]
31. Gerik, K.J.; Bhimireddy, S.R.; Ryerse, J.S.; Specht, C.A.; Lodge, J.K. PKC1 is essential for protection against both oxidative and nitrosative stresses, cell integrity, and normal manifestation of virulence factors in the pathogenic fungus Cryptococcus neoformans. *Eukaryot. Cell* **2008**, *7*, 1685–1698. [CrossRef]
32. Kraus, P.R.; Fox, D.S.; Cox, G.M.; Heitman, J. The Cryptococcus neoformans MAP kinase Mpk1 regulates cell integrity in response to antifungal drugs and loss of calcineurin function. *Mol. Microbiol.* **2003**, *48*, 1377–1387. [CrossRef] [PubMed]
33. Jung, U.S.; Levin, D.E. Genome-wide analysis of gene expression regulated by the yeast cell wall integrity signalling pathway. *Mol. Microbiol.* **1999**, *34*, 1049–1057. [CrossRef] [PubMed]
34. Lev, S.; Desmarini, D.; Chayakulkeeree, M.; Sorrell, T.C.; Djordjevic, J.T. The Crz1/Sp1 transcription factor of Cryptococcus neoformans is activated by calcineurin and regulates cell wall integrity. *PLoS ONE* **2012**, *7*, e51403. [CrossRef]
35. Adler, A.; Park, Y.D.; Larsen, P.; Nagarajan, V.; Wollenberg, K.; Qiu, J.; Myers, T.G.; Williamson, P.R. A novel specificity protein 1 (SP1)-like gene regulating protein kinase C-1 (Pkc1)-dependent cell wall integrity and virulence factors in Cryptococcus neoformans. *J. Biol. Chem.* **2011**, *286*, 20977–20990. [CrossRef] [PubMed]
36. Perfect, J.R. Cryptococcus neoformans: The yeast that likes it hot. *FEMS Yeast Res.* **2006**, *6*, 463–468. [CrossRef]
37. Ballou, E.R.; Nichols, C.B.; Miglia, K.J.; Kozubowski, L.; Alspaugh, J.A. Two CDC42 paralogues modulate Cryptococcus neoformans thermotolerance and morphogenesis under host physiological conditions. *Mol. Microbiol.* **2010**, *75*, 763–780. [CrossRef] [PubMed]
38. Jung, W.H.; Son, Y.E.; Oh, S.H.; Fu, C.; Kim, H.S.; Kwak, J.H.; Cardenas, M.E.; Heitman, J.; Park, H.S. Had1 Is Required for Cell Wall Integrity and Fungal Virulence in Cryptococcus neoformans. *G3* **2018**, *8*, 643–652. [CrossRef] [PubMed]
39. Raschke, W.C.; Baird, S.; Ralph, P.; Nakoinz, I. Functional macrophage cell lines transformed by Abelson leukemia virus. *Cell* **1978**, *15*, 261–267. [CrossRef]
40. Zaragoza, O.; Chrisman, C.J.; Castelli, M.V.; Frases, S.; Cuenca-Estrella, M.; Rodriguez-Tudela, J.L.; Casadevall, A. Capsule enlargement in *Cryptococcus neoformans* confers resistance to oxidative stress suggesting a mechanism for intracellular survival. *Cell. Microbiol.* **2008**, *10*, 2043–2057. [CrossRef]
41. Rodrigues, M.L.; Nakayasu, E.S.; Oliveira, D.L.; Nimrichter, L.; Nosanchuk, J.D.; Almeida, I.C.; Casadevall, A. Extracellular vesicles produced by *Cryptococcus neoformans* contain protein components associated with virulence. *Eukaryot. Cell* **2008**, *7*, 58–67. [CrossRef]
42. Rodrigues, M.L.; Nimrichter, L.; Oliveira, D.L.; Frases, S.; Miranda, K.; Zaragoza, O.; Alvarez, M.; Nakouzi, A.; Feldmesser, M.; Casadevall, A. Vesicular polysaccharide export in *Cryptococcus neoformans* is a eukaryotic solution to the problem of fungal trans-cell wall transport. *Eukaryot. Cell* **2007**, *6*, 48–59. [CrossRef]
43. Donlin, M.J.; Upadhya, R.; Gerik, K.J.; Lam, W.; VanArendonk, L.G.; Specht, C.A.; Sharma, N.K.; Lodge, J.K. Cross talk between the cell wall integrity and cyclic AMP/protein kinase A pathways in Cryptococcus neoformans. *mBio* **2014**, *5*. [CrossRef] [PubMed]
44. Pukkila-Worley, R.; Gerrald, Q.D.; Kraus, P.R.; Boily, M.J.; Davis, M.J.; Giles, S.S.; Cox, G.M.; Heitman, J.; Alspaugh, J.A. Transcriptional network of multiple capsule and melanin genes governed by the Cryptococcus neoformans cyclic AMP cascade. *Eukaryot. Cell* **2005**, *4*, 190–201. [CrossRef] [PubMed]
45. Zhu, X.; Gibbons, J.; Garcia-Rivera, J.; Casadevall, A.; Williamson, P.R. Laccase of *Cryptococcus neoformans* is a cell wall-associated virulence factor. *Infect. Immun.* **2001**, *69*, 5589–5596. [CrossRef]
46. Wang, Y.; Casadevall, A. Decreased susceptibility of melanized *Cryptococcus neoformans* to UV light. *Appl. Environ. Microbiol.* **1994**, *60*, 3864–3866. [CrossRef] [PubMed]
47. Rosas, A.L.; Casadevall, A. Melanization affects susceptibility of *Cryptococcus neoformans* to heat and cold. *FEMS Microbiol. Lett.* **1997**, *153*, 265–272. [CrossRef]
48. Noverr, M.C.; Williamson, P.R.; Fajardo, R.S.; Huffnagle, G.B. CNLAC1 is required for extrapulmonary dissemination of *Cryptococcus neoformans* but not pulmonary persistence. *Infect. Immun.* **2004**, *72*, 1693–1699. [CrossRef]
49. Liu, L.; Tewari, R.P.; Williamson, P.R. Laccase protects *Cryptococcus neoformans* from antifungal activity of alveolar macrophages. *Infect. Immun.* **1999**, *67*, 6034–6039. [CrossRef]
50. Casadevall, A.; Rosas, A.L.; Nosanchuk, J.D. Melanin and virulence in *Cryptococcus neoformans*. *Curr. Opin. Microbiol.* **2000**, *3*, 354–358. [CrossRef]
51. Salas, S.D.; Bennett, J.E.; Kwon-Chung, K.J.; Perfect, J.R.; Williamson, P.R. Effect of the laccase gene *CNLAC1*, on virulence of *Cryptococcus neoformans*. *J. Exp. Med.* **1996**, *184*, 377–386. [CrossRef]
52. Heung, L.J.; Luberto, C.; Plowden, A.; Hannun, Y.A.; Del Poeta, M. The sphingolipid pathway regulates Pkc1 through the formation of diacylglycerol in Cryptococcus neoformans. *J. Biol. Chem.* **2004**, *279*, 21144–21153. [CrossRef]
53. Heung, L.J.; Kaiser, A.E.; Luberto, C.; Del Poeta, M. The role and mechanism of diacylglycerol-protein kinase C1 signaling in melanogenesis by Cryptococcus neoformans. *J. Biol. Chem.* **2005**, *280*, 28547–28555. [CrossRef]
54. De Oliveira, H.; Castelli, R.F.; Kuczera, D.; Souza, T.N.; Marcos, C.M.; Scorzoni, L.; Nimrichter, L.; Rodrigues, M. Fungal Infections of the Central Nervous System. In *Encyclopedia of Mycology*; Zaragoza, O., Casadevall, A., Eds.; Elsevier: Amsterdam, The Netherlands, 2021; Volume 1, pp. 736–748.

55. Denham, S.T.; Brown, J.C.S. Mechanisms of Pulmonary Escape and Dissemination by Cryptococcus neoformans. *J. Fungi* **2018**, *4*. [CrossRef]
56. De Oliveira, H.C.; Trevijano-Contador, N.; Garcia-Rodas, R. Cryptococcal Pathogenicity and Morphogenesis. *Curr. Fungal Infect. Rep.* **2019**, *13*, 67–76. [CrossRef]
57. Casadevall, A.; Coelho, C.; Cordero, R.J.B.; Dragotakes, Q.; Jung, E.; Vij, R.; Wear, M.P. The capsule of Cryptococcus neoformans. *Virulence* **2019**, *10*, 822–831. [CrossRef]
58. Feldmesser, M.; Kress, Y.; Casadevall, A. Dynamic changes in the morphology of *Cryptococcus neoformans* during murine pulmonary infection. *Microbiology* **2001**, *147*, 2355–2365. [CrossRef] [PubMed]
59. Zaragoza, O.; Telzak, A.; Bryan, R.A.; Dadachova, E.; Casadevall, A. The polysaccharide capsule of the pathogenic fungus *Cryptococcus neoformans* enlarges by distal growth and is rearranged during budding. *Mol. Microbiol.* **2006**, *59*, 67–83. [CrossRef]
60. Zaragoza, O.; Garcia-Rodas, R.; Nosanchuk, J.D.; Cuenca-Estrella, M.; Rodriguez-Tudela, J.L.; Casadevall, A. Fungal cell gigantism during mammalian infection. *PLoS Pathog.* **2010**, *6*, e1000945. [CrossRef]
61. Okagaki, L.H.; Strain, A.K.; Nielsen, J.N.; Charlier, C.; Baltes, N.J.; Chretien, F.; Heitman, J.; Dromer, F.; Nielsen, K. Cryptococcal cell morphology affects host cell interactions and pathogenicity. *PLoS Pathog.* **2010**, *6*, e1000953. [CrossRef]
62. Gerstein, A.C.; Fu, M.S.; Mukaremera, L.; Li, Z.; Ormerod, K.L.; Fraser, J.A.; Berman, J.; Nielsen, K. Polyploid titan cells produce haploid and aneuploid progeny to promote stress adaptation. *mBio* **2015**, *6*, e01340-15. [CrossRef] [PubMed]
63. Okagaki, L.H.; Nielsen, K. Titan cells confer protection from phagocytosis in *Cryptococcus neoformans* infections. *Eukaryot. Cell* **2012**, *11*, 820–826. [CrossRef]
64. Garcia-Barbazan, I.; Trevijano-Contador, N.; Rueda, C.; de Andres, B.; Perez-Tavarez, R.; Herrero-Fernandez, I.; Gaspar, M.L.; Zaragoza, O. The formation of titan cells in *Cryptococcus neoformans* depends on the mouse strain and correlates with induction of Th2-type responses. *Cell. Microbiol.* **2016**, *18*, 111–124. [CrossRef] [PubMed]
65. Trevijano-Contador, N.; de Oliveira, H.C.; Garcia-Rodas, R.; Rossi, S.A.; Llorente, I.; Zaballos, A.; Janbon, G.; Arino, J.; Zaragoza, O. Cryptococcus neoformans can form titan-like cells in vitro in response to multiple signals. *PLoS Pathog.* **2018**, *14*, e1007007. [CrossRef]
66. Wang, Y.; Liu, T.B.; Patel, S.; Jiang, L.; Xue, C. The casein kinase I protein Cck1 regulates multiple signaling pathways and is essential for cell integrity and fungal virulence in Cryptococcus neoformans. *Eukaryot. Cell* **2011**, *10*, 1455–1464. [CrossRef] [PubMed]
67. Bahn, Y.S.; Kojima, K.; Cox, G.M.; Heitman, J. Specialization of the HOG pathway and its impact on differentiation and virulence of Cryptococcus neoformans. *Mol. Biol. Cell* **2005**, *16*, 2285–2300. [CrossRef] [PubMed]
68. Huang, Y.M.; Tao, X.H.; Xu, D.F.; Yu, Y.; Teng, Y.; Xie, W.Q.; Fan, Y.B. HOG1 has an essential role in the stress response, virulence and pathogenicity of Cryptococcus gattii. *Exp. Ther. Med.* **2021**, *21*, 476. [CrossRef]
69. Meyers, G.L.; Jung, K.W.; Bang, S.; Kim, J.; Kim, S.; Hong, J.; Cheong, E.; Kim, K.H.; Bahn, Y.S. The water channel protein aquaporin 1 regulates cellular metabolism and competitive fitness in a global fungal pathogen Cryptococcus neoformans. *Environ. Microbiol. Rep.* **2017**, *9*, 268–278. [CrossRef] [PubMed]
70. Bahn, Y.S.; Kojima, K.; Cox, G.M.; Heitman, J. A unique fungal two-component system regulates stress responses, drug sensitivity, sexual development, and virulence of Cryptococcus neoformans. *Mol. Biol. Cell* **2006**, *17*, 3122–3135. [CrossRef]
71. Song, M.H.; Lee, J.W.; Kim, M.S.; Yoon, J.K.; White, T.C.; Floyd, A.; Heitman, J.; Strain, A.K.; Nielsen, J.N.; Nielsen, K.; et al. A flucytosine-responsive Mbp1/Swi4-like protein, Mbs1, plays pleiotropic roles in antifungal drug resistance, stress response, and virulence of Cryptococcus neoformans. *Eukaryot. Cell* **2012**, *11*, 53–67. [CrossRef]
72. Kim, M.S.; Ko, Y.J.; Maeng, S.; Floyd, A.; Heitman, J.; Bahn, Y.S. Comparative transcriptome analysis of the CO2 sensing pathway via differential expression of carbonic anhydrase in Cryptococcus neoformans. *Genetics* **2010**, *185*, 1207–1219. [CrossRef]
73. Ko, Y.J.; Yu, Y.M.; Kim, G.B.; Lee, G.W.; Maeng, P.J.; Kim, S.; Floyd, A.; Heitman, J.; Bahn, Y.S. Remodeling of global transcription patterns of Cryptococcus neoformans genes mediated by the stress-activated HOG signaling pathways. *Eukaryot. Cell* **2009**, *8*, 1197–1217. [CrossRef]
74. Garcia-Rodas, R.; de Oliveira, H.C.; Trevijano-Contador, N.; Zaragoza, O. Cryptococcal Titan Cells: When Yeast Cells Are All Grown up. *Curr. Top. Microbiol. Immunol.* **2019**, *422*, 101–120. [CrossRef]
75. Dambuza, I.M.; Drake, T.; Chapuis, A.; Zhou, X.; Correia, J.; Taylor-Smith, L.; LeGrave, N.; Rasmussen, T.; Fisher, M.C.; Bicanic, T.; et al. The Cryptococcus neoformans Titan cell is an inducible and regulated morphotype underlying pathogenesis. *PLoS Pathog.* **2018**, *14*, e1006978. [CrossRef] [PubMed]
76. Hommel, B.; Mukaremera, L.; Cordero, R.J.B.; Coelho, C.; Desjardins, C.A.; Sturny-Leclere, A.; Janbon, G.; Perfect, J.R.; Fraser, J.A.; Casadevall, A.; et al. Titan cells formation in Cryptococcus neoformans is finely tuned by environmental conditions and modulated by positive and negative genetic regulators. *PLoS Pathog.* **2018**, *14*, e1006982. [CrossRef] [PubMed]
77. Lee, Y.; Puumala, E.; Robbins, N.; Cowen, L.E. Antifungal Drug Resistance: Molecular Mechanisms in Candida albicans and Beyond. *Chem Rev* **2021**, *121*, 3390–3411. [CrossRef] [PubMed]
78. Perfect, J.R.; Bicanic, T. Cryptococcosis diagnosis and treatment: What do we know now. *Fungal. Genet. Biol.* **2015**, *78*, 49–54. [CrossRef]
79. Miceli, M.H.; Lee, S.A. Emerging moulds: Epidemiological trends and antifungal resistance. *Mycoses* **2008**, *54*, e666–e678. [CrossRef]

80. Scorzoni, L.; de Paula, E.S.A.C.; Marcos, C.M.; Assato, P.A.; de Melo, W.C.; de Oliveira, H.C.; Costa-Orlandi, C.B.; Mendes-Giannini, M.J.; Fusco-Almeida, A.M. Antifungal Therapy: New Advances in the Understanding and Treatment of Mycosis. *Front. Microbiol.* **2017**, *8*, 36. [CrossRef]
81. Iyer, K.R.; Revie, N.M.; Fu, C.; Robbins, N.; Cowen, L.E. Treatment strategies for cryptococcal infection: Challenges, advances and future outlook. *Nat. Rev. Microbiol.* **2021**, *19*, 454–466. [CrossRef]
82. Reinoso-Martin, C.; Schuller, C.; Schuetzer-Muehlbauer, M.; Kuchler, K. The yeast protein kinase C cell integrity pathway mediates tolerance to the antifungal drug caspofungin through activation of Slt2p mitogen-activated protein kinase signaling. *Eukaryot. Cell* **2003**, *2*, 1200–1210. [CrossRef]
83. Markovich, S.; Yekutiel, A.; Shalit, I.; Shadkchan, Y.; Osherov, N. Genomic approach to identification of mutations affecting caspofungin susceptibility in Saccharomyces cerevisiae. *Antimicrob. Agents Chemother.* **2004**, *48*, 3871–3876. [CrossRef]
84. Perfect, J.R.; Dismukes, W.E.; Dromer, F.; Goldman, D.L.; Graybill, J.R.; Hamill, R.J.; Harrison, T.S.; Larsen, R.A.; Lortholary, O.; Nguyen, M.H.; et al. Clinical practice guidelines for the management of cryptococcal disease: 2010 update by the infectious diseases society of america. *Clin. Infect. Dis. Off. Publ. Infect. Dis. Soc. Am.* **2010**, *50*, 291–322. [CrossRef] [PubMed]
85. Bicanic, T.; Muzoora, C.; Brouwer, A.E.; Meintjes, G.; Longley, N.; Taseera, K.; Rebe, K.; Loyse, A.; Jarvis, J.; Bekker, L.G.; et al. Independent association between rate of clearance of infection and clinical outcome of HIV-associated cryptococcal meningitis: Analysis of a combined cohort of 262 patients. *Clin. Infect. Dis. Off. Publ. Infect. Dis. Soc. Am.* **2009**, *49*, 702–709. [CrossRef] [PubMed]
86. Sionov, E.; Chang, Y.C.; Garraffo, H.M.; Kwon-Chung, K.J. Heteroresistance to fluconazole in Cryptococcus neoformans is intrinsic and associated with virulence. *Antimicrob. Agents Chemother.* **2009**, *53*, 2804–2815. [CrossRef]
87. LaFayette, S.L.; Collins, C.; Zaas, A.K.; Schell, W.A.; Betancourt-Quiroz, M.; Gunatilaka, A.A.; Perfect, J.R.; Cowen, L.E. PKC signaling regulates drug resistance of the fungal pathogen Candida albicans via circuitry comprised of Mkc1, calcineurin, and Hsp90. *PLoS Pathog.* **2010**, *6*, e1001069. [CrossRef]
88. Parsons, A.B.; Brost, R.L.; Ding, H.; Li, Z.; Zhang, C.; Sheikh, B.; Brown, G.W.; Kane, P.M.; Hughes, T.R.; Boone, C. Integration of chemical-genetic and genetic interaction data links bioactive compounds to cellular target pathways. *Nat. Biotechnol.* **2004**, *22*, 62–69. [CrossRef] [PubMed]
89. Lee, H.; Khanal Lamichhane, A.; Garraffo, H.M.; Kwon-Chung, K.J.; Chang, Y.C. Involvement of PDK1, PKC and TOR signalling pathways in basal fluconazole tolerance in Cryptococcus neoformans. *Mol. Microbiol.* **2012**, *84*, 130–146. [CrossRef]
90. Sellers-Moya, A.; Nuevalos, M.; Molina, M.; Martin, H. Clotrimazole-Induced Oxidative Stress Triggers Novel Yeast Pkc1-Independent Cell Wall Integrity MAPK Pathway Circuitry. *J. Fungi* **2021**, *7*. [CrossRef]
91. Banerjee, D.; Umland, T.C.; Panepinto, J.C. De Novo Pyrimidine Biosynthesis Connects Cell Integrity to Amphotericin B Susceptibility in Cryptococcus neoformans. *mSphere* **2016**, *1*. [CrossRef]
92. Hartland, K.; Pu, J.; Palmer, M.; Dandapani, S.; Moquist, P.N.; Munoz, B.; DiDone, L.; Schreiber, S.L.; Krysan, D.J. High-Throughput Screen in Cryptococcus neoformans Identifies a Novel Molecular Scaffold That Inhibits Cell Wall Integrity Pathway Signaling. *ACS Infect. Dis.* **2016**, *2*, 93–102. [CrossRef]

Journal of *Fungi*

MDPI

Review

Substrates of the MAPK Slt2: Shaping Yeast Cell Integrity

Gema González-Rubio, Lucía Sastre-Vergara, María Molina, Humberto Martín * and Teresa Fernández-Acero *

Departamento de Microbiología y Parasitología, Facultad de Farmacia, Universidad Complutense de Madrid, and Instituto Ramón y Cajal de Investigaciones Sanitarias (IRYCIS), 28040 Madrid, Spain; gemagonzalezrubio@ucm.es (G.G.-R.); lusastre@ucm.es (L.S.-V.); molmifa@farm.ucm.es (M.M.)
* Correspondence: humberto@ucm.es (H.M.); teresafe@ucm.es (T.F.-A.); Tel.: +34-91-3941888 (H.M.)

Abstract: The cell wall integrity (CWI) MAPK pathway of budding yeast *Saccharomyces cerevisiae* is specialized in responding to cell wall damage, but ongoing research shows that it participates in many other stressful conditions, suggesting that it has functional diversity. The output of this pathway is mainly driven by the activity of the MAPK Slt2, which regulates important processes for yeast physiology such as fine-tuning of signaling through the CWI and other pathways, transcriptional activation in response to cell wall damage, cell cycle, or determination of the fate of some organelles. To this end, Slt2 precisely phosphorylates protein substrates, modulating their activity, stability, protein interaction, and subcellular localization. Here, after recapitulating the methods that have been employed in the discovery of proteins phosphorylated by Slt2, we review the bona fide substrates of this MAPK and the growing set of candidates still to be confirmed. In the context of the complexity of MAPK signaling regulation, we discuss how Slt2 determines yeast cell integrity through phosphorylation of these substrates. Increasing data from large-scale analyses and the available methodological approaches pave the road to early identification of new Slt2 substrates and functions.

Keywords: yeast; phosphorylation; cell wall integrity pathway; MAPK substrate; Slt2; kinase assay

Citation: González-Rubio, G.; Sastre-Vergara, L.; Molina, M.; Martín, H.; Fernández-Acero, T. Substrates of the MAPK Slt2: Shaping Yeast Cell Integrity. *J. Fungi* **2022**, *8*, 368. https://doi.org/10.3390/jof8040368

Academic Editor: Aaron Neiman

Received: 16 March 2022
Accepted: 31 March 2022
Published: 4 April 2022

1. Cell Wall Integrity Pathway: An Introductory View

All eukaryotes use signal transduction pathways mediated by mitogen-activated protein kinases (MAPKs) to adequately respond and adapt to distinct stresses and environmental changes. Upon stimulation, these pathways engage a three-tiered hierarchical phosphorylation cascade, involving the sequential activation of a MAP kinase kinase kinase (MAPKKK), a MAP kinase kinase (MAPKK), and finally a MAP kinase (MAPK). Studies over the last three decades performed by our group and others have revealed many details about the architecture, regulation, stimulation, and functionality of the cell wall integrity (CWI) MAPK pathway of budding yeast *Saccharomyces cerevisiae*. The structure of this pathway displays the typical arrangement of signaling components operating in MAPK pathways, in which a set of membrane-spanning mechanosensors (Wsc1, Wsc2, Wsc3, Mid2, and Mtl1) detect stimuli and relay signals to a GTPase (Rho1), mainly through the GDP/GTP exchange factor (GEF) Rom2 (Figure 1). One of the main effectors of Rho1 is the protein kinase Pkc1, which triggers a phosphorylation cascade composed of the MAPKKK Bck1; two redundant MAPKKs, Mkk1 and Mkk2; and the downstream MAPK Slt2 (also known as Mpk1) [1–3].

The concerted action of specific regulatory components and the precise molecular mechanisms of activation and inactivation ensure proper spatiotemporal control of the pathway, providing the right timing, tuning, and teaming in the signaling process. These additional components include positive regulators, such as the GEFs for Rho1, Rom1, and Tus1 [4]; and mammalian 3-phosphoinositide-dependent protein kinase 1 (PDK1) homologue kinases Pkh1 and Pkh2, which act on the protein kinase Pkc1 [5–7]; as well as some negative regulatory elements such as Rho1 GTPase-activating proteins (GAPs)

Sac7, Bem2, and Lrg1 [2,5] and down-modulatory phosphatases. The latter consists of the serin/threonine phosphatase Ptc1, tyrosine phosphatases Ptp2 and Ptp3, and dual-specificity phosphatases (DSPs) Msg5 and Sdp1. Phosphatases ultimately modulate the phosphorylation level and thereby the activation status of Slt2, either directly or through the downregulation of upstream kinases, as occurs with Ptc1 [2,8–10].

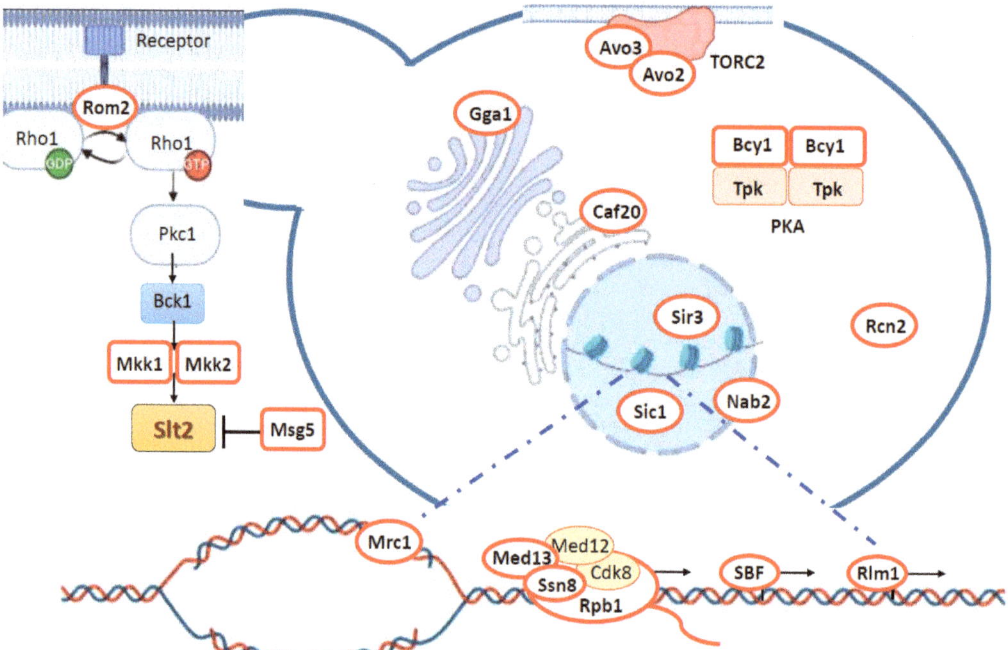

Figure 1. Schematic representation of CWI pathway and subcellular localization of Slt2 substrates within yeast cell. Substrates are depicted in white and encircled by a red line. The CWI pathway and most Slt2 substrates implicated in gene expression regulation can be found at the emergent bud and in an enlarged view at the bottom of the image, respectively.

Regarding the regulatory mechanisms, post-translational modifications such as phosphorylation of signaling components are fundamental for their governance [11]. Moreover, there are additional regulatory layers operating on the pathway. Among them, cell membrane surface metabolism deeply influences the functionality of the pathway; complex sphingolipids and phosphoinositol phosphates (PIPs) are essential for signaling, since interactions with these lipid species ensure the correct membrane localization of some CWI pathway components [5]. For example, Rom2 interacts through its PH domain with PI(4,5)P_2 [12]. The chaperoning activity of Hsp90 can also be considered as an example of a fundamental regulatory level for signaling because it is essential for dually phosphorylated Slt2 to activate its downstream target Rlm1 [13].

As studies progressed, the range of stimuli found to engage the CWI pathway expanded beyond the initially identified heat shock and cell wall-specific stresses to also include plasma membrane, oxidative, and genotoxic stresses, stationary phase, protein unfolding, and low or high pH, among other conditions [2,14,15]. This means that a great variety of physical changes and chemical compounds that induce those stresses trigger the activation of the CWI pathway [15], resulting in the stimulation of Slt2-specific MAPKKs Mkk1 and Mkk2, which phosphorylate threonine (T190) and tyrosine (Y192) residues within the T-E-Y motif of the activation loop of Slt2 [16]. An alternative mode of activation has been described for Slt2 driven by genotoxic stress, which does not involve the stimulation of these MAPKKs. Instead, DNA damage activates Slt2 through the induction of proteasomal degradation of the Msg5 phosphatase [17]. Since it is widely accepted that full activation of Slt2 occurs in parallel with its dual phosphorylation, stimulation of the pathway is commonly and easily monitored using commercial antibodies that detect the dually phosphorylated form of Slt2 [16]. However, some of these antibodies are also able to detect the monophosphorylated species at either T190 or T192. The fact that some Slt2 mutants with low catalytic activity have recently been shown to display Y192 hyperphosphorylation highlights the convenience of discerning among dual and each monophosphorylated form [18].

Regarding the functional outputs, Slt2 is known to be involved in various cellular functions, including the control of cell wall biogenesis, actin cytoskeleton dynamics [2], osmolyte and metalloid transport [19], iron homeostasis [20], cell cycle progression [21], control of sphingolipid synthesis [5], mitophagy and pexophagy [22], and endoplasmic reticulum and mitochondrial inheritance [23,24]. As detailed below, Slt2 is also implicated in the regulation of its own CWI pathway by mediating feedback mechanisms [11,25,26]. Interestingly, the introduction of synthetic genetic feedback circuits that alter this autocontrol upon stimulating conditions could lead to lethality due to detrimental hyperactivation of the pathway [27]. Slt2 also modulates other signaling routes, such as the TORC2 and PKA pathways [28,29]. Most of these functions require the catalytic activity of Slt2 on its target proteins, although non-catalytic mechanisms have also been described for Slt2 that promote the induction of a subset of cell wall stress-activated genes through the SBF transcription factor [30,31].

By taking advantage of the versatility and reversibility of phosphorylation, MAPKs can modify the activity, stability, interacting properties, or subcellular localization of their substrates. MAPK substrates are therefore essential links between stimulus-triggered signal transmission and the functional outputs elicited by MAPK pathways. MAPKs are considered serine/threonine protein kinases with a requirement for proline at the +1 position of the phosphorylation residue. The phosphorylation motif recognized by mammalian MAPKs ERK1/2 can be generalized as P-X-S/T-P [32], although proline at −2 seems to be dispensable for phosphorylation and proline at +1 may be absent in particular cases [33]. Interestingly, a peptide library screening approach suggested that in addition to proline at the +1 position, and in contrast to the other yeast MAPKs, Slt2 was especially selective at the −3 position, in which the preferred residue was arginine [34]. Although the cyclin-dependent kinases (CDKs) are also proline-directed serine/threonine kinases, their consensus phosphorylation signature is S/T-P-X-K [35].

In addition to the presence of an S/T-P phosphorylation motif in the primary sequence of the substrate, which is very common in all proteins, the existence of interaction motifs between MAPK and its targets also contributes to increasing the specificity and efficiency of the phosphorylation process. Substrate binding is commonly mediated by short MAPK docking sites present in different MAPK partners. D-motifs are the most used docking sites [36], which are frequently located N-terminally from the targeted phosphorylation site [37]. The consensus sequence of D-motifs is $\psi_{1-3}X_{3-7}\phi X\phi$ (where ψ is a positively charged residue, X is any residue, and ϕ is a hydrophobic residue) [37,38]. D-motifs interact with the negatively charged common docking (CD) domain and hydrophobic docking grooves at MAPKs [39,40]. Another binding motif to anchor MAPKs is the F-motif or DEF motif (docking site for ERK, FXF) [41]. While Slt2 displays a CD domain, it lacks an FXF-binding site [42].

The wide variety of stimuli and the numerous functions performed by the CWI pathway predict a high number of phosphorylation targets for Slt2. However, the complete repertoire of cellular substrates of this MAPK remains far from being elaborated. This may be why there are several reviews covering multiple aspects of this key signaling pathway, but an analysis focused on Slt2 substrates has not yet been accomplished. In this review, we aim to compile the bona fide Slt2 substrates found to date and attempt to both understand the kinase–substrate relationship and provide links between Slt2 activity and cellular functions of the CWI pathway.

2. Substrate Fishing: Methods Used for Searching Slt2 Targets

The order of action of genes in a regulatory hierarchy such as that governing MAPK signaling pathways can be determined by an epistasis analysis. Loss-of-function mutations in upstream components can be suppressed by either gain-of-function mutations or overexpression of downstream elements of the pathway. Conversely, loss-of-function mutations in downstream components can ameliorate defects provoked by upstream hyperactive alleles. Based on this concept, the transcription factor Rlm1 (resistance to lethality of *MKK1*^{P386} overexpression) was the first Slt2 target identified as a mutant suppressor of the toxicity caused by Slt2 hyperactivation triggered by overexpression of a constitutively active mutant of *MKK1* [43]. This study positioned Rlm1 downstream of Slt2, but the confirmation of direct phosphorylation required an additional biochemical analysis [44].

Over the years, several methods for identifying MAPK substrates based on phosphorylation assays have been used in yeast studies, ranging from traditional radioactive in vitro kinase assays to a large-scale in vivo phosphoproteomic analysis. Kinase enzymes catalyze the transfer of a phosphate group from a donor ATP molecule to a substrate. Therefore, in vitro kinase assays provide a direct measurement of catalytic activity by detecting and quantifying the formation of the phosphorylated product. The possibility of using purified proteins in these assays allows the addition of large amounts of both kinase and substrate, increasing the sensitivity and avoiding spurious reactions that may occur due to the eventual presence of contaminant kinases. Typically, activated polyhistidine- or glutathione S-transferase (GST)-tagged MAPKs are purified by immunoprecipitation or affinity chromatography from yeast cells. The MAPK can be obtained in its active state by either treating the cells with the adequate stimulus or overexpressing hyperactive MAPK-activating proteins. Tagged substrates are usually expressed and purified from either *Escherichia coli* or yeast cells. Protein yield is much higher with bacterial expression, but solubility or folding problems can arise, preventing effective protein recovery. When the phosphorylation site in the substrate is known and specific antibodies against the phosphorylated form are available, immunoblotting can be performed to detect the activity of the kinase on the substrate [45]. Alternatively, the use of radiolabeled [Y-^{32}P] ATP in the kinase reaction allows tracking of the transfer of radioisotope ^{32}P to the kinase substrate by either radioactive scintillation methods or SDS-PAGE followed by autoradiography analysis [46]. This type of assay can be scaled up for global kinase substrate identification by using protein microarrays representing the whole proteome [47]. Both single and large-scale radioactive in vitro kinase assays have been effectively used to identify Slt2 substrates.

Although radioactive kinase assays are very sensitive, other detection methods are usually preferred to avoid potential adverse health effects. Among them, analog-sensitive (AS) kinase technology provides a very powerful tool for non-radioactive selective labeling of substrates in in vitro kinase assays [48]. In this approach, the active site of the kinase is engineered by mutating the normal gatekeeper to a smaller residue, such as glycine, to create a larger active site pocket that allows the enzyme to accept a bulky ATP analog, in which the γ-phosphate is replaced with a thiophosphate moiety. Then, proteins thiophosphorylated by AS kinase are reacted with the thiol-specific alkylating agent p-nitrobenzyl mesylate (PNBM) to generate a thiophosphate ester, which can be detected by immunoblotting with commercial thiophosphate ester-specific antibodies [49]. The advantage of this methodology is that it can be used to confirm direct phosphorylation by the engineered AS kinase of candidate substrates either purified or in complex samples, such as cell extracts. The bulky ATP analog is exclusively recognized by the modified AS kinase, which transfers the thiophosphate group to its target proteins in the sample. Therefore, only the labeled AS kinase substrates are detected by immunoblotting. The identity of the tagged substrate is then confirmed with tag-specific antibodies. This strategy has been successfully used to confirm several candidate targets of Slt2 [50].

The above-mentioned kinase assays are designed to be performed in vitro. Among the methods available for analyzing phosphorylation in vivo, a classical one is an electrophoretic mobility shift analysis, based on the change in band migration usually displayed by phosphorylated proteins. When antibodies or tagged versions of proteins are available, it is possible to detect such changes, which may reflect a modification in the phosphorylation status [51]. Eliminating the shift by phosphatase treatment confirms that it is caused by phosphorylation. If the shift is not clearly observable by conventional SDS-PAGE, phosphate-affinity technology (Phos-tag) can be used to induce slower migration of phosphoproteins by reversible binding of the Phos-tag reagent to phosphate moieties on them [52]. Candidate substrates of a given kinase can be identified by assessing the loss of protein phosphorylation in mutant cells lacking the active kinase. Mutagenesis of putative phosphorylation residues in the substrate leading to the loss of mobility shift allows identification of the actual phosphosites.

On a large-scale basis, MS-based phosphoproteomics enable global studies of dynamic protein phosphorylation in vivo [53]. Phosphoproteomic data, including phosphosite identification and phosphoprotein quantification, obtained under MAPK-stimulated vs. unstimulated conditions, or in mutants lacking the MAPK vs. wild-type cells, provide a map of potential kinase substrates. This strategy has been applied by our group to yeast cells overexpressing a constitutively active version of Pkc1 to unveil putative Slt2 targets [54].

Both in vitro and in vivo approaches have advantages and limitations. The former can provide substrate identification in a highly selective and sensitive manner, but the alteration of the stoichiometry of the reaction and the loss of cellular compartmentalization may result in the identification of false kinase substrates. In vivo assays can associate a kinase with authentic phosphorylation events, but direct phosphorylation cannot be inferred without additional experimentation [55]. Therefore, a combination of different strategies is required to designate a protein as a genuine MAPK substrate; therefore, it is the best way to obtain a complete and accurate picture of the MAPK-dependent signaling network.

3. Targeting Different Yeast Processes: Genuine Slt2 Substrates

Several lines of evidence from both in vivo and in vitro assays provide increased confidence to consider a protein as a kinase substrate. Thus, we selected as bona fide Slt2 substrates those proteins that have been found to be directly phosphorylated by Slt2 in vitro, but they have also been strongly linked to Slt2 in vivo, either genetically or biochemically, for example, by being phosphorylated under CWI pathway activating conditions and/or displaying a genetic or physical interaction with Slt2 (Table 1, Figure 1). In the next section of this review, we recapitulate the Slt2 substrates that meet these criteria and discuss the relevance of this phosphorylation to yeast physiology described to date. In addition, we include information on potential Slt2 targets that have not yet been fully validated as authentic substrates.

Although some non-catalytic roles have been described for Slt2, most of its functions are carried out through direct phosphorylation of protein targets. Even though Slt2 transiently localizes in polarity sites, such as the tip of emergent buds and the bud neck during cytokinesis, this MAPK is particularly enriched at the nucleus [56]. The predominant nuclear localization of Slt2 reflects its preeminent role in regulating gene expression through the phosphorylation of nuclear targets. However, as new Slt2 cytosolic substrates are discovered, the implication of this MAPK in the control of additional cellular processes, including the cell cycle and signaling pathways, becomes more evident (Figure 1).

Table 1. Genuine Slt2 substrates, indicating in vitro and in vivo phosphorylation assays used, physiological role and consequences of Slt2-dependent phosphorylation, and precise phosphosites. Substrates are grouped by color based on their functional implication: modulation of signaling through CWI pathway (blue), regulation of other signaling pathways (green), regulation of gene transcription and mRNA transport (yellow), cell cycle control (gray), and undetermined Slt2-dependent phosphorylation role (pink).

Substrate	In Vitro Kinase Assay			In Vivo Kinase Assay		Effects on Protein Function/Cell Physiology	P-Site	Ref.
	Method	Stimulus [a]	Substrate Expressed in	Method [b]	Stimulus [a]			
Rom2	Radioactive	Heat stress	Yeast	Mobility shift	Heat stress	Cellular redistribution/ Inactivation of Rho1-mediated CWI signaling	ND	[25]
Mkk1	Radioactive AS kinase	Heat stress	Bacteria	Mobility shift	Vanadate	Downregulation /CWI signaling attenuation	ND	[50,57]
Mkk2	Radioactive	Heat stress	Bacteria	Mobility shift	Vanadate	Downregulation /CWI signaling attenuation	S^{50}	[57]
Msg5	Radioactive AS kinase	Heat stress	Bacteria [c]	Mobility shift	Heat stress Cell wall stress (CFW)	Reduced binding to Slt2 /Increased Slt2 activation	ND	[50,58]
Bcy1	Radioactive	Rapa	Bacteria	Anti-p-Bcy1 (T^{129})	Rapa, SDS, CR, vanadate, Bck1 [d]	PKA inhibition	T^{129}	[28]
Avo2	Radioactive	Pkc1 [e]	Bacteria	Phos-tag	Pkc1 [e]	Negative regulation of TORC2 activity	(T^{157}, T^{232}, S^{246}, S^{253}, S^{262}, T^{323}, S^{328}, T^{343}, S^{348}) [f]	[29]
Avo3	Radioactive	Pkc1 [e]	Bacteria [g]	Phos-tag [h]	Pkc1 [e]	ND	(S^{12}, T^{25}, T^{29}, S^{51}, S^{85}, S^{88}) [f]	[29]
Rlm1	Radioactive	Heat stress	Bacteria [g]	Mobility shift	Heat stress	Activation/ Transcriptional induction of cell wall genes	S^{374}, S^{427}, T^{439} (S^{234}, S^{261}, T^{276}, S^{299}, S^{518}, T^{646}, T^{654}) [f]	[44,59]
Swi6	Radioactive	Heat stress	Bacteria	Mobility shift	Heat stress	Nuclear exit/Downregulation of transcriptional activity	S^{238}	[60,61]

Table 1. *Cont.*

Substrate	In Vitro Kinase Assay			In Vivo Kinase Assay		Effects on Protein Function/Cell Physiology	P-Site	Ref.
	Method	Stimulus [a]	Substrate Expressed in	Method [b]	Stimulus [a]			
Swi4	Radioactive	Heat stress	Insect cells	Epistatic and physical interaction with Slt2	Heat stress	ND	ND	[60]
Ssn8 (Cyclin C)	Radioactive	Oxidative stress (H_2O_2)	Bacteria	Epistatic and physical interaction with Slt2	Oxidative stress (H_20_2)	Cytoplasmic release, mitochondrial targeting and degradation/ Mitochondrial fission and transcriptional activation of stress genes	S^{266}	[62,63]
Med13	Radioactive	Vanadate	Bacteria [g]	Epistatic interaction with Slt2	Oxidative stress (H_20_2)	Degradation/ Cytoplasmic release of cyclin C	T^{835}, T^{837}	[64]
Rpb1	Anti-p-Y	Heat stress	Bacteria [g]	Anti-p-Y	Cell wall stress (CFW)	Activation/ Transcriptional induction of stress genes	Y_1 of heptad repeats YSPTSPS	[65]
Sir3	Radioactive	Rapa, Heat stress	Bacteria, Bacteria [g]	Mobility shift	Rapa, Non-stimulated cells	Reduced association with subtelomeric sequences/ Derepression of *PAU* genes, Chronological lifespanshortening	S^{275}, S^{282} S^{289}, S^{295}	[66,67]
Nab2	Radioactive	Non-stimulated	Bacteria	Mobility shift	Heat stress	Non-hsp RNA retention in the nucleus/ Recovery of heat-stressed cells	T^{178}, S^{180}	[68]

Table 1. *Cont.*

Substrate	In Vitro Kinase Assay			In Vivo Kinase Assay		Effects on Protein Function/Cell Physiology	P-Site	Ref.
	Method	Stimulus [a]	Substrate Expressed in	Method [b]	Stimulus [a]			
Mrc1	Radioactive	NS	Bacteria	Anti-p-S/T	Heat stress	Delay in DNA replication to avoid transcription-replication conflicts	$T^{169}, S^{215}, S^{229}$	[69]
Sic1	Anti-p-Sic1 (T^{163})	Rapa	Bacteria	Anti-p-Sic1 (T^{163}) Phos-tag	Rapa	Stabilization/G1-S arrest	T^{163}	[70]
Rcn2	AS kinase	Cell wall stress (CR)	Bacteria Yeast	Phospho-peptide increase [h]	Pkc1 [e]	ND	$S^{152}, S^{160}, S^{255}$	[50,54]
Gga1	AS kinase	Cell wall stress (CR)	Bacteria Yeast	Phospho-peptide Increase [h]	Pkc1 [e]	ND	ND	[50,54]
Caf20	AS kinase	Cell wall stress (CR)	Bacteria Yeast	Phospho-peptide increase [h]	Pkc1 [e]	ND	T^{102}	[50,54]

ND (not determined), NS (not specified), CR (Congo red), CFW (calcofluor white), Rapa (rapamycin), Vanadate (sodium orthovanadate). [a]: Stimulus used for Slt2 activation. [b]: In vivo phosphorylation assay or alternative evidence (epistatic or physical interaction). [c]: Phosphorylation assay on a phosphatase dead version of Msg5 ($Msg5^{C319A}$) [58]. [d]: Expression of a constitutively active allele of Bck1 (*BCK1*-20) [71]. [e]: Overexpression of a constitutively active allele of Pkc1 ($PKC1^{R398A, R405A, K406K}$) [16]. [f]: Putative S/T-P sites. [g]: Phosphorylation assay on a protein fragment ($Avo3^{1-100}$, $Rlm1^{329-445}$, $Med13^{571-906}$, $Rpb1^{1556-1718}$, $Sir3^{1-439}$). [h]: Slt2-dependence was not tested.

3.1. Feedback Regulation of the CWI Pathway: Rom2, Mkk1/2, and Msg5

With the aim of ensuring the best fitness or even cell survival, Slt2 adjusts the signal flow through the CWI pathway by phosphorylating several of the components that constitute this signaling cascade. While some of these phosphorylation events enhance MAPK activation, others negatively regulate it. For example, under stress conditions that require full activation of the CWI pathway, Slt2 guarantees the induction of an appropriate adaptive cellular response by inhibiting the action of its main negative regulator, the phosphatase Msg5. However, when excessive activation of the CWI pathway is detrimental to the cell, Slt2 inhibits its own activation through a negative feedback loop acting on the GEF Rom2 and MAPK activating proteins Mkk1/2 (Table 1).

Rom2, Rom1, and Tus1 are the GEFs that activate the Rho1 GTPase in *S. cerevisiae*, acting upstream of the CWI MAPK module (Figure 1) [4,72]. Under normal growth conditions, basal activity of the CWI pathway is required to maintain the polarized distribution of actin and control cell wall synthesis [2]. In this situation, Rom2 localizes to the growing bud surface during bud emergence and to the bud neck during cytokinesis in a cell cycle-dependent manner [73,74], similarly to Rho1 [75]. However, in response to heat stress, which activates the CWI pathway, Rom2 becomes depolarized [12] and undergoes an Slt2-dependent mobility shift [25]. In vitro phosphorylation assays confirmed that Slt2 directly phosphorylates Rom2. This retrophosphorylation event would lead to the redistribution of Rom2 from the bud to the cell periphery and to subsequent inactivation of Rho1 activity, suggesting the existence of an Slt2-dependent feedback control that downregulates CWI pathway signaling by depriving Rho1 of its GEF when cells are exposed to adverse conditions [25]. Even though the exact position of this phosphorylation has not been described, several studies have revealed that Rom2 is phosphorylated at S171 and T216 by the CDK Cdc28 [76], and at five additional S/T-P sites (S126, S233, S391, T398, S494), which appear to be phosphorylated in several phosphoproteomic analyses, pointing to these sites as potential Slt2 targets (for details, see the *Saccharomyces* Genome Database at [77]).

Mkk1 and Mkk2 interact with Slt2 [78], leading to its phosphorylation at T190 and Y192 and its subsequent activation [79]. These MAPKKs were first described to be functionally redundant [80]. However, later studies attributed a preeminent role of Mkk1 over Mkk2 in the transmission of signals through the CWI pathway [9,16]. Specifically, it has been suggested that priming phosphorylation at Y192 is mainly carried out by Mkk1, and that this modification is necessary for subsequent phosphorylation at T190 [18]. On the other hand, Slt2 has been shown to phosphorylate Mkk1 and Mkk2 in vivo and in vitro by mobility shift and kinase assays, respectively [57]. In particular, Mkk2 retrophosphorylation at S50 does not affect its localization, stability, or ability to interact with Slt2, but it appears to downregulate its function, constituting a negative feedback regulatory mechanism of the CWI pathway [57]. Intriguingly, it has been demonstrated that, in contrast to the wild-type version of Slt2, the threonine monophosphorylatable mutant, Slt2^{Y192F}, is able to retrophosphorylate Mkk1 but not Mkk2 [18], suggesting that this feedback phosphorylation mechanism can be different in each MAPKK, providing high versatility to this pathway. Retrophosphorylation of upstream components by MAPKs has also been shown to regulate signaling specificity and intensity in other yeast [81,82] and mammalian [83] MAPK pathways.

Msg5 is a dual-specificity phosphatase (DSP) that negatively regulates the mating and CWI MAPK pathways by dephosphorylating threonine and tyrosine residues located at the activation loop of Fus3 and Slt2, respectively [10]. Importantly, following heat stress, Msg5 is phosphorylated in vivo in an Slt2 kinase activity-dependent manner. Moreover, Slt2 directly phosphorylates the phosphatase-dead version Mgs5^{C319A} in in vitro kinase assays. It is likely that this phosphorylation negatively regulates the interaction between Msg5 and Slt2, and has been thus proposed to serve as a mechanism by which Slt2 ensures its proper phosphorylation state and the subsequent cell wall remodeling response, as long as cell surface stress is present [58]. Since Msg5 also negatively regulates Fus3, it has been suggested that retrophosphorylation of Msg5 could prevent it from acting on Slt2, but not

Fus3, providing substrate specificity [58]. The precise residues phosphorylated by Slt2 have not yet been identified. Different phosphoproteomics studies have revealed that Msg5 contains 10 S/T-P phosphorylated sites, six of them in the N-terminal regulatory domain (S22, S62, S85, S115, S135, and T178) and four in the C-terminal half of the protein (S377, S422, T434, and T437). Although both domains are phosphorylated in vitro, only the latter seemed to be causative of the electrophoretic mobility shift displayed by Msg5 upon CWI activation [50].

Reciprocal regulation between MAPKs and DSPs is a conserved modulatory mechanism that is also found in mammalian cells [84–86]. Considering that Slt2 activation leads to a decrease in the overall amount of Msg5 [58], it is quite possible that Msg5 phosphorylation by Slt2 also negatively regulates the stability of this DSP, sustaining Slt2 activation by Msg5 degradation.

3.2. *Slt2 Impinges on Central Yeast Signaling Pathways via Bcy1 and Avo2/3*

Besides regulating its own pathway, Slt2 controls the activity of cAMP-dependent protein kinase (PKA) and target of rapamycin (TOR) complex 2 (TORC2) through phosphorylation of their subunits Bcy1 and Avo2/3, respectively (Table 1).

Bcy1 is the negative regulatory subunit of PKA [87]. Working together with the TOR complex 1 (TORC1) pathway, the PKA pathway regulates central processes for yeast growth, such as translation, ribosome biogenesis, autophagy, stress response, glucose metabolism, and life span [88]. In the absence of glucose, Bcy1 forms an inactive heterotetrameric complex composed by a Bcy1 dimer and two catalytic subunits, which are encoded by three homologs, *TPK1*, *TPK2*, and *TPK3*. However, in the presence of fermentable sugars, cAMP is synthetized, then it binds to Bcy1, causing its dissociation from the complex and the subsequent release of the Tpk active catalytic subunits that promote cell growth in these conditions [88,89].

The phosphorylation status of Bcy1 affects its affinity for catalytic subunits Tpk1–3 [90] and the localization of PKA. In yeast growing rapidly on glucose, PKA is almost exclusively localized in the nucleus. However, under stress conditions such as growth on a non-fermentable carbon source or an increase in temperature, Bcy1 is phosphorylated by kinases Yak1 and Mck1, causing its cytoplasmic localization via interaction with the protein Zds1 [91,92]. Bcy1 is also phosphorylated by Mck1 in response to DNA damage, restraining mitosis under these conditions [93]. In addition, inhibition of TORC1 with rapamycin leads to Bcy1 phosphorylation on several sites, including T129. This phosphorylation is totally abolished in cells lacking *SLT2*, and recombinant Bcy1 is phosphorylated in vitro on T129 by Slt2 purified from rapamycin-treated cells, indicating that Slt2 directly phosphorylates Bcy1. Upon rapamycin-dependent TORC1 inhibition, Sch9 becomes inactivated, promoting Slt2 hyperphosphorylation in vivo. This allows the subsequent Slt2-mediated phosphorylation of Bcy1 and the resulting inhibition of PKA catalytic activity. However, it is important to note that Bcy1 T129 phosphorylation is not exclusively dependent on TORC1 inhibition, but is also induced in vivo under different stresses, leading to activation of the CWI pathway, such as by cell wall disrupting agents or a genetically activated Bck1 version [28]. Strikingly, T129 is not a canonical proline-directed MAPK phosphosite, but rather a TS site. Considering that Bcy1 contains a TP site at T131 and that phosphoproteomic analyses have revealed that this residue is phosphorylated [77]), it is possible that T131 phosphoresidue participates in T129 recognition by Slt2.

The evolutionarily conserved TORC2 complex is an essential regulator of plasma membrane homeostasis in *S. cerevisiae*. TORC2 contains four essential core subunits (Avo1, Avo3, Lst8, and Tor2), two classes of peripherally located non-essential subunits (Avo2 and Bit61 or its paralog Bit2), and two essential ancillary subunits (Slm1 and Slm2) that shuttle from eisosomes to TORC2 following plasma membrane stress [94]. Phosphoproteomic analyses have detected in vivo phosphorylation of the MAPK S/T-P target motifs in all proteins comprising TORC2 except Bit2 [77].

Particularly, Avo2 and Avo3 are phosphorylated at several sites in vivo upon overexpression of a constitutively active version of Pkc1, which leads to the activation of Slt2 and Hog1 MAPKs. However, Avo2 hyperphosphorylation was markedly reduced in *slt2Δ* cells but not in *hog1Δ* cells, as detected by Phos-tag SDS-PAGE, confirming that Slt2 is the major MAPK responsible for this phosphorylation. In addition, Avo2 and the N-terminal fragment of Avo3 (1–100), which contain nine and six S/T-P sites, respectively, are robustly phosphorylated in vitro by Slt2. Mutation of these residues to non-phosphorylatable alanine eliminates and markedly reduces Avo2^{9A} and Avo3(1–100)^{6A} Slt2-dependent phosphorylation, respectively, confirming that both proteins are direct targets of Slt2 [29]. Slt2-dependent phosphorylation of Avo2 downregulates TORC2 activity on its primary downstream effector, Ypk1. Because the expression of the phosphomimetic Avo2^{9E} version renders cells sensitive to myriocin-induced sphingolipid depletion, showing significant displacement from the plasma membrane, it has been proposed that phosphorylation of Avo2 by Slt2 promotes its dissociation from TORC2 [29]. Thus, the Slt2-mediated phosphorylation of Avo2 constitutes the first evidence that an MAPK pathway regulates TORC2 function and reveals the regulatory circuitry by which *S. cerevisiae* controls the growth-promoting functions of TORC2 depending on cell wall stress.

3.3. Cell Wall Stress-Related Gene Transcription: Rlm1 and SBF Complex

In addition to its important role in regulating signaling events, the main function of Slt2 is to regulate transcription factors, with the aim of adjusting the transcriptional response to the environmental context [95]. CWI activation triggers the expression of a characteristic pattern of stress-related genes that allows the yeast to cope with cell wall or plasma membrane insults. This transcriptional program is regulated by Slt2 through the phosphorylation of two transcription factors, Rlm1 and SBF (Table 1) [96].

Rlm1 was the first identified Slt2 substrate, and its ability to rescue the growth inhibition caused by overexpression of a hyperactive version of Mkk1 [43] prompted additional investigations. Rlm1 drives the main CWI transcriptional reprogramming response to cell wall stress [96]. Hence, upon CWI activation, a particular cluster of genes is induced, including *SLT2* and *RLM1* [97]. This way, Rlm1 phosphorylation by Slt2 leads to a positive feedback circuit that ensures a high signaling flow through the pathway [26].

In vitro phosphorylation assays indicated that upon heat stress, Slt2 phosphorylates a fragment of Rlm1 comprising amino acids 329 to 445, containing three S/T-P motifs, S374, S427, and S439 [44]. A later study demonstrated that S427 and S439 residues are responsible for the majority of the transcriptional activation function of Rlm1 [98]. Neither of these two residues appears as phosphorylated in a phosphoproteomic analysis [77]. Further studies showed that the triple *rlm1* mutant in the aforementioned sites is still phosphorylated by Slt2 and retains transcriptional activity. Mutation of the seven additional S/T-P sites of Rlm1 outside of the DNA-binding domain, S234, S261, T276, S299, S518, T646, and T654, totally abolishes the Slt2-dependent phosphorylation of Rlm1 and its transcriptional activity, indicating that together with S374, S427, and S439, other residues of Rlm1 are phosphorylated by Slt2 and required for full transcriptional activation [99]. In vitro kinase assays with individual non-phosphorylatable Rlm1 mutants in these positions would make it possible to identify all the regulatory phosphorylation sites and to clarify the mechanism underlying Slt2-dependent Rlm1 activation.

Besides phosphorylation, activation by Slt2 requires the integrity of a MAPK docking site in Rlm1 [98]. However, a recent study demonstrated that the lack of a functional CD domain in Slt2 does not completely abolish signaling to Rlm1, suggesting that additional sites are involved in the interaction between Slt2 and Rlm1 [18]. This study also showed that Rlm1 phosphorylation depends on the presence of both T190 and Y192 residues within the activation loop of Slt2, since monophosphorylatable mutants are as ineffective as the catalytically inactive Slt2 version inactivating Rlm1 [18].

In a less prominent position than Rlm1, the CWI pathway also organizes its adaptive transcriptional response through activation of the SBF complex. SBF and MBF are transcription factor complexes that regulate the activation of the transcriptional program that mediates the G1/S transition [100]. SBF is a heterodimeric protein composed of DNA binding factor Swi4 and transcriptional activator Swi6. The interaction of Swi6 with Swi4 relieves an autoinhibitory intramolecular association of the Swi4 C-terminus with its DNA binding domain, allowing binding to its target promoters [101]. Although the CWI pathway has been related to the SBF transcriptional complex at multiple levels, its role in the regulation of the G1/S transcriptional program remains to be fully elucidated [21]. However, the role of SBF in regulating cell wall stress-induced gene transcription is better understood. It was demonstrated that overexpression of *SWI4* restored the viability of *slt2Δ* cells exposed to cell wall damage, and that SBF controlled the transcription of several cell wall-related genes, linking this complex to Slt2 function [102]. Although in vitro phosphorylation of Swi4 by Slt2 was reported quite some time ago [60], subsequent experiments showed that Slt2 activates SBF by a mechanism in which its kinase activity is dispensable. Slt2 relieves autoinhibitory Swi4 interaction by binding to an MAPK docking site near the C-terminal Swi6 binding site, leading to the association of Swi4 with the promoter region of a subset of genes (*FKS2, CHA1, YLR042C, YKR031w*) to enhance their expression upon cell wall stress [30,31,103].

On the other hand, in vivo and in vitro phosphorylation assays have demonstrated that Slt2 directly phosphorylates Swi6 upon heat shock. However, Swi6 is initially directed to the promoters of its stress-related target genes, where it associates with Swi4 and Slt2 for transcription to initiate in a Slt2-catalytic activity independent manner. It is the phosphorylation of Swi6 by Slt2 on Ser238 that subsequently drives this transcription factor out of the nucleus, since this modification interferes with the function of an adjacent nuclear localization signal (NLS). By this means, Slt2 exerts a negative regulation on the transcriptional function of Swi6 [61]. Given that Swi6 acts as a transcriptional activator, it is not clear whether the Slt2-mediated phosphorylation of Swi6 could affect the expression of the other target genes it regulates. In addition, the Slt2-mediated cytoplasmic re-localization of Swi6 could promote its association with other proteins.

3.4. Regulation of RNA Polymerase Holoenzyme Complex: Cyclin C, Med13, and Rbp1

Along with activating transcription factors Rlm1 and SBF, Slt2 also modulates gene expression by directly controlling the transcription machinery. Three components of the RNA polymerase holoenzyme complex are phosphorylated by Slt2 upon stress conditions: the cyclin C (Ssn8), the Mediator component Med13 (Ssn2), and the RNA polymerase II (Pol II) catalytic subunit (Rpb1) (Table 1). Pol II is a complex 12-subunit enzyme responsible for mRNA transcription in eukaryotes. Despite its complexity, Pol II itself lacks the ability to initiate transcription and needs to interact with different proteins and complexes to regulate its activity. For example, the association of Pol II with the Mediator is required for the initiation of transcription of some eukaryotic genes [104] and the repression of a subset of others [105]. The Mediator is a multi-subunit transcriptional coactivator complex of proteins, highly conserved among eukaryotes. Structurally, Mediator proteins are assembled separately into a core Mediator and a dissociable subcomplex called Cdk8 kinase module (CKM), containing Med12-Med13-Cdk8-Ssn8 [106]. The CKM plays an important function as a transcriptional repressor by modulating the activity of Rpb1 through phosphorylation of its carboxy-terminal domain (CTD) [107].

Yeast Ssn8 is generally degraded upon exposure to some stresses, which relieves the transcriptional machinery from its repression function and allows stress-induced gene expression [108,109]. Particularly, exposure to reactive oxygen species (ROS) promotes the Slt2-dependent phosphorylation of Ssn8 in the only MAPK target site S266, and subsequent proteasomal degradation. Interestingly, this is a specific role for Slt2 in response to ROS, but not to other types of cellular stress, such as thermal stress [62,63,110,111]. Upon Slt2 phosphorylation, Ssn8 translocates from the nucleus to the cytoplasm, where it promotes mitochondrial fission via a transcription-independent mechanism and induces programmed cell death before its complete degradation. In addition, the Mediator component Med13 is also degraded by the proteasome in response to ROS. This is a sequential process, in which the first step consists of priming phosphorylation by Cdk8-Ssn8, and the second step involves Slt2-mediated phosphorylation on S266 of Ssn8, facilitating Slt2-dependent Med13 phosphorylation, which in turn leads to its degradation [64]. This parallel regulation of two different targets within the same complex ensures Ssn8 nuclear release and subsequent activation of transcriptional and mitochondrial responses when facing oxidative stress, to finally promote cell survival.

Rpb1 CTD is subjected to extensive regulatory phosphorylation not only by the Ssn8-Cdk8 complex, as mentioned above, but also by many other protein kinases. Usually, these modifications are carried out in the CTD consensus repeated heptad $Y_1S_2P_3T_4S_5P_6S_7$. Even though the most common modifications take place in S2 and S5, Y1 phosphorylation has been shown to play a role in the regulation of gene expression. Slt2 is the kinase responsible for Rbp1 Y1 phosphorylation, and this event is important for upregulating the transcription of stress-induced genes, especially those related to cell wall stress, iron homeostasis, and processes related to oxide-reductive stress. Rbp1 Y1 phosphorylation by Slt2 seems to control the function of the Nrd1-Nab3-Sen1 (NSS) complex during the stress response, preventing this transcription termination complex from prematurely binding to the CTD [65]. So far, this is the only reported tyrosine phosphorylation executed by Slt2, except for the auto-phosphorylation on its TEY activation domain after removal of the C-terminal tail [112]. It is important to note that a non-phosphorylatable Rpb1 Y1F mutant exhibits increased Ssn8-Cdk8 promoter occupancy on several stress genes. Given the fact that Ssn8-Cdk8 is a transcriptional repressor, Slt2-mediated Rbp1 Y1 phosphorylation may facilitate Ssn8-Cdk8 recognition by Slt2, favoring stress gene expression.

3.5. Epigenetic Control of Gene Expression and Yeast Life Span Extension: Sir3 Phosphorylation

Chromatin consists of packaged genomic DNA interacting with histones. The dynamic control of chromatin is an important layer in the regulation of gene expression. Besides the direct modification of DNA and histones, remodeling protein complexes play a key role in chromatin regulation [113]. The silent information regulator (SIR) complex, which consists of Sir2, Sir3, and Sir4, is involved in gene silencing on mating-type loci *HML* and *HMR* on chromosome III and regions surrounding the telomeres [114,115].

Previous evidence indicated that Sir3 is a phosphoprotein that becomes hyperphosphorylated upon exposure to several stresses [116]. Further work carried out independently by two groups confirmed, by in vitro kinase assays, that Sir3 is a direct substrate of Slt2 (Table 1) [66,67]. Slt2 phosphorylates Sir3 upon rapamycin and chlorpromazine treatment, but not in nutrient starvation conditions [66]. This phosphorylation event impedes Sir3 from exerting its silencing function on subtelomeric regions and causes the derepression of certain cell wall stress-related genes [66], such as seripauperin (*PAU*) genes, which are located at these sites [117]. Sir3 phosphorylation by Slt2 therefore constitutes an additional mechanism by which the CWI MAPK pathway acts as an enzymatic regulator of gene expression. Mutants affected in cohesin, a key architectural chromosomal protein complex, have been shown to display telomere silencing defects as well as Slt2-dependent Sir3 hyperphosphorylation. However, in this case, Slt2 activity contributed to derepression only to a very limited extent, suggesting the existence of a Sir-independent mode of repression mediated by cohesin [118]. Several studies with different organism models, including *S. cerevisiae*, have shown that sirtuins are linked to aging [119]. Given that Sir3 is required for the recruitment of sirtuin Sir2 to telomeres [120], and that Sir2 activity has been linked to telomere silencing and a subsequent increase in life span [121], Sir3 phosphorylation by Slt2 could promote dissociation of Sir2 from telomeres, with an impact on life span and aging in *S. cerevisiae*. Indeed, Sir3 phosphorylation on S275 and S282 by Slt2 has been shown to shorten life span [67].

3.6. Regulation of mRNA Nuclear Export: Nab2

Restriction of the mRNA nuclear exit constitutes an important mechanism by which eukaryotic cells control protein expression by preventing transcripts from entering the cytoplasm in stress situations. In *S. cerevisiae*, exporting of mRNA is blocked when defects in splicing are detected [122] or in stress circumstances, such as heat stress [123]. However, there is also selective exporting of transcripts that is induced under such stress conditions, so that only those transcripts required in a particular context are exported to the cytoplasm. For example, heat shock protein transcripts are selectively exported from the nucleus in response to heat stress [123]. The selective mRNA nuclear export is controlled by post-translational modifications of export adaptors, which dissociate them from mRNA [124].

The CWI MAPK Slt2 controls the export of transcripts through phosphorylation of the polyadenylated RNA-binding protein Nab2, which mediates the nuclear mRNA export. Nab2 is phosphorylated upon heat shock, displaying an Slt2-dependent mobility shift. In vitro kinase assays have demonstrated that Slt2 directly phosphorylates Nab2 on residues T178 and S180 (Table 1) [68]. Moreover, the phosphorylation of bacterially produced Nab2 by an AS version of Slt2 (Slt2-AS) has been observed in our laboratory [50]. The Slt2-dependent phosphorylation of Nab2 reduces its binding to the export receptor Mex67 [68]. In this work, a model was proposed in which activated Slt2 promoted the nuclear retention of non-heat-shock mRNAs by uncoupling Mex67 from Nab2 and possibly other components of mRNA export machinery during heat shock. Coincident with Nab2 phosphorylation, this protein and the mRNA binding protein Yra1 co-localize in nuclear foci with the nuclear pore-associated myosin-like protein 1 (Mlp1), a protein involved in mRNA retention. However, Nab2 nuclear focus formation and Nab2 phosphorylation are independent, suggesting that several mechanisms are implicated in mRNA transport during heat stress. In addition to mRNA nuclear export, Nab2 has been implicated in the previous proper 3′ end processing as well as splicing of nascent RNA, serving as a checkpoint for the fidelity of pre-mRNA processing [122,125]. Thus, it is tempting to speculate that Slt2 may also be implicated in the regulation of these processes through Nab2 phosphorylation.

3.7. Control of Cell Cycle: Sic1 and Mrc1

MAPKs are key players in the control of the cell cycle in eukaryotes. In response to plasma membrane and cell wall damage, the CWI pathway regulates cell cycle progression at different points [21]. In line with this, the Slt2-dependent phosphorylation of Sic1 and Mrc1 (Table 1) has been implicated in cell cycle arrest at Start point and S-phase, respectively.

When the Start transcriptional program is activated, the cyclin dependent kinase inhibitor (CKI) Sic1 is sequentially phosphorylated on multiple sites within its N-terminal region, first by the G1/S CDK–cyclin complexes Cln1,2-Cdc28, and later by Clb-Cdc28. These phosphorylation events target Sic1 for degradation mediated by the ubiquitin ligase SCF^{Cdc4}, thus allowing progression from the G1 to S phase. TORC1 inhibition by rapamycin treatment or nitrogen limitation promotes the activation of both Greatwall/Rim15 kinase and Slt2. As a result, whereas Slt2 phosphorylates Sic1 in T173, Rim15 inhibits the phosphatase PP2A-Cdc55, which dephosphorylates Sic1 in this site. These events greatly stabilize Sic1, since T173 phosphorylation is critical for Sic1 stability in rapamycin-treated cells [126] by preventing its association with SCF^{Cdc4} [127], thus avoiding Sic1 degradation. Consequently, the stabilized pool of Sic1 inhibits the G1 to S transition, and thus promotes cell cycle arrest in a stress context [128]. Such Slt2-dependent Sic1 phosphorylation at T173 has been demonstrated both in vitro and in vivo (Table 1). Even though this phosphorylation has been well characterized in rapamycin-treated cells, an increase in Sic1 T173 phosphorylation is also observed during the G1 phase in proliferating cells [128]. Sic1 is also stabilized after SDS-induced plasma membrane damage [129], a condition that triggers Slt2 activation. Thus, Slt2 may restrain cell cycle progression through Sic1 stabilization in other situations that have not yet been studied.

Another substrate through which Slt2 exerts cell cycle control is Mrc1, a component of the DNA replication complex. Mrc1, an evolutionarily conserved replisome-associated factor required for efficient DNA replication [130], couples DNA polymerase and helicase activities [131] and is involved in establishing the S-phase checkpoint to prevent genome instability [132]. Upon heat shock, Mrc1 is phosphorylated in its N-terminus at positions T169, S215, and S229 by the MAPK Slt2 (Table 1), causing a delay in DNA replication and favoring widespread transcriptional reprogramming. This way, Slt2 prevents transcription–replication conflicts that may arise in situations in which a massive transcriptional response and DNA replication occur simultaneously [69]. A recent study showed that a sublethal concentration of ethanol causes DNA replication stress and relocalization of Mrc1 from the replication fork to a perinuclear compartment, affecting replisome stability, replication rate, and genome stability. However, whether ethanol triggers Mrc1 phosphorylation is still unknown [133]. Future studies analyzing the effect of Slt2-dependent phosphorylation on Mrc1 localization will help to elucidate the mechanism underlying this S-phase checkpoint.

Interestingly, both Sic1 and Mrc1 are phosphorylated on the same sites by Hog1, the MAPK of the yeast osmolarity pathway [127,134]. This suggests that these proteins contain regulatory sites sensitive to changes in the environment, reflecting the variety of mechanisms that operate in cell cycle control.

3.8. Orphan Slt2 Substrates: Caf20, Rcn2, and Gga1

Several Slt2 substrates have been found in proteomic studies and confirmed through direct kinase assays, but the function of Slt2-exerted phosphorylation remains to be discovered.

As mentioned above, our group performed a phosphoproteomic analysis of *S. cerevisiae* upon overexpression of a constitutively active version of Pkc1, which led to hyperactivation of the CWI pathway. This assay generated a list of potential Slt2 target proteins, which contained peptides that appeared hyperphosphorylated at S/T-P sites in this condition [54]. Further experiments with Slt2-AS led us to confirm that among these candidates, Caf20, Rcn2, and Gga1 were direct targets of Slt2 phosphorylation in vitro (Table 1) [50].

The translation initiation repressor Caf20 is phosphorylated by Slt2 on the only consensus S/T-P site at T102 [50]. This residue has also been found to be phosphorylated in other phosphoproteomic studies [77]. Since Caf20 represses the translation of its target mRNAs, and mRNAs linked to processes such as cell cycle, intracellular signaling cascades, and cell morphogenesis were found to be enriched in the pool of Caf20-associated mRNAs [135], it is very possible that the CWI pathway specifically regulates the translation of these subsets of proteins through phosphorylation of this protein.

The calcium/calmodulin-dependent protein phosphatase regulator Rcn2 [136] was phosphorylated by Slt2 on three target sites: S152, S160, and S255 [50]. Phosphorylation of this protein by Slt2-AS was totally lost only in the mutant protein with all three residues substituted by alanine. Each individual mutation resulted in a reduced phosphorylation, but the most intense effect was observed when S255, the only one of these Rcn2 phosphosites detected as hyperphosphorylated in our previous phosphoproteomic analysis [54], was mutated (Table 1). These three residues also appeared to be phosphorylated in a global phosphoproteomic analysis using the endoplasmic reticulum (ER) stress-inducing agent DTT [137]. Taking into account that calcineurin has an essential role in response to ER stress [138], and that DTT triggers activation of the CWI pathway [15,139], Slt2 phosphorylation of Rcn2 may be part of a regulatory mechanism by which the CWI pathway modulates the calcineurin response to ER stress.

The Golgi-associated protein Gga1, involved in protein trafficking, contains four S/T-P target sites, which have been shown to be phosphorylated in a different phosphoproteomic analysis [77]. Although the specific Slt2-dependent phosphoresidue remains undetermined, the strong negative genetic interaction found between *SLT2* and either *GGA1* or its paralog *GGA2* under cell wall stress conditions suggests a connection of the CWI pathway with trafficking from the Golgi complex to the vacuole [50].

3.9. Is There a Specific Slt2 Phosphorylation Signature?

In searching for an Slt2 consensus signature, we compared the S/T phosphorylation motifs of the 14 genuine Slt2 targets whose phosphorylation sites are known (Table 1, Table S1). We excluded Rpb1 because it is atypically hyperphosphorylated at tyrosine. As shown in Figure 2, most of the peptides analyzed are phosphorylated in the consensus MAPK site S/T-P [32], except Bcy1 which shows serine in the +1 position instead of proline. The expected Slt2 preference for an R at the P−3 position [34] was not found within the Slt2 phosphorylation consensus signature. Our analysis reveals the difficulties in establishing the rules that determine Slt2 specificity and limits the feasibility of predicting its phosphorylation sites. It is also interesting to note the importance of knowing the precise biological relevance of Slt2-mediated phosphorylation. However, while this is easily achieved in cases of single phosphorylation sites, the existence of multiple phosphorylation sites, which is common among Slt2 substrates, hinders this task.

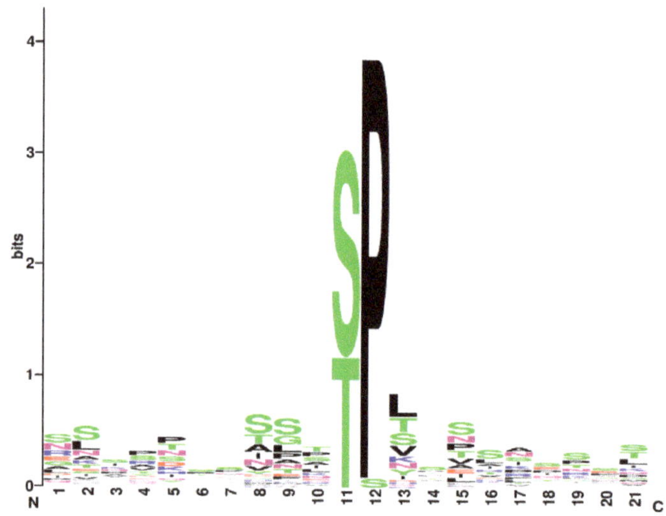

Figure 2. Slt2 consensus phosphorylation site. Motif logo representing Slt2 phosphorylation signature, obtained with WebLogo, a program for alignment and motif enrichment [140]. The 11th position at the logo corresponds to serine or threonine phosphorylated by Slt2, and the rest to 10 upstream/downstream amino acids surrounding this position. Complete list of sequences corresponding to phosphorylated peptides can be found in Table S1 and correspond to proteins included in Table 1 whose S/T phosphorylation sites are known.

4. Candidate Slt2 Substrates: A Growing List

Apart from the proteins listed above, which we have considered to be genuine Slt2 substrates, other proteins have been described that are phosphorylated in vitro by Slt2, or in vivo upon CWI-activating conditions, suggesting that they are likely Slt2 targets. They have been found through either systematic approaches, such as protein array screenings and phosphoproteomics, or targeted experiments. However, they do not meet the criteria initially indicated and more evidence is needed to consider them as bona fide Slt2 substrates. They are briefly reviewed below.

A global analysis of yeast protein phosphorylation by in vitro kinase assays performed with 87 kinases over a microarray representing the yeast proteome found direct phosphorylation of Slt2 on Brx1 and Cmk2 [141]. In spite of this, no further work has confirmed either these observations by individual in vitro or in vivo assays, or the precise phosphorylated positions. Brx1 is a ribosomal assembly factor [142] that contains a unique MAPK target site (S244), which was found to be phosphorylated in phosphoproteomic analyses [77]. Therefore, S244 is a very plausible target site for Slt2 on Brx1. On the other hand, Cmk2 is a calmodulin-dependent protein kinase that acts as a negative feedback controller within the calcium/calmodulin signaling pathway [143]. Although Cmk2 contains two MAPK target sites, neither of them has been reported to be phosphorylated [77].

Eisosome, also known as the membrane compartment of Can1 (MCC), is a protein complex distributed throughout the plasma membrane of *S. cerevisiae*. Eisosome formation is promoted by two paralogous proteins, Pil1 and Lsp1 [144]. Our quantitative phosphoproteomic analysis of yeast cells under Pkc1 hyperactivation revealed that Pil1 and Lsp1 displayed increased phosphorylation at MAPK target motif T233. Both proteins showed electrophoretic mobility promotion corresponding to their phosphorylation status, which was dependent on the presence of Slt2 in yeast cells [54]. Interestingly Pil1 T233 residue has also been described to be phosphorylated by Pkh1 and Pkh2 [145,146], the two redundant PDK1 homologs required for Pkc1 activation [7]. Considering that the sequence surrounding the T233 phosphorylation site in Pil1 does not match the consensus site for PDK kinases [147] and that Slt2 is downstream of Pkh1 and Pkh2 [2], T233 may be targeted by Slt2, thus regulating eisosome assembly and organization. However, we could not detect direct kinase activity on Pil1 by the Slt2-AS-based kinase assay in vitro [50], pointing to the possibility that an Slt2-dependent intermediate kinase is responsible for Pil1 T233 phosphorylation. In any case, the functional implication of Pil1 and Lsp1 phosphorylation remains controversial, given that Pil1 phosphorylation has been shown to promote eisosome assembly or disassembly depending on the phosphorylated residue [148].

Slt2 is mainly localized in the nucleus, where it controls, among other processes, transcription, epigenetic modification, and mRNA nuclear export through phosphorylation of the different nuclear substrates described above. In addition, this MAPK also localizes at the tip of small buds and at the mother-bud neck region in late mitosis, promoting the expansion of the daughter cell by new cell wall synthesis and stimulating septum construction for cell separation, respectively. Thus, even though no direct substrates at sites of polarized growth have been found, it is plausible that Slt2 phosphorylates proteins implicated in the regulation of these morphogenetic events. Among them, Bni4, a phosphoprotein with an important role in septum formation during cytokinesis [149], is a potential substrate of Slt2 within the yeast bud neck. In favor of this idea, it has been shown that Slt2 physically interacts with Bni4 and regulates its localization. Moreover, Bni4 contains several S/T-P sites with important roles in its function at the bud neck, and *slt2Δ* mutants exhibit a decrease in Bni4 phosphorylation [150]. An interesting phosphoproteomic analysis of the transcriptional response to the ER stressor DTT revealed Mkk1/2-dependent phosphorylation of 28 proteins, mainly involved in budding, polarity, cytoskeleton, and endocytosis (Spa2, Exo84, Bps1, Kin1, Mon2, Myo5, Vrp1, Yck1, Bbc1, Sec31, Vrp1, Twf1, and Bud6), which uncovers new potential Slt2 substrates related to these functions. Because the vast majority of phosphorylation was observed in long-term treatment with DTT, these protein modifications may be involved in the adaptive response to DTT-induced ER stress [137].

As mentioned above, a large-scale phosphoproteomic analysis under conditions that lead to activation of the CWI pathway has yielded potential Slt2 substrates that remain to be individually assayed. A recent example of this approach is the rapamycin-induced phosphoproteome, which has shown the Slt2-dependent phosphorylation of seven proteins, including the transcriptional repressor Mig1 and the calcineurin-activated transcription factor Crz1 [151]. The Slt2-dependent phosphorylation of Cmk2, Rcn2, and Crz1, all proteins related to calcium homeostasis [136,143], together with the reported negative epistatic interactions between Slt2 and calcineurin [152], provide evidence of the involvement of the MAPK Slt2 in the control of the calcium/calcineurin signaling pathway.

5. Concluding Remarks

Work over the last 30 years has provided evidence that Slt2, the MAPK of the CWI pathway, not only plays a key role in cell wall remodeling, but also has an important function in the control of cell signaling through its own and other pathways, and in coordinating essential physiological processes such as cell cycle, morphogenesis, and responses to different stress situations (Figure 3). The mechanism of action of Slt2 goes beyond its role as a protein kinase, as it has been shown to have kinase-independent functions. However, most of its wide cellular effects rely on its kinase activity. Thus, it is expected that the number of Slt2 substrates is much larger than that found to date. Furthermore, only a few of the already described substrates have been thoroughly characterized.

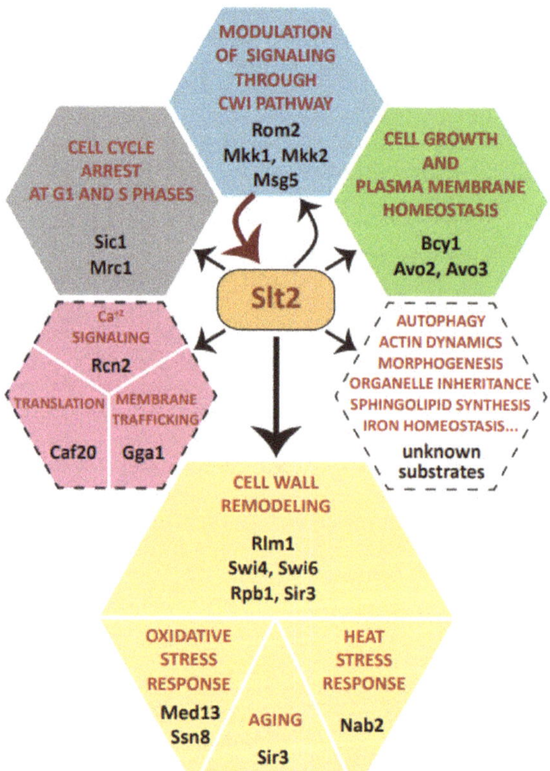

Figure 3. Graphical scheme showing cellular processes that are affected by Slt2 through the phosphorylation of the indicated substrates. Cellular processes are grouped by color based on the functional implication of substrate phosphorylation, as in Table 1: modulation of signaling through CWI pathway (blue), regulation of other signaling pathways (green), regulation of gene transcription and mRNA transport (yellow), cell cycle control (gray), and undetermined Slt2-dependent phosphorylation role (pink). The hexagons delimited by dashed lines show suggested but not formally proven Slt2-dependent functions (pink), or Slt2-regulated processes for which the specific substrates involved have not yet been identified (white).

Exhaustive exploitation of genetic screening and proteomic approaches will reveal a good number of novel putative substrates of Slt2 in the near future. These efforts should further include biological validation experiments with the large number of tools available for the rapid confirmation of genuine Slt2 substrates, including hyperactive, kinase-dead, and analog-sensitive versions of this kinase [18,50]. The biological interpretation of the data should bring important information to link signaling through the CWI MAPK module to Slt2 functional roles.

Many currently known substrates of Slt2 are transcription factors or proteins related to transcriptional regulation, which underlines both the preeminent role of the CWI pathway as a regulator of gene expression under stress conditions and the induction of specific mRNAs as a hallmark of the response to cell wall insults. However, there is an important gap in the knowledge regarding other outputs of this key MAPK. For example, it would be important to unveil the unknown substrates responsible for key functions of Slt2, such as autophagy and inheritance of organelles, the control of sphingolipid synthesis, and actin cytoskeleton dynamics, among others (Figure 3). In addition, there are still some interesting issues to be addressed. For example, keeping in mind that Slt2 is transiently localized in polarity sites, it is surprising that no genuine substrate has been identified at this cellular localization. Additionally, as occurs with other MAPKs, such as ERK1/2, Slt2 dimerizes [153]. While it has been shown that dimerization impacts the activation of different pools of ERK substrates [154], the role of dimerization on Slt2 functionality and substrate targeting is still unknown.

In sum, over the coming years, new insights into how Slt2 shapes yeast cell integrity are likely to emerge. Identifying new Slt2 substrates will also help us to understand how pathogenic fungi use this widely conserved CWI pathway for virulence and adaptation to antifungal-induced stress. Thus, we should keep our eyes wide open, as this important new information could translate into applied medical research.

Supplementary Materials: The following supporting information can be downloaded at: https://www.mdpi.com/article/10.3390/jof8040368/s1, Table S1: Peptides phosphorylated by Slt2 within the substrates listed in Table 1 and whose alignment has been performed in search for the consensus phosphorylation site shown in Figure 2.

Author Contributions: Conceptualization, G.G.-R., M.M., H.M., and T.F.-A.; data curation, G.G.-R., L.S.-V., and T.F.-A.; software analysis, G.G.-R., L.S.-V., and T.F.-A.; original draft preparation, G.G.-R., M.M., H.M., and T.F.-A.; review and editing, G.G.-R., M.M., H.M., and T.F.-A.; supervision, M.M. and H.M.; project administration, M.M. and H.M.; funding acquisition, M.M. All authors have read and agreed to the published version of the manuscript.

Funding: This research was funded by Ministerio de Ciencia e Innovación (Spain), grant number PID2019-105342GB-I00; Comunidad de Madrid and European Structural and Investment Funds, grant number S2017/BMD-3691 (InGEMICS-CM); G.G-R. was recipient of a predoctoral fellowship from the Universidad Complutense de Madrid and is currently funded by the grant PID2019-105342GB-I00; L.S-V. was recipient of a contract funded by Fondo de Garantía Juvenil de la Comunidad de Madrid and is currently funded by the grant PID2019-105342GB-I00.

Institutional Review Board Statement: Not applicable.

Informed Consent Statement: Not applicable.

Data Availability Statement: The data presented in this study are available on request from the corresponding authors.

Acknowledgments: We acknowledge our colleagues from Unit 3 of the Departamento de Microbiología y Parasitología at UCM for their useful comments and discussions throughout our work.

Conflicts of Interest: The authors declare no conflict of interest.

References

1. Kock, C.; Dufrêne, Y.F.; Heinisch, J.J. Up against the wall: Is yeast cell wall integrity ensured by mechanosensing in plasma membrane microdomains? *Appl. Environ. Microbiol.* **2015**, *81*, 806–811. [CrossRef]
2. Levin, D.E. Regulation of cell wall biogenesis in *Saccharomyces cerevisiae*: The cell wall integrity signaling pathway. *Genetics* **2011**, *189*, 1145–1175. [CrossRef]
3. Chen, R.E.; Thorner, J. Function and regulation in MAPK signaling pathways: Lessons learned from the yeast *Saccharomyces cerevisiae*. *Biochim. Biophys. Acta* **2007**, *1773*, 1311–1340. [CrossRef]

4. Schmelzle, T.; Helliwell, S.B.; Hall, M.N. Yeast protein kinases and the RHO1 exchange factor TUS1 are novel components of the cell integrity pathway in yeast. *Mol. Cell Biol.* **2002**, *22*, 1329–1339. [CrossRef]
5. Heinisch, J.J.; Rodicio, R. Protein kinase C in fungi-more than just cell wall integrity. *FEMS Microbiol. Rev.* **2018**, *42*, fux051. [CrossRef]
6. Roelants, F.M.; Leskoske, K.L.; Martinez Marshall, M.N.; Locke, M.N.; Thorner, J. The TORC2-Dependent Signaling Network in the Yeast *Saccharomyces cerevisiae*. *Biomolecules* **2017**, *7*, 66. [CrossRef]
7. Inagaki, M.; Schmelzle, T.; Yamaguchi, K.; Irie, K.; Hall, M.N.; Matsumoto, K. PDK1 homologs activate the Pkc1-mitogen-activated protein kinase pathway in yeast. *Mol. Cell Biol.* **1999**, *19*, 8344–8352. [CrossRef]
8. Martín, H.; Flandez, M.; Nombela, C.; Molina, M. Protein phosphatases in MAPK signalling: We keep learning from yeast. *Mol. Microbiol.* **2005**, *58*, 6–16. [CrossRef]
9. Tatjer, L.; Sacristán-Reviriego, A.; Casado, C.; González, A.; Rodríguez-Porrata, B.; Palacios, L.; Canadell, D.; Serra-Cardona, A.; Martín, H.; Molina, M.; et al. Wide-Ranging Effects of the Yeast Ptc1 Protein Phosphatase Acting Through the MAPK Kinase Mkk1. *Genetics* **2016**, *202*, 141–156. [CrossRef]
10. González-Rubio, G.; Fernández-Acero, T.; Martín, H.; Molina, M. Mitogen-Activated Protein Kinase Phosphatases (MKPs) in Fungal Signaling: Conservation, Function, and Regulation. *Int. J. Mol. Sci.* **2019**, *20*, 1709. [CrossRef]
11. Molina, M.; Cid, V.J.; Martín, H. Fine regulation of *Saccharomyces cerevisiae* MAPK pathways by post-translational modifications. *Yeast* **2010**, *27*, 503–511. [CrossRef]
12. Audhya, A.; Emr, S.D. Stt4 PI 4-kinase localizes to the plasma membrane and functions in the Pkc1-mediated MAP kinase cascade. *Dev. Cell* **2002**, *2*, 593–605. [CrossRef]
13. Millson, S.H.; Truman, A.W.; King, V.; Prodromou, C.; Pearl, L.H.; Piper, P.W. A two-hybrid screen of the yeast proteome for Hsp90 interactors uncovers a novel Hsp90 chaperone requirement in the activity of a stress-activated mitogen-activated protein kinase, Slt2p (Mpk1p). *Eukaryot. Cell* **2005**, *4*, 849–860. [CrossRef]
14. Lee, J.; Liu, L.; Levin, D.E. Stressing out or stressing in: Intracellular pathways for SAPK activation. *Curr. Genet.* **2019**, *65*, 417–421. [CrossRef]
15. Jiménez-Gutiérrez, E.; Alegría-Carrasco, E.; Sellers-Moya, A.; Molina, M.; Martín, H. Not just the wall: The other ways to turn the yeast CWI pathway on. *Int. Microbiol.* **2020**, *23*, 107–119. [CrossRef]
16. Martín, H.; Rodríguez-Pachón, J.M.; Ruiz, C.; Nombela, C.; Molina, M. Regulatory mechanisms for modulation of signaling through the cell integrity Slt2-mediated pathway in *Saccharomyces cerevisiae*. *J. Biol. Chem.* **2000**, *275*, 1511–1519. [CrossRef]
17. Liu, L.; Levin, D.E. Intracellular mechanism by which genotoxic stress activates yeast SAPK Mpk1. *Mol. Biol. Cell* **2018**, *29*, 2898–2909. [CrossRef]
18. González-Rubio, G.; Sellers-Moya, Á.; Martín, H.; Molina, M. Differential Role of Threonine and Tyrosine Phosphorylation in the Activation and Activity of the Yeast MAPK Slt2. *Int. J. Mol. Sci.* **2021**, *22*, 1110. [CrossRef]
19. Ahmadpour, D.; Maciaszczyk-Dziubinska, E.; Babazadeh, R.; Dahal, S.; Migocka, M.; Andersson, M.; Wysocki, R.; Tamás, M.J.; Hohmann, S. The mitogen-activated protein kinase Slt2 modulates arsenite transport through the aquaglyceroporin Fps1. *FEBS Lett.* **2016**, *590*, 3649–3659. [CrossRef]
20. Pujol-Carrion, N.; Pavón-Vergés, M.; Arroyo, J.; de la Torre-Ruiz, M.A. The MAPK Slt2/Mpk1 plays a role in iron homeostasis through direct regulation of the transcription factor Aft1. *Biochim. Biophys. Acta Mol. Cell Res.* **2021**, *1868*, 118974. [CrossRef]
21. Quilis, I.; Gomar-Alba, M.; Igual, J.C. The CWI Pathway: A Versatile Toolbox to Arrest Cell-Cycle Progression. *J. Fungi* **2021**, *7*, 1041. [CrossRef] [PubMed]
22. Mao, K.; Klionsky, D.J. MAPKs regulate mitophagy in *Saccharomyces cerevisiae*. *Autophagy* **2011**, *7*, 1564–1565. [CrossRef] [PubMed]
23. Du, Y.; Walker, L.; Novick, P.; Ferro-Novick, S. Ptc1p regulates cortical ER inheritance via Slt2p. *Embo J.* **2006**, *25*, 4413–4422. [CrossRef]
24. Li, X.; Du, Y.; Siegel, S.; Ferro-Novick, S.; Novick, P. Activation of the mitogen-activated protein kinase, Slt2p, at bud tips blocks a late stage of endoplasmic reticulum inheritance in *Saccharomyces cerevisiae*. *Mol. Biol. Cell* **2010**, *21*, 1772–1782. [CrossRef]
25. Guo, S.; Shen, X.; Yan, G.; Ma, D.; Bai, X.; Li, S.; Jiang, Y. A MAP kinase dependent feedback mechanism controls Rho1 GTPase and actin distribution in yeast. *PLoS ONE* **2009**, *4*, e6089. [CrossRef] [PubMed]
26. García, R.; Sanz, A.B.; Rodríguez-Peña, J.M.; Nombela, C.; Arroyo, J. Rlm1 mediates positive autoregulatory transcriptional feedback that is essential for Slt2-dependent gene expression. *J. Cell Sci.* **2016**, *129*, 1649–1660. [CrossRef] [PubMed]
27. Jiménez-Gutiérrez, E.; Alegría-Carrasco, E.; Alonso-Rodríguez, E.; Fernández-Acero, T.; Molina, M.; Martín, H. Rewiring the yeast cell wall integrity (CWI) pathway through a synthetic positive feedback circuit unveils a novel role for the MAPKKK Ssk2 in CWI pathway activation. *FEBS J.* **2020**, *287*, 4881–4901. [CrossRef]
28. Soulard, A.; Cremonesi, A.; Moes, S.; Schütz, F.; Jenö, P.; Hall, M.N. The rapamycin-sensitive phosphoproteome reveals that TOR controls protein kinase A toward some but not all substrates. *Mol. Biol. Cell* **2010**, *21*, 3475–3486. [CrossRef]
29. Leskoske, K.L.; Roelants, F.M.; Emmerstorfer-Augustin, A.; Augustin, C.M.; Si, E.P.; Hill, J.M.; Thorner, J. Phosphorylation by the stress-activated MAPK Slt2 down-regulates the yeast TOR complex 2. *Genes Dev.* **2018**, *32*, 1576–1590. [CrossRef]
30. Kim, K.Y.; Levin, D.E. Transcriptional reporters for genes activated by cell wall stress through a non-catalytic mechanism involving Mpk1 and SBF. *Yeast* **2010**, *27*, 541–548. [CrossRef]
31. Kim, K.Y.; Truman, A.W.; Levin, D.E. Yeast Mpk1 mitogen-activated protein kinase activates transcription through Swi4/Swi6 by a noncatalytic mechanism that requires upstream signal. *Mol. Cell Biol.* **2008**, *28*, 2579–2589. [CrossRef] [PubMed]

32. Roskoski, R., Jr. ERK1/2 MAP kinases: Structure, function, and regulation. *Pharm. Res.* **2012**, *66*, 105–143. [CrossRef] [PubMed]
33. Carlson, S.M.; Chouinard, C.R.; Labadorf, A.; Lam, C.J.; Schmelzle, K.; Fraenkel, E.; White, F.M. Large-scale discovery of ERK2 substrates identifies ERK-mediated transcriptional regulation by ETV3. *Sci. Signal.* **2011**, *4*, rs11. [CrossRef]
34. Mok, J.; Kim, P.M.; Lam, H.Y.; Piccirillo, S.; Zhou, X.; Jeschke, G.R.; Sheridan, D.L.; Parker, S.A.; Desai, V.; Jwa, M.; et al. Deciphering protein kinase specificity through large-scale analysis of yeast phosphorylation site motifs. *Sci. Signal.* **2010**, *3*, ra12. [CrossRef] [PubMed]
35. Bradley, D.; Beltrao, P. Evolution of protein kinase substrate recognition at the active site. *PLoS Biol.* **2019**, *17*, e3000341. [CrossRef]
36. Tanoue, T.; Nishida, E. Docking interactions in the mitogen-activated protein kinase cascades. *Pharmacol. Ther.* **2002**, *93*, 193–202. [CrossRef]
37. Garai, Á.; Zeke, A.; Gógl, G.; Törő, I.; Fördős, F.; Blankenburg, H.; Bárkai, T.; Varga, J.; Alexa, A.; Emig, D.; et al. Specificity of linear motifs that bind to a common mitogen-activated protein kinase docking groove. *Sci. Signal.* **2012**, *5*, ra74. [CrossRef]
38. Tanoue, T.; Adachi, M.; Moriguchi, T.; Nishida, E. A conserved docking motif in MAP kinases common to substrates, activators and regulators. *Nat. Cell Biol.* **2000**, *2*, 110–116. [CrossRef]
39. Chang, C.I.; Xu, B.E.; Akella, R.; Cobb, M.H.; Goldsmith, E.J. Crystal structures of MAP kinase p38 complexed to the docking sites on its nuclear substrate MEF2A and activator MKK3b. *Mol. Cell* **2002**, *9*, 1241–1249. [CrossRef]
40. Zeke, A.; Bastys, T.; Alexa, A.; Garai, Á.; Mészáros, B.; Kirsch, K.; Dosztányi, Z.; Kalinina, O.V.; Reményi, A. Systematic discovery of linear binding motifs targeting an ancient protein interaction surface on MAP kinases. *Mol. Syst. Biol.* **2015**, *11*, 837. [CrossRef]
41. Akella, R.; Moon, T.M.; Goldsmith, E.J. Unique MAP Kinase binding sites. *Biochim. Biophys. Acta* **2008**, *1784*, 48–55. [CrossRef] [PubMed]
42. González-Rubio, G.; Sellers-Moya, Á.; Martín, H.; Molina, M. A walk-through MAPK structure and functionality with the 30-year-old yeast MAPK Slt2. *Int. Microbiol.* **2021**, *24*, 531–543. [CrossRef] [PubMed]
43. Watanabe, Y.; Irie, K.; Matsumoto, K. Yeast RLM1 encodes a serum response factor-like protein that may function downstream of the Mpk1 (Slt2) mitogen-activated protein kinase pathway. *Mol. Cell Biol.* **1995**, *15*, 5740–5749. [CrossRef] [PubMed]
44. Watanabe, Y.; Takaesu, G.; Hagiwara, M.; Irie, K.; Matsumoto, K. Characterization of a serum response factor-like protein in *Saccharomyces cerevisiae*, Rlm1, which has transcriptional activity regulated by the Mpk1 (Slt2) mitogen-activated protein kinase pathway. *Mol. Cell Biol.* **1997**, *17*, 2615–2623. [CrossRef] [PubMed]
45. Jia, Y.; Quinn, C.M.; Kwak, S.; Talanian, R.V. Current in vitro kinase assay technologies: The quest for a universal format. *Curr. Drug Discov. Technol.* **2008**, *5*, 59–69. [CrossRef]
46. Hastie, C.J.; McLauchlan, H.J.; Cohen, P. Assay of protein kinases using radiolabeled ATP: A protocol. *Nat. Protoc.* **2006**, *1*, 968–971. [CrossRef]
47. Mok, J.; Im, H.; Snyder, M. Global identification of protein kinase substrates by protein microarray analysis. *Nat. Protoc.* **2009**, *4*, 1820–1827. [CrossRef]
48. Shah, K.; Shokat, K.M. A chemical genetic approach for the identification of direct substrates of protein kinases. *Methods Mol. Biol.* **2003**, *233*, 253–271.
49. Hertz, N.T.; Wang, B.T.; Allen, J.J.; Zhang, C.; Dar, A.C.; Burlingame, A.L.; Shokat, K.M. Chemical genetic approach for kinase-substrate mapping by covalent capture of thiophosphopeptides and analysis by mass spectrometry. *Curr. Protoc. Chem. Biol.* **2010**, *2*, 15–36. [CrossRef]
50. Alonso-Rodríguez, E.; Fernández-Pinar, P.; Sacristán-Reviriego, A.; Molina, M.; Martín, H. An Analog-sensitive Version of the Protein Kinase Slt2 Allows Identification of Novel Targets of the Yeast Cell Wall Integrity Pathway. *J. Biol. Chem.* **2016**, *291*, 5461–5472. [CrossRef]
51. Koch, A.; Hauf, S. Strategies for the identification of kinase substrates using analog-sensitive kinases. *Eur J. Cell Biol.* **2010**, *89*, 184–193. [CrossRef] [PubMed]
52. Sugiyama, Y.; Uezato, Y. Analysis of protein kinases by Phos-tag SDS-PAGE. *J. Proteom.* **2022**, *255*, 104485. [CrossRef] [PubMed]
53. von Stechow, L.; Francavilla, C.; Olsen, J.V. Recent findings and technological advances in phosphoproteomics for cells and tissues. *Expert Rev. Proteom.* **2015**, *12*, 469–487. [CrossRef] [PubMed]
54. Mascaraque, V.; Hernaez, M.L.; Jimenez-Sanchez, M.; Hansen, R.; Gil, C.; Martin, H.; Cid, V.J.; Molina, M. Phosphoproteomic analysis of protein kinase C signaling in *Saccharomyces cerevisiae* reveals Slt2 mitogen-activated protein kinase (MAPK)-dependent phosphorylation of eisosome core components. *Mol. Cell Proteom.* **2013**, *12*, 557–574. [CrossRef] [PubMed]
55. Knight, J.D.; Tian, R.; Lee, R.E.; Wang, F.; Beauvais, A.; Zou, H.; Megeney, L.A.; Gingras, A.C.; Pawson, T.; Figeys, D.; et al. A novel whole-cell lysate kinase assay identifies substrates of the p38 MAPK in differentiating myoblasts. *Skelet Muscle* **2012**, *2*, 5. [CrossRef]
56. van Drogen, F.; Peter, M. Spa2p functions as a scaffold-like protein to recruit the Mpk1p MAP kinase module to sites of polarized growth. *Curr. Biol.* **2002**, *12*, 1698–1703. [CrossRef]
57. Jiménez-Sánchez, M.; Cid, V.J.; Molina, M. Retrophosphorylation of Mkk1 and Mkk2 MAPKKs by the Slt2 MAPK in the yeast cell integrity pathway. *J. Biol. Chem.* **2007**, *282*, 31174–31185. [CrossRef]
58. Flández, M.; Cosano, I.C.; Nombela, C.; Martín, H.; Molina, M. Reciprocal regulation between Slt2 MAPK and isoforms of Msg5 dual-specificity protein phosphatase modulates the yeast cell integrity pathway. *J. Biol. Chem.* **2004**, *279*, 11027–11034. [CrossRef]

59. Marín, M.J.; Flández, M.; Bermejo, C.; Arroyo, J.; Martín, H.; Molina, M. Different modulation of the outputs of yeast MAPK-mediated pathways by distinct stimuli and isoforms of the dual-specificity phosphatase Msg5. *Mol. Genet. Genom.* **2009**, *281*, 345–359. [CrossRef]
60. Madden, K.; Sheu, Y.J.; Baetz, K.; Andrews, B.; Snyder, M. SBF cell cycle regulator as a target of the yeast PKC-MAP kinase pathway. *Science* **1997**, *275*, 1781–1784. [CrossRef]
61. Kim, K.Y.; Truman, A.W.; Caesar, S.; Schlenstedt, G.; Levin, D.E. Yeast Mpk1 cell wall integrity mitogen-activated protein kinase regulates nucleocytoplasmic shuttling of the Swi6 transcriptional regulator. *Mol. Biol. Cell* **2010**, *21*, 1609–1619. [CrossRef] [PubMed]
62. Jin, C.; Strich, R.; Cooper, K.F. Slt2p phosphorylation induces cyclin C nuclear-to-cytoplasmic translocation in response to oxidative stress. *Mol. Biol. Cell* **2014**, *25*, 1396–1407. [CrossRef] [PubMed]
63. Krasley, E.; Cooper, K.F.; Mallory, M.J.; Dunbrack, R.; Strich, R. Regulation of the oxidative stress response through Slt2p-dependent destruction of cyclin C in *Saccharomyces cerevisiae*. *Genetics* **2006**, *172*, 1477–1486. [CrossRef] [PubMed]
64. Stieg, D.C.; Willis, S.D.; Ganesan, V.; Ong, K.L.; Scuorzo, J.; Song, M.; Grose, J.; Strich, R.; Cooper, K.F. A complex molecular switch directs stress-induced cyclin C nuclear release through SCF(Grr1)-mediated degradation of Med13. *Mol. Biol. Cell* **2018**, *29*, 363–375. [CrossRef] [PubMed]
65. Yurko, N.; Liu, X.; Yamazaki, T.; Hoque, M.; Tian, B.; Manley, J.L. MPK1/SLT2 Links Multiple Stress Responses with Gene Expression in Budding Yeast by Phosphorylating Tyr1 of the RNAP II CTD. *Mol. Cell* **2017**, *68*, 913–925.e3. [CrossRef]
66. Ai, W.; Bertram, P.G.; Tsang, C.K.; Chan, T.F.; Zheng, X.F. Regulation of subtelomeric silencing during stress response. *Mol. Cell* **2002**, *10*, 1295–1305. [CrossRef]
67. Ray, A.; Hector, R.E.; Roy, N.; Song, J.H.; Berkner, K.L.; Runge, K.W. Sir3p phosphorylation by the Slt2p pathway effects redistribution of silencing function and shortened lifespan. *Nat. Genet.* **2003**, *33*, 522–526. [CrossRef]
68. Carmody, S.R.; Tran, E.J.; Apponi, L.H.; Corbett, A.H.; Wente, S.R. The mitogen-activated protein kinase Slt2 regulates nuclear retention of non-heat shock mRNAs during heat shock-induced stress. *Mol. Cell Biol.* **2010**, *30*, 5168–5179. [CrossRef]
69. Duch, A.; Canal, B.; Barroso, S.I.; García-Rubio, M.; Seisenbacher, G.; Aguilera, A.; de Nadal, E.; Posas, F. Multiple signaling kinases target Mrc1 to prevent genomic instability triggered by transcription-replication conflicts. *Nat. Commun.* **2018**, *9*, 379. [CrossRef]
70. Moreno-Torres, M.; Jaquenoud, M.; De Virgilio, C. TORC1 controls G1-S cell cycle transition in yeast via Mpk1 and the greatwall kinase pathway. *Nat. Commun.* **2015**, *6*, 8256. [CrossRef]
71. Lee, K.S.; Levin, D.E. Dominant mutations in a gene encoding a putative protein kinase (BCK1) bypass the requirement for a *Saccharomyces cerevisiae* protein kinase C homolog. *Mol. Cell Biol.* **1992**, *12*, 172–182.
72. Ozaki, K.; Tanaka, K.; Imamura, H.; Hihara, T.; Kameyama, T.; Nonaka, H.; Hirano, H.; Matsuura, Y.; Takai, Y. Rom1p and Rom2p are GDP/GTP exchange proteins (GEPs) for the Rho1p small GTP binding protein in *Saccharomyces cerevisiae*. *EMBO J.* **1996**, *15*, 2196–2207. [CrossRef] [PubMed]
73. Manning, B.D.; Padmanabha, R.; Snyder, M. The Rho-GEF Rom2p localizes to sites of polarized cell growth and participates in cytoskeletal functions in *Saccharomyces cerevisiae*. *Mol. Biol. Cell* **1997**, *8*, 1829–1844. [CrossRef] [PubMed]
74. Kobayashi, T.; Takematsu, H.; Yamaji, T.; Hiramoto, S.; Kozutsumi, Y. Disturbance of sphingolipid biosynthesis abrogates the signaling of Mss4, phosphatidylinositol-4-phosphate 5-kinase, in yeast. *J. Biol. Chem.* **2005**, *280*, 18087–18094. [CrossRef] [PubMed]
75. Yamochi, W.; Tanaka, K.; Nonaka, H.; Maeda, A.; Musha, T.; Takai, Y. Growth site localization of Rho1 small GTP-binding protein and its involvement in bud formation in *Saccharomyces cerevisiae*. *J. Cell Biol.* **1994**, *125*, 1077–1093. [CrossRef] [PubMed]
76. Holt, L.J.; Tuch, B.B.; Villén, J.; Johnson, A.D.; Gygi, S.P.; Morgan, D.O. Global analysis of Cdk1 substrate phosphorylation sites provides insights into evolution. *Science* **2009**, *325*, 1682–1686. [CrossRef] [PubMed]
77. Saccharomyces Genome Database. Available online: https://www.yeastgenome.org/ (accessed on 14 February 2022).
78. Soler, M.; Plovins, A.; Martín, H.; Molina, M.; Nombela, C. Characterization of domains in the yeast MAP kinase Slt2 (Mpk1) required for functional activity and in vivo interaction with protein kinases Mkk1 and Mkk2. *Mol. Microbiol.* **1995**, *17*, 833–842. [CrossRef]
79. Lee, K.S.; Irie, K.; Gotoh, Y.; Watanabe, Y.; Araki, H.; Nishida, E.; Matsumoto, K.; Levin, D.E. A yeast mitogen-activated protein kinase homolog (Mpk1p) mediates signalling by protein kinase C. *Mol. Cell Biol.* **1993**, *13*, 3067–3075.
80. Irie, K.; Takase, M.; Lee, K.S.; Levin, D.E.; Araki, H.; Matsumoto, K.; Oshima, Y. MKK1 and MKK2, which encode *Saccharomyces cerevisiae* mitogen-activated protein kinase-kinase homologs, function in the pathway mediated by protein kinase C. *Mol. Cell Biol.* **1993**, *13*, 3076–3083.
81. Errede, B.; Gartner, A.; Zhou, Z.; Nasmyth, K.; Ammerer, G. MAP kinase-related FUS3 from S. cerevisiae is activated by STE7 in vitro. *Nature* **1993**, *362*, 261–264. [CrossRef]
82. Zhou, Z.; Gartner, A.; Cade, R.; Ammerer, G.; Errede, B. Pheromone-induced signal transduction in *Saccharomyces cerevisiae* requires the sequential function of three protein kinases. *Mol. Cell Biol.* **1993**, *13*, 2069–2080. [PubMed]
83. Buscà, R.; Pouysségur, J.; Lenormand, P. ERK1 and ERK2 Map Kinases: Specific Roles or Functional Redundancy? *Front. Cell Dev. Biol.* **2016**, *4*, 53. [CrossRef]
84. Katagiri, C.; Masuda, K.; Urano, T.; Yamashita, K.; Araki, Y.; Kikuchi, K.; Shima, H. Phosphorylation of Ser-446 determines stability of MKP-7. *J. Biol. Chem.* **2005**, *280*, 14716–14722. [CrossRef]

85. Brondello, J.M.; Pouysségur, J.; McKenzie, F.R. Reduced MAP kinase phosphatase-1 degradation after p42/p44MAPK-dependent phosphorylation. *Science* **1999**, *286*, 2514–2517. [CrossRef] [PubMed]
86. Sohaskey, M.L.; Ferrell, J.E., Jr. Activation of p42 mitogen-activated protein kinase (MAPK), but not c-Jun NH(2)-terminal kinase, induces phosphorylation and stabilization of MAPK phosphatase XCL100 in Xenopus oocytes. *Mol. Biol. Cell* **2002**, *13*, 454–468. [CrossRef] [PubMed]
87. Portela, P.; Rossi, S. cAMP-PKA signal transduction specificity in *Saccharomyces cerevisiae*. *Curr. Genet.* **2020**, *66*, 1093–1099. [CrossRef]
88. Santangelo, G.M. Glucose signaling in *Saccharomyces cerevisiae*. *Microbiol. Mol. Biol. Rev.* **2006**, *70*, 253–282. [CrossRef]
89. Johnson, K.E.; Cameron, S.; Toda, T.; Wigler, M.; Zoller, M.J. Expression in Escherichia coli of BCY1, the regulatory subunit of cyclic AMP-dependent protein kinase from Saccharomyces cerevisiae. Purification and characterization. *J. Biol. Chem.* **1987**, *262*, 8636–8642. [CrossRef]
90. Kuret, J.; Johnson, K.E.; Nicolette, C.; Zoller, M.J. Mutagenesis of the regulatory subunit of yeast cAMP-dependent protein kinase. Isolation of site-directed mutants with altered binding affinity for catalytic subunit. *J. Biol. Chem.* **1988**, *263*, 9149–9154. [CrossRef]
91. Griffioen, G.; Branduardi, P.; Ballarini, A.; Anghileri, P.; Norbeck, J.; Baroni, M.D.; Ruis, H. Nucleocytoplasmic distribution of budding yeast protein kinase A regulatory subunit Bcy1 requires Zds1 and is regulated by Yak1-dependent phosphorylation of its targeting domain. *Mol. Cell Biol.* **2001**, *21*, 511–523. [CrossRef]
92. Griffioen, G.; Swinnen, S.; Thevelein, J.M. Feedback inhibition on cell wall integrity signaling by Zds1 involves Gsk3 phosphorylation of a cAMP-dependent protein kinase regulatory subunit. *J. Biol. Chem.* **2003**, *278*, 23460–23471. [CrossRef] [PubMed]
93. Searle, J.S.; Wood, M.D.; Kaur, M.; Tobin, D.V.; Sanchez, Y. Proteins in the nutrient-sensing and DNA damage checkpoint pathways cooperate to restrain mitotic progression following DNA damage. *PLoS Genet.* **2011**, *7*, e1002176. [CrossRef] [PubMed]
94. Locke, M.N.; Thorner, J. Regulation of TORC2 function and localization by Rab5 GTPases in *Saccharomyces cerevisiae*. *Cell Cycle* **2019**, *18*, 1084–1094. [CrossRef] [PubMed]
95. Turjanski, A.G.; Vaqué, J.P.; Gutkind, J.S. MAP kinases and the control of nuclear events. *Oncogene* **2007**, *26*, 3240–3253. [CrossRef]
96. Sanz, A.B.; García, R.; Pavón-Vergés, M.; Rodríguez-Peña, J.M.; Arroyo, J. Control of Gene Expression via the Yeast CWI Pathway. *Int. J. Mol. Sci.* **2022**, *23*, 1791. [CrossRef]
97. García, R.; Bermejo, C.; Grau, C.; Pérez, R.; Rodríguez-Peña, J.M.; Francois, J.; Nombela, C.; Arroyo, J. The global transcriptional response to transient cell wall damage in *Saccharomyces cerevisiae* and its regulation by the cell integrity signaling pathway. *J. Biol. Chem.* **2004**, *279*, 15183–15195. [CrossRef]
98. Jung, U.S.; Sobering, A.K.; Romeo, M.J.; Levin, D.E. Regulation of the yeast Rlm1 transcription factor by the Mpk1 cell wall integrity MAP kinase. *Mol. Microbiol.* **2002**, *46*, 781–789. [CrossRef]
99. Sanz, A.B.; Garcia, R.; Rodriguez-Pena, J.M.; Nombela, C.; Arroyo, J. Slt2 MAPK association with chromatin is required for transcriptional activation of Rlm1 dependent genes upon cell wall stress. *Biochim. Biophys. Acta Gene Regul. Mech.* **2018**, *1861*, 1029–1039. [CrossRef]
100. Haase, S.B.; Wittenberg, C. Topology and control of the cell-cycle-regulated transcriptional circuitry. *Genetics* **2014**, *196*, 65–90. [CrossRef]
101. Baetz, K.; Andrews, B. Regulation of cell cycle transcription factor Swi4 through auto-inhibition of DNA binding. *Mol. Cell Biol.* **1999**, *19*, 6729–6741. [CrossRef]
102. Igual, J.C.; Johnson, A.L.; Johnston, L.H. Coordinated regulation of gene expression by the cell cycle transcription factor Swi4 and the protein kinase C MAP kinase pathway for yeast cell integrity. *EMBO J.* **1996**, *15*, 5001–5013. [CrossRef] [PubMed]
103. Truman, A.W.; Kim, K.Y.; Levin, D.E. Mechanism of Mpk1 mitogen-activated protein kinase binding to the Swi4 transcription factor and its regulation by a novel caffeine-induced phosphorylation. *Mol. Cell Biol.* **2009**, *29*, 6449–6461. [CrossRef]
104. Schier, A.C.; Taatjes, D.J. Structure and mechanism of the RNA polymerase II transcription machinery. *Genes Dev.* **2020**, *34*, 465–488. [CrossRef] [PubMed]
105. Kornberg, R.D. Mediator and the mechanism of transcriptional activation. *Trends BioChem. Sci.* **2005**, *30*, 235–239. [CrossRef] [PubMed]
106. Harper, T.M.; Taatjes, D.J. The complex structure and function of Mediator. *J. Biol. Chem.* **2018**, *293*, 13778–13785. [CrossRef] [PubMed]
107. Ježek, J.; Smethurst, D.G.J.; Stieg, D.C.; Kiss, Z.A.C.; Hanley, S.E.; Ganesan, V.; Chang, K.T.; Cooper, K.F.; Strich, R. Cyclin C: The Story of a Non-Cycling Cyclin. *Biology* **2019**, *8*, 3. [CrossRef] [PubMed]
108. Cooper, K.F.; Mallory, M.J.; Smith, J.B.; Strich, R. Stress and developmental regulation of the yeast C-type cyclin Ume3p (Srb11p/Ssn8p). *EMBO J.* **1997**, *16*, 4665–4675. [CrossRef]
109. Cooper, K.F.; Mallory, M.J.; Strich, R. Oxidative stress-induced destruction of the yeast C-type cyclin Ume3p requires phosphatidylinositol-specific phospholipase C and the 26S proteasome. *Mol. Cell Biol.* **1999**, *19*, 3338–3348. [CrossRef]
110. Cooper, K.F.; Khakhina, S.; Kim, S.K.; Strich, R. Stress-induced nuclear-to-cytoplasmic translocation of cyclin C promotes mitochondrial fission in yeast. *Dev. Cell* **2014**, *28*, 161–173. [CrossRef]
111. Cooper, K.F.; Scarnati, M.S.; Krasley, E.; Mallory, M.J.; Jin, C.; Law, M.J.; Strich, R. Oxidative-stress-induced nuclear to cytoplasmic relocalization is required for Not4-dependent cyclin C destruction. *J. Cell Sci.* **2012**, *125*, 1015–1026. [CrossRef]

112. Levin-Salomon, V.; Maayan, I.; Avrahami-Moyal, L.; Marbach, I.; Livnah, O.; Engelberg, D. When expressed in yeast, mammalian mitogen-activated protein kinases lose proper regulation and become spontaneously phosphorylated. *BioChem. J.* **2009**, *417*, 331–340. [CrossRef] [PubMed]
113. Valencia, A.M.; Kadoch, C. Chromatin regulatory mechanisms and therapeutic opportunities in cancer. *Nat. Cell Biol.* **2019**, *21*, 152–161. [CrossRef] [PubMed]
114. Gartenberg, M.R.; Smith, J.S. The Nuts and Bolts of Transcriptionally Silent Chromatin in *Saccharomyces cerevisiae*. *Genetics* **2016**, *203*, 1563–1599. [CrossRef] [PubMed]
115. Sauty, S.M.; Shaban, K.; Yankulov, K. Gene repression in S. cerevisiae-looking beyond Sir-dependent gene silencing. *Curr. Genet.* **2021**, *67*, 3–17. [CrossRef]
116. Stone, E.M.; Pillus, L. Activation of an MAP kinase cascade leads to Sir3p hyperphosphorylation and strengthens transcriptional silencing. *J. Cell Biol.* **1996**, *135*, 571–583. [CrossRef]
117. Viswanathan, M.; Muthukumar, G.; Cong, Y.S.; Lenard, J. Seripauperins of *Saccharomyces cerevisiae*: A new multigene family encoding serine-poor relatives of serine-rich proteins. *Gene* **1994**, *148*, 149–153. [CrossRef]
118. Kothiwal, D.; Laloraya, S. A SIR-independent role for cohesin in subtelomeric silencing and organization. *Proc. Natl. Acad. Sci. USA* **2019**, *116*, 5659–5664. [CrossRef]
119. Blasl, A.T.; Schulze, S.; Qin, C.; Graf, L.G.; Vogt, R.; Lammers, M. Post-translational lysine ac(et)ylation in health, ageing and disease. *Biol. Chem.* **2022**, *403*, 151–194. [CrossRef]
120. Strahl-Bolsinger, S.; Hecht, A.; Luo, K.; Grunstein, M. SIR2 and SIR4 interactions differ in core and extended telomeric heterochromatin in yeast. *Genes Dev.* **1997**, *11*, 83–93. [CrossRef]
121. Lin, S.J.; Kaeberlein, M.; Andalis, A.A.; Sturtz, L.A.; Defossez, P.A.; Culotta, V.C.; Fink, G.R.; Guarente, L. Calorie restriction extends *Saccharomyces cerevisiae* lifespan by increasing respiration. *Nature* **2002**, *418*, 344–348. [CrossRef]
122. Lu, Y.Y.; Krebber, H. Nuclear mRNA Quality Control and Cytoplasmic NMD Are Linked by the Guard Proteins Gbp2 and Hrb1. *Int. J. Mol. Sci.* **2021**, *22*, 11275. [CrossRef] [PubMed]
123. Saavedra, C.; Tung, K.S.; Amberg, D.C.; Hopper, A.K.; Cole, C.N. Regulation of mRNA export in response to stress in *Saccharomyces cerevisiae*. *Genes Dev.* **1996**, *10*, 1608–1620. [CrossRef] [PubMed]
124. Zarnack, K.; Balasubramanian, S.; Gantier, M.P.; Kunetsky, V.; Kracht, M.; Schmitz, M.L.; Sträßer, K. Dynamic mRNP Remodeling in Response to Internal and External Stimuli. *Biomolecules* **2020**, *10*, 1310. [CrossRef] [PubMed]
125. Alpert, T.; Straube, K.; Carrillo Oesterreich, F.; Herzel, L.; Neugebauer, K.M. Widespread Transcriptional Readthrough Caused by Nab2 Depletion Leads to Chimeric Transcripts with Retained Introns. *Cell Rep.* **2020**, *33*, 108324. [CrossRef]
126. Zinzalla, V.; Graziola, M.; Mastriani, A.; Vanoni, M.; Alberghina, L. Rapamycin-mediated G1 arrest involves regulation of the Cdk inhibitor Sic1 in *Saccharomyces cerevisiae*. *Mol. Microbiol.* **2007**, *63*, 1482–1494. [CrossRef]
127. Escoté, X.; Zapater, M.; Clotet, J.; Posas, F. Hog1 mediates cell-cycle arrest in G1 phase by the dual targeting of Sic1. *Nat. Cell Biol.* **2004**, *6*, 997–1002. [CrossRef]
128. Moreno-Torres, M.; Jaquenoud, M.; Péli-Gulli, M.P.; Nicastro, R.; De Virgilio, C. TORC1 coordinates the conversion of Sic1 from a target to an inhibitor of cyclin-CDK-Cks1. *Cell Discov.* **2017**, *3*, 17012. [CrossRef]
129. Kono, K.; Al-Zain, A.; Schroeder, L.; Nakanishi, M.; Ikui, A.E. Plasma membrane/cell wall perturbation activates a novel cell cycle checkpoint during G1 in *Saccharomyces cerevisiae*. *Proc. Natl. Acad. Sci. USA* **2016**, *113*, 6910–6915. [CrossRef]
130. Yeeles, J.T.P.; Janska, A.; Early, A.; Diffley, J.F.X. How the Eukaryotic Replisome Achieves Rapid and Efficient DNA Replication. *Mol. Cell* **2017**, *65*, 105–116. [CrossRef]
131. Katou, Y.; Kanoh, Y.; Bando, M.; Noguchi, H.; Tanaka, H.; Ashikari, T.; Sugimoto, K.; Shirahige, K. S-phase checkpoint proteins Tof1 and Mrc1 form a stable replication-pausing complex. *Nature* **2003**, *424*, 1078–1083. [CrossRef]
132. Uzunova, S.D.; Zarkov, A.S.; Ivanova, A.M.; Stoynov, S.S.; Nedelcheva-Veleva, M.N. The subunits of the S-phase checkpoint complex Mrc1/Tof1/Csm3: Dynamics and interdependence. *Cell Div.* **2014**, *9*, 4. [CrossRef] [PubMed]
133. Voordeckers, K.; Colding, C.; Grasso, L.; Pardo, B.; Hoes, L.; Kominek, J.; Gielens, K.; Dekoster, K.; Gordon, J.; Van der Zande, E.; et al. Ethanol exposure increases mutation rate through error-prone polymerases. *Nat. Commun.* **2020**, *11*, 3664. [CrossRef] [PubMed]
134. Duch, A.; Felipe-Abrio, I.; Barroso, S.; Yaakov, G.; García-Rubio, M.; Aguilera, A.; de Nadal, E.; Posas, F. Coordinated control of replication and transcription by a SAPK protects genomic integrity. *Nature* **2013**, *493*, 116–119. [CrossRef] [PubMed]
135. Castelli, L.M.; Talavera, D.; Kershaw, C.J.; Mohammad-Qureshi, S.S.; Costello, J.L.; Rowe, W.; Sims, P.F.; Grant, C.M.; Hubbard, S.J.; Ashe, M.P.; et al. The 4E-BP Caf20p Mediates Both eIF4E-Dependent and Independent Repression of Translation. *PLoS Genet.* **2015**, *11*, e1005233. [CrossRef]
136. Mehta, S.; Li, H.; Hogan, P.G.; Cunningham, K.W. Domain architecture of the regulators of calcineurin (RCANs) and identification of a divergent RCAN in yeast. *Mol. Cell Biol.* **2009**, *29*, 2777–2793. [CrossRef]
137. MacGilvray, M.E.; Shishkova, E.; Place, M.; Wagner, E.R.; Coon, J.J.; Gasch, A.P. Phosphoproteome Response to Dithiothreitol Reveals Unique Versus Shared Features of *Saccharomyces cerevisiae* Stress Responses. *J. Proteome Res.* **2020**, *19*, 3405–3417. [CrossRef]
138. Bonilla, M.; Nastase, K.K.; Cunningham, K.W. Essential role of calcineurin in response to endoplasmic reticulum stress. *EMBO J.* **2002**, *21*, 2343–2353. [CrossRef]

139. Chen, Y.; Feldman, D.E.; Deng, C.; Brown, J.A.; De Giacomo, A.F.; Gaw, A.F.; Shi, G.; Le, Q.T.; Brown, J.M.; Koong, A.C. Identification of mitogen-activated protein kinase signaling pathways that confer resistance to endoplasmic reticulum stress in *Saccharomyces cerevisiae*. *Mol. Cancer Res.* **2005**, *3*, 669–677. [CrossRef]
140. Crooks, G.E.; Hon, G.; Chandonia, J.M.; Brenner, S.E. WebLogo: A sequence logo generator. *Genome Res.* **2004**, *14*, 1188–1190. [CrossRef]
141. Ptacek, J.; Devgan, G.; Michaud, G.; Zhu, H.; Zhu, X.; Fasolo, J.; Guo, H.; Jona, G.; Breitkreutz, A.; Sopko, R.; et al. Global analysis of protein phosphorylation in yeast. *Nature* **2005**, *438*, 679–684. [CrossRef]
142. Shimoji, K.; Jakovljevic, J.; Tsuchihashi, K.; Umeki, Y.; Wan, K.; Kawasaki, S.; Talkish, J.; Woolford, J.L., Jr.; Mizuta, K. Ebp2 and Brx1 function cooperatively in 60S ribosomal subunit assembly in *Saccharomyces cerevisiae*. *Nucleic Acids Res.* **2012**, *40*, 4574–4588. [CrossRef] [PubMed]
143. Xu, H.; Fang, T.; Yan, H.; Jiang, L. The protein kinase Cmk2 negatively regulates the calcium/calcineurin signalling pathway and expression of calcium pump genes PMR1 and PMC1 in budding yeast. *Cell Commun. Signal.* **2019**, *17*, 7. [CrossRef] [PubMed]
144. Lanze, C.E.; Gandra, R.M.; Foderaro, J.E.; Swenson, K.A.; Douglas, L.M.; Konopka, J.B. Plasma Membrane MCC/Eisosome Domains Promote Stress Resistance in Fungi. *Microbiol. Mol. Biol. Rev.* **2020**, *84*, e00063-19. [CrossRef] [PubMed]
145. Luo, G.; Gruhler, A.; Liu, Y.; Jensen, O.N.; Dickson, R.C. The sphingolipid long-chain base-Pkh1/2-Ypk1/2 signaling pathway regulates eisosome assembly and turnover. *J. Biol. Chem.* **2008**, *283*, 10433–10444. [CrossRef]
146. Walther, T.C.; Aguilar, P.S.; Fröhlich, F.; Chu, F.; Moreira, K.; Burlingame, A.L.; Walter, P. Pkh-kinases control eisosome assembly and organization. *EMBO J.* **2007**, *26*, 4946–4955. [CrossRef]
147. Pearce, L.R.; Komander, D.; Alessi, D.R. The nuts and bolts of AGC protein kinases. *Nat. Rev. Mol. Cell Biol.* **2010**, *11*, 9–22. [CrossRef]
148. Foderaro, J.E.; Douglas, L.M.; Konopka, J.B. MCC/Eisosomes Regulate Cell Wall Synthesis and Stress Responses in Fungi. *J. Fungi* **2017**, *3*, 61. [CrossRef]
149. Kozubowski, L.; Panek, H.; Rosenthal, A.; Bloecher, A.; DeMarini, D.J.; Tatchell, K. A Bni4-Glc7 phosphatase complex that recruits chitin synthase to the site of bud emergence. *Mol. Biol. Cell* **2003**, *14*, 26–39. [CrossRef]
150. Pérez, J.; Arcones, I.; Gómez, A.; Casquero, V.; Roncero, C. Phosphorylation of Bni4 by MAP kinases contributes to septum assembly during yeast cytokinesis. *FEMS Yeast Res.* **2016**, *16*, fow060. [CrossRef]
151. Dokládal, L.; Stumpe, M.; Hu, Z.; Jaquenoud, M.; Dengjel, J.; De Virgilio, C. Phosphoproteomic responses of TORC1 target kinases reveal discrete and convergent mechanisms that orchestrate the quiescence program in yeast. *Cell Rep.* **2021**, *37*, 110149. [CrossRef]
152. Garrett-Engele, P.; Moilanen, B.; Cyert, M.S. Calcineurin, the Ca2+/calmodulin-dependent protein phosphatase, is essential in yeast mutants with cell integrity defects and in mutants that lack a functional vacuolar H(+)-ATPase. *Mol. Cell Biol.* **1995**, *15*, 4103–4114. [CrossRef] [PubMed]
153. Kim, K.Y.; Cosano, I.C.; Levin, D.E.; Molina, M.; Martin, H. Dissecting the transcriptional activation function of the cell wall integrity MAP kinase. *Yeast* **2007**, *24*, 335–342. [CrossRef] [PubMed]
154. Casar, B.; Pinto, A.; Crespo, P. Essential role of ERK dimers in the activation of cytoplasmic but not nuclear substrates by ERK-scaffold complexes. *Mol. Cell* **2008**, *31*, 708–721. [CrossRef] [PubMed]

Journal of *Fungi*

MDPI

Review

The Fission Yeast Cell Integrity Pathway: A Functional Hub for Cell Survival upon Stress and Beyond

José Cansado *, Teresa Soto, Alejandro Franco, Jero Vicente-Soler and Marisa Madrid *

Yeast Physiology Group, Department of Genetics and Microbiology, Campus de Excelencia Internacional de Ambito Regional (CEIR)—Campus Mare Nostrum, Universidad de Murcia, 30071 Murcia, Spain; teresaso@um.es (T.S.); afranco@um.es (A.F.); jerovic@um.es (J.V.-S.)

* Correspondence: jcansado@um.es (J.C.); marisa@um.es (M.M.)

Abstract: The survival of eukaryotic organisms during environmental changes is largely dependent on the adaptive responses elicited by signal transduction cascades, including those regulated by the Mitogen-Activated Protein Kinase (MAPK) pathways. The Cell Integrity Pathway (CIP), one of the three MAPK pathways found in the simple eukaryote fission of yeast *Schizosaccharomyces pombe*, shows strong homology with mammalian Extracellular signal-Regulated Kinases (ERKs). Remarkably, studies over the last few decades have gradually positioned the CIP as a multi-faceted pathway that impacts multiple functional aspects of the fission yeast life cycle during unperturbed growth and in response to stress. They include the control of mRNA-stability through RNA binding proteins, regulation of calcium homeostasis, and modulation of cell wall integrity and cytokinesis. Moreover, distinct evidence has disclosed the existence of sophisticated interplay between the CIP and other environmentally regulated pathways, including Stress-Activated MAP Kinase signaling (SAPK) and the Target of Rapamycin (TOR). In this review we present a current overview of the organization and underlying regulatory mechanisms of the CIP in *S. pombe*, describe its most prominent functions, and discuss possible targets of and roles for this pathway. The evolutionary conservation of CIP signaling in the dimorphic fission yeast *S. japonicus* will also be addressed.

Keywords: fission yeast; MAPK; Cell Integrity Pathway; stress

Citation: Cansado, J.; Soto, T.; Franco, A.; Vicente-Soler, J.; Madrid, M. The Fission Yeast Cell Integrity Pathway: A Functional Hub for Cell Survival upon Stress and Beyond. *J. Fungi* **2022**, *8*, 32. https://doi.org/10.3390/jof8010032

Academic Editors: María Molina and Humberto Martín

Received: 3 December 2021
Accepted: 27 December 2021
Published: 30 December 2021

1. Introduction

The ability to sense and respond to external and internal stimuli is critical for the survival of living organisms in an ever-changing environment. In eukaryotic cells, the highly conserved Mitogen Activated Protein Kinase (MAPK) pathways are specialized signal transduction cascades that detect environmental signals at the cell surface, which are transmitted to a set of downstream cytoplasmic and nuclear effectors to control multiple aspects of cell function, including metabolism, division, death, differentiation, and movement [1–3]. The organization of MAPK pathways is based in a modular structure that maintains a remarkable degree of evolutionary conservation [2,4,5]. The first module of the pathway includes the sensor/s and downstream effectors that recognize the stimulus, while the second transmission module comprises the three-tiered and highly conserved kinases known as the MAPK kinase kinase (MAPKKK), the MAPK kinase (MAPKK), and the MAP kinase (MAPK), which is the core component of the pathway. MAPKKKs are usually activated through phosphorylation in response to environmental changes by upstream signaling cascades composed by GTPases of the Rho or Ras families together with different kinases, although they can also become activated through oligomerization or changes in subcellular location. Once activated, MAPKKKs phosphorylate and activate the downstream MAPKK at the Thr and Ser residues. In turn, the MAPKK associates and dually phosphorylates the MAPK at two strongly conserved T and Y residues of a T-X-Y motif located within the activation loop [6,7]. Among the most biologically relevant downstream effectors phosphorylated by MAPKs are transcription factors, cell cycle regulators, chaperones, cytoskeletal proteins, RNA-binding proteins, and different cytoplasmic substrates [2,6,8].

The number of MAPKs varies depending on the eukaryotic organism. Mammalian cells carry five types of conventional MAPKs, which comprise the extracellular signal-regulated kinases Erk1 and Erk2, JUN amino-terminal kinase (JNK), p38, and Erk5. JNK and p38 are also known as "Stress-Activated Protein Kinases" (SAPK), because they become preferentially activated in response to multiple environmental cues and are involved in the regulation of apoptosis and stress responses [9]. On the other hand, MAPKs of the ERK group are activated in response to mitogenic signals elicited by growth factors and phorbol esters and regulate several aspects of cellular functions including proliferation, survival, growth, metabolism, migration, and differentiation in response to extracellular cues [10]. While JNKs have only been described in mammals, orthologs to ERK and p38 MAPKs are also present in yeasts, including the rod-shaped fission yeast *Schizosaccharomyces pombe*, which, during the last few decades, has been positioned as an excellent model organism to depict functional mechanisms of general significance in eukaryotes, particularly in the field of fungal biology. *S. pombe* has three MAPK-signaling cascades known as the mating-pheromone, the stress-activated (SAPK), and the cell integrity (CIP) pathways. The SAPK pathway is homologous to mammalian p38, while the CIP pathway shows strong homology with mammalian ERKs [11–15]. In this review we present an exhaustive and updated description of the organization, components, and multiple biological roles of the CIP in *S. pombe*, from its discovery almost 30 years ago to the present day.

2. Architecture and Organization of the Fission Yeast CIP

The cell wall (CW) is an essential structure for the viability of fungal cells. It is responsible for the maintenance of their morphology and provides protection against osmotic lysis and mechanical damage, as well as being a key target for various antifungals. The main components of the CW in *S. pombe* are polysaccharides, such as β-glucans (54–60%) and α-glucans (28–32%), which confer rigidity and are essential for cell integrity [16]. In this organism, the CIP regulates multiple processes such as CW synthesis and maintenance during stress, morphogenesis, cytokinesis, mRNA stabilization, vacuole fusion, and ionic homeostasis [15,17–19]. The core effector of the CIP is MAPK Pmk1, an extracellular signal-regulated kinase ERK ortholog, which becomes cyclically activated during cytokinesis [20] and also in response to multiple environmental changes including heat, osmotic and oxidative stresses, CW damage, and glucose depletion [15,18,21]. Accordingly, Pmk1-less mutants display cytokinetic defects as well as growth sensitivity to osmotic stress and CW-damaging agents. While Pmk1 activity is necessary to preserve cellular integrity under multiple circumstances, its constitutive hyperactivation can be harmful [15,17–19,22–24]. Pioneering studies demonstrated that the Ca^{2+}/calmodulin-dependent protein phosphatase calcineurin (Ppb1) and Pmk1 pathway play antagonistic roles during chloride homeostasis [25]. Based on this negative interaction between calcineurin and Pmk1, a genetic screen was developed to isolate mutants that were *v*iable in the presence of *i*mmunosuppressant and *c*hloride ion (a phenotype called *vic*), which allowed the identification of key members of the CIP. Indeed, the *vic* phenotype is a characteristic feature of CIP null mutants, since the absence of Pmk1 activity suppresses the strong chloride sensitivity of *ppb1*Δ mutants or of wild-type cells subjected to pharmacological inhibition of calcineurin activity with the immunosuppressant FK506 [25,26] (further details on this issue will be described in the "*Calcium homeostasis*" section below).

The main components of the CIP have been extensively characterized in *S. pombe* and will be described thereafter (Figure 1A). A short description of their biological roles and the corresponding *S. cerevisiae* orthologs are also shown in Table 1. Its overall architecture, as in every MAPK-signaling cascade, is based on a modular organization. The first and most upstream module comprises one or several sensors that specifically perceive the stimuli, plus a number of immediate downstream regulatory components. The second module comprises the signature MAPK transmission module, which is composed of three highly conserved kinases: MAPKKK, MAPKK, and MAPK. The nature of last module is heterogeneous depending on the signaling cascade and includes biologically distinct targets and effectors that become phosphorylated by the active MAPK (Figure 1A) [2,6,27]. Of note, very recently, we have characterized the main components of the CIP and their role during hyphal development in *S. japonicus*, a dimorphic fission yeast species. Our findings are presented in an accompanying research paper in this collection [28].

2.1. Upstream of the CIP MAPK Module

In *S. pombe*, the upstream elements of the CIP MAPK include, among others, several putative sensors, Rho-GTPases and their regulators (GEFs and GAPs), the phospholipid-dependent kinase (PDK), and two PKC orthologs (Figure 1A). Their structure, organization, and functional roles during CIP activation are described below.

2.1.1. Sensors

Sensors are key components of cell integrity-signaling pathways, as they have the primary role for detecting changes in the CW and/or membrane structure and integrity in response to external stressors. In *S. cerevisiae*, as in other fungi, the transmembrane CW sensors belong to two different families, the WSC-type (Wsc1, Wsc2, and Wsc3) and MID-type (Mid2-Mtl1 pair) families, which function upstream and trigger CIP activation [29]. CW sensors show a cytoplasmic C-terminal tail that mediates downstream signaling, a transmembrane domain (TMD), and an N-terminal serine- and threonine-rich region (STR) that is highly mannosylated. In addition, the WSC-type sensors carry a cysteine-rich domain (CRD or WSC domain), while MID-type sensors have an N-glycosylated asparagine residue near the N-terminus, which mediates its interaction with CW polysaccharides [29,30]. Their characteristic conformation and ability to anchor both the plasma membrane and CW allow *S. cerevisiae* WSC and MID proteins to behave as mechanosensors that detect the perturbations at the CW and plasma membrane, which are transduced through their cytoplasmic tail to the CIP through Rom2 [29,31]. Rom2 is a Guanine Exchange Factor (GEF) that activates the GTPase Rho1, which in turn activates Pkc1, the only Protein Kinase C (PKC) ortholog present in budding yeast, which acts upstream of the CIP MAPK cascade [29–31].

In *S. pombe*, the two putative sensors Wsc1 and Mlt2 have been described as structural homologs of *S. cerevisiae* Wsc1 and Mlt1, respectively (Figure 1A) [32]. Wsc1 is located at active growth sites and the division septum, while Mlt2 is located at the cellular periphery. Mlt2 is necessary for fission yeast survival in response to different CW stresses, whereas Wsc1 overexpression activates CW biosynthesis [32]. Wsc1 and Mtl2 physically couple the CW with the plasma membrane and transmit signals mainly through Rgf1, a GEF that activates the essential GTPase Rho1, which in turn regulates β-(1,3)-glucan synthase activity and stabilizes the two PKC orthologs Pck1 and Pck2 [33]. Indeed, both Wsc1 and Mtl2 are necessary to maintain the physiological levels of active Rho1-GTP upon CW stress, and the overexpression of Rho1 or its GEFs suppresses the lethality caused by simultaneous deletion of *wsc1*$^+$ and *mlt2*$^+$ genes [32]. However, in contrast to the *S. cerevisiae* CWI sensors, CIP activity upon stress is not significantly compromised in the absence of Wsc1 and/or Mlt2, suggesting that they may regulate *S. pombe* CW integrity independently of this MAPK cascade [32]. Recent studies have proposed that both proteins perform CW mechanosensing activities that detect cell growth as a mechanical stress, i.e., as the thickness of the CW at the growing tips is very dynamic in fission yeast [34–36]. Wsc1 may represent an autonomous sensing module that detects mechanical stress at the CW and forms stable clusters that act as signaling platforms to recruit and activate downstream signaling elements to mechano-perception points [35,37]. This mechanosensing mechanism mediated by Wsc1 could represent a basal homeostatic module that, together with the CIP pathway, may contribute to fine-tuning of CW thickness and homeostasis [37].

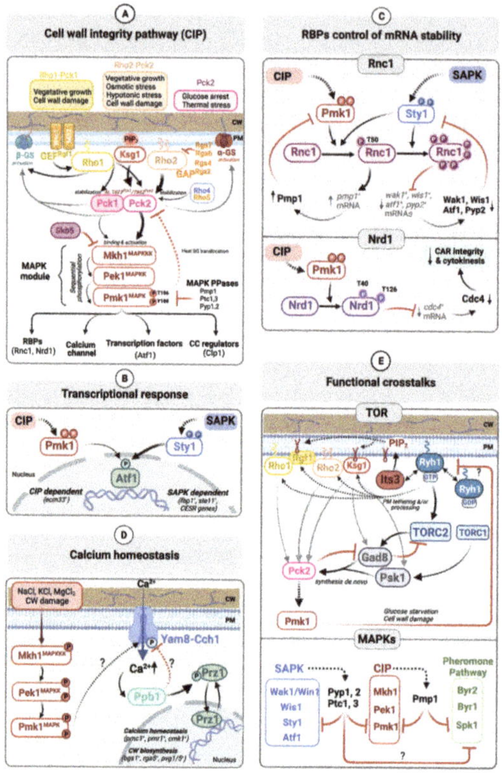

Figure 1. Organization and biological roles of the fission yeast Cell Integrity Pathway (CIP). (**A**) The overall architecture and main components of the CIP in *S. pombe*. (**B**) Transcriptional regulation. Atf1, the main downstream transcription factor regulated by the stress activated MAPK pathway (SAPK) via Sty1 MAPK, can be also targeted by Pmk1 under specific stimuli, such as cell wall stress. The possibility that Pmk1 may, together with Sty1, regulate the expression of other Atf1-dependent genes besides *ecm33*$^{+}$ is currently unknown. (**C**) Control of mRNA-stability through RNA binding proteins (RBPs). The RBPs Nrd1 and Rnc1 become phosphorylated by Pmk1 during growth and stress to modulate the stability of mRNAs encoding the dual specificity phosphatase Pmp1 (Rnc1), and the myosin II essential light chain Cdc4 (Nrd1), thus allowing an accurate control of the Pmk1 activity threshold and cytokinesis, respectively. Sty1 also phosphorylates Rnc1 in vivo to prompt destabilization of mRNAs encoding SAPK components, resulting in reduced MAPK activity. (**D**) Regulation of calcium homeostasis. Pmk1 and calcineurin antagonistically regulate the activity of the Yam8-Cch1 channel complex to modulate the calcium influx in response to saline and cell wall stresses. (**E**) Functional cross-talks. Upper: Functional interaction between CIP and target of rapamycin (TOR) signaling pathways. Tor complex-2 (TORC2) and its main activator, the Rab GTPase Ryh1, positively control de novo synthesis of Pck2 and trigger Pmk1 activation in response to cell wall (CW) damage or upon glucose limitation. Ryh1 also promotes the plasma membrane targeting of several CIP upstream components and Pmk1 activation independently of TORC2 signaling. Conversely, activated Pmk1 down-regulates Ryh1 activity. Lower: The SAPK pathway negatively controls CIP signaling through the transcriptional induction of the MAPK phosphatases, Pyp1, Pyp2, Ptc1, and Ptc3, that dephosphorylate both Sty1 and Pmk1. Pmk1 dual specificity phosphatase Pmp1 also dephosphorylates Spk1, the MAPK of the pheromone response pathway. Dotted lines indicate that the molecular links and/or mechanisms have not been fully established. CW: cell wall; PM: plasma membrane; CC: cell cycle; RBP: RNA binding protein; and PPases: phosphatases. Please see the main text for specific nomenclature and details.

Table 1. Main functional roles of fission yeast CIP components.

	Protein	Function	*S. cerevisiae* Ortholog
Cell-surface sensors	Wsc1	- Plasma membrane-associated serine-rich cell wall mechanosensor located at active growth sites and the division septum - Detects perturbations at the CW and plasma membrane and transmits signals through Rgf1 (Rho1 GEF)	WSC1
	Mtl2	- Plasma membrane-associated serine-rich cell wall mechanosensor located at cellular periphery - Detects perturbations at the CW and plasma membrane and transmits signals through Rgf1	MLT1
Regulators of GTPases	Rgf1	- Guanine nucleotide exchange factor (GEF) for Rho1 GTPase - Activates the CIP via Rho1 and Pck2	ROM1
	Rga2	- GTPase activating protein (GAP) for Rho1 and Rho2. - Negatively regulates the Rho2–Pck2 interaction with the CIP	BEM3
	Rga4	Rho2 GAP that negatively regulates the activity of the CIP	RGA2
	Rga6	Rho2 GAP	
	Rga7	Rho2 GAP that negatively regulates the activity of the CIP	RGD1
Rho GTPases	Rho1	- Regulation of cell wall biosynthesis by β-glucan synthases Bgs1-4 - Stabilizes Pck1 and Pck2 - Regulates the CIP in response to cell wall damage, independently of Wsc1 and Mlt2 - Cytokinesis checkpoint to cell wall damage through Pmk1	RHO1
	Rho2	- Regulation of cell wall biosynthesis by α-glucan synthase Mok1/Ags1 - Major CIP regulator together with Pck2	RHO2
	Rho4	- Minor role in CIP signaling	RHO4
	Rho5	- Functional paralogue of Rho1 - Minor role in CIP signaling	RHO1
Phosphoinositide metabolism	Ksg1	Serine/threonine protein kinase PDK-1ortholog involved in CIP signaling through the activation of Pck1 and Pck2	PKH1/PKH2
PKCs and the CIP MAPK cascade	Pck1	Rho1 target involved in CIP signaling in response to CW damage	PKC1
	Pck2	-Rho1 and Rho2 target - Main upstream activator of the CIP MAPK module during growth and stress	PKC2
	Mkh1	MAPK kinase kinase	BCK1
	Pek1	MAPK kinase	MKK1/MKK2
	Pmk1	MAPK	MPK1/SLT2
Negative regulators	Skb5	- Shk1 kinase binding protein - Inhibits Pmk1 by downregulating Mkh1 localization to cell tips	NBP2
	Pmp1	- Dual-specificity MAP kinase phosphatase - Binds and dephosphorylates Pmk1 during vegetative growth	SDP1/MSG5
	Pyp1	Tyrosine phosphatase: - Transcriptionally regulated by the SAPK pathway (Sty1-Atf1) - Binds and dephosphorylates both Sty1 and Pmk1 during vegetative growth	PTP3, PTP2
	Pyp2	Tyrosine phosphatase: - Transcriptionally regulated by the SAPK pathway (Sty1-Atf1). - Binds and dephosphorylates both Sty1 and Pmk1 during stress	PTP3, PTP2
	Ptc1	Serine/threonine phosphatase: - Transcriptionally regulated by the SAPK pathway (Sty1-Atf1). - Binds and dephosphorylates both Sty1 and Pmk1 during vegetative growth	PTC1
	Ptc3	Serine/threonine phosphatase: - Transcriptionally regulated by the SAPK pathway (Sty1-Atf1). - Binds and dephosphorylates both Sty1 and Pmk1 during stress	PTC3, PTC2
Downstream targets	Atf1	- Atf-CREB family bZIP domain transcription factor - Downstream effector of Sty1 MAPK(SAPK pathway) in response to environmental cues and Pmk1 in response to CW damage	—
	Nrd1	RNA-binding protein (RBP): - Phosphorylated by Pmk1 during growth and stress - Negatively modules myosin II essential light chain mRNA stability - Regulation of actomyosin ring integrity during cytokinesis	MRN1

Table 1. *Cont.*

Protein	Function	*S. cerevisiae* Ortholog
Rnc1	KH domain RBP: - Phosphorylated by Pmk1 during growth and stress - Phosphorylated by Sty1 during growth and stress - Stabilizes mRNA encoding Pmk1 dual specificity phosphatase Pmp1 - Negative regulation of Pmk1 activity	PBP2
Clp1	Cdc14-related serine/threonine protein phosphatase: - Phosphorylated by Pmk1 - Pmk1-dependent phosphorylation promotes its nucleoplasmic accumulation during genotoxic stress	CDC14
Cch1-Yam8	Plasma-membrane channel complex: - Putative substrate for Pmk1 phosphorylation - Putative substrate for dephosphorylation by Calcineurin (Ppb1) - Positively regulates import of calcium ions in response to salt stress	CCH1-MID1

2.1.2. Rho-GTPases and Their Regulators

The Rho family GTPases are highly conserved and fundamental regulators in eukaryotes of multiple processes such as actin cytoskeleton dynamics, cell cycle, gene expression, vesicle trafficking, and cell polarity [38,39]. They perform their biological functions bound to biological membranes, for which they must be post-translationally modified [40]. Rho-GTPases oscillate between an active form bound to GTP that interacts with the downstream effectors, and an inactive form coupled with GDP. The exchange of GDP to GTP induced by Guanine Exchange Factors (GEFs) promotes their activation, while the hydrolysis of GTP to GDP is mediated by GTPase Activator Proteins (GAPs). Rho-GTPases are also negatively regulated by cytoplasmic Guanine nucleotide Dissociation Inhibitors (GDIs), which prompt GTPase removal from membranes [38]. *S. pombe* has six Rho GTPases (Rho1 to Rho5 and Cdc42), two of which, Rho1 and Cdc42, are essential [41,42]. Rho2 and, to a lesser extent, Rho1 are recognized as the key activating regulators of fission yeast CIP signaling (Figure 1A) [33].

Rho2 is a non-essential GTPase that localizes to the sites of active growth, and its deletion produces slightly rounded cells that are sensitive to glucanase treatment or staurosporine, a potent PKC inhibitor [6,43]. Fission yeast Rho2 regulates cell polarity, actin cytoskeleton organization, and CW biosynthesis by controlling the activity of α-glucan synthase Mok1 through the PKC orthologue Pck2 [6,44,45] (see below). In addition, Rho2 and Pck2 are the core upstream components that activate the CIP MAPK module (Figure 1A). Accordingly, fission yeast mutants lacking both proteins display a sharp *vic* phenotype [26]. Rho2 positively regulates the activation of the CIP MAPK module, fundamentally through Pck2, since the lethality associated with the Rho2 overexpression is totally suppressed in the absence of Pck2 or any of the MAPK module components [18,26]. Post-translational lipid modification of Rho2 mediates its proper membrane tethering and activation of the CIP [46–48]. This process takes place in vivo in three sequential steps. Rho2 becomes first farnesylated at C-197 by farnesyltransferase Cpp1 [26], and then the free carboxyl group of the isoprenylated cysteine is methylated by Mam4, a specific isoprenylcysteine-O-carboxylmethyltransferase (ICMT) [47]. Curiously, while Mam4 absence reduces Rho2-targeting to the plasma membrane, this effect is not observed with other methylated GTPases such as Rho1 or Cdc42 [47]. Finally, farnesylated and methylated Rho2 undergoes in vivo palmitoylation at C-196 by the DHHC palmitoyltransferase Erf2 [46]. This three-step lipidation process is essential for Rho2 full plasma membrane localization and the ensuing morphogenesis control and signaling to the CIP during unperturbed growth and stress [26,46,47,49].

The control of phosphoinositide turnover by eisosomes, which are multiprotein structures that generate linear invaginations at the plasma membrane of yeast cells, has also been proposed to modulate Rho2 localization and CIP activity [48]. This is based on the observation that Pmk1 activation during osmostress, which is Rho2-dependent, is impaired in a fission yeast hypomorphic mutant in the phosphatidylinositol (PI) 5-kinase Its3, and that this defect is suppressed in absence of eisosomes [48].

The four Rho2 GAPs identified to date, Rga2, Rga4, Rga6, and Rga7, have been shown to be involved in the control of CIP activity, since their single deletion prompts a significant, Rho2-dependent increase in Pmk1 MAPK activity during growth and stress (Figure 1A) [50–52]. The identity of the putative Rho2 GEFs, and their possible role during CIP signaling, is currently unknown. Rdi1, the only known Rho GDI in *S. pombe*, has been shown to negatively regulate the CIP, but it does so in a Rho2-independent fashion [46]. The fission yeast essential GTPase Rho1 is a functional homolog of human RhoA and *S. cerevisiae* Rho1. Rho1 localizes at active growth sites and is a direct activator of the (1,3)-β-D-glucan synthase necessary for CW glucan synthesis and cytokinesis [41,44,53,54]. Other prominent effectors of Rho1 are the PKC orthologues Pck1 and Pck2 [53,55,56]. Subsequently, it was shown that Rho1 regulates CIP signaling in an additive and/or alternative way to Rho2 (Figure 1A) [21,57]. By using a mutant that expresses a hypomorphic allele of Rho1 (Rho1-596), it was demonstrated that Rho1 and Rho2 independently regulate Pck2 to maintain Pmk1 basal activity during vegetative growth [23,57]. According to the current model, Pck1, a second PKC ortholog, additionally regulates Pmk1 activity by acting as a specific target for Rho1, fundamentally in response to CW damage [23,57]. Rho1 GTPase activity is upregulated in vivo by three GEFs, Rgf1, Rgf2 and Rgf3, and is downregulated by three GAPs, Rga1, Rga5, and Rga8, together with Rdi1 GDI [58–62]. Rgf1 is the main GEF responsible for the maintenance of basal Rho1 GTPase activity in fission yeast cells during unperturbed growth, since its deletion strongly reduces the amount of the available Rho1-GTP pool. Activation of Rho1 by Rgf1 is necessary for the correct reorganization of actin cytoskeletons during the transition from monopolar to bipolar growth that occurs during *S. pombe* cell growth [63]. Later, Rgf1 was identified as a canonical member of the CIP, since its knockout blocked CIP activation upon different stresses, including osmotic stress. It was proposed that the Rgf1 positively regulates the CIP during this specific situation through the activation of Rho1 and Pck2 (Figure 1A) [21]. However, as described above, a Rho2–Pck2 branch positively regulates Pmk1 activation during vegetative growth and in response to many environmental cues including osmotic stress, which is totally Rho2-dependent [64]. It might be possible that Rgf1 could act as a Rho2 GEF under this specific situation, although this hypothesis has not been proven. On the other hand, the two redundant Rho1 GEFs, Rgf2 and Rgf3, do not seem to play a significant role in the regulation of CIP signaling, because Pmk1 activation upon stress was largely unaffected by their respective deletion [21].

Rho3, Rho4, and Rho5 GTPases have also been proposed as upstream regulators of the CIP besides Rho1 and Rho2, although their biological relevance is less clear. Rho3 has a general role in the secretion, exocytosis, and Golgi to endosome trafficking [38] and is a putative negative regulator of the CIP, as *rho3*Δ cells show a mild increase in Pmk1 phosphorylation and chloride sensitivity, which are typical features of CIP hyperactivation [46]. However, negative control of the CIP by Rho3 is likely an indirect effect and is independent of Rho2 function [46]. In fission yeast, Rho4 regulates cell morphology, septation and CW integrity [65–67], whereas Rho5 is a functional parologue of Rho1 (86% amino acids identity) that regulates cellular morphology and septation and also the survival of both vegetative cells and ascospores during the stationary phase [68,69]. Interestingly, *rho4*Δ and *rho5*Δ mutants display a *vic* phenotype and are growth-sensitive to micafungin, a specific inhibitor of CW beta-glucan synthase [70]. Pmk1 activation was lower in *rho4*Δ and *rho5*Δ mutants than in wild-type cells in response to heat shock, and their respective over-expression resulted in the stimulation of CIP signaling, mainly through Pck2 [70]. However, besides the plasma membrane, a notable portion of both GTPases localizes to internal membranes, suggesting that they may positively control the CIP through as yet unidentified target(s) [70]. In conclusion, Rho2, and to a lesser extent Rho1, are the main activating GTPases for the CIP (Figure 1A), while Rho3, Rho4, and/or Rho5 might contribute positively or negatively to modulate the activity of the pathway under very specific conditions.

2.1.3. Phospholipid-Dependent Kinase-1 (PDK-1) and Protein Kinase C (PKC) Orthologs

The protein kinase C (PKC) family is a group of serine/threonine AGC kinases with essential roles in signaling pathways that participate in many different processes including cell proliferation, apoptosis, actin cytoskeleton remodeling, and ion channel modulation, among others [71]. All PKC isoforms share a basic structure that includes an N-terminal regulatory domain linked by a hinge region to a highly conserved C-terminal kinase domain [71]. The regulatory domain contains an autoinhibitory pseudosubstrate segment that is a key molecular switch during PCK activation, plus two putative membrane-binding modules named C1 and C2. The C1 module is conserved in all PKC isozymes, while the C2 module is found only in the conventional group of kinases [71]. The C-terminal kinase domain contains three conserved phosphorylation sites known as the activation loop (AL), turn motif (TM), and hydrophobic motif (HM), which are essential for catalytic activity [71–74]. *S. pombe* has two PKC orthologs known as Pck1 (111.78 kDa) and Pck2 (116 kDa), that share broad homology (70% identity in amino acid sequences within the catalytic domain), and have overlapping and essential roles in cell viability, since their simultaneous deletion is lethal [55,75]. Deletion of Pck1 does not cause obvious phenotypes, except for a moderate sensitivity to CW damage agents, and the appearance of a small percentage of cells showing mispositioned septa [55]. The *pck2Δ* mutants show clear defects in cell polarity and thin CWs and are hypersensitive to high temperatures or treatment with lytic enzymes [55,75]. While Pck1 overexpression prompts cellular elongation and a slight cell cycle delay, the overexpression of Pck2 is lethal and results in cells with very thick CWs and strong cytokinetic defects [55]. Both Pck1 and Pck2 contain all the typical domains of PKCs and, similar to the mammalian Rho family-responsive protein N kinases (PKNs), also show two polybasic coiled-coil HR1 motifs at their N-terminus that are responsible for binding to Rho1 and Rho2 GTPases [52,55]. Furthermore, both kinases are short half-life proteins that contain N-terminal PEST sequences (i.e., proline-, glutamic acid-, serine-, and threonine-rich sequences), which are involved in their degradation by the proteasome, and the interaction with activated Rho1 and Rho2 prompts their stabilization [52,55,74]. Pck1 and Pck2 co-localize with Rho1 and Rho2 in zones of polarized growth and relocate to the division zone before ring constriction [56,76,77]. Pck1 and Pck2 activation is controlled by several mechanisms, including autophosphorylation and phosphorylation by 3-phosphoinositide-dependent kinase-1 (PDK-1). Both modifications require prior binding and stabilization by Rho1 and/or Rho2 (see below) [55,74,78].

Rho1 and Rho2 regulate the biosynthesis of CW polymers (α- and β-D-glucan) through Pck1 and Pck2 and are also upstream activators of the CIP (Figure 1A) [18,26,46,52,55,57]. Rho1 regulates the synthesis of β-D-glucan, either directly or indirectly through Pck1 and Pck2 [55], while the synthesis of α-glucan is modulated by Rho2 strictly via Pck2 [45]. Moreover, the Rho2–Pck2 branch is the major upstream activator of the CIP upon stress, whereas both Rho1 and Rho2 target Pck2 to control basal Pmk1 activity during unperturbed vegetative growth [18,26,57]. It has been recently shown that Pmk1 promotes Pck2 translocation from the cell tips into stress granules upon heat stress, and this response that could contribute to prevent MAPK hyperactivation under such specific circumstances [79]. Instead, Pck1 plays a less meaningful role than Pck2 as a positive regulator of CIP, and its signaling activity seems biologically relevant only in response to CW damage by acting as a Rho1 effector (Figure 1A) [23,57].

PDK-1 is an essential AGC kinase family member that phosphorylates the AL of many AGC kinases including most PKC isozymes [71,80]. Ksg1 (65.66 kDa), one of the two *S. pombe* PDK-1 orthologs, plays an essential role in cell growth, conjugation, and sporulation [81,82]. It was initially shown that Ksg1 interacts with Pck1 and Pck2 in two hybrid assays [82]. Later, we demonstrated that Ksg1 is involved in CIP signaling through the activation of Pck1 and Pck2 [74,78]. In the current model, both Ksg1 and an autophosphorylation mechanism are responsible for the in vivo phosphorylation of Pck2 at the conserved T-842 within the AL during vegetative growth and in response to stress. AL phosphorylation, together with turn TM autophosphorylation at T-984 and binding to Rho1 and/or Rho2, prompt Pck2 stabilization and downstream activation of the CIP (Figure 1A) [74,78]. The equivalent AL phosphorylation site of Pck1, T-823, is also phosphorylated in vivo by Ksg1, and this modification is critical for Pck1 stabilization, catalytic activity, and biological function. Contrariwise, Pck2 is still stable and partially functional in the absence of AL phosphorylation [74,78]. Furthermore, the de novo synthesis of Pck1 and Pck2 is positively regulated by the TORC2 complex during vegetative growth and in response to several stresses (please see "*CIP and TOR signaling*") [72,78]. Thus, despite their strong structural similarity and functional redundancy, the mechanisms that regulate the maturation, activation, and stabilization of Pck1 and Pck2 have a very different impact on their biological function as CIP activators [74]. In summary, a molecular network composed of Rho1, Rho2, Ksg1, Pck1, and Pck2 alternatively modulates CIP MAPK cascade signaling in *S. pombe* during vegetative growth and depending on the type of stress, as discussed below (Figure 1A) [57,64,83].

Recently, we have described that, in the dimorphic fission yeast *S. japonicus*, Pck1 and Pck2 are also key upstream regulators of the CIP pathway [28]. As in *S. pombe*, *S. japonicus* Pck2 is the major activator of Pmk1 during unperturbed growth. However, MAPK activation during CW damage, which is Pck1- and Pck2-co-dependent in *S. pombe*, is exclusively transmitted by Pck2 in *S. japonicus* [28]. Moreover, Pck1 and Pck2 antagonistically regulate hyphal differentiation through fine-tuning control of CIP activity. The putative conserved role of Rho1 and Rho2 GTPases and Ksg1 orthologs during Pck1/2 stabilization and activation is currently unknown [28].

2.2. The CIP MAPK Module

The core CIP MAPK module is composed of MAPKKK Mkh1, MAPKK Pek1, and MAPK Pmk1/Spm1 (Figure 1A). As with other MAPK module components, their structural integrity is essential for proper CIP signaling, and the absence of any of the three kinase components causes the typical features associated with loss of function of CIP signaling, including the *vic* phenotype, multiseptation, defective vacuole fusion, altered CW organization, and growth sensitivity to beta-glucan synthase inhibitors stress treatments [15,17,18,22,26].

2.2.1. MAPKKK: Mkh1

Mkh1 ("Mek Kinase Homolog1"), is a 125.1 kDa protein, which was identified as the *S. pombe* CIP MAPKKK and is a structural homolog to the budding yeast CIP MAPKKK Bck1 [24]. Mkh1 associates in vivo with Pck1 and Pck2. However, while the Mkh1–Pck2 association is detected during vegetative growth and in response to stress, the Mkh1–Pck1 interaction has only been documented during CW stress [57]. This agrees with Pck2 being the major activator of CIP signaling during unperturbed growth. Once activated by Pck2, and occasionally by Pck1, Mkh1 transmits the activation signal to Pek1 and Pmk1, the remaining components of the MAPK module, through an in vivo interaction by forming a ternary complex (Figure 1A) [18,19,57]. Although Mkh1 is essentially a cytoplasmic protein, it can also be located at the septum area during cytokinesis and at the cell tips during the G1/S phase of the cell cycle [18,84]. Mkh1 targeting to the cell tips is regulated by Skb5 (Shk1 kinase binding protein) [84], an SH3 domain protein which is a direct activator of the p21-activated kinase (PAK) ortholog Pak1/Shk1 in fission yeast [85]. Skb5 inhibits Pmk1 MAPK signaling by downregulating Mkh1 localization to the cell tips, since its deletion results in the increased cell-tip accumulation of Mkh1 and the downstream activation of the CIP MAPK Pmk1 [84]. Similar to Mkh1, Pck2 is also tethered to the cell tips at the G1/S phase, suggesting that the Skb5–Mkh1 interaction may inhibit Pck2-dependent downstream signaling to the MAPK module [84].

2.2.2. MAPKK: Pek1

Pek1 ("pombe mEK 1"), also named Skh1, is a 40.7 kDa protein encoded by the *pek1*$^{+}$ gene, which shows a strong affinity with *S. cerevisiae* Mkk1 and Mkk2 MAPKKs [19,22]. Pek1, like Mkh1, is mostly a cytoplasmic protein that also localizes to the septum during cytokinesis [18]. Pek1 is phosphorylated and activated by Mkh1 in response to multiple stimuli (Figure 1A) [22]. Interestingly, in the absence of stress, inactive Pek1 binds to and inhibits Pmk1 activity. The dual activating/non-activating role of Pek1 allows this MAPKK to act as a molecular "all or nothing" switch during downstream signaling to Pmk1 MAPK depending on its phosphorylation status [22].

2.2.3. MAPK: Pmk1

The MAPK Pmk1/Spm1 ("*S. pombe* MAP kinase 1"), a 48.26 kDa protein, was initially identified as a structural counterpart of Mpk1/Slt2 MAPK of *S. cerevisiae* and is ortologous to human ERK1/ERK2 and ERK5 [11,12,15,17]. Pmk1 is the core component of the CIP in *S. pombe* (Figure 1A), and its activation is fully dependent on the signaling transduced exclusively by Mkh1 and Pek1, which supports the fact that the CIP comprises an unbranched transmission MAPK module [17,18,24]. Similar to other MAPKs of the ERK-type, Pmk1 is activated by the upstream MAPKK (Pek1) through dual phosphorylation at T-186 and Y-188 located within the strongly conserved -TEY- activation motif [18,19,22]. In spite of these observations, it was later demonstrated that a threonine-monophosphorylated Pmk1 isoform performs, to a notable extent, many of the biological functions of the dually phosphorylated kinase [86]. On the contrary, tyrosine-monophosphorylated Pmk1 is not biologically active, despite showing an increased level of phosphorylation in this residue, which likely results from a trapping effect by Pek1 when the phosphorylation at the threonine residue is not available [86]. These observations support the fact that tyrosine phosphorylation is a prerequisite for the subsequent phosphorylation of Pmk1 at the threonine residue and that the dual phosphorylation of the MAPK is attained by an ordered and processive mechanism [86].

Although Pmk1 is constitutively localized in both the cytoplasm and the nucleus, it has also been detected in the mitotic spindle and in the septum during cytokinesis [18]. Contrary to mammalian ERK1/2, the sub-cellular targeting of Pmk1 is not significantly altered during stress or in the absence of the upstream module components Mkh1 or Pek1, suggesting that it becomes activated at the cytoplasm and/or septum, and that either the active or the inactive MAPK isoforms can freely translocate to the nucleus [18]. Detailed analysis of the biological consequences of the forced nuclear exclusion of Pmk1 has revealed that nuclear localization of the MAPK may be relevant for the full regulation of CW integrity at a transcriptional level. Conversely, many of the biological functions of Pmk1, including the regulation of chloride homeostasis and cellular separation, are executed by the cytoplasmic pool of the MAPK [87].

Pmk1 is cyclically activated during the unperturbed cell cycle, peaking at the G1-S transition during cytokinesis [20]. It also becomes activated in response to multiple environmental changes including heat, osmotic and oxidative stresses, CW damage, or glucose withdrawal [15,18,21]. Pmk1-less mutants display cytokinesis defects and are growth-sensitive to most of the above stresses, suggesting that CIP/Pmk1 activity is necessary to preserve cellular integrity in fission yeast under different biological scenarios [18,25,26,64,88]. Pmk1 activation must also be tightly controlled as its constitutive hyperactivation is detrimental, leading to notable morphological alterations that result in strong cell-growth defects [15,17–19,22–24].

The kinetics and magnitude of Pmk1 activation in response to environmental stimuli varies considerably depending on the stressor [18,64]. Some of the upstream regulatory components of the MAPK module may or may not function during signal transduction to the CIP according to the nature of the stimulus, suggesting the existence of alternative architectures for this signaling cascade [57,64]. For example, Pmk1 activation induced by hypertonic and hypotonic stresses is quick and transitory and relies completely on the Rho2–

Pck2 branch, whereas MAPK activation induced during CW damage occurs progressively, reaches its maximum at long incubation times, and depends upon the activity of Rho1, Rho2, Pck1, and Pck2 [18,26,64,74]. Pmk1 activation in the absence of glucose occurs only after the total depletion of the sugar, requires de novo protein synthesis, is Rho2 independent, and involves Pck2 [83]. Contrariwise, Pck2 function is required to maintain a basal Pmk1 activity to allow further CIP activation during heat shock [64]. Rho4 and Rho5 might also play a positive role during this specific response [70].

S. japonicus Pmk1 shares a strong amino-acid sequence identity with *S. pombe* Pmk1 at the N- and C-lobes of the kinase domain, which includes the ERK-type TEY activation loop. However, a unique characteristic of the *S. japonicus* Pmk1 secondary structure is the absence of a 24 amino acids motif within the N-lobe, which includes the putative gatekeeper residue [28]. While *S. japonicus* Pmk1 also shows a nucleo-cytoplasmic location, it is not targeted to the septum during cytokinesis like its *S. pombe* counterpart. Despite these differences, comparative phenotypic analysis of *pmk1Δ* mutants revealed that regulation by the CIP of ionic homeostasis and CW integrity is conserved in both fission yeast species. Thereby, *S. japonicus* Pmk1 suppressed the phenotypes associated with a lack of CIP signaling when expressed in *S. pombe pmk1Δ* cells [28]. *S. japonicus* Pmk1 becomes activated by identical stressors to the *S. pombe* MAPK, except in response to glucose withdrawal, which does not occur in *S. japonicus*. Lack of MAPK activation upon this specific stimulus in *S. japonicus* might occur as the result of evolutionary loss, since this species utilizes glucose exclusively via fermentation and is respiration defective [28,89].

2.3. *Downstream Targets of Pmk1*

The main role of CIP is to ensure cellular adaptation in response to environmental changes through the phosphorylation of its targets. Only a few of direct Pmk1′s targets have been identified to date. The short list includes the transcription factor Atf1, the RNA binding proteins Rnc1 and Nrd1, and the cell cycle protein Clp1 (Figure 1A).

2.3.1. Transcription Factor Atf1

The Atf-CREB family bZIP domain transcription factor Atf1 (59.71 kDa), is regarded as the key downstream effector of Sty1, which is the core MAPK of the SAPK pathway in *S. pombe*. Activated Sty1 associates with, phosphorylates, and stabilizes Atf1, which binds to promoters and ORFs of multiple stress-responsive genes, including the CESR (Core Environmental Stress Response), and modulates their expression in response to environmental cues [90–95]. However, some evidence supports the fact that Atf1 may also be targeted by Pmk1 under specific biological stimuli (Figure 1A). For instance, Pmk1 associates with and phosphorylates Atf1 specifically in response to CW damage, and it has been proposed that this transcription factor might enable the induced expression of genes required to maintain CW integrity [96]. One transcriptional target for the Pmk1–Atf1 branch is the *ecm33*$^{+}$ gene, encoding a glycosyl-phosphatidylinositol (GPI)-anchored cell surface protein that negatively modulates Pmk1 signaling [97] (Figure 1B). The *ecm33*$^{+}$ promoter shows a putative ATF/CREB-binding site, and its transcriptional induction also depends on Mbx1, a MADS-box transcription factor that controls gene expression during M-G1 cell cycle transition [98]. Mbx1 could also be a direct downstream target of Pmk1, but the in vivo association and phosphorylation by the MAPK have not been yet demonstrated. Aside from Ecm33, no additional CW gene target/s regulated via Pmk1–Atf1 transcription have been identified to date. As compared to wild-type cells, the growth sensitivity of cells lacking Atf1 to CW-damaging compounds (i.e., caspofungin) is rather modest. Moreover, a nuclear excluded version of Pmk1, which does not bind Atf1, is able to partially support growth in the presence of CW-damaging compounds [96]. Therefore, in contrast to *S. cerevisiae*, the biological significance of the transcriptional control of fission yeast CW integrity by the CIP seems rather modest.

Treatment of fission yeast cells with the pro-oxidant terc-buthyl hydroperoxide (TBH), induced the expression of a number of genes through Pmk1–Atf1 control in a Sty1-

independent fashion [94], suggesting that this transcriptional branch is important for *S. pombe* survival in the presence of the oxidant. However, in strong contrast to the *pmk1Δ* mutant, *atf1Δ* cells are not growth-sensitive in the presence of TBH, supporting the fact that the CIP may promote fission yeast cellular adaptation to this stimulus more at a post-translational level [94]. Finally, it has been proposed that Pmk1 might work through Atf1 in addition to the canonical Sty1–Atf1 branch to promote the expression of genes involved in the adaptation of *S. pombe* from a fermentative to a respiratory metabolism such as fbp1+, encoding fructose-1,6-bisphosphatase [83]. Atf1 may thus act as a common transcriptional nexus between the SAPK and CIP MAPKs cascades under certain environmental scenarios.

2.3.2. RNA Binding Proteins (RBPs) Nrd1 and Rnc1

RNA-binding proteins (RBPs) assemble into different mRNA-protein complexes and play key post-transcriptional roles in eukaryotes during splicing, mRNA transport, and the modulation of mRNA translation and decay, thus regulating essential cellular processes [99]. Phosphorylation of RBPs by MAPKs impacts their ability to bind and stabilize targeted mRNAs and influence cellular functions [100,101]. Indeed, Nrd1 and Rnc1, two of the ~140 RBPs present in the *S. pombe* proteome [102], have been described as direct downstream targets of Pmk1 during CIP signaling (Figure 1A). Their structural and functional characteristics, together with the biological significance of their functional control by Pmk1-dependent phosphorylation will be addressed exhaustively in a section below (*"Control of mRNA-stability through RBPs"*).

2.3.3. Other Downstream Targets

The Cdc14-related serine/threonine protein phosphatase Clp1, also known as Flp1, regulates several aspects of mitotic exit in *S. pombe*. Clp1 locates at the nucleolus during vegetative growth and leaves this organelle to access its substrates when fission yeast cells face DNA damage agents or upon oxidative stress [103]. It has been shown that Pmk1 directly phosphorylates Clp1 at conserved -TP- MAPK phosphosites to promote its nucleoplasmic accumulation during genotoxic stress (Figure 1A). Accordingly, a Clp1 mutant lacking TP phosphosites showed a delayed nucleolus release during H_2O_2 treatment, whereas maintained phosphorylation at these sites prompted constitutive nucleoplasm targeting [104].

The plasma membrane Cch1–Yam8 channel complex that positively regulates the Ca^{+2} influx when *S. pombe* cells are subjected to salt stress, has been proposed as a direct phosphorylated substrate of Pmk1. However, the possibility that this is indeed the case remains to be confirmed. The antagonistic role in the fission yeast of calcineurin and CIP signaling during the control of Cch1–Yam8 activity and calcium homeostasis is addressed below (*"Calcium homeostasis"*).

2.4. *Downregulation of Pmk1 Signaling by MAPK Phosphatases*

Besides its activation, a tight control of Pmk1's deactivation status is critical for the precise control of CIP signaling during unperturbed growth and environmental stresses. The primary negative regulator of CIPs during vegetative growth is Pmp1, a dual-specificity MAPK phosphatase that associates in vivo to Pmk1 and dephosphorylates the T-186 and Y-188 residues within the conserved -TEY- activation motif (Figure 1A) [25]. Accordingly, Pmp1 overexpression results in a strong *vic* phenotype in the absence of the calcineurin function, whereas its deletion increases Pmk1 dual phosphorylation in vivo [25]. The stability of *pmp1*$^+$ mRNA is positively regulated by the RBP Rnc1, which is, in turn, phosphorylated and activated by Pmk1 to settle a negative feedback loop of MAPK signaling [105] (see *"Control of mRNA-stability through RBPs"* below). Later work demonstrated that the tyrosine phosphatases Pyp1 and Pyp2 and the threonine phosphatases Ptc1 and Ptc3, whose expression is mostly regulated by the SAPK pathway through Sty1 MAPK and the Atf1 transcription factor, also bind and dephosphorylate Pmk1 in vivo (Figure 1A). While Pyp1 and Ptc1 dephosphorylate Pmk1 during vegetative growth together with Pmp1, Pyp2 is only involved in MAP deactivation during stress [20]. Thus, SAPK-regulated MAPK phosphatases are involved in a crosstalk mechanism between the SAPK and CIP pathways (see *"Interplay with other fission yeast MAPK signaling cascades"* for specific details).

3. Main Regulatory Functions of the CIP

3.1. Control of Mrna-Stability through Rbps

3.1.1. Nrd1

Nrd1 (Negative regulator of differentiation-1), also named Msa2 (Multicopy suppressor of sporulation abnormal mutant 2), is a 529 aa (57.76 kDa) RBP which was originally identified by its ability to repress sexual differentiation in *S. pombe* [106–108]. Conversely, deletion of the *nrd1*$^+$ gene causes fission yeast cells to initiate sexual development in the absence of nutrient starvation [106]. Further analysis revealed that Nrd1 blocks the onset of sexual development until cells reach a critical level of starvation by repressing the expression of genes induced by the transcription factor Ste11, which is a master transcriptional regulator of early sexual differentiation in fission yeast [106,109]. Nrd1 contains four copies of a putative RNA-recognition motif (RRM) distributed along its amino-acid sequence and was initially found to bind poly(U)-rich sequences with high affinity [106]. However, it is currently unknown if Nrd1 binds in vivo to mRNAs of either Ste11 or Ste11-dependent genes.

Later work identified Nrd1 as a multicopy suppressor of the thermo-sensitive and multiseptation phenotypes of cells expressing a hypomorphic version of the myosin II essential light chain Cdc4 (*cdc4-8* allele) [110]. Cdc4 plays an essential function to regulate fission yeast myosin II activity for the assembly and constriction of a contractile actin- and myosin-based ring (CAR) during cytokinesis, which enables the physical separation of daughter cells once mitosis has been completed [111]. While *cdc4*$^+$ mRNA and protein levels decrease upon Nrd1 deletion, both increase in response to Nrd1 overexpression. Furthermore, Nrd1 binds and stabilizes *cdc4*$^+$ mRNA in vivo [110]. This mechanism is exerted at least in part through its association with two UCUU motifs present in *cdc4*$^+$ mRNA, since their mutation results in a significant reduction in Nrd1′s binding affinity and lack of suppression of both the thermo-sensitive and multiseptation defects elicited upon Nrd1 overexpression [110]. Furthermore, Nrd1 is a direct target of Pmk1 MAPK, and Pmk1-dependent phosphorylation negatively regulates its ability to bind and stabilize *cdc4*$^+$ mRNA (Figure 1C). Pmk1 phosphorylates Nrd1 in vivo in at least at two threonine residues, T-40 and T-126, which lie at a two "perfect" MAPK phosphorylation sites (consensus sequence -PNTP-). These residues are located at the intrinsically disordered N-terminus and the RNA recognition motif-1, respectively [110,112]. The Pmk1 phosphorylation of Nrd1 at T-40 and T-126 is regulated in a cell cycle-dependent manner, being low at the M and G1-S phases and increasing sharply during G2. Conversely, *cdc4*$^+$ mRNA levels are maximal during the M to G1-S phases and strongly decrease during the G2 phase [110]. Taken together, these observations draw a scenario where Nrd1 binding and stabilization of the mRNA encoding the myosin II essential light chain are controlled by Pmk1-dependent phosphorylation during the cell cycle, thus promoting an accurate regulation of myosin-II function for CAR assembly and constriction during cytokinesis (Figure 1C).

Nrd1 localizes at the cytoplasm during unperturbed growth and is recruited to stress granules (SGs) in response to various stressful situations including heat shock, arsenite treatment, oxidative stress, and glucose starvation. This is accompanied by enhanced phosphorylation at MAPK phospho-sites T-40 and T-126 [113,114]. Accordingly, as compared to an unphosphorylatable Nrd1 mutant (Nrd1-T40A T126A), a phosphorylation-mimicked version of the RBP (Nrd1-T40D T126D mutant) translocates to SGs more quickly in response to stress, suggesting that MAPK-dependent phosphorylation at both residues promotes its recruitment to SGs [113]. Phenotypically, Nrd1 deletion induced resistance to sustained stresses and enhanced sensitivity to transient stresses [113]. However, the observation that the Nrd1 localization dynamics at SGs in response to stress do not change significantly in cells lacking Pmk1 [113] suggests that other MAPKs besides Pmk1 might regulate

Nrd1 phosphorylation and subcellular localization at SGs. In support of this hypothesis, Nrd1 becomes phosphorylated in response to nitrogen starvation by Spk1, the MAPK of the pheromone-signaling pathway, and phosphorylation at T-126 plays an important role during the negative control of mating efficiency [114].

It has been proposed that Nrd1 adopts a beads-on-a-string structure, where each of its four RRM motifs, which are connected by flexible hinges, may undergo phosphorylation-induced conformational changes through these linkers [112]. Besides T-40 and T-126, the Nrd1 amino acid sequence has seven additional putative MAPK phosphorylation sites, three of which (S-5, S-93, and S-309) have been shown to become phosphorylated in vivo in global phosphoproteomic studies [115–117]. Altogether, this evidence suggests that alternative Nrd1 phosphorylation at multiple residues by the three *S. pombe* MAPK cascades might allow for the exquisite regulation of its biological functions as a RBP during unperturbed growth and in response to specific environmental cues. This control might somehow be exerted through its interaction with Cpc2, the ortholog of the receptor of activated C kinase (RACK1), that, in fission yeast, controls the extent of the activation of MAPK cascades from the ribosome, the cellular defense against oxidative stress, and the progression of the cell cycle [118]. Cpc2 associates with Nrd1 in vivo [108,113,114], and its deletion affects the formation of Nrd1 SGs upon specific stresses [113].

3.1.2. Rnc1

Rnc1 (RNA-binding protein that suppresses calcineurin deletion 1), a 398 aa (43.38 kDa) RBP, was isolated in a genetic screen for regulators of cell integrity signaling based on the functional interaction between calcineurin and Pmk1. Rnc1 overexpression suppressed the strong chloride sensitivity of fission yeast cells lacking calcineurin (Ppb1) by repressing Pmk1 activity [105]. The Rnc1 amino-acid sequence contains three K-homology (KH) domains, which are also found as multiple copies in many eukaryotic RBPs that coordinate the different steps of RNA synthesis, metabolism, and localization [119]. Further analysis identified the mRNA encoding Pmp1, the dual specificity phosphatase that dephosphorylates Pmk1 during vegetative growth [25], as a direct target of Rnc1 (Figure 1C) [105]. Rnc1 directly binds *pmp1*$^{+}$ mRNA in vivo and in vitro through its three KH domains. Accordingly, the respective mutants in each of the conserved GXXG loops that are essential for RNA recognition and docking in KH-containing RBPs [120] failed to suppress the chloride sensitivity of calcineurin deletion [105]. Rnc1 stabilizes *pmp1*$^{+}$ mRNA by binding to several UCAU repeats located at its 3′ UTR [105]. This control is biologically relevant, since *pmp1*$^{+}$ mRNA levels decrease upon Rnc1 deletion, resulting in the increased basal phosphorylation of Pmk1 MAPK, whereas Rnc1 overexpression strongly reduces MAPK activity (Figure 1C) [105]. Importantly, activated Pmk1 binds and phosphorylates Rnc1 in vivo at a MAPK consensus phospho-site located at a threonine residue at position 50, and this post-translational modification enhances the Rnc1 binding and stabilization of *pmp1*$^{+}$ mRNA, thus posing Rnc1 as key component of a negative feedback loop of cell integrity MAPK signaling (Figure 1C) [105,121]. In agreement with this model, expression of an unphosphorylatable Rnc1 mutant at T-50 (T50A) failed to suppress the chloride sensitivity of *ppb1*Δ cells, whereas phosphorylation-mimic mutants (T50D or T50E), were more potent suppressors of this phenotype than the wild-type RBP [105,121].

Further studies revealed that Rnc1 localization at the cytoplasm is mediated by a putative Nuclear Export Signal (NES), present at the N-terminus, that mediates its nuclear to cytoplasm shuttling. Curiously, this mechanism is not mediated by the exportin-1 ortholog Crm1 and relies on unknown exportin(s) [122,123]. Rnc1 nucleo-cytoplasmic export shuttling is also exerted by the nucleoporin RNA export factor Rae1 [122,123]. An Rnc1 mutant version lacking NES accumulates in the nucleus and fails to bind *pmp1*$^{+}$ mRNA, thus prompting its destabilization and the enhancement of Pmk1 activity. Similarly, Rnc1 mutants at the KH domains do not bind *pmp1*$^{+}$ mRNA and accumulate in the nucleus due to the loss of nuclear exports of the mRNA-protein complex. Thus, nuclear export of Rnc1 requires an mRNA-binding ability and the mRNA export factor Rae1 [122,123].

Similar to Nrd1, Rnc1 is a component of the SGs that relocates to these structures as a result of different stimuli [124]. Rnc1 deletion results in decreased SGs, whereas Rnc1 overproduction induced a strong accumulation and aggregation of SGs even in the absence of stress, indicating that it plays a positive role in their formation. Rnc1 prompts SG assembly in response to stress dependently or independently of its RNA-binding activity, suggesting that different signaling pathways may target Rnc1 during the regulation of this process [124].

Recent work has shown that, besides Pmk1, the MAPK of the SAPK pathway Sty1 phosphorylates Rnc1 in vivo at multiple phosphosites (T-45, T-50, T-171, T-177, S-278, and S-286) during vegetative growth and stress (Figure 1C). These modifications trigger Rnc1 binding through its KH domains and destabilization of mRNA targets encoding both the activators (Wak1 MAPKKK and Wis1 MAPKK) and downregulators (Atf1 transcription factor and Pyp1 and Pyp2 phosphatases) of Sty1 phosphorylation (Figure 1C) [125]. Biologically, this control results in an overall reduction of Sty1 activity through a negative feedback mechanism via Rnc1 that ensures a precise control of fission yeast cell-size progression during vegetative growth and in response to acute stress [125]. Accordingly, cells expressing a mutated version of the RBP at the KH-domains, which are unable to bind mRNAs (Rnc1-m3KH) or lack phosphorylatable MAPK sites (Rnc1-S/T6A), phenocopy *rnc1Δ* cells and show increased Sty1 activity, reduced cell length at division, and enhanced tolerance to heat shock [125].

Altogether, this evidence presents an intriguing situation, where alternative phosphorylation of Rnc1 by Pmk1 and Sty1 MAPKs may prompt either mRNA stabilization (Pmp1 mRNA) or destabilization (Wak1, Wis1, Atf1, and Pyp1 and Pyp2 mRNAs), during unperturbed growth and stress (Figure 1C). The differences in Rnc1 binding affinity towards the target mRNAs might rely upon strong conformational changes induced by the alternative phosphorylation of MAPK phospho-sites, which are located outside of the KH domains and lie within intrinsically disordered regions of the protein [125]. The shared modulation of Rnc1 mRNA binding ability by the cell integrity and stress-activated MAP kinases Pmk1 and Sty1 is, to a certain point, an elegant example of MAPK crosstalk that allows for precise self-regulation during growth and in response to environmental changes.

3.2. Calcium Homeostasis

The first evidence suggesting that CIP signaling is involved in the regulation of ionic homeostasis in fission yeast came from the observation that mutants lacking Pmk1 exhibit differential growth sensitivity in the presence of cations. Indeed, as compared to wild-type cells, *pmk1Δ* cells are tolerant to sodium chloride, whereas they are sensitive to the presence of potassium and calcium salts in the growth media [17,18]. Later work established a putative functional link during the control of ionic homeostasis between the CIP and the Ca^{2+}/calmodulin-dependent protein phosphatase Calcineurin (Ppb1). The strong chloride sensitivity of fission yeast cells lacking calcineurin activity (i.e., in *ppb1Δ* cells or through specific pharmacological inhibition with Tracolimus/FK506) could be totally abrogated by the simultaneous deletion of Pmk1. This phenotype, named *vic* (viable in the presence of immunosuppressant and chloride ions), set the experimental ground for the identification of key core components of the CIP such as Pmp1, Pek1, and Rnc1 [22,25,105] and for the precise assessment of the biological relevance of upstream regulators of the MAPK module including Rho1, Rho2, Pck1, and Pck2 [57,64]. The *vic* phenotype is also present in the lack of function mutants in CIP signaling in *S. japonicus*, thus supporting its conservation within the *Schizosaccharomyces* genus [28].

Calcium homeostasis is critical for the regulation of multiple physiological processes in eukaryotes. Accordingly, minimal changes in its cytoplasmic levels lead to the activation of several calcium-sensing proteins, including calcineurin, which results in the induction of various downstream transduction pathways [126]. In *S. pombe*, activation of the calcineurin catalytic subunit Ppb1 in response to calcium accumulation triggers the dephosphorylation and nuclear translocation of the nuclear factor Prz1, an activated T-cell family (NFAT) transcription factor that regulates the expression of genes showing Calcium Dependent Response Element (CDRE)-like motifs at their regulatory sequences [126–128]. Prz1 regulates the expression of genes involved in calcium homeostasis, such as *pmc1*$^+$, *pmr1*$^+$, *ncs1*$^+$, *cmk1*$^+$, and *prz1*$^+$ [126–130], and also CW biosynthesis genes such as *bgs1*$^+$, *rga5*$^+$, *omh1*$^+$, *pvg1*$^+$, and *pvg5*$^+$ [131]. While mutants lacking Prz1 are hypersensitive to calcium, *ppb1*Δ cells exhibit additional phenotypes including cytokinesis and cell polarity defects [126,132]. Importantly, the overexpression of Pmp1 phosphatase, which negatively regulates Pmk1 activity, fully suppressed the chloride hypersensitivity, but not the calcium hypersensitivity, of *ppb1*Δ cells. From these observations, it was hypothesized that Prz1 is not the sole target of calcineurin during the control of ionic homeostasis, and that this control is exerted through at least two different signaling pathways instead. One requires Prz1-dependent transcriptional regulation that regulates calcium homeostasis, whereas a second and unknown mechanism would function antagonistically with the CIP to modulate cellular morphogenesis and chloride homeostasis [126].

The transient receptor potential (TRP) calcium channels, Trp1322 and Pkd2, play significant functional roles during fission yeast-CW biosynthesis and membrane trafficking [23,133]. Similarly, the Yam8 channel (also known as Ehs1), has also been involved in CW integrity control and calcium import [134]. By performing live measurements of cellular cytoplasmic Ca^{2+} levels, it was found that Trp1322 and Pkd2 directly regulate calcium influx in *S. pombe*, whereas exposure of fission yeast cells to KCl, NaCl, and $MgCl_2$, elicits calcium import and cytoplasmic accumulation through the Yam8/Cch1 channel complex, thus resulting in calcineurin activation [127,135]. Those treatments prompted a further increase in cytoplasmic calcium content in either *ppb1*Δ cells or FK506-treated cells as compared to wild-type cells, and this feature was suppressed in the absence of Yam8 and/or Cch1. Therefore, calcineurin may directly downregulate the Ca^{2+} influx activity of the Yam8–Cch1 channel complex [135].

Remarkably, fission yeast mutants lacking either Pmk1 or upstream CIP elements exhibit extremely low cytoplasmic calcium levels during unperturbed growth, and their deletion suppresses the synergistic increase in calcium content elicited in the absence of Ppb1 activity. Contrariwise, the enhancement of CIP activity achieved by the overexpression of a constitutively active version of Pek1 MAPKK further increased the cytoplasmic calcium levels in a Yam8/Cch1-dependent fashion [135]. As a whole, these findings suggest that both the calcineurin and the CIP pathways may antagonistically regulate calcium uptake in fission yeast through the fine-tuning of the dephosphorylation/phosphorylation status of the Yam8–Chh1 channel (Figure 1D) [135]. Hence, the *vic* phenotype, which was initially associated with altered chloride homeostasis, is actually due to the suppression by CIP inhibition of the lethal effect resulting from the persistent uptake of calcium via Yam8–Chh1 during salt stress and in the absence of calcineurin activity. This model supports that calcineurin activity is dominant over CIP signaling during unperturbed growth, and, therefore, the Yam8–Cch1 channel remains basically inactive. Conversely, in response to salt stress, activated Pmk1 may phosphorylate and promote the Cch1–Yam8 opening and Ca^{+2} influx. In turn, the increase in intra-cytoplasmic calcium levels triggers calcineurin activation, resulting in the dephosphorylation and closure of the channel (Figure 1D) [135]. Formal confirmation of this regulatory mechanism will require the demonstration of a direct physical association of Pmk1 and Ppb1 with Yam8–Cch1 and the identification of the putative conserved residues that become phosphorylated or dephosphorylated by the kinase and phosphatase, respectively, to modulate the activity of the channel depending on the environmental context.

3.3. Cell-Wall Integrity

The fungal CW, largely composed of glucans, mannans, and chitin, is an essential barrier that protects and maintains cellular integrity during polarized vegetative growth and other morphogenetic processes such as mating, sporulation, or pseudohyphal growth [136]. The CW is a highly dynamic structure that adjusts its composition and structure in response to environmental changes (pH, temperature shifts, carbon source, etc.), or during drug exposure and immune responses [136,137]. The CW must be assembled to become both malleable and mechanically robust, in order to withstand the elevated cytoplasmic turgor pressure and adequately perform its biological functions. These adaptive and compensatory mechanisms involve different strategies depending on the fungi, but their ultimate goal is to repair the damaged/altered CW and guarantee cell survival. Although the CIP lies at the core of such responses, both in budding and fission yeasts, the molecular processes involved are known with greater detail in the former species [31,33].

Most fungal CWs are composed of an innermost and relatively conserved layer that functionally represents the load-bearing component of the wall (glucans and chitin), while the outer layers are more variable and specifically tailored to suit the needs of each particular species [136]. In *S. pombe*, most of the CW is composed of branched β-1,3-glucan synthesized by the membrane-bound β-1,3-glucan synthase (β-GS) complex. During division, a three-layered septum is formed with the central area being made of linear β-1,3-glucan surrounded by two outer secondary septa [33]. The CW behaves differently at the tips during polarized growth as compared to the cell sides and exhibits a fluctuating behavior with thickening phases followed by thinning phases. On the contrary, the cell sides remain stable [35]. Therefore, the tips are twice as soft and thin with respect to the remainder of the cell. Importantly, CW thickness is positively regulated by synthesis and negatively regulated by growth [35].

In *S. pombe*, the essential GTPase Rho1 is the β-GS regulatory subunit responsible for the activation of GS catalytic subunits Bgs1 to Bgs4 [41,53], likely through conformational changes dependent on their direct physical binding, as described in *S. cerevisiae* [138]. In addition, Rho1 activates β-1,3-glucan biosynthesis indirectly through the CIP components and PKC orthologs, Pck1 and Pck2 [55,56]. On the other hand, Rho2 activates Mok1 [45], which is the essential enzyme responsible for CW α-1,3-glucan synthesis, via Pck2 [33]. How Rho1 and Rho2 activities are precisely controlled in response to internal or external cues to dynamically modulate CW structure remains an open question, but it likely involves the regulation of the activity of their specific GEFs and/or GAPs [38]. As previously noted, the fission yeast sensors Wsc1 and Mtl2 physically couple the CW with the plasma membrane to activate Rho1 and CW biogenesis through its interaction with Rgf1 GEF and, to a lesser extent, Rgf2 [32,35]. However, the data obtained so far suggest that Wsc1 and Mlt2 may perform this function independently of the CIP [32]. These findings support that mechanosensing through these sensors might not be the major mechanism responsible for Rho- and PKC-dependent CIP activation, which mainly relies on the Rho2–Pck2 branch. Besides, Wsc1 clustering at the membrane does not depend on any members of CIP, including Pmk1 MAPK [37]. Even so, Pmk1 has been involved in the homeostatic maintenance of CW thickness during polarized cell growth by adjusting synthesis-to-surface growth in order to avoid excessive thinning [35]. However, the molecular mechanisms involved in this control are far from being understood. As mentioned earlier, the only transcriptional CW-related target for the Pmk1–Atf1 branch is the *ecm33*$^+$ gene, encoding a GPI-anchored cell-surface protein [97]. This makes it rather unlikely that the CIP may play a direct and significant role during the control of CW homeostasis at a transcriptional level. Nevertheless, Pmk1 might affect CW integrity through alternative pathways, such as regulating calcium influx in response to CW damage through the Yam8–Cch1 channel and involving the calmodulin/Prz1-dependent transcriptional response or by exerting regulatory feedbacks on the Rho-PKC module to elicit CW synthesis and/or CIP signaling. In budding yeast, it has been suggested that the MAPK Mpk1/Slt2 activates a negative feedback loop that downregulates CIP signaling by detaching Rho1 from its GEF Rom2 [139]. A recent study has shown that constitutive activation of the CIP in cells grown in heavy water (D_2O), causes gross morphological changes, thickened CWs, and aberrant septa that limit cell growth [140]. Remarkably, blocking CIP activity partially alleviates those phenotypes, while its activation increases cellular sensitivity to D_2O [140]. Moreover, it has been recently described that moderate cell-wall stress extends chronological lifespan in fission yeast in a

Pmk1-dependent manner [141]. Theis evidence supports the fact that the activity of the CIP must be precisely regulated in time and space to preserve CW integrity and cellular viability.

Another remarkable difference between budding and fission yeasts with regard to the functional role of the CIP in CW biogenesis relates to the control of the actin cytoskeleton. In *S. cerevisiae*, Rho1 is important for proper actin cytoskeleton organization through the control of the CIP MAPK Mpk1/Slt2 [31]. However, such a functional link has not been established so far in fission yeast, and the regulatory mechanisms underlying actin cytoskeleton reorganization and integrity in response to environmental stimuli are mostly unknown. Future studies in *S. pombe* will help to elucidate the details of the possible relevance of Rho GTPases and the CIP during actin cytoskeleton organization.

As discussed earlier, the CIP plays a homeostatic role during the control of polarized cell growth [35]. In fission yeast, the Cdc42 GTPase module is a major regulator of polarized growth, which involves intertwined connections between surface material synthesis by exocytosis and endocytosis, the actin cortex, and the CW [38]. Considering that polarized growth may be counteracted by CW stiffness, the existence of functional crosstalk between the Cdc42 polarity network and the CIP should be expected. In this regard, it was suggested in early studies that Cdc42 might signal through the CIP pathway via the redundant PAK kinase Pak2, since the expression of a dominant-activated *cdc42G12V* allele prompted a strong growth defect that was rescued by both *pak2*Δ and *mkh1*Δ deletions [142]. Unexpectedly, neither *pak2*$^+$ deletion nor *cdc42G12V* overexpression had a noticeable effect on Pmk1 activity during growth and stress [18]. Cdc42 could regulate a minimal subset of CIP components localized at the growing poles, which might hamper their identification in the studies performed so far. Another possibility is that the CIP Rho GTPases and/or PKCs might be targeted by Cdc42 to modulate CW homeostasis independently of their role during CIP signaling [38]. In agreement with this hypothesis, a functional relationship between the polarity machinery and the Rho1–Pkc1 function has been described in *S. cerevisiae* [143].

Altogether, these findings support that the CIP plays a determinant role in maintaining and modulating cell-wall integrity in fission yeast during vegetative growth and in response to different stimuli, although further studies will be required to shed light onto the detailed molecular mechanisms and processes responsible for such control.

3.4. *Cytokinesis*

Cytokinesis, the final stage of cell division, involves the assembly and constriction of a contractile actomyosin ring (CAR) that physically divides the mother cell into two daughter cells. Contrary to most animal and fungal cells, including fission yeast *S. japonicus*, which assemble the CAR when chromosomes are fully segregated, CAR assembly in *S. pombe* is a rather counterintuitive biological phenomenon that starts early in anaphase before chromosomal segregation takes place [144]. Fission yeast cytokinesis can be divided into four distinct stages: (i) node condensation and CAR assembly; (ii) CAR maturation; (iii) CAR constriction and septum synthesis; and iv) septum digestion and cell separation. First, pre-nodes containing the anillin-like protein Mid1 and cell-cycle kinases evolve into cytokinetic nodes through the addition of myosin II Myo2, IQGAP Rng2, F-BAR protein Cdc15, and the formin Cdc12, which together with For3, another formin, stimulate the nucleation of actin filaments [145]. The subsequent and dynamic interactions between myosin II and the actin filaments promote node condensation for assembly of the CAR. The CAR maturation stage involves the recruitment to the CAR of Cdc15 and several other proteins, including Rga7 (Rho2 GAP) and Rgf3 (Rho1 GEF) [145]. Later, the CAR constricts by the action of the motor type II myosin Myp2 and cofilins, which cut actin fragments stochastically to shorten the ring [145]. CAR constriction and the synthesis of a new CW via the formation of a three-layered septum is triggered by the Septation Initiation Network (SIN) cascade, which is related to the Hippo and MEN signaling pathways in mammals and *S. cerevisiae*, respectively [111]. Finally, cell separation liberates the two daughter cells upon dissolution of the primary septum through the specific action of the glucanases Eng1 and Agn1 localized at the division site [146].

In *S. cerevisiae* the CIP regulates mechanistically distinct aspects of cytokinesis both at the transcriptional and post-transcriptional levels. During late mitosis, Polo-activated Rho1 mediates the assembly of the actomyosin ring through the formin Bni1, a fission yeast Cdc12 ortholog [147], which leads to deposition of the primary septum by the chitin synthase Chs2 [148]. Right after mitosis, Rho1 also activates the chitin synthase Chs3 and the β-glucan synthases Fks1/2 that participate in secondary septum biogenesis via Pkc1 [149]. Activation of the CIP and Mpk1/Slt2 MAPK by Rho1 is also biologically relevant for the transcriptional induction of several cytokinetic genes, such as Fks2 (glucan synthase), Gfa1 (required for the synthesis of a metabolic precursor of chitin), and Chr1/2 (transglycosidases) [150,151]. Contrariwise, the CIP-dependent transcriptional response seems rather irrelevant during cytokinesis control in fission yeast. However, similar to budding yeast, several studies have highlighted the key role of fission yeast Rho1, and Rho2, in the activation of glucan synthases during primary and secondary septum deposition (reviewed in [38]).

The assembly of a septin ring, which is one of the first cytokinetic events in budding yeast, is modulated by the CIP under specific circumstances [152]. In this regard, it has been shown that the ER Stress Surveillance (ERSU) pathway, a mechanism that safeguards the inheritance of a functional endoplasmic reticulum (ER) by the daughter cell, is regulated by Mpk1/Slt2 signaling [152]. Upon ER stress, both ER inheritance and cytokinesis are inhibited due to the relocalization of the septin complex away from the bud neck [153]. However, in the absence of Mpk1/Slt2, cells do not block ER inheritance, and the septin ring remains positioned at the bud neck [152]. The putative functional connection between ER stress, the CIP, and septum assembly has not been explored so far in fission yeast. Nevertheless, several studies point to a functional role of Pmk1 during CAR assembly and maturation that influences septal dynamics in response to several stressors, although the precise molecular details remain obscure [110,154]. As described earlier, the Pmk1-dependent phosphorylation of the RNA-binding protein Nrd1 negatively regulates its ability to bind and stabilize *cdc4⁺* mRNA, thus allowing for the accurate regulation of myosin II function during CAR assembly and constriction (Figure 1C) [110]. In addition, recent data suggest that the CIP may participate in a checkpoint that imposes a delay during the maturation phase of CAR assembly upon CW-integrity stress [154]. This inhibition is exerted through Rgf1–Rho1 activation of the CIP components Pck2 and Pmk1 and ensures that cytokinesis is completed after cells have adapted to the new growth conditions [154].

S. pombe mutants lacking CIP components display distinct phenotypic penetrance during cytokinesis. The upstream CIP GTPase members, Rho1 and Rho2, and PKC orthologs, Pck1 and Pck2, play an essential role in this process, as either Rho1 downregulation or simultaneous deletion of Pck1 and Pck2 produce strong cytokinetic defects that result in cell death [6]. Rho1 (and Rho2) plays a prominent role during septum synthesis, whereas Rho3 and Rho4 participate in the dissolution step [38]. It has been proposed that the SIN pathway may target the Rho1 GEF Rgf3 to promote septation [155], although this has not been formally proven. Rho1 activation at the CAR by Rgf3 is essential for proper ring maturation and constriction [59,156]. In addition to this, low levels of Rgf3 at the division site cause septum defects and cell lysis due to poor β-GS activity [156]. On the other hand, CIP mutants lacking the MAPK module components Mkh1, Pek1, and Pmk1 show mild separation defects, probably due to late defects in CW-remodeling that become exacerbated under stress. Furthermore, Pmk1 hyperactivation also leads to cell separation defects [6,74]. These phenotypes might be partially related, as previously mentioned, to the activity of the Pmk1–Nrd1 regulatory axis [110]. Alternatively, Pmk1 might be necessary for the activation at the CAR of calcineurin, which participates in the dephosphorylation of the F-BAR protein Cdc15 during cytokinesis [157].

Activation of the CIP pathway in budding yeast induces the phosphorylation of eisosome core components such as Pil1 and Lsp1 [158], whereas Pmk1 activation in fission yeast is inhibited by eisosomes [48]. These observations raise the possibility that Pmk1-dependent regulation of eisosome dynamics might alter plasma membrane compartmentalization, thus affecting the endocytosis of proteins involved in septum assembly. In this regard, the CIP components Rgf1, Pck1, Pck2, and Pmk1 cooperate to localize the polarity landmarks Tea1, Tea4, and Pom1 at the cell cortex; it has been suggested that the CIP likely regulates membrane–lipid homeostasis [159]. The possibility that CIP signaling impacts protein turnover at the membrane, suggests that its role during polarized growth and cytokinesis will surely be investigated further in the future. Finally, *S. cerevisiae* mutants with defects in Chs2 function, or in the actomyosin and septin rings, are able to assemble abnormal and engrossed septa (known as "remedial septa"), to prevent lysis during cytokinesis. Importantly, these situations trigger a strong CIP activation that results in the up-regulated synthesis of Gfa1, Fkh1/2, and Chr1/2 [150]. Remedial septa have also been described in *S. pombe* mutants that are defective in proper septum assembly [160], but the involvement of the CIP pathway during their formation has not been depicted.

The evidence obtained so far suggests that the CIP has a significant impact on fission yeast cytokinesis, where the cell wall provides the strength to generate the invaginating furrow. However, this contribution might have been overlooked, as the CAR and septal dynamics have been studied during unperturbed growth in most studies.

4. Functional Crosstalks of the CIP

4.1. Interplay with Other Fission Yeast MAPK-Signaling Cascades

S. pombe only has two conserved MAPK cascades besides the CIP, the mating pheromone-responsive pathway, and the stress-activated MAP kinase (SAPK) pathway [14,15,17,90,161–163]. While MAPK cascades are evolutionarily organized in a strict and modular fashion to achieve signal specificity in response to external cues, some core components (i.e., kinases, scaffolds) are sometimes shared by different MAPK cascades, resulting in an enhanced signaling plasticity and a more precise coordination of biological responses. This situation does not seem to occur in fission yeast, as the main components of its three MAPK modules and their corresponding upstream activators are pathway specific. However, some downstream elements of the CIP are shared by the stress- and pheromone-signaling cascades [18,91,164]. These include the Atf1 transcription factor and several MAPK phosphatases (Pyp1, Pyp2, Ptc1, Ptc2, and Ptc3), together with RNA-binding proteins (Rnc1 and Nrd1).

Early observations pointed out the existence of a strong functional crosstalk between the CIP and the SAPK pathways during the stress response in fission yeast. Sty1 MAPK, which is the ortholog of budding yeast Hog1 and mammalian p38, plays a crucial role during fission yeast survival in response to multiple forms of stress [88,90,91,162,163,165–169]. Similar to the SAPK pathway, the CIP becomes activated to multiple stressful conditions [18]. However, while some stimuli specifically activate Pmk1 (CW stress) or Sty1 (nitrogen starvation), many others, such as oxidative stress, hyperosmotic stress, thermal stress, glucose depletion, and hypergravity, prompt the simultaneous activation of both pathways [6].

Various pro-oxidants and oxidizing agents such as hydrogen peroxide, diamide, diethyl maleate (DEM), Paraquat, and TBH, induce activation in both Sty1 and Pmk1 [18,167,170]. However, whereas Sty1- or Atf1-deleted mutants are strongly growth-sensitive for growth in the presence of these compounds [167], the sensitivity of Pmk1-less mutants is minimal [18]. Sty1 is activated in response to both low and high concentrations of hydrogen peroxide [167], but Pmk1 activation only occurs at high concentrations of the oxidant (>5 mM) [18]. These observations support the fact that the biological significance of the CIP pathway during the defensive response against oxidative stress is rather low compared with the SAPK pathway. From a general perspective, the differences in the respective activation patterns of both MAPKs and the fact that the overall stress-sensitivity of mutants devoid of Sty1 is higher than in *pmk1Δ* cells have led to the suggestion that the CIP pathway may reinforce the SAPK pathway to control cell survival and adaptation to sub-lethal stressful conditions [6,18].

Deletion of Pmk1 does not affect Sty1 activation in response to hyperosmotic stress [18]. However, a lack of the SAPK MAPK module components, including Sty1, increased both the basal and the stress-induced levels of Pmk1 phosphorylation [20]. Moreover, accurate deactivation of Pmk1 following osmostress required a functional Sty1–Atf1 transcriptional branch [18], suggesting that one or more MAPK phosphatases induced through the SAPK pathway might be responsible for Pmk1 deactivation during growth and stress. Negative regulation of Sty1 activity is carried out by tyrosine phosphatases Pyp1 and Pyp2 and serine/threonine phosphatases Ptc1 and Ptc3 [171]. Pyp1 is the main phosphatase deactivating Sty1 during unperturbed growth [90,172], and both Pyp1 and Pyp2 dephosphorylate Sty1 activated under osmotic or oxidative stress [162,163]. During heat shock, Pyp1 binding to Sty1 is abolished, resulting in a quick and strong Sty1 activation that is subsequently attenuated by Ptc1 and Ptc3 [173]. The basal expression of *pyp1*$^{+}$ and the stress-induced expression of *pyp1*$^{+}$, *pyp2*$^{+}$, and *ptc1*$^{+}$ are regulated by Sty1 and Atf1 through negative feedback loops [90–92,174]. Significantly, these three phosphatases have been shown to associate with and dephosphorylate Pmk1 in vivo during growth and stress (Figure 1E) [20]. It has been reported that Ptc3 is also involved in Pmk1 down-regulation [96]. Pyp1 and Ptc1 control Pmk1 basal activity during unperturbed vegetative growth and Pmk1 hyperactivation during cellular adaptation to osmostress is reduced by Pyp1, Pyp2, and Ptc1 [20]. In support of this model, cells devoid of the Receptor of Activated C Kinase (RACK1) ortholog Cpc2, which binds to the ribosomal 40S subunit and is essential for efficient up-regulation of Pyp1 and Pyp2 translation, show enhanced Pmk1 activation as a result of a decrease in the protein levels of both phosphatases [118].

Finally, the dual-specificity phosphatase Pmp1/Dsp1, which is the main negative regulator of Pmk1 activity during vegetative growth [25], also downregulates the MAPK of the pheromone response pathway Spk1 (Figure 1E) [164]. Pmp1 is closely related to the budding yeast dual-specificity phosphatase Msg5, which was isolated as a suppressor of the pheromone-signaling pathway that dephosphorylates Fus3 MAPK [175]. Msg5 also negatively regulates Mpk1/Sst2 and Kss1, which are the respective MAP kinases of the cell-integrity and filamentation/invasion pathways [176–178]. Hence, the substrate promiscuity of dual-specificity MAPK phosphatases is likely a common theme in the group of yeast MAPK cascades.

4.2. CIP and TOR Signaling Crosstalk

The Target of Rapamycin (TOR) is a conserved Ser/Thr kinase that regulates cell growth and metabolism in response to environmental cues [179]. TOR is found as two evolutionary conserved complexes, TORC1 and TORC2 [180–182]. In fission yeast, TORC1 is essential for vegetative growth and regulates the switch between cell proliferation and differentiation by sensing nutrient availability [183–185]. In contrast to TORC1, TORC2 is dispensable for proliferation under optimal growth conditions but is required for cellular adaptation during starvation and in response to stress [11,186]. The AGC-family kinases Psk1 and Gad8 are two direct targets for TORC1 and TORC2, respectively [184,187,188]. Notably, similar to Pck2, which is the main upstream kinase activator of the CIP MAPK module [78], the activity of both Gad8 and Psk1 is dependent upon AL phosphorylation mediated by the PDK1 ortholog Ksg1 [184,187].

For budding and fission yeasts, TOR and MAPK pathways play essential cellular functions in response to environmental changes in budding and fission yeasts; therefore, it is not surprising that both pathways are functionally interconnected to a large extent [11,72,189–193]. In *S. pombe*, a highly intricate cross-regulatory relationship between the TORC2 complex and the CIP secures cell adaptation and survival in response to environmental cues (Figure 1E) [72]. Increased de novo synthesis of Pck2 is essential to trigger Pmk1 activation in response to CW damage or upon glucose limitation. TORC2 positively controls this process through its main activator, the Rab GTPase Ryh1 [72]. GTP-Ryh1 associates with TORC2 to induce the phosphorylation and activation of Gad8 in glucose-rich media, and its activity is strongly reduced under glucose deprivation [194]. Moreover, Ryh1 modulates the CIP via TORC2-dependent and -independent mechanisms. In the TORC2-dependent mechanism Gad8, with the participation of TORC1 target Psk1, there is an increase in Pck2 protein levels that contributes to Pmk1 activation during CW stress and glucose limitation (Figure 1E) [72]. Interestingly, the inhibition of TORC1 in *S. pombe* by nutrient deprivation induces the phosphorylation of the translation initiation factor eIF2α [185], which negatively regulates translation in response to different environmental challenges [195–197]. Psk1 might reinforce Pck2 synthesis during stress by maintaining low eIF2α phosphorylation levels. Nevertheless, the upregulation of Pck2 synthesis by Gad8–Psk1 kinases does not operate through the phosphorylation of Rps6, an evolutionary conserved ribosomal protein, which is phosphorylated in vivo by both Psk1 and Gad8 in response to nitrogen and glucose sources [72,198,199]. Further, Ryh1 may prompt Pmk1 activation, regardless of its role as an upstream regulator of TORC2 signaling, by promoting proper plasma membrane tethering and/or processing of several key activators of the CIP, including Ksg1, Rho1, Rho2, Rgf1, and Pck2 (Figure 1A,E). This alternative mechanism is particularly relevant during MAPK activation in response to osmotic stress. The novel role of Ryh1 as elicitor of MAPK signaling does not seem to be conserved in budding yeast, since the deletion of Ypt6 (Ryh1 ortholog) did not avoid stress-activation of the respective CWIs and p38 MAPKs Slt2 or Hog1 [72]. Moreover, the phosphatidylinositol (PI) 4-phosphate 5-kinase orthologs Its3 and PI(4,5)P2, which promote Ryh1-TORC2 signaling, also act downstream Ryh1 to regulate trafficking and localization of several CIP activators of the CIP cited above (Figure 1E) [72]. The dual functional role of PI kinases might be evolutionarily conserved, since the budding yeast PI kinases also regulate PDK orthologs Pkh1/2 and TORC2 at the plasma membrane [200].

Glucose starvation results in the activation of Pmk1 through TORC2-dependent and -independent mechanisms. TORC2 activity, as measured using the in vivo phosphorylation status of Gad8 at S-546 as a readout, decreases quickly but recovers to half of the initial levels after 30 min of incubation and remains unchanged for longer times [11,72,194]. Importantly, CIP signaling inhibits TORC2-dependent Gad8 phosphorylation and activation during glucose starvation and CW stress. Accordingly, Gad8 phosphorylation at S-546 was quickly reduced in wild-type and *pmk1*Δ cells upon both stresses and increased significantly in the *pmk1*Δ mutant at longer times [11,72]. It has been reported that Pck2 phosphorylates Gad8 in vivo to downregulate TORC2–Gad8 signaling [201]. However, the Pmk1 negative regulation of TORC2 activity does not rely on Pck2 and/or Gad8. The reduction of Ryh1 GTPase activity and Gad8-S546 phosphorylation levels displayed by fission yeast cells during prolonged incubation without glucose became significantly attenuated in a *pmk1*Δ background. This led to the hypothesis that Pmk1 may negatively modulate TORC2–Gad8 signaling in response to glucose deprivation and likely CW stress, by downregulating the Ryh1 GTPase activation cycle (Figure 1E) [72].

Hyperosmotic stress transiently inhibits TORC2 activity independent of the function of the SAPK MAPK Sty1 [191,202]. However, Sty1 promotes TORC2 reactivation after osmostress through the transcription factor Atf1 and the glycerol-3-phosphate dehydrogenase Gpd1, whose expression was induced by the SAPK pathway [191]. Although CIP signaling was involved in the negative regulation of the TORC2–Gad8 pathway under osmotic saline stress conditions [11,72], the evidence obtained so far suggests that Pmk1 is not required for the osmo-inhibition of TORC2 signaling [191].

4.3. Interaction between the CIP and Camp/PKA Pathways

Evidence points to the existence of a functional crosstalk between the CIP and cAMP/PKA pathways in different fungal models. For example, alterations of the β-1,3-glucan network in budding yeast induce both CIP activation via Wsc1 and the parallel inhibition of cAMP/PKA signaling [203]. Although cAMP levels were reduced during CW damage induced using caspofungin independent of Slt2 activation, both responses are required to counteract the detrimental effects of the drug [203]. Further, deletion of *Cryptococcus neoformans* genes encoding the core kinases in the CIP pathway (namely, Pkc1, Bck1, Mkk2, and Mpk1) results in severe cell wall defects, including a loss of surface capsule and decreased virulence in a mouse model [204]. Interestingly, in these mutants, capsule formation is impaired due to reduced cAMP levels, and this defect is suppressed by the addition of exogenous cAMP [205]. These data suggest the existence of an intricate functional network between the CIP and the cAMP/PKA pathways, which is relevant for the appropriate regulation of fungal cell wall homeostasis.

The fission yeast CIP has also been shown to functionally interact with cAMP/PKA signaling [11,83]. In this organism, the cAMP/PKA pathway is involved in the control of relevant cellular events, including glucose sensing, sexual differentiation, spore germination, and osmotic stress response [206,207]. The pathway starts with the detection of extracellular glucose by the G-protein coupled receptor Git3. The signal is then transmitted to a membrane-bound adenylate cyclase Cyr1/Git2 through a heterotrimeric G protein composed of the Gpa2 Gα, the Git5 Gβ, and the Git11 Gγ subunits. Cyr1 triggers an increase in cAMP levels, which prompts the activation of the Pka1 catalytic subunit through its dissociation from the regulatory subunit Cgs1 [208]. Pka1 phosphorylates and negatively regulates the activity of Rst2, a transcription factor responsible for the induced expression of genes, such as *fbp1*$^{+}$, encoding the gluconeogenic enzyme fructose-1,6-bisphosphatase, whose activity is critical for adaptive growth to non-fermentable carbon sources (i.e., in the absence of glucose) [206]. Pmk1 becomes activated upon glucose deprivation, but the magnitude of this response is reduced in the absence of a cAMP/PKA function, as observed in *git3*Δ, *gpa2*Δ, *cyr1*Δ, and *pka1*Δ mutants. Contrariwise, Pmk1 activation in response to glucose starvation remained unchanged in cells lacking the Pka1-dependent transcriptional regulator Rst2 [83]. These findings suggest that an operative cAMP/PKA pathway is necessary for the full activation of the CIP under glucose limitation, but this control is exerted in a Rst2-independent way. However, the detailed responsible molecular mechanisms involved in this crosstalk wait to be elucidated.

In summary, although the SAPK has arguably been considered the most prominent and physiologically relevant MAPK pathway in fission yeast, it is becoming increasingly evident that the functional interactions of the CIP with additional MAPKs and TOR and PKA signaling cascades are also essential for this organism to integrate multiple adaptive responses to the changing environment.

5. Concluding Remarks and Perspectives

CIP signaling is involved in the regulation of essential cellular processes in fission yeast, such as the RBPs-mediated control of mRNA-stability, ionic homeostasis, CW integrity, and cytokinesis. Studies completed during the last 30 years have depicted the complexity and pleiotropism of environmentally controlled CIP signaling during the fission yeast life cycle, which is likely due to the existence of a wide number of effector substrates. However, very few of these putative targets have been identified to date. Therefore, their discovery will help to clarify and expand the repertoire of cellular processes regulated by this MAPK cascade. Another area of future investigation that surely deserves further deepening relates to the elucidation of the precise molecular mechanisms responsible for the functional interaction of the CIP with other environmentally regulated signaling pathways. Ultimately, comprehension of how CIP activity modulates fission yeast cellular responses might be compared to conserved pathways present in other fungal models and also in higher eukaryotes.

Author Contributions: Conceptualization: J.C., T.S., A.F., J.V.-S. and M.M.; original draft preparation: J.C., T.S., A.F., J.V.-S. and M.M.; review and editing J.C., T.S., A.F., J.V.-S. and M.M. All authors have read and agreed to the published version of the manuscript.

Funding: This research was funded by the Agencia Estatal de Investigación and Ministerio de Ciencia e Innovación, Spain, grant number PID2020-112569GB-I00; Fundación Séneca de la Región de Murcia,

Spain, grant number 20856/PI/18; and the European Regional Development Fund (ERDF) co-funding from the European Union.

Acknowledgments: The authors wish to thank all the past and current members of the Yeast Physiology Group for many helpful discussions and their continuous support. A special thanks to Pilar Pérez for her inspiring ideas and unconditional support during our shared scientific journey along these years.

Conflicts of Interest: The authors declare no conflict of interest.

References

1. Errede, B.; Levin, D.E. A conserved kinase cascade for MAP kinase activation in yeast. *Curr. Opin. Cell Biol.* **1993**, *5*, 254–260. [CrossRef]
2. Marshall, C.J. Specificity of receptor tyrosine kinase signaling: Transient versus sustained extracellular signal-regulated kinase activation. *Cell* **1995**, *80*, 179–185. [CrossRef]
3. Bluthgen, N.; Legewie, S. Systems analysis of MAPK signal transduction. *Essays Biochem.* **2008**, *45*, 95–107. [CrossRef]
4. Waskiewicz, A.J.; Cooper, J.A. Mitogen and stress response pathways: MAP kinase cascades and phosphatase regulation in mammals and yeast. *Curr. Opin. Cell Biol.* **1995**, *7*, 798–805. [CrossRef]
5. Marshall, C.J. MAP kinase kinase kinase, MAP kinase kinase and MAP kinase. *Curr. Opin. Genet. Dev.* **1994**, *4*, 82–89. [CrossRef]
6. Pérez, P.; Cansado, J. Cell integrity signaling and response to stress in fission yeast. *Curr. Protein Pept. Sci.* **2010**, *11*, 680–692. [CrossRef]
7. Kyriakis, J.M.; Avruch, J. Mammalian MAPK signal transduction pathways activated by stress and inflammation: A 10-year update. *Physiol. Rev.* **2012**, *92*, 689–737. [CrossRef]
8. Hill, C.S.; Treisman, R. Transcriptional Regulation by Extracellular signals: Mechanisms and Specificity. *Cell* **1995**, *80*, 199–211. [CrossRef]
9. Krishna, M.; Narang, H. The complexity of mitogen-activated protein kinases (MAPKs) made simple. *Cell. Mol. Life Sci.* **2008**, *65*, 3525–3544. [CrossRef]
10. Lavoie, H.; Gagnon, J.; Therrien, M. ERK signalling: A master regulator of cell behaviour, life and fate. *Nat. Rev. Mol. Cell Biol.* **2020**, *21*, 607–632. [CrossRef]
11. Cohen, A.; Kupiec, M.; Weisman, R. Glucose activates TORC2-Gad8 protein via positive regulation of the cAMP/cAMP-dependent protein kinase A (PKA) pathway and negative regulation of the Pmk1 protein-mitogen-activated protein kinase pathway. *J. Biol. Chem.* **2014**, *289*, 21727–21737. [CrossRef]
12. Schaeffer, H.J.; Weber, M.J. Mitogen-activated protein kinases: Specific messages from ubiquitous messengers. *Mol. Cell. Biol.* **1999**, *19*, 2435–2444. [CrossRef]
13. Shiozaki, K.; Russell, P. Cell-cycle control linked to extracellular environment by MAP kinase pathway in fission yeast. *Nature* **1995**, *378*, 739–743. [CrossRef]
14. Toda, T.; Shimanuki, M.; Yanagida, M. Fission yeast genes that confer resistance to staurosporine encode an AP-1-like transcription factor and a protein kinase related to the mammalian ERK1/MAP2 and budding yeast FUS3 and KSS1 kinases. *Genes Dev.* **1991**, *5*, 60–73. [CrossRef]
15. Zaitsevskaya-Carter, T.; Cooper, J.A. Spm1, a stress-activated MAP kinase that regulates morphogenesis in *S. pombe*. *Embo. J.* **1997**, *16*, 1318–1331. [CrossRef]
16. Sugawara, T.; Sato, M.; Takagi, T.; Kamasaki, T.; Ohno, N.; Osumi, M. In situ localization of cell wall alpha-1,3-glucan in the fission yeast *Schizosaccharomyces pombe*. *J. Electron. Microsc.* **2003**, *52*, 237–242. [CrossRef]
17. Toda, T.; Dhut, S.; Superti-Furga, G.; Gotoh, Y.; Nishida, E.; Sugiura, R.; Kuno, T. The fission yeast pmk1+ gene encodes a novel mitogen-activated protein kinase homolog which regulates cell integrity and functions coordinately with the protein kinase C pathway. *Mol. Cell. Biol.* **1996**, *16*, 6752–6764. [CrossRef]
18. Madrid, M.; Soto, T.; Khong, H.K.; Franco, A.; Vicente, J.; Pérez, P.; Gacto, M.; Cansado, J. Stress-induced response, localization, and regulation of the Pmk1 cell integrity pathway in Schizosaccharomyces pombe. *J. Biol. Chem.* **2006**, *281*, 2033–2043. [CrossRef]
19. Loewith, R.; Hubberstey, A.; Young, D. Skh1, the MEK component of the mkh1 signaling pathway in Schizosaccharomyces pombe. *J. Cell Sci.* **2000**, *113 Pt 1*, 153–160. [CrossRef]
20. Madrid, M.; Núñez, A.; Soto, T.; Vicente-Soler, J.; Gacto, M.; Cansado, J. Stress-activated protein kinase-mediated down-regulation of the cell integrity pathway mitogen-activated protein kinase Pmk1p by protein phosphatases. *Mol. Biol. Cell* **2007**, *18*, 4405–4419. [CrossRef]
21. Garcia, P.; Tajadura, V.; Sanchez, Y. The Rho1p exchange factor Rgf1p signals upstream from the Pmk1 mitogen-activated protein kinase pathway in fission yeast. *Mol. Biol. Cell* **2009**, *20*, 721–731. [CrossRef]
22. Sugiura, R.; Toda, T.; Dhut, S.; Shuntoh, H.; Kuno, T. The MAPK kinase Pek1 acts as a phosphorylation-dependent molecular switch. *Nature* **1999**, *399*, 479–483. [CrossRef]
23. Viana, R.A.; Pinar, M.; Soto, T.; Coll, P.M.; Cansado, J.; Pérez, P. Negative functional interaction between cell integrity MAPK pathway and Rho1 GTPase in fission yeast. *Genetics* **2013**, *195*, 421–432. [CrossRef]

24. Sengar, A.S.; Markley, N.A.; Marini, N.J.; Young, D. Mkh1, a MEK kinase required for cell wall integrity and proper response to osmotic and temperature stress in Schizosaccharomyces pombe. *Mol. Cell. Biol.* **1997**, *17*, 3508–3519. [CrossRef]
25. Sugiura, R.; Toda, T.; Shuntoh, H.; Yanagida, M.; Kuno, T. pmp1+, a suppressor of calcineurin deficiency, encodes a novel MAP kinase phosphatase in fission yeast. *EMBO J.* **1998**, *17*, 140–148. [CrossRef]
26. Ma, Y.; Kuno, T.; Kita, A.; Asayama, Y.; Sugiura, R. Rho2 is a target of the farnesyltransferase Cpp1 and acts upstream of Pmk1 mitogen-activated protein kinase signaling in fission yeast. *Mol. Biol. Cell* **2006**, *17*, 5028–5037. [CrossRef]
27. Qi, M.; Elion, E.A. MAP kinase pathways. *J. Cell Sci.* **2005**, *118*, 3569–3572. [CrossRef]
28. Gómez-Gil, E.; Franco, A.; Vázquez-Marín, B.; Prieto-Ruiz, F.; Pérez-Díaz, A.; Vicente-Soler, J.; Madrid, M.; Soto, T.; Cansado, J. Specific Functional Features of the Cell Integrity MAP Kinase Pathway in the Dimorphic Fission Yeast. *J. Fungi* **2021**, *7*, 482. [CrossRef]
29. Kock, C.; Dufrêne, Y.F.; Heinisch, J.J. Up against the wall: Is yeast cell wall integrity ensured by mechanosensing in plasma membrane microdomains? *Appl. Environ. Microbiol.* **2015**, *81*, 806–811. [CrossRef]
30. Heinisch, J.J.; Rodicio, R. Protein kinase C in fungi-more than just cell wall integrity. *FEMS Microbiol. Rev.* **2018**, *42*, 22–39. [CrossRef]
31. Levin, D.E. Regulation of cell wall biogenesis in Saccharomyces cerevisiae: The cell wall integrity signaling pathway. *Genetics* **2011**, *189*, 1145–1175. [CrossRef]
32. Cruz, S.; Muñoz, S.; Manjón, E.; García, P.; Sanchez, Y. The fission yeast cell wall stress sensor-like proteins Mtl2 and Wsc1 act by turning on the GTPase Rho1p but act independently of the cell wall integrity pathway. *Microbiologyopen* **2013**, *2*, 778–794. [CrossRef]
33. Pérez, P.; Cortés, J.C.G.; Cansado, J.; Ribas, J.C. Fission yeast cell wall biosynthesis and cell integrity signalling. *Cell Surf.* **2018**, *4*, 1–9. [CrossRef]
34. Banavar, S.P.; Gomez, C.; Trogdon, M.; Petzold, L.R.; Yi, T.M.; Campàs, O. Mechanical feedback coordinates cell wall expansion and assembly in yeast mating morphogenesis. *PLoS Comput. Biol.* **2018**, *14*, e1005940. [CrossRef]
35. Davì, V.; Tanimoto, H.; Ershov, D.; Haupt, A.; De Belly, H.; Le Borgne, R.; Couturier, E.; Boudaoud, A.; Minc, N. Mechanosensation Dynamically Coordinates Polar Growth and Cell Wall Assembly to Promote Cell Survival. *Dev. Cell* **2018**, *45*, 170–182.e7. [CrossRef]
36. Mishra, R.; van Drogen, F.; Dechant, R.; Oh, S.; Jeon, N.L.; Lee, S.S.; Peter, M. Protein kinase C and calcineurin cooperatively mediate cell survival under compressive mechanical stress. *Proc. Natl. Acad. Sci. USA* **2017**, *114*, 13471–13476. [CrossRef]
37. Neeli-Venkata, R.; Diaz, C.M.; Celador, R.; Sanchez, Y.; Minc, N. Detection of surface forces by the cell-wall mechanosensor Wsc1 in yeast. *Dev. Cell* **2021**, *56*, 2856–2870.e2857. [CrossRef]
38. Vicente-Soler, J.; Soto, T.; Franco, A.; Cansado, J.; Madrid, M. The Multiple Functions of Rho GTPases in Fission Yeasts. *Cells* **2021**, *10*, 1422. [CrossRef]
39. Hodge, R.G.; Ridley, A.J. Regulating Rho GTPases and their regulators. *Nat. Rev. Mol. Cell Biol.* **2016**, *17*, 496–510. [CrossRef]
40. Pérez, P.; Soto, T.; Gómez-Gil, E.; Cansado, J. Functional interaction between Cdc42 and the stress MAPK signaling pathway during the regulation of fission yeast polarized growth. *Int. Microbiol.* **2020**, *23*, 31–41. [CrossRef]
41. Arellano, M.; Durán, A.; Pérez, P. Rho 1 GTPase activates the (1-3)beta-D-glucan synthase and is involved in *Schizosaccharomyces pombe* morphogenesis. *EMBO J.* **1996**, *15*, 4584–4591. [CrossRef]
42. Miller, P.J.; Johnson, D.I. Cdc42p GTPase is involved in controlling polarized cell growth in Schizosaccharomyces pombe. *Mol. Cell. Biol.* **1994**, *14*, 1075–1083. [CrossRef]
43. Hirata, D.; Nakano, K.; Fukui, M.; Takenaka, H.; Miyakawa, T.; Mabuchi, I. Genes that cause aberrant cell morphology by overexpression in fission yeast: A role of a small GTP-binding protein Rho2 in cell morphogenesis. *J. Cell Sci.* **1998**, *111 Pt 2*, 149–159. [CrossRef]
44. Arellano, M.; Duran, A.; Perez, P. Localisation of the *Schizosaccharomyces pombe* rho1p GTPase and its involvement in the organisation of the actin cytoskeleton. *J. Cell Sci.* **1997**, *110 Pt 20*, 2547–2555. [CrossRef]
45. Calonge, T.M.; Nakano, K.; Arellano, M.; Arai, R.; Katayama, S.; Toda, T.; Mabuchi, I.; Perez, P. Schizosaccharomyces pombe rho2p GTPase regulates cell wall alpha-glucan biosynthesis through the protein kinase pck2p. *Mol. Biol. Cell* **2000**, *11*, 4393–4401. [CrossRef]
46. Sánchez-Mir, L.; Franco, A.; Martín-García, R.; Madrid, M.; Vicente-Soler, J.; Soto, T.; Gacto, M.; Pérez, P.; Cansado, J. Rho2 palmitoylation is required for plasma membrane localization and proper signaling to the fission yeast cell integrity mitogen-activated protein kinase pathway. *Mol. Cell. Biol.* **2014**, *34*, 2745–2759. [CrossRef]
47. Franco, A.; Soto, T.; Martín-García, R.; Madrid, M.; Vázquez-Marín, B.; Vicente-Soler, J.; Coll, P.M.; Gacto, M.; Pérez, P.; Cansado, J. Distinct functional relevance of dynamic GTPase cysteine methylation in fission yeast. *Sci. Rep.* **2017**, *7*, 6057. [CrossRef]
48. Kabeche, R.; Madrid, M.; Cansado, J.; Moseley, J.B. Eisosomes Regulate Phosphatidylinositol 4,5-Bisphosphate (PI(4,5)P2) Cortical Clusters and Mitogen-activated Protein (MAP) Kinase Signaling upon Osmotic Stress. *J. Biol. Chem.* **2015**, *290*, 25960–25973. [CrossRef]
49. Cansado, J. To finish things well: Cysteine methylation ensures selective GTPase membrane localization and signalling. *Curr. Genet.* **2018**, *64*, 341–344. [CrossRef]

50. Soto, T.; Villar-Tajadura, M.A.; Madrid, M.; Vicente, J.; Gacto, M.; Pérez, P.; Cansado, J. Rga4 modulates the activity of the fission yeast cell integrity MAPK pathway by acting as a Rho2 GTPase-activating protein. *J. Biol. Chem.* **2010**, *285*, 11516–11525. [CrossRef]
51. Cansado, J.; Soto, T.; Gacto, M.; Pérez, P. Rga4, a Rho-GAP from fission yeast: Finding specificity within promiscuity. *Commun. Integr. Biol.* **2010**, *3*, 436–439. [CrossRef]
52. Villar-Tajadura, M.A.; Coll, P.M.; Madrid, M.; Cansado, J.; Santos, B.; Pérez, P. Rga2 is a Rho2 GAP that regulates morphogenesis and cell integrity in S. pombe. *Mol. Microbiol.* **2008**, *70*, 867–881. [CrossRef]
53. Nakano, K.; Arai, R.; Mabuchi, I. The small GTP-binding protein Rho1 is a multifunctional protein that regulates actin localization, cell polarity, and septum formation in the fission yeast Schizosaccharomyces pombe. *Genes Cells* **1997**, *2*, 679–694. [CrossRef]
54. Arellano, M.; Coll, P.M.; Pérez, P. RHO GTPases in the control of cell morphology, cell polarity, and actin localization in fission yeast. *Microsc. Res. Technol.* **1999**, *47*, 51–60. [CrossRef]
55. Arellano, M.; Valdivieso, M.H.; Calonge, T.M.; Coll, P.M.; Duran, A.; Perez, P. Schizosaccharomyces pombe protein kinase C homologues, pck1p and pck2p, are targets of rho1p and rho2p and differentially regulate cell integrity. *J. Cell Sci.* **1999**, *112 Pt 20*, 3569–3578. [CrossRef]
56. Sayers, L.G.; Katayama, S.; Nakano, K.; Mellor, H.; Mabuchi, I.; Toda, T.; Parker, P.J. Rho-dependence of *Schizosaccharomyces pombe* Pck2. *Genes Cells* **2000**, *5*, 17–27. [CrossRef]
57. Sánchez-Mir, L.; Soto, T.; Franco, A.; Madrid, M.; Viana, R.A.; Vicente, J.; Gacto, M.; Pérez, P.; Cansado, J. Rho1 GTPase and PKC ortholog Pck1 are upstream activators of the cell integrity MAPK pathway in fission yeast. *PLoS ONE* **2014**, *9*, e88020. [CrossRef]
58. Nakano, K.; Mutoh, T.; Mabuchi, I. Characterization of GTPase-activating proteins for the function of the Rho-family small GTPases in the fission yeast Schizosaccharomyces pombe. *Genes Cells* **2001**, *6*, 1031–1042. [CrossRef]
59. Mutoh, T.; Nakano, K.; Mabuchi, I. Rho1-GEFs Rgf1 and Rgf2 are involved in formation of cell wall and septum, while Rgf3 is involved in cytokinesis in fission yeast. *Genes Cells* **2005**, *10*, 1189–1202. [CrossRef]
60. García, P.; Tajadura, V.; García, I.; Sánchez, Y. Role of Rho GTPases and Rho-GEFs in the regulation of cell shape and integrity in fission yeast. *Yeast* **2006**, *23*, 1031–1043. [CrossRef]
61. Tajadura, V.; García, B.; García, I.; García, P.; Sánchez, Y. Schizosaccharomyces pombe Rgf3p is a specific Rho1 GEF that regulates cell wall beta-glucan biosynthesis through the GTPase Rho1p. *J. Cell Sci.* **2004**, *117*, 6163–6174. [CrossRef]
62. Calonge, T.M.; Arellano, M.; Coll, P.M.; Perez, P. Rga5p is a specific Rho1p GTPase-activating protein that regulates cell integrity in Schizosaccharomyces pombe. *Mol. Microbiol.* **2003**, *47*, 507–518. [CrossRef]
63. García, P.; Tajadura, V.; García, I.; Sánchez, Y. Rgf1p is a specific Rho1-GEF that coordinates cell polarization with cell wall biogenesis in fission yeast. *Mol. Biol. Cell* **2006**, *17*, 1620–1631. [CrossRef]
64. Barba, G.; Soto, T.; Madrid, M.; Núñez, A.; Vicente, J.; Gacto, M.; Cansado, J.; Group, Y.P. Activation of the cell integrity pathway is channelled through diverse signalling elements in fission yeast. *Cell. Signal.* **2008**, *20*, 748–757. [CrossRef]
65. Nakano, K.; Mutoh, T.; Arai, R.; Mabuchi, I. The small GTPase Rho4 is involved in controlling cell morphology and septation in fission yeast. *Genes Cells* **2003**, *8*, 357–370. [CrossRef]
66. Santos, B.; Martín-Cuadrado, A.B.; Vázquez de Aldana, C.R.; del Rey, F.; Pérez, P. Rho4 GTPase is involved in secretion of glucanases during fission yeast cytokinesis. *Eukaryot. Cell* **2005**, *4*, 1639–1645. [CrossRef]
67. Santos, B.; Gutiérrez, J.; Calonge, T.M.; Pérez, P. Novel Rho GTPase involved in cytokinesis and cell wall integrity in the fission yeast Schizosaccharomyces pombe. *Eukaryot. Cell* **2003**, *2*, 521–533. [CrossRef]
68. Nakano, K.; Arai, R.; Mabuchi, I. Small GTPase Rho5 is a functional homologue of Rho1, which controls cell shape and septation in fission yeast. *FEBS Lett.* **2005**, *579*, 5181–5186. [CrossRef]
69. Rincón, S.A.; Santos, B.; Pérez, P. Fission yeast Rho5p GTPase is a functional paralogue of Rho1p that plays a role in survival of spores and stationary-phase cells. *Eukaryot. Cell* **2006**, *5*, 435–446. [CrossRef]
70. Doi, A.; Kita, A.; Kanda, Y.; Uno, T.; Asami, K.; Satoh, R.; Nakano, K.; Sugiura, R. Geranylgeranyltransferase Cwg2-Rho4/Rho5 module is implicated in the Pmk1 MAP kinase-mediated cell wall integrity pathway in fission yeast. *Genes Cells* **2015**, *20*, 310–323. [CrossRef]
71. Newton, A.C. Protein kinase C: Perfectly balanced. *Crit. Rev. Biochem. Mol. Biol.* **2018**, *53*, 208–230. [CrossRef]
72. Madrid, M.; Vazquez-Marin, B.; Franco, A.; Soto, T.; Vicente-Soler, J.; Gacto, M.; Cansado, J. Multiple crosstalk between TOR and the cell integrity MAPK signaling pathway in fission yeast. *Sci. Rep.* **2016**, *6*, 37515. [CrossRef]
73. Freeley, M.; Kelleher, D.; Long, A. Regulation of Protein Kinase C function by phosphorylation on conserved and non-conserved sites. *Cell. Signal.* **2011**, *23*, 753–762. [CrossRef]
74. Madrid, M.; Vázquez-Marín, B.; Soto, T.; Franco, A.; Gómez-Gil, E.; Vicente-Soler, J.; Gacto, M.; Pérez, P.; Cansado, J. Differential functional regulation of protein kinase C (PKC) orthologs in fission yeast. *J. Biol. Chem.* **2017**, *292*, 11374–11387. [CrossRef]
75. Toda, T.; Shimanuki, M.; Yanagida, M. Two novel protein kinase C-related genes of fission yeast are essential for cell viability and implicated in cell shape control. *EMBO J.* **1993**, *12*, 1987–1995. [CrossRef]
76. Matsuyama, A.; Arai, R.; Yashiroda, Y.; Shirai, A.; Kamata, A.; Sekido, S.; Kobayashi, Y.; Hashimoto, A.; Hamamoto, M.; Hiraoka, Y.; et al. ORFeome cloning and global analysis of protein localization in the fission yeast Schizosaccharomyces pombe. *Nat. Biotechnol.* **2006**, *24*, 841–847. [CrossRef]
77. Magliozzi, J.O.; Sears, J.; Cressey, L.; Brady, M.; Opalko, H.E.; Kettenbach, A.N.; Moseley, J.B. Fission yeast Pak1 phosphorylates anillin-like Mid1 for spatial control of cytokinesis. *J. Cell Biol.* **2020**, *219*, e201908017. [CrossRef]

78. Madrid, M.; Jiménez, R.; Sánchez-Mir, L.; Soto, T.; Franco, A.; Vicente-Soler, J.; Gacto, M.; Pérez, P.; Cansado, J. Multiple layers of regulation influence cell integrity control by the PKC ortholog Pck2 in fission yeast. *J. Cell Sci.* **2015**, *128*, 266–280. [CrossRef]
79. Kanda, Y.; Satoh, R.; Takasaki, T.; Tomimoto, N.; Tsuchiya, K.; Tsai, C.A.; Tanaka, T.; Kyomoto, S.; Hamada, K.; Fujiwara, T.; et al. Sequestration of the PKC ortholog Pck2 in stress granules as a feedback mechanism of MAPK signaling in fission yeast. *J. Cell Sci.* **2021**, *134*, jcs250191. [CrossRef]
80. Newton, A.C. Protein kinase C: Poised to signal. *Am. J. Physiol. Endocrinol. Metab.* **2010**, *298*, E395–E402. [CrossRef]
81. Niederberger, C.; Schweingruber, M.E. A *Schizosaccharomyces pombe* gene, ksg1, that shows structural homology to the human phosphoinositide-dependent protein kinase PDK1, is essential for growth, mating and sporulation. *Mol. Gen. Genet.* **1999**, *261*, 177–183. [CrossRef]
82. Gräub, R.; Hilti, N.; Niederberger, C.; Schweingruber, M.E. Ksg1, a homologue of the phosphoinositide-dependent protein kinase 1, controls cell wall integrity in Schizosaccharomyces pombe. *J. Basic Microbiol.* **2003**, *43*, 473–482. [CrossRef]
83. Madrid, M.; Fernández-Zapata, J.; Sánchez-Mir, L.; Soto, T.; Franco, A.; Vicente-Soler, J.; Gacto, M.; Cansado, J. Role of the fission yeast cell integrity MAPK pathway in response to glucose limitation. *BMC Microbiol.* **2013**, *13*, 34. [CrossRef]
84. Kanda, Y.; Satoh, R.; Matsumoto, S.; Ikeda, C.; Inutsuka, N.; Hagihara, K.; Matzno, S.; Tsujimoto, S.; Kita, A.; Sugiura, R. Skb5, an SH3 adaptor protein, regulates Pmk1 MAPK signaling by controlling the intracellular localization of the MAPKKK Mkh1. *J. Cell Sci.* **2016**, *129*, 3189–3202. [CrossRef]
85. Yang, P.; Pimental, R.; Lai, H.; Marcus, S. Direct activation of the fission yeast PAK Shk1 by the novel SH3 domain protein, Skb5. *J. Biol. Chem.* **1999**, *274*, 36052–36057. [CrossRef]
86. Vázquez, B.; Soto, T.; del Dedo, J.E.; Franco, A.; Vicente, J.; Hidalgo, E.; Gacto, M.; Cansado, J.; Madrid, M. Distinct biological activity of threonine monophosphorylated MAPK isoforms during the stress response in fission yeast. *Cell. Signal.* **2015**, *27*, 2534–2542. [CrossRef]
87. Sánchez-Mir, L.; Franco, A.; Madrid, M.; Vicente-Soler, J.; Villar-Tajadura, M.A.; Soto, T.; Pérez, P.; Gacto, M.; Cansado, J. Biological significance of nuclear localization of mitogen-activated protein kinase Pmk1 in fission yeast. *J. Biol. Chem.* **2012**, *287*, 26038–26051. [CrossRef]
88. Soto, T.; Núñez, A.; Madrid, M.; Vicente, J.; Gacto, M.; Cansado, J. Transduction of centrifugation-induced gravity forces through mitogen-activated protein kinase pathways in the fission yeast Schizosaccharomyces pombe. *Microbiology* **2007**, *153*, 1519–1529. [CrossRef]
89. Kaino, T.; Tonoko, K.; Mochizuki, S.; Takashima, Y.; Kawamukai, M. Schizosaccharomyces japonicus has low levels of CoQ(10) synthesis, respiration deficiency, and efficient ethanol production. *Biosci. Biotechnol. Biochem.* **2018**, *82*, 1031–1042. [CrossRef]
90. Degols, G.; Shiozaki, K.; Russell, P. Activation and regulation of the Spc1 stress-activated protein kinase in Schizosaccharomyces pombe. *Mol. Cell. Biol.* **1996**, *16*, 2870–2877. [CrossRef]
91. Shiozaki, K.; Russell, P. Conjugation, meiosis, and the osmotic stress response are regulated by Spc1 kinase through Atf1 transcription factor in fission yeast. *Genes Dev.* **1996**, *10*, 2276–2288. [CrossRef]
92. Wilkinson, M.G.; Samuels, M.; Takeda, T.; Toone, W.M.; Shieh, J.C.; Toda, T.; Millar, J.B.; Jones, N. The Atf1 transcription factor is a target for the Sty1 stress-activated MAP kinase pathway in fission yeast. *Genes Dev.* **1996**, *10*, 2289–2301. [CrossRef]
93. Lawrence, C.L.; Maekawa, H.; Worthington, J.L.; Reiter, W.; Wilkinson, C.R.; Jones, N. Regulation of *Schizosaccharomyces pombe* Atf1 protein levels by Sty1-mediated phosphorylation and heterodimerization with Pcr1. *J. Biol. Chem.* **2007**, *282*, 5160–5170. [CrossRef]
94. Chen, D.; Wilkinson, C.R.; Watt, S.; Penkett, C.J.; Toone, W.M.; Jones, N.; Bähler, J. Multiple pathways differentially regulate global oxidative stress responses in fission yeast. *Mol. Biol. Cell* **2008**, *19*, 308–317. [CrossRef]
95. Reiter, W.; Watt, S.; Dawson, K.; Lawrence, C.L.; Bahler, J.; Jones, N.; Wilkinson, C.R. Fission yeast MAP kinase Sty1 is recruited to stress-induced genes. *J. Biol. Chem.* **2008**, *283*, 9945–9956. [CrossRef]
96. Takada, H.; Nishimura, M.; Asayama, Y.; Mannse, Y.; Ishiwata, S.; Kita, A.; Doi, A.; Nishida, A.; Kai, N.; Moriuchi, S.; et al. Atf1 is a target of the mitogen-activated protein kinase Pmk1 and regulates cell integrity in fission yeast. *Mol. Biol. Cell* **2007**, *18*, 4794–4802. [CrossRef]
97. Takada, H.; Nishida, A.; Domae, M.; Kita, A.; Yamano, Y.; Uchida, A.; Ishiwata, S.; Fang, Y.; Zhou, X.; Masuko, T.; et al. The cell surface protein gene ecm33+ is a target of the two transcription factors Atf1 and Mbx1 and negatively regulates Pmk1 MAPK cell integrity signaling in fission yeast. *Mol. Biol. Cell* **2010**, *21*, 674–685. [CrossRef]
98. Papadopoulou, K.; Ng, S.S.; Ohkura, H.; Geymonat, M.; Sedgwick, S.G.; McInerny, C.J. Regulation of gene expression during M-G1-phase in fission yeast through Plo1p and forkhead transcription factors. *J. Cell Sci.* **2008**, *121*, 38–47. [CrossRef]
99. Oliveira, C.; Faoro, H.; Alves, L.R.; Goldenberg, S. RNA-binding proteins and their role in the regulation of gene expression in Trypanosoma cruzi and Saccharomyces cerevisiae. *Genet. Mol. Biol.* **2017**, *40*, 22–30. [CrossRef]
100. Sugiura, R.; Satoh, R.; Ishiwata, S.; Umeda, N.; Kita, A. Role of RNA-Binding Proteins in MAPK Signal Transduction Pathway. *J. Signal Transduct.* **2011**, *2011*, 109746. [CrossRef]
101. Venigalla, R.K.; Turner, M. RNA-binding proteins as a point of convergence of the PI3K and p38 MAPK pathways. *Front. Immunol.* **2012**, *3*, 398. [CrossRef]
102. Hasan, A.; Cotobal, C.; Duncan, C.D.; Mata, J. Systematic analysis of the role of RNA-binding proteins in the regulation of RNA stability. *PLoS Genet.* **2014**, *10*, e1004684. [CrossRef]

103. Díaz-Cuervo, H.; Bueno, A. Cds1 controls the release of Cdc14-like phosphatase Flp1 from the nucleolus to drive full activation of the checkpoint response to replication stress in fission yeast. *Mol. Biol. Cell* **2008**, *19*, 2488–2499. [CrossRef]
104. Broadus, M.R.; Gould, K.L. Multiple protein kinases influence the redistribution of fission yeast Clp1/Cdc14 phosphatase upon genotoxic stress. *Mol. Biol. Cell* **2012**, *23*, 4118–4128. [CrossRef]
105. Sugiura, R.; Kita, A.; Shimizu, Y.; Shuntoh, H.; Sio, S.O.; Kuno, T. Feedback regulation of MAPK signalling by an RNA-binding protein. *Nature* **2003**, *424*, 961–965. [CrossRef]
106. Tsukahara, K.; Yamamoto, H.; Okayama, H. An RNA binding protein negatively controlling differentiation in fission yeast. *Mol. Cell. Biol.* **1998**, *18*, 4488–4498. [CrossRef]
107. Jeong, H.T.; Ozoe, F.; Tanaka, K.; Nakagawa, T.; Matsuda, H.; Kawamukai, M. A novel gene, msa1, inhibits sexual differentiation in Schizosaccharomyces pombe. *Genetics* **2004**, *167*, 77–91. [CrossRef]
108. Jeong, H.T.; Oowatari, Y.; Abe, M.; Tanaka, K.; Matsuda, H.; Kawamukai, M. Interaction between a negative regulator (Msa2/Nrd1) and a positive regulator (Cpc2) of sexual differentiation in Schizosaccharomyces pombe. *Biosci. Biotechnol. Biochem.* **2004**, *68*, 1621–1626. [CrossRef]
109. Mata, J.; Bähler, J. Global roles of Ste11p, cell type, and pheromone in the control of gene expression during early sexual differentiation in fission yeast. *Proc. Natl. Acad. Sci. USA* **2006**, *103*, 15517–15522. [CrossRef]
110. Satoh, R.; Morita, T.; Takada, H.; Kita, A.; Ishiwata, S.; Doi, A.; Hagihara, K.; Taga, A.; Matsumura, Y.; Tohda, H.; et al. Role of the RNA-binding Protein Nrd1 and Pmk1 Mitogen-activated Protein Kinase in the Regulation of Myosin mRNA Stability in Fission Yeast. *Mol. Biol. Cell* **2009**, *20*, 2473–2485. [CrossRef]
111. Pollard, T.D.; Wu, J.Q. Understanding cytokinesis: Lessons from fission yeast. *Nat. Rev. Mol. Cell Biol.* **2010**, *11*, 149–155. [CrossRef]
112. Kobayashi, A.; Kanaba, T.; Satoh, R.; Fujiwara, T.; Ito, Y.; Sugiura, R.; Mishima, M. Structure of the second RRM domain of Nrd1, a fission yeast MAPK target RNA binding protein, and implication for its RNA recognition and regulation. *Biochem. Biophys. Res. Commun.* **2013**, *437*, 12–17. [CrossRef]
113. Satoh, R.; Tanaka, A.; Kita, A.; Morita, T.; Matsumura, Y.; Umeda, N.; Takada, M.; Hayashi, S.; Tani, T.; Shinmyozu, K.; et al. Role of the RNA-Binding Protein Nrd1 in Stress Granule Formation and Its Implication in the Stress Response in Fission Yeast. *PLoS ONE* **2012**, *7*, e29683. [CrossRef]
114. Oowatari, Y.; Jeong, H.; Tanae, K.; Nakagawa, T.; Kawamukai, M. Regulation and role of an RNA-binding protein Msa2 in controlling the sexual differentiation of fission yeast. *Curr. Genet.* **2011**, *57*, 191–200. [CrossRef]
115. Kettenbach, A.N.; Deng, L.; Wu, Y.; Baldissard, S.; Adamo, M.E.; Gerber, S.A.; Moseley, J.B. Quantitative phosphoproteomics reveals pathways for coordination of cell growth and division by the conserved fission yeast kinase pom1. *Mol. Cell. Proteom.* **2015**, *14*, 1275–1287. [CrossRef]
116. Swaffer, M.P.; Jones, A.W.; Flynn, H.R.; Snijders, A.P.; Nurse, P. Quantitative Phosphoproteomics Reveals the Signaling Dynamics of Cell-Cycle Kinases in the Fission Yeast Schizosaccharomyces pombe. *Cell Rep.* **2018**, *24*, 503–514. [CrossRef]
117. Tay, Y.D.; Leda, M.; Spanos, C.; Rappsilber, J.; Goryachev, A.B.; Sawin, K.E. Fission Yeast NDR/LATS Kinase Orb6 Regulates Exocytosis via Phosphorylation of the Exocyst Complex. *Cell Rep.* **2019**, *26*, 1654–1667.e1657. [CrossRef]
118. Nunez, A.; Franco, A.; Madrid, M.; Soto, T.; Vicente, J.; Gacto, M.; Cansado, J. Role for RACK1 Orthologue Cpc2 in the Modulation of Stress Response in Fission Yeast. *Mol. Biol. Cell* **2009**, *20*, 3996–4009. [CrossRef]
119. Hollingworth, D.; Candel, A.M.; Nicastro, G.; Martin, S.R.; Briata, P.; Gherzi, R.; Ramos, A. KH domains with impaired nucleic acid binding as a tool for functional analysis. *Nucleic Acids Res.* **2012**, *40*, 6873–6886. [CrossRef]
120. Grishin, N.V. KH domain: One motif, two folds. *Nucleic Acids Res.* **2001**, *29*, 638–643. [CrossRef]
121. Sugiura, R.; Kita, A.; Kuno, T. Upregulation of mRNA in MAPK signaling—Transcriptional activation or mRNA stabilization? *Cell Cycle* **2004**, *3*, 286–288. [CrossRef]
122. Satoh, R.; Matsumura, Y.; Tanaka, A.; Takada, M.; Ito, Y.; Hagihara, K.; Inari, M.; Kita, A.; Fukao, A.; Fujiwara, T.; et al. Spatial regulation of the KH domain RNA-binding protein Rnc1 mediated by a Crm1-independent nuclear export system in Schizosaccharomyces pombe. *Mol. Microbiol.* **2017**, *104*, 428–448. [CrossRef]
123. Satoh, R.; Hagihara, K.; Sugiura, R. Rae1-mediated nuclear export of Rnc1 is an important determinant in controlling MAPK signaling. *Curr. Genet.* **2018**, *64*, 103–108. [CrossRef]
124. Satoh, R.; Hara, N.; Kawasaki, A.; Takasaki, T.; Sugiura, R. Distinct modes of stress granule assembly mediated by the KH-type RNA-binding protein Rnc1. *Genes Cells* **2018**, *23*, 778–785. [CrossRef]
125. Prieto-Ruiz, F.; Vicente-Soler, J.; Franco, A.; Gomez-Gil, E.; Sanchez-Marinas, M.; Vazquez-Marin, B.; Aligue, R.; Madrid, M.; Moreno, S.; Soto, T.; et al. RNA-Binding Protein Rnc1 Regulates Cell Length at Division and Acute Stress Response in Fission Yeast through Negative Feedback Modulation of the Stress-Activated Mitogen-Activated Protein Kinase Pathway. *mBio* **2020**, *11*, e02815-19. [CrossRef]
126. Hirayama, S.; Sugiura, R.; Lu, Y.; Maeda, T.; Kawagishi, K.; Yokoyama, M.; Tohda, H.; Giga-Hama, Y.; Shuntoh, H.; Kuno, T. Zinc finger protein Prz1 regulates Ca^{2+} but not Cl^{-} homeostasis in fission yeast. Identification of distinct branches of calcineurin signaling pathway in fission yeast. *J. Biol. Chem.* **2003**, *278*, 18078–18084. [CrossRef]
127. Deng, L.; Sugiura, R.; Takeuchi, M.; Suzuki, M.; Ebina, H.; Takami, T.; Koike, A.; Iba, S.; Kuno, T. Real-time monitoring of calcineurin activity in living cells: Evidence for two distinct Ca^{2+}-dependent pathways in fission yeast. *Mol. Biol. Cell* **2006**, *17*, 4790–4800. [CrossRef]

128. Hamasaki-Katagiri, N.; Ames, J.B. Neuronal calcium sensor-1 (Ncs1p) is up-regulated by calcineurin to promote Ca^{2+} tolerance in fission yeast. *J. Biol. Chem.* **2010**, *285*, 4405–4414. [CrossRef]
129. Maeda, T.; Sugiura, R.; Kita, A.; Saito, M.; Deng, L.; He, Y.; Yabin, L.; Fujita, Y.; Takegawa, K.; Shuntoh, H.; et al. Pmr1, a P-type ATPase, and Pdt1, an Nramp homologue, cooperatively regulate cell morphogenesis in fission yeast: The importance of Mn^{2+} homeostasis. *Genes Cells* **2004**, *9*, 71–82. [CrossRef]
130. Cisneros-Barroso, E.; Yance-Chávez, T.; Kito, A.; Sugiura, R.; Gómez-Hierro, A.; Giménez-Zaragoza, D.; Aligue, R. Negative feedback regulation of calcineurin-dependent Prz1 transcription factor by the CaMKK-CaMK1 axis in fission yeast. *Nucleic Acids Res.* **2014**, *42*, 9573–9587. [CrossRef]
131. Chatfield-Reed, K.; Vachon, L.; Kwon, E.J.; Chua, G. Conserved and Diverged Functions of the Calcineurin-Activated Prz1 Transcription Factor in Fission Yeast. *Genetics* **2016**, *202*, 1365–1375. [CrossRef]
132. Yoshida, T.; Toda, T.; Yanagida, M. A calcineurin-like gene ppb1+ in fission yeast: Mutant defects in cytokinesis, cell polarity, mating and spindle pole body positioning. *J. Cell Sci.* **1994**, *107 Pt 7*, 1725–1735. [CrossRef]
133. Aydar, E.; Palmer, C.P. Polycystic kidney disease channel and synaptotagmin homologues play roles in schizosaccharomyces pombe cell wall synthesis/repair and membrane protein trafficking. *J. Membr. Biol.* **2009**, *229*, 141–152. [CrossRef]
134. Carnero, E.; Ribas, J.C.; García, B.; Durán, A.; Sánchez, Y. *Schizosaccharomyces pombe* ehs1p is involved in maintaining cell wall integrity and in calcium uptake. *Mol. Gen. Genet.* **2000**, *264*, 173–183. [CrossRef]
135. Ma, Y.; Sugiura, R.; Koike, A.; Ebina, H.; Sio, S.O.; Kuno, T. Transient receptor potential (TRP) and Cch1-Yam8 channels play key roles in the regulation of cytoplasmic Ca^{2+} in fission yeast. *PLoS ONE* **2011**, *6*, e22421. [CrossRef]
136. Gow, N.A.R.; Latge, J.P.; Munro, C.A. The Fungal Cell Wall: Structure, Biosynthesis, and Function. *Microbiol. Spectr.* **2017**, *5*, FUNK-0035-2016. [CrossRef]
137. Hopke, A.; Brown, A.J.P.; Hall, R.A.; Wheeler, R.T. Dynamic Fungal Cell Wall Architecture in Stress Adaptation and Immune Evasion. *Trends Microbiol.* **2018**, *26*, 284–295. [CrossRef]
138. Mazur, P.; Baginsky, W. In vitro activity of 1,3-beta-D-glucan synthase requires the GTP-binding protein Rho1. *J. Biol. Chem.* **1996**, *271*, 14604–14609. [CrossRef]
139. Guo, S.; Shen, X.; Yan, G.; Ma, D.; Bai, X.; Li, S.; Jiang, Y. A MAP Kinase Dependent Feedback Mechanism Controls Rho1 GTPase and Actin Distribution in Yeast. *PLoS ONE* **2009**, *4*, e6089. [CrossRef]
140. Kampmeyer, C.; Johansen, J.V.; Holmberg, C.; Karlson, M.; Gersing, S.K.; Bordallo, H.N.; Kragelund, B.B.; Lerche, M.H.; Jourdain, I.; Winther, J.R.; et al. Mutations in a Single Signaling Pathway Allow Cell Growth in Heavy Water. *ACS Synth. Biol.* **2020**, *9*, 733–748. [CrossRef]
141. Imai, Y.; Shimasaki, T.; Enokimura, C.; Ohtsuka, H.; Tsubouchi, S.; Ihara, K.; Aiba, H. gas1 mutation extends chronological lifespan via Pmk1 and Sty1 MAPKs in Schizosaccharomyces pombe. *Biosci. Biotechnol. Biochem.* **2020**, *84*, 330–337. [CrossRef]
142. Merla, A.; Johnson, D.I. The *Schizosaccharomyces pombe* Cdc42p GTPase signals through Pak2p and the Mkh1p-Pek1p-Spm1p MAP kinase pathway. *Curr. Genet.* **2001**, *39*, 205–209. [CrossRef]
143. Kono, K.; Saeki, Y.; Yoshida, S.; Tanaka, K.; Pellman, D. Proteasomal degradation resolves competition between cell polarization and cellular wound healing. *Cell* **2012**, *150*, 151–164. [CrossRef]
144. Gu, Y.; Oliferenko, S. Comparative biology of cell division in the fission yeast clade. *Curr. Opin. Microbiol.* **2015**, *28*, 18–25. [CrossRef]
145. Rincon, S.A.; Paoletti, A. Molecular control of fission yeast cytokinesis. *Semin. Cell Dev. Biol.* **2016**, *53*, 28–38. [CrossRef]
146. Garcia Cortes, J.C.; Ramos, M.; Osumi, M.; Perez, P.; Ribas, J.C. The Cell Biology of Fission Yeast Septation. *Microbiol. Mol. Biol. Rev.* **2016**, *80*, 779–791. [CrossRef]
147. Tolliday, N.; VerPlank, L.; Li, R. Rho1 directs formin-mediated actin ring assembly during budding yeast cytokinesis. *Curr. Biol.* **2002**, *12*, 1864–1870. [CrossRef]
148. Chin, C.F.; Bennett, A.M.; Ma, W.K.; Hall, M.C.; Yeong, F.M. Dependence of Chs2 ER export on dephosphorylation by cytoplasmic Cdc14 ensures that septum formation follows mitosis. *Mol. Biol. Cell* **2012**, *23*, 45–58. [CrossRef]
149. Yoshida, S.; Bartolini, S.; Pellman, D. Mechanisms for concentrating Rho1 during cytokinesis. *Genes Dev.* **2009**, *23*, 810–823. [CrossRef]
150. García, R.; Bermejo, C.; Grau, C.; Pérez, R.; Rodríguez-Peña, J.M.; Francois, J.; Nombela, C.; Arroyo, J. The global transcriptional response to transient cell wall damage in *Saccharomyces cerevisiae* and its regulation by the cell integrity signaling pathway. *J. Biol. Chem.* **2004**, *279*, 15183–15195. [CrossRef]
151. Cabib, E.; Arroyo, J. How carbohydrates sculpt cells: Chemical control of morphogenesis in the yeast cell wall. *Nat. Rev. Microbiol.* **2013**, *11*, 648–655. [CrossRef]
152. Piña, F.J.; Niwa, M. The ER Stress Surveillance (ERSU) pathway regulates daughter cell ER protein aggregate inheritance. *Elife* **2015**, *4*, e06970. [CrossRef]
153. Babour, A.; Bicknell, A.A.; Tourtellotte, J.; Niwa, M. A surveillance pathway monitors the fitness of the endoplasmic reticulum to control its inheritance. *Cell* **2010**, *142*, 256–269. [CrossRef]
154. Edreira, T.; Celador, R.; Manjón, E.; Sánchez, Y. A novel checkpoint pathway controls actomyosin ring constriction trigger in fission yeast. *eLife* **2020**, *9*, e59333. [CrossRef]
155. Jin, Q.-W.; Zhou, M.; Bimbo, A.; Balasubramanian, M.K.; McCollum, D. A role for the septation initiation network in septum assembly revealed by genetic analysis of sid2-250 suppressors. *Genetics* **2006**, *172*, 2101–2112. [CrossRef]

156. Davidson, R.; Laporte, D.; Wu, J.-Q. Regulation of Rho-GEF Rgf3 by the arrestin Art1 in fission yeast cytokinesis. *Mol. Biol. Cell* **2014**, *26*, 453–466. [CrossRef]
157. Martín-García, R.; Arribas, V.; Coll, P.M.; Pinar, M.; Viana, R.A.; Rincón, S.A.; Correa-Bordes, J.; Ribas, J.C.; Pérez, P. Paxillin-Mediated Recruitment of Calcineurin to the Contractile Ring Is Required for the Correct Progression of Cytokinesis in Fission Yeast. *Cell Rep.* **2018**, *25*, 772–783.e774. [CrossRef]
158. Mascaraque, V.; Hernáez, M.L.; Jiménez-Sánchez, M.; Hansen, R.; Gil, C.; Martín, H.; Cid, V.J.; Molina, M. Phosphoproteomic analysis of protein kinase C signaling in *Saccharomyces cerevisiae* reveals Slt2 mitogen-activated protein kinase (MAPK)-dependent phosphorylation of eisosome core components. *Mol. Cell Proteom.* **2013**, *12*, 557–574. [CrossRef]
159. Dodgson, J.; Chessel, A.; Vaggi, F.; Giordan, M.; Yamamoto, M.; Arai, K.; Madrid, M.; Geymonat, M.; Abenza, J.F.; Cansado, J.; et al. Reconstructing regulatory pathways by systematically mapping protein localization interdependency networks. *bioRxiv* **2017**. [CrossRef]
160. Cortés, J.C.; Konomi, M.; Martins, I.M.; Muñoz, J.; Moreno, M.B.; Osumi, M.; Durán, A.; Ribas, J.C. The (1,3)beta-D-glucan synthase subunit Bgs1p is responsible for the fission yeast primary septum formation. *Mol. Microbiol.* **2007**, *65*, 201–217. [CrossRef]
161. Gotoh, Y.; Nishida, E.; Shimanuki, M.; Toda, T.; Imai, Y.; Yamamoto, M. Schizosaccharomyces pombe Spk1 is a tyrosine-phosphorylated protein functionally related to Xenopus mitogen-activated protein kinase. *Mol. Cell. Biol.* **1993**, *13*, 6427–6434. [CrossRef]
162. Millar, J.B.; Buck, V.; Wilkinson, M.G. Pyp1 and Pyp2 PTPases dephosphorylate an osmosensing MAP kinase controlling cell size at division in fission yeast. *Genes Dev.* **1995**, *9*, 2117–2130. [CrossRef]
163. Shiozaki, K.; Russell, P. Counteractive roles of protein phosphatase 2C (PP2C) and a MAP kinase kinase homolog in the osmoregulation of fission yeast. *EMBO J.* **1995**, *14*, 492–502. [CrossRef]
164. Didmon, M.; Davis, K.; Watson, P.; Ladds, G.; Broad, P.; Davey, J. Identifying regulators of pheromone signalling in the fission yeast Schizosaccharomyces pombe. *Curr. Genet.* **2002**, *41*, 241–253. [CrossRef]
165. Degols, G.; Russell, P. Discrete roles of the Spc1 kinase and the Atf1 transcription factor in the UV response of Schizosaccharomyces pombe. *Mol. Cell. Biol.* **1997**, *17*, 3356–3363. [CrossRef]
166. Soto, T.; Beltrán, F.F.; Paredes, V.; Madrid, M.; Millar, J.B.; Vicente-Soler, J.; Cansado, J.; Gacto, M. Cold induces stress-activated protein kinase-mediated response in the fission yeast Schizosaccharomyces pombe. *Eur. J. Biochem.* **2002**, *269*, 5056–5065. [CrossRef]
167. Quinn, J.; Findlay, V.J.; Dawson, K.; Millar, J.B.; Jones, N.; Morgan, B.A.; Toone, W.M. Distinct regulatory proteins control the graded transcriptional response to increasing H(2)O(2) levels in fission yeast Schizosaccharomyces pombe. *Mol. Biol. Cell* **2002**, *13*, 805–816. [CrossRef]
168. Rodríguez-Gabriel, M.A.; Russell, P. Distinct signaling pathways respond to arsenite and reactive oxygen species in Schizosaccharomyces pombe. *Eukaryot. Cell* **2005**, *4*, 1396–1402. [CrossRef]
169. George, V.T.; Brooks, G.; Humphrey, T.C. Regulation of cell cycle and stress responses to hydrostatic pressure in fission yeast. *Mol. Biol. Cell* **2007**, *18*, 4168–4179. [CrossRef]
170. Buck, V.; Quinn, J.; Soto Pino, T.; Martin, H.; Saldanha, J.; Makino, K.; Morgan, B.A.; Millar, J.B. Peroxide sensors for the fission yeast stress-activated mitogen-activated protein kinase pathway. *Mol. Biol. Cell* **2001**, *12*, 407–419. [CrossRef]
171. Hohmann, S. Osmotic stress signaling and osmoadaptation in yeasts. *Microbiol. Mol. Biol. Rev. MMBR* **2002**, *66*, 300–372. [CrossRef]
172. Samejima, I.; Mackie, S.; Fantes, P.A. Multiple modes of activation of the stress-responsive MAP kinase pathway in fission yeast. *EMBO J.* **1997**, *16*, 6162–6170. [CrossRef]
173. Nguyen, A.N.; Shiozaki, K. Heat-shock-induced activation of stress MAP kinase is regulated by threonine- and tyrosine-specific phosphatases. *Genes Dev.* **1999**, *13*, 1653–1663. [CrossRef]
174. Gaits, F.; Russell, P. Active nucleocytoplasmic shuttling required for function and regulation of stress-activated kinase Spc1/StyI in fission yeast. *Mol. Biol. Cell* **1999**, *10*, 1395–1407. [CrossRef]
175. Doi, K.; Gartner, A.; Ammerer, G.; Errede, B.; Shinkawa, H.; Sugimoto, K.; Matsumoto, K. MSG5, a novel protein phosphatase promotes adaptation to pheromone response in S. cerevisiae. *EMBO J.* **1994**, *13*, 61–70. [CrossRef]
176. Watanabe, Y.; Irie, K.; Matsumoto, K. Yeast RLM1 encodes a serum response factor-like protein that may function downstream of the Mpk1 (Slt2) mitogen-activated protein kinase pathway. *Mol. Cell. Biol.* **1995**, *15*, 5740–5749. [CrossRef]
177. Davenport, K.D.; Williams, K.E.; Ullmann, B.D.; Gustin, M.C. Activation of the *Saccharomyces cerevisiae* filamentation/invasion pathway by osmotic stress in high-osmolarity glycogen pathway mutants. *Genetics* **1999**, *153*, 1091–1103. [CrossRef]
178. Martín, H.; Rodríguez-Pachón, J.M.; Ruiz, C.; Nombela, C.; Molina, M. Regulatory mechanisms for modulation of signaling through the cell integrity Slt2-mediated pathway in Saccharomyces cerevisiae. *J. Biol. Chem.* **2000**, *275*, 1511–1519. [CrossRef]
179. González, A.; Hall, M.N. Nutrient sensing and TOR signaling in yeast and mammals. *EMBO J.* **2017**, *36*, 397–408. [CrossRef]
180. Wullschleger, S.; Loewith, R.; Hall, M.N. TOR signaling in growth and metabolism. *Cell* **2006**, *124*, 471–484. [CrossRef]
181. Cybulski, N.; Hall, M.N. TOR complex 2: A signaling pathway of its own. *Trends Biochem. Sci.* **2009**, *34*, 620–627. [CrossRef]
182. Loewith, R. A brief history of TOR. *Biochem. Soc. Trans.* **2011**, *39*, 437–442. [CrossRef]
183. Alvarez, B.; Moreno, S. Fission yeast Tor2 promotes cell growth and represses cell differentiation. *J. Cell Sci.* **2006**, *119*, 4475–4485. [CrossRef]

184. Nakashima, A.; Otsubo, Y.; Yamashita, A.; Sato, T.; Yamamoto, M.; Tamanoi, F. Psk1, an AGC kinase family member in fission yeast, is directly phosphorylated and controlled by TORC1 and functions as S6 kinase. *J. Cell Sci.* **2012**, *125*, 5840–5849. [CrossRef]
185. Valbuena, N.; Rozalén, A.E.; Moreno, S. Fission yeast TORC1 prevents eIF2α phosphorylation in response to nitrogen and amino acids via Gcn2 kinase. *J. Cell Sci.* **2012**, *125*, 5955–5959. [CrossRef]
186. Weisman, R.; Choder, M. The fission yeast TOR homolog, tor1+, is required for the response to starvation and other stresses via a conserved serine. *J. Biol. Chem.* **2001**, *276*, 7027–7032. [CrossRef]
187. Matsuo, T.; Kubo, Y.; Watanabe, Y.; Yamamoto, M. Schizosaccharomyces pombe AGC family kinase Gad8p forms a conserved signaling module with TOR and PDK1-like kinases. *EMBO J.* **2003**, *22*, 3073–3083. [CrossRef]
188. Ikeda, K.; Morigasaki, S.; Tatebe, H.; Tamanoi, F.; Shiozaki, K. Fission yeast TOR complex 2 activates the AGC-family Gad8 kinase essential for stress resistance and cell cycle control. *Cell Cycle* **2008**, *7*, 358–364. [CrossRef]
189. Duran, R.V.; Hall, M.N. Regulation of TOR by small GTPases. *EMBO Rep.* **2012**, *13*, 121–128. [CrossRef]
190. Mendoza, M.C.; Er, E.E.; Blenis, J. The Ras-ERK and PI3K-mTOR pathways: Cross-talk and compensation. *Trends Biochem. Sci.* **2011**, *36*, 320–328. [CrossRef]
191. Morigasaki, S.; Chin, L.C.; Hatano, T.; Emori, M.; Iwamoto, M.; Tatebe, H.; Shiozaki, K. Modulation of TOR complex 2 signaling by the stress-activated MAPK pathway in fission yeast. *J. Cell Sci.* **2019**, *132*, jcs236133. [CrossRef]
192. Petersen, J.; Hagan, I.M. Polo kinase links the stress pathway to cell cycle control and tip growth in fission yeast. *Nature* **2005**, *435*, 507–512. [CrossRef]
193. Torres, J.; Di Como, C.J.; Herrero, E.; De La Torre-Ruiz, M.A. Regulation of the cell integrity pathway by rapamycin-sensitive TOR function in budding yeast. *J. Biol. Chem.* **2002**, *277*, 43495–43504. [CrossRef]
194. Hatano, T.; Morigasaki, S.; Tatebe, H.; Ikeda, K.; Shiozaki, K. Fission yeast Ryh1 GTPase activates TOR Complex 2 in response to glucose. *Cell Cycle* **2015**, *14*, 848–856. [CrossRef]
195. Zhan, K.; Narasimhan, J.; Wek, R.C. Differential activation of eIF2 kinases in response to cellular stresses in Schizosaccharomyces pombe. *Genetics* **2004**, *168*, 1867–1875. [CrossRef]
196. Berlanga, J.J.; Rivero, D.; Martín, R.; Herrero, S.; Moreno, S.; de Haro, C. Role of mitogen-activated protein kinase Sty1 in regulation of eukaryotic initiation factor 2alpha kinases in response to environmental stress in Schizosaccharomyces pombe. *Eukaryot. Cell* **2010**, *9*, 194–207. [CrossRef]
197. Asano, K. Origin of translational control by eIF2α phosphorylation: Insights from genome-wide translational profiling studies in fission yeast. *Curr. Genet.* **2021**, *67*, 359–368. [CrossRef]
198. Nakashima, A.; Sato, T.; Tamanoi, F. Fission yeast TORC1 regulates phosphorylation of ribosomal S6 proteins in response to nutrients and its activity is inhibited by rapamycin. *J. Cell Sci.* **2010**, *123*, 777–786. [CrossRef]
199. Du, W.; Hálová, L.; Kirkham, S.; Atkin, J.; Petersen, J. TORC2 and the AGC kinase Gad8 regulate phosphorylation of the ribosomal protein S6 in fission yeast. *Biol. Open* **2012**, *1*, 884–888. [CrossRef]
200. Omnus, D.J.; Manford, A.G.; Bader, J.M.; Emr, S.D.; Stefan, C.J. Phosphoinositide kinase signaling controls ER-PM cross-talk. *Mol. Biol. Cell* **2016**, *27*, 1170–1180. [CrossRef]
201. Du, W.; Forte, G.M.; Smith, D.; Petersen, J. Phosphorylation of the amino-terminus of the AGC kinase Gad8 prevents its interaction with TORC2. *Open Biol.* **2016**, *6*, 150189. [CrossRef]
202. Tatebe, H.; Murayama, S.; Yonekura, T.; Hatano, T.; Richter, D.; Furuya, T.; Kataoka, S.; Furuita, K.; Kojima, C.; Shiozaki, K. Substrate specificity of TOR complex 2 is determined by a ubiquitin-fold domain of the Sin1 subunit. *eLife* **2017**, *6*, e19594. [CrossRef]
203. García, R.; Bravo, E.; Diez-Muñiz, S.; Nombela, C.; Rodríguez-Peña, J.M.; Arroyo, J. A novel connection between the Cell Wall Integrity and the PKA pathways regulates cell wall stress response in yeast. *Sci. Rep.* **2017**, *7*, 5703. [CrossRef]
204. Gerik, K.J.; Donlin, M.J.; Soto, C.E.; Banks, A.M.; Banks, I.R.; Maligie, M.A.; Selitrennikoff, C.P.; Lodge, J.K. Cell wall integrity is dependent on the PKC1 signal transduction pathway in Cryptococcus neoformans. *Mol. Microbiol.* **2005**, *58*, 393–408. [CrossRef]
205. Donlin, M.J.; Upadhya, R.; Gerik, K.J.; Lam, W.; VanArendonk, L.G.; Specht, C.A.; Sharma, N.K.; Lodge, J.K. Cross talk between the cell wall integrity and cyclic AMP/protein kinase A pathways in Cryptococcus neoformans. *mBio* **2014**, *5*, e01573-14. [CrossRef]
206. Higuchi, T.; Watanabe, Y.; Yamamoto, M. Protein kinase A regulates sexual development and gluconeogenesis through phosphorylation of the Zn finger transcriptional activator Rst2p in fission yeast. *Mol. Cell. Biol.* **2002**, *22*, 1–11. [CrossRef]
207. Hoffman, C.S. Glucose sensing via the protein kinase A pathway in Schizosaccharomyces pombe. *Biochem. Soc. Trans.* **2005**, *33*, 257–260. [CrossRef]
208. Hoffman, C.S. Except in every detail: Comparing and contrasting G-protein signaling in *Saccharomyces cerevisiae* and *Schizosaccharomyces pombe*. *Eukaryot. Cell* **2005**, *4*, 495–503. [CrossRef]

Journal of *Fungi*

MDPI

Review

The CWI Pathway: A Versatile Toolbox to Arrest Cell-Cycle Progression

Inma Quilis †, **Mercè Gomar-Alba** † and **Juan Carlos Igual** *

Departament de Bioquímica i Biologia Molecular, Institut Universitari de Biotecnologia i Biomedicina (BIOTECMED), Universitat de València, 46100 València, Spain; inmaculada.quilis@uv.es (I.Q.); merce.gomar@uv.es (M.G.-A.)

* Correspondence: jcigual@uv.es

† These authors contributed equally to this work.

Abstract: Cell-signaling pathways are essential for cells to respond and adapt to changes in their environmental conditions. The cell-wall integrity (CWI) pathway of *Saccharomyces cerevisiae* is activated by environmental stresses, compounds, and morphogenetic processes that compromise the cell wall, orchestrating the appropriate cellular response to cope with these adverse conditions. During cell-cycle progression, the CWI pathway is activated in periods of polarized growth, such as budding or cytokinesis, regulating cell-wall biosynthesis and the actin cytoskeleton. Importantly, accumulated evidence has indicated a reciprocal regulation of the cell-cycle regulatory system by the CWI pathway. In this paper, we describe how the CWI pathway regulates the main cell-cycle transitions in response to cell-surface perturbance to delay cell-cycle progression. In particular, it affects the Start transcriptional program and the initiation of DNA replication at the G1/S transition, and entry and progression through mitosis. We also describe the involvement of the CWI pathway in the response to genotoxic stress and its connection with the DNA integrity checkpoint, the mechanism that ensures the correct transmission of genetic material and cell survival. Thus, the CWI pathway emerges as a master brake that stops cell-cycle progression when cells are coping with distinct unfavorable conditions.

Keywords: cell cycle; cell wall integrity; Pkc1; Slt2; checkpoint; DNA damage

Citation: Quilis, I.; Gomar-Alba, M.; Igual, J.C. The CWI Pathway: A Versatile Toolbox to Arrest Cell-Cycle Progression. *J. Fungi* **2021**, *7*, 1041. https://doi.org/10.3390/jof7121041

Academic Editors: María Molina and Humberto Martín

Received: 8 November 2021
Accepted: 2 December 2021
Published: 4 December 2021

Cell proliferation is the basis of life propagation. For this reason, the cell cycle is a crucial process subject to sophisticated molecular control. At the heart of this regulatory system is a family of kinases called CDKs (cyclin-dependent kinases), whose activity oscillates throughout the cell cycle depending on their association with activator proteins (cyclins), inhibitory proteins (CDK inhibitor proteins or CKIs), and the phosphorylation status of certain Thr and Tyr residues. The existing levels of cyclins and CKIs, as well as many other proteins involved in cell-cycle processes, oscillate along the cell cycle. Two main molecular mechanisms are responsible for this: regulation of gene transcription, with the existence of successive waves of gene expression along the different cell-cycle stages, and cell-cycle regulation of protein degradation by the ubiquitin–proteosome pathway [1].

Major control points in cell-cycle regulation occur at the G1/S transition (a key control point in which cells commit to a new round of cell division, initiating DNA replication), the G2/M transition (or entry into mitosis), and the metaphase–anaphase transition, coupled with the exit of mitosis. Errors in cell-cycle processes or cell-cycle progression under inadequate conditions can have catastrophic consequences for the cells. For this reason, eukaryotic cells have developed complex surveillance mechanisms, referred to as checkpoints, to ensure faithful cell division [2]. Checkpoints monitor that internal and external conditions are favorable. Once a deleterious condition is detected, checkpoints transiently arrest cell-cycle progression until the situation is amended. The components of the checkpoint mechanism are revealed by mutations that fail to stop the cell cycle, putting cell viability at risk.

Several signaling pathways transduce external and internal signals to the cell machinery to adapt the cellular physiology. One of the main signaling pathways in *Saccharomyces cerevisiae* is the cell-wall integrity (CWI) pathway. The major functions of the pathway are control of cell-wall biosynthesis and actin cytoskeleton dynamics, responding to environmental stresses and morphogenetic processes that affect the cell surface. Briefly, Pkc1 is activated by GTPase Rho1 after a perturbation in the cell wall is detected by membrane-sensor proteins. Pkc1, in turn, phosphorylates and activates a mitogen-activated protein kinase (MAPK) module comprising the MAPKKK Bck1, the redundant pair MAPKK Mkk1 and Mkk2, and the MAPK Slt2/Mpk1, which ultimately activates the transcriptional factor Rlm1 in order to induce the expression of genes involved in cell-wall biogenesis [3].

1. *PKC1*: A Cell-Cycle Gene?

The relationship between the CWI pathway and the cell cycle has been a remarkable aspect since the beginning of research into this pathway. Protein kinase C (Pkc1) was the first discovered component of the CWI signaling pathway, presented as an essential protein for the yeast cell cycle [4]. As is characteristic for cell division cycle (*cdc*) mutants, *pkc1* mutant cells arrested with a uniform terminal phenotype, which indicates a defect in a stage-specific function of the cell cycle. In particular, *pkc1* mutant cells accumulate in G2 as small budded cells, with a single nucleus and replicated DNA. At that time, it was suggested that *PKC1* might control an unknown checkpoint in the cell cycle. Immediately, subsequent studies reported that the lethality of *pkc1* mutant cells was due to important cell-wall defects, causing cell lysis during budding [5,6]. This raised the question of whether Pkc1 was really a relevant gene for cell-cycle control, or whether its terminal phenotype simply reflected that its defect in cell-wall biosynthesis was critical at a specific point in the cell cycle. During the following years, however, accumulated evidence has supported that the connection of Pkc1 and the CWI pathway with the cell cycle goes beyond the expected cell-cycle regulation of the pathway, given that cell-wall biosynthesis and actin cytoskeleton dynamics are cell-cycle-regulated processes. In fact, the CWI pathway acts as a bridge to interweave the morphogenetic processes that it regulates with the cell-cycle control machinery. Furthermore, it also acts on other aspects of the cell cycle, which are apparently unrelated to the control of the cell wall.

The characteristics and functions of the CWI pathway in cell-wall biogenesis and morphogenesis are very well-known [3]. More recently, other authors have revised other functions of the CWI beyond the cell wall [7,8]. In this review, we focus on the direct mechanisms by which the CWI pathway affects the cell-cycle control system, considering the latest advances made.

2. The CWI Pathway and Start

The G1/S transition, called Start in budding yeast, is a fundamental point in cell-cycle control, in which cells irreversibly commit to a new round of cell division [9]. Start comprises the activation of a transcriptional program which involves the periodic expression of hundreds of genes, including genes coding for cell-cycle regulators and the cellular machinery needed for the S-phase or cell-wall biosynthesis. Two transcription factors—SBF and MBF—regulate this transcriptional wave at G1/S [10]. Both are heterodimeric proteins composed of a DNA-binding factor, Swi4 (SBF) or Mbp1 (MBF), and a common protein, Swi6. In the case of SBF, it has been characterized that Swi6 acts, in part, by alleviating an auto-inhibitory intramolecular interaction in Swi4 that blocks its DNA-binding domain [11]. Although SBF and MBF preferentially regulate a subset of specific genes, there is an important functional redundancy between them [12–14]. The importance of the Start transcriptional program has been evidenced by the lethality of *swi4 swi6* and *swi4 mbp1* double mutants [15]. Strikingly, the *mbp1 swi6* cells are viable, indicating that Swi4 could activate transcription in the absence of Swi6 [16–18].

The CWI pathway has been related to SBF at multiple levels. However, we still lack a complete picture of how it directly regulates the Start transcriptional machinery and the implications that this has for the control of cell-cycle entry.

2.1. SBF and Slt2: Co-Regulation of Cell-Wall Gene Transcription

Genetic and biochemical studies have long connected the CWI pathway with the transcription factor SBF, supporting the idea that SBF is involved in the maintenance of cell integrity. First, the cell lysis defect of an *slt2Δ* mutant was suppressed by *SWI4* overexpression [19], and mutations in both Slt2 and Pkc1 showed synthetic lethality with Swi4 and Swi6 mutations [20]. Second, *swi4Δ* and *swi6Δ* mutants have been shown to be hypersensitive to the cell-wall-perturbing compounds CFW and SDS [20], and the lethality of *swi4Δ* in some backgrounds was suppressed by an osmotic stabilizer [19]. In addition, SBF is responsible for the periodic expression (peaking in the G1/S transition) of cell-wall protein genes, genes which are also regulated by the CWI pathway [20] through its main transcriptional effector, Rlm1 [21]. Interestingly, inactivation of SBF eliminated cell-cycle periodic expression, but did not affect the global mRNA level, whereas CWI pathway mutants showed reduced expression of cell-wall genes without affecting their periodic expression [20]. Therefore, there exists a co-regulation of cell-wall genes, in which the CWI pathway and the cell-cycle transcription factor SBF act in parallel: Slt2 activates Rlm1 to control the global level of expression, whereas SBF modulates this expression throughout the cell cycle. Genomic studies have further confirmed the existence of a group of genes that are coregulated by both SBF and Slt2 [22].

In addition to collaborate in the expression of common target genes, direct relationships between SBF and the CWI pathway have been described. Biochemical studies have shown that Slt2 phosphorylates Swi6 and Swi4 and associates with SBF [19,22]. In fact, it has been proposed that, in the absence of Swi6, Slt2 could control the association of Swi4 with the promoters of a subset of specific genes [22]. Subsequent studies have described that, in response to cell-wall stress, Slt2 and its pseudokinase paralog Mlp1 activate SBF through a noncatalytic mechanism [23–25]. In this mechanism, Slt2 and Mlp1 must be in their active form (phosphorylated), but their kinase activity is not required. Then, Slt2 and Mlp1, in complex with Swi4, associate with the *FKS2* promoter independently of Swi6 [24]. Slt2 associates through a D-motif binding site common in MAPK with the D-domain present in the C-terminal autoinhibitory region of Swi4. As a consequence, the Swi4 DNA-binding domain is released to direct promoter association [25]. Although, in this mechanism, Slt2 replaces Swi6 to allow for Swi4 binding to DNA, Swi6 must be incorporated into the Slt2–Swi4 complex to allow subsequent recruitment of RNA pol II and activation of *FKS2* transcription in response to cell-wall stress [24].

Remarkably, this mechanism is not general for SBF-regulated genes. On the contrary, it only affects a very small number of genes: *FKS2*, *CHA1*, *YLR042c*, and *YKR013* [23,24]. Note that the *FKS2* gene is an exceptional cell-wall gene, as it is only expressed under cell-wall stress conditions and its expression—although Slt2-dependent—is not mediated by Rlm1, as it occurs for the vast majority of cell-wall genes [21], but by the SBF transcription factor [24]. In spite of this, the expression of *FKS2* by SBF is not cell-cycle regulated [26]. Interestingly, in the case of *YKR013w*, this mechanism acts specifically through only one of the three SBF binding sites present in its promoter, the other sites likely being responsible for the periodic expression of the gene at Start [23]. Moreover, although Slt2 and Swi6 both alleviate the Swi4 self-inhibition mechanism for DNA binding, they act through specific and distinct interactions with Swi4, as preventing the interaction of Slt2 with Swi4—which inhibits *FKS2* expression—does not prevent either the association of Swi4 with Swi6, nor SBF-dependent activation of cell-cycle genes [25]. In conclusion, this direct regulation of Swi4 by Slt2 is independent of the role of SBF in cell-cycle-regulated transcription. How this mechanism of Swi4 activation by Slt2 is restricted to a handful of SBF-regulated genes (and SBF binding sites) remains an open question.

2.2. Regulation of the Subcellular Localization of Swi6 by Slt2

In addition to the connections mentioned above, two other mechanisms involving the direct regulation of mediators of the Start transcriptional program by the CWI pathway have functional implications for cell-cycle control. One of them involves controlling the subcellular localization of the transcription factor Swi6. The localization of Swi6 changes throughout the cell cycle: it is nuclear in G1, relocates to the cytosol at G2/M, and re-enters the nucleus at the end of mitosis [27,28]. This regulation along the cell cycle is controlled by the phosphorylation of Ser^{160} by Clb6-Cdc28 kinase [29], which acts by inhibiting a nearby nuclear localization signal (NLS) [30]. Later, it was identified that Swi6 localization also undergoes changes in response to cell-wall stress and that these changes are mediated by Slt2. Under cell-wall stress conditions, there is a transient accumulation of Swi6 inside the nucleus: it rapidly enters the nucleus through its interaction with the Slt2-Swi4 complex discussed above, and then relocates to the cytosol due to phosphorylation of Ser^{238} by Slt2 [31]. This phosphorylation acts by inhibiting a nearby second NLS present in Swi6. As expected, this phosphorylation and the consequent nuclear import blockage has an inhibitory function on the sustained expression of *FKS2* under cell-wall stress conditions [31]. Thus, Swi6 has one NLS that is regulated throughout the cell cycle by Cdc28, and another NLS which is regulated in response to cell-wall stress by Slt2. Slt2-dependent phosphorylation can be detected by Western blot, as the appearance of slow migrating bands [31]. It is interesting to note that this band-shift has also been observed—although to a lesser extent—under basal conditions ([19] and our unpublished results), suggesting that regulation of the subcellular localization of Swi6 by Slt2 may have implications beyond cell-wall stress conditions. In support of this, inactivation of the Slt2-regulated NLS of Swi6 also reduced the expression of the *CLN2*: *lacZ* gene under basal conditions [31]. It is foreseeable that this mechanism could have a more global impact, affecting other genes regulated by Swi6. In short, the MAPK Slt2, by phosphorylating Swi6 to restrain its nuclear import, may act as an inhibitor of the Start transcriptional program. However, this does not seem to be a limiting step in the case of *CLN2* and *CLN1* genes, as inactivation of Slt2 does not affect their expression [19]. Global expression analysis with an Swi6 mutant protein lacking the Slt2-regulated NLS may help in identifying target genes whose expression is affected by this regulatory mechanism, both in basal and cell-wall stress conditions.

2.3. Regulation of the Start Repressor Whi7 by the CWI Pathway

A new direct mechanism that links the CWI pathway with Start has been revealed with the discovery of the Start repressor Whi7. Although SBF is bound to the G1/S gene promoters in early G1, Start transcription is inhibited by the association of transcriptional repressors Whi5 and Whi7, which are functional homologs of Rb in mammalian cells [32–34]. Whi7 also negatively regulates Start through the retention of the Cln3 cyclin in the endoplasmic reticulum membrane [35]. Whi5 and Whi7 show sequence homology and similar cell-cycle-dependent localization, but Whi7 is a less-abundant protein than Whi5, which could explain why Whi5 is the main transcriptional repressor under normal conditions [32,36]. Moreover, both repressors show different preferences for promoter binding: Whi5 is preferentially associated with the G1/S cyclin genes *CLN1* and *CLN2*—key regulators of Start—whereas Whi7 has a preferred association with cell-wall genes [36]; however, Whi7 could substitute for Whi5 in the control of Start when overexpressed [32].

WHI7 expression is upregulated under various stress conditions [37–39]. Under cell-wall stress, an eightfold induction of *WHI7* transcription has been observed. This induction is mediated by Slt2 and the Rlm1 transcription factor, which binds to the *WHI7* promoter [40,41]. Strikingly, inactivation of Slt2 or Rlm1 not only affects cell-wall stress induction, but *WHI7* mRNA levels are also almost depleted under both normal and stress conditions [36]. Consistently, Whi7 protein levels were also decreased in a *pck1* mutant. Thus, *WHI7* is a gene whose expression is totally dependent on the CWI pathway; in particular, on the MAPK Slt2 and Rlm1. This suggests that Whi7 must be an important mediator of CWI pathway functions, specifically in connection with cell-cycle regulation.

As mentioned above, in normal growth conditions, Whi7 plays a modest role as a Start repressor, compared to Whi5. Importantly, this situation changes under cell-wall stress conditions. In response to this stress, *cln3* mutant cells arrest in G1. This arrest is mediated mostly by Whi7 and not by Whi5, as *whi7* deletion is more efficient than *whi5* deletion to override the Start blockage. Similarly, Whi7 and not Whi5 mediates the partial G1 arrest observed in wild-type cells. Thus, under cell-wall stress conditions, Whi7, and not Whi5, becomes the main Start repressor to regulate cell-cycle entry [36].

The gain-of-function of Whi7 under cell-wall stress conditions is totally coherent with the fact that *WHI7* expression is dependent on the CWI integrity pathway and the preference of Whi7, compared to Whi5, for binding to cell-wall gene promoters. Is this gain-of-function simply the result of the stress-dependent accumulation of Whi7 protein? The fact that Whi7 protein levels become similar to those of Whi5 protein (which remain constant during stress) after cell-wall stress treatments provides evidence against a merely quantitative explanation. Alternatively, qualitative differences between the Start repressors probably explain their different abilities to inhibit cell-cycle progression after cell-wall stress. It is possible that, in response to stress, the CWI pathway could increase the intrinsic functionality of Whi7. Furthermore, it is interesting to remark that each repressor has different determinants for binding to promoters. Whi5 binding relies on an integral SBF complex, as binding to any of the Swi4 or Swi6 proteins, as well as association to promoters, is totally destroyed in the absence of the other member of the complex [33,34]. On the contrary, Whi7 can associate with promoters and interact with monomeric Swi4 in the absence of Swi6 [36]. The inhibition of Swi6 import in response to cell-wall stress may result in an enrichment of monomeric Swi4 in SBF-regulated promoters, making regulation more dependent on Whi7 than Whi5. The functional specialization of Whi7 in stress conditions opens the possibility that stress-specific Start repressors may be key elements linking signaling pathways with cell-cycle arrest and subsequent cell-cycle re-entry under adverse conditions.

2.4. The CW/START (Cell-Wall/Start) Checkpoint

A new link between the CWI pathway and the G1/S control, called the cell-wall/Start checkpoint (CW/START) or the Rlm1-dependent checkpoint, has been recently described [42]. The authors noticed that *rlm1* mutant cells growing in a nonfermentable carbon source with low osmolarity underwent the formation of cell aggregates composed of a large mother cell surrounded by several small satellite daughter cells. These aggregates formed when the mother cell progressed through several rounds of cell division, whereas the daughter cells were unviable due to defects in cell integrity and the actin cytoskeleton appearing after cytokinesis, causing their rapid shrinkage and death. In such culture conditions, a strong activation of the CWI pathway, including Rlm1, occurs. This activation is consistent with cell-wall defects and the role of Rlm1-mediated transcription of cell-wall biosynthesis genes in the repair of the damage. It is worth noting that phenotypic analysis of the mother cells revealed that Rlm1 plays another fundamental role as, in its absence, the G1 phase was significantly shortened compared to that of wild-type cells. This observation led the authors to propose that Rlm1 negatively regulates Start when grown under a nonfermentable carbon source and low osmolarity conditions. This environment-specific G1 delay would act as a checkpoint that responds to cellular damage blocking Start, thus allowing for optimal biosynthesis of the cell wall in the poor media before embarking on a new division, in order to guarantee the viability of the future daughter cell.

The mechanism by which Rlm1 regulates Start is unknown, but it does not involve inhibition of the CDK Cdc28 by Tyr^{19} phosphorylation [42]. A good possibility is that Rlm1 might be acting through Whi7, which, as commented above, is a repressor of the Start transcriptional program whose expression depends on Rlm1 [36]. However, *slt2* mutant cells, which also showed depletion of the Whi7 protein [36], lacked the satellite daughter phenotype observed in *rlm1* cells [42]. This is intriguing, given that the activation of Rlm1 is dependent on Slt2, although it cannot be ruled out that, under these particular conditions, Rlm1 can be activated by mechanisms independent of Slt2 and that Whi7 could be involved.

More work is needed to unveil the exact nature of the cellular damage caused by respiration under low osmolarity conditions that triggers this checkpoint, and how that signal reaches Rlm1 in an Slt2-independent way, as well as the Rlm1 targets delaying the G1/S transition.

3. The CWI Pathway and DNA Replication

3.1. The CWI Pathway Regulates the CDK Inhibitor Sic1

A key event in the G1/S transition of the cell cycle is degradation of the CDK inhibitor Sic1. Sic1 blocks the activity of the S-phase CDKs Clb5,6-Cdc28, preventing the activation of DNA replication [43,44]. When the Start transcriptional program is activated, the G1/S CDKs Cln1,2-Cdc28 accumulate and phosphorylate Sic1 at multiple sites, targeting it for degradation mediated by the ubiquitin ligase SCF^{Cdc4} [45,46]. Clb5,6-Cdc28 further phosphorylate Sic1, establishing positive feedback that ensures their abrupt and irreversible activation, and the consequent initiation of DNA replication [47].

The TORC1 complex is the major controller of growth in eukaryotes. The inhibition of TORC1 by rapamycin, which mimics a nutrient deficiency, causes a blockage of cell-cycle progression in G1. This arrest is due—in addition to the decrease in the cellular levels of Cln cyclins—to the increase in Sic1 protein levels. This increase occurs through stabilization of the Sic1 protein caused by its phosphorylation at the Ser^{173} residue [48]. The kinase responsible for such phosphorylation is the MAPK Slt2 of the CWI pathway, this phosphorylation being reversed by the phosphatase $PP2A^{Cdc55}$ [49]. Under normal conditions, TORC1 indirectly inhibits Slt2 [50,51] and activates phosphatase $PP2A^{Cdc55}$ by downregulation of the inhibitory Greatwall (Rim15) endosulfin (Igo1,2) pathway. This maintains Sic1 Ser^{173} in a dephosphorylated state. However, when TORC1 is downregulated by rapamycin or an absence of nutrients, Slt2 kinase activity is induced and the Greatwall–endosulfin pathway is activated to inhibit $PP2A^{Cdc55}$, resulting in the phosphorylation of Ser^{173} and consequent stabilization of Sic1 [49,52]. Accordingly, an *slt2* mutant showed defects in the rapamycin-induced G1 arrest.

The phosphorylation of Sic1 in Ser^{173} could be also relevant in cell-cycle control in proliferating cells under normal conditions. It has been proposed that Rim15 is negatively regulated by Cln-Cdc28 [52,53]. This could facilitate Ser^{173} dephosphorylation by $PP2A^{Cdc55}$, contributing to Sic1 degradation at the G1/S transition. Some results have shown that cells expressing the $Sic1^{T173A}$ mutant protein slightly advance the accumulation of CDK activity [52] and DNA replication [54] after release from an α-factor arrest in the absence of stress. To the best of our knowledge, no effect in DNA replication initiation has been associated with Slt2 inactivation under normal growth conditions, calling into question whether this mechanism—or, at least, the role of Slt2—is critical for the control of Start under optimal conditions. Note that another MAPK, Hog1, phosphorylates Sic1 Ser^{173} to impose a G1 arrest in response to hyperosmotic stress [54]. In any case, by stabilizing Sic1, Slt2 plays an inhibitory role in the initiation of DNA replication, acting as a brake for the G1/S transition when nutrient availability is not adequate.

3.2. The PM/CW (Plasma Membrane/Cell Wall) Damage Checkpoint

The existence of a cell-cycle checkpoint by which cells block the initiation of DNA replication in response to PM/CW damage caused by SDS has been suggested [55]. Two mechanisms that contribute to this response have been characterized. The first one plays a more important role in cell-cycle arrest and involves degradation of the Cdc6 protein. Cdc6 is an essential protein required for DNA replication, which acts in the formation of pre-replicative complexes at origins [56]. Its degradation in response to cell-surface damage is mediated through phosphorylation by the yeast GSK-3 Mck1 at Thr^{39} and Thr^{368}. The second mechanism involves stabilization of the CDK-inhibitor Sic1. In fact, S-phase CDKs' activity is greatly reduced after SDS treatment [55]. These two processes help to block DNA replication and ensure cell survival while membrane damage is under repair.

It is not surprising that the CWI pathway is directly involved in plasma membrane damage repair, through reorganization of the actin cytoskeleton from the growth point toward the area of damage [57] and the transcriptional activation of genes involved in cell-wall biosynthesis [21]. However, more importantly, different observations strongly suggest that the CWI pathway also plays an important role in this checkpoint response. The molecular mechanism underlying Sic1 stabilization in this condition has not been described. However, it is very likely that, as occurs with nutritional [48] or osmotic stress [54], Sic1 is stabilized by phosphorylation in Ser^{173}. Taking into account that treatment with SDS activates the CWI pathway, Slt2 might be the kinase responsible for this phosphorylation, as occurs in the response to nutritional stress. Regarding Cdc6 degradation, phosphorylation of Thr^{368} by Mck1 is dependent on a priming phosphorylation at nearby Ser^{372} by Cdc28 [58]; however, it has been suggested that Cdc28 kinase is not required for Cdc6 degradation in response to plasma membrane damage, and that other kinases can replace Cdc28 to prime Cdc6 phosphorylation by Mck1 in this condition [55]. Slt2 is a good candidate as the phosphorylation sites of CDK and MAPK are similar and, as indicated, Slt2 is activated under this condition. On the other hand, Slt2 positively regulates Mck1 in Ca^{2+} signaling [59]. Thus, it is possible that the CWI pathway could also regulate the degradation of Cdc6 in response to SDS.

It is interesting to note that the cell-cycle arrest after SDS treatment is only partially alleviated by a nonphosphorylatable Cdc6 protein, or is not affected by *sic1* deletion [55]. This could be due to the redundancy of both mechanisms or, alternatively, may point to the existence of additional mechanisms acting to restrain the G1/S transition in response to SDS treatment.

4. The CWI Pathway and the G2/M Transcriptional Program

4.1. Pkc1 Negatively Regulates the G2/M Transcriptional Program

The cell-cycle-regulated transcriptional wave that occurs in late G2 involves a large variety of genes encoding proteins important for mitosis, among them the gene of the major mitotic cyclin, *CLB2*. This transcriptional program is regulated by the transcriptional complex constituted by Mcm1–Fkh2–Ndd1 [10]. The treatment of cells with the Pkc1 inhibitor cercosporamide resulted in increased and advanced expression of the *CLB2* gene cluster, suggesting that Pkc1 could act to negatively regulate this transcriptional program. Pkc1 associates with the promoter of genes of the *CLB2* cluster in a cell-cycle-regulated manner: it is recruited to target promoters during S-phase, while binding declines as cells progress into G2/M when the transcriptional program is turned on. In vitro kinase assays have demonstrated that Pkc1 phosphorylates Ndd1 in two residues: Ser^{520} and Ser^{527}. Interestingly, mutation of these Ser residues to Ala caused an increased and advanced association of Ndd1 to promoters of the *CLB2* cluster genes, and a higher level of gene expression was consistently detected [60]. All of these observations are consistent with Pkc1 inhibiting the expression of *CLB2* cluster genes by phosphorylating Ndd1 transcription factor. Cells expressing the $Ndd1^{S520A,S527A}$ mutant protein progressed through the G2/M transition faster after release from an HU-induced arrest and had severe growth defects in the presence of HU or at high temperature [60]. Thus, Pkc1-dependent phosphorylation of Ndd1 must be important for cell physiology under certain conditions.

4.2. The CWI (Cell-Wall Integrity) Checkpoint

The study of mutant strains in enzymes involved in the biosynthesis of cell-wall components have revealed that, in addition to the expected defect in bud growth, there is a blockage in cell-cycle progression at the G2 phase: cells accumulate with replicated DNA and duplicated but not separated SPB. These observations have led to the proposal of the existence of the CWI checkpoint by which cells respond to alterations in cell-wall structure by inhibiting entry into mitosis [61]. This arrest is associated with an inhibition of mitotic CDK activity caused by a drop in the levels of the major mitotic cyclin, Clb2. In fact, *CLB2* overexpression partially alleviates the cell-cycle arrest induced by cell-wall damage.

The regulation of Clb2 by the CWI checkpoint occurs at the transcriptional level. As commented above, the cell-cycle transcriptional wave in G2 phase that includes the *CLB2* gene is controlled by the Mcm1–Fkh2–Ndd1 transcription complex. Increased levels of Fkh2 [61] and Ndd1 [62] override the checkpoint, indicating that inhibition of Fkh2–Ndd1 mediated transcription is critical for the induced arrest. However, the regulation of Fkh2–Ndd1 by cell-wall damage is not a direct event; rather, the CWI checkpoint regulates another cell-cycle transcription factor: Hcm1. Hcm1 is responsible for the cell-cycle transcriptional wave in S-phase, in which many genes involved in spindle assembly and dynamics, chromosome segregation, or budding are expressed, including the genes for the transcriptional factors *FKH2* and *NDD1* required for the next transcriptional wave [63]. When the CWI checkpoint is triggered by cell-wall damage, the cellular levels of Hcm1 drop dramatically. This regulation occurs through a post-transcriptional mechanism involving the phosphorylation of at least Ser^{61}, Ser^{65}, and Ser^{66}, which likely destabilizes the protein. As a consequence, the transcription of many genes regulated by Hcm1 (including the *FKH2* and *NDD1* genes) is impaired. This leads to inhibition of transcription in G2, resulting in cell-cycle arrest [64]. In this context, the inhibition of Ndd1 by Pkc1 phosphorylation described above is expected to contribute to the checkpoint arrest.

The fact that conditions that activate the CWI checkpoint also activate the CWI pathway, and that these Hcm1 phosphorylation sites are Ser–Pro sites, suggest that the MAPK Slt2 could be involved in these phosphorylation events. In support of this hypothesis, Hcm1 protein levels increase when Slt2 is inactivated and decrease when Slt2 is constitutively activated. Importantly, in the absence of Slt2, Hcm1 is partially stabilized, similar to that observed in the nonphosphorylatable Hcm1 mutant [64]. All these observations have led to the suggestion that MAPK Slt2 triggers Hcm1 degradation by phosphorylation at Ser^{61}, Ser^{65}, and Ser^{66} in response to cell-wall damage, eventually inducing the blockage of cell-cycle progression in G2/M. Supporting this, more recently, it has been reported that mutant strains in all components of the MAPK cascade are defective in blocking cell-cycle progression, as deduced from the presence of SPB separation under cell-wall stress conditions [65]. Future work is necessary to demonstrate whether Hcm1 is indeed a substrate for Slt2 kinase, and to evaluate the relevance of this regulation in the response to cell-wall damage.

5. The CWI Pathway and Mitosis

The key step in mitotic entry is the activation of mitotic CDK through the removal of inhibitory phosphorylation in a specific Tyr residue (Tyr^{19} of Cdc28 in *S. cerevisiae*). This phosphorylation is controlled by the Wee1 kinase (Swe1 in *S. cerevisiae*) and the Cdc25 phosphatase (Mih1 in *S. cerevisiae*). A sophisticated system-level mechanism for the control of mitotic entry in both yeast and mammals ensures step-wise activation of CDK: it is first activated at a low level, in order to enter mitosis, and then strongly activated to progress through mitosis [66–68]. Both Swe1 and Mih1 are hyperphosphorylated at multiple sites in a cell-cycle-regulated manner. Cdc28 partially phosphorylates Swe1 to activate it, thus limiting Clb2-Cdc28 activity to a basal level to initiate entry into mitosis [67,69]. Subsequently, complete hyperphosphorylation of Swe1 by Cdc28 and other kinases will inactivate Swe1, inducing its degradation and, thus, allowing the burst of Cdc28 activity necessary for mitotic progression [69–71]. On the other hand, Mih1 is hyperphosphorylated at the beginning of the cell cycle by casein kinase I. Although the role of this phosphorylation is not clear, it has been suggested that it may have an inhibitory role [72]. In mitotic entry, Mih1 is partially dephosphorylated and undergoes activating phosphorylation by Cdc28 [73]. As can be seen, the regulation of Swe1 and Mih1 is complex, with both proteins undertaking both inhibitory and activating phosphorylations.

5.1. Pkc1 Negatively Regulates the Mih1/Cdc25 Phosphatase

In cells that contains a single mitotic cyclin, Clb2, Mih1 phosphatase is essential for entry into mitosis at elevated temperature [74]. The Mih1 sequence presents four possible Pkc1 phosphorylation consensus sites. Substitution of these Ser by Asp or Glu (Mih1^{3E1D}) reduces the ability of Mih1 to trigger mitotic entry. This is due to the nuclear exclusion of the protein, as evidenced by the fact that the functionality of Mih1^{3E1D} is recovered by forcing its nuclear localization with the SV40 NLS. It could be considered that this mechanism prevents Mih1 from activating Cdc28 by dephosphorylating Tyr^{19}. However, the suppression of the mitotic arrest of the *clb1,3,4 mih1* strain by Cdc28^{Y19F} is very slight, and the difference in the extent of Tyr^{19} phosphorylation between Mih1^{3E1D} and Mih1 is very small, indicating that the blockage of mitotic entry associated with Mih1 inactivation involves other targets besides Cdc28.

Two observations suggest that Pkc1 is the kinase responsible for the phosphorylation of these four Ser residues: first, bands of lower electrophoretic mobility associated with in vivo phosphorylation of these residues (absent in Mih1^{4A}, a mutant protein with the four Ser substituted by Ala) disappear when Pkc1 is inactivated in the cell. Second, *PKC1* overexpression causes Mih1 nuclear exclusion, but has no effect on the Mih1^{4A} mutant protein. These observations suggest that Mih1 is downregulated by Pkc1-dependent phosphorylation, which excludes Mih1 from the nucleus under certain conditions. An attractive hypothesis is that this mechanism could help to prevent premature entry into mitosis until bud formation is complete, as the CWI pathway is activated during the budding process and inactivated in the G2/M transition [75–77]. However, there is no definitive evidence that Mih1 is a substrate for Pkc1 and, in fact, other authors have reported that Pkc1 does not phosphorylate Mih1 in vitro [73]. Therefore, it is necessary to clarify the role of Pkc1 in the phosphorylation of Mih1 in further research.

5.2. The Morphogenesis Checkpoint

The morphogenesis checkpoint has been proposed as a mechanism that blocks the cell cycle in G2/M in response to defects in the actin cytoskeleton caused by various stresses, allowing time for bud formation and growth before cell division [78]. More recently, it has been proposed that the signal that activates the checkpoint is actually a defect in membrane trafficking at sites of polarized growth [79], and that this mechanism could function as a cell sizer to block the G2/M transition until the bud has grown sufficiently [80], although the latter is a matter of controversy [81].

Blockage of cell-cycle progression occurs when Cdc28 is inhibited by phosphorylation of Tyr^{19}, a mechanism regulated by Swe1 tyrosine kinase and Mih1 phosphatase. In the morphogenesis checkpoint, in response to budding defects or alterations of the actin cytoskeleton, Swe1 degradation is deregulated and the protein accumulates to inhibit Cdc28 [78,82]. Stabilization of Swe1 is insufficient to generate a long mitotic delay, but simultaneous inactivation of Mih1 can cause a lethal G2/M blockage. These observations have led to the suggestion that the morphogenesis checkpoint operates both through Swe1 stabilization and Mih1 inhibition [78,83].

The CWI pathway plays an important role in the morphogenesis checkpoint. Mutants in Rho1, Pkc1, and the MAPK Slt2 cascade have been reported to override the G2/M arrest [84]. High Ca^{2+} concentrations also induce Slt2-dependent G2/M arrest [85]. Specifically, acting through GSK-3 Mrc1, Slt2 downregulates Hsl1 kinase [59], which is involved in the control of Swe1 degradation [83]. This mechanism may also be acting in response to disturbances in the actin cytoskeleton. Furthermore, as Slt2 inactivation caused checkpoint defects in an *hsl1* mutant but not in a *mih1* mutant strain, it has been suggested that Slt2 could act by inhibiting Mih1 phosphatase [84].

On the other hand, in the context of the studies connecting bud growth with the control of mitosis, Pkc1 phosphorylates the endosulphin proteins Igo1/Igo2 and Cdc55, in order to prevent their interaction and the consequent inhibition of the PP2A^{Cdc55} phosphatase [86]. PP2A^{Cdc55} regulates the phosphorylation state of both Swe1 and Mih1, playing an ambivalent role in mitosis. PP2A^{Cdc55} partially dephosphorylates Mih1 and counteracts the activating phosphorylation of Swe1 by Cdc28 upon entering mitosis [67,73]. Thus, PP2A^{Cdc55} plays a positive role in mitotic entry, by helping to establish a basal threshold for Cdc28 activity. Therefore, mutants in PP2A^{Cdc55} show a significantly delayed entry into mitosis, inhibiting Cdc28 through phosphorylation at Tyr^{19} [73]. However, in addition to this positive role, PP2A^{Cdc55} negatively regulates mitotic progression. Various studies have indicated that PP2A^{Cdc55} counteracts Cdc28-dependent phosphorylation in many substrates [87,88], which implies a general function as an opposing activity to CDKs in mitosis—a role well-characterized in other systems [89]. Furthermore, PP2A^{Cdc55}, by reverting Cdc28-dependent phosphorylation of Swe1, could avoid the full hyperphosphorylation of Swe1 necessary for its degradation, thus negatively regulating the full activation of Cdc28 in mitosis. In fact, the progression through mitosis depends on PP2A^{Cdc55} inactivation [90]. More important, PP2A^{Cdc55} activity is necessary to inhibit mitotic progress in response to morphogenetic defects that activate the checkpoint [91]. In this context, the activation of PP2A^{Cdc55} by Pkc1 mentioned above may reflect the role of Pkc1 in retaining mitosis through the morphogenesis checkpoint. As PP2A^{Cdc55} has also a positive role in initiating mitosis, it is tempting to speculate that the abovementioned mechanism of Pkc1 downregulating Mih1 might counteract this positive role of PP2A^{Cdc55} when Pkc1 is activated. In fact, restricting the nuclear import of Mih1 causes an extended delay during the morphogenesis checkpoint response [92]. The mutation of Pkc1-phosphorylation sites on Cdc55 and Igo2 had no effect in mitotic progression in nonstressed cells [86]. It would be interesting to test the physiological relevance of Pkc1-dependent regulation of PP2A^{Cdc55} when the morphogenesis checkpoint is in action.

6. A Collection of Checkpoints Related to Cell-Surface Perturbations: A Unified Vision

Checkpoints are surveillance mechanisms that prevent the initiation of a cell-cycle process if abnormalities exist, or when a previous process has not been successfully completed. As described above, in recent years, various checkpoints have been proposed to stop cell-cycle progression in response to different perturbations affecting the cell surface, in which the CWI pathway has been involved. Furthermore, additional mechanisms have been described that support a negative role for the CWI pathway in cell-cycle progression, which are clearly complementary or overlapping with some of the proposed checkpoints, although they have not been named as such.

We propose a unified vision of all these checkpoints and mechanisms. For example, the DNA damage checkpoint responds to different signals depending on the damage, and the DNA replication checkpoint responds to problems in replicative forks. These responses affect different cell-cycle regulators, depending on the stage in the cycle. Strikingly, all of these responses converge through a common core mechanism, composed of a signaling pathway that involves apical kinases ATR/ATM and effector kinases Chk1/Chk2 [93]. Similarly, the various proposed checkpoints related to cell-surface alterations may respond to different signals: actin cytoskeletal alterations or membrane trafficking (morphogenesis checkpoint), alteration of the plasma membrane by SDS (PM/CW checkpoint), or defects in cell-wall biosynthesis (CWI checkpoint or CW/START checkpoint); however, all of them converge in the activation of the CWI pathway, which acts on different effectors to block the cell cycle at different stages (Figure 1). Thus, the CWI pathway emerges as a major brake in the control of the cell cycle when the cell surface is compromised.

The differences observed between the different checkpoints could be due to various causes. A first possibility is that the different treatments used are sensed by different sensors generating different outputs. For instance, in the case of lesions due to *fks1* mutation (CWI checkpoint), the damage is detected by the Sho1-branch of the HOG pathway, which activates Hog1 and, acting through dynactin and the Las17 complex, transduces the signal to the MAPK Slt2 cascade, without requiring the classical Wsc1-3, Mid2, and Mtl1 sensors of the CWI pathway [65]. Wsc1-3 and Mid2 also do not participate in the morphogenesis checkpoint [84]. Classical pathway sensors can act on other stresses, but it has been described that compounds that damage the cell wall use pathway-specific sensors [21]. In this sense, it would be interesting to characterize the molecular basis of the signals generated under different conditions, the components required to transduce them to the CWI pathway, and at which level in the pathway the signal impinges.

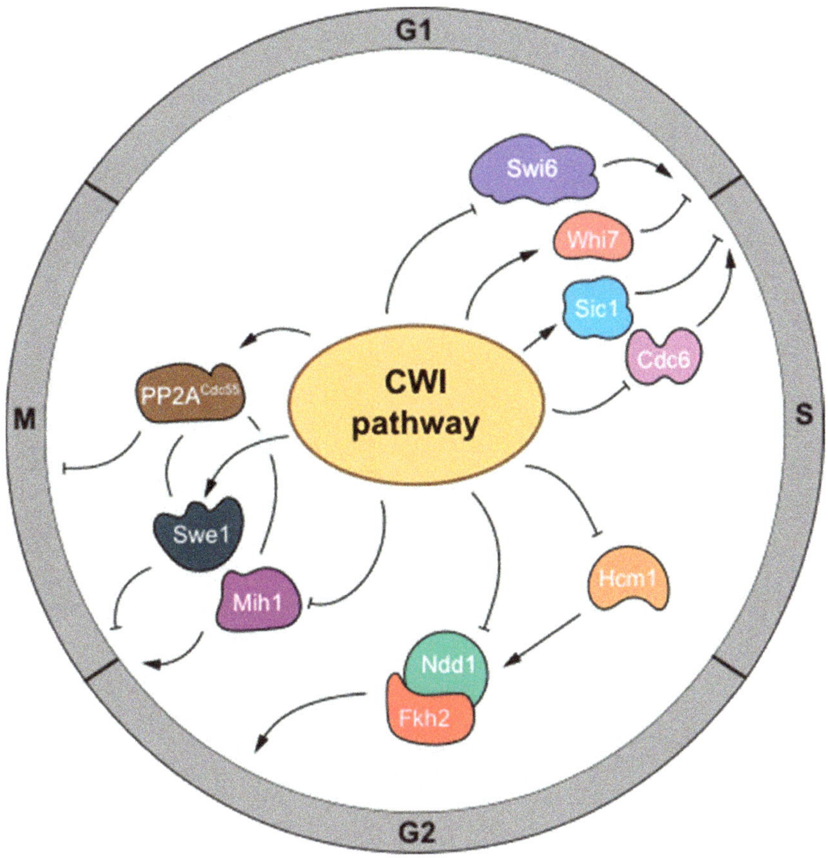

Figure 1. Cell-cycle regulation by the CWI pathway in response to perturbations in the cell surface. This scheme represents the mechanism by which the CWI pathway impinges on cell-cycle regulators to respond to stresses in the cell surface. The CWI pathway has a negative role in the Start transcriptional program, through inducing the expression of the transcriptional repressor Whi7 and promoting the nuclear export of the transcription factor Swi6. It also negatively regulates the initiation of DNA replication, by stabilizing the CKI Sic1 and inhibiting the DNA replication factor Cdc6. The CWI pathway plays a negative role in the G2 transcriptional program, inhibiting the Hcm1 and Ndd1 transcription factors, which results in the impaired expression of mitotic cyclins, among many other mitotic genes. Finally, it affects other aspects of mitotic entry and progression, acting by regulating Swe1, and probably Mih1, to control the inhibition of CDK Cdc28 through Tyr-19 phosphorylation, and by the activation of phosphatase $PP2A^{Cdc55}$, which positively and negatively affects Swe1 and Mih1 and inhibits mitotic progression. Although some molecular details and the specific relevance to cell-cycle regulation of some of these mechanisms are yet to be fully characterized, a scenario has emerged in which the CWI pathway is an important player mediating the arrest of the cell cycle in response to cell-wall/plasma membrane stresses.

Another alternative is that the response would depend on the intensity of the generated signal. For instance, Hog1 activation is induced by cell-wall perturbation in the CWI checkpoint, but well-known downstream effectors of Hog1 are not activated. This is probably due to the fact that Hog1 activation by cell-wall perturbation is very weak, when compared to Hog1 activation by hyperosmotic stress [65]. Similarly, it is possible that, depending on the intensity of the signal generated by a specific cell-surface damage, the response is directed to one or another cell-cycle effector. In this sense, it would be interesting to compare, in parallel, the intensities of the MAPK Slt2 activation in response to the treatments used in the different checkpoints, in order to evaluate this possibility.

Finally, it cannot be ruled out that, in some cases, the differences were caused by the different experimental approaches used to generate the damage. For example, using G1-synchronized cells, it has been reported that SDS (PM/CW checkpoint) blocks DNA replication, while *fks1* mutation (CWI checkpoint) blocks mitosis. This could reflect that the SDS-induced damage was sensed earlier than *fks1*-induced damage. Maybe the inactivation of glucan synthase requires time to generate damage in the cell-wall structure and this damage occurs in the period of bud growth, once the cells have passed Start and activated DNA replication. It would be interesting to carry out an experiment with SDS on post-replicative cells, in order to see whether SDS acts on the G2/M transition or on the next Start, and an experiment with the *fks1* mutation maintaining cells in G1 for enough time to generate cell-wall damage before Start, in order to see whether it could act on DNA replication. Or, is there a homogeneous arrested phenotype when the stress is induced in asynchronous cultures?

It can be expected that future works focused on the connections between the CWI pathway and the control of cell proliferation in response to alterations in the cell surface and morphogenesis will clarify all of these aspects.

7. The CWI Pathway in the Response to DNA Damage

Genetic material is constantly exposed to insults caused by physiological processes, such as replication errors, defective activities of enzymes or reactive oxygen species, or by external physical and chemical agents. To cope with these threats, cells have developed surveillance mechanisms—namely, DNA damage and replication checkpoints, collectively referred to as the DNA integrity checkpoint hereafter—to arrest cell-cycle progression and initiate damage repair [93,94]. The DNA integrity checkpoint has evolved to become a complex signaling network which is evolutionarily well-conserved. The major regulators of this network are apical kinases Mec1 and Tel1 (ATR and ATM in mammals) that, through the adaptor proteins Rad9 and Mrc1 (53BP1 and claspin in mammals), activate the downstream effector kinases Rad53 (Chk2 in mammals) and Chk1. Once activated, DNA damage signaling kinases mediate hallmark responses, such as cell-cycle arrest, inhibition of origin firing, protection and restart of stalled replication forks, induction of a transcriptional response, induction of DNA repair, control of dNTP levels, and the induction of apoptosis (in vertebrates) [93,95].

Several studies have provided evidence supporting the existence of cross-talk between the DNA integrity checkpoint and cellular morphogenesis in budding yeast. DNA integrity checkpoint mutants have aberrant cell morphology, cell wall, and polarized growth [96–98], and there exist proteins that have separable functions in both the cell wall and genome integrity pathways [98,99]. Therefore, it is not surprising that the CWI pathway has been related to the DNA integrity checkpoint in the last decade.

7.1. Slt2 and the DNA Damage Response

First reports connecting Slt2 and the DNA damage response described synthetic genetic interactions between mutations in *SLT2* and DNA damage genes [100,101]. More important, Slt2 is activated in response to different genotoxic stresses, such as treatment with the mutagen methylmetanosulfonate (MMS), hydroxyurea (HU), phleomycin, or UV irradiation [102], suggesting that Slt2 plays an important role in the cellular response to

DNA damage. Slt2 activation involves the phosphatase Msg5, which dephosphorylates Slt2 to maintain it in a low-activity state in the absence of stress. When DNA damage occurs, Msg5 is degraded, leading to Slt2 activation. Although components of the CKI pathway upstream of Slt2 are not directly involved in the mechanism by which DNA damage activates Slt2, a functional pathway is needed. The authors propose that the pathway must have a basal level of activity that will be modulated in response to DNA damage directly at the level of Slt2 by Msg5 [103].

On the other hand, Slt2 is a direct substrate for checkpoint kinases. A genomic study has indicated that Slt2 is phosphorylated at Ser^{423} and Ser^{428} in response to MMS [104]. These same sites were phosphorylated by Rad53 (Ser^{423}) and Mec1/Tel1 (Ser^{428}) when cells were treated with caffeine [25]. As described above, the phosphorylation of Ser^{423} blocks the interaction of Slt2 with Swi4. Although the functional relevance of this phosphorylation in the context of DNA damage is unknown, it has been suggested that it could help redirect Slt2 to targets involved in the DNA damage response.

What are the targets of Slt2 in response to genotoxic stress? A protein candidate to mediate the role of Slt2 in response to DNA damage is Swe1. A morphogenetic function of the DNA integrity checkpoint consists of switching from apical to isotropic bud growth after DNA damage by the degradation of Swe1, mediated by Rad53 [96]. The *slt2* mutant cells have shown hyperpolarized bud morphology and defects in Swe1 degradation under replicative stress, indicating that DNA-damage-induced Swe1 degradation is mediated by Slt2 [102]. This role is apparently contradictory to the results previously described in the morphogenesis checkpoint context as, in that case, Slt2 is supposed to act by stabilizing Swe1. However, it must be taken into account that Swe1 also acts as a downstream effector in the DNA replication checkpoint, blocking cell-cycle progression through the inactivation of mitotic CDK [105]. Apparently, it is difficult to reconcile this role as an important mediator of the checkpoint response with the mentioned checkpoint-induced Swe1 degradation. One hypothesis is that these two aspects could respond to a different timing and/or signal intensity. It is possible that Swe1 is necessary at the beginning of the response to stop proliferation and only later on, when switching off apical growth is required, is Swe1 degraded. This scenario opens the possibility that the positive regulation of Swe1 by Slt2 in the context of the morphogenesis checkpoint, could be also relevant in the DNA integrity checkpoint. Thus, Slt2, as the DNA integrity checkpoint, could play a positive or negative role on Swe1, to stop proliferation or to switch from apical to isotropic bud growth, respectively. In any case, although defects in the control of Swe1 stability have been related to HU sensitivity [106], *slt2* sensitivity to genotoxic agents was not suppressed by *swe1* deletion [102]. This indicates that Slt2 must affect cell survival to genotoxic stress through an Swe1-independent mechanism.

In this sense, our group has observed that *slt2* hypersensitivity to genotoxic stress is, in fact, suppressed by Cyclin C (also known as Cnc1/Ssn8/Ume3/Srb11) deletion (unpublished results). This result mimics what occurs under oxidative stress: cyclin C deletion suppresses the hypersensitivity of *slt2* mutants to oxidative damage [107]. In response to oxidative stress, Slt2 phosphorylates cyclin C, leading to its translocation from the nucleus to the cytoplasm, where it promotes programmed cell death through extensive mitochondrial fragmentation before its degradation [108–110]. It is tempting to speculate that this same mechanism could also occur in the response to genotoxic stress. In support of this idea, an autophagy pathway that is specific to the DNA damage response, called genotoxin-induced targeted autophagy (GTA), has been described [111]; furthermore, Slt2 is required for the optimal induction of autophagy in response to DNA damage, although no specific targets have been identified to date [112].

Another target of Slt2 is the checkpoint adaptor Mrc1. Mrc1 is also a basic regulatory component of the replication complex responsible for maintaining the replication fork progression rate [113] and replication initiation [114]. Slt2 is capable of delaying DNA replication through Mrc1 phosphorylation in response to increased transcription under heat-stress conditions [115]. The authors have proposed the existence of a general S-phase control mechanism mediated by Mrc1, that serves to prevent genomic instability when outbursts of transcription or unscheduled transcription occur during the S-phase [115]. Redundancy between checkpoint adaptor proteins has been proposed recently, suggesting that Mrc1 could also mediate the activation of Rad53 outside of the S-phase [116,117]. This opens the possibility that Mrc1 regulation by Slt2 may be important beyond the S-phase.

Finally, another possible candidate to mediate the role of Slt2 in the DNA damage response could be the Start repressor Whi7. Interestingly, the *WHI7/SRL3* gene has been identified as a suppressor of the *rad53* checkpoint mutant [118] and Whi7 has been localized in gene promoters after HU treatment [119]. This indicates that Whi7 is somehow part of the cellular response to genotoxic stress. As previously mentioned, expression of the *WHI7* gene is almost totally dependent on Slt2 [36,40]. It is reasonable to propose that the regulation of *WHI7* expression by Slt2 could have relevance in the DNA damage response, although this aspect has not yet been tested.

7.2. Pkc1 and the DNA Damage Response

The first CWI pathway component that was related to DNA metabolism was Pkc1 [120]. In particular, a *pkc1* mutant strain showed an elevated rate of mitotic recombination. The recombination increase was not rescued by osmotic stabilizing agents, suggesting that Pkc1 regulates DNA metabolism by an alternative pathway not related to the cell wall. This hyper-recombination phenotype was not observed in an *slt2* mutant, which led to the suggestion that this is a specific function of Pkc1, independent of the MAPK cascade. However, recent works have reported that the inactivation of Slt2 [115] or Bck1 [121] results in a significant increase in foci of Rad52 protein, which is considered a marker for homologue recombination [121]. This suggests that the MAPK cascade is involved in some aspect of recombination. In fact, in the case of the *bck1* mutant, although the appearance of Rad52 foci was not associated with a general hyper-recombination phenotype, an increased recombination rate was indeed observed in rDNA, supporting a region-specific role of Bck1 [121]. This could likely be the case for Slt2 as well, which could explain why the *slt2* mutant was not reported in the first study, which assayed the recombination rate using an *ADE2*-locus sectoring assay.

Another relevant connection of Pkc1 with DNA metabolism is that Pkc1 phosphorylates and activates the CTP synthetase [122,123], an enzyme involved in nucleotide biosynthesis. However, CTP synthetase regulation by Pkc1 phosphorylation may be more complex. CTP synthetase contains four putative Pkc1 phosphorylation consensus sites. The mutation of three of them (Ser^{36}, Ser^{354}, and Ser^{454}) to Ala resulted in reduced enzymatic activity and reduced cellular levels of CTP. On the contrary, mutation of the fourth site (Ser^{330}) to Ala resulted in increased enzymatic activity and increased cellular levels of CTP [124]. Consistent with a role of Pkc1 in nucleotide biosynthesis, a proteomic study detected phosphorylation of the Rnr2 and Rnr4 subunits of ribonucleotide reductase when a hyperactive Pkc1 is overexpressed [125]. It would be interesting to analyze which residues of CTP synthetase are phosphorylated in vivo by Pkc1 and to determine the effect of this phosphorylation, as well as Rnr2/Rnr4 phosphorylation, and Pkc1 inactivation, on the cellular levels of nucleotide pools. This will help to obtain a better picture of how Pkc1 is implied in nucleotide biosynthesis regulation, an essential issue in the cellular response and survival to genotoxic stress.

On the other hand, the *pkc1* mutant is hypersensitive to different compounds that cause DNA damage, such as methyl metanosulfonate (MMS), 4-nitroquinoline 1-oxide (4NQO) [126], bleomycin [127], and hydroxyurea (HU) [101]. This clearly supports the role of Pkc1 in the cellular response to genotoxic stress. Further supporting this, Pkc1 is phosphorylated in response to DNA damage in a manner dependent on the checkpoint kinases Tel1/Mec1 [128,129]. Under genotoxic stress conditions, Mec1/Tel1 mediate the interaction between Pkc1 and the casein kinase 1 Hrr25. Inactivation of Hrr25 or mutation of Hrr25 consensus sites in Pkc1 abolish Pkc1 phosphorylation, suggesting that Hrr25 is the bona fide kinase responsible for Pkc1 modification [128]. Cells expressing a mutant

Pkc1 protein with all the Hrr25 consensus sites mutated showed the same viability as wild-type cells under HU and UV treatments, questioning the physiological relevance of Pkc1 phosphorylation to survive under stress conditions. However, these cells and cells depleted for Pkc1 manifested a defect in the expression of the *RNR3* gene (coding for a large subunit of ribonucleotide reductase), both under basal and genotoxic stress conditions [128]. This result points to a role of Pkc1 in the DNA damage transcriptional response, although the nature of this connection is unknown at present.

Importantly, Pkc1 has also been related to the activation of the DNA integrity checkpoint. Our group has described that Pkc1 activity is required for the activation of the DNA integrity checkpoint in response to different genotoxic stresses in conditional and deleted *pkc1* mutant strains [129]. This result has been questioned by the group of Dr. Levin, who described checkpoint activation in distinct *pkc1*Δ strains [103]. We do not know the exact reason for this apparent discrepancy, but it could be due to the different experimental conditions used. Liu and Levin used harder stress conditions involving longer incubation times or higher doses of the genotoxic agents than normally used. We have reproduced their conditions with a *pkc1*Δ strain and observed that, although checkpoint activation could be detected, it was severely impaired compared to the wild-type strain. In fact, accurate inspection of the Western blots in [103] showed a mild (but detectable) effect of Pkc1 deletion on checkpoint activation when compared to wild-type strains. A way to reconciliate both results is that Pkc1 was not strictly necessary, but may be required for optimal activation of the checkpoint. Further work is needed to definitively clarify this point.

A fact that supports DNA integrity checkpoint control by PKC is that this mechanism is not restricted to yeast, but must be a general trait of eukaryotic cells. Humanized yeast expressing PKCδ revealed that this mammalian isoform is able to rescue the checkpoint defect of a *pkc1* mutant strain [129]. Moreover, PKCδ activity is relevant for checkpoint activation by DNA damage in human HeLa cells [129], as well as in mouse NSC (neural stem cells) and ES (embryonic stem) cells, as deduced from the diminished activation of checkpoint effector kinases in the absence of PKCδ (our unpublished results). Other observations support a link between PKCδ and the DNA damage checkpoint. For instance, in the presence of DNA damage, the ATM checkpoint kinase functions upstream of PKCδ activation, inducing PKCδ translocation into the nucleus where it targets the hRad9 protein, a component of the 9-1-1 checkpoint clamp complex, in order to activate apoptosis [130]. PKCδ has also been situated upstream of nuclear DNA-PK and ATM [131,132], and blocking its activity inhibited the phosphorylation of ATM and histone H2AX [131,133]. Moreover, overexpression of PKCδ induced S-phase arrest and activation of the DNA integrity checkpoint [134]. Finally, a recent functional proteomic analysis has proposed additional targets of PKCδ involved in the DNA damage response, DNA repair, and cell-cycle checkpoint activation [135]. The role of PKCδ in regulating cell survival or death has been under discussion for years, but an increasing collection of studies is now illustrating its ability to induce cell-cycle arrest in response to DNA damage at the G1 phase, through the induction of p53 and p21 [136–138], S-phase [134], or G2/M phase [139]. All of these observations reinforce the idea that PKC acts in the DNA integrity checkpoint response, in order to brake cell-cycle progression.

In conclusion, the CWI pathway and, more specifically, Pkc1 and the MAPK Slt2 are new players in the cellular response to DNA damage, although their roles are far from being fully understood. In this study, we collected evidence and insights on the possible functions of both proteins, as summarized in Figure 2. Future work will clarify the molecular basis and the relevance for cell-cycle regulation and cell survival of Pkc1, as well as the action of Slt2 in response to DNA damage.

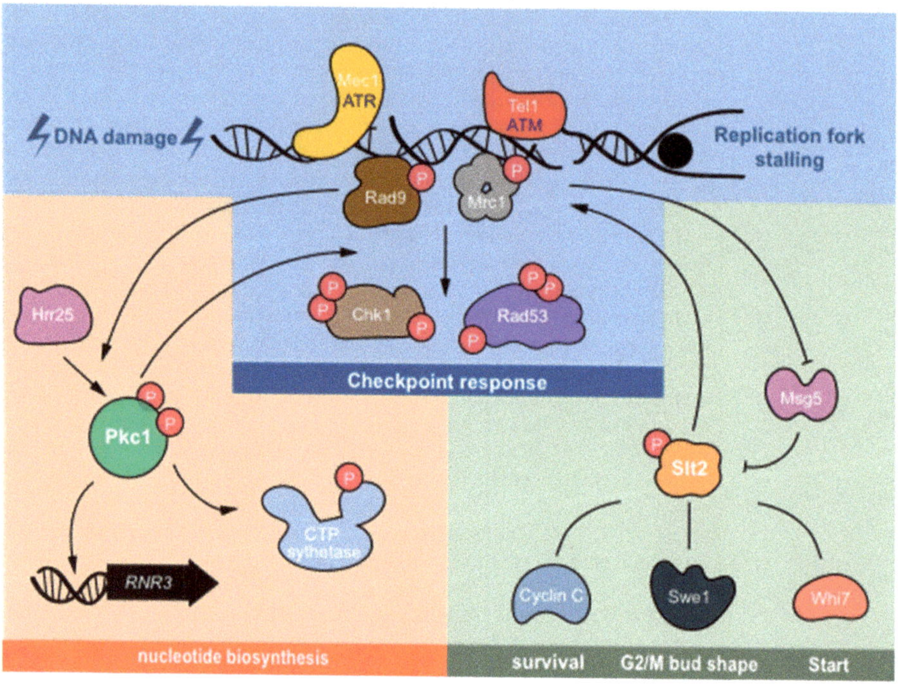

Figure 2. Pkc1 and Slt2 connections with the DNA damage response. This scheme represents the possible roles of Pkc1 and Slt2 related to the response to genotoxic stress. Pkc1 participates in the optimal activation of the DNA integrity checkpoint upstream or downstream of the Mec1/Tel1 apical kinases. Mec1/Tel1, in turn, mediate phosphorylation of Pkc1 by Hrr25 kinase after DNA damage. This regulation is involved in the expression of the *RNR3* gene. Pkc1 is also linked to nucleotide biosynthesis, by regulating the CTP synthetase. Slt2 is activated by genotoxic stress through the inactivation of phosphatase Msg5. Possible Slt2 effectors include: Swe1, which is important for bud morphogenesis and G2/M arrest, whose DNA-damage induced degradation depends on Slt2; Cyclin C, an inducer of programmed cell death responsible for the hypersensitivity of *slt2* mutants to genotoxic treatments; Whi7, a Start repressor regulated by Slt2 whose overexpression suppresses the checkpoint mutation; and Mrc1, a DNA integrity checkpoint adaptor protein that is regulated by Slt2 to stop DNA replication under certain conditions. Although the details of most of these connections are unknown, a scenario is beginning to emerge in which Pkc1 and Slt2 are important players in the response to genotoxic stress.

8. Concluding Remarks

Cells use signaling pathways to detect and adapt to adverse conditions. The CWI pathway is activated in response to a wide variety of stimuli coming from the outermost structure (the cell wall) to the heart of the nucleus (the genetic material). In response to signals as diverse as plasma membrane/cell-wall damage caused by different compounds, defects in enzymatic activities involved in the biosynthesis of cell-wall components, defects in the actin cytoskeleton or membrane trafficking at the sites of polarized growth, growth in poor medium and low osmolarity, nutrient deficiencies, high temperature, or damage to the DNA and replicative stress, the CWI pathway is able to act on the cell cycle as a canonical checkpoint regulator: it arrests cell-cycle progression until the damage is repaired and the cell can resume cell division with all the guarantees for cell survival. In this scenario, the CWI pathway has been revealed as a versatile toolbox for the cell, with different mechanisms acting on a diverse group of cell-cycle regulators at specific cell-cycle transitions.

Author Contributions: Conceptualization: I.Q., M.G.-A. and J.C.I.; Writing—Original Draft Preparation, I.Q., M.G.-A. and J.C.I.; Writing—Review & Editing, I.Q., M.G.-A. and J.C.I.; Visualization, J.C.I.; Supervision, J.C.I.; Project Administration, J.C.I.; Funding Acquisition, J.C.I. All authors have read and agreed to the published version of the manuscript.

Funding: This work was funded by the Spanish Government and co-financed by ERDF from the European Union, grant numbers BFU2017-88692 and PID2020-119793GB-I00.

Institutional Review Board Statement: Not applicable.

Informed Consent Statement: Not applicable.

Acknowledgments: We thank M. Carmen Bañó for her useful comments, Ana Domingo for her collaboration in the elaboration of the figures, and all members of our research group.

Conflicts of Interest: The authors declare no conflict of interest.

References

1. Morgan, D.O. The Cell-Cycle Control System. In *The Cell Cycle: Principles of Control*; New Science Press Ltd.: London, UK, 2007; pp. 28–54.
2. Hartwell, L.H.; Weinert, T.A. Checkpoints: Controls that ensure the order of cell cycle events. *Science* **1989**, *246*, 629–634. [CrossRef] [PubMed]
3. Levin, D.E. Regulation of cell wall biogenesis in *Saccharomyces Cerevisiae*: The cell wall integrity signaling pathway. *Genetics* **2011**, *189*, 1145–1175. [CrossRef] [PubMed]
4. Levin, D.E.; Fields, F.O.; Kunisawa, R.; Bishop, J.M.; Thorner, J. A candidate protein kinase C gene, *PKC1*, is required for the *S. cerevisiae* cell cycle. *Cell* **1990**, *62*, 213–224. [CrossRef]
5. Levin, D.E.; Bartlett-Heubusch, E. Mutants in the *S. cerevisiae PKC1* gene display a cell cycle-specific osmotic stability defect. *J. Cell Biol.* **1992**, *116*, 1221–1229. [CrossRef] [PubMed]
6. Paravicini, G.; Cooper, M.; Friedli, L.; Smith, D.J.; Carpentier, J.L.; Klig, L.S.; Payton, M.A. The osmotic integrity of the yeast cell requires a functional *PKC1* gene product. *Mol. Cell. Biol.* **1992**, *12*, 4896–4905.
7. Heinisch, J.J.; Rodicio, R. Protein kinase C in fungi-more than just cell wall integrity. *FEMS Microbiol. Rev.* **2018**, *42*, 22–39. [CrossRef]
8. Jimenez-Gutierrez, E.; Alegria-Carrasco, E.; Sellers-Moya, A.; Molina, M.; Martin, H. Not just the wall: The other ways to turn the yeast CWI pathway on. *Int. Microbiol.* **2020**, *23*, 107–119. [CrossRef]
9. Johnson, A.; Skotheim, J.M. Start and the restriction point. *Curr. Opin. Cell Biol.* **2013**, *25*, 717–723. [CrossRef]
10. Haase, S.B.; Wittenberg, C. Topology and control of the cell-cycle-regulated transcriptional circuitry. *Genetics* **2014**, *196*, 65–90. [CrossRef]
11. Baetz, K.; Andrews, B. Regulation of cell cycle transcription factor Swi4 through auto-inhibition of DNA binding. *Mol. Cell. Biol.* **1999**, *19*, 6729–6741. [CrossRef]
12. Bastos de Oliveira, F.M.; Harris, M.R.; Brazauskas, P.; de Bruin, R.A.; Smolka, M.B. Linking DNA replication checkpoint to MBF cell-cycle transcription reveals a distinct class of G1/S genes. *EMBO J.* **2012**, *31*, 1798–1810. [CrossRef]
13. Bean, J.M.; Siggia, E.D.; Cross, F.R. High functional overlap between Mlu1 cell-cycle box binding factor and Swi4/6 cell-cycle box binding factor in the G1/S transcriptional program in *Saccharomyces cerevisiae*. *Genetics* **2005**, *171*, 49–61. [CrossRef] [PubMed]
14. Ferrezuelo, F.; Colomina, N.; Futcher, B.; Aldea, M. The transcriptional network activated by Cln3 cyclin at the G1-to-S transition of the yeast cell cycle. *Genome Biol.* **2010**, *11*, R67. [CrossRef] [PubMed]
15. Koch, C.; Moll, T.; Neuberg, M.; Ahorn, H.; Nasmyth, K. A role for the transcription factors Mbp1 and Swi4 in progression from G1 to S phase. *Science* **1993**, *261*, 1551–1557. [CrossRef] [PubMed]
16. Adames, N.R.; Schuck, P.L.; Chen, K.C.; Murali, T.M.; Tyson, J.J.; Peccoud, J. Experimental testing of a new integrated model of the budding yeast Start transition. *Mol. Biol. Cell* **2015**, *26*, 3966–3984. [CrossRef]
17. Nasmyth, K.; Dirick, L. The role of Swi4 and Swi6 in the activity of G1 cyclins in yeast. *Cell* **1991**, *66*, 995–1013. [CrossRef]
18. Wijnen, H.; Landman, A.; Futcher, B. The G(1) cyclin Cln3 promotes cell cycle entry via the transcription factor Swi6. *Mol. Cell. Biol.* **2002**, *22*, 4402–4418. [CrossRef]
19. Madden, K.; Sheu, Y.J.; Baetz, K.; Andrews, B.; Snyder, M. SBF cell cycle regulator as a target of the yeast PKC-MAP kinase pathway. *Science* **1997**, *275*, 1781–1784. [CrossRef]
20. Igual, J.C.; Johnson, A.L.; Johnston, L.H. Coordinated regulation of gene expression by the cell cycle transcription factor Swi4 and the protein kinase C MAP kinase pathway for yeast cell integrity. *EMBO J.* **1996**, *15*, 5001–5013. [CrossRef]
21. Sanz, A.B.; Garcia, R.; Rodriguez-Pena, J.M.; Arroyo, J. The CWI pathway: Regulation of the transcriptional adaptive response to cell wall stress in yeast. *J. Fungi* **2017**, *4*, 1. [CrossRef] [PubMed]
22. Baetz, K.; Moffat, J.; Haynes, J.; Chang, M.; Andrews, B. Transcriptional coregulation by the cell integrity mitogen-activated protein kinase Slt2 and the cell cycle regulator Swi4. *Mol. Cell. Biol.* **2001**, *21*, 6515–6528. [CrossRef]

23. Kim, K.Y.; Levin, D.E. Transcriptional reporters for genes activated by cell wall stress through a non-catalytic mechanism involving Mpk1 and SBF. *Yeast* **2010**, *27*, 541–548. [CrossRef] [PubMed]
24. Kim, K.Y.; Truman, A.W.; Levin, D.E. Yeast Mpk1 mitogen-activated protein kinase activates transcription through Swi4/Swi6 by a noncatalytic mechanism that requires upstream signal. *Mol. Cell. Biol.* **2008**, *28*, 2579–2589. [CrossRef] [PubMed]
25. Truman, A.W.; Kim, K.Y.; Levin, D.E. Mechanism of Mpk1 mitogen-activated protein kinase binding to the Swi4 transcription factor and its regulation by a novel caffeine-induced phosphorylation. *Mol. Cell. Biol.* **2009**, *29*, 6449–6461. [CrossRef]
26. Mazur, P.; Morin, N.; Baginsky, W.; el-Sherbeini, M.; Clemas, J.A.; Nielsen, J.B.; Foor, F. Differential expression and function of two homologous subunits of yeast 1,3-beta-D-glucan synthase. *Mol. Cell. Biol.* **1995**, *15*, 5671–5681. [CrossRef] [PubMed]
27. Sidorova, J.M.; Mikesell, G.E.; Breeden, L.L. Cell cycle-regulated phosphorylation of Swi6 controls its nuclear localization. *Mol. Biol. Cell* **1995**, *6*, 1641–1658. [CrossRef]
28. Queralt, E.; Igual, J.C. Cell cycle activation of the Swi6p transcription factor is linked to nucleocytoplasmic shuttling. *Mol. Cell. Biol.* **2003**, *23*, 3126–3140. [CrossRef]
29. Geymonat, M.; Spanos, A.; Wells, G.P.; Smerdon, S.J.; Sedgwick, S.G. Clb6/Cdc28 and Cdc14 regulate phosphorylation status and cellular localization of Swi6. *Mol. Cell. Biol.* **2004**, *24*, 2277–2285. [CrossRef] [PubMed]
30. Harreman, M.T.; Kline, T.M.; Milford, H.G.; Harben, M.B.; Hodel, A.E.; Corbett, A.H. Regulation of nuclear import by phosphorylation adjacent to nuclear localization signals. *J. Biol. Chem.* **2004**, *279*, 20613–20621. [CrossRef]
31. Kim, K.Y.; Truman, A.W.; Caesar, S.; Schlenstedt, G.; Levin, D.E. Yeast Mpk1 cell wall integrity mitogen-activated protein kinase regulates nucleocytoplasmic shuttling of the Swi6 transcriptional regulator. *Mol. Biol. Cell* **2010**, *21*, 1609–1619. [CrossRef] [PubMed]
32. Gomar-Alba, M.; Mendez, E.; Quilis, I.; Bano, M.C.; Igual, J.C. Whi7 is an unstable cell-cycle repressor of the Start transcriptional program. *Nat. Commun.* **2017**, *8*, 329. [CrossRef] [PubMed]
33. de Bruin, R.A.; McDonald, W.H.; Kalashnikova, T.I.; Yates, J., 3rd; Wittenberg, C. Cln3 activates G1-specific transcription via phosphorylation of the SBF bound repressor Whi5. *Cell* **2004**, *117*, 887–898. [CrossRef]
34. Costanzo, M.; Nishikawa, J.L.; Tang, X.; Millman, J.S.; Schub, O.; Breitkreuz, K.; Dewar, D.; Rupes, I.; Andrews, B.; Tyers, M. Cdk activity antagonizes Whi5, an inhibitor of G1/S transcription in yeast. *Cell* **2004**, *117*, 899–913. [CrossRef] [PubMed]
35. Yahya, G.; Parisi, E.; Flores, A.; Gallego, C.; Aldea, M. A Whi7-anchored loop controls the G1 cdk-cyclin complex at Start. *Mol. Cell* **2014**, *53*, 115–126. [CrossRef]
36. Méndez, E.; Gomar-Alba, M.; Bañó, M.C.; Mendoza, M.; Quilis, I.; Igual, J.C. The budding yeast start repressor Whi7 differs in regulation from Whi5 emerging as a major cell cycle brake in response to stress. *J. Cell Sci.* **2020**, *133*, jcs251413. [CrossRef] [PubMed]
37. Waern, K.; Snyder, M. Extensive transcript diversity and novel upstream open reading frame regulation in yeast. *G3 Genes | Genomes | Genet.* **2013**, *3*, 343–352. [CrossRef] [PubMed]
38. Gasch, A.P.; Spellman, P.T.; Kao, C.M.; Carmel-Harel, O.; Eisen, M.B.; Storz, G.; Botstein, D.; Brown, P.O. Genomic expression programs in the response of yeast cells to environmental changes. *Mol. Biol. Cell* **2000**, *11*, 4241–4257. [CrossRef]
39. Garcia, R.; Bermejo, C.; Grau, C.; Perez, R.; Rodriguez-Pena, J.M.; Francois, J.; Nombela, C.; Arroyo, J. The global transcriptional response to transient cell wall damage in *Saccharomyces cerevisiae* and its regulation by the cell integrity signaling pathway. *J. Biol. Chem.* **2004**, *279*, 15183–15195. [CrossRef]
40. Sanz, A.B.; Garcia, R.; Rodriguez-Pena, J.M.; Diez-Muniz, S.; Nombela, C.; Peterson, C.L.; Arroyo, J. Chromatin remodeling by the SWI/SNF complex is essential for transcription mediated by the yeast cell wall integrity MAPK pathway. *Mol. Biol. Cell* **2012**, *23*, 2805–2817. [CrossRef]
41. Sanz, A.B.; García, R.; Rodríguez-Peña, J.M.; Nombela, C.; Arroyo, J. Slt2 MAPK association with chromatin is required for transcriptional activation of Rlm1 dependent genes upon cell wall stress. *BBA Gene Regul. Mech.* **2018**, *1861*, 1029–1039. [CrossRef] [PubMed]
42. Piccirillo, S.; Neog, D.; Spade, D.; Van Horn, J.D.; Tiede-Lewis, L.M.; Dallas, S.L.; Kapros, T.; Honigberg, S.M. Shrinking daughters: Rlm1-dependent G1/S checkpoint maintains *Saccharomyces cerevisiae* daughter cell size and viability. *Genetics* **2017**, *206*, 1923–1938. [CrossRef] [PubMed]
43. Schneider, B.L.; Yang, Q.H.; Futcher, A.B. Linkage of replication to Start by the CDK inhibitor Sic1. *Science* **1996**, *272*, 560–562. [CrossRef]
44. Schwob, E.; Böhm, T.; Mendenhall, M.D.; Nasmyth, K. The B-type cyclin kinase inhibitor p40SIC1 controls the G1 to S transition in *S. cerevisiae*. *Cell* **1994**, *79*, 233–244. [CrossRef]
45. Nash, P.; Tang, X.; Orlicky, S.; Chen, Q.; Gertler, F.B.; Mendenhall, M.D.; Sicheri, F.; Pawson, T.; Tyers, M. Multisite phosphorylation of a CDK inhibitor sets a threshold for the onset of DNA replication. *Nature* **2001**, *414*, 514–521. [CrossRef]
46. Verma, R.; Annan, R.S.; Huddleston, M.J.; Carr, S.A.; Reynard, G.; Deshaies, R.J. Phosphorylation of Sic1p by G1 CDK required for its degradation and entry into S phase. *Science* **1997**, *278*, 455–460. [CrossRef]
47. Kõivomägi, M.; Valk, E.; Venta, R.; Iofik, A.; Lepiku, M.; Balog, E.R.; Rubin, S.M.; Morgan, D.O.; Loog, M. Cascades of multisite phosphorylation control Sic1 destruction at the onset of S phase. *Nature* **2011**, *480*, 128–131. [CrossRef]
48. Zinzalla, V.; Graziola, M.; Mastriani, A.; Vanoni, M.; Alberghina, L. Rapamycin-mediated G1 arrest involves regulation of the CDK inhibitor Sic1 in *Saccharomyces cerevisiae*. *Mol. Microbiol.* **2007**, *63*, 1482–1494. [CrossRef] [PubMed]

49. Moreno-Torres, M.; Jaquenoud, M.; De Virgilio, C. TORC1 controls G1-S cell cycle transition in yeast via Mpk1 and the greatwall kinase pathway. *Nat. Commun.* **2015**, *6*, 8256. [CrossRef]
50. Soulard, A.; Cremonesi, A.; Moes, S.; Schütz, F.; Jenö, P.; Hall, M.N. The rapamycin-sensitive phosphoproteome reveals that TOR controls protein kinase A toward some but not all substrates. *Mol. Biol. Cell* **2010**, *21*, 3475–3486. [CrossRef] [PubMed]
51. Torres, J.; Di Como, C.J.; Herrero, E.; De La Torre-Ruiz, M.A. Regulation of the cell integrity pathway by rapamycin-sensitive TOR function in budding yeast. *J. Biol. Chem.* **2002**, *277*, 43495–43504. [CrossRef]
52. Moreno-Torres, M.; Jaquenoud, M.; Peli-Gulli, M.P.; Nicastro, R.; De Virgilio, C. TORC1 coordinates the conversion of Sic1 from a target to an inhibitor of cyclin-CDK-Cks1. *Cell Discov.* **2017**, *3*, 17012. [CrossRef] [PubMed]
53. Talarek, N.; Gueydon, E.; Schwob, E. Homeostatic control of Start through negative feedback between Cln3-CDK1 and Rim15/Greatwall kinase in budding yeast. *eLife* **2017**, *6*, e26233. [CrossRef]
54. Escoté, X.; Zapater, M.; Clotet, J.; Posas, F. Hog1 mediates cell-cycle arrest in G1 phase by the dual targeting of Sic1. *Nat. Cell Biol.* **2004**, *6*, 997–1002. [CrossRef] [PubMed]
55. Kono, K.; Al-Zain, A.; Schroeder, L.; Nakanishi, M.; Ikui, A.E. Plasma membrane/cell wall perturbation activates a novel cell cycle checkpoint during G1 in *Saccharomyces cerevisiae*. *Proc. Natl. Acad. Sci. USA* **2016**, *113*, 6910–6915. [CrossRef]
56. Bell, S.P.; Dutta, A. DNA replication in eukaryotic cells. *Annu. Rev. Biochem.* **2002**, *71*, 333–374. [CrossRef]
57. Kono, K.; Saeki, Y.; Yoshida, S.; Tanaka, K.; Pellman, D. Proteasomal degradation resolves competition between cell polarization and cellular wound healing. *Cell* **2012**, *150*, 151–164. [CrossRef]
58. Al-Zain, A.; Schroeder, L.; Sheglov, A.; Ikui, A.E. Cdc6 degradation requires phosphodegron created by GSK-3 and CDK1 for SCF^{Cdc4} recognition in *Saccharomyces cerevisiae*. *Mol. Biol. Cell* **2015**, *26*, 2609–2619. [CrossRef]
59. Mizunuma, M.; Hirata, D.; Miyaoka, R.; Miyakawa, T. GSK-3 kinase Mck1 and calcineurin coordinately mediate Hsl1 down-regulation by Ca^{2+} in budding yeast. *EMBO J.* **2001**, *20*, 1074–1085. [CrossRef]
60. Darieva, Z.; Han, N.; Warwood, S.; Doris, K.S.; Morgan, B.A.; Sharrocks, A.D. Protein kinase C regulates late cell cycle-dependent gene expression. *Mol. Cell. Biol.* **2012**, *32*, 4651–4661. [CrossRef]
61. Suzuki, M.; Igarashi, R.; Sekiya, M.; Utsugi, T.; Morishita, S.; Yukawa, M.; Ohya, Y. Dynactin is involved in a checkpoint to monitor cell wall synthesis in *Saccharomyces cerevisiae*. *Nat. Cell Biol.* **2004**, *6*, 861–871. [CrossRef] [PubMed]
62. Sekiya, M.; Nogami, S.; Ohya, Y. Transcription factors of M-phase cyclin Clb2 in the yeast cell wall integrity checkpoint. *Genes Genet. Syst.* **2009**, *84*, 269–276. [CrossRef] [PubMed]
63. Pramila, T.; Wu, W.; Miles, S.; Noble, W.S.; Breeden, L.L. The forkhead transcription factor Hcm1 regulates chromosome segregation genes and fills the S-phase gap in the transcriptional circuitry of the cell cycle. *Genes Dev.* **2006**, *20*, 2266–2278. [CrossRef]
64. Negishi, T.; Veis, J.; Hollenstein, D.; Sekiya, M.; Ammerer, G.; Ohya, Y. The late S-phase transcription factor Hcm1 is regulated through phosphorylation by the cell wall integrity checkpoint. *Mol. Cell. Biol.* **2016**, *36*, 941–953. [CrossRef]
65. Sukegawa, Y.; Negishi, T.; Kikuchi, Y.; Ishii, K.; Imanari, M.; Ghanegolmohammadi, F.; Nogami, S.; Ohya, Y. Genetic dissection of the signaling pathway required for the cell wall integrity checkpoint. *J. Cell Sci.* **2018**, *131*, jcs219063. [CrossRef]
66. Deibler, R.W.; Kirschner, M.W. Quantitative reconstitution of mitotic CDK1 activation in somatic cell extracts. *Mol. Cell* **2010**, *37*, 753–767. [CrossRef]
67. Harvey, S.L.; Enciso, G.; Dephoure, N.; Gygi, S.P.; Gunawardena, J.; Kellogg, D.R. A phosphatase threshold sets the level of CDK1 activity in early mitosis in budding yeast. *Mol. Biol. Cell* **2011**, *22*, 3595–3608. [CrossRef]
68. Lindqvist, A.; van Zon, W.; Karlsson Rosenthal, C.; Wolthuis, R.M. Cyclin B1-CDK1 activation continues after centrosome separation to control mitotic progression. *PLoS Biol.* **2007**, *5*, e123. [CrossRef] [PubMed]
69. Harvey, S.L.; Charlet, A.; Haas, W.; Gygi, S.P.; Kellogg, D.R. Cdk1-dependent regulation of the mitotic inhibitor Wee1. *Cell* **2005**, *122*, 407–420. [CrossRef]
70. Asano, S.; Park, J.E.; Sakchaisri, K.; Yu, L.R.; Song, S.; Supavilai, P.; Veenstra, T.D.; Lee, K.S. Concerted mechanism of Swe1/Wee1 regulation by multiple kinases in budding yeast. *EMBO J.* **2005**, *24*, 2194–2204. [CrossRef] [PubMed]
71. McMillan, J.N.; Theesfeld, C.L.; Harrison, J.C.; Bardes, E.S.; Lew, D.J. Determinants of Swe1p degradation in *Saccharomyces cerevisiae*. *Mol. Biol. Cell* **2002**, *13*, 3560–3575. [CrossRef] [PubMed]
72. Wicky, S.; Tjandra, H.; Schieltz, D.; Yates, J., 3rd; Kellogg, D.R. The Zds proteins control entry into mitosis and target protein phosphatase 2A to the Cdc25 phosphatase. *Mol. Biol. Cell* **2011**, *22*, 20–32. [CrossRef]
73. Pal, G.; Paraz, M.T.; Kellogg, D.R. Regulation of Mih1/Cdc25 by protein phosphatase 2A and casein kinase 1. *J. Cell Biol.* **2008**, *180*, 931–945. [CrossRef] [PubMed]
74. Yano, K.; Uesono, Y.; Yoshida, S.; Kikuchi, A.; Kashiwazaki, J.; Mabuchi, I.; Kikuchi, Y. Mih1/Cdc25 is negatively regulated by Pkc1 in *Saccharomyces cerevisiae*. *Genes Cells* **2013**, *18*, 425–441. [CrossRef]
75. Gray, J.V.; Ogas, J.P.; Kamada, Y.; Stone, M.; Levin, D.E.; Herskowitz, I. A role for the Pkc1 MAP kinase pathway of *Saccharomyces cerevisiae* in bud emergence and identification of a putative upstream regulator. *EMBO J.* **1997**, *16*, 4924–4937. [CrossRef] [PubMed]
76. Kono, K.; Nogami, S.; Abe, M.; Nishizawa, M.; Morishita, S.; Pellman, D.; Ohya, Y. G1/S cyclin-dependent kinase regulates small GTPase Rho1p through phosphorylation of RhoGEF Tus1p in *Saccharomyces cerevisiae*. *Mol. Biol. Cell* **2008**, *19*, 1763–1771. [CrossRef]
77. Zarzov, P.; Mazzoni, C.; Mann, C. The Slt2(Mpk1) MAP kinase is activated during periods of polarized cell growth in yeast. *EMBO J.* **1996**, *15*, 83–91. [CrossRef] [PubMed]

78. Lew, D.J. The morphogenesis checkpoint: How yeast cells watch their figures. *Curr. Opin. Cell Biol.* **2003**, *15*, 648–653. [CrossRef]
79. Anastasia, S.D.; Nguyen, D.L.; Thai, V.; Meloy, M.; MacDonough, T.; Kellogg, D.R. A link between mitotic entry and membrane growth suggests a novel model for cell size control. *J. Cell Biol.* **2012**, *197*, 89–104. [CrossRef]
80. Harvey, S.L.; Kellogg, D.R. Conservation of mechanisms controlling entry into mitosis: Budding yeast wee1 delays entry into mitosis and is required for cell size control. *Curr. Biol.* **2003**, *13*, 264–275. [CrossRef]
81. McNulty, J.J.; Lew, D.J. Swe1p responds to cytoskeletal perturbation, not bud size, in *S. cerevisiae*. *Curr. Biol.* **2005**, *15*, 2190–2198. [CrossRef]
82. Keaton, M.A.; Lew, D.J. Eavesdropping on the cytoskeleton: Progress and controversy in the yeast morphogenesis checkpoint. *Curr. Opin. Microbiol.* **2006**, *9*, 540–546. [CrossRef]
83. McMillan, J.N.; Longtine, M.S.; Sia, R.A.; Theesfeld, C.L.; Bardes, E.S.; Pringle, J.R.; Lew, D.J. The morphogenesis checkpoint in *Saccharomyces cerevisiae*: Cell cycle control of Swe1p degradation by Hsl1p and Hsl7p. *Mol. Cell. Biol.* **1999**, *19*, 6929–6939. [CrossRef]
84. Harrison, J.C.; Bardes, E.S.; Ohya, Y.; Lew, D.J. A role for the Pkc1p/Mpk1p kinase cascade in the morphogenesis checkpoint. *Nat. Cell Biol.* **2001**, *3*, 417–420. [CrossRef]
85. Mizunuma, M.; Hirata, D.; Miyahara, K.; Tsuchiya, E.; Miyakawa, T. Role of calcineurin and Mpk1 in regulating the onset of mitosis in budding yeast. *Nature* **1998**, *392*, 303–306. [CrossRef] [PubMed]
86. Thai, V.; Dephoure, N.; Weiss, A.; Ferguson, J.; Leitao, R.; Gygi, S.P.; Kellogg, D.R. Protein kinase C controls binding of Igo/ENSA proteins to protein phosphatase 2A in budding yeast. *J. Biol. Chem.* **2017**, *292*, 4925–4941. [CrossRef]
87. Baro, B.; Játiva, S.; Calabria, I.; Vinaixa, J.; Bech-Serra, J.J.; de LaTorre, C.; Rodrigues, J.; Hernáez, M.L.; Gil, C.; Barceló-Batllori, S.; et al. SILAC-based phosphoproteomics reveals new PP2A-Cdc55-regulated processes in budding yeast. *GigaScience* **2018**, *7*, 1–18. [CrossRef] [PubMed]
88. Godfrey, M.; Touati, S.A.; Kataria, M.; Jones, A.; Snijders, A.P.; Uhlmann, F. $PP2A^{Cdc55}$ phosphatase imposes ordered cell-cycle phosphorylation by opposing threonine phosphorylation. *Mol. Cell* **2017**, *65*, 393–402. [CrossRef]
89. Domingo-Sananes, M.R.; Kapuy, O.; Hunt, T.; Novak, B. Switches and latches: A biochemical tug-of-war between the kinases and phosphatases that control mitosis. *Philos. Trans. R. Soc. Lond. B Biol. Sci.* **2011**, *366*, 3584–3594. [CrossRef]
90. Queralt, E.; Lehane, C.; Novak, B.; Uhlmann, F. Downregulation of $PP2A^{Cdc55}$ phosphatase by separase initiates mitotic exit in budding yeast. *Cell* **2006**, *125*, 719–732. [CrossRef] [PubMed]
91. Chiroli, E.; Rossio, V.; Lucchini, G.; Piatti, S. The budding yeast $PP2A^{Cdc55}$ protein phosphatase prevents the onset of anaphase in response to morphogenetic defects. *J. Cell Biol.* **2007**, *177*, 599–611. [CrossRef]
92. Keaton, M.A.; Szkotnicki, L.; Marquitz, A.R.; Harrison, J.; Zyla, T.R.; Lew, D.J. Nucleocytoplasmic trafficking of G2/M regulators in yeast. *Mol. Biol. Cell* **2008**, *19*, 4006–4018. [CrossRef] [PubMed]
93. Lanz, M.C.; Dibitetto, D.; Smolka, M.B. DNA damage kinase signaling: Checkpoint and repair at 30 years. *EMBO J.* **2019**, *38*, e101801. [CrossRef]
94. Waterman, D.P.; Haber, J.E.; Smolka, M.B. Checkpoint responses to DNA double-strand breaks. *Annu. Rev. Biochem.* **2020**, *89*, 103–133. [CrossRef] [PubMed]
95. Putnam, C.D.; Jaehnig, E.J.; Kolodner, R.D. Perspectives on the DNA damage and replication checkpoint responses in *Saccharomyces cerevisiae*. *DNA Repair* **2009**, *8*, 974–982. [CrossRef]
96. Enserink, J.M.; Smolka, M.B.; Zhou, H.; Kolodner, R.D. Checkpoint proteins control morphogenetic events during DNA replication stress in *Saccharomyces cerevisiae*. *J. Cell Biol.* **2006**, *175*, 729–741. [CrossRef]
97. Smolka, M.B.; Chen, S.H.; Maddox, P.S.; Enserink, J.M.; Albuquerque, C.P.; Wei, X.X.; Desai, A.; Kolodner, R.D.; Zhou, H. An FHA domain-mediated protein interaction network of Rad53 reveals its role in polarized cell growth. *J. Cell Biol.* **2006**, *175*, 743–753. [CrossRef]
98. Diani, L.; Colombelli, C.; Nachimuthu, B.T.; Donnianni, R.; Plevani, P.; Muzi-Falconi, M.; Pellicioli, A. *Saccharomyces* CDK1 phosphorylates Rad53 kinase in metaphase, influencing cellular morphogenesis. *J. Biol. Chem.* **2009**, *284*, 32627–32634. [CrossRef]
99. Traven, A.; Lo, T.L.; Pike, B.L.; Friesen, H.; Guzzo, J.; Andrews, B.; Heierhorst, J. Dual functions of Mdt1 in genome maintenance and cell integrity pathways in *Saccharomyces cerevisiae*. *Yeast* **2010**, *27*, 41–52. [PubMed]
100. Bandyopadhyay, S.; Mehta, M.; Kuo, D.; Sung, M.K.; Chuang, R.; Jaehnig, E.J.; Bodenmiller, B.; Licon, K.; Copeland, W.; Shales, M.; et al. Rewiring of genetic networks in response to DNA damage. *Science* **2010**, *330*, 1385–1389. [CrossRef]
101. Queralt, E.; Igual, J.C. Functional connection between the Clb5 cyclin, the protein kinase C pathway and the Swi4 transcription factor in *Saccharomyces cerevisiae*. *Genetics* **2005**, *171*, 1485–1498. [CrossRef]
102. Soriano-Carot, M.; Bano, M.C.; Igual, J.C. The yeast mitogen-activated protein kinase Slt2 is involved in the cellular response to genotoxic stress. *Cell Div.* **2012**, *7*, 1. [CrossRef]
103. Liu, L.; Levin, D.E. Intracellular mechanism by which genotoxic stress activates yeast SAPK Mpk1. *Mol. Biol. Cell* **2018**, *29*, 2898–2909. [CrossRef] [PubMed]
104. Albuquerque, C.P.; Smolka, M.B.; Payne, S.H.; Bafna, V.; Eng, J.; Zhou, H. A multidimensional chromatography technology for in-depth phosphoproteome analysis. *Mol. Cell. Proteom.* **2008**, *7*, 1389–1396. [CrossRef] [PubMed]
105. Palou, G.; Palou, R.; Zeng, F.; Vashisht, A.A.; Wohlschlegel, J.A.; Quintana, D.G. Three different pathways prevent chromosome segregation in the presence of DNA damage or replication stress in budding yeast. *PLoS Genet.* **2015**, *11*, e1005468. [CrossRef] [PubMed]

106. Liu, H.; Wang, Y. The function and regulation of budding yeast Swe1 in response to interrupted DNA synthesis. *Mol. Biol. Cell* **2006**, *17*, 2746–2756. [CrossRef]
107. Krasley, E.; Cooper, K.F.; Mallory, M.J.; Dunbrack, R.; Strich, R. Regulation of the oxidative stress response through Slt2p-dependent destruction of cyclin C in *Saccharomyces cerevisiae*. *Genetics* **2006**, *172*, 1477–1486. [CrossRef]
108. Cooper, K.F.; Khakhina, S.; Kim, S.K.; Strich, R. Stress-induced nuclear-to-cytoplasmic translocation of cyclin C promotes mitochondrial fission in yeast. *Dev. Cell* **2014**, *28*, 161–173. [CrossRef]
109. Ježek, J.; Smethurst, D.G.J.; Stieg, D.C.; Kiss, Z.A.C.; Hanley, S.E.; Ganesan, V.; Chang, K.T.; Cooper, K.F.; Strich, R. Cyclin C: The story of a non-cycling cyclin. *Biology* **2019**, *8*, 3. [CrossRef]
110. Jin, C.; Strich, R.; Cooper, K.F. Slt2p phosphorylation induces cyclin C nuclear-to-cytoplasmic translocation in response to oxidative stress. *Mol. Biol. Cell* **2014**, *25*, 1396–1407. [CrossRef]
111. Eapen, V.V.; Waterman, D.P.; Bernard, A.; Schiffmann, N.; Sayas, E.; Kamber, R.; Lemos, B.; Memisoglu, G.; Ang, J.; Mazella, A.; et al. A pathway of targeted autophagy is induced by DNA damage in budding yeast. *Proc. Natl. Acad. Sci. USA* **2017**, *114*, E1158–E1167. [CrossRef] [PubMed]
112. Ueda, S.; Ozaki, R.; Kaneko, A.; Akizuki, R.; Katsuta, H.; Miura, A.; Matsuura, A.; Ushimaru, T. TORC1, Tel1/Mec1, and Mpk1 regulate autophagy induction after DNA damage in budding yeast. *Cell. Signal.* **2019**, *62*, 109344. [CrossRef]
113. Yeeles, J.T.P.; Janska, A.; Early, A.; Diffley, J.F.X. How the eukaryotic replisome achieves rapid and efficient DNA replication. *Mol. Cell* **2017**, *65*, 105–116. [CrossRef] [PubMed]
114. Masai, H.; Yang, C.C.; Matsumoto, S. Mrc1/Claspin: A new role for regulation of origin firing. *Curr. Genet.* **2017**, *63*, 813–818. [CrossRef]
115. Duch, A.; Canal, B.; Barroso, S.I.; García-Rubio, M.; Seisenbacher, G.; Aguilera, A.; de Nadal, E.; Posas, F. Multiple signaling kinases target Mrc1 to prevent genomic instability triggered by transcription-replication conflicts. *Nat. Commun.* **2018**, *9*, 379. [CrossRef]
116. Bacal, J.; Moriel-Carretero, M.; Pardo, B.; Barthe, A.; Sharma, S.; Chabes, A.; Lengronne, A.; Pasero, P. Mrc1 and Rad9 cooperate to regulate initiation and elongation of DNA replication in response to DNA damage. *EMBO J.* **2018**, *37*, e99319. [CrossRef] [PubMed]
117. Moriel-Carretero, M.; Pasero, P.; Pardo, B. DDR Inc., one business, two associates. *Curr. Genet.* **2019**, *65*, 445–451. [CrossRef] [PubMed]
118. Desany, B.A.; Alcasabas, A.A.; Bachant, J.B.; Elledge, S.J. Recovery from DNA replicational stress is the essential function of the S-phase checkpoint pathway. *Genes Dev.* **1998**, *12*, 2956–2970. [CrossRef] [PubMed]
119. Travesa, A.; Kalashnikova, T.I.; de Bruin, R.A.; Cass, S.R.; Chahwan, C.; Lee, D.E.; Lowndes, N.F.; Wittenberg, C. Repression of G1/S transcription is mediated via interaction of the GTB motifs of Nrm1 and Whi5 with Swi6. *Mol. Cell. Biol.* **2013**, *33*, 1476–1486. [CrossRef]
120. Huang, K.N.; Symington, L.S. Mutation of the gene encoding protein kinase C 1 stimulates mitotic recombination in *Saccharomyces cerevisiae*. *Mol. Cell. Biol.* **1994**, *14*, 6039–6045.
121. Alvaro, D.; Lisby, M.; Rothstein, R. Genome-wide analysis of Rad52 foci reveals diverse mechanisms impacting recombination. *PLoS Genet.* **2007**, *3*, e228. [CrossRef]
122. Yang, W.L.; Bruno, M.E.; Carman, G.M. Regulation of yeast CTP synthetase activity by protein kinase C. *J. Biol. Chem.* **1996**, *271*, 11113–11119. [CrossRef]
123. Yang, W.L.; Carman, G.M. Phosphorylation of CTP synthetase from *Saccharomyces cerevisiae* by protein kinase C. *J. Biol. Chem.* **1995**, *270*, 14983–14988. [CrossRef]
124. Park, T.S.; O'Brien, D.J.; Carman, G.M. Phosphorylation of CTP synthetase on Ser^{36}, Ser^{330}, Ser^{354}, and Ser^{454} regulates the levels of CTP and phosphatidylcholine synthesis in *Saccharomyces cerevisiae*. *J. Biol. Chem.* **2003**, *278*, 20785–20794. [CrossRef]
125. Mascaraque, V.; Hernaez, M.L.; Jimenez-Sanchez, M.; Hansen, R.; Gil, C.; Martin, H.; Cid, V.J.; Molina, M. Phosphoproteomic analysis of protein kinase C signaling in *Saccharomyces cerevisiae* reveals Slt2 mitogen-activated protein kinase (MAPK)-dependent phosphorylation of eisosome core components. *Mol. Cell. Proteom.* **2013**, *12*, 557–574. [CrossRef]
126. Zu, T.; Verna, J.; Ballester, R. Mutations in WSC genes for putative stress receptors result in sensitivity to multiple stress conditions and impairment of Rlm1-dependent gene expression in *Saccharomyces cerevisiae*. *Mol. Genet. Genom.* **2001**, *266*, 142–155. [CrossRef]
127. Leduc, A.; He, C.H.; Ramotar, D. Disruption of the *Saccharomyces cerevisiae* cell-wall pathway gene Slg1 causes hypersensitivity to the antitumor drug bleomycin. *Mol. Genet. Genom.* **2003**, *269*, 78–89. [CrossRef]
128. Liu, L.; Veis, J.; Reiter, W.; Motari, E.; Costello, C.E.; Samuelson, J.C.; Ammerer, G.; Levin, D.E. Regulation of Pkc1 hyper-phosphorylation by genotoxic stress. *J. Fungi* **2021**, *7*, 874. [CrossRef]
129. Soriano-Carot, M.; Quilis, I.; Bañó, M.C.; Igual, J.C. Protein kinase C controls activation of the DNA integrity checkpoint. *Nucleic Acids Res.* **2014**, *42*, 7084–7095. [CrossRef]
130. Yoshida, K.; Wang, H.G.; Miki, Y.; Kufe, D. Protein kinase Cδ is responsible for constitutive and DNA damage-induced phosphorylation of Rad9. *EMBO J.* **2003**, *22*, 1431–1441. [CrossRef]
131. Arango, D.; Parihar, A.; Villamena, F.A.; Wang, L.; Freitas, M.A.; Grotewold, E.; Doseff, A.I. Apigenin induces DNA damage through the PKCδ-dependent activation of ATM and H2AX causing down-regulation of genes involved in cell cycle control and DNA repair. *Biochem. Pharmacol.* **2012**, *84*, 1571–1580. [CrossRef] [PubMed]

132. Bharti, A.; Kraeft, S.K.; Gounder, M.; Pandey, P.; Jin, S.; Yuan, Z.M.; Lees-Miller, S.P.; Weichselbaum, R.; Weaver, D.; Chen, L.B.; et al. Inactivation of DNA-dependent protein kinase by protein kinase Cδ: Implications for apoptosis. *Mol. Cell. Biol.* **1998**, *18*, 6719–6728. [CrossRef] [PubMed]
133. Li, B.; Wang, X.; Rasheed, N.; Hu, Y.; Boast, S.; Ishii, T.; Nakayama, K.; Nakayama, K.I.; Goff, S.P. Distinct roles of c-Abl and Atm in oxidative stress response are mediated by protein kinase Cδ. *Genes Dev.* **2004**, *18*, 1824–1837. [CrossRef]
134. Santiago-Walker, A.E.; Fikaris, A.J.; Kao, G.D.; Brown, E.J.; Kazanietz, M.G.; Meinkoth, J.L. Protein kinase Cδ stimulates apoptosis by initiating G1 phase cell cycle progression and S phase arrest. *J. Biol. Chem.* **2005**, *280*, 32107–32114. [CrossRef] [PubMed]
135. Speidel, J.T.; Affandi, T.; Jones, D.N.M.; Ferrara, S.E.; Reyland, M.E. Functional proteomic analysis reveals roles for PKCδ in regulation of cell survival and cell death: Implications for cancer pathogenesis and therapy. *Adv. Biol. Regul.* **2020**, *78*, 100757. [CrossRef]
136. Liu, H.; Lu, Z.G.; Miki, Y.; Yoshida, K. Protein kinase Cδ induces transcription of the TP53 tumor suppressor gene by controlling death-promoting factor Btf in the apoptotic response to DNA damage. *Mol. Cell. Biol.* **2007**, *27*, 8480–8491. [CrossRef]
137. Nakagawa, M.; Oliva, J.L.; Kothapalli, D.; Fournier, A.; Assoian, R.K.; Kazanietz, M.G. Phorbol ester-induced G1 phase arrest selectively mediated by protein kinase Cδ-dependent induction of p21. *J. Biol. Chem.* **2005**, *280*, 33926–33934. [CrossRef]
138. Saha, K.; Adhikary, G.; Kanade, S.R.; Rorke, E.A.; Eckert, R.L. p38δ regulates p53 to control $p21^{Cip1}$ expression in human epidermal keratinocytes. *J. Biol. Chem.* **2014**, *289*, 11443–11453. [CrossRef] [PubMed]
139. LaGory, E.L.; Sitailo, L.A.; Denning, M.F. The protein kinase Cδ catalytic fragment is critical for maintenance of the G2/M DNA damage checkpoint. *J. Biol. Chem.* **2010**, *285*, 1879–1887. [CrossRef] [PubMed]

Journal of *Fungi*

MDPI

Review

The Role of the Cell Integrity Pathway in Septum Assembly in Yeast

Cesar Roncero *, Rubén Celador, Noelia Sánchez, Patricia García and Yolanda Sánchez *

Departamento de Microbiología y Genética, Instituto de Biología Funcional y Genómica, CSIC/Universidad de Salamanca, C/Zacarías González, s/n, 37007 Salamanca, Spain; rubencelador@usal.es (R.C.); nsn@usal.es (N.S.); pgr@usal.es (P.G.)

* Correspondence: crm@usal.es (C.R.); ysm@usal.es (Y.S.)

Abstract: Cytokinesis divides a mother cell into two daughter cells at the end of each cell cycle and proceeds via the assembly and constriction of a contractile actomyosin ring (CAR). Ring constriction promotes division furrow ingression, after sister chromatids are segregated to opposing sides of the cleavage plane. Cytokinesis contributes to genome integrity because the cells that fail to complete cytokinesis often reduplicate their chromosomes. While in animal cells, the last steps of cytokinesis involve extracellular matrix remodelling and mid-body abscission, in yeast, CAR constriction is coupled to the synthesis of a polysaccharide septum. To preserve cell integrity during cytokinesis, fungal cells remodel their cell wall through signalling pathways that connect receptors to downstream effectors, initiating a cascade of biological signals. One of the best-studied signalling pathways is the cell wall integrity pathway (CWI) of the budding yeast *Saccharomyces cerevisiae* and its counterpart in the fission yeast *Schizosaccharomyces pombe*, the cell integrity pathway (CIP). Both are signal transduction pathways relying upon a cascade of MAP kinases. However, despite strong similarities in the assembly of the septa in both yeasts, there are significant mechanistic differences, including the relationship of this process with the cell integrity signalling pathways.

Keywords: yeast; cytokinesis; actomyosin ring; septum; cell integrity

Citation: Roncero, C.; Celador, R.; Sánchez, N.; García, P.; Sánchez, Y. The Role of the Cell Integrity Pathway in Septum Assembly in Yeast. *J. Fungi* **2021**, *7*, 729. https://doi.org/10.3390/jof7090729

Academic Editors: María Molina and Humberto Martín

Received: 30 July 2021
Accepted: 31 August 2021
Published: 6 September 2021

1. Maintaining the Shape: The Cell Integrity Signaling Pathways

The cell integrity signalling pathways are usually described as fairly linear, they channel the signal from the cell surface to the nucleus without significant branching (Figure 1). Cell wall stress is detected by two conserved families of single-pass transmembrane cell wall sensors of the WSC and MID types. Their role is well documented in *S. cerevisiae*, where these sensors can detect the mechanical tension between the cell wall (CW) and the plasma membrane (PM) [1,2]. They function as upstream triggers of the cell integrity pathway and activate membrane-associated RhoA type GTPases through specific GEFs, the ScRom2p and SpRgf1p [3,4]. In *S. cerevisiae*, signalling is also associated with Tus1p, another Rho1p-GEF and a functional homologue of SpRgf3p, whose relevance in CIP activation is unknown. Signalling in *S. cerevisiae* is directly translated to the MAP kinase cascade through the GTPase Rho1p and its effector the Pkc1p kinase (Figure 1B). In *S. pombe*, the situation is a bit more complex (Figure 1A), there are two different Rho A homologues (Rho1p and Rho2p) and two Pkc kinases (Pck1p and Pck2p), participating in signalling. The major signal input is channelled through Pkc2p, which receives different inputs, the main one through the Rho2p GTPase and a minor one through Rho1p, which also transmits a minor signal through the Pkc1p branch [5,6]. In both yeasts, the signal from the PKC kinase is transmitted to a MAP kinase cascade that ends with the MAP kinases ScSlt2p and SpPmk1p. These kinases participate in phosphorylation events of specific nuclear transcription factors. The *S. pombe* Pmk1p phosphorylates Atf1p and Mbx2p transcription factors, but to date, only a few downstream targets have been characterised. In *S. cerevisiae*, the MAP kinase Slt2p phosphorylates the Rlm1p transcription factor that mediates a strong transcriptional response. This response encompasses genes involved in chitin and glucan synthesis as well as genes encoding cell wall remodelling activities, among others [7,8]. ScSlt2p also phosphorylates the heterodimeric SBF transcription factor that promotes the G1/S transition [9].

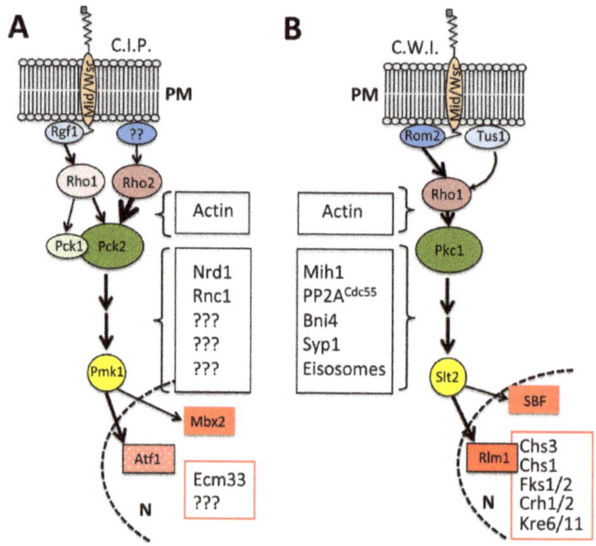

Figure 1. The cell integrity cascade in yeast. (**A**) The cell integrity pathway (CIP) in *Schizosaccharomyces pombe*. (**B**) The cell wall integrity pathway (CWI) in *Saccharomyces cerevisiae*. Both cascades signal from the mechanosensors in the PM (Mid/Wsc) to the transcription factor in the nucleus (N). Red boxed proteins are those for which expression is transcriptionally regulated by the cascade and are involved in the cell wall and/or septum assembly. Black boxed proteins represent cytosolic targets of the cascade at different levels that are related to septum assembly. For additional description of the cascade, see text.

In addition to their signalling through transcriptional regulation, these pathways act through cytosolic targets. The Rho1/2p GTPases are directly involved in the biological cycles of actin, regulating patches and filaments turnover that affect endocytosis, cell polarisation and therefore the assembly of the yeast cell wall. Additional targets of the signal transduction in *S. pombe* have been elusive, but are numerous in *S. cerevisiae*. The potential effect of such targets will be described later in the context of septum assembly.

2. *S. pombe*, the Fission Yeast

In *S. pombe*, the cells are cylindrical and grow by elongation at their tips. Cell division is accomplished by medial fission using a contractile actomyosin ring (CAR), which guides the formation of the cell wall septum. Cell separation also involves cell wall degradation between the two halves of the division septum that will constitute the new ends of the daughter cells. The overall process has been deeply studied, and there are excellent reviews on cytokinesis [10–13], septation [14,15] and cell separation [16]. In this review, we will first summarise key steps of fission yeast cytokinesis, including ring assembly, constriction and septum formation, before discussing emerging mechanisms that involve the cell integrity pathway (CIP) in the regulation of cytokinesis and their biological implications.

2.1. *Before Septum Assembly: Actomyosin Ring Positioning and Assembly*

The CAR in *S. pombe* is composed of short actin filaments assembled by formins and myosin motors (actomyosin) that produce the force to arrange the actin filaments (revised in [12,17]). CAR dynamics are tightly regulated in space and time and can be divided into several steps including positioning, assembly, maintenance, constriction and disassembly (Figure 2A, upper panel). Ring position is determined during interphase by a broad band of cortical cytokinetic precursor nodes located at the equator of the cell [18].

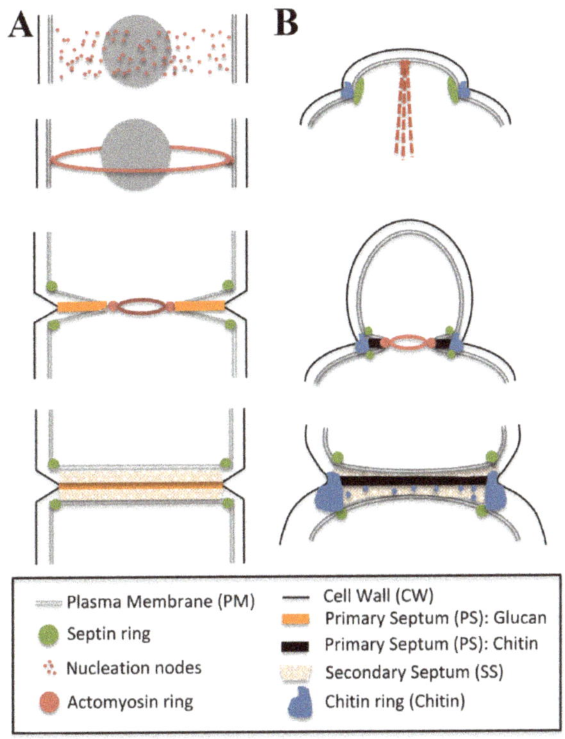

Figure 2. Septum assembly in yeasts: *S. pombe* (**A**) and *S. cerevisiae* (**B**). The schemes represent the temporal sequence of the process, from the early stages involved in the selection of the septation site (upper panels) to the assembly of the secondary septa (lower panels). The final stages in cell separation are not depicted. For additional details in the process, see text.

The anillin Mid1p and the DYRK kinase Pom1p dictate CAR positioning in the cell middle [13]. At mitosis, Plo1p kinase activate Mid1p allowing its exit from the nucleus [19,20] and Pak1p kinase promotes its association to the cortical nodes [21]. Then, Mid1p initiates the recruitment of cytokinetic factors, including the IQGAP Rngp2, Myosin II heavy and light chains, the F-BAR protein Cdc15p and the formin Cdc12p (Figure 2A) [13,22].

After spindle pole bodies (SPBs) separation, the cytokinesis nodes condense, and the contractile ring is formed through dynamic interactions between the actin filaments assembled by Cdc12p and Myo2p in adjacent nodes. Afterwards, the ring is maintained until the completion of anaphase in an interval known as maturation that lasts ~10 min until the onset of ring constriction. In maturation, more proteins are recruited from the cytoplasmic pool, while others leave the ring. Mid1p disappears, and the ring adds more polymerised actin, Cdc15p with its partners (Imp2p, Pxl1p, Fic1p and Rgf3p) [23], unconventional myosin-II (Myp2p) [24] and the glucan synthases Bgs1p and Bgs4p [25,26] among others. In the mature ring, proteins are ordered roughly in a three-layered structure. Starting from the inside, the first layer contains membrane-bound proteins that anchor the ring and act as scaffolds, an intermediate layer that contains signalling components that influence cell division and a distal layer that contains F-actin filaments, myosin motors and F-actin cross-linkers [27].

2.2. Anchoring the Ring to the Plasma Membrane and the Growing Septum

To perform its function, the CAR needs to be anchored to the PM, a linkage defined by at least three types of attachments: protein–protein and protein–lipid interactions, the cell wall and the arrangement of microtubules at the cell equator.

Upon mitotic entry, more Mid1p binds to the PM, anchoring ring proteins (and then the ring itself) to this structure. It has been shown that Mid1p dimerisation favours its interaction with membrane phospholipids, in particular phosphatidylinositol 4,5-bisphosphate [PI(4,5)P2] [28], preventing the sliding of the CAR [29]. An additional link of the CAR to the PM is provided by Cdc15p (*S. cerevisiae* Hof1p), a CAR scaffold that binds to the membrane through its BAR domain. When *cdc15* expression is repressed, the CAR slides along the PM and disassembles [30,31]. Cdc15p helps to deliver Bgs1p to the PM [30] and binds to paxillin, Pxl1p, Fic1p, Rgf3p (Rho1p-GEF) that also play a role in CAR anchoring [23]. Pxl1p mediates the interaction between the β-glucan synthase Bgs1p and the contractile ring contributing to the initiation of septum synthesis [32].

A different alteration in the cell wall structure also leads to CAR sliding. This has been shown in spherical protoplasts deprived of the wall [33], as well as in cells depleted for Bgs4p that bear low levels of branched-β(1,3)glucan [34], suggesting an additional level of linkage between the CAR and the cell wall. Finally, certain evidence also links the CAR and the cytoskeleton. For instance, microtubule depolymerisation in the β-GS mutant *cps1-191*, which arrests with a stable CAR, leads to CAR sliding [35], and in the absence of the microtubule nucleator Mto2p, the cells also fail to anchor of the CAR in the medial region under conditions that mildly perturb actin structures [36].

2.3. Triggering Septation: The Role of the Septation Initiation Network (SIN) in CAR Maintenance and Constriction

Once the CAR is assembled at the division site, it must constrict to guide and power membrane ingression and cell wall synthesis (Figure 2A). The regulation of CAR assembly/maintenance and constriction coupled to septum synthesis depends on a signalling cascade of the septation initiation network (SIN) [37–39]. The SIN induces cytokinesis only after the decrease in CDK activity in anaphase, guaranteeing that cytokinesis occurs after chromosome segregation. A pathway similar to the SIN, termed the mitotic exit network (MEN), exists in *S. cerevisiae* [40]. In addition, SIN/MEN orthologues also exist in mammals that conform to the HIPPO pathway [41]. In yeasts, these networks monitor the position of the spindle pole bodies (SPBs), the yeasts equivalent to centrosomes, throughout the cell cycle to coordinate cytokinesis with other cell cycle phases.

The SIN signal begins with the activity of the GTPase Spg1p and involves a regulatory GAP complex, a scaffold complex that anchors the pathway to the SPBs and a linear cascade of three kinases (Cdc7p, Sid1p and Sid2p), in order of their activation [39]. Insufficient SIN results in improper assembly of the contractile ring and failure of cytokinesis, generating multinucleated cells without rings or septa [42]. On the opposite, ectopic activation of the SIN triggers formation of a contractile ring and septum at any point in the cell cycle [43].

How does the SIN achieve CAR assembly and maintenance? What are the SIN substrates that direct CAR constriction and septum synthesis? Most of these questions are still not solved. Among the SIN components, only Sid2p kinase (a nuclear Dbf2-related (NDR) kinase) and its counterpart Mob1p associate with the CAR in mid-late anaphase [44]. Sid2p phosphorylation of Clp1p (*S. cerevisiae* Cdc14p) keeps the phosphatase out of the nucleolus, allowing the protein to operate on cytoplasmic targets. Clp1p reverses Cdk1p phosphorylation of itself, Cdc25p and other Cdk1p substrates antagonising CDK [45]. Clp1p also dephosphorylates Cdc15p, inducing its oligomerisation and the scaffolding activity necessary to recruit CAR components [46]. In addition, Sid2p targets CAR components such as the essential formin Cdc12p, triggering an oligomeric switch that positively modulates formin function [47]. The SIN pathway directly targets the SAD kinase Cdr2p, promoting its dissociation from the cortex [48]. Similarly, Sid2p phosphorylation of Mid1p disrupts Mid1p interaction with membrane [49]. In both situations, removal of these landmark proteins from the cortex cytokinesis resets the division plane for the next cell cycle.

Finally, the SIN should coordinate many aspects of CAR and septum regulation during late cytokinesis, although the molecular mechanism of this regulation remains unexplored. It has been shown that upregulation of the GTPase Rho1p partially rescues the lethality of *sid2* mutants at a low-restrictive temperature [50,51]. Based on these results, it has been proposed that the SIN activates Rho1p, which in turn activates the Bgs enzymes. However, the SIN target(s) involved in septum assembly remain unknown. Therefore, identifying SIN targets [52] and elucidating the consequence of known phosphorylation events at the ring will be central tasks in advancing our understanding of *S. pombe* cytokinesis.

2.4. Furrow Ingression and Septum Deposition

Initiation of furrow ingression by *S. pombe* depends on an intact CAR, a signal from the cell cycle clock and septum synthesis (Figure 2A). CAR constriction provides the pulling force for membrane and primary septum deposition, although septum synthesis also contributes to the membrane ingression [17]. The actin and myosin-dependent forces from the ring promote septum deposition and maintain the circularity of the pore [53,54]; in that sense, mutations compromising contractility slow furrow ingression, suggesting that the ring may positively modulate the rate of ingression [24,55]. One of the problems of the CAR being the principal force of ingression is the huge turgor pressure inside the cell that forces the ring to work against this pressure [56]. In addition, furrow ingression and septum growth can still occur in the absence of F-actin. These findings suggest that cell-wall assembly pushing from the outside of the membrane could afford the force for furrow ingression [57]. Moreover, the delivery of exocytic vesicles and membrane edge expansion [58–60] could provide stream-like forces, as has been shown in animal and plant cytokinesis [61,62]. A conciliatory model for septum synthesis proposes a two-phase motion. In the first phase, the septum ingresses slowly, remains immature and depends on the CAR integrity; in the second phase, the ingression rate increases, and the CAR becomes dispensable [63].

The fission yeast septum is made of β and α-glucans and lacks chitin. Under electron microscopy, it looks like a three-layered structure with a primary septum (PS) in the middle of two secondary septa (SS). The PS is mainly composed of linear-β(1,3)glucan synthesised by glucan synthase Bgs1p/Cps1p and contains branched-β(1,3)glucan [64,65]. This linear-β(1,3)glucan would play a role similar to chitin in budding yeasts (see below and Figure 2B) and other fungi in the synthesis of the PS. The secondary septum (SS) forms the cell wall once the cells are separated and consists of 1,6 branched β-1,3-glucans synthesised by Bgs4p, α-1,3-glucans synthesised by Ags1p and β-1,6-glucans [15].

One of the most studied aspects of septum synthesis at the PM is the regulation and deposition of the β-glucans. The enzyme responsible is the β–GS complex that consists of four catalytic subunits, Bgs1p-4p, and a unique regulatory subunit, the GTPase Rho1p [66]. Bgs1p, Bgs3p and Bgs4p are essential transmembrane proteins. Bgs1p localises as a ring, tightly associated with the CAR, at the edge of the septum membrane (PM) during ingression and is responsible for the synthesis of the PS. Bgs3p and Bgs4p follow the CAR but remain localised as a disk along the invaginated PM and are required for the synthesis and assembly of the SS [66].

Rho1p activates all three β-glucan synthases and is positively regulated by three guanine nucleotide exchange factors (GEFs): Rgf1p, Rgf2p and Rgf3p [67–69]. Rgf3p-GFP is a CAR component and is the main candidate to regulate Rho1p function in septum synthesis. Moreover, *rgf3* depleted cells lyse as couples during cell separation, mimicking the characteristic phenotype of *rho1* depleted mutants and mutants of Pck1p and Pck2p (*S. cerevisiae* Pkc1p homologues). In addition, it is possible that Rgf3p acts as a physical link between components of the CAR and the membrane–bound Bgs-mediated septum growth [27]. CAR-localised proteins, such as Cdc15p, Imp2p and Art1p, recruit Rgf3p, probably activating the regulatory subunit of the β-GS [70,71]. As mentioned before, Cdc15p participates in the transport of Bgs1p to the septum membrane and Rga7p (a Rho GAP) also contributes to the transfer of the Bgs4p to the same area [25,30]; therefore, it is possible that the concerted action of these proteins also regulate traffic of the Bgs1p to the PM. To date, it remains unclear whether Rgf3p regulates only Bgs1p activity or if it is involved in the regulation of all β-glucan synthases.

By contrast, Rgf1p follows the ring from outside, leaving behind a trail as division proceeds [72], making it a likely candidate for regulating the β-glucan synthesis forming the SS. Interestingly, Rgf1p is the most abundant GEF and activates the cell integrity pathway (CIP) in response to cell wall damage and osmotic stress [73,74].

The enzyme responsible for the synthesis of α–glucans in the SS septum is Ags1p, whose activity is regulated by the GTPase Rho2p. Ags1p localization to the CAR is very similar to that of Bgs4p and together with Bgs4p grants to the PS the robustness needed to counteract the turgor pressure for a gradual cell separation [66].

In *S. pombe*, the septin ring apparently does not have a function to recruit proteins for the assembly of the CAR, as occurs in *S. cerevisiae* (see below). This difference is probably associated with the mechanism used by each yeast for the selection of septum position. *S. pombe* septins form a ring structure at the septum during the constriction of the CAR, which serves as a scaffold to recruit the GTPases and glucanases that ultimately mediate daughter cell separation [16]. The glucanases are funnelled by the exocyst and concentrated by the double septin ring. Cell wall degradation starts with the erosion of the wall material that surrounds the septum driven by the α-1,3 glucanase activity of Agn1p [75,76]. After that, cell turgor pressure and the action of the endoglucanase Eng1p [77] finish the dissolution of the PS. Interestingly, the lysis phenotype characteristic of the *rgf3* (*ehs2-1*) mutants is suppressed by elimination of Eng1p but not Agn1p (unpublished results). This result suggests that Rgf3p is involved in PS synthesis and that a balance between cell wall synthesis and degradation is necessary to accomplish cell separation safely.

2.5. Integrating Septum Assembly and the Cell Integrity Pathway during Cytokinesis in Fission Yeast

What is the role of the CIP in cytokinesis? Are there cytoplasmic targets of Pmk1p with a potential role in cytokinesis?

In *S. pombe*, the cell integrity pathway (CIP) is composed of a module of three MAP kinases, the MAPKKK Pek1p, the MAPKK Mkh1p and the MAPK Pmk1p, which are regulated by upstream activators, as Rho GTPases and PKC homologues (Figure 1A) [4]. In general, the CIP null mutants display lysis and multiseptated phenotypes characteristic of cell separation defects; however, there are significant differences between mutants in the upper part of the pathways from others in the lower parts. We have described before how Rho1/2p and Pck1/2p regulate the synthesis of the main components of the cell wall, thereby organising the synthesis and/or assembly of the primary and secondary septa. Not surprisingly, most of these mutants shrink as doublets during cytokinesis. In addition, cells of *rgf1*, *pck1* and *pck2* null mutants show monopolar growth [68,78]; suggesting problems in the recognition of a faulty disassembled end [79], a factor that could also affect septum assembly.

Mutants lacking Mkh1p, Skh1p/Pek1p and Pmk1p show separation defects when grown in nutrient-limiting conditions, at high temperature and in hyperosmotic medium [80–82]. Apparently, these cells had almost finished cytokinesis but had not completely lysed the external wall, suggesting late defects in cell wall remodelling. Specifically, the phenotype of Pmk1p mutants may reflect fine-tuning of septation, as expected for mutations in CIP repairers of cell wall damage caused under environmental stresses. The differences in the phenotypic penetrance of the different mutants resembled that observed in *S. cerevisiae*.

2.6. The Role of the CIP beyond Septum Assembly

In *S. pombe*, contrary to *S. cerevisiae*, the transcriptional response mediated by the CIP seemed rather weak. Upon activation of the route, Pmk1p phosphorylates at least two transcription factors: Atf1p and Mbx2p. Atf1p is also phosphorylated by the SAP MAPK Sty1p in response to various stresses [83]. Atf1p mutants show sensitivity to antifungal agents and the *vic* (*v*iable in the presence of *i*mmunosuppressant and *c*hloride ion) phenotype characteristic of null mutants in bonafide components of the CIP [84,85]. Mbx2p is the closest homologue to Rlm1p in fission yeasts; *mbx*Δ mutants displayed only a modest sensitivity to cell wall-damaging agents, suggesting that Mbx2p plays a minor role in this process. While in *S. cerevisiae*, it is known that Rlm1p regulates the expression of tens of genes, most of which have been implicated in cell wall biogenesis [86], in *S. pombe*, the only well-known target of Atf1p and Mbx2p is Ecm33p. Ecm33p is a glycosylphosphatidylinositol (GPI)-anchored cell surface protein; Ecm33p deletion mutants display abnormal morphology and hypersensitivity to antifungal agents, although the relationship between the CIP, Ecm33 and cell wall assembly remains uncertain.

Pmk1p phosphorylation varies periodically during the cell cycle, reaching its maximum activity during cytokinesis. In fact, 15–20% of a population of *pmk1*Δ synchronised cells were unable to complete cytokinesis, suggesting that the Pmk1p pathway is activated to control septum formation and/or dissolution [87]. However, to date, we do not know how this is accomplished. Pmk1p localises to the mitotic spindle and the septum during cytokinesis and constitutively resides in both cytoplasm and the nucleus [88]; however, the role of Pmk1p in cell separation seems mostly independent of its nuclear localisation [89]. Among the few known cytoplasmic targets for Pmk1p in vivo are the RNA-binding proteins Nrd1p and Rnc1p. Nrd1p binds and stabilises the essential myosin II light chain mRNA, thereby playing an important role in the regulation of CAR synthesis and contraction [90]. The potential participation of Rnc1p is unknown.

It has been known for a long time that the MAP kinase Pmk1p becomes activated within minutes by cell wall stress [5,87]; however, it is still unclear how the CIP integrates this stress input with successful cell separation [91]. A quick response is required when the cell's genome has already split, and the cell becomes ready to separate its cytoplasm. Recently, it has been shown that cell wall damage inflicted during cytokinesis triggers a checkpoint-like response, promoting a delay right before CAR constriction [72]. This delay depends on Rgf1p/Rho1p and Pck2p and was also abolished in the absence of the MAP kinase of the CIP. Because inactivation of this pathway in stressed cells causes defects in septation [80–82], it is possible that the CIP signalling delays CAR constriction in response to cell wall perturbations to ensure that cytokinesis reaches completion only after the cell has adjusted to the new conditions.

Finally, there is a connection between the checkpoint response to cell wall damage and the SIN [72]. It has been shown that the cell wall cytokinesis checkpoint depends on the SIN to be achieved. Moreover, the cell wall delay correlates with a prolonged SIN signal. Given that Sid2p is required for CAR maintenance when the cytokinesis checkpoint is active [50], it is very likely that the prolonged SIN activity serves to maintain the CAR in a competent state to achieve constriction safely.

3. *S. cerevisiae*, the Budding Yeast

In *S. cerevisiae*, the building of a septum is initiated very early in the cell cycle. Bud site selection is mediated by the landmark proteins inherited during cell division [92]. Then a cascade of GTP-GDP-bound proteins recruits and activates the Rho family GTPase Cdc42p, which in turn stimulates actin cable polarisation, targeted exocytosis and septin ring formation [93]. Here, we shall focus on the assembly of the septin ring that in *S. cerevisiae* acts as a scaffold structure for the sequential recruitment of the components that build up a septum.

The initial septin ring recruits the major chitin synthase, Chs3p, and its activator Chs4p, to the mother site of the neck. There, they will promote the synthesis of the chitin ring that serves as a scaffold for the septum, although it is not strictly a part of it. The initial septin ring undergoes significant structural modifications monitored by the morphogenesis checkpoint that finally result in the split of the septin hourglass structure into a double ring. Formation of this structure is not essential for cytokinesis, but it could set the limits of the area for septum assembly [94], favoring the correct positioning of the ingression progression complexes (IPCs) to trigger the synthesis of the primary septum. In addition, the double septin ring acts as a landmark to redirect the polarised secretion in the daughter cell to the neck region, facilitating the assembly of the septum [95,96].

3.1. Before Septum Assembly: Shaping the Yeast Cell through the Synthesis and Assembly of the Cell Wall

Although this review is mainly focused on the role of CWI in septum assembly, we could not ignore the general roles of CWI in the synthesis of the yeast cell wall. Activation of Cdc42p at the site of bud formation triggers the synthesis of new cellular material forming the growing bud (Figure 2B) [97]. This material includes β-glucans synthesised by the β-glucan synthases Fks1p and Fks2p. The FKS1/2 are activated by the Rho1p GTPase [98,99] and its GEF Rom2p recruited to the site of bud emergence by the CWI sensors, Mid2p and Wsc1p [100]. At the same time, the main CWI Kinase PKC1 is recruited to the site of polarised growth [101], allowing the synthesis of the major component of the yeast cell wall. Later on, the polarisation machinery is displaced to the growing tip (Figure 2B, upper panel), where the synthesis of the cell wall continues along the cell cycle. Interestingly, it has been elegantly shown that the CWI is engaged in the localisation of the cellular machinery after the local cell wall is damaged through proteasomal degradation of critical components previously assembled at the site of cell division [102].

The structure of the septin ring is externally reinforced by a chitin ring synthesized by Chs3p, defining the width of the yeast neck [103]. The absence of the chitin ring, which is not essential on its own, exacerbates minor cytokinesis defects leading to severe synthetic lethal effects [104,105]; moreover, upregulation of the chitin synthesis mediated by Chs3p relieves multiple strong cell wall defects mediated by antifungal therapies [106]. In this scenario, it is therefore, not surprising to find chitin synthesis under the control of the CWI. How is this control achieved?

It has been shown that phosphorylation of Chs3p depends on PKC1 activity [107], although there is no direct evidence that the CWI activates chitin synthesis. However, there are multiple circumstantial pieces of evidence indicating that this could be indeed the case. Chs3p activity depends on its polarised delivery to the neck where it binds to the septins [108]. This localisation strictly depends on the endocytic turnover of Chs3p [109,110]; therefore, the potential effects of the CWI on actin patch localisation (see below) will affect Chs3p endocytosis. This effect can be direct or could be mediated through Chs4p that links Chs3p to the septins [109,111]. A more important action of CWI in chitin synthesis is through the transcriptional regulation of the *GFA1* gene encoding one of the enzymes required for the synthesis of the UDP N-acetylglucosamine (UDP-NAGA), the metabolic precursor of chitin. Gfa1p is normally synthesised in limited amounts and acts as a bottleneck in chitin synthesis [112]. Cell wall damage triggers a compensatory response accompanied by a significant increase in chitin synthesis that relays in higher levels of Gfa1p, promoted by the activation of the CWI response through its transcriptional program [7]. Not surprisingly, overexpression of GFA1 by other means also alleviated many cell wall defects. The regulation of the Chs3p related to the synthesis of chitin in the daughter cell and in the secondary septa, will be describe later.

Finally, a potential role of the CWI in septum assembly through the ER stress surveillance (ERSU) cell cycle checkpoint that ensures that cells inherit functional ER into the daughter cell cannot be ignored; the role of CWI in this process is well documented, but the molecular mechanisms that underlie this function are still poorly defined (reviewed in [113]).

3.2. Starting the Separation: The Building of a Primary Septum

While yeast cells progress along the cell cycle, cell growth moves from apical to isodiametric, and the septin ring undergoes strong structural modifications that eventually end with splitting in a double ring that would mark the position of the septa [95,96]. This progression is monitored by the morphogenesis checkpoint through the function of the Swe1p kinase [93].

Progression along the cell cycle triggers the so-called mitotic exit system, which involves the Cdc14p early anaphase release (FEAR) and the mitotic exit network (MEN) which is the start point for cytokinesis. This signal triggers the destruction of the mitotic kinases allowing the arrival of the chitin synthase Chs2p to the septation site [114,115]. In addition, the mitotic exit also promotes the Rho1p-mediated assembly of the actomyosin ring though the Bni1p formin [116,117]. Concomitantly, the symmetric relocalisation of the polarisation machinery to both sites of the division site funnels the secretion machinery in order to provide the building blocks for the synthesis of the new membrane units and the septum. The primary septum (PS) is then assembled (Figure 2B) by the coordinated action of two independent but interconnected mechanisms: the centripetal synthesis of the chitin disk and the contraction of the actomyosin ring [118].

The chitinous nature of the primary septa in *S. cerevisiae* represents a strong difference with *S. pombe* PS, which is mainly made of linear β-(1,3)-glucan [14,64]. In fission yeast, the CWI could control the synthesis of PS by modulating Rho1p and/or Bgs1p, the regulatory and the catalytic subunit, respectively, of the β-glucan synthase complex [119]. In *S. cerevisiae,* the activation of Chs2p orchestrated by the CWI pathway has not been described to date.

The pioneering work in Cabib and Li labs established that chitin synthesis mediated by Chs2p and the actomyosin ring contraction are two interdependent, but interconnected processes that led to the synthesis of the primary septa [120,121]. Chs2p dephosphorylation promoted by the MEN substrate Cdc14p triggers its exit from the ER and its delivery to the division site, which ensures that septum formation takes place only after the completion of mitotic events [114,115]. Then, Chs2p interacts with the SH3 domains of Hof1p (SpCdc15p) as well as with Cyk3p (SpCyk3p), Inn1p (SpFic1p) and with the scaffold Spa2p, favouring its incorporation to the IPC complexes, where is activated to synthesise the chitin disk [122,123]. It has been proposed that the C2 domain of Inn1p participates in Chs2p activation [123]. Accordingly, some hypermorphic alleles of Chs2p can bypass the cell division defect seen in Inn1p mutants and mutants of other IPC components [124]. However, the molecular mechanism of Chs2p activation at the neck remains uncertain.

It is known that the chitin synthases stay competent for chitin synthesis after endocytosis blockade [109]; therefore, modification of the endocytic turnover by the CWI pathway through Rho1p could have a direct impact on PS synthesis [125,126]. In this sense, it has been shown that Rho1p and Pkc1p modulate the neck localisation of Syp1p, a protein involved in the negative regulation of actin patch assembly [127,128]. In addition, activation of the CWI induces phosphorylation of the eisosome core components such as Pil1p and Lsp1p [129], which could participate in PM compartmentalisation [130]. Altogether, this evidence suggests a direct role of CWI in regulating the endocytic turnover of proteins involved in septum assembly. Finally, the MEN kinase Dbf2 directly phosphorylated Chs2p, triggers its dissociation from the neck [131].

3.3. At the End of the Process, Secondary Septa Synthesis and Cell Separation

At the end of the PS assembly, there begins the synthesis of the SS, which in *S. cerevisiae* is formed by β-glucans with minor quantities of chitin. Chitin is synthesised by the chitin synthase Chs3p [132], and its localisation at the neck is dependent on Rho1p [133], although the details are unknown. One possibility is that the relocalisation of secretory and endocytic machinery on both sides of the neck increased the endocytic turnover of Chs3p/Chs4p, favouring chitin synthesis at the SS. The regulation of the β-glucan synthesis at the SS by the CWI is achieved through Rho1p. During septation Rho1p is recruited and activated by a distinct mechanism that involves its binding to membrane phosphoinositide's in order to activate the β-glucan synthases Fks1/2p [133]. In addition, SS synthesis could be also achieved through transcriptional activation of FKS2 and GFA1 genes by the CWI pathway.

Moreover, chitin and β-glucans are linked together at the septa by the action of Chr1/2p transglycosidases, which are localised at the neck [134]. In the neck region, cross-linkage occurs between the β-(1-3)-glucans and the Chs3p-made chitin, therefore linking the SS β-glucan material to the chitin ring [135]. However, additional linkages between chitin and β(1-6) glucans cannot be discarded as part of the SS layered out in the mother cells of the septa, as it has been reported for the lateral cell walls [135]. Interestingly, Chr1/2p expression is under the transcriptional control of the CWI, which could also contribute to the strength of the septa through increased levels of chitin-glucans cross-linkages.

The last step in septum dynamics is its dissolution to achieve cell separation. This process is triggered by the RAM pathway and involves the Ace2p mediated expression of several hydrolytic enzymes, specifically in the daughter cell [136]. In *S. cerevisiae*, the major role in cell separation is performed by the Cts1p chitinase [137] together with the Eng1p endoglucanase [138]. These proteins are secreted in a polarised manner to the periplasmic space surrounding the neck region of the daughter cell [138,139]. Chitinase acts centripetally, first on the chitin ring and later on the chitin disk formed by Chs2p, allowing the separation of chitin from the SS material mainly formed by β-glucans. The partial degradation of β-glucans by Eng1p would contribute to the process. Interestingly, these actions, performed only from the daughter side, leave most of the primary septa at the mother cell as the bud scar.

Cell separation involves degradation coupled to the repair of the cell wall when degradation takes place in an excess that could compromise cell integrity. Therefore, the CWI performs a critical role by regulating the expression of the repair enzyme, Chs1p [7]. Interestingly, while the transcriptional response mediated by the CWI pathway is not relevant under normal circumstances, the transcriptional regulation of Chs1p becomes critical in the presence of an excess of chitinase activity [104,140]. More recently, it has been described a sort of cell separation checkpoint-like named ECO (enforcement of cytokinesis order) that down-regulates directly Cts1p secretion upon cytokinesis defects. The ECO detects cytokinesis defects and signals through the Cbk1p kinase in order to prevent Cts1p secretion independently of the transcriptional regulation exerted by Cbk1p trough the RAM pathway [141]. The precise mechanism used by this pathway is uncertain and, so far, its potential relationship with the CWI is untested.

3.4. Beyond Septal Assembly: The Generation of a Remedial Septum

Besides the roles of CWI in septum assembly, the CWI participates in the synthesis of the remedial septa [118]. These septa are assembled upon catastrophic events caused by defects in Chs2p or in the actomyosin and septin rings. The failure in separating mother and daughter cells compromises cell integrity and therefore triggers a strong CWI response. This response is directed in part by the chitin synthase Chs3p, which promotes an abundant synthesis of chitin at the neck region independently of its role in the assembly of the chitin ring. Synthesis is triggered by up-regulation of the *GFA1* gene, although the collapse on the actomyosin ring contraction very likely favoured Chs3p accumulation at the neck by preventing its endocytosis. In addition, FKS1/2 and CHR1/2 genes are also upregulated [7,112], increasing chitin-glucans cross-linkages and in general favouring the strength of the remedial septa. Moreover, the activation of CWI triggers the increased expression of genes involved in the synthesis of β-(1,6)-glucans and mannoproteins that could also contribute to its assembly. Altogether, these actions promote the synthesis of abnormally engrossed septa that lack the typical layered structure but prevent cell lysis during cytokinesis.

As stated above, the upregulation of chitin synthesis mediated by the CWI seemed to be a cellular general response against cell wall damage that can be mimicked simply by the addition of glucosamine to the media as a direct precursor in chitin synthesis [104].

3.5. The Action of CWI beyond Septum Assembly

The CWI pathway has been described extensively based on its strong transcriptional in response to cell injuries that compromised cell integrity. Due to the different phenotypic penetrance of mutations in their components, very soon it was apparent that the CWI pathway was not linear. The phenotypes associated with the absence of the Rlm1p transcription factor were milder than those associated with the upper part of the route, while the phenotypes seen in the absence of the Slt2p MAP kinase were clearly intermediate, suggesting significant branching in the functional signalling along the route. We have highlighted before some of the most relevant non-transcriptional effects of the CWI response in the assembly of a functional septum, but many other aspects have not yet been addressed.

In recent years, and through different approaches, it has been shown that the CWI mediates, directly or indirectly, the phosphorylation of multiple proteins, potentially influencing all aspects of cell physiology, including carbohydrate metabolism, protein synthesis and DNA repair, among others [100]. There is also a close relationship between cell wall synthesis and cell cycle progression; therefore, a tight link between CWI and cell cycle progression is expected, exemplified by early reports that established the SBF complex as a direct target of the CWI [142]. This linkage is established at multiple levels that have been reviewed recently [100]. However, we shall highlight here the close relationship between the PP2A^{Cdc55}-Zds1/Zds2 complex and the CWI. This PP2A^{Cdc55} complex is an effector of Rho1p that in unperturbed growth conditions favours polarised cell growth, inactivating the Rho1p GAP Lrg1, while preventing CWI activation by stabilising the other Rho1p GAP, Lrg7. Upon cell wall stress, Rho1p rapidly activates the CWI, which downregulates the PP2A^{Cdc55} complex at multiple levels, reducing polarised growth in favour of the stress response [143].

An additional link between the CWI and cell cycle progression has been recently proposed through the protein Bni4p, a direct target of the MAP kinase Slt2p [144] and the cyclin kinase Pho85p [145]. Bni4p is involved in septum assembly [146] and its location at the neck region is cell-cycle-regulated [147], allowing a new level of the crosstalk between CWI and cell cycle progression. It is tempting to speculate a potential functional relationship between this level of control and the new check-point-like response associated with the cell wall damage inflicted during cytokinesis recently described in *S. pombe* [72]. However, the answer will have to wait until the identification of the potential target/s of the CIP at the neck.

4. Concluding Remarks and Future Perspectives

In this review, we have highlighted the multiple interconnections between the cell integrity signalling responses and the assembly of yeast septa using two model yeasts with different modes of growth. We believe that though the main rules governing the physiological relationship between cell integrity signalling and septum assembly are similar, sometimes the molecular mechanisms underlying both processes are different.

Some of the differences described in the text may reflect the levels of knowledge accumulated for both yeasts. However, in many others, the differences are probably associated with how both yeasts choose the site of septum synthesis, the different structure and composition of their cell walls and so on. In this context, it would be very interesting to know how the CWI influences septum assembly in other fungi. The CWI response seems conserved across fungi and participates in the response to several stresses, including antifungal therapies and in pathogeny [148]. In addition, although some members of the cascade have been linked with the synthesis of α and β-glucans, the precise implications of this cascade in septum assembly have been poorly explored.

In summary, it would be interesting to gain a deeper understanding of the relationship between CWI and septum assembly in these yeasts as well as in other fungi to define the general rules governing them. This knowledge eventually will allow the identification of new targets useful in the design of efficient antifungal therapies.

Funding: MEIC, Spain (BFU2017-84508-P), (PID2020-115111GB-I00) and Regional Government of Castile and Leon [SA116G19]. Y.S. and C.R. also thanks the financial support awarded to the Institute of Biological and Functional Genomics (IBFG) (CLU-2017-03) provided by the Junta de Castilla y León through the program "Escalera de Excelencia," co-financed by the PO FEDER of Castilla y León 14-20. N.S. was supported by an FPU fellowship from the Spanish Ministry of Education. P.G. was supported by a postdoctoral contract granted by the University of Salamanca and R.C. was financially supported by predoc grant from the University of Salamanca.

Acknowledgments: Thanks to the proof-reading-service for English revision of the manuscript.

Conflicts of Interest: The authors declare no conflict of interest.

References

1. Elhasi, T.; Blomberg, A. Integrins in disguise—Mechanosensors in Saccharomyces cerevisiae as functional integrin analogues. *Microb. Cell* **2019**, *6*, 335–355. [CrossRef] [PubMed]
2. Kock, C.; Dufrene, Y.F.; Heinisch, J.J. Up against the wall: Is yeast cell wall integrity ensured by mechanosensing in plasma membrane microdomains? *Appl. Environ. Microbiol.* **2015**, *81*, 806–811. [CrossRef] [PubMed]
3. Levin, D.E. Regulation of cell wall biogenesis in *Saccharomyces cerevisiae*: The cell wall integrity signaling pathway. *Genetics* **2011**, *189*, 1145–1175. [CrossRef]
4. Perez, P.; Cortes, J.C.G.; Cansado, J.; Ribas, J.C. Fission yeast cell wall biosynthesis and cell integrity signalling. *Cell Surf.* **2018**, *4*, 1–9. [CrossRef]
5. Barba-Espín, G.; Soto, T.; Madrid, M.; Núñez, A.; Vicente, J.; Gacto, M.; Cansado, J. Activation of the cell integrity pathway is channelled through diverse signalling elements in fission yeast. *Cell. Signal.* **2008**, *20*, 748–757. [CrossRef]
6. Sánchez-Mir, L.; Soto, T.; Franco, A.; Madrid, M.; Viana, R.A.; Vicente, J.; Gacto, M.; Perez, P.; Cansado, J. Rho1 GTPase and PKC Ortholog Pck1 Are Upstream Activators of the Cell Integrity MAPK Pathway in Fission Yeast. *PLoS ONE* **2014**, *9*, e88020. [CrossRef]
7. Garcia, R.; Bermejo, C.; Grau, C.; Pérez, R.; Peña, J.M.R.; François, J.M.; Nombela, C.; Arroyo, J. The Global Transcriptional Response to Transient Cell Wall Damage in Saccharomyces cerevisiae and Its Regulation by the Cell Integrity Signaling Pathway. *J. Biol. Chem.* **2004**, *279*, 15183–15195. [CrossRef]
8. Lagorce, A.; Hauser, N.C.; Labourdette, D.; Rodriguez, C.; Martin-Yken, H.; Arroyo, J.; Hoheisel, J.D.; François, J.M. Genome-wide Analysis of the Response to Cell Wall Mutations in the Yeast Saccharomyces cerevisiae. *J. Biol. Chem.* **2003**, *278*, 20345–20357. [CrossRef]
9. Madden, K.; Snyder, M. Cell polarity and morphogenesis in budding yeast. *Annu. Rev. Microbiol.* **1998**, *52*, 687–744. [CrossRef]
10. Pollard, T.D.; O'Shaughnessy, B. Molecular Mechanism of Cytokinesis. *Annu. Rev. Biochem.* **2019**, *88*, 661–689. [CrossRef]
11. Gerien, K.S.; Wu, J.Q. Molecular mechanisms of contractile-ring constriction and membrane trafficking in cytokinesis. *Biophys. Rev.* **2018**, *10*, 1649–1666. [CrossRef] [PubMed]
12. Mangione, M.C.; Gould, K.L. Molecular form and function of the cytokinetic ring. *J. Cell Sci.* **2019**, *132*, 12. [CrossRef]
13. Rincon, S.A.; Paoletti, A. Molecular control of fission yeast cytokinesis. *Semin. Cell Dev. Biol.* **2016**, *53*, 28–38. [CrossRef]
14. Cortés, J.C.G.; Ramos, M.; Osumi, M.; Pérez, P.; Ribas, J.C. Fission yeast septation. *Commun. Integr. Biol.* **2016**, *9*, e1189045. [CrossRef]
15. Pérez, P.; Cortés, J.C.G.; Martín-García, R.; Ribas, J.C. Overview of fission yeast septation. *Cell. Microbiol.* **2016**, *18*, 1201–1207. [CrossRef]
16. Martín-García, R.; Santos, B. The price of independence: Cell separation in fission yeast. *World J. Microbiol. Biotechnol.* **2016**, *32*, 1–7. [CrossRef]
17. Cheffings, T.H.; Burroughs, N.J.; Balasubramanian, M.K. Actomyosin Ring Formation and Tension Generation in Eukaryotic Cytokinesis. *Curr. Biol.* **2016**, *26*, R719–R737. [CrossRef]
18. Akamatsu, M.; Berro, J.; Pu, K.-M.; Tebbs, I.R.; Pollard, T.D. Cytokinetic nodes in fission yeast arise from two distinct types of nodes that merge during interphase. *J. Cell Biol.* **2014**, *204*, 977–988. [CrossRef]
19. Steever, A.; Pringle, J.; Bähler, J.; Wang, Y.-L.; Gould, K.; McCollum, D.; Wheatley, S. Role of Polo Kinase and Mid1p in Determining the Site of Cell Division in Fission Yeast. *J. Cell Biol.* **1998**, *143*, 1603–1616. [CrossRef]
20. Mccollum, D. Cytokinesis: ER keeps Mid1 in the middle. *Curr. Biol.* **2010**, *20*, R484–R486. [CrossRef]
21. Magliozzi, J.O.; Sears, J.; Cressey, L.; Brady, M.; Opalko, H.E.; Kettenbach, A.N.; Moseley, J.B. Fission yeast Pak1 phosphorylates anillin-like Mid1 for spatial control of cytokinesis. *J. Cell Biol.* **2020**, *219*, 8. [CrossRef]
22. Willet, A.H.; McDonald, N.A.; Gould, K.L. Regulation of contractile ring formation and septation in Schizosaccharomyces pombe. *Curr. Opin. Microbiol.* **2015**, *28*, 46–52. [CrossRef]
23. Roberts-Galbraith, R.H.; Chen, J.S.; Wang, J.; Gould, K.L. The SH3 domains of two PCH family members cooperate in assembly of the Schizosaccharomyces pombe contractile ring. *J. Cell Biol.* **2009**, *184*, 113–127. [CrossRef] [PubMed]
24. Laplante, C.; Berro, J.; Karatekin, E.; Hernandez-Leyva, A.; Lee, R.; Pollard, T.D. Three Myosins Contribute Uniquely to the Assembly and Constriction of the Fission Yeast Cytokinetic Contractile Ring. *Curr. Biol.* **2015**, *25*, 1955–1965. [CrossRef]

25. Arasada, R.; Pollard, T.D. A role for F-BAR protein Rga7p during cytokinesis in *S. pombe*. *J. Cell Sci.* **2015**, *128*, 2259–2268. [CrossRef] [PubMed]
26. Goss, J.W.; Kim, S.; Bledsoe, H.; Pollard, T.D. Characterization of the roles of Blt1p in fission yeast cytokinesis. *Mol. Biol. Cell* **2014**, *25*, 1946–1957. [CrossRef]
27. McDonald, N.A.; Lind, A.L.; Smith, S.E.; Li, R.; Gould, K.L. Nanoscale architecture of the Schizosaccharomyces pombe contractile ring. *eLife* **2017**, *6*, e28865. [CrossRef]
28. Sun, L.; Guan, R.; Lee, I.-J.; Liu, Y.; Chen, M.; Wang, J.; Wu, J.-Q.; Chen, Z. Mechanistic Insights into the Anchorage of the Contractile Ring by Anillin and Mid1. *Dev. Cell* **2015**, *33*, 413–426. [CrossRef]
29. Snider, C.E.; Willet, A.H.; Chen, J.-S.; Arpağ, G.; Zanic, M.; Gould, K.L. Phosphoinositide-mediated ring anchoring resists perpendicular forces to promote medial cytokinesis. *J. Cell Biol.* **2017**, *216*, 3041–3050. [CrossRef]
30. Arasada, R.; Pollard, T.D. Contractile Ring Stability in *S. pombe* Depends on F-BAR Protein Cdc15p and Bgs1p Transport from the Golgi Complex. *Cell Rep.* **2014**, *8*, 1533–1544. [CrossRef]
31. McDonald, N.A.; Kooi, C.W.V.; Ohi, M.D.; Gould, K.L. Oligomerization but Not Membrane Bending Underlies the Function of Certain F-BAR Proteins in Cell Motility and Cytokinesis. *Dev. Cell* **2015**, *35*, 725–736. [CrossRef] [PubMed]
32. Cortés, J.C.G.; Pujol, N.; Sato, M.; Pinar, M.; Ramos, M.; Moreno, M.B.; Osumi, M.; Ribas, J.; Perez, P. Cooperation between Paxillin-like Protein Pxl1 and Glucan Synthase Bgs1 Is Essential for Actomyosin Ring Stability and Septum Formation in Fission Yeast. *PLoS Genet.* **2015**, *11*, e1005358. [CrossRef]
33. Mishra, M.; Huang, Y.; Srivastava, P.; Srinivasan, R.; Sevugan, M.; Shlomovitz, R.; Gov, N.; Rao, M.; Balasubramanian, M. Cylindrical Cellular Geometry Ensures Fidelity of Division Site Placement in Fission Yeast. *J. Cell Sci.* **2012**, *125*, 3850–3857. [CrossRef]
34. Muñoz, J.; Cortes, J.G.; Sipiczki, M.; Ramos, M.; Clemente-Ramos, J. Ángel; Moreno, M.B.; Martins, I.; Perez, P.; Ribas, J.C. Extracellular cell wall β(1,3)glucan is required to couple septation to actomyosin ring contraction. *J. Cell Biol.* **2013**, *203*, 265–282. [CrossRef] [PubMed]
35. Pardo, M.; Nurse, P. Equatorial retention of the contractile actin ring by microtubules during cytokinesis. *Science* **2003**, *300*, 1569–1574. [CrossRef] [PubMed]
36. Venkatram, S.; Jennings, J.L.; Link, A.; Gould, K.L. Mto2p, a novel fission yeast protein required for cyto-plasmic microtubule organization and anchoring of the cytokinetic actin ring. *Mol. Biol. Cell* **2005**, *16*, 3052–3063. [CrossRef]
37. Goyal, A.; Takaine, M.; Simanis, V.; Nakano, K. Dividing the spoils of growth and the cell cycle: The fission yeast as a model for the study of cytokinesis. *Cytoskeleton* **2011**, *68*, 69–88. [CrossRef] [PubMed]
38. Johnson, A.E.; Mccollum, D.; Gould, K.L. Polar opposites: Fine-tuning cytokinesis through SIN asymmetry. *Cytoskeleton* **2012**, *69*, 686–699. [CrossRef]
39. Simanis, V. Pombe's thirteen—Control of fission yeast cell division by the septation initiation network. *J. Cell Sci.* **2015**, *128*, 1465–1474. [CrossRef]
40. Baro, B.; Queralt, E.; Monje-Casas, F. Regulation of Mitotic Exit in Saccharomyces cerevisiae. *Methods Mol. Biol.* **2017**, *1505*, 3–17. [CrossRef]
41. Ma, S.; Meng, Z.; Chen, R.; Guan, K.-L. The Hippo Pathway: Biology and Pathophysiology. *Annu. Rev. Biochem.* **2019**, *88*, 577–604. [CrossRef]
42. Hachet, O.; Simanis, V. Mid1p/anillin and the septation initiation network orchestrate contractile ring assembly for cytokinesis. *Genes Dev.* **2008**, *22*, 3205–3216. [CrossRef]
43. Schmidt, S.; Sohrmann, M.; Hofmann, K.; Woollard, A.; Simanis, V. The Spg1p GTPase is an essential, dosage-dependent inducer of septum formation in Schizosaccharomyces pombe. *Genes Dev.* **1997**, *11*, 1519–1534. [CrossRef]
44. Sparks, C.A.; Morphew, M.; McCollum, D. Sid2p, a spindle pole body kinase that regulates the onset of cytokinesis. *J. Cell Biol.* **1999**, *146*, 777–790. [CrossRef] [PubMed]
45. Lu, L.X.; Domingo-Sananes, M.R.; Huzarska, M.; Novak, B.; Gould, K.L. Multisite phosphoregulation of Cdc25 activity refines the mitotic entrance and exit switches. *Proc. Natl. Acad. Sci. USA* **2012**, *109*, 9899–9904. [CrossRef]
46. Roberts-Galbraith, R.H.; Ohi, M.D.; Ballif, B.A.; Chen, J.S.; McLeod, I.; McDonald, W.H.; Gygi, S.P.; Yates, J.R., 3rd; Gould, K.L. Dephosphorylation of F-BAR protein Cdc15 modulates its conformation and stimulates its scaffolding activity at the cell division site. *Mol. Cell* **2010**, *39*, 86–99. [CrossRef]
47. Bohnert, K.A.; Grzegorzewska, A.P.; Willet, A.; Kooi, C.W.V.; Kovar, D.R.; Gould, K.L. SIN-dependent phosphoinhibition of formin multimerization controls fission yeast cytokinesis. *Genes Dev.* **2013**, *27*, 2164–2177. [CrossRef]
48. Rincon, S.A.; Estravis, M.; Dingli, F.; Loew, D.; Tran, P.T.; Paoletti, A. SIN-Dependent Dissociation of the SAD Kinase Cdr2 from the Cell Cortex Resets the Division Plane. *Curr. Biol.* **2017**, *27*, 534–542. [CrossRef] [PubMed]
49. Willet, A.; DeWitt, A.K.; Beckley, J.R.; Clifford, D.M.; Gould, K.L. NDR Kinase Sid2 Drives Anillin-like Mid1 from the Membrane to Promote Cytokinesis and Medial Division Site Placement. *Curr. Biol.* **2019**, *29*, 1055–1063. [CrossRef]
50. Gavilan, M.A.; Lahoz, A.; Daga, R.; Jimenez, J. Feedback Regulation of SIN by Etd1 and Rho1 in Fission Yeast. *Genetics* **2014**, *196*, 455–470. [CrossRef]
51. Jin, Q.-W.; Zhou, M.; Bimbo, A.; Balasubramanian, M.; McCollum, D. A Role for the Septation Initiation Network in Septum Assembly Revealed by Genetic Analysis of sid2-250 Suppressors. *Genetics* **2006**, *172*, 2101–2112. [CrossRef]

52. Gupta, S.; Mana-Capelli, S.; McLean, J.R.; Chen, C.-T.; Ray, S.; Gould, K.; McCollum, D. Identification of SIN Pathway Targets Reveals Mechanisms of Crosstalk between NDR Kinase Pathways. *Curr. Biol.* **2013**, *23*, 333–338. [CrossRef] [PubMed]
53. Thiyagarajan, S.; Munteanu, E.L.; Arasada, R.; Pollard, T.D.; O'Shaughnessy, B. The fission yeast cytokinetic contractile ring regulates septum shape and closure. *J. Cell Sci.* **2015**, *128*, 3672–3681. [CrossRef]
54. Zhou, Z.; Munteanu, E.L.; He, J.; Ursell, T.; Bathe, M.; Huang, K.C.; Chang, F. The contractile ring coordinates curvature-dependent septum assembly during fission yeast cytokinesis. *Mol. Biol. Cell* **2015**, *26*, 78–90. [CrossRef] [PubMed]
55. Tebbs, I.R.; Pollard, T.D. Separate roles of IQGAP Rng2p in forming and constricting the Schizosaccharomyces pombe cytokinetic contractile ring. *Mol. Biol. Cell* **2013**, *24*, 1904–1917. [CrossRef]
56. Chang, F. Forces that shape fission yeast cells. *Mol. Biol. Cell* **2017**, *28*, 1819–1824. [CrossRef]
57. Proctor, S.A.; Minc, N.; Boudaoud, A.; Chang, F. Contributions of Turgor Pressure, the Contractile Ring, and Septum Assembly to Forces in Cytokinesis in Fission Yeast. *Curr. Biol.* **2012**, *22*, 1601–1608. [CrossRef]
58. Vjestica, A.; Tang, X.Z.; Oliferenko, S. The actomyosin ring recruits early secretory compartments to the division site in fission yeast. *Mol. Biol. Cell* **2008**, *19*, 1125–1138. [CrossRef]
59. Wang, N.; Lee, I.J.; Rask, G.; Wu, J.Q. Roles of the TRAPP-II Complex and the Exocyst in Membrane Deposition during Fission Yeast Cytokinesis. *PLoS Biol.* **2016**, *14*, e1002437. [CrossRef]
60. Zhu, Y.; Hyun, J.; Pan, Y.-Z.; Hopper, J.E.; Rizo, J.; Wu, J.-Q. Roles of the fission yeast UNC-13/Munc13 protein Ync13 in late stages of cytokinesis. *Mol. Biol. Cell* **2018**, *29*, 2259–2279. [CrossRef]
61. Frémont, S.; Echard, A. Membrane Traffic in the Late Steps of Cytokinesis. *Curr. Biol.* **2018**, *28*, R458–R470. [CrossRef] [PubMed]
62. Smertenko, A. Phragmoplast expansion: The four-stroke engine that powers plant cytokinesis. *Curr. Opin. Plant Biol.* **2018**, *46*, 130–137. [CrossRef] [PubMed]
63. Ramos, M.; Cortés, J.C.G.; Sato, M.; Rincón, S.A.; Moreno, M.B.; Clemente-Ramos, J.Á.; Osumi, M.; Pérez, P.; Ribas, J.C. Two *S. pombe* septation phases differ in ingression rate, septum structure, and response to F-actin loss. *J. Cell Biol.* **2019**, *218*, 4171–4194. [CrossRef] [PubMed]
64. Cortes, J.C.; Konomi, M.; Martins, I.M.; Munoz, J.; Moreno, M.B.; Osumi, M.; Duran, A.; Ribas, J.C. The (1,3)beta-D-glucan synthase subunit Bgs1p is responsible for the fission yeast primary septum formation. *Mol. Microbiol.* **2007**, *65*, 201–217. [CrossRef] [PubMed]
65. Humbel, B.M.; Konomi, M.; Takagi, T.; Kamasawa, N.; Ishijima, S.A.; Osumi, M. In situ localization of beta-glucans in the cell wall of *Schizosaccharomyces pombe*. *Yeast* **2001**, *18*, 433–444. [CrossRef] [PubMed]
66. Garcia Cortes, J.C.; Ramos, M.; Osumi, M.; Perez, P.; Ribas, J.C. The Cell Biology of Fission Yeast Septation. *Microbiol. Mol. Biol. Rev.* **2016**, *80*, 779–791. [CrossRef]
67. Garcia, P.; Garcia, I.; Marcos, F.; de Garibay, G.R.; Sanchez, Y. Fission yeast rgf2p is a rho1p guanine nucleotide exchange factor required for spore wall maturation and for the maintenance of cell integrity in the absence of rgf1p. *Genetics* **2009**, *181*, 1321–1334. [CrossRef]
68. García, P.; Tajadura, V.; García, I.; Sanchez, Y. Role of Rho GTPases and Rho-GEFs in the regulation of cell shape and integrity in fission yeast. *Yeast* **2006**, *23*, 1031–1043. [CrossRef]
69. Tajadura, V.; Garcia, B.; Garcia, I.; Garcia, P.; Sanchez, Y. *Schizosaccharomyces pombe* Rgf3p is a specific Rho1 GEF that regulates cell wall beta-glucan biosynthesis through the GTPase Rho1p. *J. Cell Sci.* **2004**, *117 Pt 25*, 6163–6174. [CrossRef]
70. Davidson, R.; Laporte, D.; Wu, J.-Q. Regulation of Rho-GEF Rgf3 by the arrestin Art1 in fission yeast cytokinesis. *Mol. Biol. Cell* **2015**, *26*, 453–466. [CrossRef]
71. Ren, L.; Willet, A.H.; Roberts-Galbraith, R.H.; McDonald, N.; Feoktistova, A.; Chen, J.-S.; Huang, H.; Guillen, R.; Boone, C.; Sidhu, S.S.; et al. The Cdc15 and Imp2 SH3 domains cooperatively scaffold a network of proteins that redundantly ensure efficient cell division in fission yeast. *Mol. Biol. Cell* **2015**, *26*, 256–269. [CrossRef] [PubMed]
72. Edreira, T.; Celador, R.; Manjon, E.; Sanchez, Y. A novel checkpoint pathway controls actomyosin ring constriction trigger in fission yeast. *eLife* **2020**, *9*, e59333. [CrossRef]
73. Davi, V.; Tanimoto, H.; Ershov, D.; Haupt, A.; De Belly, H.; Le Borgne, R.; Couturier, E.; Boudaoud, A.; Minc, N. Mechanosensation Dynamically Coordinates Polar Growth and Cell Wall Assembly to Promote Cell Survival. *Dev. Cell* **2018**, *45*, 170–182. [CrossRef]
74. Garcia, P.; Tajadura, V.; Sanchez, Y. The Rho1p exchange factor Rgf1p signals upstream from the Pmk1 mitogen-activated protein kinase pathway in fission yeast. *Mol. Biol. Cell* **2009**, *20*, 721–731. [CrossRef]
75. Dekker, N.; Speijer, D.; Grun, C.H.; van den Berg, M.; de Haan, A.; Hochstenbach, F. Role of the alpha-glucanase Agn1p in fission-yeast cell separation. *Mol. Biol. Cell* **2004**, *15*, 3903–3914. [CrossRef]
76. García, I.; Jiménez, D.; Martín, V.; Durán, A.; Sánchez, Y. The α-glucanase Agn1p is required for cell separation in *Schizosaccharomyces pombe*. *Biol. Cell* **2005**, *97*, 569–576. [CrossRef]
77. Martin-Cuadrado, A.B.; Duenas, E.; Sipiczki, M.; Vazquez de Aldana, C.R.; del Rey, F. The en-do-beta-1,3-glucanase eng1p is required for dissolution of the primary septum during cell separation in *Schizosaccharomyces pombe*. *J. Cell Sci.* **2003**, *116 Pt 9*, 1689–1698. [CrossRef]
78. Koyano, T.; Kume, K.; Konishi, M.; Toda, T.; Hirata, D. Search for Kinases Related to Transition of Growth Polarity in Fission Yeast. *Biosci. Biotechnol. Biochem.* **2010**, *74*, 1129–1133. [CrossRef]
79. Bohnert, K.A.; Gould, K.L. Cytokinesis-Based Constraints on Polarized Cell Growth in Fission Yeast. *PLoS Genet.* **2012**, *8*, e1003004. [CrossRef]

80. Sengar, A.; Markley, N.A.; Marini, N.J.; Young, D. Mkh1, a MEK kinase required for cell wall integrity and proper response to osmotic and temperature stress in *Schizosaccharomyces pombe*. *Mol. Cell. Biol.* **1997**, *17*, 3508–3519. [CrossRef]
81. Toda, T.; Dhut, S.; Superti-Furga, G.; Gotoh, Y.; Nishida, E.; Sugiura, R.; Kuno, T. The fission yeast pmk1+ gene encodes a novel mitogen-activated protein kinase homolog which regulates cell integrity and functions coordinately with the Protein Kinase C pathway. *Mol. Cell. Biol.* **1996**, *16*, 6752–6764. [CrossRef]
82. Zaitsevskaya-Carter, T.; Cooper, J.A. Spm1, a stress-activated MAP kinase that regulates morphogenesis in *S.pombe*. *EMBO J.* **1997**, *16*, 1318–1331. [CrossRef]
83. Wilkinson, M.G.; Samuels, M.; Takeda, T.; Toone, W.M.; Shieh, J.C.; Toda, T.; Millar, J.; Jones, N. The Atf1 transcription factor is a target for the Sty1 stress-activated MAP kinase pathway in fission yeast. *Genes Dev.* **1996**, *10*, 2289–2301. [CrossRef]
84. Takada, H.; Nishida, A.; Domae, M.; Kita, A.; Yamano, Y.; Uchida, A.; Ishiwata, S.; Fang, Y.; Zhou, X.; Masuko, T.; et al. The cell surface protein gene ecm33+ is a target of the two transcription factors Atf1 and Mbx1 and negatively regulates Pmk1 MAPK cell integrity signaling in fission yeast. *Mol. Biol. Cell* **2010**, *21*, 674–685. [CrossRef]
85. Takada, H.; Nishimura, M.; Asayama, Y.; Mannse, Y.; Ishiwata, S.; Kita, A.; Doi, A.; Nishida, A.; Kai, N.; Moriuchi, S.; et al. Atf1 Is a Target of the MAP Kinase Pmk1 and Regulates Cell Integrity in Fission Yeast. *Mol. Biol. Cell* **2007**, *18*, 4794–4802. [CrossRef]
86. Levin, D.E. Cell Wall Integrity Signaling in *Saccharomyces cerevisiae*. *Microbiol. Mol. Biol. Rev.* **2005**, *69*, 262–291. [CrossRef]
87. Madrid, M.; Nunez, A.; Soto, T.; Vicente-Soler, J.; Gacto, M.; Cansado, J. Stress-activated protein kinase-mediated down-regulation of the cell integrity pathway mitogen-activated protein kinase Pmk1p by protein phosphatases. *Mol. Biol. Cell* **2007**, *18*, 4405–4419. [CrossRef]
88. Madrid, M.; Soto, T.; Khong, H.K.; Franco, A.; Vicente, J.; Perez, P.; Gacto, M.; Cansado, J. Stress-induced response, localization, and regulation of the Pmk1 cell integrity pathway in *Schizosaccharomyces pombe*. *J. Biol. Chem.* **2006**, *281*, 2033–2043. [CrossRef]
89. Sanchez-Mir, L.; Franco, A.; Martin-Garcia, R.; Madrid, M.; Vicente-Soler, J.; Soto, T.; Gacto, M.; Perez, P.; Cansado, J. Rho2 Palmitoylation Is Required for Plasma Membrane Localization and Proper Signaling to the Fission Yeast Cell Integrity Mitogen-Activated Protein Kinase Pathway. *Mol. Cell. Biol.* **2014**, *34*, 2745–2759. [CrossRef]
90. Satoh, R.; Morita, T.; Takada, H.; Kita, A.; Ishiwata, S.; Doi, A.; Hagihara, K.; Taga, A.; Matsumura, Y.; Tohda, H.; et al. Role of the RNA-binding Protein Nrd1 and Pmk1 Mitogen-activated Protein Kinase in the Regulation of Myosin mRNA Stability in Fission Yeast. *Mol. Biol. Cell* **2009**, *20*, 2473–2485. [CrossRef] [PubMed]
91. Bähler, J. A Transcriptional Pathway for Cell Separation in Fission Yeast. *Cell Cycle* **2004**, *4*, 39–41. [CrossRef]
92. Bi, E.; Park, H.-O. Cell Polarization and Cytokinesis in Budding Yeast. *Genetics* **2012**, *191*, 347–387. [CrossRef]
93. Howell, A.S.; Lew, D.J. Morphogenesis and the Cell Cycle. *Genetics* **2012**, *190*, 51–77. [CrossRef] [PubMed]
94. Wloka, C.; Nishihama, R.; Onishi, M.; Oh, Y.; Hanna, J.; Pringle, J.R.; Krauss, M.; Bi, E. Evidence that a septin diffusion barrier is dispensable for cytokinesis in budding yeast. *Biol. Chem.* **2011**, *392*, 813–829. [CrossRef]
95. Farkašovský, M. Septin architecture and function in budding yeast. *Biol. Chem.* **2020**, *401*, 903–919. [CrossRef]
96. Spiliotis, E.T.; McMurray, M.A. Masters of asymmetry—Lessons and perspectives from 50 years of septins. *Mol. Biol. Cell* **2020**, *31*, 2289–2297. [CrossRef]
97. Miller, K.E.; Kang, P.J.; Park, H.-O. Regulation of Cdc42 for polarized growth in budding yeast. *Microb. Cell* **2020**, *7*, 175–189. [CrossRef]
98. Drgonova, J.; Drgon, T.; Tanaka, K.; Kollár, R.; Chen, G.-C.; Ford, R.A.; Chan, C.S.M.; Takai, Y.; Cabib, E. Rho1p, a Yeast Protein at the Interface between Cell Polarization and Morphogenesis. *Science* **1996**, *272*, 277–279. [CrossRef]
99. Qadota, H.; Ishii, I.; Fujiyama, A.; Ohya, Y.; Anraku, Y. RHO gene products, putative small GTP-binding proteins, are important for activation of the CAL1/CDC43 gene product, a protein geranylgeranyltransferase in *Saccharomyces cerevisiae*. *Yeast* **1992**, *8*, 735–741. [CrossRef]
100. Heinisch, J.J.; Rodicio, R. Protein kinase C in fungi—More than just cell wall integrity. *FEMS Microbiol. Rev.* **2017**, *42*, fux051. [CrossRef] [PubMed]
101. Andrews, P.D.; Stark, M.J. Dynamic, Rho1p-dependent localization of Pkc1p to sites of polarized growth. *J. Cell Sci.* **2000**, *113*, 2685–2693. [CrossRef] [PubMed]
102. Kono, K.; Saeki, Y.; Yoshida, S.; Tanaka, K.; Pellman, D. Proteasomal degradation resolves competition between cell polarization and cellular wound healing. *Cell* **2012**, *150*, 151–164. [CrossRef] [PubMed]
103. Schmidt, M.; Varma, A.; Drgon, T.; Bowers, B.; Cabib, E. Septins, under Cla4p Regulation, and the Chitin Ring Are Required for Neck Integrity in Budding Yeast. *Mol. Biol. Cell* **2003**, *14*, 2128–2141. [CrossRef]
104. Gomez, A.; Perez, J.; Reyes, A.; Duran, A.; Roncero, C. Slt2 and Rim101 Contribute Independently to the Correct Assembly of the Chitin Ring at the Budding Yeast Neck in Saccharomyces cerevisiae. *Eukaryot. Cell* **2009**, *8*, 1449–1459. [CrossRef]
105. Lesage, G.; Shapiro, J.; Specht, C.A.; Sdicu, A.-M.; Ménard, P.; Hussein, S.; Tong, A.H.Y.; Boone, C.; Bussey, H. An interactional network of genes involved in chitin synthesis in *Saccharomyces cerevisiae*. *BMC Genet.* **2005**, *6*, 8. [CrossRef] [PubMed]
106. Lenardon, M.D.; Munro, C.; Gow, N.A. Chitin synthesis and fungal pathogenesis. *Curr. Opin. Microbiol.* **2010**, *13*, 416–423. [CrossRef]
107. Valdivia, R.H.; Schekman, R. The yeasts Rho1p and Pkc1p regulate the transport of chitin synthase III (Chs3p) from internal stores to the plasma membrane. *Proc. Natl. Acad. Sci. USA* **2003**, *100*, 10287–10292. [CrossRef]
108. DeMarini, D.J.; Adams, A.E.; Fares, H.; De Virgilio, C.; Valle, G.; Chuang, J.S.; Pringle, J.R. A septin-based hierarchy of proteins required for localized deposition of chitin in the Saccharomyces cerevisiae cell wall. *J. Cell Biol.* **1997**, *139*, 75–93. [CrossRef]

109. Reyes, A.; Sanz, M.; Duran, A.; Roncero, C. Chitin synthase III requires Chs4p-dependent translocation of Chs3p into the plasma membrane. *J. Cell Sci.* **2007**, *120*, 1998–2009. [CrossRef]
110. Sacristan, C.; Reyes, A.; Roncero, C. Neck compartmentalization as the molecular basis for the different endocytic behaviour of Chs3 during budding or hyperpolarized growth in yeast cells. *Mol. Microbiol.* **2012**, *83*, 1124–1135. [CrossRef]
111. Oh, Y.; Schreiter, J.H.; Okada, H.; Wloka, C.; Okada, S.; Yan, D.; Duan, X.; Bi, E. Hof1 and Chs4 Interact via F-BAR Domain and Sel1-like Repeats to Control Extracellular Matrix Deposition during Cytokinesis. *Curr. Biol.* **2017**, *27*, 2878–2886. [CrossRef]
112. Lagorce, A.; Le Berre-Anton, V.; Aguilar-Uscanga, B.; Martin-Yken, H.; Dagkessamanskaia, A.; François, J.M. Involvement ofGFA1, which encodes glutamine-fructose-6-phosphate amidotransferase, in the activation of the chitin synthesis pathway in response to cell-wall defects in *Saccharomyces cerevisiae. JBIC J. Biol. Inorg. Chem.* **2002**, *269*, 1697–1707. [CrossRef]
113. Niwa, M. A cell cycle checkpoint for the endoplasmic reticulum. *Biochim. Biophys. Acta Mol. Cell. Res.* **2020**, *1867*, 118825. [CrossRef]
114. Chin, C.F.; Bennett, A.M.; Ma, W.K.; Hall, M.; Yeong, F.M. Dependence of Chs2 ER export on dephosphorylation by cytoplasmic Cdc14 ensures that septum formation follows mitosis. *Mol. Biol. Cell* **2012**, *23*, 45–58. [CrossRef]
115. Zhang, G.; Kashimshetty, R.; Ng, K.E.; Tan, H.B.; Yeong, F.M. Exit from mitosis triggers Chs2p transport from the endoplasmic reticulum to mother–daughter neck via the secretory pathway in budding yeast. *J. Cell Biol.* **2006**, *174*, 207–220. [CrossRef]
116. Tolliday, N.; VerPlank, L.; Li, R. Rho1 directs formin-mediated actin ring assembly during budding yeast cytokinesis. *Curr. Biol.* **2002**, *12*, 1864–1870. [CrossRef]
117. Yoshida, S.; Kono, K.; Lowery, D.M.; Bartolini, S.; Yaffe, M.B.; Ohya, Y.; Pellman, D. Polo-Like Kinase Cdc5 Controls the Local Activation of Rho1 to Promote Cytokinesis. *Science* **2006**, *313*, 108–111. [CrossRef]
118. Cabib, E. The septation apparatus, a chitin-requiring machine in budding yeast. *Arch. Biochem. Biophys.* **2004**, *426*, 201–207. [CrossRef]
119. Arellano, M.; Duran, A.; Perez, P. Rho 1 GTPase activates the (1-3)beta-D-glucan synthase and is involved in *Schizosaccharomyces pombe* morphogenesis. *EMBO J.* **1996**, *15*, 4584–4591. [CrossRef]
120. Schmidt, M.; Bowers, B.; Varma, A.; Roh, D.-H.; Cabib, E. In budding yeast, contraction of the actomyosin ring and formation of the primary septum at cytokinesis depend on each other. *J. Cell Sci.* **2002**, *115*, 293–302. [CrossRef]
121. Ver Plank, L.; Li, R. Cell Cycle-regulated Trafficking of Chs2 Controls Actomyosin Ring Stability during Cytokinesis. *Mol. Biol. Cell* **2005**, *16*, 2529–2543. [CrossRef] [PubMed]
122. Foltman, M.; Filali-Mouncef, Y.; Crespo, D.; Sanchez-Diaz, A. Cell polarity protein Spa2 coordinates Chs2 incorporation at the division site in budding yeast. *PLoS Genet.* **2018**, *14*, e1007299. [CrossRef] [PubMed]
123. Foltman, M.; Molist, I.; Arcones, I.; Sacristan, C.; Filali-Mouncef, Y.; Roncero, C.; Sanchez-Diaz, A. Ingression Progression Complexes Control Extracellular Matrix Remodelling during Cytokinesis in Budding Yeast. *PLoS Genet.* **2016**, *12*, e1005864. [CrossRef]
124. Devrekanli, A.; Foltman, M.; Roncero, C.; Sanchez-Diaz, A.; Labib, K. Inn1 and Cyk3 regulate chitin synthase during cytokinesis in budding yeasts. *J. Cell Sci.* **2012**, *125*, 5453–5466. [CrossRef]
125. Guo, S.; Shen, X.; Yan, G.; Ma, D.; Bai, X.; Li, S.; Jiang, Y. A MAP Kinase Dependent Feedback Mechanism Controls Rho1 GTPase and Actin Distribution in Yeast. *PLoS ONE* **2009**, *4*, e6089. [CrossRef]
126. Prosser, D.C.; Drivas, T.G.; Maldonado-Báez, L.; Wendland, B. Existence of a novel clathrin-independent endocytic pathway in yeast that depends on Rho1 and formin. *J. Cell Biol.* **2011**, *195*, 657–671. [CrossRef]
127. Boettner, D.R.; D'Agostino, J.L.; Torres, O.T.; Daugherty-Clarke, K.; Uygur, A.; Reider, A.; Wendland, B.; Lemmon, S.K.; Goode, B.L. The F-BAR protein Syp1 negatively regulates WASp-Arp2/3 complex activity during endocytic patch formation. *Curr. Biol.* **2009**, *19*, 1979–1987. [CrossRef]
128. Merlini, L.; Bolognesi, A.; Juanes, M.A.; Vandermoere, F.; Courtellemont, T.; Pascolutti, R.; Séveno, M.; Barral, Y.; Piatti, S. Rho1- and Pkc1-dependent phosphorylation of the F-BAR protein Syp1 contributes to septin ring assembly. *Mol. Biol. Cell* **2015**, *26*, 3245–3262. [CrossRef]
129. Mascaraque, V.; Hernaez, M.L.; Jimenez-Sanchez, M.; Hansen, R.; Gil, C.; Martin, H.; Cid, V.J.; Molina, M. Phosphoproteomic analysis of protein kinase C signaling in Saccharomyces cerevisiae reveals Slt2 mitogen-activated protein kinase (MAPK)-dependent phosphorylation of eisosome core components. *Mol. Cell Proteom.* **2013**, *12*, 557–574. [CrossRef] [PubMed]
130. Ziolkowska, N.E.; Karotki, L.; Rehman, M.; Huiskonen, J.T.; Walther, T.C. Eisosome-driven plasma membrane organization is mediated by BAR domains. *Nat. Struct. Mol. Biol.* **2011**, *18*, 854–856. [CrossRef] [PubMed]
131. Oh, Y.; Chang, K.J.; Orlean, P.; Wloka, C.; Deshaies, R.; Bi, E. Mitotic exit kinase Dbf2 directly phosphorylates chitin synthase Chs2 to regulate cytokinesis in budding yeast. *Mol. Biol. Cell* **2012**, *23*, 2445–2456. [CrossRef]
132. Shaw, J.A.; Mol, P.C.; Bowers, B.; Silverman, S.J.; Valdivieso, M.-H.; Durán, A.; Cabib, E. The function of chitin synthases 2 and 3 in the Saccharomyces cerevisiae cell cycle. *J. Cell Biol.* **1991**, *114*, 111–123. [CrossRef]
133. Yoshida, S.; Bartolini, S.; Pellman, D. Mechanisms for concentrating Rho1 during cytokinesis. *Genes Dev.* **2009**, *23*, 810–823. [CrossRef] [PubMed]
134. Cabib, E.; Blanco, N.; Grau, C.; Rodriguez-Pena, J.M.; Arroyo, J. Crh1p and Crh2p are required for the cross-linking of chitin to beta(1-6)glucan in the Saccharomyces cerevisiae cell wall. *Mol. Microbiol.* **2007**, *63*, 921–935. [CrossRef]
135. Cabib, E.; Arroyo, J. How carbohydrates sculpt cells: Chemical control of morphogenesis in the yeast cell wall. *Nat. Rev. Genet.* **2013**, *11*, 648–655. [CrossRef] [PubMed]

136. Dohrmann, P.R.; Butler, G.; Tamai, K.; Dorland, S.; Greene, J.R.; Thiele, D.J.; Stillman, D.J. Parallel pathways of gene regulation: Homologous regulators SWI5 and ACE2 differentially control transcription of HO and chitinase. *Genes Dev.* **1992**, *6*, 93–104. [CrossRef]
137. Kuranda, M.J.; Robbins, P.W. Chitinase is required for cell separation during growth of *Saccharomyces cerevisiae*. *J. Biol. Chem.* **1991**, *266*, 19758–19767. [CrossRef]
138. Baladron, V.; Ufano, S.; Duenas, E.; Martin-Cuadrado, A.B.; del Rey, F.; Vazquez de Aldana, C.R. Eng1p, an endo-1,3-beta-glucanase localized at the daughter side of the septum, is involved in cell separation in Saccharomyces cerevisiae. *Eukaryot. Cell* **2002**, *1*, 774–786. [CrossRef]
139. Colman-Lerner, A.; Chin, T.E.; Brent, R. Yeast Cbk1 and Mob2 Activate Daughter-Specific Genetic Programs to Induce Asymmetric Cell Fates. *Cell* **2001**, *107*, 739–750. [CrossRef]
140. Cabib, E.; Silverman, S.J.; Shaw, J.A. Chitinase and chitin synthase 1: Counterbalancing activities in cell separation of Saccharomyces cerevisiae. *J. Gen. Microbiol.* **1992**, *138*, 97–102. [CrossRef]
141. Brace, J.L.; Doerfler, M.; Weiss, E.L. A cell separation checkpoint that enforces the proper order of late cytokinetic events. *J. Cell Biol.* **2019**, *218*, 150–170. [CrossRef]
142. Madden, K.; Sheu, Y.-J.; Baetz, K.; Andrews, B.; Snyder, M. SBF Cell Cycle Regulator as a Target of the Yeast PKC-MAP Kinase Pathway. *Science* **1997**, *275*, 1781–1784. [CrossRef]
143. Jonasson, E.M.; Rossio, V.; Hatakeyama, R.; Abe, M.; Ohya, Y.; Yoshida, S. Zds1/Zds2-PP2ACdc55 com-plex specifies signaling output from Rho1 GTPase. *J. Cell Biol.* **2016**, *212*, 51–61. [CrossRef] [PubMed]
144. Perez, J.; Arcones, I.; Gomez, A.; Casquero, V.; Roncero, C. Phosphorylation of Bni4 by MAP kinases contributes to septum assembly during yeast cytokinesis. *FEMS Yeast Res.* **2016**, *16*, fow060. [CrossRef] [PubMed]
145. Zou, J.; Friesen, H.; Larson, J.; Huang, D.; Cox, M.; Tatchell, K.; Andrews, B. Regulation of Cell Polarity through Phosphorylation of Bni4 by Pho85 G1 Cyclin-dependent Kinases in Saccharomyces cerevisiae. *Mol. Biol. Cell* **2009**, *20*, 3239–3250. [CrossRef] [PubMed]
146. Sanz, M.; Castrejón, F.; Durán, A.; Roncero, C. *Saccharomyces cerevisiae* Bni4p directs the formation of the chitin ring and also participates in the correct assembly of the septum structure. *Microbiology* **2004**, *150*, 3229–3241. [CrossRef]
147. Kozubowski, L.; Larson, J.; Tatchell, K. Role of the Septin Ring in the Asymmetric Localization of Proteins at the Mother-Bud Neck in Saccharomyces cerevisiae. *Mol. Biol. Cell* **2005**, *16*, 3455–3466. [CrossRef]
148. Dichtl, K.; Samantaray, S.; Wagener, J. Cell wall integrity signalling in human pathogenic fungi. *Cell. Microbiol.* **2016**, *18*, 1228–1238. [CrossRef] [PubMed]

Journal of *Fungi*

MDPI

Article

Structure of the Yeast Cell Wall Integrity Sensor Wsc1 Reveals an Essential Role of Surface-Exposed Aromatic Clusters

Philipp Schöppner [1], Anne Pia Lutz [1], Bernard Johannes Lutterbach [1], Stefan Brückner [1], Lars-Oliver Essen [2,*] and Hans-Ulrich Mösch [1,*]

1 Department of Genetics, Philipps-Universität, Karl-von-Frisch-Strasse 8, 35043 Marburg, Germany; schoeppl@staff.uni-marburg.de (P.S.); anne.lutz@hotmail.de (A.P.L.); bernard.lutterbach@gmx.net (B.J.L.); brueckns@staff.uni-marburg.de (S.B.)

2 Department of Biochemistry, Philipps-Universität, Hans-Meerwein-Strasse 4, 35043 Marburg, Germany

* Correspondence: essenl@staff.uni-marburg.de (L.-O.E.); moesch@staff.uni-marburg.de (H.-U.M.); Tel.: +49-6421-282-2032 (L.-O.E.); +49-6421-282-3013 (H.-U.M.)

Abstract: In the yeast *Saccharomyces cerevisiae* and other ascomycetes, the maintenance of cell wall integrity is governed by a family of plasma-membrane spanning sensors that include the Wsc-type proteins. These cell wall proteins apparently sense stress-induced mechanical forces at the cell surface and target the cell wall integrity (CWI) signaling pathway, but the structural base for their sensor function is yet unknown. Here, we solved a high-resolution crystal structure of the extracellular cysteine-rich domain (CRD) of yeast Wsc1, which shows the characteristic PAN/Apple domain fold with two of the four Wsc1 disulfide bridges being conserved in other PAN domain cores. Given the general function of PAN domains in mediating protein–protein and protein–carbohydrate interactions, this finding underpins the importance of Wsc domains in conferring sensing and localization functions. Our Wsc1 CRD structure reveals an unusually high number of surface-exposed aromatic residues that are conserved in other fungal CRDs, and can be arranged into three solvent-exposed clusters. Mutational analysis demonstrates that two of the aromatic clusters are required for conferring *S. cerevisiae* Wsc1-dependent resistance to the glucan synthase inhibitor caspofungin, and the chitin-binding agents Congo red and Calcofluor white. These findings suggest an essential role of surface-exposed aromatic clusters in fungal Wsc-type sensors that might include an involvement in stress-induced sensor-clustering required to elicit appropriate cellular responses via the downstream CWI pathway.

Keywords: fungal cell wall; *Saccharomyces cerevisiae*; membrane sensor; cysteine-rich domain; PAN domain; aromatic clusters

Citation: Schöppner, P.; Lutz, A.P.; Lutterbach, B.J.; Brückner, S.; Essen, L.-O.; Mösch, H.-U. Structure of the Yeast Cell Wall Integrity Sensor Wsc1 Reveals an Essential Role of Surface-Exposed Aromatic Clusters. *J. Fungi* **2022**, *8*, 379. https://doi.org/10.3390/jof8040379

Academic Editors: María Molina and Humberto Martín

Received: 25 March 2022
Accepted: 7 April 2022
Published: 8 April 2022

1. Introduction

The fungal cell wall is a highly specialized and dynamic cellular compartment that plays a key role in the maintenance of cellular integrity and protection of cells against physical stress or mechanical damage [1,2]. Genome-wide studies employing the budding yeast *Saccharomyces cerevisiae* and other fungi have revealed that approximately 20% of fungal genomes are involved in cell wall biosynthesis, function, and regulation [3–5]. Comprehensive studies in *S. cerevisiae* have shown that sensing of structural alterations within the cell wall during cell growth, and in response to chemical or mechanical stress, is essential for the viability of fungal cells [6]. These studies have also provided detailed insights into the signaling systems that confer appropriate physiological responses to counteract cell wall stress, such as the treatment of fungal cells with cell-wall-perturbing agents [7–13]. Together, the sensors and signaling components of this system are named the cell wall integrity (CWI) pathway [6], which can be found not only in *S. cerevisiae*, but, in homologous form, also in other fungi [14]. At its core, the CWI pathway comprises a small number of membrane-spanning sensors, protein kinase C (PKC), a mitogen-activated protein kinase (MAPK) cascade, and a number of nuclear transcription factors that confer expression of cell wall biogenesis genes [15,16].

In *S. cerevisiae*, five plasma membrane-spanning proteins have been identified as sensors of the CWI pathway [13,17]. Three of these sensors (*Sc*Wsc1, *Sc*Wsc2, and *Sc*Wsc3) are characterized by an N-terminal, extracellular cysteine-rich domain (CRD), or WSC domain (PFAM entry PF01822; INTERPRO entry: IPR002889), which is followed by an extracellular, glycosylated serine/threonine-rich region (STR), a single transmembrane domain (TMD), and a C-terminal cytoplasmic tail (Figure 1). The best studied variant is *Sc*Wsc1, which has been characterized by genetic, biophysical, and structural methods. Genetic studies have shown that *Sc*Wsc1 is required to confer resistance against numerous types of cell wall stress, including treatment of cells with the β-1,3-glucan synthase inhibitor caspofungin, the chitin-binding agents Congo red and Calcofluor white, or caffeine [12,18]. Further studies have demonstrated that *Sc*Wsc1 function requires the presence of the N-terminal CRD [19], which can also be inactivated by mutation of its eight cysteine residues that are a hallmark of this highly conserved domain [20]. Experiments employing atomic-force microscopy (AFM) further indicate that *Sc*Wsc1 sensors form nanosprings, and act as mechanosensors [21], which can be localized in specific microdomains within the plasma membrane [22]. These studies also suggest that the CRD of *Sc*Wsc proteins is required for sensor clustering observed in response to cell wall stress [20]. Purification of *Sc*Wsc1 from yeast cells in lipid nanoparticles and transmission electron microscopy have recently provided the first broad three-dimensional model, which confirms the proposed tripartite organization of the sensor [23]. These studies suggest that *Sc*Wsc proteins may act as mechanosensors that detect structural alterations of the cell wall by their extracellular CRD and STR domains, and, upon stress, undergo extensive conformational changes, which are transferred to the cytoplasmic tail, and trigger activation of downstream signaling components of the CWI pathway [17,23]. In addition, Wsc1 may have structural functions at the yeast cell wall independently of the CWI pathway.

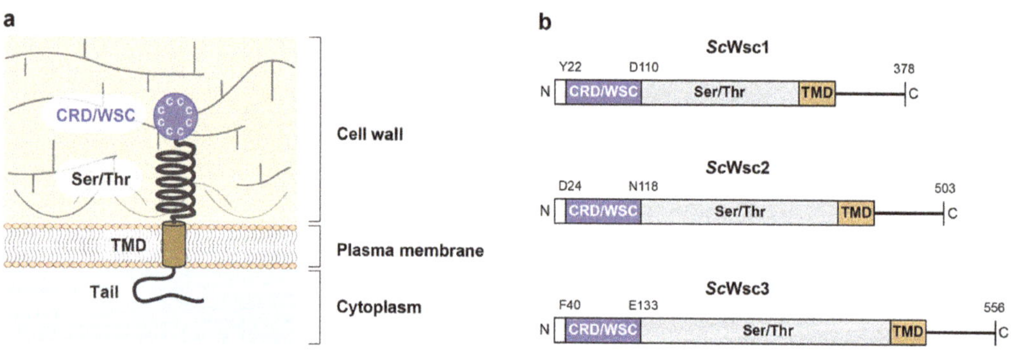

Figure 1. Structural model and architecture of fungal WSC family sensors. (**a**) Model showing the topology and proposed functional domains of WSC domain sensors, including the N-terminal cysteine-rich domain (CRD) or WSC domain, the glycosylated serine/threonine-rich domain (Ser/Thr), the transmembrane domain (TMD), and the C-terminal cytoplasmic tail [17]. (**b**) Domain architecture of the *S. cerevisiae* Wsc family mechanosensors Wsc1 (*Sc*Wsc1, UniProt entry P54867), Wsc2 (*Sc*Wsc2, UniProt entry P53832), and Wsc3 (*Sc*Wsc3, UniProt entry Q12215). Residues corresponding to the borders of the CRD/WSC domains and the C-terminal ends are indicated.

Although the CRD of yeast Wsc proteins is essential for sensor function in vivo, its precise molecular function is unknown, and no high-resolution crystal structure of any fungal WSC domain is available so far. In the PFAM 35.0 database (state: 11/2021), the CRD/WSC domain (PF01822) can be found in 7660 eukaryotic sequences, 2672 of which are present in a total of 331 different fungal species. Despite their wide distribution, only a few fungal proteins with a WSC domain have been attributed a molecular function. These include the *Sc*Wsc sensors discussed above and a number of fungal orthologs, which also contain a CRD together with a transmembrane domain, and are involved in activation of the CWI pathway [24–27]. WSC domains have also been found in other domain contexts, e.g., in the β-1-3-exoglucanase *Th*Crd2 of the mycoparasitic fungus *Trichoderma harzianum* [28], and the β-glucan-binding lectin *Si*Wsc3 from the root endophyte *Serendipita indica* [29], indicating a carbohydrate-binding function of CRD/WSC domains. In *Homo sapiens*, the WSC-domain-containing Kremen receptor proteins Krm1/2 have been described to form ternary complexes with the Wnt co-receptors Lrp5/6 and the Wnt antagonistic Dickkopf proteins Dkk1/2 in the vertebrate Wnt/β-catenin signaling pathway [30,31]. Importantly, crystal structures are available for Krm1 (PDB ID 5FWS) and the ternary Krm1–Lrp6–Dkk1 complex (PDB ID 5FWW), which show that the WSC domain of Krm1 forms direct polar contacts with Dkk1 [32]. These studies indicate a second function of CRD/WSC domains in conferring specific protein–protein interactions.

In this study, we aimed at characterizing the structural and functional properties of *Sc*Wsc sensors in more detail. For this purpose, we determined the high-resolution crystal structure of the CRD of *Sc*Wsc1, revealing a tightly packed globular core domain stabilized by four disulfide bonds. The protein surface of the CRD contains three aromatic clusters that are conserved in the CRDs of both *Sc*Wsc2 and *Sc*Wsc3. Our mutational analysis reveals that these surface-exposed aromatic clusters play essential roles in conferring sensor function under cell wall stress conditions, indicating that they might be involved in eliciting appropriate cellular responses via the downstream CWI pathway.

2. Materials and Methods

2.1. Yeast Strains and Growth Tests

Yeast strain YHUM2959 carrying a *wsc1Δ* mutation was obtained by deletion of the *WSC1* gene in yeast strain ESM356-1 (S288c, *MATα*, *ura3-52*, *leu2Δ1*, *his3Δ200*, *trp1Δ63*), [33] following a previously described protocol, and plasmid pFA6a-natNT2 [34].

For growth tests under cell wall stress conditions, yeast strains ESM356-1 or YHUM2959 were freshly transformed with appropriate plasmids (Table 1). Transformants were grown to logarithmic phase in liquid SC-Trp medium, and fivefold serial dilutions of the cultures were spotted onto solid YPD medium supplemented with either 0.4 μg/mL caspofungin (Sigma-Aldrich, Taufkirchen, Germany), 30 μg/mL Congo red (Carl-Roth, Karlsruhe, Germany), 100 μg/mL calcofluor (fluorescent brightener 28, Sigma-Aldrich, Taufkirchen, Germany), or 2 mg/mL caffeine (Sigma-Aldrich, Taufkirchen, Germany). Growth on plates was documented by photography after incubation at 30 °C for 3–5 days. Standard methods for yeast culture medium and transformation were used as described previously [35].

Table 1. Plasmids used in this study.

Plasmid	Relevant Genotype	Source
pET-28(a)+	*P_{T7}-6xHis lacI Kan^R*	Merck, Germany
BHUM3120	*WSC1*$^{(22-118)}$ in pET-28(a)+	this study
BHUM3121	*WSC2*$^{(24-118)}$ in pET-28(a)+	this study
BHUM3122	*WSC3*$^{(40-133)}$ in pET-28(a)+	this study
pFA6a-natNT2	*PFA6a-NatNT2 Amp^R*	[34]
pCR95	*mNeonGreen::NatNT2 Amp^R*	[36]
pRS314	*TRP1 ARS CEN2 Amp^R*	[37]
BHUM3291	*WSC1* in pRS314	this study
BHUM3293	*WSC1*$^{Y22A\ Y24A\ Y107A}$ in pRS314	this study
BHUM3295	*WSC1*$^{Y64A\ Y70A\ Y104A}$ in pRS314	this study
BHUM3297	*WSC1*$^{Y41A\ W43A\ Y89A\ F91A\ Y93A}$ in pRS314	this study
BHUM3301	*WSC1*$^{\Delta CRD}$ in pRS314	this study
BHUM3303	*WSC1-mNeonGreen* in pRS314	this study
BHUM3304	*WSC1*$^{Y22A\ Y24A\ Y107A}$-mNeonGreen in pRS314	this study
BHUM3305	*WSC1*$^{Y64A\ Y70A\ Y104A}$*-mNeonGreen* in pRS314	this study
BHUM3306	*WSC1*$^{Y41A\ W43A\ Y89A\ F91A\ Y93A}$*-mNeonGreen* in pRS314	this study
BHUM3308	*WSC1*$^{\Delta CRD}$*-mNeonGreen* in pRS314	this study

2.2. Plasmids

All plasmids used in this study are listed in Table 1. For bacterial production of CRDs from *Sc*Wsc1, *Sc*Wsc2, and *Sc*Wsc3, the pET-28a(+) expression system was used (Merck, Darmstadt, Germany). Construction of expression plasmids BHUM3120, BHUM3121, and BUM3122, respectively, was carried out by the cloning of codon-optimized sequences (BioCat, Heidelberg, Germany) from pMA-based cloning vectors into the expression vector pET-28a(+) by *Nde*I/*Xho*I restriction and ligation. For expression of different *mNeonGreen*-tagged *WSC1* gene variants in *S. cerevisiae*, plasmids BHUM3303, BHMU3304, BHMU3305, BHMU3306, and BHUM3308 were constructed using several steps. (i) In a first step, a 2 kb fragment carrying the *WSC1* locus (−711 to +181 relative to the translational start site ATG at +1) was isolated from the yeast genome by PCR, integrated into the yeast vector pRS314 to obtain plasmid BHUM 3291 (pRS314-WSC1; map shown in Figure S1), and verified by sequencing. (ii) In a second step, plasmids BHUM3293 (pRS314-WSC1$^{Y22A\ Y24A\ Y107A}$), BHUM3295 (pRS314-WSC1$^{Y64A\ Y70A\ Y104A}$), BHUM3297 (pRS314-WSC1$^{Y41A\ W43A\ Y89A\ F91A\ Y93A}$), and BHUM3301 (pRS314-WSC1$^{\Delta CRD}$) were constructed by site-directed mutagenesis using BHUM3291 as a template. All mutations were verified by DNA sequencing. (iii) In a third step, plasmids BHUM 3303 (pRS314-WSC1-mNeonGreen), BHUM3304 (pRS314-WSC1$^{Y22A\ Y24A\ Y107A}$-mNeonGreen), BHUM3305 (pRS314-WSC1$^{Y64A\ Y70A\ Y104A}$-mNeonGreen), BHUM3306 (pRS314-WSC1$^{Y41A\ W43A\ Y89A\ F91A\ Y93A}$-mNeonGreen), and BHUM3308 (pRS314-WSC1$^{\Delta CRD}$-mNeon Green) were constructed by homologous recombination cloning in yeast using three fragments with overlapping ends. Two of the fragments, which carry the vector pRS314 and the *WSC1* gene or mutated variants, respectively, were obtained by PCR using either one of the plasmids BHUM3291, BHUM3293, BHUM3295, BHUM3297, or BHUM 3301 as templates, and appropriate primers adding 30–40 bp long homologous ends. A third fragment, coding for the *mNeonGreen* gene, was obtained by PCR using plasmid pCR95 [36] as template, and two primers adding appropriate homologous areas. Each three appropriate fragments were then simultaneously transformed into competent yeast cells to obtain the desired plasmids by tripartite homologous recombination. Upon growth of transformed yeast on solid SC-Trp selection medium, resulting plasmids were isolated from transformants, amplified in *E. coli*, and verified by DNA sequence analysis. Maps of the resulting plasmids are shown in Figure S1.

2.3. Recombinant Production of CRD Proteins

CRDs of *Sc*Wsc1, *Sc*Wsc2, and *Sc*Wsc3 were produced following a low temperature protocol [38] using the expression plasmids BHUM3120, BHUM3121, and BUM3122, and the *E. coli* strain SHuffle T7 express (New England Biolabs GmbH, Frankfurt, Germany).

Proteins were purified by Ni-NTA affinity chromatography (Macherey-Nagel, Düren, Germany) and subsequent size exclusion chromatography using a HiLoad Superdex 75 pg column (GE Healthcare, Munich, Germany). All steps were carried out in AML buffer (20 mM Tris/HCl, pH 8.0, 350 mM NaCl).

2.4. CD Spectroscopy and Thermal Shift Assays

CD spectroscopy with recombinant CRDs was performed with a JASCO J-810 spectropolarimeter (JASCO, Pfungstadt, Germany) in quartz cuvettes with a 1 mm gap. Proteins were measured at a concentration of 100 µg/mL in a 50 mM $NaHCO_3$ buffer [pH 8.3] in the far UV range (190–260 nm) at 20 °C. All spectra were measured three times against the buffer spectrum. Collected data were converted to molar ellipticity per amino acid. The *Jasco Secondary Structure Estimation* software (JASCO, Pfungstadt, Germany) was used for comparison with a reference spectrum to estimate secondary structure compositions.

Thermal shift assays with recombinant CRDs were performed in a *Rotor-Gene Q* real-time PCR cycler (Qiagen, Hilden, Germany) in volumes of 40 µL, using UV-permeable PCR cups. The temperature was increased linearly by 2 °C per minute. Proteins were measured at a concentration of 1 mg/mL in the presence of *SYPRO*™ Orange dye (7.8×) in either one of the following buffers: 20 mM sodium citrate (pH 3); 20 mM sodium acetate (pH 4); 20 mM sodium acetate (pH 5); 20 mM MES (pH 6); 20 mM HEPES (pH 7); 20 mM Tris-Cl (pH 8); 20 mM CHES (pH 9); 20 mM CHES (pH 10).

2.5. Crystallization

Crystal screening with recombinant CRDs was performed with commercially available screens (Qiagen, Hilden, Germany) in a 600 nl sitting drop setup using a Digilab Honeybee 963 dispensing system (Genomic Solutions, Huntingdon Cambridgeshire, UK). Crystals for the CRD of *Sc*Wsc1 belonging to the space group $P12_11$ could be obtained at a protein concentration of 70 mg/mL at 18 °C in conditions containing 80 mM sodium acetate (pH 4.5), 1.6 M ammonium sulfate, and 20% (*v*/*v*) glycerol. All crystals were frozen in mother liquor.

2.6. X-ray Data Collection, Structure Solution, and Analysis

Datasets were recorded at the ESRF (Grenoble), beamline ID29. Data integration was performed with *XDS* [39]. *XSCALE* [40] or *SCALA* [41] were used for scaling, both run within the CCP4 [42] software suite. The structure of the *Sc*Wsc1 CRD was solved via molecular replacement with *PHASER* [43] using a carefully trimmed model of the WSC domain from the human transmembrane sensor Krm1 (PDB ID 5FWS). Refinement was done with alternating rounds of *phenix.refine* [44] and *Coot* [45].

2.7. Western-Blot Analysis

Yeast strains expressing different *mNeonGreen*-tagged *WSC1* gene variants were analyzed by Western blot analysis after growth to logarithmic phase and preparation of protein extracts as previously described [46]. Equal amounts of protein extracts were separated by 12% SDS-PAGE, and transferred to nitrocellulose membranes. Wsc1-mNeonGreen proteins were detected using enhanced chemiluminescence (ECL) technology after incubation of membranes with monoclonal mouse anti-mNeonGreen antibodies (ChromoTek GmbH, Munich, Germany) and peroxidase-coupled goat anti-mouse secondary antibodies (Dianova GmbH, Hamburg, Germany). Membranes were then stripped by treatment with SDS and β-mercaptoethanol, and tubulin was detected by incubation of membranes with monoclonal rabbit anti-tubulin antibodies (Abcam plc, Cambridge, MA, UK) and peroxidase-coupled goat anti-rabbit secondary antibodies (Cayman Chemical, Ann Arbor, MI, USA).

2.8. Microscopy

Yeast strains expressing different *mNeonGreen*-tagged *WSC1* gene variants were analyzed by microscopy after growth to logarithmic phase under a Zeiss Axiovert 200M microscope using (i) transmission light microscopy (TM) and (ii) fluorescence microscopy with a GFP filter set (AHF Analysentechnik AG, Tübingen, Germany). Cells were photographed with a Hamamatsu Orca ER digital camera, and pictures were processed and analyzed using the Improvision Volocity software (Improvision, Coventry, UK).

2.9. Bioinformatic Analysis

Figures of protein structures were generated with the Molecular Graphics Software PyMOL v2.3.0 (Schrödinger, LLC, New York, NY, USA). Sequence alignments were obtained by using Clustal Omega [47].

2.10. Data Availability

The atomic coordinates and structure factors for the *Sc*Wsc1 CRD obtained in this study have been deposited in the Protein Data Bank (www.rcsb.org) under the accession code 7PZ2.

3. Results

3.1. Characterization of Recombinant CRDs from S. cerevisiae Wsc1, Wsc2, and Wsc3

In order to biochemically and structurally characterize the CRDs from *S. cerevisiae* Wsc1, Wsc2, and Wsc3, appropriate proteins (Figure 2a) were produced by heterologous expression in *E. coli* and purification employing the hexahistidine (*6xHis*) affinity tag, followed by size-exclusion chromatography. Because removal of the *6xHis* tag profoundly reduced the solubility of the three purified CRDs, further experiments were performed only with the tagged variants. Secondary structure elements of purified proteins were determined by CD spectroscopy, revealing a comparable distribution of α-helical- and β-sheet-containing regions (Figure 2b). In addition, we found a significant stability of the CRD of *Sc*Wsc1 against denaturation by heat treatment up to 95 °C (Figure 2b,c). Further characterization of the recombinant CRDs by thermal shift assays corroborated the high thermal stability of the CRD from *Sc*Wsc1, as no reproducible melting curves could be obtained at different pH values ranging from pH 3 to pH 10 (data not shown). Similarly, no significant melting was observed for the *Sc*Wsc2 and *Sc*Wsc3 CRDs at neutral pH conditions. In the case of *Sc*Wsc2, however, melting curves for the CRD could be obtained either at very acidic conditions with a melting temperature (T_m) of 66 °C at pH 3, or under alkaline conditions, with a T_m of 73 °C at pH 10 (Figure 2d). Similarly, melting was observed for the *Sc*Wsc3 CRD, with melting temperatures of 65 °C at pH 3, and 69 °C at pH 6 (Figure 2e). We also found that addition of the reducing agent DTT significantly decreases the melting temperature of CRDs (Figure 2e, yellow curve), indicating a markedly stabilizing role of the intramolecular disulfide bridges formed by the eight cysteines.

Together, characterization of recombinant CRDs from *Sc*Wsc1, *Sc*Wsc2, and *Sc*Wsc3 reveals a significant pH-dependent stability of WSC domains, and further emphasizes the essential role of their conserved cysteine residues [20,48].

a

CRD	Residues	Mw [Da]	pI	calc. ext. coefficient [M^{-1} cm^{-1}]
ScWsc1	Y22-I118	12878	5.20	20900
ScWsc2	D24-N118	12525	7.01	19410
ScWsc3	F40-E133	12429	6.34	19410

b

	ScWsc1	*ScWsc1*^{heat}	*ScWsc2*	*ScWsc3*
α-helix	12%	5 %	10%	16%
β-sheet	46%	36%	60%	41%
turn	<0.1%	<0.1%	<0.1%	17%
random	42%	59%	30%	26%

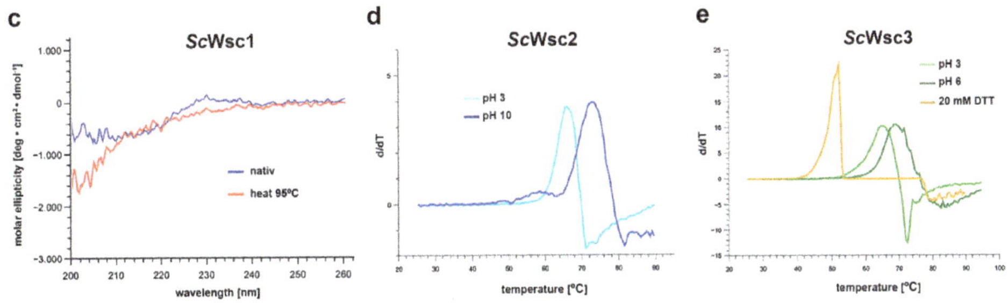

Figure 2. Characterization of recombinant CRDs from *Sc*Wsc1, *Sc*Wsc2, and *Sc*Wsc3. (**a**) Theoretical physiochemical properties of His-tagged proteins. Values were determined with the *ProtParam* tool (https://www.expasy.org; 24 March 2022) using the amino acid sequences encompassing the indicated residues. (**b**) Quantification of secondary structure elements of recombinant CRDs from *Sc*Wsc1, *Sc*Wsc2, and *Sc*Wsc3. Values were determined by CD spectroscopy at 20 °C. For *Sc*Wsc1, values were also determined after heat treatment at 95 °C (heat). (**c**) CD spectroscopy of the recombinant *Sc*Wsc1 CRD before (native; blue line) and after heating to 95 °C (heat; red line) for 1 h. (**d**,**e**) Thermal stability of recombinant CRDs as determined by thermal shift assays (TSA). For *Sc*Wsc2, TSA was performed at pH 3 (light blue line) and pH 10 (dark blue line), respectively (**d**). For *Sc*Wsc3, thermal stability was measured at pH 3 (light green line), pH 6 (dark green line), and at pH 6 in the presence of additional 20 mM DTT (yellow line), respectively (**e**).

3.2. High-Resolution Structure of the ScWsc1 CRD

In order to solve the 3D structure of fungal CRDs, and to gain insights into the structural properties of *S. cerevisiae* Wsc family sensors, X-ray diffraction analysis was performed with purified CRD domains. For this purpose, large-scale crystallization screenings were performed with protein concentrations at the highest achievable solubilities. The *Sc*Wsc1 CRD showed a solubility of 70 mg/mL, followed by *Sc*Wsc3 (30 mg/mL) and *Sc*Wsc2 (15 mg/mL). To date, we were only able to obtain crystals of the *Sc*Wsc1 CRD, with first crystals growing after three weeks of incubation. In the cases of *Sc*Wsc2 and *Sc*Wsc3 CRDs, crystallization attempts so far resulted in either proto-crystalline material or precipitation. A crystallographic dataset collected for the *Sc*Wsc1 CRD allowed structure determination by molecular replacement, using as template the structure of the WSC domain from the human transmembrane sensor Krm1 (PDB ID 5FWS) (Figure 3a,b) despite a low sequence identity of only ~20% between the two CRDs. The final structure of the *Sc*Wsc1 CRD was refined at a resolution of 1.6 Å (Table 2).

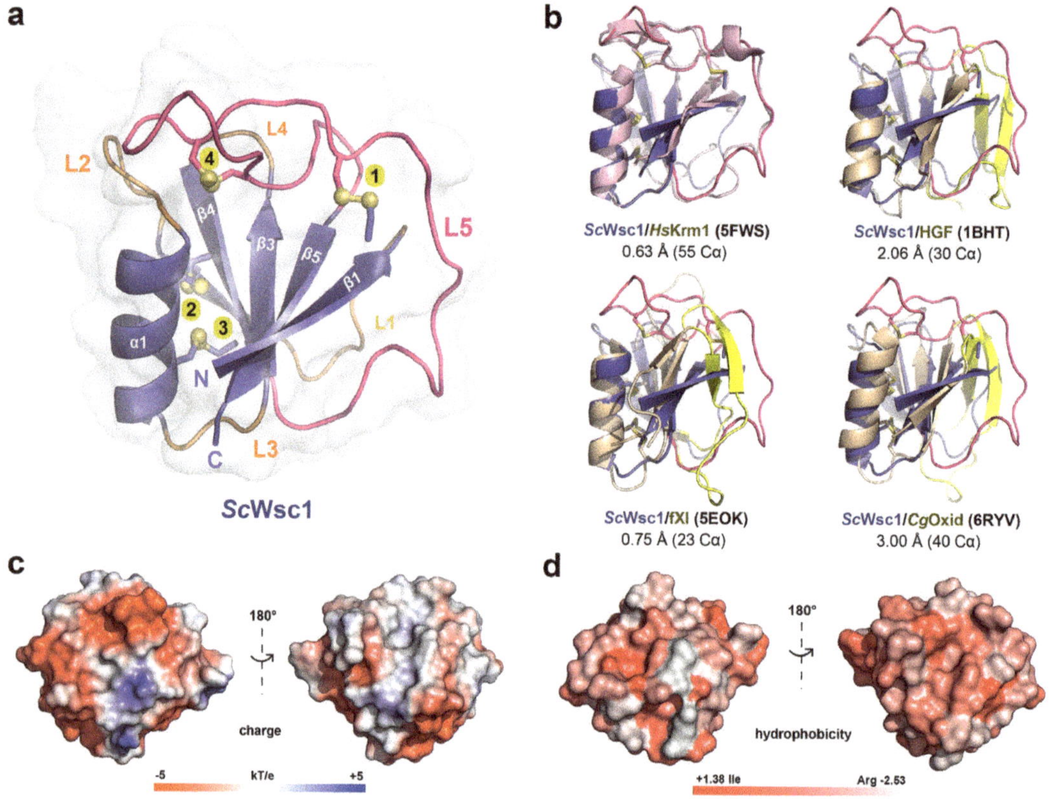

Figure 3. Structural features of the *Sc*Wsc1 cysteine-rich domain. (**a**) The overall structure of the *Sc*Wsc1 CRD (PDB ID 7PZ2; UniProt entry P54867) is shown as a cartoon with the secondary structure elements and the termini indicated. The eight cysteines forming four disulfide bonds (numbered from 1 to 4) are shown as sticks. (**b**) Pairwise superposition between the CRD structures from *Sc*Wsc1 (blue) and human Kremen1 (*Hs*Krm1; PDB ID 5FWS; UniProt entry Q96MU8; light purple), human hepatocytic growth factor (HGF; PDB ID 1BHT; UniProt entry P14210; light orange), human factor XI (fXI; PDB ID 5EOK; UniProt entry P03951; light orange), and a fungal copper oxidase from *Colletotrichum graminearum* (*Cg*Oxid; PDB ID 7PZ2; UniProt entry E3Q9X3; light orange). The PAN/Apple-domain-specific insertions of HGF, fXI, and *Cg*Oxid are highlighted in yellow, the L5 loop of *Sc*Wsc1 in pink. Root mean square standard deviations for the superpositions are indicated in Å, the number of superposed C_α-positions in parentheses. (**c**) Surface charge and (**d**) surface hydrophobicity predictions for *Sc*Wsc1.

Table 2. Data collection and refinement statistics of *Sc*Wsc1^{CRD}.

PDB Code	7PZ2
X-ray source	ESRF ID29
Wavelength (Å)	0.979
Resolution range (Å) [1]	26.75–1.58 (1.64–1.58)
Space group	$P\,1\,2_1\,1$
Unit cell	a = 31.82 Å, b = 53.5 Å, c = 51.93 Å α = 90°, β = 95.46°, γ = 90°
Total reflections [1]	70,058 (6136)
Unique reflections [1]	23,125 (2133)
Multiplicity [1]	3.0 (2.8)
Completeness (%) [1]	97.06 (91.42)
Mean I/sigma(I) [1]	10.35 (2.17)
Wilson B-factor (Å^2)	16.72
R_{merge} [1]	0.06339 (0.4653)
$CC_{1/2}$ [1]	0.997 (0.756)
CC* [1]	0.999 (0.928)
Reflections used in refinement [1]	23,052 (2132)
Reflections used for R_{free} [1]	1113 (126)
R_{work} [1]	0.150 (0.258)
R_{free} [1]	0.184 (0.251)
CC (work, free) [1]	0.973 (0.819), 0.965 (0.824)
Number of non-hydrogen atoms	1731
Macromolecules	1527
Ligands, solvent	7197
Protein residues	197
r.m.s.d. bonds (Å)	0.005
r.m.s.d. angles (°)	0.99
Ramachandran favored (%)	92.23
Ramachandran allowed (%)	7.77
Ramachandran outliers (%)	0.00
Rotamer outliers (%)	0.56
Clashscore [2]	1.06
Average B-factor (Å^2)	20.88
Macromolecules (Å^2)	19.53
Ligands, Solvent (Å^2)	25.84, 31.15

[1] Values in parentheses correspond to highest resolution shell. [2] As calculated by MolProbity.

The overall structure of the *Sc*Wsc1 CRD is characterized by a central β-sheet, composed of five antiparallel β-strands (β1–β5) which form the core of a tightly packed globular domain (Figure 3a). This β-sheet engulfs an α-helix (α1, S46-G57), and is connected by five loops, L1–L5. Besides the WSC domain of human Krm1, this fold with its β1β5β3β4β2(α1) topology, also corresponds to the PAN/Apple domain fold that has been characterized before for human hepatocytic growth factor HGF [49], human plasma coagulation factor XI [50], and a fungal copper oxidase from *Colletotrichum graminearum* [51] (Figure 3b). The *Sc*Wsc1 CRD is stabilized by the formation of four disulfide bonds from eight highly conserved cysteines (Figure 4), resulting in a highly rigidized structure (Figure 3a). Helix α1 is attached by two disulfide bridges, C49–C69 and C53–C71, to the β4 strand of the β-sheet (Figure 3a; yellow bonds 3 and 4).

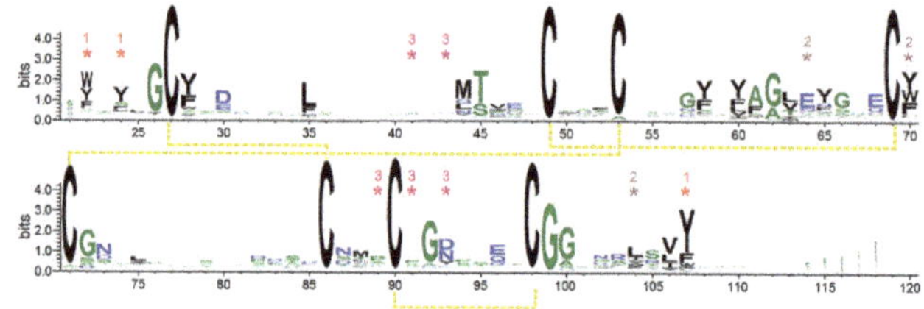

Figure 4. WebLogo [52] generated for 2933 Wsc-type CRD sequences, which have been collected by GREMLIN [53] using the *Sc*Wsc1 sequence A21-S120 as query sequence, and default parameters. Residues belonging to aromatic clusters (Figure 6) have been marked by asterisks. The numbering scheme is derived from *Sc*Wsc1.

Both of these intramolecular cross-links are likewise conserved in the PAN/Apple domains, which have been assigned to a different protein family (PFAM entry PF00024) than the Wsc-like domains (PF01822) due to their low, crosswise sequence similarity. The other two disulfide bridges shown by *Sc*Wsc1 (C27–C86, C90–C98) and the Krm1 CRD are a specific feature of Wsc-like domains due to their strict conservation (Figure 4). These disulfide bridges stabilize the unusually long, but distinctly folded, loop L5 (G72-A103), which enwraps almost the whole PAN/Apple domain (Figures 3a and 5).

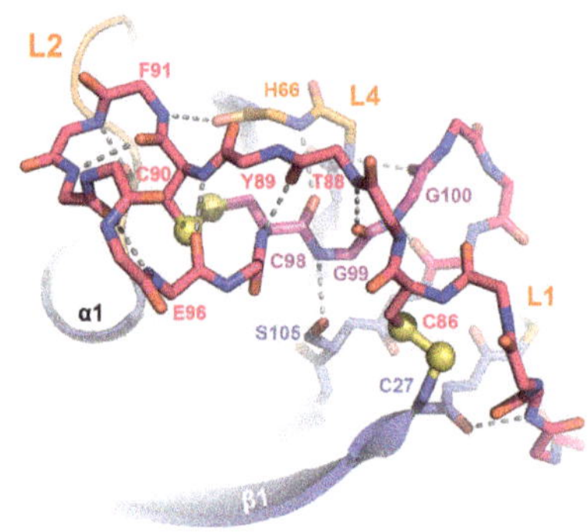

Figure 5. The L5 loop of *Sc*Wsc1. The L5 loop (pink, only main chain shown for S81-A103) covers one end of the PAN/Apple domain together with the accompanying loops L1, L2, and L4. By making numerous interactions, the CGG motif of Wsc-like CRDs (highlighted in purple) is centrally located between the two disulfide bridges C27-C86 and C90-C98, the L4 loop, the N-terminal end of β5, and the crossing L5 stretch N87-C90. Main chain hydrogen bonds of L5 are marked as dashed lines (grey).

Here, one disulfide bridge (Figure 3a; yellow bond 1, C27-C86) connects the β1-strand to the L5 loop, whose conformation is further stabilized by the intra-loop disulfide bridge C90-C98. A comparison of the *Sc*Wsc1 and *Hs*Krm1 CRDs shows that the C-terminal part of this long L5 stretch adopts a conserved conformation, which is stabilized by intra-loop hydrogen bonds, as well as interactions with the L1, L2, and L4 loops (Figure 5). A second determinant for this conformation besides the two disulfide bridges is given by the highly conserved CGG motif of Wsc-like CRDs (*Sc*Wsc1: C98-G100, Figure 4). These two glycines do not only form a hydrogen bonding network from L5 to L4 and β5 (Figure 5), but also allow main chain crossing of L5 between its two conserved disulfide bridges.

The surface of the Wsc1 CRD lacks an apparent site or pocket for the binding of metal ions or ligands, such as glycans. However, calculation of the electrostatic surface potential reveals a mainly negative charge on one side of the domain, whereas the opposite side has mainly non-polar characteristics (Figure 3c). The negative surface potential results from eleven surface-exposed residues of aspartic and glutamic acids, which explains the acidic pI of 5.20 calculated for the *Sc*Wsc1 CRD (Figure 2a). Calculation of the surface hydrophobicity characteristics does not reveal particularly hydrophobic areas (Figure 3d). However, a further analysis of the surface properties of the *Sc*Wsc1 CRD reveals an unusually high number of surface-exposed aromatic residues (Figure 6), which is reminiscent of the surface-exposed aromatic stretches of yeast adhesins such as Flo11 [54] or Awp1 [55].

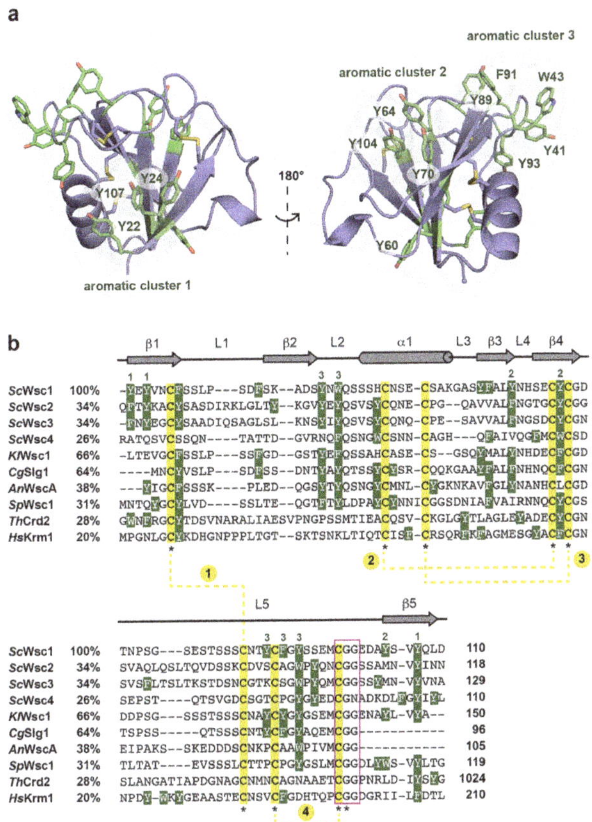

Figure 6. Surface exposure of aromatic residues in fungal CRD/WSC domains. (**a**) Surface-exposed aromatic clusters in the *Sc*Wsc1 CRD. The 12 surface-exposed aromatic residues are shown in green. Three residues form both the aromatic clusters 1 (Y22, Y24, Y107) and 2 (Y64, Y70, Y104), respectively, whereas aromatic cluster 3 encompasses five residues (Y41, W43, Y89, F91, Y93). (**b**) Multiple sequence alignment of selected CRD/WSC domains from *S. cerevisiae* (*Sc*Wsc1, UniProt entry P54867; *Sc*Wsc2, UniProt entry P53832; *Sc*Wsc3, UniProt entry Q12215; *Sc*Wsc4, UniProt entry P38739), *Kluyveromyes lactis* (*Kl*Wsc1, GenBank entry CAH00797.1), *Candida glabrata* (*Cg*Slg1, UniProt entry Q6FUQ7), *Aspergillus nidulans* (*An*WscA, GenBank entry XP_663264.1), *Schizosaccharomyces pombe* (*Sp*Wsc1, UniProt entry P87179), *Trichoderma harzianum* (*Th*Crd2, UniProt entry O14402), and *Homo sapiens* (*Hs*Krm1, UniProt entry Q96MU8). The secondary structure elements of the *Sc*Wsc1 CRD are shown on top (grey). The highly conserved cysteine residues are underlaid in yellow and marked by asterisks. The four disulfide bonds found in the structures of the *Sc*Wsc1 and *Hs*Krm1 CRDs are indicated by yellow dotted lines, and consecutively numbered. Aromatic residues (Phe, Tyr, and Trp) are indicated by white letters, underlaid in green. Green numbers on top indicate the residues of *Sc*Wsc1 forming the three surface-exposed aromatic clusters shown in (**a**). Numbers after protein names indicate percentage identity to *Sc*Wsc1. Numbers on the right indicate position of residues relative to the translational start site. The highly conserved CGG motif is marked by a purple box.

Specifically, twelve of the fifteen aromatic amino acids found in the domain, mostly tyrosines, are surface-exposed, and form three distinct aromatic clusters. The three residues from aromatic cluster 1 (Y22, Y24, Y107) are highly conserved within Wsc-like CRDs, with notable exchanges such as tryptophan or phenylalanine (Figure 4). Likewise, cluster 2 (Y64, Y70, Y104) shows a preference for aromatic residues at least at positions 70 and 104, whereas the larger cluster 3, with its five residues (Y41, W43, Y89, F91, Y93), lacks any apparent conservation pattern (Figures 4 and 6). Nevertheless, a sequence alignment reveals that the CRDs of *Sc*Wsc2 and *Sc*Wsc3 contain aromatic residues at a total of eight positions as well, which correspond to the surface-exposed aromatic sidechains of *Sc*Wsc1 (Figure 6b). Moreover, homology modelling shows that these conserved residues together with two additional non-conserved aromatic residues are also arranged into three aromatic clusters at the CRD surfaces of *Sc*Wsc2 and *Sc*Wsc3 (Figure 7).

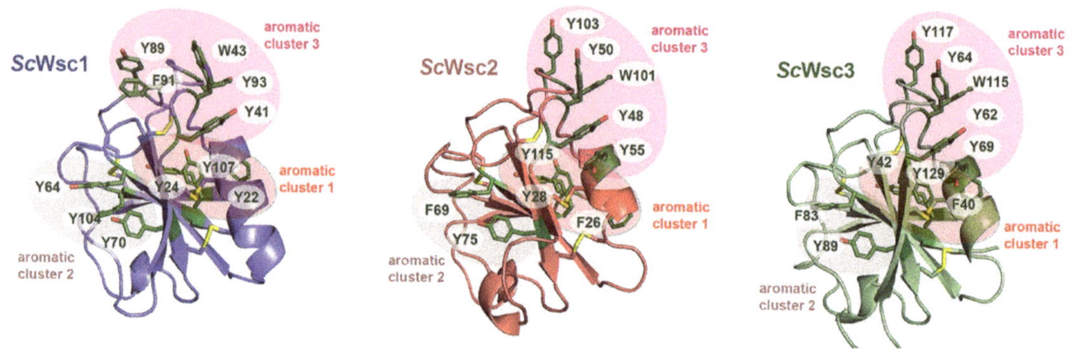

Figure 7. Homology models of *Sc*Wsc2 and *Sc*Wsc3 CRDs. Modelling was performed with *SWISS-MODEL* (https://www.expasy.org; 10 March 2022) based on the crystal structure of the *Sc*Wsc1 CRD (PDB ID 7PZ2) shown on the left. Surface-exposed aromatic residues are shown in green with numbering referring to sequences shown in Figure 5b. Aromatic clusters are indicated by circled areas.

In summary, X-ray crystallography of the *Sc*Wsc1 CRD reveals a conserved fold that is characterized by a tightly packed globular core domain enwrapped by the long L5 region, exposing three aromatic clusters which are conserved at least in the CRDs of *Sc*Wsc2 and *Sc*Wsc3.

3.3. Functional Characterization of Wsc1 Surface-Exposed Aromatic Clusters

Our finding that CRDs from *S. cerevisiae* Wsc-type mechanosensors carry an unusually high number of surface-exposed aromatic residues prompted us to further analyze their involvement in sensor function, as aromatic clusters have been described to be involved in optimizing both affinity, as well as specificity, of protein–protein and protein–ligand interactions [56,57]. For this purpose, we constructed a set of yeast strains lacking the chromosomal *WSC1* gene (*wsc1*Δ), and individually expressing different *WSC1* variants from plasmids. Specifically, we included a WT version, a variant lacking the complete CRD, and three variants each carrying alanine mutations in the residues that form one of the three aromatic clusters. All *WSC1* variants on plasmids were additionally fused to the green fluorescent protein mNeonGreen at their C-terminus, in order to monitor expression levels. Yeast strains were then assayed for viability by serial dilution growth tests in the presence of the β-1,3-glucan synthase inhibitor caspofungin, the chitin-binding agents Congo red and Calcofluor white, and caffeine. As expected, a *wsc1*Δ yeast strain carrying a control plasmid without a *WSC1* gene is highly sensitive to the presence of all four agents, whereas strains expressing a genomic or a plasmid borne WT variant of *WSC1* are resistant (Figure 8a).

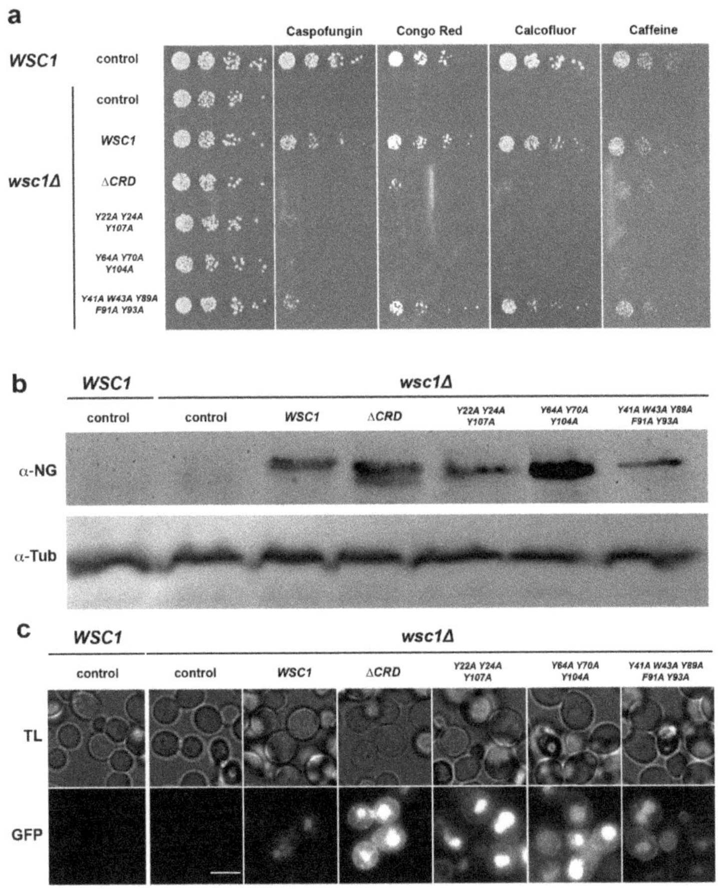

Figure 8. Functional analysis of surface-exposed aromatic residues in the CRD of *Sc*Wsc1. (**a**) A yeast strain carrying a chromosomal deletion of the *WSC1* gene (*wsc1*Δ; YHUM2959) was transformed with an empty control plasmid (control; pRS314) or plasmids carrying either WT *WSC1* (BHUM3303) or *WSC1* lacking the CRD (Δ*CRD*; BHUM3308) or *WSC1* with mutations in one of the aromatic clusters 1 (Y22A Y24A Y107A; BHUM3304), 2 (Y64A Y70A, Y104A; BHUM3305), or 3 (Y41A W43A Y89A F91A Y93A; BHUM3306). All *WSC1* variants on the plasmids carry a C-terminal translational fusion to the green fluorescent protein mNeonGreen. As an additional control, a yeast strain carrying an untagged chromosomal *WSC1* gene (*WSC1*; ESM356-1) and an empty control plasmid (pRS314) was used (shown on the left). Plasmid carrying strains were grown to logarithmic phase in liquid SC-Trp medium, and fivefold serial dilutions of the cultures were spotted onto solid YPD medium or YPD medium supplemented with either 0.4 µg/mL caspofungin (Caspofungin), 30 µg/mL Congo red (Congo Red), 100 µg/mL Calcofluor white (Calcofluor), or 2 mg/mL caffeine (Caffeine). Plates were photographed after incubation for 3–5 days at 30 °C. The pictures shown are representative for at least two independent transformants. (**b**) Western blot analysis. Strains described in (**a**) were grown to logarithmic phase, and full-length Wsc1-mNeonGreen proteins were detected in cell extracts by anti-mNeonGreen antibodies (α-NG). As an internal loading control, tubulin was detected using anti-tubulin antibodies (α-Tub). The pictures shown are representative for at least two independent transformants. (**c**) Microscopy. All strains were further analyzed for expression of *WSC1* variants by fluorescence microscopy after growth to logarithmic phase using the channels for transmission light (TL) or for mNeonGreen (GFP). The white bar corresponds to 5 µm.

Expression of the variants either lacking the complete CRD or carrying points mutations in aromatic cluster 1 or 2, respectively, led to enhanced sensitivity towards Caspofungin, Congo red, and Calcofluor white when compared to the strains expressing WT *WSC1*. Somewhat weaker effects were observed on the medium containing caffeine. In contrast, expression of the *WSC1* variant carrying mutations in the aromatic cluster 3 did not confer enhanced sensitivity towards three of the four cell-wall-perturbing agents. These results corroborate the previously described importance of the CRD for sensor function, and reveal an important functional role of surface-exposed aromatic clusters 1 and 2. Importantly, all mutant Wsc1-mNeonGreen fusion proteins were expressed at levels comparable to or above the levels of the WT variant, as detected by Western-blot analysis (Figure 8b). Similarly, GFP signals of mutant Wsc1-mNeonGreen fusion proteins were at levels comparable to or above the levels of the WT variant, as observed by fluorescence microscopy (Figure 8c). However, a noticeable increase of intracellular GFP signals was observed for the variants lacking the CRD or carrying mutations in the aromatic clusters 1 or 2.

In summary, our mutational analysis reveals that the surface-exposed aromatic clusters 1 and 2 play essential roles in conferring *Sc*Wsc1 sensor function under different cell wall stress conditions.

4. Discussion

In this study, we have solved the first crystal structure of a fungal cell wall integrity sensor domain: the CRD of *Sc*Wsc1. Together with the WSC domain structure from the human type I transmembrane protein Krm1 [32], our structure shows the typical fold of a PAN/Apple domain that is enwrapped by the long and twisted L5 loop region. This structural determinant apparently replaces the two-stranded antiparallel β-hairpin-like insertions found in the L1 and L5 regions of canonical PAN/Apple domains, such as the human hepatocyte growth factor or coagulation factor XI (Figure 3b). Krm1, together with its paralog Krm2, have been identified as co-receptors for Dickkopf (Dkk) proteins that act as secreted antagonists of vertebrate Wnt-signaling pathways [30]. In the presence of Dkk, Krm paralogs form a ternary complex with the Wnt co-receptors Lrp5/6, leading to endocytosis and amplification of the Wnt-antagonistic activity of Dkk [31]. Reminiscent of the Wsc mechanosensors of *S. cerevisiae*, Krm proteins have a tripartite architecture, and consist of a WSC-domain-containing extracellular region, followed by a transmembrane domain and a short cytoplasmic tail. In contrast to fungal Wsc sensors, however, the extracellular WSC domain of Krm proteins is sandwiched between a Kringle domain (PFAM entry PF00051) and a CUB domain (PFAM entry PF00431). Our study reveals that despite a sequence identity of only 20%, the folds of the *Hs*Krm1 and *Sc*Wsc1 WSC domains are highly similar with their common β1β5β3β4β2(α) topology and stabilization by four disulfide bridges. In addition, both structures show very similar packing arrangements for their L1, L2, L4, and L5 loop regions, and lack an obvious cavity for the binding of ligands. These features are apparently shared by most other known Wsc-like CRD regions, a conclusion that is corroborated by the recently released structural predictions of numerous WSC domains as found in the AlphaFold Protein Structure Database (https://alphafold.ebi.ac.uk). Our study further shows that larger structural differences between the WSC domains of *Hs*Krm1 and *Sc*Wsc1 are mainly local, e.g., within the L3 loop or the N-terminal third of L5, and that the two CRDs have distinct surface properties. Given the low sequence identity of 20%, this finding is not unexpected, but it is noteworthy that the *Sc*Wsc1 CRD carries roughly twice as many surface-exposed aromatic residues than the *Hs*Krm1 WSC domain. One of these aromatic residues of *Hs*Krm1, Y165 (*Sc*Wsc1: S67), has been found to be directly contacting *Hs*Dkk1 [32]. Interestingly, *Hs*Krm1 further contacts *Hs*Dkk1 via a hydrophobic pocket formed by three aromatic residues present on the surface of its Kringle domain, which is adjacent to the WSC domain, and not found in fungal WSC domain sensors. It thus appears that mammalian WSC-type domain receptors have undergone an expansion in their domain architecture and—in comparison to fungal cell wall integrity sensors—exert their sensing functions by employing their CRDs in combination with additional domains.

A central finding of our study is the presence of functionally essential aromatic clusters on the surface of *S. cerevisiae* CRDs. Though we do not yet know the precise function of these aromatic clusters, they could be potentially involved in protein–protein interactions such as the *Hs*Krm proteins [56,57] Our previous work on yeast adhesins demonstrated the crucial role of aromatic stretches on the surfaces of adhesin domains in (i) mediating the formation of cell aggregates by homotypic interactions of neighboring adhesin molecules [54,55], and (ii) conferring kin discrimination [58]. Previous work has also demonstrated that (i) the CRDs of *Sc*Wsc proteins are required for sensor clustering in response to cell wall stress, an event that is thought to trigger the downstream CWI signaling cascade [20]; and (ii) that homodimeric sensor interactions in membrane microdomains (MCMs) depend on CRDs [22]. Thus, aromatic clusters of Wsc CRDs might mediate sensor clustering in MCMs via hydrophobic interactions, and thereby initiate downstream signaling. Our finding that aromatic clusters are present on the CRD surface of three different *Sc*Wsc sensors further indicates that homotypic, as well as heterotypic, sensor interactions are conceivable. This conclusion is supported by the previous demonstration that MCMs formed by *Sc*Wsc1 and *Sc*Wsc2 can overlap [22]. In that regard, CRD-mediated clustering to MCMs resembles models describing cell–cell and cell–substrate interactions mediated by unspecific adhesins such as yeast Flo11 [54,58]. In alternate scenarios, aromatic clusters of Wsc CRDs might of course also mediate bindings to other, yet unknown cell wall components, β-glucans, or proteins, thereby conferring sensor clustering, anchoring in MCMs and/or CWI-pathway-independent structural functions at the yeast cell wall. Finally, our data do not rule out the possibility that aromatic clusters might be involved in intracellular sensor distribution by, e.g., affecting efficiency of their endocytosis (reviewed in [13,17]), and thereby affecting their functionality. A previous study has shown that non-clustering Wsc1 sensors can accumulate in vacuoles [20]. Interestingly, we also observe stronger intracellular GFP signals for some of the non-functional Wsc1 variants (Figure 8c), indicating vacuolar accumulation of such sensors that could be triggered by reduced CRD-mediated clustering.

The CRDs of Wsc sensors have previously been proposed to be involved in binding to glycan structures of the cell wall [20]. However, no direct glycan binding has been experimentally demonstrated so far. Our structural analysis of the *Sc*WSc1 CRD does not reveal an obvious glycan-binding cavity comparable to, e.g., the binding pockets for high-affinity binding of terminal glycans found in fungal adhesins [59–63]. Thus, single WSC domains of fungal mechanosensors alone might not be prone for high-affinity binding of cell wall glycans. However, a recent study has demonstrated that the secreted lectin *Si*Wsc3 (UNiProt entry G4TKP7) from the root endophyte *Serendipita indica* binds to linear long-chain β-1,3-glucans with affinities in the low micromolar range [29]. In contrast to fungal WSC-type sensors, the ~39 kDA integral cell wall component *Si*Wsc3 merely consists of three repeated WSC domains. Interestingly, the individual WSC domains of *Si*Wsc3 also lack an obvious binding pocket for glycan binding, but together appear to form a carbohydrate-binding site [29]. This opens the possibility that high-affinity glycan binding by fungal WSC domain sensors could be enabled by multimerization of individual WSC domains, e.g., upon sensor clustering. Such a mechanism might, for instance, confer anchoring of sensor clusters and MCMs at specific sites of the cell wall, and influence the interaction with downstream signaling components. It might, therefore, be interesting to test whether synthetic multimers of, e.g., *Sc*Wsc domains, are able to efficiently bind to linear long-chain β-1,3-glucans, and if such a binding involves any of the aromatic surface clusters, even though the precise spatial configuration of clustered WSC sensor domains is currently unknown.

Taken together, our study provides novel insights into the structure and function of WSC domains in general, and pinpoints aromatic clusters as functional hotspots of *S. cerevisiae* mechanosensors. As such, our data provide an important basis for future structure-based functional analysis of Wsc-type mechanosensors, and the elucidation of the precise role of CRDs in sensor clustering, downstream signaling, and interactions with other cell wall proteins and/or glycans.

Supplementary Materials: The following are available online at https://www.mdpi.com/article/10.3390/jof8040379/s1, Figure S1: Plasmid maps.

Author Contributions: Conceptualization, L.-O.E. and H.-U.M.; methodology, P.S., A.P.L., B.J.L., S.B., L.-O.E. and H.-U.M.; software, P.S., A.P.L., B.J.L., L.-O.E. and H.-U.M.; validation, P.S., A.P.L., B.J.L., S.B., L.-O.E. and H.-U.M.; formal analysis, P.S., A.P.L., B.J.L., S.B., L.-O.E. and H.-U.M.; investigation, P.S., A.P.L., B.J.L., S.B.; resources, P.S., A.P.L. and B.J.L.; data curation, P.S., A.P.L., B.J.L. and S.B.; writing—original draft preparation, L.-O.E. and H.-U.M.; writing—review and editing, P.S., A.P.L., B.J.L., S.B., L.-O.E. and H.-U.M.; visualization, P.S., A.P.L., B.J.L., S.B., L.-O.E. and H.-U.M.; supervision, L.-O.E. and H.-U.M.; project administration, L.-O.E. and H.-U.M.; funding acquisition, L.-O.E. and H.-U.M. All authors have read and agreed to the published version of the manuscript.

Funding: This research was funded by the DEUTSCHE FORSCHUNGSGEMEINSCHAFT (DFG) to L.-O.E., (SFB 987) and H.-U.M. (SFB 987).

Institutional Review Board Statement: Not applicable.

Informed Consent Statement: Not applicable.

Data Availability Statement: Original datasets, strains, and constructs employed in this work are freely available upon request to the academic community for research purposes only.

Acknowledgments: The authors thank Daniela Störmer for technical support, and the staff of Beamline ID29 at the European Synchrotron Radiation Facility (ESRF) in Grenoble (France).

Conflicts of Interest: The authors declare no conflict of interest. The funders had no role in the design of the study; in the collection, analyses, or interpretation of data; in the writing of the manuscript; or in the decision to publish the results.

References

1. Garcia-Rubio, R.; de Oliveira, H.C.; Rivera, J.; Trevijano-Contador, N. The Fungal Cell Wall: Candida, Cryptococcus, and Aspergillus Species. *Front. Microbiol.* **2019**, *10*, 2993. [CrossRef] [PubMed]
2. Gow, N.A.R.; Latge, J.P.; Munro, C.A. The Fungal Cell Wall: Structure, Biosynthesis, and Function. *Microbiol. Spectr.* **2017**, *5*. [CrossRef] [PubMed]
3. De Groot, P.W.; Hellingwerf, K.J.; Klis, F.M. Genome-wide identification of fungal GPI proteins. *Yeast* **2003**, *20*, 781–796. [CrossRef] [PubMed]
4. Lesage, G.; Bussey, H. Cell wall assembly in *Saccharomyces cerevisiae. Microbiol. Mol. Biol. Rev.* **2006**, *70*, 317–343. [CrossRef]
5. Orlean, P. Architecture and biosynthesis of the Saccharomyces cerevisiae cell wall. *Genetics* **2012**, *192*, 775–818. [CrossRef]
6. Levin, D.E. Cell wall integrity signaling in Saccharomyces cerevisiae. *Microbiol. Mol. Biol. Rev.* **2005**, *69*, 262–291. [CrossRef]
7. Elorza, M.V.; Rico, H.; Sentandreu, R. Calcofluor white alters the assembly of chitin fibrils in Saccharomyces cerevisiae and Candida albicans cells. *J. Gen. Microbiol.* **1983**, *129*, 1577–1582. [CrossRef]
8. Roncero, C.; Duran, A. Effect of Calcofluor white and Congo red on fungal cell wall morphogenesis: In vivo activation of chitin polymerization. *J. Bacteriol.* **1985**, *163*, 1180–1185. [CrossRef]
9. Ketela, T.; Green, R.; Bussey, H. Saccharomyces cerevisiae mid2p is a potential cell wall stress sensor and upstream activator of the PKC1-MPK1 cell integrity pathway. *J. Bacteriol.* **1999**, *181*, 3330–3340. [CrossRef]
10. Garcia, R.; Bermejo, C.; Grau, C.; Perez, R.; Rodriguez-Pena, J.M.; Francois, J.; Nombela, C.; Arroyo, J. The global transcriptional response to transient cell wall damage in Saccharomyces cerevisiae and its regulation by the cell integrity signaling pathway. *J. Biol. Chem.* **2004**, *279*, 15183–15195. [CrossRef]
11. Imai, K.; Noda, Y.; Adachi, H.; Yoda, K. A novel endoplasmic reticulum membrane protein Rcr1 regulates chitin deposition in the cell wall of Saccharomyces cerevisiae. *J. Biol. Chem.* **2005**, *280*, 8275–8284. [CrossRef] [PubMed]
12. Bermejo, C.; Garcia, R.; Straede, A.; Rodriguez-Pena, J.M.; Nombela, C.; Heinisch, J.J.; Arroyo, J. Characterization of sensor-specific stress response by transcriptional profiling of wsc1 and mid2 deletion strains and chimeric sensors in Saccharomyces cerevisiae. *Omics J. Integr. Biol.* **2010**, *14*, 679–688. [CrossRef] [PubMed]
13. Rodicio, R.; Heinisch, J.J. Together we are strong—Cell wall integrity sensors in yeasts. *Yeast* **2010**, *27*, 531–540. [CrossRef]
14. Dichtl, K.; Samantaray, S.; Wagener, J. Cell wall integrity signalling in human pathogenic fungi. *Cell Microbiol.* **2016**, *18*, 1228–1238. [CrossRef] [PubMed]
15. Levin, D.E. Regulation of cell wall biogenesis in Saccharomyces cerevisiae: The cell wall integrity signaling pathway. *Genetics* **2011**, *189*, 1145–1175. [CrossRef]

16. Heinisch, J.J.; Rodicio, R. Protein kinase C in fungi-more than just cell wall integrity. *FEMS Microbiol. Rev.* **2018**, *42*, fux051. [CrossRef] [PubMed]
17. Kock, C.; Dufrene, Y.F.; Heinisch, J.J. Up against the wall: Is yeast cell wall integrity ensured by mechanosensing in plasma membrane microdomains? *Appl. Environ. Microbiol.* **2015**, *81*, 806–811. [CrossRef] [PubMed]
18. Jacoby, J.J.; Nilius, S.M.; Heinisch, J.J. A screen for upstream components of the yeast protein kinase C signal transduction pathway identifies the product of the SLG1 gene. *Mol. Gen. Genet.* **1998**, *258*, 148–155. [CrossRef]
19. Lodder, A.L.; Lee, T.K.; Ballester, R. Characterization of the Wsc1 protein, a putative receptor in the stress response of Saccharomyces cerevisiae. *Genetics* **1999**, *152*, 1487–1499. [CrossRef]
20. Heinisch, J.J.; Dupres, V.; Wilk, S.; Jendretzki, A.; Dufrene, Y.F. Single-molecule atomic force microscopy reveals clustering of the yeast plasma-membrane sensor Wsc1. *PLoS ONE* **2010**, *5*, e11104. [CrossRef]
21. Dupres, V.; Alsteens, D.; Wilk, S.; Hansen, B.; Heinisch, J.J.; Dufrene, Y.F. The yeast Wsc1 cell surface sensor behaves like a nanospring in vivo. *Nat. Chem. Biol.* **2009**, *5*, 857–862. [CrossRef] [PubMed]
22. Kock, C.; Arlt, H.; Ungermann, C.; Heinisch, J.J. Yeast cell wall integrity sensors form specific plasma membrane microdomains important for signalling. *Cell Microbiol.* **2016**, *18*, 1251–1267. [CrossRef] [PubMed]
23. Voskoboynikova, N.; Karlova, M.; Kurre, R.; Mulkidjanian, A.Y.; Shaitan, K.V.; Sokolova, O.S.; Steinhoff, H.J.; Heinisch, J.J. A Three-Dimensional Model of the Yeast Transmembrane Sensor Wsc1 Obtained by SMA-Based Detergent-Free Purification and Transmission Electron Microscopy. *J. Fungi* **2021**, *7*, 118. [CrossRef] [PubMed]
24. Rodicio, R.; Buchwald, U.; Schmitz, H.P.; Heinisch, J.J. Dissecting sensor functions in cell wall integrity signaling in Kluyveromyces lactis. *Fungal Genet. Biol.* **2008**, *45*, 422–435. [CrossRef] [PubMed]
25. Futagami, T.; Nakao, S.; Kido, Y.; Oka, T.; Kajiwara, Y.; Takashita, H.; Omori, T.; Furukawa, K.; Goto, M. Putative stress sensors WscA and WscB are involved in hypo-osmotic and acidic pH stress tolerance in Aspergillus nidulans. *Eukaryot. Cell* **2011**, *10*, 1504–1515. [CrossRef]
26. Maddi, A.; Dettman, A.; Fu, C.; Seiler, S.; Free, S.J. WSC-1 and HAM-7 are MAK-1 MAP kinase pathway sensors required for cell wall integrity and hyphal fusion in Neurospora crassa. *PLoS ONE* **2012**, *7*, e42374. [CrossRef]
27. Tong, S.M.; Chen, Y.; Zhu, J.; Ying, S.H.; Feng, M.G. Subcellular localization of five singular WSC domain-containing proteins and their roles in Beauveria bassiana responses to stress cues and metal ions. *Environ. Microbiol. Rep.* **2016**, *8*, 295–304. [CrossRef]
28. Cohen-Kupiec, R.; Broglie, K.E.; Friesem, D.; Broglie, R.M.; Chet, I. Molecular characterization of a novel beta-1,3-exoglucanase related to mycoparasitism of Trichoderma harzianum. *Gene* **1999**, *226*, 147–154. [CrossRef]
29. Wawra, S.; Fesel, P.; Widmer, H.; Neumann, U.; Lahrmann, U.; Becker, S.; Hehemann, J.H.; Langen, G.; Zuccaro, A. FGB1 and WSC3 are in planta-induced beta-glucan-binding fungal lectins with different functions. *New Phytol.* **2019**, *222*, 1493–1506. [CrossRef]
30. Mao, B.; Wu, W.; Davidson, G.; Marhold, J.; Li, M.; Mechler, B.M.; Delius, H.; Hoppe, D.; Stannek, P.; Walter, C.; et al. Kremen proteins are Dickkopf receptors that regulate Wnt/beta-catenin signalling. *Nature* **2002**, *417*, 664–667. [CrossRef]
31. Nakamura, T.; Nakamura, T.; Matsumoto, K. The functions and possible significance of Kremen as the gatekeeper of Wnt signalling in development and pathology. *J. Cell Mol. Med.* **2008**, *12*, 391–408. [CrossRef] [PubMed]
32. Zebisch, M.; Jackson, V.A.; Zhao, Y.; Jones, E.Y. Structure of the Dual-Mode Wnt Regulator Kremen1 and Insight into Ternary Complex Formation with LRP6 and Dickkopf. *Structure* **2016**, *24*, 1599–1605. [CrossRef] [PubMed]
33. Pereira, G.; Tanaka, T.U.; Nasmyth, K.; Schiebel, E. Modes of spindle pole body inheritance and segregation of the Bfa1p-Bub2p checkpoint protein complex. *EMBO J.* **2001**, *20*, 6359–6370. [CrossRef] [PubMed]
34. Janke, C.; Magiera, M.M.; Rathfelder, N.; Taxis, C.; Reber, S.; Maekawa, H.; Moreno-Borchart, A.; Doenges, G.; Schwob, E.; Schiebel, E.; et al. A versatile toolbox for PCR-based tagging of yeast genes: New fluorescent proteins, more markers and promoter substitution cassettes. *Yeast* **2004**, *21*, 947–962. [CrossRef] [PubMed]
35. Guthrie, C.; Fink, G.R. (Eds.) *Guide to Yeast Genetics and Molecular Biology*; Academic Press: San Diego, CA, USA, 1991; Volume 194, pp. 1–863.
36. Lutz, A.P.; Schladebeck, S.; Renicke, C.; Spadaccini, R.; Mosch, H.U.; Taxis, C. Proteasome Activity Is Influenced by the HECT_2 Protein Ipa1 in Budding Yeast. *Genetics* **2018**, *209*, 157–171. [CrossRef] [PubMed]
37. Sikorski, R.S.; Hieter, P. A system of shuttle vectors and yeast host strains designed for efficient manipulation of DNA in *Saccharomyces cerevisiae. Genetics* **1989**, *122*, 19–27. [CrossRef]
38. Veelders, M.; Brückner, S.; Ott, D.; Unverzagt, C.; Mösch, H.U.; Essen, L.O. Structural basis of flocculin-mediated social behavior in yeast. *Proc. Natl. Acad. Sci. USA* **2010**, *107*, 22511–22516. [CrossRef]
39. Kabsch, W. Xds. *Acta Crystallogr.* **2010**, *66*, 125–132. [CrossRef]
40. Kabsch, W. Integration, scaling, space-group assignment and post-refinement. *Acta Crystallogr.* **2010**, *66*, 133–144. [CrossRef]
41. Evans, P.R. An introduction to data reduction: Space-group determination, scaling and intensity statistics. *Acta Crystallogr.* **2011**, *67*, 282–292. [CrossRef]
42. Winn, M.D.; Ballard, C.C.; Cowtan, K.D.; Dodson, E.J.; Emsley, P.; Evans, P.R.; Keegan, R.M.; Krissinel, E.B.; Leslie, A.G.; McCoy, A.; et al. Overview of the CCP4 suite and current developments. *Acta Crystallogr.* **2011**, *67*, 235–242. [CrossRef]
43. McCoy, A.J.; Grosse-Kunstleve, R.W.; Adams, P.D.; Winn, M.D.; Storoni, L.C.; Read, R.J. Phaser crystallographic software. *J. Appl. Crystallogr.* **2007**, *40*, 658–674. [CrossRef] [PubMed]

44. Adams, P.D.; Afonine, P.V.; Bunkoczi, G.; Chen, V.B.; Davis, I.W.; Echols, N.; Headd, J.J.; Hung, L.W.; Kapral, G.J.; Grosse-Kunstleve, R.W.; et al. PHENIX: A comprehensive Python-based system for macromolecular structure solution. *Acta Crystallogr.* **2010**, *66*, 213–221. [CrossRef] [PubMed]
45. Emsley, P.; Lohkamp, B.; Scott, W.G.; Cowtan, K. Features and development of Coot. *Acta Crystallogr.* **2010**, *66*, 486–501. [CrossRef]
46. Yaffe, M.P.; Schatz, G. Two nuclear mutations that block mitochondrial protein import in yeast. *Proc. Natl. Acad. Sci. USA* **1984**, *81*, 4819–4823. [CrossRef]
47. Madeira, F.; Park, Y.M.; Lee, J.; Buso, N.; Gur, T.; Madhusoodanan, N.; Basutkar, P.; Tivey, A.R.N.; Potter, S.C.; Finn, R.D.; et al. The EMBL-EBI search and sequence analysis tools APIs in 2019. *Nucleic Acids Res.* **2019**, *47*, W636–W641. [CrossRef]
48. Dupres, V.; Heinisch, J.J.; Dufrene, Y.F. Atomic force microscopy demonstrates that disulfide bridges are required for clustering of the yeast cell wall integrity sensor Wsc1. *Langmuir* **2011**, *27*, 15129–15134. [CrossRef]
49. Ultsch, M.; Lokker, N.A.; Godowski, P.J.; de Vos, A.M. Crystal structure of the NK1 fragment of human hepatocyte growth factor at 2.0 A resolution. *Structure* **1998**, *6*, 1383–1393. [CrossRef]
50. Wong, S.S.; Ostergaard, S.; Hall, G.; Li, C.; Williams, P.M.; Stennicke, H.; Emsley, J. A novel DFP tripeptide motif interacts with the coagulation factor XI apple 2 domain. *Blood* **2016**, *127*, 2915–2923. [CrossRef]
51. Mathieu, Y.; Offen, W.A.; Forget, S.M.; Ciano, L.; Viborg, A.H.; Blagova, E.; Henrissat, B.; Walton, P.H.; Davies, G.J.; Brumer, H. Discovery of a Fungal Copper Radical Oxidase with High Catalytic Efficiency toward 5-Hydroxymethylfurfural and Benzyl Alcohols for Bioprocessing. *ACS Catal.* **2020**, *10*, 3042–3058. [CrossRef]
52. Crooks, G.E.; Hon, G.; Chandonia, J.M.; Brenner, S.E. WebLogo: A sequence logo generator. *Genome Res.* **2004**, *14*, 1188–1190. [CrossRef] [PubMed]
53. Ovchinnikov, S.; Kamisetty, H.; Baker, D. Robust and accurate prediction of residue-residue interactions across protein interfaces using evolutionary information. *eLife* **2014**, *3*, e02030. [CrossRef] [PubMed]
54. Kraushaar, T.; Brückner, S.; Veelders, M.; Rhinow, D.; Schreiner, F.; Birke, R.; Pagenstecher, A.; Mösch, H.-U.; Essen, L.-O. Interactions by the Fungal Flo11 Adhesin Depend on a Fibronectin Type III-like Adhesin Domain Girdled by Aromatic Bands. *Structure* **2015**, *23*, 1005–1017. [CrossRef] [PubMed]
55. Reithofer, V.; Fernandez-Pereira, J.; Alvarado, M.; de Groot, P.; Essen, L.O. A novel class of Candida glabrata cell wall proteins with beta-helix fold mediates adhesion in clinical isolates. *PLoS Pathog.* **2021**, *17*, e1009980. [CrossRef] [PubMed]
56. Lanzarotti, E.; Biekofsky, R.R.; Estrin, D.A.; Marti, M.A.; Turjanski, A.G. Aromatic-aromatic interactions in proteins: Beyond the dimer. *J. Chem. Inf. Model* **2011**, *51*, 1623–1633. [CrossRef] [PubMed]
57. Lanzarotti, E.; Defelipe, L.A.; Marti, M.A.; Turjanski, A.G. Aromatic clusters in protein-protein and protein-drug complexes. *J. Cheminform.* **2020**, *12*, 30. [CrossRef] [PubMed]
58. Brückner, S.; Schubert, R.; Kraushaar, T.; Hartmann, R.; Hoffmann, D.; Jelli, E.; Drescher, K.; Müller, D.J.; Oliver Essen, L.O.; Mösch, H.U. Kin discrimination in social yeast is mediated by cell surface receptors of the Flo11 adhesin family. *eLife* **2020**, *9*, e55587. [CrossRef]
59. Maestre-Reyna, M.; Diderrich, R.; Veelders, M.S.; Eulenburg, G.; Kalugin, V.; Brückner, S.; Keller, P.; Rupp, S.; Mösch, H.U.; Essen, L.O. Structural basis for promiscuity and specificity during *Candida glabrata* invasion of host epithelia. *Proc. Natl. Acad. Sci. USA* **2012**, *109*, 16864–16869. [CrossRef]
60. Diderrich, R.; Kock, M.; Maestre-Reyna, M.; Keller, P.; Steuber, H.; Rupp, S.; Essen, L.O.; Mösch, H.U. Structural Hot Spots Determine Functional Diversity of the Candida glabrata Epithelial Adhesin Family. *J. Biol. Chem.* **2015**, *290*, 19597–19613. [CrossRef]
61. Kock, M.; Brückner, S.; Wozniak, N.; Maestre-Reyna, M.; Veelders, M.; Schlereth, J.; Mösch, H.U.; Essen, L.O. Structural and Functional Characterization of PA14/Flo5-Like Adhesins From Komagataella pastoris. *Front. Microbiol.* **2018**, *9*, 2581. [CrossRef]
62. Hoffmann, D.; Diderrich, R.; Reithofer, V.; Friederichs, S.; Kock, M.; Essen, L.O.; Mösch, H.U. Functional reprogramming of Candida glabrata epithelial adhesins: The role of conserved and variable structural motifs in ligand binding. *J. Biol. Chem.* **2020**, *295*, 12512–12524. [CrossRef] [PubMed]
63. Essen, L.O.; Vogt, M.S.; Mösch, H.U. Diversity of GPI-anchored fungal adhesins. *Biol. Chem.* **2020**, *401*, 1389–1405. [CrossRef] [PubMed]

Journal of *Fungi*

MDPI

Article

A Three-Dimensional Model of the Yeast Transmembrane Sensor Wsc1 Obtained by SMA-Based Detergent-Free Purification and Transmission Electron Microscopy

Natalia Voskoboynikova [1,*], Maria Karlova [2], Rainer Kurre [3], Armen Y. Mulkidjanian [1,4], Konstantin V. Shaitan [2,5], Olga S. Sokolova [2,6], Heinz-Jürgen Steinhoff [1] and Jürgen J. Heinisch [7,*]

1 Faculty of Physics, University of Osnabrück, Barbarastrasse 7, D-49076 Osnabrück, Germany; armen.mulkidjanian@uni-osnabrueck.de (A.Y.M.); hsteinho@uni-osnabrueck.de (H.-J.S.)
2 Department of Bioengineering, Faculty of Biology, M. V. Lomonosov Moscow State University, 119234 Moscow, Russia; mkarlova@yandex.ru (M.K.); shaytan49@yandex.ru (K.V.S.); sokolova@mail.bio.msu.ru (O.S.S.)
3 Center of Cellular Nanoanalytics, Integrated Bioimaging Facility iBiOs, Faculty of Biology/Chemistry, University of Osnabrück, Barbarastrasse 11, D-49076 Osnabrück, Germany; rainer.kurre@uos.de
4 Faculty of Bioengineering and Bioinformatics, Belozersky Institute of Physico-Chemical Biology, M. V. Lomonosov Moscow State University, 119234 Moscow, Russia
5 N.N. Semenov Federal Research Center for Chemical Physics, Russian Academy of Sciences, 119991 Moscow, Russia
6 Biology Department, Shenzhen MSU-BIT University, Shenzhen 518172, China
7 Department of Genetics, Faculty of Biology/Chemistry, University of Osnabrück, Barbarastrasse 11, D-49076 Osnabrück, Germany
* Correspondence: nvoskobo@uni-osnabrueck.de (N.V.); jheinisc@uni-osnabrueck.de (J.J.H.); Tel.: +49-541-969-2668 (N.V.); +49-541-969-2290 (J.J.H.)

Citation: Voskoboynikova, N.; Karlova, M.; Kurre, R.; Mulkidjanian, A.Y.; Shaitan, K.V.; Sokolova, O.S.; Steinhoff, H.-J.; Heinisch, J.J. A Three-Dimensional Model of the Yeast Transmembrane Sensor Wsc1 Obtained by SMA-Based Detergent-Free Purification and Transmission Electron Microscopy. *J. Fungi* **2021**, *7*, 118. https://doi.org/10.3390/jof7020118

Academic Editors: María Molina and Humberto Martín

Received: 18 January 2021
Accepted: 1 February 2021
Published: 5 February 2021

Abstract: The cell wall sensor Wsc1 belongs to a small family of transmembrane proteins, which are crucial to sustain cell integrity in yeast and other fungi. Wsc1 acts as a mechanosensor of the cell wall integrity (CWI) signal transduction pathway which responds to external stresses. Here we report on the purification of Wsc1 by its trapping in water-soluble polymer-stabilized lipid nanoparticles, obtained with an amphipathic styrene-maleic acid (SMA) copolymer. The latter was employed to transfer tagged sensors from their native yeast membranes into SMA/lipid particles (SMALPs), which allows their purification in a functional state, i.e., avoiding denaturation. The SMALPs composition was characterized by fluorescence correlation spectroscopy, followed by two-dimensional image acquisition from single particle transmission electron microscopy to build a three-dimensional model of the sensor. The latter confirms that Wsc1 consists of a large extracellular domain connected to a smaller intracellular part by a single transmembrane domain, which is embedded within the hydrophobic moiety of the lipid bilayer. The successful extraction of a sensor from the yeast plasma membrane by a detergent-free procedure into a native-like membrane environment provides new prospects for in vitro structural and functional studies of yeast plasma proteins which are likely to be applicable to other fungi, including plant and human pathogens.

Keywords: Wsc1; membrane sensor; SMALP; detergent-free extraction; fluorescence correlation spectroscopy; transmission electron microscopy; 3D reconstruction

1. Introduction

Fungal cell walls are essential structures that serve as a first barrier against adverse environmental conditions and ensure form and stability of yeast cells and hyphae in filamentous fungi. As such, they are a prime choice for the development of antifungal drugs and of utmost importance in medicine and agriculture [1]. In this context, cell wall biogenesis and its regulation has been extensively studied in the yeast *Saccharomyces cerevisiae* [2]. In short, cell surface stress is detected by a small family of membrane spanning sensors, which, mediated by the yeast protein kinase C (Pkc1), signal to a mitogen-activated protein kinase (MAPK) cascade to nuclear transcription factors that promote cell wall reconstruction (reviewed in [3]). Taken together, the term cell wall integrity (CWI) pathway has been coined for this signal transduction cascade in yeast and has been adopted for homologous systems in other fungi [4,5]. Amongst the five sensors of *S. cerevisiae*, Wsc1 has best been studied both by genetic and biophysical methods (reviewed in [6]). Atomic-force microscopy (AFM) experiments suggest that individual Wsc1 proteins act as mechanosensors in the form of a nanospring [7]. AFM studies also showed that the clustering of sensors in the plasma membrane in response to cell surface stress is controlled by the cysteine-rich extracellular domain (CRD, also known as the Wsc-domain; [3,5]). Quantitative live cell fluorescence microscopy showed that, in cells growing under standard conditions, Wsc1 is localized in specific microdomains of the plasma membrane, termed Membrane Compartment carrying Wsc1 (MCW) [8].

In all studies performed so far, Wsc1 functions were assessed either in living yeast cells [7,9,10] or in crude cell extracts [5,11,12]. No studies on the purification of the sensor, much less in a native membrane environment, were available, until now. The routine isolation and purification of membrane proteins in most studies involves the use of detergents. While such approaches allow structural assessments of membrane proteins (MPs) in solution, they may negatively affect their stability and activity of the protein in question, thus leading to modifications and even complete inactivation [13].

To overcome these technical barriers, novel membrane mimetic systems for MPs on the nanometer scale have been developed, which avoid the use of detergents [14]. Amphipathic styrene-maleic acid (SMA) and SMA-related copolymers enable the direct, detergent-free extraction of membrane proteins from both artificial membranes with a given composition of lipids, and natural lipid bilayers, providing stable membrane patches with incorporated MPs (reviewed in [11,15–18]). The SMA copolymer-driven lipid solubilization also has the advantage of being non-selective with regard to the lipid-type of the donor membrane [19,20]. SMA-encased lipoprotein particles comprise lipid/protein cores surrounded by a SMA copolymer belt and usually have diameters of approximately 10 nm, depending on the preparation routine [21]. The SMA copolymer with a 3:1 styrene-to-maleic acid molar ratio (3:1 SMA) has been shown to extract both α-helical bundle proteins, such as bacteriorhodopsin [14,22], and β-barrel proteins, such as lipid A palmitoyltransferase PagP [14], from 1,2-dimyristoyl-sn-glycero-3-phosphocholine (DMPC) liposomes. The non-selective nature of the method was also demonstrated by the extraction of transmembrane protein complexes from the archaeal photoreceptor–transducer complex (*Np*SRII/*Np*HtrII) into a 3:1 SMA copolymer, with a photoreceptor: transducer stoichiometry of 2:2, from *Escherichia coli* polar lipid liposomes [23]. Detergent-free solubilization of MPs by SMA copolymers has been demonstrated for cellular membranes from yeast [24–28] and mammalian cells [29].

An extensive set of biophysical methods is compatible with the SMA lipid particle (SMALP) system [30]. Thus, the small size and single-particle character of SMALPs facilitated the use of solution-based structural investigation by electron microscopy (EM; [31–33]).

In this work, we applied the 3:1 SMA copolymer to isolate Wsc1-GFP (green fluorescent protein) fusions into SMALPs directly from native yeast membranes. We purified Wsc1-GFP-containing SMALPs by affinity chromatography and characterized SMALP preparations using dynamic light scattering (DLS), fluorescence correlation spectroscopy (FCS) and single particle transmission electron microscopy (TEM). This serves as a proof-of-principle for further structural and functional studies of Wsc1 and other fungal MPs in vitro.

2. Materials and Methods

2.1. Media, Growth Conditions, Strains and Genetic Manipulations

Growth media and genetic manipulations followed previously described standard protocols [34]. The *S. cerevisiae* strain HOD356-1A (*MATa ura3-52 leu2-3,112 his3-11,15 WSC1-GFP-TEV-3xHA::SpHIS5*) was obtained in this work as a source of the Wsc1 sensor. It was constructed in the background of the HD56-5A strain by homologous recombination with PCR-generated tags at the native *WSC1* locus, based on constructs used in previous works [7,8,10]. Note that in addition to the green fluorescent protein (GFP) used for detection, the encoded fusion protein carries a recognition sequence for the TEV protease and a triple hemagglutinin tag (3 × HA) to facilitate purification. Details on the construction and the sequence of the genomic construct are available upon request.

The *WSC1* gene was expressed under the control of its native promoter. Yeast cells were grown to late logarithmic phase (~5 × 10^7 cells/mL) in 50 mL of rich medium (1% *w/v* yeast extract, 2% *w/v* Bacto peptone; Difco Laboratories Inc., Detroit, MI, USA) with 2% *w/v* glucose as a carbon source (YEPD) at 30 °C with constant rotary shaking at 180 rpm.

2.2. Spheroplast Preparation

Yeast cells were harvested from 10 mL aliquots of the overnight culture by centrifugation to obtain approximately130 mg of wet cells washed once with 3 mL of sterile water and transferred to an Eppendorf tube. The wet cell pellet after centrifugation was resuspended in 0.5 mL of spheroplast buffer (0.9 M sorbitol, 0.1 M EDTA, 20 mM dithiothreitol). To digest the cell wall glucans, 2 µL Zymolyase 20T (an enzyme preparation from a submerged culture of *Arthrobacter luteus* provided by MP Biomedicals, Eschwege/Germany; applied from a stock solution of 10 mg/mL in sterile water and stored at −20 °C) were added and the suspension was incubated for 30 min at 37 °C. After centrifugation (13,500 rpm; 1 min; at room temperature), pellets containing the spheroplasts were either processed immediately or flash frozen and kept at −80 °C until further use.

2.3. Cell Lysis and Fractionation

Spheroplasts were washed twice and resuspended at 20 mg/mL of wet cells in buffer A: 10 mM Tris-HCl, pH 8.0, 150 mM NaCl supplemented with one tablet per 50 mL of protease inhibitor cocktail (Sigma-Aldrich, Merck KGaA, Darmstadt/Germany). The cell suspension was sonicated on ice (Branson Ultrasonic Corporation, USA; duty cycle 20; output control 2; 5 × 1 min with 1 min break in between). After centrifugation (134,000× *g*; 20 min; 4 °C), the resulting pellet was resuspended at 80 mg/mL of wet cells in buffer A and directly used for protein extraction by SMA or flash frozen and kept at −80 °C until further use.

2.4. Preparation of Styrene-Maleic Acid Copolymer Solution

The styrene maleic acid (SMA) copolymer with a styrene-to-maleic acid molar ratio of 3:1 (MW 9500 Da, supplied as an aqueous sodium salt solution SMA 3000 HNa) was kindly provided as a gift by Cray Valley (Exton; PA; USA). A 5% (*w/v*) solution of SMA, which was extensively dialyzed against buffer A, was used for the membrane solubilization.

2.5. Membrane Solubilization by SMA

A 5% (*w/v*) solution of SMA copolymer in buffer A was added dropwise to the membrane suspension to get a final cell-to-SMA weight ratio of 1:2.5. The suspension was incubated for 30 min at RT and then for 16 hrs at 4 °C. All incubations were performed with gentle shaking. The suspension was then centrifuged for 20 min at 134,000× *g* at 4 °C. Before the affinity purification, aliquots of the supernatant were taken for Dynamic Light Scattering and Fluorescence Correlation Spectroscopy analyses. The pellet and the supernatant were also analyzed by sodium dodecyl sulfate/polyacrylamide gel electrophoresis (SDS-PAGE) and immunoblotting. The supernatant was subsequently purified on affinity resin.

2.6. Affinity Chromatography

Anti-HA (hemagglutinin) agarose (Sigma-Aldrich, Munich, Germany) was pre-equilibrated with PBS buffer. SMA-solubilized membranes were added to the resin and incubated overnight at 4 °C with gentle mixing. The resin was then pelleted by brief centrifugation, and the supernatant (column flow through) was discarded. The resin was washed with 40 column volumes of PBS buffer and protein was eluted with four column volumes of the same buffer, supplemented with 100 μg/mL HA peptide (Sigma-Aldrich, Munich, Germany).

2.7. Electrophoresis and Immunoblot

All stages of cell fractionation, SMA solubilization and affinity purification were applied to 12% SDS-PAGE. For immunodetection proteins from gels were transferred onto a 0.2 μm nitrocellulose membrane (GE Healthcare, Chalfont St. Giles, UK), blocked with 5% non-fat milk and probed with rabbit polyclonal HA-antiserum (Abcam, Cambridge, UK) as primary antibodies. The secondary antibodies were anti-rabbit (H+L) HRP-conjugated (BioRad, Hercules, CA, USA). Chemiluminescent signals were developed with ECL Super-Signal West Pico substrate (Thermo, Bartlesville, OK, USA) and registered on a ChemiDoc XRS+ imager with ImageLab software (BioRad, Hercules, CA, USA).

2.8. Dynamic Light Scattering

Dynamic light scattering (DLS) measurements were performed on a Zetasizer Nano ZS (Malvern Instruments, Worcestershire, UK) at 550 nm and 25 °C. Data represent the average of three sets of 14 runs of 10 s each. The particle size distribution was obtained by using the ZETASIZER software package v7.02. under the assumption that the analyzed particles would have a spherical shape.

2.9. Fluorescence Correlation Spectroscopy

Fluorescence Correlation Spectroscopy (FCS) was conducted on a confocal laser scanning microscope (FluoView 1000, Olympus, Tokyo, Japan) equipped with a FLIM/FCS upgrade kit (Picoquant, Berlin, Germany). For excitation of GFP, a picosecond pulsed 485 nm laser diode (LDH-D-C-485, Picoquant) at a repetition rate of 20–40 MHz was used. Fluorescence was detected by a single photon avalanche detector (Picoquant) using a bandpass filter from 500–550 nm (BrightLine HC 525/50, Semrock, Tübingen, Germany). All measurements were performed in a 100 μL droplet placed on a high-precision coverslip at room temperature (24–26 °C). Exact temperature at sample position was determined for each measurement. Fluorescence was collected with a 60× water immersion objective (UPLSAPO 60×, NA 1.2, Olympus) at 20 μm above coverslip inside the droplet. Autocorrelation functions (ACF) were analyzed by the software SymPhoTime 64 (v2.5, Picoquant). This software allowed for correction of background and afterpulsing artefacts using fluorescence lifetime information. A diffusion model assuming a single population of particles showing three-dimensional free diffusion and considering triplet state blinking of GFP was fitted to the resulting ACFs [35]. FCS fitting requires the determination of the effective confocal volume, which can be calibrated with a dye solution with known diffusion constant measured at exact same acquisition conditions. We used a 1 nM fluorescein solution with a diffusion constant of (425 ± 0.01) $\mu m^2/s$ in water at 25 °C [36]. Diffusion coefficients D can be related to temperature T, viscosity $\eta(T)$ and hydrodynamic radius r by the Stokes-Einstein equation $D(T) = k_B T/(6\pi\eta(T)r)$. Viscosity was estimated by $\eta(T) = a \cdot \exp(-b/(T-c))$ with $a = 2.414 \cdot 10^{-5}$ Pas, b = 247.8 K, and c = 140 K (Picoquant application note). Calibration and SMALP measurements at different temperatures were corrected by these equations. The final hydrodynamic radius was estimated by the measured diffusion coefficient and the Stokes–Einstein equation.

2.10. Transmission Electron Microscopy

Copper grids (300 mesh formvar/carbon-coated) (Ted Pella, Redding, CA, USA) were hydrophilized by glow discharge (−20 mA, 45 s) with a Pelco EasyGlow apparatus (Ted Pella, USA). Fresh protein samples (3 μL) were placed onto the grid and incubated at room temperature (RT) for 30 s. Excess of the sample was removed with filter paper. Grids were stained twice, using 1% aquatic uranyl acetate solution for 30 s at RT, and air-dried.

Micrographs were acquired using an analytical transmission electron microscope Jem-2100 (Jeol, Akishima City, Japan) equipped with a 2K × 2K CCD camera Ultrascan 1000XP (Gatan, Pleasanton, CA, USA). The microscope operated at 200 kV in a low dose mode, with a magnification ×40,000 (2.5 Å/pix) and defocus 0.5–1.9 μm. Images were acquired automatically with SerialEM3.8.0 software [37].

2.11. Image Processing

To obtain the 2D projections of purified Wsc1 protein molecules, 107712 particles were selected automatically from the 870 EM images using crYOLO neural network [38]. Particles were windowed into 80 × 80 pixel images and analyzed in the EMAN2.3 suite [39,40]. The 3D reconstruction of Wsc1 was accomplished with RELION 2.0 [41]. Briefly, 2D class averages were produced by classification images of Wsc1 particles. For each complex, 50 classes were produced, and 25 iterations were used to increase the signal-to-noise ratio. After classification, all classes were ranked according to quality. The worst 2D classes (representing approximately 5% of all particles) were discarded. For the 3D reconstruction only those particles were selected, that represent elongated Wisc1-GFP structure. In total, 46,500 particles were subjected to 3D classification in RELION 2.0. It produced three 3D classes with the largest number of particles and highest signal-to-noise, that were chosen to build 3D models. C1 symmetry was used for all reconstructions, including the final one from the class 1 (16818 images) obtained at a resolution of 18 Å (Figure S2). The atomic structures of the GFP (Protein Data Bank ID: 4ogs) and the N-terminal homologue-$KRM1_{WSC}$ domain (Protein Data Bank ID: 5fws) were fit into the density map using UCSF Chimera [42].

2.12. Bioinformatics Analysis

For modelling the 3D structure of WSC1, we searched for its structural homologs by performing searches with the Position-Specific Iterative Basic Local Alignment Search Tool (PSI-BLAST) (https://blast.ncbi.nlm.nih.gov/Blast.cgi [43]) against all the structures in Protein Data Bank PDB (https://www.rcsb.org/ [44]). The validity of the search results was checked by comparison with the PROSITE database (https://prosite.expasy.org/ [45]).

3. Results

3.1. Construction of a Yeast Strain for Stable Expression of a Tagged WSC1 Construct

In previous works we employed a yeast strain carrying a *WSC1* gene tagged at its native locus with the coding sequence for the green fluorescent protein (GFP) to produce a C-terminally tagged, fully functional sensor fusion protein [46]. While preliminary experiments demonstrated that this construct was suitable to detect the detergent-free incorporation of the sensor from the yeast plasma membrane into a styrene maleic acid (SMA) copolymer, it raised the problem that the preparation of SMA lipoproteins (SMALPs) contained a mixture of membrane proteins, which only in a small fraction represented the Wsc1 sensor. In order to provide a means to enrich the latter fraction, we decided to additionally tag the sensor GFP-fusion at the C-terminal end with the recognition sequence for the TEV protease followed by three copies of the hemagglutin antigen (Figure 1; [10]; note that the TEV sequence was inserted to further facilitate purification, which turned out not be necessary in the course of this work). Spheroplasts prepared from this yeast donor strain were then used as a source of membrane proteins to obtain nanosized SMALPs to be characterized by a set of biophysical approaches, as outlined below.

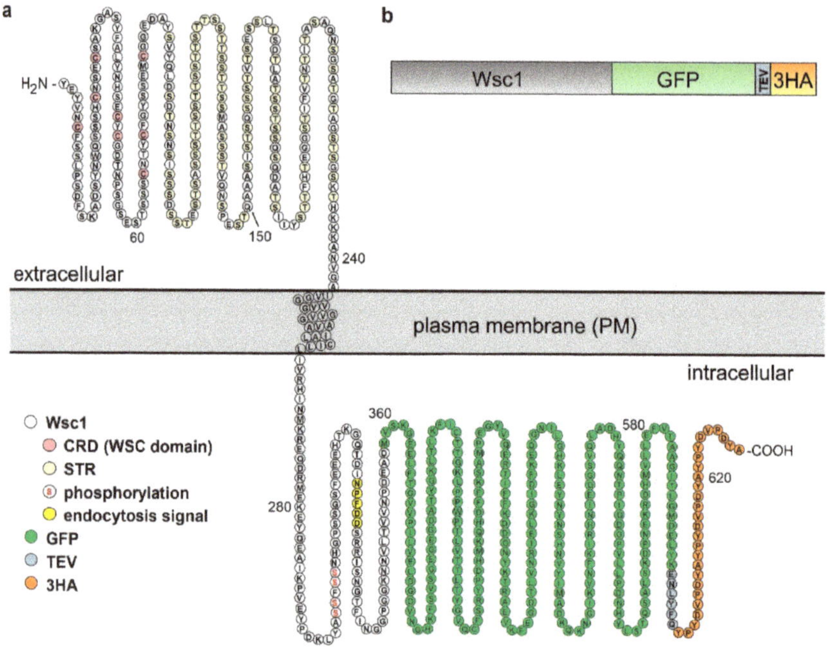

Figure 1. Schematic representations of the Wsc1 fusion construct employed in this work. (**a**) Primary structure of the sensor fusion protein indicating its orientation within the yeast plasma membrane, designed with Protter according to [47] and modified manually. Colour codes within the Wsc1 sequence highlight the eight cysteine residues characteristic of the WSC- or cysteine-rich domain (CRD), the residues believed to be glycosylated in the serine/threonine-rich region (STR; [6]), the serine residues within the cytoplasmic tail which could be phosphorylated [48], and the signal motif for endocytosis [46]. (**b**) Simplified overview of different protein sequences fused to the C-terminal end of Wsc1.

3.2. Enrichment of Wsc1-GFP in SMALPs

In order to purify the tagged Wsc1 protein for further characterization, the general scheme depicted in Figure 2a was followed. Thus, we first prepared yeast spheroplasts from logarithmically growing cells by digesting the cell wall with the Zymolyase glucanase preparation. This gentle treatment prevents the liberation of endogenous intracellular proteases caused by mechanical damage, which should also help to protect the Wsc1 membrane sensor from proteolytic degradation [49]. After disruption of the spheroplasts with ultrasound, the membrane fraction was harvested by centrifugation and resuspended in Tris-HCl buffer with sodium chloride. For extraction of the tagged sensor, the membrane suspension was then incubated with a 2.5% SMA copolymer solution and subsequently clarified by centrifugation. Finally, the supernatant containing SMALP-preparation was loaded onto an anti-HA tag affinity column and eluted to obtain Wsc1-GFP purified in SMALPs.

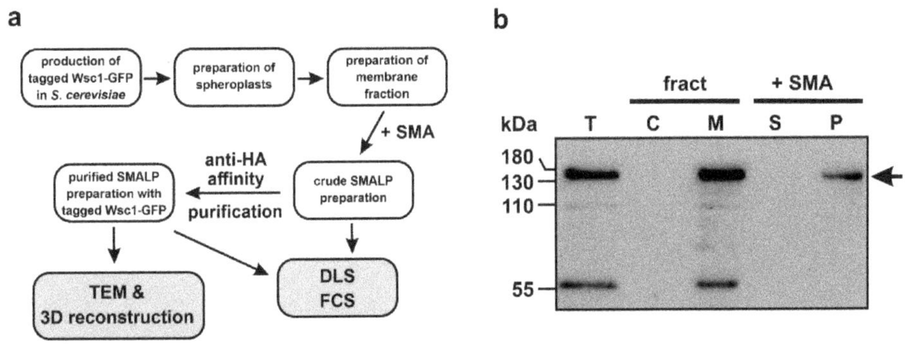

Figure 2. Purification and examination of styrene maleic acid lipoproteins (SMALPs) containing Wsc1-green fluorescent protein (GFP). (**a**) General scheme of the experimental strategy followed in this work. (**b**) Immunoblot analysis of cell fractionation and solubilization of the sensor fusion in the styrene maleic acid (SMA) polymer. T = total cell lysate; fract = cell fractionation with C = soluble cytosolic proteins, M = membrane-bound, pelleted fraction; +SMA = solubilization with SMA with S = supernatant (i.e., SMALPs), P = pellet with debris. The arrow indicates the signal of the presumably highly glycosylated sensor, according to [50].

Western blotting and immunological detection (Figure 2b) verified the presence of the Wsc1-GFP protein in the whole cell membranes, after cell fractionation and membrane solubilization. Wsc1-GFP was exclusively retained in the membrane fraction (Figure 2b, lane M), as demonstrated by comparison to the total cell lysate (Figure 2b, lane T). The protein appeared in a major band at approximately 130 kDa on the immunoblot probed with anti-HA (Figure 2b), which presumably represents the highly glycosylated sensor [50]. A minor signal at 55 kDa probably reflects a proteolytic cleavage product, which was eliminated by the copolymer treatment (note that the Wsc1 fusion protein has a calculated molecular mass of 73 kDa). The smaller band is unlikely to represent a GFP dimer, as this would dissociate in the denaturing gel conditions and is not likely to form in the first place with the GFP employed in our construct [51]. With regard to the glycosylated sensor, a significant portion remained in the original membrane fraction (Figure 2b, lane P) after treatment with the SMA copolymer, while the amount of Wsc1-GFP solubilized was below the sensitivity threshold of the immunoblot (Figure 2b, lane S), it could be detected by fluorescence spectroscopy (as explained below).

3.3. Biophysical Characterization of Crude SMALP Preparations

Dynamic light scattering (DLS) was employed to analyze the size distribution of the SMA copolymer particles in the supernatant fraction obtained after membrane solubilization (further referred to as soluble membrane fraction). These data confirmed the presence of monodisperse nano-sized particles (Figure 3a). Supporting data from previous works, their intensity-weighted diameter was estimated to be in the 10-nm range [14,15,22,52]. More importantly, we thus demonstrated that incubation of the yeast cellular membrane fraction with the SMA copolymer in a ratio of 3:1 indeed yielded nano-sized SMALPs.

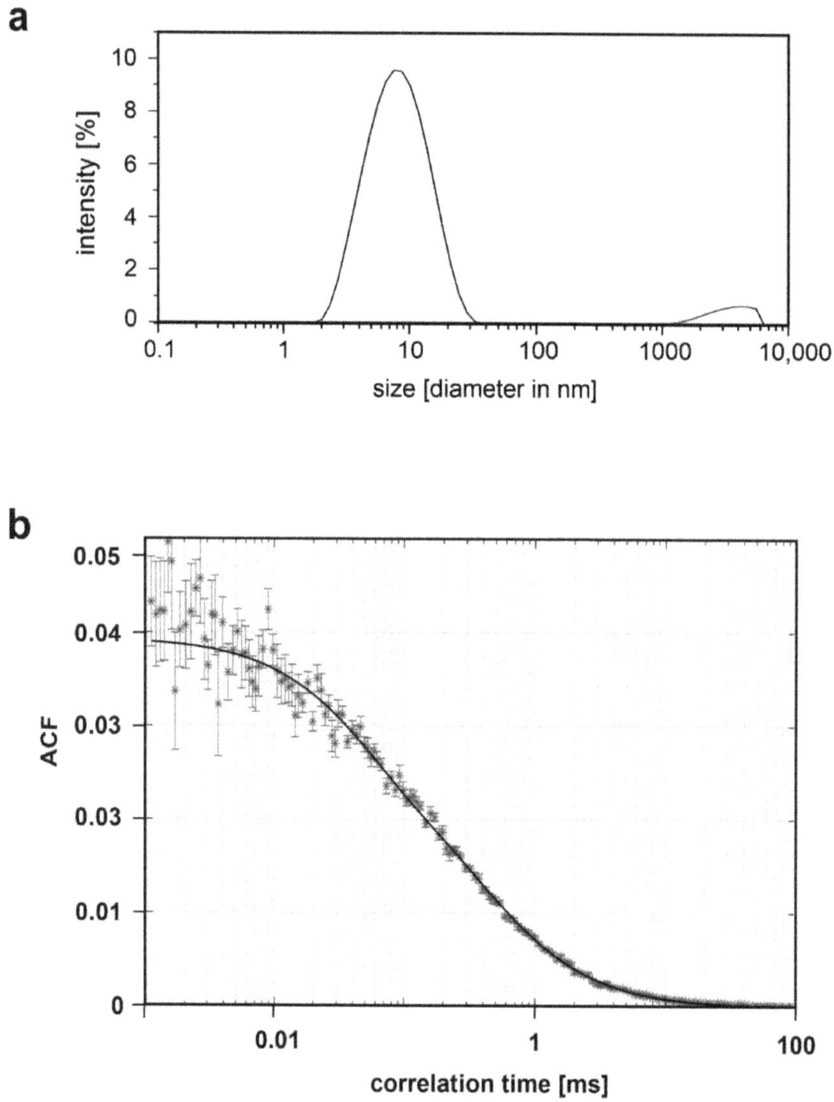

Figure 3. Quality assessment of crude SMALP preparations. (**a**) Representative image of size distribution of a SMALP preparation after membrane solubilization based on dynamic light scattering (DLS) data. (**b**) Representative example of a Fluorescence Correlation Spectroscopy (FCS) analysis. The curve was fitted by a model considering triplet-state blinking (black line). Calculations given in the text were based on at least three such curves obtained for four different SMALP preparations, each, which displayed similar kinetics. ACF = autocorrelation function.

These SMALPs were then subjected to fluorescence correlation spectroscopy (FCS), which allows the robust determination of diffusion constants at nanomolar concentrations following fluorescent labels, in order to determine their content of Wsc1-GFP. First, the identity of the GFP signals was verified by determination of the fluorescence emission spectrum. From the diffusion constants obtained in the FCS, hydrodynamic radii were estimated using the Stokes-Einstein equation, assuming a spherical shape of the particles. Samples were then subjected to robust FCS measurements based on their GFP fluorescence, using appropriate dilutions to nanomolar concentrations if necessary. Typical FCS autocorrelation results are shown in Figure 3b. The fit indicates a monodisperse population of SMALP particles undergoing three-dimensional free diffusion. Furthermore, fitting the experimental autocorrelation function revealed an average particle diffusion coefficient of 53.0 ± 3.0 $\mu m^2/s$ (± standard error). The average hydrodynamic particle radius, r, was 4.9 ± 0.3 nm and the Wsc1-GFP concentration was approximately 30 nM. We took these data as strong evidence that the Wsc1-GFP protein was indeed present in the SMALP preparation after membrane solubilization, basically carrying a single sensor molecule per particle. The latter assumption was confirmed by the TEM pictures taken of the purified SMALPs as described below. Moreover, the FCS values obtained for the size of the nanoparticles correlated well with the DLS data discussed above.

3.4. Biophysical Characterization of Affinity-Purified SMALPs with Wsc1-GFP

We assumed that the soluble membrane fraction, i.e., the supernatant fraction obtained after membrane solubilization by incubation with the SMA copolymer, contained both SMALPs loaded with Wsc1-GFP and sensor-free SMALPs carrying other yeast membrane proteins. The fraction containing the GFP-tagged sensor was further purified by application of the supernatant to an anti-HA affinity column. Similar to the soluble membrane fraction preparation shown in Figure 2b, Wsc1-GFP concentrations in the purified preparations remained below the detection limit in immunoblotting and were also too low for reliable DLS measurements (data not shown). However, a GFP signal could be detected and characterized in the purified SMALPs by FCS. The diffusion coefficient was close to the one determined above for the non-purified SMALPs, with 50.0 ± 4.6 $\mu m^2/s$ as compared to 53.0 ± 3.0 $\mu m^2/s$, respectively. Likewise, the corresponding hydrodynamic radius, r, of 4.9 ± 0.3 nm remained unaltered. This confirmed that affinity purification did not affect the physicochemical properties of the SMALP-embedded sensor.

3.5. Three-Dimensional Modelling of the Tagged Wsc1 Sensor Based on Negative-Stain Transmission Electron Microscopy and Bioinformatic Analyses

Finally, affinity-purified SMALPs loaded with Wsc1-GFP were analyzed by negative-stain transmission electron microscopy (nsTEM). As shown in the exemplary image in Figure 4a, the particles observed were mostly single, rather homogeneous, and did not form larger aggregates. A total of 107,000 single particle images were then acquired and first classified for their appearance in two dimensions (2D). Some of the class images displayed elongated structures with a length of approximately 10–15 nm, with three clearly distinguishable segments (Figure 4b). Nevertheless, the particle geometry was highly dynamic, with some showing an extended conformation (top row in Figure 4b), while others adopted a 'folded' shape (bottom row in Figure 4b).

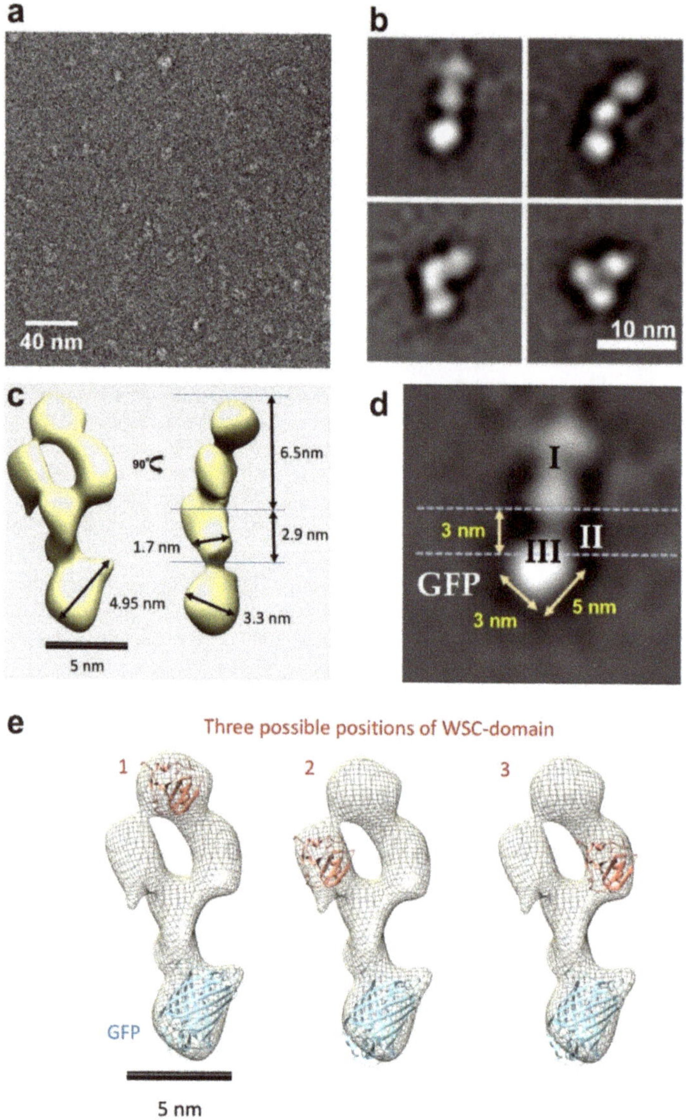

Figure 4. Construction of a three-dimensional model of the tagged Wsc1 sensor. (**a**) Images of the purified Wsc1-carrying SMALP preparation obtained by negative-stain transmission electron microscopy. (**b**) Four representative examples of two-dimensional class images of the Wsc1-GFP sensor fusion. 2D classes that are representative of those used for the 3D reconstruction are shown in the two images of the upper row. (**c**) Three-dimensional reconstruction of the sensor structure. Measurements were performed in ImageJ [53]. The two views depicted in were obtained by successive 90° rotations around the vertical axis. (**d**) 2D class-average of the extended Wsc1-GFP construct with designated domains, depicting their dimensions and the proposed position of the membrane. I, II, and III designate the proposed extracellular, transmembrane, and cytosolic domains, respectively. (**e**) Possible locations of homologous domains with a resolved crystal structure. The structure of GFP (pdb ID 4OGS) was fitted into the presumed cytosolic extension of the protein (blue ribbon) and the crystal structure homologous to the WSC domain of KRM1 (pdb ID 5FWS) would fit into any of the three cytosolic densities shown (salmon ribbon).

A PSI-BLAST search immediately revealed sequence similarity between residues 28–80 of the mature Wsc1 protein and a sequence stretch of the human KREMEN1 receptor. All eight functionally important cysteine residues matched (see alignments from run 1 presented in supplementary materials File S1, with the first four KREMEN1 receptor sequences). This region corresponds to the WSC domain in the KREMEN1 receptor structure (PDB ID 5FWS [54]) named after the homologous region near the N-terminus of Wsc1 (PROSITE entry PS51212; WSC). This first analysis revealed no further obvious candidates that could provide structural information for other parts of Wsc1.

However, the second iteration of the PSI-BLAST search (supplementary File S2) detected a sequence similarity between the following 200 residues of Wsc1 with the Ebola virus membrane-bound glycoprotein (GP), for which the structure of its soluble variant has been resolved [55]. Although the overall sequence similarity is fairly low, both proteins are highly enriched in serine and threonine residues in this region, which are highly glycosylated and typically inherent to intrinsically disordered proteins [56].

Indeed, the structure of Ebola virus GP (PDB ID 5KEL) displays short, disordered α-helices and β-strands [55]. It should be noted that the folding of these regions depends on the interaction of the respective proteins with physiological partners and can be determined only in their presence (see [57,58] for recent reviews). Thus, the Ebola virus GP structure was resolved only in a complex with specific antibodies [55], which impeded its use in the modelling of Wsc1. In any case, the periplasmic part of Wsc1 is built of two structural segments: the N-terminal domain stabilized by disulfide bridges forming the known WSC domain (further described in the next paragraph), and the following segment, which probably has a variable, disordered structure sensitive to the degree of its glycosylation [56]. This could cause the second segment to look somewhat fuzzier on processed EM images.

Since the position of the membrane-crossing α-helix could only be identified in images with the extended conformation (Figure 4c,d), only those were considered for extraction by RELION 2.0 and employed to calculate the three-dimensional (3D) structure of the protein (Figure 4c). The remaining 46,500 single particles which met this criterion then served to generate the 3D reconstructions of Wsc1-GFP in an extended conformation with C1 symmetry, as depicted in Figure 4c.The resulting model reflected the three putative functional parts of the Wsc1 sensor (Figure 4d). We used the 3D structure of the WSC-homologous domain of KREMEN1 (residues Gly129-Thr220; PDB ID 5FWS [54]) for modelling the shape of the N-terminal Wsc1 part. The rest of the extracellular part was assumed to be entropically unstable and to lack a defined structure, while its exact architecture is unknown (note that the overall size of this domain is compatible with any of three reconstructed extracellular periplasmic densities). The WSC-like domain of KRM1 may fit in three different ways (Figure 4e). The entire extracellular part then has a clasp-like appearance and is connected to the transmembrane domain (TMD) via a hinge, which confers flexibility relative to the TMD. With approximately 1.7 nm $\times$ 2.9 nm the TMD appears to be slightly broader than expected from the primary sequence of a single alpha-helix (Figure 4c,d). This could be attributed to the varying position of GFP and the extracellular part relative to the transmembrane helix within different lipodiscs and/or to a distortion from lipodisc surrounding the membrane domain.

The size of the third, bulkier domain fits the short cytosolic sequence of Wsc1 fused to the GFP ß-barrel. The crystal structure of GFP [59] was thus modelled into this domain, leaving space for the cytosolic tail of the sensor. As seen in Figure 4e, the GFP does not fit perfectly, leaving an additional thumb-like projection on the right-hand side close to the inner membrane surface, which may correspond to the folded cytosolic tail of Wsc1 (60 amino acids in length).

4. Discussion

Recently, the amphipathic styrene-maleic acid (SMA) and SMA-related copolymers became a promising tool for membrane protein purification. Here, we have reported the application of SMA in the solubilization of the membrane spanning sensor protein Wsc1 as an important constituent of the native plasma membrane from the model yeast *S. cerevisiae*. Wsc1 is a major player in the activation of the cell wall integrity (CWI) pathway, which regulates cell wall remodeling during periods of extracellular stress, mating and normal cell growth [5]. We took advantage of the GFP-moiety of a tagged functional protein employed in previous works [10,46] and equipped it with an additional 3xHA-tag for affinity purification. By combining the ease of genetic engineering in yeast cells for specific experimental purposes with the use of advanced biophysical methods, we were able to successfully isolate and purify the tagged sensor in a detergent-free manner and in a semi-native lipid environment. This allowed a compilation of images obtained by negative-stain electron microscopy (nsTEM), from which a first three-dimensional structural model of Wsc1 was deduced.

The model is consistent with a tripartite organization of Wsc1, previously suggested based on its primary sequence and mechanical features revealed by single-molecule atomic force microscopy (AFM, [7]). These three parts correspond to the relatively large extracellular region, a single transmembrane domain, and a relatively short cytoplasmic tail at the C-terminus. Due to its high flexibility in the nsTEM images, modelling of the extracellular region is most problematic and probably does not closely reflect the in vivo conformation. The models deduced from our 3D class images (Figure 4c and Figure S1) exhibit a relatively large extracellular domain (approximately 6.5×6.4 nm) forming a horseshoe, attached to the smaller TMD density of 1.7×2.9 nm. The N-terminal cysteine-rich WSC-domain seems to be the only folded part of the whole extracellular domain of Wsc1, which is expected to form disulfide bridges and interact with the carbohydrate chains of the cell wall glucans (Figure 5; [9,10]). It is tempting to speculate that the polysaccharide chains could go through the hairpin structure and thus pull the sensor if they are distorted. Consistent with the proposed function as a mechanosensor [6], the force could be generated by dislocation of the polysaccharide chains, as a consequence of stretching caused by changes in turgor in reaction to medium osmolarity or by cell wall damage. The hairpin then could be easily imagined to break open under extensive mechanical stress thus allowing the sensor to be internalized by the rapid endocytosis observed in live yeast cells [46,60] solving the puzzle of how mechanical interaction with the cell wall is compatible with sensor internalization. Clearly, evidence for such a mechanism would require visualization of the protein associated with the cell wall, for which the methodology does not yet exist. The serine/threonine-rich region (STR), that is likely to be intrinsically disordered [56], has been proposed to penetrate the yeast cell wall in the form of a nanospring, which could stretch for 42 nm as shown by AFM measurements [7]. In fact, the disordered arrangement of the serine/threonine rich segment, which is held together by entropic forces, would be consistent with a spring-like behavior. Consistent with this view, our estimates based on this model now would predict that this 200 aa-long segment would cover some 50 nm distance, if completely stretched (Figure 5).

We assume that the shrinking and reduction of the external length to about 6–7 nm, as observed for the EM images, is a manifestation of an entropically collapsed structure of the nanospring, which, otherwise, could be considerably stretched by interaction of the N-terminal WSC-domain with the cell wall in living yeast cells under standard conditions, and even more though upon perturbances of the cell surface (Figure 5). As stated above, the compressed conformation(s) of the periplasmic domain may be a consequence of the extraction of full-length Wsc1 from the plasma membrane and foremost of preventing its interaction with the yeast cell wall, due to the Zymolyase treatment and transfer to SMA. The ability for stretching in live yeast cells greatly enhances the sensitivity of the sensor towards perturbations in either the cell wall or the plasma membrane, which would be transmitted through the transmembrane part to the cytosolic tail to trigger CWI signaling, supporting the role of Wsc1 as a mechanosensor [7]. It should be noted that the external domain also confers the species-specific properties of CWI signaling in different yeasts, as deduced from heterologous expression studies of the related Mid2 sensor [61]. It would thus be interesting to resolve the structures of Wsc1 homologues from other yeasts and/or those of Mid2, which lack a WSC domain, in future studies.

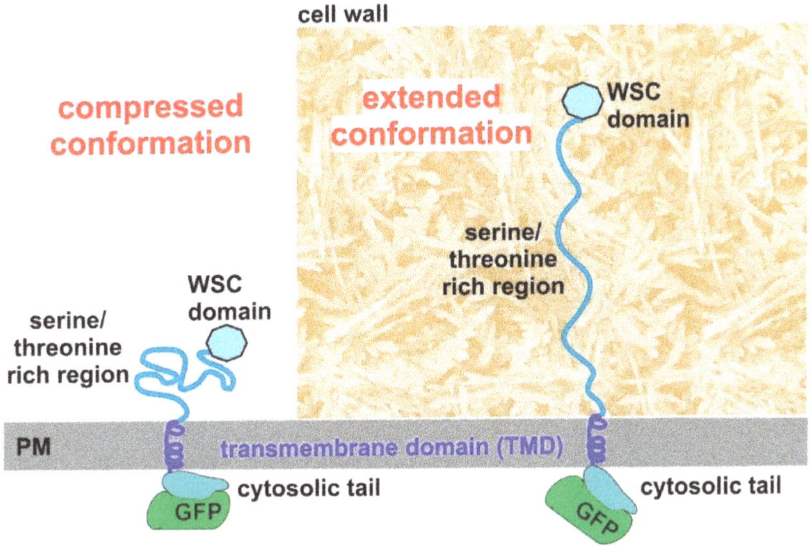

Figure 5. Proposed conformations of the Wsc1 sensor and position of its cysteine-rich domain (WSC domain) in the SMALP preparations (left) and in living yeast cells (right). PM = plasma membrane.

The cytosolic tail of Wsc1 is of utmost importance in signal transduction to the CWI pathway. It has been shown to interact with the guanine nucleotide exchange factor (GEF) Rom2 triggering the downstream signaling cascade [48]. Interesting, a tyrosine residue (Y282 in Figure 1a) was shown in the latter work to mediate this interaction, which is counteracted by phosphorylation of downstream serine residues, which could lead to an electrostatic repulsion from the positively charged inner leaf of the PM. The cytosolic domain also mediates the highly dynamic intracellular distribution of Wsc1, which is constantly subjected to endocytosis, depending on an NPFXD motif [46,60,62]. Endocytosis of the sensor also depends on its lipid environment within the PM, as depletion of phosphoinositol-4,5-bisphosphate results in a massive accumulation of Wsc1 and other CWI signaling components in recycling endosomes [63]. Moreover, early studies indicated that the Wsc-type sensors reside in so-called lipid rafts, i.e., membrane compartments defined by their resistance towards solubilization by Triton X100 [50]. However, apart from these two reports indicating their functional influence, the exact lipid composition of the membrane locally surrounding Wsc1 sensors is not known.

In our 3D reconstruction, the cytosolic part probably best reflects the natural conformation in living yeast cells, as it is in the natural aqueous environment and relatively short. As evident from Figures 4 and 5, the thumb-like structure of the cytosolic tail seems to align with the inner leaf of the membrane. Its location is consistent with previous observations, that Wsc1 function is not affected by fusion to the GFP tag [10,60]. Moreover, as indicated at the left hand side of Figure 5, it could be dislocated from this position by external stress and/or by electrostatic repulsion due to the observed phosphorylation of serine residues [48]. This then would enable the sensor to interact with the downstream components of the CWI pathway or other modification factors. The model also explains how slight modifications in the angle of the hinge region connecting the TMD to the intracellular part, e.g., by pulling it towards the lipid bilayer, could result in dramatic changes in the exposure of the cytosolic tail. Obviously, such conformational changes would be brought about by the interaction of the extracellular part of Wsc1 with the yeast cell wall, thus transmitting perturbations through the TMD to the interior of the cell.

This work provides interesting insights into the structure of the Wsc1 sensor and confirms the importance of studying TM receptors and sensors in lipid-based, native-like environments. For instance, the monodisperse purified SMALPs containing the sensor had an average diameter of only 10 nm, accommodating only single sensor molecules. However, Wsc1 has been shown to cluster in large complexes within the plasma membrane of living yeast cells and was suggested to form at least dimers, both homodimers and heterodimers with other Wsc-type sensors, as deduced from bimolecular fluorescence interactions [8]. An extraction method for sensor multimers would thus be the next challenge.

Diisobutylene/maleic acid (DIBMA) copolymers, which differ from SMA copolymers in the non-aromatic nature of their hydrophobic groups, can also solubilize membrane proteins directly from a lipid bilayer. However, the size of DIBMA/lipid particles (DIBMALPs) can be varied in a broader range than that of SMALPs, thus reaching dimensions of up to 50 nm [64–66]. In future studies it would be interesting to employ DIBMA copolymers for the extraction of dimers and larger sensor complexes from the yeast plasma membrane into large sized polymer/lipid particles. Furthermore, the interaction between yeast cell wall extracts and SMALP-embedded sensors should be studied, to determine whether Wsc1 can access its extended conformation in the presence of cell wall polysaccharides.

Nevertheless, this work demonstrated the applicability of a detergent-free extraction and purification procedure, which is suitable not only for other membrane proteins of *S. cerevisiae*, but also of pathogenic yeasts and other fungi. Choosing essential members of such membrane proteins to study their 3D structure could thus provide a new angle for the design of antifungal agents, urgently needed in medicine and agriculture.

Supplementary Materials: The following materials are available online at https://www.mdpi.com/2309-608X/7/2/118/s1, File S1: Report on the initial PSI-BLAST search with the Wsc1 sequence as query. File S2: Report on the second iteration PSI-BLAST search with the Wsc1 sequence as query. The Ebola virus GP alignment is marked yellow. Figure S1: Additional 3D classes of Wsc1-GFP construct, calculated in RELION2.0 [43]. For each reconstruction two orientations are shown, that differ by rotation for 90 degrees around vertical axis. Bar–5 nm; Figure S2: Fourier shell correlation plot for the 3D reconstruction in Figure 4c.

Author Contributions: Conceptualization, N.V. and J.J.H.; methodology, N.V., M.K., R.K.; software, M.K. and R.K.; validation, M.K. and R.K.; formal analysis, N.V.,M.K., K.V.S., O.S.S., A.Y.M., and R.K.; investigation, N.V., M.K., R.K., K.V.S., and O.S.S.; resources, O.S.S., K.V.S., H.-J.S., and J.J.H.; data curation, M.K. and R.K.; writing—original draft preparation, N.V. and J.J.H.; writing—review and editing, N.V., A.Y.M., O.S.S., H.-J.S., and J.J.H.; visualization, M.K., K.V.S., O.S.S., and R.K.; supervision, K.V.S., O.S.S., H.-J.S. and J.J.H.; project administration, N.V. and J.J.H.; funding acquisition, H.-J.S. and K.V.S. All authors have read and agreed to the published version of the manuscript.

Funding: This research was funded by German Research Foundation (DFG, STE640/15) to H.-J.S., and the Ostpartnerschaftenprogramm of DAAD and RFBR grant no. 18-504-12045 to K.V.S.

Institutional Review Board Statement: Not applicable.

Informed Consent Statement: Not applicable.

Data Availability Statement: Original datasets, strains and constructs employed in this work are freely available upon request to the academic community for research purposes, only.

Acknowledgments: We thank Andrea Murra for excellent technical support. FCS experiments were performed at the Integrated Bioimaging Facility (iBiOs) at the Center of Cellular Nanoanalytics (CellNanOs), University of Osnabrück. We thank rer. nat. Katherina Psathaki, head of EM unit of iBiOs, for the possibility to prepare and to screen EM grids. Electron microscopy was performed at the scientific facility "3D Electron Microscopy and Spectroscopy" of Biology department, M.V. Lomonosov Moscow State University. We thank Andrey Moiseenko for expert help with SerialEM. O.S.S. and K.V.S. acknowledge the support from the Interdisciplinary Scientific and Educational School of Moscow Lomonosov University «Molecular Technologies of the Living Systems and

Synthetic Biology». We acknowledge support by Deutsche Forschungsgemeinschaft (DFG) and Open Access Publishing Fund of Osnabrück University.

Conflicts of Interest: The authors declare no conflict of interest. The funders had no role in the design of the study; in the collection, analyses, or interpretation of data; in the writing of the manuscript, or in the decision to publish the results.

References

1. Odds, F.C.; Brown, A.J.; Gow, N.A. Antifungal agents: Mechanisms of action. *Trends Microbiol.* **2003**, *11*, 272–279. [CrossRef]
2. Orlean, P. Architecture and biosynthesis of the *Saccharomyces cerevisiae* cell wall. *Genetics* **2012**, *192*, 775–818. [CrossRef] [PubMed]
3. Heinisch, J.J.; Rodicio, R. Protein kinase C in fungi-more than just cell wall integrity. *FEMS Microbiol. Rev.* **2018**, *42*. [CrossRef] [PubMed]
4. Dichtl, K.; Samantaray, S.; Wagener, J. Cell wall integrity signalling in human pathogenic fungi. *Cell. Microbiol.* **2016**, *18*, 1228–1238. [CrossRef]
5. Levin, D.E. Cell wall integrity signaling in *Saccharomyces cerevisiae*. *Microbiol. Mol. Biol. Rev.* **2005**, *69*, 262–291. [CrossRef] [PubMed]
6. Kock, C.; Dufrene, Y.F.; Heinisch, J.J. Up against the wall: Is yeast cell wall integrity ensured by mechanosensing in plasma membrane microdomains? *Appl. Environ. Microbiol.* **2015**, *81*, 806–811. [CrossRef]
7. Dupres, V.; Alsteens, D.; Wilk, S.; Hansen, B.; Heinisch, J.J.; Dufrene, Y.F. The yeast Wsc1 cell surface sensor behaves like a nanospring in vivo. *Nat. Chem. Biol.* **2009**, *5*, 857–862. [CrossRef]
8. Kock, C.; Arlt, H.; Ungermann, C.; Heinisch, J.J. Yeast cell wall integrity sensors form specific plasma membrane microdomains important for signalling. *Cell. Microbiol.* **2016**, *18*, 1251–1267. [CrossRef]
9. Dupres, V.; Heinisch, J.J.; Dufrene, Y.F. Atomic force microscopy demonstrates that disulfide bridges are required for clustering of the yeast cell wall integrity sensor Wsc1. *Langmuir* **2011**, *27*, 15129–15134. [CrossRef]
10. Heinisch, J.J.; Dupres, V.; Wilk, S.; Jendretzki, A.; Dufrene, Y.F. Single-molecule atomic force microscopy reveals clustering of the yeast plasma-membrane sensor Wsc1. *PLoS ONE* **2010**, *5*, e11104. [CrossRef]
11. Lemieux, M.J.; Overduin, M. Structure and function of proteins in membranes and nanodiscs. *Biochim. Biophys. Acta Biomembr.* **2021**, *1863*, 183445. [CrossRef] [PubMed]
12. Rodriguez-Pena, J.M.; Diez-Muniz, S.; Bermejo, C.; Nombela, C.; Arroyo, J. Activation of the yeast cell wall integrity MAPK pathway by zymolyase depends on protease and glucanase activities and requires the mucin-like protein Hkr1 but not Msb2. *FEBS Lett.* **2013**, *587*, 3675–3680. [CrossRef]
13. De Zorzi, R.; Mi, W.; Liao, M.; Walz, T. Single-particle electron microscopy in the study of membrane protein structure. *Microscopy* **2016**, *65*, 81–96. [CrossRef] [PubMed]
14. Knowles, T.J.; Finka, R.; Smith, C.; Lin, Y.P.; Dafforn, T.; Overduin, M. Membrane proteins solubilized intact in lipid containing nanoparticles bounded by styrene maleic acid copolymer. *J. Am. Chem. Soc.* **2009**, *131*, 7484–7485. [CrossRef] [PubMed]
15. Dorr, J.M.; Scheidelaar, S.; Koorengevel, M.C.; Dominguez, J.J.; Schafer, M.; van Walree, C.A.; Killian, J.A. The styrene-maleic acid copolymer: A versatile tool in membrane research. *Eur. Biophys. J.* **2016**, *45*, 3–21. [CrossRef]
16. Lee, S.C.; Pollock, N.L. Membrane proteins: Is the future disc shaped? *Biochem. Soc. Trans.* **2016**, *44*, 1011–1018. [CrossRef] [PubMed]
17. Overduin, M.; Esmaili, M. Memtein: The fundamental unit of membrane-protein structure and function. *Chem. Phys. Lipids* **2019**, *218*, 73–84. [CrossRef]
18. Ravula, T.; Hardin, N.Z.; Ramamoorthy, A. Polymer nanodiscs: Advantages and limitations. *Chem. Phys. Lipids* **2019**, *219*, 45–49. [CrossRef]
19. Cuevas Arenas, R.; Klingler, J.; Vargas, C.; Keller, S. Influence of lipid bilayer properties on nanodisc formation mediated by styrene/maleic acid copolymers. *Nanoscale* **2016**, *8*, 15016–15026. [CrossRef]
20. Dominguez Pardo, J.J.; Dorr, J.M.; Iyer, A.; Cox, R.C.; Scheidelaar, S.; Koorengevel, M.C.; Subramaniam, V.; Killian, J.A. Solubilization of lipids and lipid phases by the styrene-maleic acid copolymer. *Eur. Biophys. J.* **2017**, *46*, 91–101. [CrossRef]
21. Craig, A.F.; Clark, E.E.; Sahu, I.D.; Zhang, R.; Frantz, N.D.; Al-Abdul-Wahid, M.S.; Dabney-Smith, C.; Konkolewicz, D.; Lorigan, G.A. Tuning the size of styrene-maleic acid copolymer-lipid nanoparticles (SMALPs) using RAFT polymerization for biophysical studies. *Biochim. Biophys. Acta* **2016**, *1858*, 2931–2939. [CrossRef]
22. Orwick-Rydmark, M.; Lovett, J.E.; Graziadei, A.; Lindholm, L.; Hicks, M.R.; Watts, A. Detergent-free incorporation of a seven-transmembrane receptor protein into nanosized bilayer Lipodisq particles for functional and biophysical studies. *Nano Lett.* **2012**, *12*, 4687–4692. [CrossRef]
23. Voskoboynikova, N.; Mosslehy, W.; Colbasevici, A.; Ismagulova, T.T.; Bagrov, D.V.; Akovantseva, A.A.; Timashev, P.S.; Mulkidjanian, A.Y.; Bagratashvili, V.N.; Schaitan, K.V.; et al. Characterization of an archaeal photoreceptor/transducer complex from *Natronomonas pharonis* assembled within styrene-maleic acid lipid particles. *RSC Adv.* **2017**, *7*, 51324–51334. [CrossRef]
24. Gulati, S.; Jamshad, M.; Knowles, T.J.; Morrison, K.A.; Downing, R.; Cant, N.; Collins, R.; Koenderink, J.B.; Ford, R.C.; Overduin, M.; et al. Detergent-free purification of ABC (ATP-binding-cassette) transporters. *Biochem. J.* **2014**, *461*, 269–278. [CrossRef] [PubMed]

25. Jamshad, M.; Charlton, J.; Lin, Y.P.; Routledge, S.J.; Bawa, Z.; Knowles, T.J.; Overduin, M.; Dekker, N.; Dafforn, T.R.; Bill, R.M.; et al. G-protein coupled receptor solubilization and purification for biophysical analysis and functional studies, in the total absence of detergent. *Biosci. Rep.* **2015**, *35*. [CrossRef] [PubMed]
26. Long, A.R.; O'Brien, C.C.; Malhotra, K.; Schwall, C.T.; Albert, A.D.; Watts, A.; Alder, N.N. A detergent-free strategy for the reconstitution of active enzyme complexes from native biological membranes into nanoscale discs. *BMC Biotechnol.* **2013**, *13*, 41. [CrossRef] [PubMed]
27. Skaar, K.; Korza, H.J.; Tarry, M.; Sekyrova, P.; Hogbom, M. Expression and subcellular distribution of GFP-tagged human tetraspanin proteins in *Saccharomyces cerevisiae*. *PLoS ONE* **2015**, *10*, e0134041. [CrossRef] [PubMed]
28. Smirnova, I.A.; Sjostrand, D.; Li, F.; Bjorck, M.; Schafer, J.; Ostbye, H.; Hogbom, M.; von Ballmoos, C.; Lander, G.C.; Adelroth, P.; et al. Isolation of yeast complex IV in native lipid nanodiscs. *Biochim. Biophys. Acta* **2016**, *1858*, 2984–2992. [CrossRef]
29. Karlova, M.G.; Voskoboynikova, N.; Gluhov, G.S.; Abramochkin, D.; Malak, O.A.; Mulkidzhanyan, A.Y.; Loussouarn, G.; Steinhoff, H.J.; Shaitan, K.V.; Sokolova, O.S. Detergent-free solubilization of human Kv channels expressed in mammalian cells. *Chem. Phys. Lipids* **2019**, *219*, 50–57. [CrossRef]
30. Overduin, M.; Esmaili, M. Structures and interactions of transmembrane targets in native nanodiscs. *SLAS Discov.* **2019**, *24*, 943–952. [CrossRef]
31. Parmar, M.; Rawson, S.; Scarff, C.A.; Goldman, A.; Dafforn, T.R.; Muench, S.P.; Postis, V.L.G. Using a SMALP platform to determine a sub-nm single particle cryo-EM membrane protein structure. *Biochim. Biophys. Acta Biomembr* **2018**, *1860*, 378–383. [CrossRef]
32. Postis, V.; Rawson, S.; Mitchell, J.K.; Lee, S.C.; Parslow, R.A.; Dafforn, T.R.; Baldwin, S.A.; Muench, S.P. The use of SMALPs as a novel membrane protein scaffold for structure study by negative stain electron microscopy. *Biochim. Biophys. Acta* **2015**, *1848*, 496–501. [CrossRef]
33. Sun, C.; Gennis, R.B. Single-particle cryo-EM studies of transmembrane proteins in SMA copolymer nanodiscs. *Chem. Phys. Lipids* **2019**, *221*, 114–119. [CrossRef]
34. Rodicio, R.; Koch, S.; Schmitz, H.P.; Heinisch, J.J. *KlRHO1* and *KlPKC1* are essential for cell integrity signalling in *Kluyveromyces lactis*. *Microbiology* **2006**, *152*, 2635–2649. [CrossRef]
35. Widengren, J.; Mets, U.; Rigler, R. Fluorescence correlation spectroscopy of triplet-states in solution—A theoretical and experimental study. *J. Phys. Chem.* **1995**, *99*, 13368–13379. [CrossRef]
36. Culbertson, C.T.; Jacobson, S.C.; Ramsey, J.M. Diffusion coefficient measurements in microfluidic devices. *Talanta* **2002**, *56*, 365–373. [CrossRef]
37. Mastronarde, D. SerialEM: A program for automated tilt series acquisition on tecnai microscopes using prediction of specimen position. *Microsc. Microanal.* **2003**, *9*, 1182–1183. [CrossRef]
38. Wagner, T.; Merino, F.; Stabrin, M.; Moriya, T.; Antoni, C.; Apelbaum, A.; Hagel, P.; Sitsel, O.; Raisch, T.; Prumbaum, D.; et al. SPHIRE-crYOLO is a fast and accurate fully automated particle picker for cryo-EM. *Commun. Biol.* **2019**, *2*, 218. [CrossRef] [PubMed]
39. Bell, J.M.; Chen, M.; Baldwin, P.R.; Ludtke, S.J. High resolution single particle refinement in EMAN2.1. *Methods* **2016**, *100*, 25–34. [CrossRef]
40. Tang, G.; Peng, L.; Baldwin, P.R.; Mann, D.S.; Jiang, W.; Rees, I.; Ludtke, S.J. EMAN2: An extensible image processing suite for electron microscopy. *J. Struct. Biol.* **2007**, *157*, 38–46. [CrossRef] [PubMed]
41. Scheres, S.H.; Valle, M.; Nunez, R.; Sorzano, C.O.; Marabini, R.; Herman, G.T.; Carazo, J.M. Maximum-likelihood multi-reference refinement for electron microscopy images. *J. Mol. Biol.* **2005**, *348*, 139–149. [CrossRef]
42. Pettersen, E.F.; Goddard, T.D.; Huang, C.C.; Couch, G.S.; Greenblatt, D.M.; Meng, E.C.; Ferrin, T.E. UCSF Chimera–a visualization system for exploratory research and analysis. *J. Comput. Chem.* **2004**, *25*, 1605–1612. [CrossRef]
43. Altschul, S.F.; Madden, T.L.; Schaffer, A.A.; Zhang, J.; Zhang, Z.; Miller, W.; Lipman, D.J. Gapped BLAST and PSI-BLAST: A new generation of protein database search programs. *Nucleic Acids Res.* **1997**, *25*, 3389–3402. [CrossRef]
44. Berman, H.M.; Westbrook, J.; Feng, Z.; Gilliland, G.; Bhat, T.N.; Weissig, H.; Shindyalov, I.N.; Bourne, P.E. The Protein Data Bank. *Nucleic Acids Res.* **2000**, *28*, 235–242. [CrossRef] [PubMed]
45. Sigrist, C.J.; de Castro, E.; Cerutti, L.; Cuche, B.A.; Hulo, N.; Bridge, A.; Bougueleret, L.; Xenarios, I. New and continuing developments at PROSITE. *Nucleic Acids Res.* **2013**, *41*, D344–D347. [CrossRef]
46. Wilk, S.; Wittland, J.; Thywissen, A.; Schmitz, H.P.; Heinisch, J.J. A block of endocytosis of the yeast cell wall integrity sensors Wsc1 and Wsc2 results in reduced fitness in vivo. *Mol. Genet. Genom.* **2010**, *284*, 217–229. [CrossRef]
47. Omasits, U.; Ahrens, C.H.; Muller, S.; Wollscheid, B. Protter: Interactive protein feature visualization and integration with experimental proteomic data. *Bioinformatics* **2014**, *30*, 884–886. [CrossRef] [PubMed]
48. Vay, H.A.; Philip, B.; Levin, D.E. Mutational analysis of the cytoplasmic domain of the Wsc1 cell wall stress sensor. *Microbiology* **2004**, *150*, 3281–3288. [CrossRef] [PubMed]
49. Salazar, O.; Asenjo, J.A. Enzymatic lysis of microbial cells. *Biotechnol. Lett.* **2007**, *29*, 985–994. [CrossRef]
50. Lodder, A.L.; Lee, T.K.; Ballester, R. Characterization of the Wsc1 protein, a putative receptor in the stress response of *Saccharomyces cerevisiae*. *Genetics* **1999**, *152*, 1487–1499. [PubMed]
51. Zacharias, D.A.; Violin, J.D.; Newton, A.C.; Tsien, R.Y. Partitioning of lipid-modified monomeric GFPs into membrane microdomains of live cells. *Science* **2002**, *296*, 913–916. [CrossRef] [PubMed]

52. Bagrov, D.V.; Voskoboynikova, N.; Armeev, G.A.; Mosslehy, W.; Gluhov, G.S.; Ismagulova, T.T.; Mulkidjanian, A.Y.; Kirpichnikov, M.P.; Steinhoff, H.J.; Shaitan, K.V. Characterization of lipodisc nanoparticles containing sensory rhodopsin ii and its cognate transducer from *Natronomonas pharaonis*. *Biophysics* **2016**, *61*, 942–949. [CrossRef]
53. Schindelin, J.; Rueden, C.T.; Hiner, M.C.; Eliceiri, K.W. The ImageJ ecosystem: An open platform for biomedical image analysis. *Mol. Reprod. Dev.* **2015**, *82*, 518–529. [CrossRef]
54. Zebisch, M.; Jackson, V.A.; Zhao, Y.; Jones, E.Y. Structure of the dual-mode Wnt regulator Kremen1 and insight into ternary complex formation with LRP6 and Dickkopf. *Structure* **2016**, *24*, 1599–1605. [CrossRef] [PubMed]
55. Pallesen, J.; Murin, C.D.; de Val, N.; Cottrell, C.A.; Hastie, K.M.; Turner, H.L.; Fusco, M.L.; Flyak, A.I.; Zeitlin, L.; Crowe, J.E., Jr.; et al. Structures of Ebola virus GP and sGP in complex with therapeutic antibodies. *Nat. Microbiol.* **2016**, *1*, 16128. [CrossRef]
56. Uversky, V.N. The intrinsic disorder alphabet. III. Dual personality of serine. *Intrinsically Disord. Proteins* **2015**, *3*, e1027032. [CrossRef]
57. Turoverov, K.K.; Kuznetsova, I.M.; Fonin, A.V.; Darling, A.L.; Zaslavsky, B.Y.; Uversky, V.N. Stochasticity of biological soft matter: Emerging concepts in intrinsically disordered proteins and biological phase separation. *Trends Biochem. Sci.* **2019**, *44*, 716–728. [CrossRef]
58. Uversky, V.N. Protein intrinsic disorder and structure-function continuum. *Prog. Mol. Biol. Transl. Sci.* **2019**, *166*, 1–17. [CrossRef]
59. Wineman-Fisher, V.; Simkovitch, R.; Shomer, S.; Gepshtein, R.; Huppert, D.; Saif, M.; Kallio, K.; Remington, S.J.; Miller, Y. Insight into the structure and the mechanism of the slow proton transfer in the GFP double mutant T203V/S205A. *Phys. Chem. Chem. Phys.* **2014**, *16*, 11196–11208. [CrossRef] [PubMed]
60. Straede, A.; Heinisch, J.J. Functional analyses of the extra- and intracellular domains of the yeast cell wall integrity sensors Mid2 and Wsc1. *FEBS Lett.* **2007**, *581*, 4495–4500. [CrossRef] [PubMed]
61. Rodicio, R.; Buchwald, U.; Schmitz, H.P.; Heinisch, J.J. Dissecting sensor functions in cell wall integrity signaling in *Kluyveromyces lactis*. *Fungal. Genet. Biol.* **2008**, *45*, 422–435. [CrossRef]
62. Piao, H.L.; Machado, I.M.; Payne, G.S. NPFXD-mediated endocytosis is required for polarity and function of a yeast cell wall stress sensor. *Mol. Biol. Cell* **2007**, *18*, 57–65. [CrossRef] [PubMed]
63. Fernandez-Acero, T.; Rodriguez-Escudero, I.; Molina, M.; Cid, V.J. The yeast cell wall integrity pathway signals from recycling endosomes upon elimination of phosphatidylinositol (4,5)-bisphosphate by mammalian phosphatidylinositol 3-kinase. *Cell. Signal.* **2015**, *27*, 2272–2284. [CrossRef] [PubMed]
64. Colbasevici, A.; Voskoboynikova, N.; Orekhov, P.S.; Bozdaganyan, M.E.; Karlova, M.G.; Sokolova, O.S.; Klare, J.P.; Mulkidjanian, A.Y.; Shaitan, K.V.; Steinhoff, H.J. Lipid dynamics in nanoparticles formed by maleic acid-containing copolymers: EPR spectroscopy and molecular dynamics simulations. *Biochim. Biophys. Acta Biomembr.* **2020**, *1862*, 183207. [CrossRef] [PubMed]
65. Oluwole, A.O.; Danielczak, B.; Meister, A.; Babalola, J.O.; Vargas, C.; Keller, S. Solubilization of membrane proteins into functional lipid-bilayer nanodiscs using a diisobutylene/maleic acid copolymer. *Angew. Chem. Int. Ed. Engl.* **2017**, *56*, 1919–1924. [CrossRef] [PubMed]
66. Oluwole, A.O.; Klingler, J.; Danielczak, B.; Babalola, J.O.; Vargas, C.; Pabst, G.; Keller, S. Formation of lipid-bilayer nanodiscs by diisobutylene/maleic acid (DIBMA) copolymer. *Langmuir* **2017**, *33*, 14378–14388. [CrossRef]

Journal of *Fungi*

MDPI

Article

Differential Requirement for the Cell Wall Integrity Sensor Wsc1p in Diploids Versus Haploids

Allison E. Hall [1,2], Miriam Lisci [1,3] and Mark D. Rose [1,4,*]

[1] Department of Molecular Biology, Princeton University, Princeton, NJ 08544, USA; Allison.Hall@nyulangone.org (A.E.H.); Ml793@cam.ac.uk (M.L.)
[2] Helen L. and Martin S. Kimmel Center for Biology and Medicine, Skirball Institute of Biomolecular Medicine, New York, NY 10016, USA
[3] Cambridge Institute for Medical Research, University of Cambridge, Cambridge Biomedical Campus, Cambridge CB2 0XY, UK
[4] Department of Biology, Georgetown University, Washington, DC 20057, USA
* Correspondence: mark.rose@georgetown.edu

Abstract: The primary role of the Cell Wall Integrity Pathway (CWI) in *Saccharomyces cerevisiae* is monitoring the state of the cell wall in response to general life cycle stresses (growth and mating) and imposed stresses (temperature changes and chemicals). Of the five mechanosensor proteins monitoring cell wall stress, Wsc1p and Mid2p are the most important. We find that *WSC1* has a stringent requirement in zygotes and diploids, unlike haploids, and differing from *MID2*'s role in shmoos. Diploids lacking *WSC1* die frequently, independent of mating type. Death is due to loss of cell wall and plasma membrane integrity, which is suppressed by osmotic support. Overexpression of several CWI pathway components suppress *wsc1*Δ zygotic death, including *WSC2*, *WSC3*, and *BEM2*, as well as the Rho-GAPS, *BEM3* and *RGD2*. Microscopic observations and suppression by *BEM2* and *BEM3* suggest that *wsc1*Δ zygotes die during bud emergence. Downstream in the CWI pathway, overexpression of a hyperactive protein kinase C (Pkc1p-R398P) causes growth arrest, and blocks the pheromone response. With moderate levels of Pkc1p-R398P, cells form zygotes and the *wsc1*Δ defect is suppressed. This work highlights functional differences in the requirement for Wsc1p in diploids Versus haploids and between Mid2p and Wsc1p during mating.

Keywords: cell wall integrity; conjugation; ploidy; yeast; lysis; protein kinase C

Citation: Hall, A.E.; Lisci, M.; Rose, M.D. Differential Requirement for the Cell Wall Integrity Sensor Wsc1p in Diploids Versus Haploids. *J. Fungi* **2021**, 7, 1049. https://doi.org/10.3390/jof7121049

Academic Editors: María Molina and Humberto Martín

Received: 13 November 2021
Accepted: 4 December 2021
Published: 8 December 2021

1. Introduction

The budding yeast, *Saccharomyces cerevisiae*, is a single-celled organism that lives in a relatively harsh environment in the wild, undergoing stresses due to changes in osmotic pressure and temperature. To survive in this environment, yeast have evolved a robust cell wall that protects against rapid osmotic changes and mechanical stresses, while providing support for shape changes during budding and polarized growth in response to pheromone [1,2].

Stresses to the cell wall are primarily detected by the cell wall integrity (CWI) pathway [3–5]. The CWI pathway is a signaling cascade that is regulated by five transmembrane sensor proteins that extend into the cell wall (Wsc1-3p, Mid2p, and Mtl1p) and ultimately regulate a series of MAP kinases [6–9] (Figure 1). Wsc1p and Mid2p appear to be the most important of the transmembrane proteins; *wsc1*Δ *mid2*Δ cells die without osmotic support (1 M sorbitol) [6,9] whereas cells containing only Wsc1p and Mid2p grow like wildtype [10]. The transmembrane proteins interact with Rom2p, a guanosine-nucleotide exchange factor (GEF) for Rho1p, a GTPase and master regulator of the CWI pathway [8]. Rho1p plays a role in regulating cell wall deposition through the 1,3-β-glucan synthase (GS) complex, made up of Fks1p and Fks2p [11–13]. Rho1p also binds to, and activates, the sole protein kinase C in yeast, Pkc1p, and has effects on the actin cytoskeleton and polarized secretion [14–18].

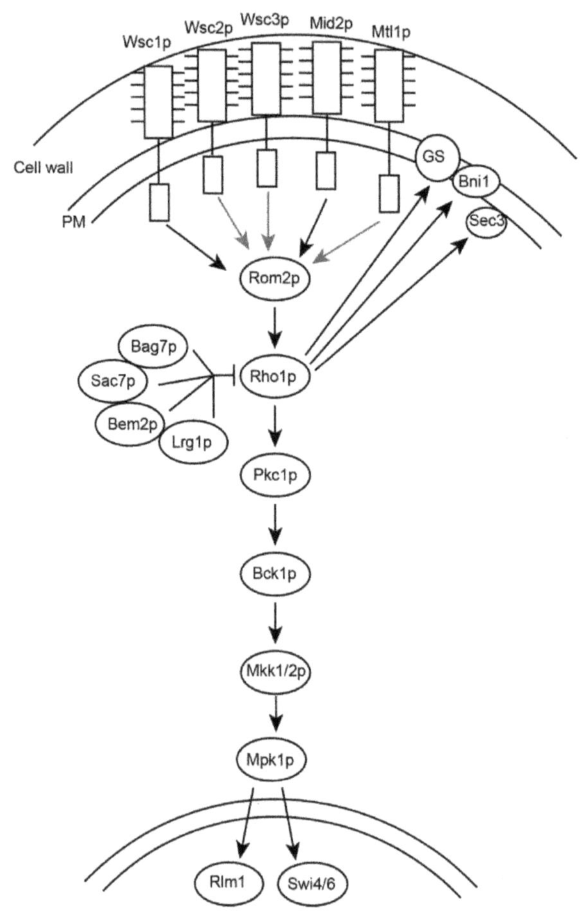

Figure 1. Overview of the Cell Wall Integrity Pathway. The CWI transmits signals through five transmembrane proteins, Wsclp, Wsc2, Wsc3, Mid2p and Mtllp. These proteins are anchored in the plasma membrane (PM) and extend into the cell wall. The transmembrane proteins signal through Rom2p to activate Rholp. Rholp is responsible for the activation of the 1,3-β-glucan synthase (GS), a formin (Bnilp), Sec3p, and Pkclp. Rholp is acted upon by four Rho-GAPs, Sac7p, Bag7, Lrg1, and Bem2p. The CWI pathway is a MAP kinase cascade that culminates in the phosphorylation of Mpklp, which activates transcription factors Rlmlp and SBF (Swi4p/Swi6p) to activate genes involved in cell cycle regulation and cell wall biogenesis. (Adapted from Levin, 2005 and Levin, 2011).

Rholp cycles between GTP and GDP bound states, which are regulated by a set of GTPase-activating proteins (GAPs) and GEFs. Romlp and Rom2p are the activating GEFs that respond to signals from the transmembrane proteins in the CWI pathway [17]. Four Rho-GAPs act on Rholp: Bag7p, Sac7p, Lrglp, and Bem2p [1,19–23] (Figure 1). The Rho-GAPs act on Rholp in a target specific manner. Lrglp primarily functions to control GS, Bem2p and Sac7p collaborate to down-regulate the CWI pathway, and Bag7p and Sac7p act on the actin cytoskeleton [19,20,22–24]. GTP-bound Rholp associates with and activates Pkclp [14,16].

Rholp activation of Pkclp allows stimulation of a downstream MAP kinase cascade, eventually culminating in the phosphorylation and activation of Mpklp [14,25–27]. Mpklp translocates to the nucleus, where it activates transcription of genes involved in cell wall biogenesis and the cell cycle [5]. Deletion of Pkclp causes cell lysis at all temperatures; loss of function of any protein downstream of Pkc1p leads to cell lysis at elevated temperature. Lysis is suppressed by high osmolarity (1 M sorbitol), indicating that death is due to defects in the cell wall [28–31].

Like *pkc1*Δ, deletion of *WSC1* leads to cell lysis, but only at elevated temperatures. The *wsc1*Δ cells also show increased sensitivity to certain drugs, such as caffeine and Caspofungin [32,33]. Deletion of *MID2*, on the other hand, causes pheromone sensitivity and sensitivity to Calcofluor White [6,10]. Both Wsc1p and Mid2p are thought to act as mechanosensors, presumably to sense changes in the cell wall for rapid response [34,35]. However, their different roles in mitotic growth and mating and the differential drug sensitivity suggest that the proteins have similar, but not completely overlapping, functions. Mid2p and Wsc1p localize to distinct microdomains within the plasma membrane, perhaps allowing for their ability to sense unique perturbations on the cell wall [36]. Wsc1p localizes to sites of polarized growth in both mitotic and mating cells [3,37,38]. While much is known about how Wsc1p contributes to temperature sensitivity in haploid cells, [39,40] its function during other stresses and life cycle stages remains less clear.

The CWI pathway is well known for its role in sensing and responding to stresses on the cell wall. Previous work has also indicated a role for the CWI pathway during mating, a situation in which cell wall degradation is actually promoted. A hyperactive allele of Pkc1p (PKC1-R398P) causes a block to cell fusion [41], and we have previously shown that *MID2* and *PKC1* negatively regulate cell fusion, which is relieved upon cell-cell contact [10]. Given the similarities between Mid2p and Wsc1p, do both proteins regulate cell fusion? During mating, zygote formation, and subsequent budding, there are dramatic changes to the cell wall and the shape of the cell. To mate, haploid cells arrest in the cell cycle and polarize growth toward one another along a pheromone gradient, forming a pear-shaped cell called a shmoo. After the mating partners come into contact they must degrade their cell wall at the zone of cell fusion (ZCF) to allow for plasma membrane fusion, cytoplasmic mixing, and nuclear fusion [42–44]. After fusion, zygotes reenter the cell cycle and usually bud from the neck connecting the two mating partners [45,46]. Each of these steps requires changes in cell shape and cell wall remodeling, which are likely sensed and regulated by the CWI pathway.

During a screen for mating phenotypes caused by loss of function mutations of the CWI pathway transmembrane proteins, we found that *WSC1* has a zygote/diploid specific role. Whereas *wsc1*Δ causes temperature sensitive growth in haploids, *WSC1* is necessary for diploid survival. The requirement is conferred by cell ploidy and not by mating type. Thus, Mid2p and Wsc1p have distinct roles; Mid2p is necessary in shmoos, whereas Wsc1p is critically important in diploids, providing a possible reason for why both proteins have been retained throughout evolution.

2. Materials and Methods

2.1. General Yeast Techniques, Strain and Plasmid Construction

Yeast media, general methods, and transformations were performed as described previously [47] with minor modifications. Strains and plasmids are listed below. Deletion strains were either created via PCR amplification of selective markers and homologous recombination at the locus of interest, or via sporulation and tetrad dissection. Where indicated cells were grown in YEPD supplemented with 1 M sorbitol.

All strains were grown at 30 °C. For pheromone induction experiments, early exponential cells growing in selective media were treated for 90 min (unless otherwise specified) with synthetic α-factor (Department of Molecular Biology Syn/Seq Facility, Princeton University, Princeton, NJ, USA) added to a final concentration of 10 µg/mL.

2.2. Yeast Mating Assays

Limited plate mating assays and quantitative filter-matings were performed as described previously with minor alterations [48]. Briefly, limited plate mating assays used a lawn of the *MAT***a** strain grown on rich media plates, and patches of the *MAT*α strains, grown on rich media or selective media for strains containing plasmids. The strains were replica plated together onto rich media, allowed to mate for 3 h at 30 °C, and then replica plated onto media selective for diploids. Mating efficiency was assessed after 2 days of growth at 30 °C. For *wsc1Δ* mating at low temperature, plates were grown, mated, and incubated at 23 °C. Matings were performed for 3 or 5 h and mating efficiency was assessed after 3 days of growth to account for slowed growth rate.

Quantitative filter-matings were performed by mixing early exponential *MAT***a** cells with *MATα* cells at a 1:1 ratio of optical density units to reach a total of ~1×10^7 cells/mL. The cells were mixed together, concentrated on 25 mm 0.45 μm nitrocellulose filter disks (Millipore Corporation, Burlington, MA, USA), and incubated on rich media plates for 2.5–3 h at 30 °C.

2.3. High Copy Suppression of *wsc1Δ*

A YEp13-based yeast genomic DNA library [49] was transformed into *MAT α wsc1Δ* strain. Approximately 24,000 transformants were mated to a *MAT***a** lawn (MY14305) as described above. Plasmids showing suppression were recovered from the cells [47] transformed into MY14306, and retested. DNA sequencing was used to identify the genes carried on the suppressing plasmids.

2.4. Microscopy

All images were acquired at 23 °C using a deconvolution microscopy system (DeltaVision; Applied Precision, LLC) equipped with an inverted microscope (TE200; Nikon) and a 100x objective with numerical aperture of 1.4. Deconvolution and image analysis were performed using Precision softWoRx and ImageJ (National Institutes of Health).

For imaging with propidium iodide, cells were prepared as described above and imaged after three hours in pheromone. Cells were spun down and resuspended in PBS, propidium iodide was added to a final concentration of 20 μm, and cells were incubated for 15 min at 23 °C in the dark prior to imaging.

Live/Dead Yeast Viability Kit (Invitrogen) was used according to protocol. Cells prepared as above after three hours in pheromone. Final concentration of 15 μm FUN-1 and 25 μm Calcofluor white were used prior to imaging.

Microscopic assays of FM4-64 stained mating mixtures and pheromone induced cells were performed as described previously [50]. Pheromone induced cells were prepared as described and then resuspended in 1 mL of TAF buffer (20mM Tris-HCl, 20mM NaN_3, 20mM NaF in water) and kept on ice. FM4-64 (Molecular Probes/Invitrogen) was added to cells to a final concentration of 4 μM and stained shmoos were imaged as above. For fusion assays, an equal OD_{600} (0.5) of each mating type in log phase was mixed, concentrated on 25 mm 0.45 μm nitrocellulose filter disks (Millipore Corporation), and incubated on rich media plates for 2.5–3 h at 30 °C. FM4-64 staining was as above.

Live imaging was performed by mixing 0.02 OD_{600} of *wsc1Δ MAT***a** and *wsc1Δ MATα* cells on a 2% agarose pad and imaged at 23 °C. Images were taken at two-minute intervals using the DeltaVision system described above.

Chi-squared statistical tests were used to determine *p*-values for microscopy data.

2.5. WSC1 Plasmid Loss and Weaning Experiments

Diploid cells obtained as described above from plate mating assays were grown in YEPD and maintained in log phase, without being allowed to reach saturation. Log phase cultures were allowed to grow for eight generations. The same population of cells were allowed to reach saturation and treated as follows. Cells were plated on either YEPD or 5-FOA plates and allowed to grow for 2 days at 30 °C, at which time colony forming units were counted and plasmid loss was determined as colonies on YEPD/colonies on FOA.

For sorbitol weaning experiments, *WSC1/wsc1Δ* or *wsc1Δ/wsc1Δ* diploid cells were streaked for single colonies on FOA + 1 M sorbitol plates to ensure the loss of URA plasmids (empty vector or *WSC1* covering plasmid). Cells were subsequently maintained in log phase in YEPD + 1 M sorbitol media and the osmolarity of the media was reduced by 0.2 M every two hours to allow for one doubling at each condition. Cells were allowed to double once in 0M sorbitol media. Another population of cells were maintained in YEPD media for the duration of the experiment. Cells were plated on YEPD or YEPD + 1 M sorbitol plates and allowed to grow for 2 days at 30 °C. Colony forming units were counted and plotted as cells on YEPD/cells on YEPD + 1 M sorbitol. One-tailed Student's *t*-tests were used to determine *p*-values for plasmid loss experiments.

2.6. Western Blotting

Cell extracts were prepared from 2 OD_{600} of each sample which had been treated with β-estradiol (for P_{Z3EV} *PKC1*) as previously described [50]. Where noted, samples were grown in YEPD or YEPD + 1 M sorbitol. Samples were lysed using NaOH/β-mercaptoethanol (1.85 M NaOH and 7.4% β-mercaptoethanol) and 50% trichloroacetic acid. Samples were run on a 10% polyacrylamide gel at 200 V for 2 h and transferred onto a nitrocellulose membrane at 100 V for 1 h. Detection of Pkc1p was done with an anti-Pkc1p antibody (Santa Cruz Biotechnology) used at 1:1000. Kar2p, used as a loading control, was detected with a custom anti-Kar2p antibody [51] used at 1:1000. HRP-conjugated anti-mouse secondary antibodies (Santa Cruz) were used at 1:2500.

3. Results

3.1. WSC1 Is Required for Diploid Survival

Wsc1p and Mid2p are thought to act as mechanosensors, detecting changes in the cell wall and transducing signals to activate the CWI pathway [34,35]. Given the evidence that the CWI pathway plays a role in cell fusion, we tested deletions of the five transmembrane sensor proteins, by mating deletions by themselves, to *fus1Δ fus2Δ*, or to wild type strains, to determine if their loss caused a mating defect. Cells exhibiting a mating defect do not form diploids efficiently when mated to *fus1Δ fus2Δ* strains [42]; we thus used this strain to initially determine if *wsc1Δ* cells showed a mating defect. Only the *wsc1Δ* mutant showed a strong apparent mating defect, specifically when mated to another *wsc1Δ* strain (bilateral mating), but not when mated to a *WSC1* wild type (Figure 2). In contrast, the *mid2Δ* mutation did not cause a detectable mating defect, whether mutants were mated to *mid2Δ*, *wsc1Δ*, or *fus1Δfus2Δ* mutants or the wild type (Figure 2). Deletions of the remaining three genes encoding transmembrane sensor proteins (*WSC2*, *WSC3* and *MTL1*) caused no apparent defects in mating.

To determine the cause of the defect in the bilateral *wsc1Δ* mating, we performed a "zygote pulling" experiment. Individual zygotes were identified microscopically and manipulated away from the mass mating to specific sites on the petri plate. If the zygotes are viable, individual colonies will form. The *wsc1Δ* x *wsc1Δ* zygotes exhibited only ~25% survival rate compared to 96% in *WSC1* x *WSC1* zygotes (*p*-value < 0.01); *WSC1* x *wsc1Δ* zygotes, however, survived at a rate comparable to the wild type crosses (*p*-value > 0.15) (Figure 3A). Given that zygotes form, these data imply that the *wsc1Δ* does not prevent mating, but rather inhibits zygote survival. To examine this further, we used FM4-64 staining and fluorescence microscopy. FM4-64 is a membrane impermeant fluorescent dye that labels the plasma membrane and is used to visualize endocytosis in living yeast [52,53]. In intact cells, FM4-64 initially stains only the plasma membrane, before entering the cell and ultimately being trafficked to the vacuole. As expected, wild-type zygotes restrict FM4-64 to the plasma membrane. However, the cytoplasmic membranes of many *wsc1Δ* x *wsc1Δ* zygotes were brightly stained by FM4-64 (Figure 3B left panels). The bright staining occurred immediately after addition of FM4-64, indicating that the plasma membranes were not intact. The *wsc1Δ* x *wsc1Δ* zygotes showed bright staining ~75% of the time, whereas almost all zygotes with at least one wild type parent showed staining only of the plasma membrane (Figure 3C).

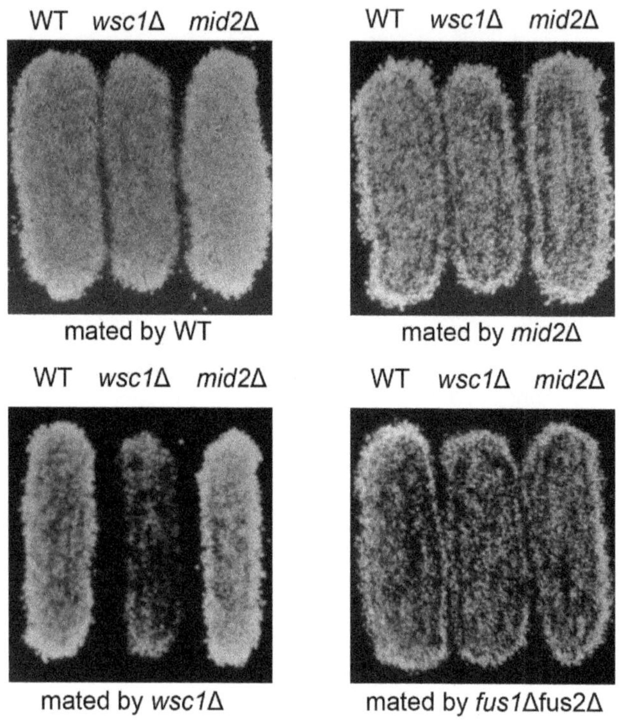

Figure 2. Bilateral loss of *WSC1* causes a defect in diploid formation. *MAT***a** patches of WT (MY8092), *wsc1*Δ (MY14305), and *mid2*Δ (MY14673) were mated to lawns of *MAT*α, WT (MY8093), *wsc1*Δ (MY14306), *mid2*Δ (MY14674) and *fus1*Δ *fus2*Δ (JY429). Cells were mated for 3 h at 30 °C and replica plated onto diploid selective media. In each case the images are of patches of cells growing on 8.5 cm petri dishes; the image panels are 4.0 cm in width.

To confirm that the FM4-64 staining of *wsc1*Δ x *wsc1*Δ zygotes did indicate cell death, both Live/Dead and Propidium Iodide staining was performed (Figure 3B). Live/Dead staining relies on FUN 1, a membrane permeable fluorescent dye that allows assessment of metabolic activity in cells [54]. In live cells, FUN 1 accumulates in the vacuoles where it is converted into a red fluorescent molecule. In wild-type zygotes, red fluorescent aggregates were seen within the vacuoles. However, the *wsc1*Δ x *wsc1*Δ zygotes were stained bright green, as expected for dead cells (Figure 3B, central panels). Propidium Iodide is also used to detect cell death [55]. Cells with intact membranes exclude the dye; cells that have lost membrane integrity stain brightly. After incubation with propidium iodide, wild type zygotes showed no fluorescent staining, whereas *wsc1*Δ x *wsc1*Δ zygotes stained brightly (Figure 3B, right panels). These results confirm that zygotes that lack both copies of *WSC1* lose cell integrity and die.

Previous work showed that deletion of *WSC1* causes haploid and homozygous diploid cells to die at high temperatures. In contrast the defect in *wsc1*Δ x *wsc1*Δ mating was observed at normal (30 °C) and low (23 °C) temperatures (Figure S1). The haploid *wsc1*Δ temperature sensitivity can be suppressed by the addition of 1 M sorbitol to the media [39,40]. When matings were performed in the presence of 1 M sorbitol, *wsc1*Δ x *wsc1*Δ zygote death was suppressed (Figure 3D). This observation indicates that the cause of death for the *wsc1*Δ x *wsc1*Δ zygotes is a cell wall defect. This is the first example of *wsc1*Δ associated cell death at normal and low temperatures.

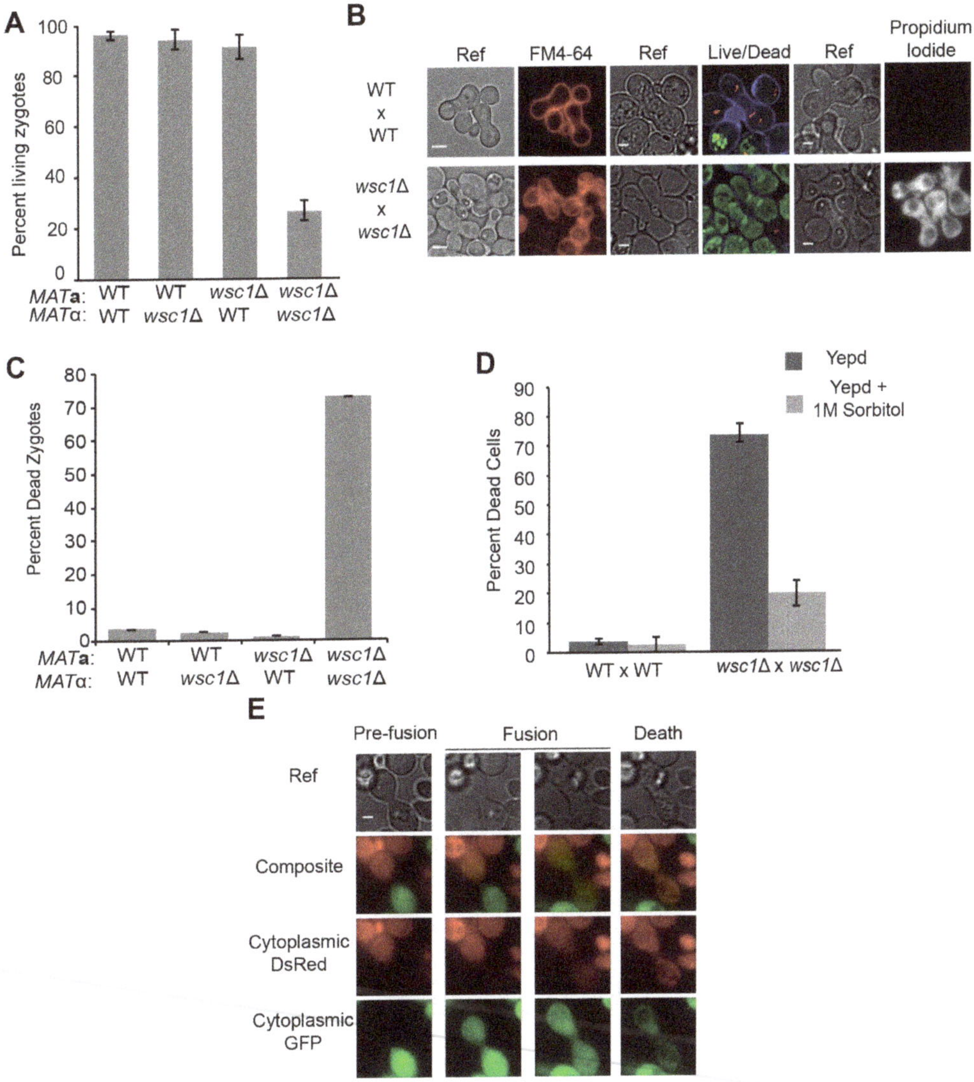

Figure 3. *wsc1Δ* causes zygote death. (**A**) Growth of zygotes formed WT and *wsc1Δ* partners. WT x WT (MY8092 xMY8093) or *wsc1Δ* x *wsc1Δ* (MY14305 xMY14306) matings were performed for 2 h. Individual zygotes were physically moved to new positions using a micromanipulator and dissecting microscope and allowed to grow for 2 days at 30 °C. Live zygotes grew to form individual colonies. n = ≥100 zygotes for each. Error bars represent the standard error. (**B**,**C**) *wsc1Δ* x *wsc1Δ* zygotes are not metabolically active and lose membrane integrity. (**B**) WT x WT (MY8092 x MY8093) or *wsc1Δ* x *wsc1Δ* (MY14305 x MY14306) cells were mated for 3 h at 30 °C. Cells were stained with FM4-64, Live/Dead staining (FUN 1 and Calcofluor White) or propidium iodide. Scale bar = 2 µm. (**C**) Microscopic quantification of zygote viability between WT and *wsc1Δ*. Filter matings of WT x WT (MY8092 x MY8093), WT *MAT***a** x *wsc1Δ MATα* (MY8092 x MY14306), WT *MATα* x *wsc1Δ MAT***a** (MY8093 x MY14305) or *wsc1Δ* x *wsc1Δ* (MY14305 x MY 14306) were performed for 3 h at 30 °C. Zygotes were resuspended in TAF buffer, stained with FM4-64 and imaged. Brightly stained zygotes were counted as dead. n > 400. Error bars represent the standard error. (**D**) Osmotic support suppresses *wsc1Δ* x *wsc1Δ* zygote death. WT (MY8092, MY8093) and *wsc1Δ* (MY14305, MY14306) cells were grown in YEPD or YEPD + 1 M sorbitol, mated for 3 h at 30 °C on the media they were grown in, stained with FM4-64 and imaged. n > 150. Error bars represent the standard error. (**E**) Live imaging

of *wsc1*Δ x *wsc1*Δ mating. *wsc1*Δ *MAT***a** (MY15101) cells transformed with an integrating plasmid containing cytoplasmic GFP (MR5909) were mated to *wsc1*Δ *MAT*α (MY15102) cells transformed with an integrating plasmid containing cytoplasmic DsRed (MR5908). Cells were mixed and imaged at 2-min intervals. Images show cells prior to fusion and cytoplasmic mixing, after fusion when both markers have transferred, and the change in fluorescence as zygotes die. Scale bar = 2 μm. Chi-squared statistical tests were used to determine significance.

The zygote pulling assay indicated that *wsc1*Δ x *wsc1*Δ zygotes form, but die before they can form a colony. To determine the timing of *wsc1*Δ x *wsc1*Δ zygote death, we used a cytoplasmic transfer assay in conjunction with live-cell imaging. One mating partner was marked with cytoplasmic dsRed, and the other was marked with cytoplasmic GFP (Figure 3E). Imaging of the mating cells showed that death occurred shortly after cytoplasmic transfer, prior to or concurrent with budding. Cell death was apparent by loss of the GFP signal and the formation of dsRed puncta, concomitant with the appearance of large vacuoles evident in the transmitted light images (Figures 3E and S2A). Therefore, cells that lack *WSC1* are able to fuse, but cannot maintain cell wall integrity. Given that death is a property of zygotes, and does not occur prior to cell fusion, we will hereafter refer to the genotype of the dying cells as *wsc1*Δ/*wsc1*Δ to indicate their diploid nature.

3.2. Overexpression of CWI Pathway Components Suppresses *wsc1*Δ Zygote Death

The temperature sensitivity of *wsc1*Δ haploid cells can be suppressed by overexpression of *WSC2*, *WSC3*, *MID2*, *ROM2*, *RHO1*, and *PKC1* [6,9,40]. To determine if the *wsc1*Δ/*wsc1*Δ zygote death phenotype was related to the temperature-dependent lysis phenotype, we performed a high-copy suppressor screen using a YEp 13-based yeast genomic library [49]. Colonies of *MAT*α *wsc1*Δ cells containing potential suppressors, along with a control patch of wild type (MY8093) cells, were mated to a *MAT***a** lawn (MY14305) and then replica plated onto diploid selective media. Large colonies, consistent with robust diploid formation, were identified as potential suppressors (Figure 4A). Approximately 24,000 transformants were screened, from which four plasmids were identified which reproducibly and significantly increased diploid formation (p-value < 0.01 for each). The genomic inserts on the plasmids contained *WSC1* (2X), *WSC2* (7X), *MID2* (5X), and *BEM2* (5X) (Figure 4B,C). In addition, we also tested high-copy *WSC3*, which provided mild suppression.

The identification of *WSC2*, *WSC3*, and *MID2* as high-copy suppressors was expected, given that they have been shown to suppress the temperature sensitivity of haploid *wsc1*Δ. Suppression by *BEM2*, however, was not expected. *BEM2* acts as a GAP for Rho1p in the CWI pathway, indicating that it should down-regulate the pathway, as opposed to the other suppressors, which would activate the pathway in the absence of *WSC1* [21,23]. However, Bem2p is also known to play a positive role in bud emergence [19,56]; cells that lack Bem2p become large and multinucleate due to a failure to bud [56,57].

Given that overexpression of one Rho-GAP was identified in our screen, we tested the other three *RHO1* GAPS, *SAC7*, *BAG7*, and *LRG1*, as well as *RGD1* and suggested CDC42p GAPS, *RGA1*, *RGA2*, and *RGD2* [5]. In addition to *BEM2*, only *BEM3* and *RGD2* showed relatively modest suppression (p-value < 0.03) (Figure 4C). Both Bem3p and Rgd2p have been suggested to act as GAPs for Cdc42p [20,58,59]. Rgd2p has also been shown to act as a GAP for Rho5p, a negative regulator of the CWI pathway [60]. It is likely that the suppression of *wsc1*Δ zygote death by overexpressed Bem2p and Bem3p relates to their regulation of cell wall deposition during budding, whereas suppression by Rgd2p over-expression is due to downregulation of a negative regulator of the CWI pathway. Overexpression of the other Rho-GAP genes did not show any suppression of the *wsc1*Δ/*wsc1*Δ zygote formation defect, as seen by plate mating, and were thus not tested further.

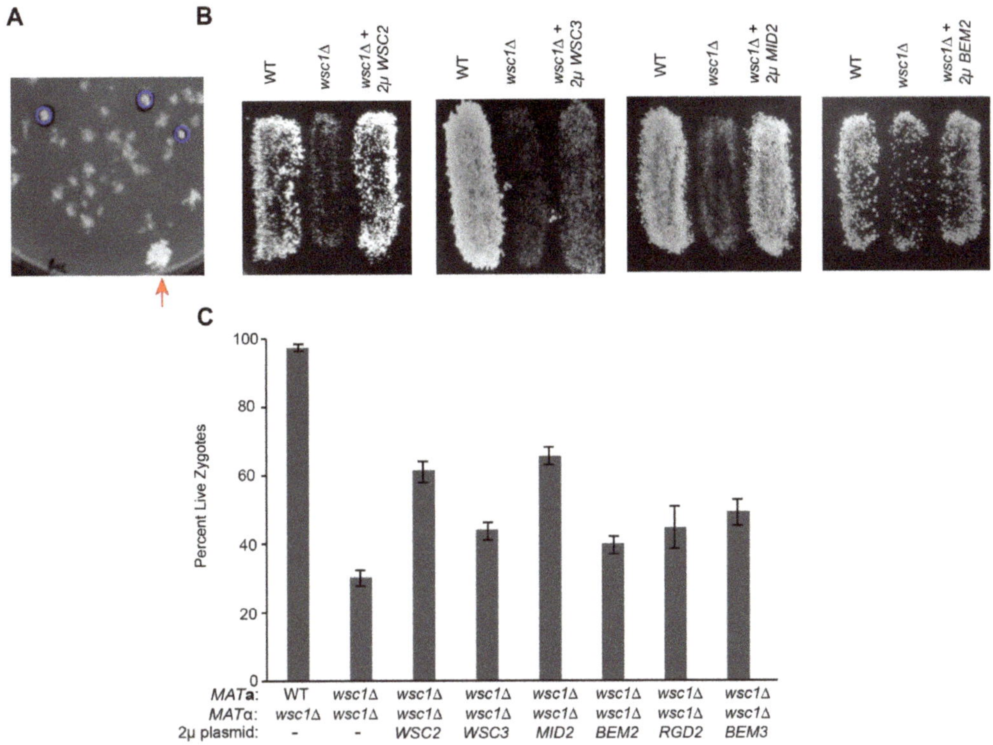

Figure 4. Overexpression of components of the CWI pathway suppress *wsc1*Δ death. (**A**) Example mating plate. A small patch of wild type (MY8093) cells were placed on the mating plate as a control (red arrow). Robust colonies, circled in blue, were picked for retesting. This image is of an 8.5 cm petri dish, the panel is 4.7 cm in width. (**B**) Overexpression suppression of *wsc1*Δ. Mating of *WSC2*, *WSC3*, *MID2*, and *BEM2* overexpression in a *wsc1*Δ background. Patches of *MAT*α WT (MY8093) and *wsc1*Δ (MY14306) strains transformed with an empty 2μ vector (pRS425) along with *wsc1*Δ transformed with 2μ *WSC2* (MY15816), 2μ *WSC3* (MY15817), 2μ *MID2* (MR6979) were mated to a *MAT***a** *wsc1*Δ lawn. For *BEM2* overexpression WT (MY8093) and *wsc1*Δ (MY14306) were transformed with empty vector (pRS426) or 2μ *BEM2* (MR6988) and mated as above. Matings were incubated for 3 h at 30 °C and then replica plated onto diploid selective media. In each case the images are of patches of cells growing on 8.5 cm petri dishes; the image panels are 4.0 cm in width. (**C**) Overexpression of potential CWI pathway regulators suppresses *wsc1*Δ x *wsc1*Δ zygote death. WT (MY8092) and *wsc1*Δ (MY14306) were transformed with an empty 2μ vector (pRS425) or *wsc1*Δ was transformed with 2μ *WSC2* (MY15816), *WSC3* (MY15817), *MID2* (MR6979), or *BEM2* (pMR6988). Each *MAT*α was mated to *MAT***a** *wsc1*Δ (MY14305) for 3 h at 30 °C, resuspended in TAF buffer and imaged with FM4-64. n ≥ 100 zygotes for each mating. Error bars represent the standard error. Chi-squared statistical tests were used to calculate significance.

3.3. *Diploid Cells Are Sensitive to Loss of WSC1*

Providing a single copy of *WSC1* allows zygotes to survive and form diploids like wild type (Figure S3). We therefore set out to determine whether the increased requirement for Wsc1p occurs only in zygotes or is also a property of mitotic diploid cells. We provided *WSC1* on a low copy, centromeric plasmid, and measured how frequently *wsc1*Δ/*wsc1*Δ zygotes lose the plasmid. After maintaining growth in log phase for eight generations, the WT/*wsc1*Δ diploids lost the *WSC1* plasmid over 100-fold more frequently than *wsc1*Δ/*wsc1*Δ diploid cells (Figure 5A). *WSC1*/*wsc1*Δ diploids were able to lose the *WSC1* plasmid at a frequency similar to the empty vector (p-value = 0.50) (Figure 5A). Thus *wsc1*Δ/*wsc1*Δ diploids also show greatly reduced viability, indicating that the defect is not restricted to zygotes. It was previously suggested that *wsc1*Δ/*wsc1*Δ diploid cells have a more severe temperature sensitivity when grown to saturation [39]. Consistent with this, the *wsc1Δ/wsc1Δ* diploid cells maintained in log phase were able to lose the plasmid at a rate 3-fold greater than those allowed to reach saturation (Figure 5B). This is likely an underrepresentation of the actual difference in plasmid loss between log phase and saturated cultures because the saturated cultures have been grown for more generations.

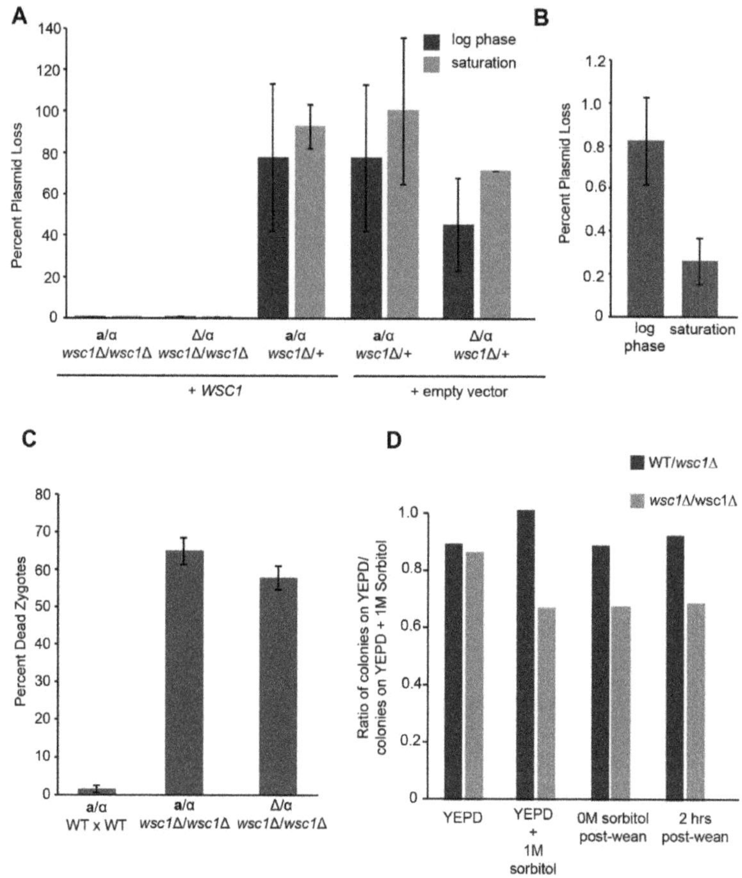

Figure 5. *WSC1* is necessary for diploid growth. (**A–C**) *WSC1* dependency is independent of mating type and exacerbated by stationary phase. (**A**) The following strains were transformed with a *URA3* centromeric empty vector plasmid or with the same plasmid bearing *WSC1*: a/α WT/*wsc1Δ* empty vector (MY16050), a/α WT/*wsc1Δ* + *WSC1* (MY16049), a/α *wsc1Δ/wsc1Δ*, Δ/α WT/*wsc1Δ* empty vector (MY16052), Δ/α WT/*wsc1Δ* + *WSC1* (MY16051), and Δ/α *wsc1Δ/wsc1Δ* + *WSC1* (MY16053). Cells were grown in YEPD liquid log phase (O.D. 0.4–0.7) or saturation and plated on both YEPD plates and 5-FOA plates to quantify plasmid loss. Colonies were counted after two days at 30 °C. (**B**) *WSC1* plasmid loss in a/α *wsc1Δ/wsc1Δ* during log phase or stationary phase. Experiments were performed three times independently. Error bars represent the standard error of the mean for both A and B. One-tailed Student's t-test was used to determine significance. (**C**) Filter matings of WT x WT (MY8092 x MY8093), *wsc1Δ* x *wsc1Δ* (MY14305 x MY14306) or *matΔ wsc1Δ* x *MATα wsc1Δ* (MY16045 x MY14306) were performed for 3 h at 30 °C. Cells were resuspended in TAF buffer, stained with FM4-64, and imaged. n ≥ 100 cells for each. Error bars represent the standard error of the mean. Chi-squared statistical tests were used to determine significance. (**D**) *wsc1Δ*/wcs1Δ diploid cells adapt to growth without osmotic support. WT/*wsc1Δ* (MY15803) or *wsc1Δ/wsc1Δ* (MY15805) diploids were

grown continuously in YEPD (left most bars) or in YEPD + 1 M sorbitol. The concentration of sorbitol in the media was dropped by 0.2 M every two hours until cells were in 0M sorbitol media when they were plated on YEPD or YEPD + 1 M sorbitol. Cells were allowed to double once without sorbitol and plated as before. Colonies were counted after 2 days at 30 °C and plotted as a ratio of cells on YEPD plates over cells on YEPD + 1 M sorbitol plates. One representative experiment is shown, based on three technical replicates.

The greater requirement for *WSC1* in diploid cells could be due to their ploidy or because they are heterozygous at the mating type locus (*MAT*). Previous work suggested that there may be a role for the *MAT* locus in maintaining cell wall integrity [60]. To determine if *wsc1Δ/wsc1Δ* death was due to diploidy or mating type, we mated a *matΔ wsc1Δ* to a *wsc1Δ MATα*. The *matΔ/α* is diploid, but, because the *matΔ* is recessive, the cell exhibits the *MATα* mating type, like a haploid cell. By both plasmid loss and FM4-64 staining of zygotes, there was no difference between the *MAT***a***/α wsc1Δ/wsc1Δ* and the *matΔ/α wsc1Δ/wsc1Δ* strains (Figure 5A,C). Thus, the *wsc1Δ/wsc1Δ* defect is diploid specific, rather than mating-type specific.

It is surprising that the severe lysis phenotype of *wsc1Δ/wsc1Δ* diploids was not previously reported. One possibility is that cells that survive a crisis of growth after *WSC1* plasmid loss are able to adapt to these conditions. To test this, we examined viability as cells were weaned off of osmotic support. *WSC1/wsc1Δ* and *wsc1Δ/wsc1Δ* diploids that had lost the *WSC1* covering plasmid were selected on media containing 1 M sorbitol. The plasmid-free strains were then grown in either YEPD or in YEPD + 1 M sorbitol. For the cells grown with 1 M sorbitol, the osmolarity of the media was reduced by 0.2 M every two hours to allow one doubling in each condition. Viability was assessed by the ratio of colony forming units on YEPD relative to YEPD + 1 M sorbitol agar plates. The *WSC1/wsc1Δ* cells grown in YEPD or YEPD + 1 M sorbitol media had equal viability on either media. When a *wsc1Δ/wsc1Δ* diploid without the covering plasmid was grown in YEPD, the cells also behaved like the *WSC1/wsc1Δ*, exhibiting equal viability on YEPD and YEPD + 1 M sorbitol. It is likely that the *wsc1Δ/wsc1Δ* cells have adapted or accumulated suppressor mutation(s) (Figure 5D). In contrast, the *wsc1Δ/wsc1Δ* strain that was grown in 1 M sorbitol showed decreased plating efficiency on YEPD, relative to YEPD + 1 M sorbitol. However, the decrease in viability was much less than expected from the frequency of plasmid loss, suggesting that the surviving cells were already largely adapted to the absence of Wsc1p. Cells that were gradually weaned off of osmotic support did not show improved viability on YEPD, relative to the starting culture (Figure 5D), suggesting that adaption takes longer than ~26 generations (generation time includes time to form a colony of FOA plates). FM4-64 staining of the weaned cells showed levels of cell death that were consistent with the reduced viability on YEPD, 27% for *wsc1Δ/wsc1Δ* versus 14% for *WSC1/wsc1Δ*. Taken together, these data suggest that *wsc1Δ/wsc1Δ* diploid cells adapt, either physiologically or by the accumulation of suppressor mutations.

3.4. Overexpression of Hyperactivated Pkc1p Causes Growth Defects

Deletion of *PKC1* causes cell lysis at any temperature, which can be suppressed by osmotic support [28–31]. It has been reported that overexpression of Pkc1p suppresses the temperature sensitivity of *wsc1Δ* haploid cells [6,9,40]. A hyperactive allele of Pkc1p, *PKC1-R398P*, has also been reported to suppress the *mid2Δ* mating-induced death and cause a mild cell fusion defect [6,41]. It was therefore surprising that high copy *PKC1* was not identified in the screen for suppression of *wsc1Δ/wsc1Δ* zygote death. However, it is possible that high copy *PKC1* is toxic in our conditions or that feedback regulation prevents the over-expressed protein from being sufficiently active. Therefore, we used the PZ_3EV promoter [61] and β-estradiol induction to express Pkc1p and Pkc1p-R398P under carefully controlled conditions (Figure 6A). To determine the best Pkc1p-R398P expression level, we induced cells with increasing concentrations of β-estradiol and plated serial dilutions to identify the point at which cells failed to grow (Figure 6B). We found that 10 nM β-estradiol was the highest concentration that could be used before significant growth inhibition occurred. At 10 nM a ~10-fold reduction in growth was observed; at 100 nM β-estradiol induction Pkc1p-R398P expression caused ~1000-fold decrease in growth (Figure 6B). In contrast, over-expression of wild-type Pkc1p had no negative effect on cell growth.

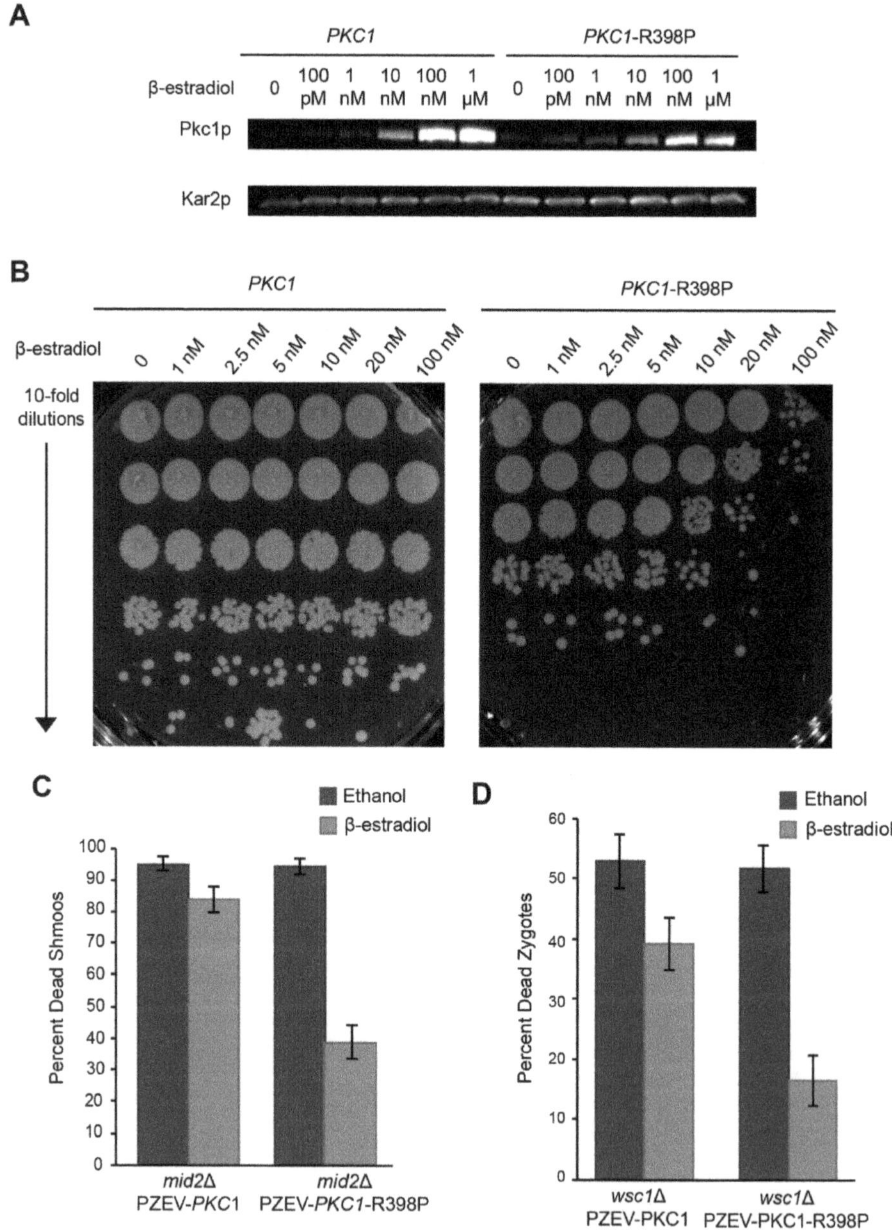

Figure 6. Hyperactivation of *PKC1*-R398P causes growth arrest and suppresses *wsc1*Δ zygote death. (**A**) P_{Z3EV} promoter allows fine control of Pkc1p and $Pkc1p^{R398P}$ protein expression. P_{Z3EV}-*PKC1* (MY15622) and P_{Z3EV}-*PKC1*-R398P (MY15623) were exposed to varying concentrations of β-estradiol for 30 min. An α-Pkc1p antibody was used to detect protein levels and

Kar2p was used as a loading control. (**B**) Hyperactivation of *PKC1*-R398P causes growth arrest. P_{Z3EV}-*PKC1* (MY15622) and P_{Z3EV}-*PKC1*-R398P (MY15623) strains induced with varying concentrations of β-estradiol for 1.5 h and plated as serial dilutions on YEPD plates. Cells were allowed to grow for 2 days at 30 °C. (**C**) *PKC1*-R398P suppresses *mid2*Δ pheromone-induced death. P_{Z3EV}-*PKC1 mid2*Δ (MY15655) and P_{Z3EV}-*PKC1*-R398P *mid2*Δ (MY15656) were induced with ethanol or 10 nM β-estradiol and induced with 8 μg/mL α-factor for 3 h and imaged with FM4-64. (**D**) *PKC1*-R398P suppresses *wsc1*Δ zygote death. P_{Z3EV}-*PKC1 wsc1*Δ (MY15651 x MY15653) and P_{Z3EV}-*PKC1*-R398P *wsc1*Δ (MY15652 x MY15654) were mated in the presence of ethanol or 10 nM β-estradiol and imaged with FM4-64. n ≥ 100 zygotes or shmoos. Error bars represent the standard error. Chi-squared statistical tests were used to determine significance.

PKC1-R398P also causes a cell fusion defect, arresting mating cells as prezygotes approximately 16% of the time [41]. Interestingly, we found that high levels of *PKC1-R398P* inhibited the pheromone response in general. Cells containing a galactose inducible *PKC1-R398P* were unable to induce expression of a *FUS1-lacZ* reporter when exposed to α-factor (Figure S5). To determine if activated Pkc1p could suppress *wsc1*Δ we identified a level of Pkc1p-R398P expression that allowed for mitotic growth and a pheromone response. As a positive control for Pkc1p-R398P suppression, we showed that hyperactivated Pkc1p-R398P was able to suppress the *mid2*Δ-dependent death of pheromone treated cells (Figure 6C). As the concentration of β-estradiol increased above 1 nM, the efficiency of zygote formation fell rapidly, reaching ~10% of wild-type at 10 nM. A concentration of 10 nM β-estradiol was chosen because this was the lowest concentration at which we saw an increase in PKC1-R398Pp production (Figure 6A), and which also provided some level of mating which would allow measurement of suppression. Among the pre-zygotes that did form, the rate of successful fusion was reduced to ~60%; the remaining 40% were blocked as prezygotes with cell wall remaining between the mating partners (Figure S4A,B). However, consistent with expectation, overexpression of hyper-activated Pkc1p suppressed the death of the *wsc1*Δ/*wsc1*Δ zygotes that did form (Figure 6D).

4. Discussion

4.1. WSC1 Is the Most Important CWI Sensor in Zygotes

Saccharomyces cerevisiae cells have a thick cell wall that changes dynamically for growth, mating and to respond to environmental stresses. The CWI pathway is responsible for monitoring the cell wall and ensuring that damage is repaired, or that adequate cell wall deposition is maintained during budding and shmoo formation [4,5]. Of the five transmembrane proteins that regulate the CWI pathway, *WSC1* and *MID2* are the most important; only *wsc1* and *mid2* mutants exhibit phenotypes, and cells containing only *WSC1* and *MID2* appear to be wild type [10]. Through this and previous work, it is apparent that *WSC1* and *MID2* play different roles; Mid2p is important during shmoo formation, whereas Wsc1p is important for maintaining cell wall integrity during mitotic growth and post fusion [10]. Deletion of *WSC1* was previously shown to cause cell lysis at high temperatures in both haploids and diploids [32,39,40]. However, using a mating assay we demonstrated that *WSC1* is actually required at all temperatures in diploids, and that previous work underestimated its role.

The *wsc1*Δ defect was initially observed during mating, as a strong zygote specific phenotype. Why might zygotes be different from haploid cells? One of the striking differences between them is their shape. Since *MID2* plays a large role in shmoos, but not in zygotes, it seems likely that Wsc1p acts as the major cell wall sensor in zygotes. However, using a plasmid loss assay, it is apparent that diploid cells also require *WSC1* to survive in log phase, and this phenotype is exacerbated when cells reach saturation. Haploid cells lacking *WSC1* have a slightly decreased growth rate, but are otherwise comparable to wild type. Why, then, would zygotes and diploid cells die without *WSC1*? Both zygotes and diploids are different in shape than haploid cells. Diploid cells are more ovoid and are twice the volume of haploid cells, on average [62]. Although, it has been shown that *MAT***a**/α diploid cells down regulate components of the cell wall synthesis machinery [62], which should exacerbate the defect of cells lacking *WSC1*. However, heterozygosity at the mating type locus is not sufficient to explain the defect, as *mat*Δ/α still require Wsc1p. Another difference between haploid and diploid cells is their surface area to volume ratio; diploids have a lower surface area to volume ratio than haploid cells. It is possible that this accounts for the increased need for *WSC1* in diploids and is independent of ploidy. Interestingly, diploid cells can adapt to growth without *WSC1*. Although *wsc1*Δ/*wsc1*Δ cells can lose the *WSC1* covering plasmid at a frequency of less than 1% (Figure 5B), diploids can survive when grown with osmotic support. The surviving *wsc1*Δ/*wsc1*Δ diploids are eventually able to grow without Wsc1p, either through adaptation or suppressor accumulation (Figure 5D). Perhaps *WSC2* or *WSC3*, both of which can suppress the zygote death of *wsc1*Δ when overexpressed, are upregulated to account for the loss of the key CWI sensor.

4.2. *WSC1 May Be Involved in Maintaining Cell Wall Integrity during Bud Emergence*

Unsurprisingly, overexpression of components of the CWI pathway suppress *wsc1*Δ zygote death. Overexpression of *BEM2*, *BEM3*, and *RGD2* were also able to suppress, which was surprising given that other GAPs for Rho1p in the CWI pathway did not. The *wsc1*Δ cells did not lyse until after fusion, and generally as a zygote with a small bud (Figures 3E and S2A). Whereas suppression by positive regulators of the CWI pathway are likely to suppress by bypassing the need for the upstream sensor, it seems unlikely that *BEM2*, *BEM3*, and *RGD2* function in a similar manner. Overexpression of *BEM2* should downregulate the CWI pathway and exacerbate the *wsc1*Δ defect. Both *BEM2* and *BEM3*, however, have been shown to play a role in bud emergence that may be independent of their function as GAPs. For example, Bem3p is a GAP for Cdc42p [19,56,63]. However, it seems unlikely that suppression by *BEM3* overexpression is through its Cdc42p GAP activity because overexpression of *RGA1* and *RGA2*, the other GAPS for Cdc42p do not suppress. In the absence of *WSC1* perhaps cells form zygotes, but are unable to regulate cell wall deposition to produce viable buds. It is also possible that *BEM2* and *BEM3* are acting as effectors rather than GAPs to suppress the zygote death phenotype. GAP proteins are generally thought to increase hydrolysis of GTP, thus inactivating the G-protein they interact with [64,65]. There is evidence, however, that GAPs act as effectors as well, potentially having a positive regulatory role downstream of the G-proteins they act upon [66]. *BEM2* may act as an effector for Rho1p in this case, activating something downstream in the CWI pathway and allowing for proper cell wall deposition and zygote survival. Suppression of the *wsc1*Δ defect must occur either though decreased cell wall removal or increased cell wall deposition, allowing for zygote survival. Though less likely, it is also possible that *BEM3* downregulation of Cdc42p activity could lead to decreased cell wall removal during budding and subsequent diploid growth, thereby suppressing the defect.

RGD2 is less well studied, but is known to act as a GAP for both Rho5p and Cdc42p [20,58–60]. Deletion of Rho5p leads to resistance to drugs such as caffeine, suggesting that its role is to down regulate the CWI pathway [60]. As a GAP, Rgd2p downregulates Rho5p activity, and would potentially lead to increased CWI pathway signaling. Thus, *RGD2* overexpression likely suppresses in a manner similar to *WSC2*, *WSC3*, and *MID2*. The fact that we did not identify specific components of the CWI pathway downstream of Rho1p and Pkc1p suggests that it is the activity of these two proteins that allows for proper cell wall deposition, rather than downstream signaling via the MAP kinase cascade. For example, Rho1p's role in regulating glucan synthase may be important for overcoming defects in the upstream components of the CWI pathway.

4.3. *Constitutive Activation of the CWI Pathway Is Detrimental to Cells*

Previous work showed that a hyperactive allele of *PKC1* (*PKC1-R398P*) causes a block in cell fusion, implicating the CWI pathway as a negative regulator of fusion [41]. Pkc1p acts downstream of *WSC1* in the CWI pathway. Surprisingly, we found that Pkc1p-R398P overexpression is toxic, leading to growth arrest and an inability to respond to pheromone (Figure 6B and Figure S5). It is likely that over-activation of Pkc1p signals constant cell wall damage that must be repaired. Pkc1p responds to cell wall damage in part by relocalizing to the site of damage [67]. Perhaps, overexpression of hyperactive Pkc1p leads to mislocalization over the entire cell cortex, signaling cell wall damage everywhere, and blocking the cell wall remodeling required for growth. Alternatively, given that the signaling output from the CWI pathway impinges on cell cycle regulation, a constant signal from Pkc1p might cause cells to arrest [5]. However, given moderate amounts of Pkc1p activity, cells can overcome the loss of either *MID2* or *WSC1*.

Although the complexity of the downstream signaling of the CWI pathway has been understood for some time, the specific functions of the transmembrane components is surprisingly complex. *MID2* and *WSC1* are the key cell wall sensors, but perform different functions in sensing the state of the cell wall. *MID2* is the primary sensor during shmoo formation, whereas *WSC1* plays a surprisingly crucial role in zygote and diploid survival, beyond that known for haploid cell growth.

Supplementary Materials: The following are available at https://www.mdpi.com/article/10.3390/jof7121049/s1. Figure S1—*wsc1*Δ/*wsc1*Δ show diploid formation defects at 23 °C. Figure S2—*wsc1*Δ/*wsc1*Δ zygote die after fusion. Figure S3—*WSC1* covering plasmid supports diploid formation. Figure S4—Induction of PZ3EV-Pkc1pR398P causes zygote formation and fusion defects. Figure S5—Hyperactivation of *PKC1-R398P* inhibits the pheromone response. Table S1—Yeast Strains. Table S2—Plasmids.

Author Contributions: Conceptualization, A.E.H. and M.D.R.; Methodology, A.E.H.; Validation, A.E.H. and M.D.R.; Formal Analysis, A.E.H.; Investigation, A.E.H. and M.L.; Resources, M.D.R.; Writing—Original Draft Preparation, A.E.H. and M.D.R.; Writing—Review and Editing, A.E.H. and M.D.R.; Visualization, A.E.H.; Supervision, M.D.R.; Project Administration, A.E.H. and M.D.R.; Funding Acquisition, M.D.R. All authors have read and agreed to the published version of the manuscript.

Funding: This work was supported by grants R01GM037739 and R35GM126998 from the National Institutes of Health to M.D.R. A.E.H. was supported by Training grant GM007388 from the National Institutes of Health to the Department of Molecular Biology, Princeton University.

Data Availability Statement: The data presented in this study are available in the main body of text or the supplementary material. All strains are available upon request.

Acknowledgments: We thank the members of the Rose and Gammie labs for general discussions and helpful suggestions. We thank Maysoun Husseini for media preparation.

Conflicts of Interest: The authors declare no conflict of interest.

References

1. Cid, V.J.; Cenamor, R.; Sanchez, M.; Nombela, C. A mutation in the Rho1-GAP-encoding gene BEM2 of *Saccharomyces cerevisiae* affects morphogenesis and cell wall functionality. *Microbiology* **1998**, *144 Pt 1*, 25–36. [CrossRef] [PubMed]
2. Klis, F.M.; Boorsma, A.; De Groot, P.W. Cell wall construction in *Saccharomyces cerevisiae*. *Yeast* **2006**, *23*, 185–202. [CrossRef] [PubMed]
3. Delley, P.A.; Hall, M.N. Cell wall stress depolarizes cell growth via hyperactivation of RHO1. *J. Cell Biol.* **1999**, *147*, 163–174. [CrossRef]
4. Levin, D.E. Cell wall integrity signaling in *Saccharomyces cerevisiae*. *Microbiol. Mol. Biol. Rev.* **2005**, *69*, 262–291. [CrossRef]
5. Levin, D.E. Regulation of cell wall biogenesis in *Saccharomyces cerevisiae*: The cell wall integrity signaling pathway. *Genetics* **2011**, *189*, 1145–1175. [CrossRef]
6. Ketela, T.; Green, R.; Bussey, H. *Saccharomyces cerevisiae* mid2p is a potential cell wall stress sensor and upstream activator of the PKC1-MPK1 cell integrity pathway. *J. Bacteriol.* **1999**, *181*, 3330–3340. [CrossRef]
7. Lodder, A.L.; Lee, T.K.; Ballester, R. Characterization of the Wsc1 protein, a putative receptor in the stress response of *Saccharomyces cerevisiae*. *Genetics* **1999**, *152*, 1487–1499. [CrossRef] [PubMed]
8. Philip, B.; Levin, D.E. Wsc1 and Mid2 are cell surface sensors for cell wall integrity signaling that act through Rom2, a guanine nucleotide exchange factor for Rho1. *Mol. Cell. Biol* **2001**, *21*, 271–280. [CrossRef]
9. Rajavel, M.; Philip, B.; Buehrer, B.M.; Errede, B.; Levin, D.E. Mid2 is a putative sensor for cell integrity signaling in *Saccharomyces cerevisiae*. *Mol. Cell. Biol.* **1999**, *19*, 3969–3976. [CrossRef]

10. Hall, A.E.; Rose, M.D. Cell fusion in yeast is negatively regulated by components of the cell wall integrity pathway. *Mol. Biol. Cell* **2019**, *30*, 441–452. [CrossRef]
11. Douglas, C.M.; Foor, F.; Marrinan, J.A.; Morin, N.; Nielsen, J.B.; Dahl, A.M.; Mazur, P.; Baginsky, W.; Li, W.; El-Sherbeini, M. The *Saccharomyces cerevisiae* FKS1 (ETG1) gene encodes an integral membrane protein which is a subunit of 1,3-beta-D-glucan synthase. *Proc. Natl. Acad. Sci. USA* **1994**, *91*, 12907–12911. [CrossRef] [PubMed]
12. Drgonova, J.; Drgon, T.; Tanaka, K.; Kollár, R.; Chen, G.-C.; Ford, R.A.; Chan, C.S.M.; Takai, Y.; Cabib, E. Rho1p, a yeast protein at the interface between cell polarization and morphogenesis. *Science* **1996**, *272*, 277–279. [CrossRef]
13. Inoue, S.B.; Takewakt, N.; Takasuka, T.; Mio, T.; Adachi, M.; Fujii, Y.; Miyamoto, C.; Arisawa, M.; Furuichi, Y.; Watanabe, T. Characterization and gene cloning of 1,3-beta-D-glucan synthase from *Saccharomyces cerevisiae*. *Eur. J. Biochem.* **1995**, *231*, 845–854. [CrossRef]
14. Kamada, Y.; Qadota, H.; Python, C.P.; Anraku, Y.; Ohya, Y.; Levin, D. Activation of yeast protein kinase C by Rho1 GTPase. *J. Biol. Chem.* **1996**, *271*, 9193–9196. [CrossRef]
15. Mazur, P.; Baginsky, W. In vitro activity of 1,3-beta-D-glucan synthase requires the GTP-binding protein Rho1. *J. Biol. Chem.* **1996**, *271*, 14604–14609. [CrossRef] [PubMed]
16. Nonaka, H.; Tanaka, K.; Hirano, H.; Fujiwara, T.; Kohno, H.; Umikawa, M.; Mino, A.; Takai, Y. A downstream target of RHO1 small GTP-binding protein is PKC1, a homolog of protein kinase C, which leads to activation of the MAP kinase cascade in *Saccharomyces cerevisiae*. *EMBO J.* **1995**, *14*, 5931–5938. [CrossRef]
17. Ozaki, K.; Tanaka, K.; Imamura, H.; Hihara, T.; Kameyama, T.; Nonaka, H.; Hirano, H.; Matsuura, Y.; Takai, Y. Rom1p and Rom2p are GDP/GTP exchange proteins (GEPs) for the Rho1p small GTP binding protein in *Saccharomyces cerevisiae*. *EMBO J.* **1996**, *15*, 2196–2207. [CrossRef]
18. Qadota, H.; Python, C.P.; Inoue, S.B.; Arisawa, M.; Anraku, Y.; Zheng, Y.; Watanabe, T.; Levin, D.E.; Ohya, Y. Identification of yeast Rho1p GTPase as a regulatory subunit of 1,3-beta-glucan synthase. *Science* **1996**, *272*, 279–281. [CrossRef] [PubMed]
19. Peterson, J.; Zheng, Y.; Bender, L.; Myers, A.; Cerione, R.; Bender, A. Interactions between the bud emergence proteins Bem1p and Bem2p and Rho-type GTPases in yeast. *J. Cell Biol.* **1994**, *127*, 1395–1406. [CrossRef] [PubMed]
20. Roumanie, O.; Weinachter, C.; Larrieu, I.; Crouzet, M.; Doignon, F. Functional characterization of the Bag7, Lrg1 and Rgd2 RhoGAP proteins from *Saccharomyces cerevisiae*. *FEBS Lett.* **2001**, *506*, 149–156. [CrossRef]
21. Schmidt, A.; Bickle, M.; Beck, T.; Hall, M.N. The yeast phosphatidylinositol kinase homolog TOR2 activates RHO1 and RHO2 via the exchange factor ROM2. *Cell* **1997**, *88*, 531–542. [CrossRef]
22. Watanabe, D.; Abe, M.; Ohya, Y. Yeast Lrg1p acts as a specialized RhoGAP regulating 1,3-beta-glucan synthesis. *Yeast* **2001**, *18*, 943–951. [CrossRef] [PubMed]
23. Schmidt, A.; Schmelzle, T.; Hall, M.N. The RHO1-GAPs SAC7, BEM2 and BAG7 control distinct RHO1 functions in *Saccharomyces cerevisiae*. *Mol. Microbiol.* **2002**, *45*, 1433–1441.
24. Martin, H.; Rodriguez-Pachon, J.M.; Ruiz, C.; Nombela, C.; Molina, M. Regulatory mechanisms for modulation of signaling through the cell integrity Slt2-mediated pathway in *Saccharomyces cerevisiae*. *J. Biol. Chem.* **2000**, *275*, 1511–1519. [CrossRef] [PubMed]
25. Levin, D.E.; Bowers, B.; Chen, C.Y.; Kamada, Y.; Watanabe, M. Dissecting the protein kinase C/MAP kinase signalling pathway of *Saccharomyces cerevisiae*. *Cell. Mol. Biol. Res.* **1994**, *40*, 229–239. [PubMed]
26. Paravicini, G.; Friedli, L. Protein-protein interactions in the yeast PKC1 pathway: Pkc1p interacts with a component of the MAP kinase cascade. *Mol. Gen. Genet.* **1996**, *251*, 682–691.
27. Kamada, Y.; Jung, U.S.; Piotrowski, J.; Levin, D.E. The protein kinase C-activated MAP kinase pathway of *Saccharomyces cerevisiae* mediates a novel aspect of the heat shock response. *Genes Dev.* **1995**, *9*, 1559–1571. [CrossRef]
28. Irie, K.; Takase, M.; Lee, K.S.; Levin, D.E.; Araki, H.; Matsumoto, K.; Oshima, Y. MKK1 and MKK2, which encode *Saccharomyces cerevisiae* mitogen-activated protein kinase-kinase homologs, function in the pathway mediated by protein kinase C. *Mol. Cell. Biol.* **1993**, *13*, 3076–3083.
29. Lee, K.S.; Irie, K.; Gotoh, Y.; Watanabe, Y.; Araki, H.; Nishida, E.; Matsumoto, K.; Levin, D.E. A yeast mitogen-activated protein kinase homolog (Mpk1p) mediates signalling by protein kinase C. *Mol. Cell. Biol.* **1993**, *13*, 3067–3075.
30. Lee, K.S.; Levin, D.E. Dominant mutations in a gene encoding a putative protein kinase (BCK1) bypass the requirement for a *Saccharomyces cerevisiae* protein kinase C homolog. *Mol. Cell. Biol.* **1992**, *12*, 172–182.
31. Torres, L.; Martín, H.; García-Saez, M.I.; Arroyo, J.; Molina, M.; Sánchez, M.; Nombela, C. A protein kinase gene complements the lytic phenotype of *Saccharomyces cerevisiae lyt2* mutants. *Mol. Microbiol.* **1991**, *5*, 2845–2854. [CrossRef]
32. Jacoby, J.J.; Nilius, S.M.; Heinisch, J.J. A screen for upstream components of the yeast protein kinase C signal transduction pathway identifies the product of the *SLG1* gene. *Mol. Gen. Genet.* **1998**, *258*, 148–155. [CrossRef] [PubMed]
33. Reinoso-Martin, C.; Schuller, C.; Schuetzer-Muehlbauer, M.; Kuchler, K. The yeast protein kinase C cell integrity pathway mediates tolerance to the antifungal drug caspofungin through activation of Slt2p mitogen-activated protein kinase signaling. *Eukaryot. Cell* **2003**, *2*, 1200–1210. [CrossRef]
34. Dupres, V.; Alsteens, D.; Wilk, S.; Hansen, B.; Heinisch, J.J.; Dufrêne, Y.F. The yeast Wsc1 cell surface sensor behaves like a nanospring in vivo. *Nat. Chem. Biol.* **2009**, *5*, 857–862. [CrossRef]
35. Heinisch, J.J.; Dufrene, Y.F. Is there anyone out there?–Single-molecule atomic force microscopy meets yeast genetics to study sensor functions. *Integr. Biol.* **2010**, *2*, 408–415. [CrossRef]

36. Kock, C.; Arlt, H.; Ungermann, C.; Heinisch, J.J. Yeast cell wall integrity sensors form specific plasma membrane microdomains important for signalling. *Cell. Microbiol.* **2016**, *18*, 1251–1267. [CrossRef] [PubMed]
37. Huh, W.-K.; Falvo, J.V.; Gerke, L.C.; Carroll, A.S.; Howson, R.W.; Weissman, J.S.; O'Shea, E.K. Global analysis of protein localization in budding yeast. *Nature* **2003**, *425*, 686–691. [CrossRef]
38. Straede, A.; Heinisch, J.J. Functional analyses of the extra- and intracellular domains of the yeast cell wall integrity sensors Mid2 and Wsc1. *FEBS Lett.* **2007**, *581*, 4495–4500. [CrossRef]
39. Gray, J.V.; Ogas, J.P.; Kamada, Y.; Stone, M.; Levin, D.; Herskowitz, I. A role for the Pkc1 MAP kinase pathway of *Saccharomyces cerevisiae* in bud emergence and identification of a putative upstream regulator. *EMBO J.* **1997**, *16*, 4924–4937. [CrossRef] [PubMed]
40. Verna, J.; Lodder, A.; Lee, K.; Vagts, A.; Ballester, R. A family of genes required for maintenance of cell wall integrity and for the stress response in *Saccharomyces cerevisiae*. *Proc. Natl. Acad. Sci. USA* **1997**, *94*, 13804–13809. [CrossRef]
41. Philips, J.; Herskowitz, I. Osmotic balance regulates cell fusion during mating in *Saccharomyces cerevisiae*. *J. Cell Biol.* **1997**, *138*, 961–974. [CrossRef] [PubMed]
42. Gammie, A.E.; Brizzio, V.; Rose, M.D. Distinct morphological phenotypes of cell fusion mutants. *Mol. Biol. Cell* **1998**, *9*, 1395–1410. [CrossRef] [PubMed]
43. Merlini, L.; Dudin, O.; Martin, S.G. Mate and fuse: How yeast cells do it. *Open Biol.* **2013**, *3*, 130008. [CrossRef]
44. Ydenberg, C.A.; Rose, M.D. Yeast mating: A model system for studying cell and nuclear fusion. *Methods Mol. Biol.* **2008**, *475*, 3–20. [CrossRef]
45. Chen, E.H.; Grote, E.; Mohler, W.; Vignery, A. Cell-cell fusion. *FEBS Lett.* **2007**, *581*, 2181–2193. [CrossRef] [PubMed]
46. Kim, J.; Bortz, E.; Zhong, H.; Leeuw, T.; Leberer, E.; Vershon, A.K.; Hirsch, J.P. Localization and signaling of G(beta) subunit Ste4p are controlled by a-factor receptor and the a-specific protein Asg7p. *Mol. Cell. Biol.* **2000**, *20*, 8826–8835. [CrossRef]
47. Amberg, D.C.; Burke, D.; Strathern, J.N.; Burke, D. Cold Spring Harbor Laboratory. In *Methods in Yeast Genetics: A Cold Spring Harbor Laboratory Course Manual*; Cold Spring Harbor Laboratory Press: ColdSpring Harbor, NY, USA, 2005.
48. Gammie, A.E.; Rose, M.D. Assays of cell and nuclear fusion. *Methods Enzymol.* **2002**, *351*, 477–498.
49. Broach, J.R.; Strathern, J.N.; Hicks, J.B. Transformation in yeast: Development of a hybrid cloning vector and isolation of the *CAN1* gene. *Gene* **1979**, *8*, 121–133. [CrossRef]
50. Grote, E. Cell fusion assays for yeast mating pairs. *Methods Mol. Biol.* **2008**, *475*, 165–196. [CrossRef] [PubMed]
51. Rose, M.D.; Misra, L.M.; Vogel, J.P. *KAR2*, a karyogamy gene, is the yeast homolog of the mammalian BiP/GRP78 gene. *Cell* **1989**, *57*, 1211–1221. [CrossRef]
52. Fischer-Parton, S.; Parton, R.M.; Hickey, P.C.; Dijksterhuis, J.; Atkinson, H.A.; Read, N.D. Confocal microscopy of FM4-64 as a tool for analysing endocytosis and vesicle trafficking in living fungal hyphae. *J. Microsc.* **2000**, *198*, 246–259. [CrossRef]
53. Yadav, V.R.; Prasad, S.; Kannappan, R.; Ravindran, J.; Chaturvedi, M.M.; Vaahtera, L.; Parkkinen, J.; Aggarwal, B.B. Cyclodextrin-complexed curcumin exhibits anti-inflammatory and antiproliferative activities superior to those of curcumin through higher cellular uptake. *Biochem. Pharmacol.* **2010**, *80*, 1021–1032. [CrossRef] [PubMed]
54. Millard, P.J.; Roth, B.L.; Thi, H.P.; Yue, S.T.; Haugland, R.P. Development of the FUN-1 family of fluorescent probes for vacuole labeling and viability testing of yeasts. *Appl. Environ. Microbiol.* **1997**, *63*, 2897–2905. [CrossRef] [PubMed]
55. Deere, D.; Shen, J.; Vesey, G.; Bell, P.; Bissinger, P.; Veal, D. Flow cytometry and cell sorting for yeast viability assessment and cell selection. *Yeast* **1998**, *14*, 147–160. [CrossRef]
56. Bender, A.; Pringle, J.R. Use of a screen for synthetic lethal and multicopy suppressee mutants to identify two new genes involved in morphogenesis in *Saccharomyces cerevisiae*. *Mol. Cell. Biol.* **1991**, *11*, 1295–1305. [PubMed]
57. Kim, Y.J.; Francisco, L.; Chen, G.C.; Marcotte, E.; Chan, C.S. Control of cellular morphogenesis by the Ip12/Bem2 GTPase-activating protein: Possible role of protein phosphorylation. *J. Cell Biol.* **1994**, *127*, 1381–1394. [CrossRef]
58. Zheng, Y.; Cerione, R.; Bender, A. Control of the yeast bud-site assembly GTPase Cdc42. Catalysis of guanine nucleotide exchange by Cdc24 and stimulation of GTPase activity by Bem3. *J. Biol. Chem.* **1994**, *269*, 2369–2372. [CrossRef]
59. Zheng, Y.; Hart, M.J.; Shinjo, K.; Evans, T.; Bender, A.; Cerione, R.A. Biochemical comparisons of the Saccharomyces cerevisiae Bem2 and Bem3 proteins. Delineation of a limit Cdc42 GTPase-activating protein domain. *J. Biol. Chem.* **1993**, *268*, 24629–24634. [CrossRef]
60. Schmitz, H.P.; Huppert, S.; Lorberg, A.; Heinisch, J.J. Rho5p downregulates the yeast cell integrity pathway. *J. Cell Sci.* **2002**, *115*, 3139–3148. [CrossRef]
61. McIsaac, R.S.; Oakes, B.L.; Wang, X.; Dummit, K.A.; Botstein, D.; Noyes, M.B. Synthetic gene expression perturbation systems with rapid, tunable, single-gene specificity in yeast. *Nucleic Acids Res.* **2013**, *41*, e57. [CrossRef]
62. De Godoy, L.M.; Olsen, J.V.; Cox, J.; Nielsen, M.L.; Hubner, N.C.; Fröhlich, F.; Walther, T.C.; Mann, M. Comprehensive mass-spectrometry-based proteome quantification of haploid versus diploid yeast. *Nature* **2008**, *455*, 1251–1254. [CrossRef]
63. Mukherjee, D.; Sen, A.; Boettner, D.R.; Fairn, G.D.; Schlam, D.; Valentin, F.J.B.; McCaffery, J.M.; Hazbun, T.; Staiger, C.J.; Grinstein, S.; et al. Bem3, a Cdc42 GTPase-activating protein, traffics to an intracellular compartment and recruits the secretory Rab GTPase Sec4 to endomembranes. *J. Cell Sci.* **2013**, *126*, 4560–4571. [CrossRef] [PubMed]
64. Bos, J.L.; Rehmann, H.; Wittinghofer, A. GEFs and GAPs: Critical elements in the control of small G proteins. *Cell* **2007**, *129*, 865–877. [CrossRef] [PubMed]
65. Sprang, S.R. G proteins, effectors and GAPs: Structure and mechanism. *Curr. Opin. Struct. Biol.* **1997**, *7*, 849–856. [CrossRef]

66. Kahn, R.A. GAPs: Terminator versus effector functions and the role(s) of ArfGAP1 in vesicle biogenesis. *Cell. Logist.* **2011**, *1*, 49–51. [CrossRef] [PubMed]
67. Ohashi, A.; Gibson, J.; Gregor, I.; Schatz, G. Import of proteins into mitochondria. The precursor of cytochrome c1 is processed in two steps, one of them heme-dependent. *J. Biol. Chem.* **1982**, *257*, 13042–13047. [CrossRef]

Journal of *Fungi*

MDPI

Article

The Cell Wall Integrity Receptor Mtl1 Contributes to Articulate Autophagic Responses When Glucose Availability Is Compromised

Sandra Montella-Manuel, Nuria Pujol-Carrion and Maria Angeles de la Torre-Ruiz *

Cell Signalling in Yeast Unit, Department of Basic Medical Sciences, Institut de Recerca Biomèdica de Lleida (IRBLleida), University of Lleida, 25198 Lleida, Spain; sandra.montella@udl.cat (S.M.-M.); nuria.pujol@udl.cat (N.P.-C.)
* Correspondence: mariaangeles.delatorre@udl.cat

Abstract: Mtl1protein is a cell wall receptor belonging to the CWI pathway. Mtl1 function is related to glucose and oxidative stress signaling. In this report, we show data demonstrating that Mtl1 plays a critical role in the detection of a descent in glucose concentration, in order to activate bulk autophagy machinery as a response to nutrient deprivation and to maintain cell survival in starvation conditions. Autophagy is a tightly regulated mechanism involving several signaling pathways. The data here show that in *Saccharomyces cerevisiae*, Mtl1 signals glucose availability to either Ras2 or Sch9 proteins converging in Atg1 phosphorylation and autophagy induction. TORC1 complex function is not involved in autophagy induction during the diauxic shift when glucose is limited. In this context, the *GCN2* gene is required to regulate autophagy activation upon amino acid starvation independent of the TORC1 complex. Mtl1 function is also involved in signaling the autophagic degradation of mitochondria during the stationary phase through both Ras2 and Sch9, in a manner dependent on either Atg33 and Atg11 proteins and independent of the Atg32 protein, the mitophagy receptor. All of the above suggest a pivotal signaling role for Mtl1 in maintaining correct cell homeostasis function in periods of glucose scarcity in budding yeast.

Keywords: cell wall integrity (CWI); Mtl1; autophagy; glucose; mitophagy; *Saccharomyces cerevisiae*

Citation: Montella-Manuel, S.; Pujol-Carrion, N.; de la Torre-Ruiz, M.A. The Cell Wall Integrity Receptor Mtl1 Contributes to Articulate Autophagic Responses When Glucose Availability Is Compromised. *J. Fungi* **2021**, 7, 903. https://doi.org/10.3390/jof7110903

Academic Editors: María Molina and Humberto Martín

Received: 24 September 2021
Accepted: 23 October 2021
Published: 26 October 2021

1. Introduction

Carbon sources have a major impact on *Saccharomyces cerevisiae* metabolism and also affect longevity. Yeast uses a fermentative metabolism when the preferred carbon source, glucose, is abundant and ethanol, organic acids, and ATP are accumulated. Glucose limitation induces slowing of growth, contributing to the switch to respiratory metabolism, the hallmark of the diauxic shift, which along with other metabolic configurations prepares cells for the stationary phase and the process of chronological ageing [1]. Budding yeast chronological life span (CLS) forced by glucose restriction is also dependent on the availability of other nutrients in the culture media, such as amino acids or nitrogen among others [2–5]. In a simple way, glucose is a promotor of ageing through the activation of the Ras-cAMP-PKA pathway and TORC1-Sch9 [6–8] pathways, whereas amino acids operate through Gcn2 protein [9–12] as well as through activation of the TORC1 complex, upstream regulators of the Sch9 protein [13].

The TORC1 and Ras/PKA pathways are both negative and independent regulators of autophagy [13–16]. Sch9 downregulates autophagy independently and coordinately with the Ras/PKA pathway [17].

Ras proteins are essential for growth in fermentable carbon sources such as glucose. In that context, Ras proteins trigger the synthesis of cAMP and activation of the PKA pathway upon binding to the inhibitor Bcy1 protein. The absence of Ras proteins does not allow growth in non-fermentative carbon sources (see reviews [18,19]).

The sucrose non-fermenting protein kinase, Snf1, is the yeast orthologue of mammalian AMPK, and is important for cell adaptation to glucose limitation. Snf1 is a key component of the main glucose repression pathway in yeast and controls genes involved in alternative carbon sources and metabolism. However, regulation of adaptation to glucose limitation is the main role of the SNF1 complex [20,21]. Snf1 also plays a role in autophagy [22]; its negative regulation is required to downregulate autophagy under certain conditions of nutritional limitation [23].

Mtl1 protein is a transmembrane protein cell wall mechano-sensor [24,25], part of the cell wall integrity pathway (CWI) with structural similarity to its paralogue Mid2 protein [26,27]. Mtl1 is also required to activate a stress response towards TORC1 and Ras/PKA signaling pathways under conditions of both oxidative stress and glucose starvation [28]. In addition, this cell wall receptor plays a role in Cyclin C localization and programmed cell death [29], as well as in the preservation of mitochondrial integrity and life span by regulating TORC1, Sch9, Slt2, and PKA [30].

The CWI is involved in sensing and transducing a wide variety of stimuli (see reviews [31,32]), including both nutritional and oxidative stresses [27,30,33]. In this pathway, a wide number of sensors are specialized in the detection of different stresses [25,34], which converge on the GTPase Rho1 to subsequently activate Pkc1 protein (see reviews [31,32]). Pkc1 in turn phosphorylates the Bck1 protein, the MAPKKK, thus activating the MAPKK module composed of Mkk1/Mkk2 proteins leading to dual phosphorylation and activation of the last member of the pathway, the MAPK Slt2 protein, whose double phosphorylation is impaired in the mutant *mtl1* [30].

Nutrient limitation strongly induces macroautophagy [23,35–37] to accomplish two main objectives; one is to detoxify and the second is to recycle components to build newly synthetized molecules.

Macroautophagy is a process in which several components (damaged or superfluous organelles, cytoplasmic elements, microorganisms ...) are engulfed within cytosolic double membrane vesicles named autophagosomes. The outer membrane of the phagosome fuses to the vacuolar membrane releasing a spherical body termed autophagic body, that is digested by hydrolases releasing the breakdown products back to the cytosol to be recycled by cells [38]. Morphological and structural characteristics of autophagy are highly conserved from yeast to humans.

The signaling network governing life span usually converges in the autophagic machinery [17,23,39]. In general, nutrient deprivation impinges on Atg13 dephosphorylation that triggers Atg1 kinase activity then leading to the formation of the complex Atg13/Atg1/Atg17/Atg29/Atg31 activating the autophagy process [40,41]. Atg7 is an E1-like enzyme essential for macroautophagy since it is part of the Atg8-Ibl conjugation system [42,43]. Mutants defective in autophagy display shorter life spans [44,45].

Autophagy entails nonselective engulfment of cytoplasmic components, but there are also other types of autophagy that selectively degrade specific cellular elements (see review [46]), such is the case of mitophagy. Mitophagy clears dysfunctional mitochondria and impinges on cellular function by promoting respiration proficiency during the process of ageing (see review [47]). Selective mitophagy requires the function of the Atg32 protein as a mitochondrial receptor and its binding to the adaptor Atg11, which interacts with the Atg8 protein in the phagophore inner surface (see review [48]). Atg33 is a yeast protein located at the outer mitochondrial membrane, its absence suppressed mitophagy in post-log cultures, however, its precise role in mitophagy is still controversial [49].

In this report, we show that bulk autophagy is highly induced during the transition to diauxic shift in a manner totally dependent on glucose and amino acids availability. The Mtl1 cell wall receptor is essential to sense glucose concentration and transmit the signal to both Ras2 and Sch9 to phosphorylate Atg1 and to activate the macroautophagic machinery. Gcn2 is the amino acid sensor. Both Mtl1 and Gcn2 operate independently of TORC1 in the signaling process leading to the activation of bulk autophagy. Moreover, Mtl1 is also relevant for mitochondrial clearance dependent on Atg33 and the inactivation of either Ras2 or Sch9 in response to glucose exhaustion.

2. Materials and Methods

2.1. Yeast Strains and Plasmids

Saccharomyces cerevisiae strains are listed in Table 1. All the strains named GSL are derivatives of the CML128 background. New null mutants described in this study were obtained by a one-step disruption method that used the *Nat*Mx4 or *Kan*Mx4 cassettes [50]. Strains GSL197, 198, 199, 200, 201, 202, 226, 265, 279, 297, 352, 370, 382, 414, and 415 were constructed upon integration of plasmid pGFP–Atg8 (original name: pHab142), previously digested with Stu1, in the *URA3* gene. Strains GSL372 and 416 were constructed upon integration of plasmid pAtg1-HA previously digested with BstEII. The plasmid pAtg1-HA was obtained upon Atg1 cloning into the Pme1 and PstI sites of the integrative vector pMM351 [51].

Table 1. Yeast strains.

Strain	Genotype	Source
CML128	*MATa leu2D3,112, ura3D52, trp1D0, his4D0*	[52]
GSL011	*MATa mtl1::NatMx4*	[53]
GSL053	*MATa ras2::Leu2MX5*	[53]
GSL054	*MATa mtl1::KanMx4 ras2::LEU2Mx5*	[53]
GSL197	*MATa leu2D3,112, ura3D52, trp1D0, his4D0 URA3::GFP-ATG8*	[23]
GSL198	*MATa mtl1::KanMx4 URA3::GFP-ATG8*	This work
GSL199	*MATa tor1::KanMx4 URA3::GFP-ATG8*	[23]
GSL200	*MATa mtl1::KanMx4 tor1::LEU2 URA3::GFP-ATG8*	This work
GSL201	*MATa ras2::Leu2MX5 URA3::GFP-ATG8*	[23]
GSL202	*MATa mtl1::KanMx4 ras2::LEU2 URA3::GFP-ATG8*	This work
GSL205	*MATa sch9::NatMx4*	[30]
GSL206	*MATa sch9::NatMx4 mtl1::KanMx4*	[30]
GSL218	*MATa atg7::NatMx4*	[23]
GSL226	*MATa atg7::NatMx4 URA3::GFP-ATG8*	[23]
GSL265	*MATa slt2::NatMx4 URA3::GFP-ATG8*	This work
GSL279	*MATa sch9::NatMx4 URA3::GFP-ATG8*	This work
GSL293	*MATa atg11::NatMx4*	[23]
GSL296	*MATa atg33::NatMX4*	This work
GSL297	*MATa atg11::NatMx4 URA3::GFP-ATG8*	[23]
GSL324	*MATa atg1::NatMx4*	[23]
GSL352	*MATa gcn2::KanMx4 URA3::GFP-ATG8*	[23]
GSL364	*MATa atg32::KanMx4*	[23]
GSL370	*MATa rho0 URA3::GFP-ATG8*	[23]
GSL372	*MATa leu2D3,112, uras3D52, trp1D0, his4D0 ATG1-HA::LEU2*	[23]
GSL382	*MATa snf1:KanMx4 URA3::GFP-ATG8*	[23]
GSL414	*MATa mtl1::KanMX4 sch9::NatMx4 URA3::GFP-ATG8*	This work
GSL415	*MATa mtl1::KanMx4 slt2::NatMx4 URA3::GFP-ATG8*	This work
GSL416	*MATa mtl1:NatMx4 ATG1-HA::LEU2*	This work
BY4741 pho8Δ	*MATa pho8 his3D1, leu2D0, met15D0, ura3D0*	[54]

Plasmid descriptions are listed in Table 2. Each particular ORF was amplified by PCR from genomic DNA to be directionally cloned in the specific plasmid.

Table 2. Plasmids employed.

Plasmid	Restriction Sites to Clone the ORF	Marker	Promoter	Epitope	Source
pGFP-Atg8	EcoRI, XhoI	*URA3*	ATG8	GFP	[55]
pSfp1-GFP	SalI, SmaI	*URA3*	MET25	GFP	[23]
pAdh1-Msn2-GFP	KspI, SalI	*LEU2*	ADH1	GFP	[56]
pRtg1-GFP	XhoI, EcoRI	*URA3*	RTG1	GFP	[57]
pIdp1-GFP	HindIII, XhoI	*URA3*	IDP1	GFP	[58]
pAtg13-HA	NotI, PstI	*URA3*	ADH1	HA	[23]
pAtg1-HA	NotI, PstI	*LEU2*	ADH1	HA	[23]
pMM351	PstI, HindIII	*LEU2*	ADH1	HA	[51]
pBcy1HA	SmaI, XhoI	*HIS3*	ADH1	HA	[30]
pPkc1*	PmeI, NotI	*LEU2*	ADH1	HA	This work
p*BCK1-20*	HindIII, PstI	*TRP1*	LAC		[28]
pYX242-cytPho8	AvrII, MluI	*LEU2*	PHO8		[54]

2.2. Media, Growth Conditions and Reagents

Yeasts were grown at 30 °C in SD medium (2% glucose, 0.67% yeast nitrogen base that lacked the corresponding amino acids for plasmid maintenance) plus amino acids [59].

Glucose depletion consisted of SD medium without glucose plus amino acids. Nitrogen depletion consisted of SD medium whose nitrogen base component was free of amino acids and ammonium sulphate plus amino acids. Amino acid depletion consisted of SD medium without adding amino acids. Glycerol medium (3% glycerol, 0.67% yeast nitrogen base that lacked the corresponding amino acids) plus amino acids. Sucrose medium (2% sucrose, 0.67% yeast nitrogen base that lacked the corresponding amino acids) plus amino acids. Fructose medium (2% fructose, 0.67% yeast nitrogen base that lacked the corresponding amino acids) plus amino acids.

Glucose was added as α-D-glucose monohydrate (Serva, Heidelberg, Germany 22720.01) at a final concentration of 2%; Amino acids were added at concentrations: 60 mg/mL Leucine, 20 mg/mL Histidine and 20 mg/mL Tryptophan. Nitrogen was added as Yeast Nitrogen Base w/o Amino Acids (Difco, Franklin Lakes, NJ, USA, 291940) at a final concentration of 0.67%. Sucrose was added as Sucrose (Sigma, Saint Louis, MI, USA, S0389) at a final concentration of 2% and Fructose was added as D-Fructose (Sigma, Saint Louis, MI, USA, 47740) at a final concentration of 2%.

We present a list of reagents detailing final concentrations in culture media and from which company they were purchased: N-Acetyl cysteine (NAC) 5 mM (Sigma, Saint Louis, MI, USA, A9165); FM4-64 30 μg/μL (Invitrogen, Waltham, MA, USA, T-3166); Rapamycin 200 ng/mL (Sigma, Saint Louis, MI, USA, R0395); ATP 200 mM (Sigma, Saint Louis, MI, USA, A1852); and Dihydroethidium (DHE) 50 μM (Sigma, Saint Louis, MI, USA, D7008). Cell cultures were exponentially grown at 600 nm [$O.D_{600}$] of 0.6. Iron was added as ammonium iron (III) sulphate hexahydrate [$NH_4Fe(SO_4)_2{\cdot}6H_2O$] (+Fe; Sigma, Saint Louis, MI, USA, F1543) at a final concentration of 10 mM.

2.3. Vacuole and Dihydroethidium Staining

For vacuole visualization, cells were stained with the fluorescent styryl dye FM4-64 (N-(3-triethylammoniumpropyl)-4-(p-diethylaminophenylhexatrienyl) pyridinium dibromide. To determine cellular oxidation, we used dihydroethidium (DHE). Both protocols were previously described by our group in [60].

2.4. Cell Survival and Chronological Life Span

To assay cell viability cells were grown to mid-log phase OD_{600}:0.6 in SD medium supplemented with the required amino acids. Viability was registered through serial dilutions and plated by triplicate onto YPD plates.

We measured the chronological life span (CLS) in the different strains based on the survival of populations of non-dividing yeast cells according to [61]. The viability was scored by counting the number of cells able to form colonies, CFU (colony-forming units). Cultures were started at an OD_{600}:0.6. The same number of cells collected from each culture were plated in triplicated into YPD plates and allowed to grow at 30 °C for 3–4 days. CLS curves were plotted with the corresponding averages and standard deviations from three independent experiments.

2.5. Protein Extraction and Immunoblot Analyses

We followed an identical procedure as described in [23]. Total yeast protein extracts were prepared as previously described in [61]. The antibodies for Western blotting were as follows: anti-HA 3F10 (No. 12158167001; Roche Applied Science, Penzberg, Germany), used at a dilution of 1:2000 in 0.25% non-fat milk, and the corresponding secondary was goat anti-Rat IgG horseradish peroxidase conjugate (No. AP136P, Millipore, Burlington, MA, USA). Anti-GFP (No. 632381; Living Colors Mountain View, CA, USA) was used at a dilution of 1:2000 and anti-Phospho-glycerate kinase 1 (anti-PGK1) (459250, Invitrogen, Waltham, MA, USA) was used at a dilution 1:1200, both with the secondary antibody anti-Mouse horseradish peroxidase conjugate (LNA931v/AG, GE Healthcare, Chicago, IL, USA) and anti-Phospho-AMPKα (Thr172) (167253S, Cell Signalling, Danvers, MA, USA) at a dilution of 1:1000 with the secondary antibody anti-rabbit horseradish peroxidase conjugate (LNA934v/AG, GE Healthcare, Chicago, IL, USA). They were used as indicated by the manufacturers.

The protein–antibody complexes were visualized by enhanced chemiluminescence, using the Supersignal substrate (Pierce, Waltham, MA, USA) in a Chemidoc (Roche Applied Science, Penzberg, Germany).

For all the figures: We used anti-PGK1 to detect PGK1, selected as a loading control in all the Western blots shown in this study. For Western blots in this paper, we have selected representative samples.

2.6. Autophagy Detection

Autophagy progression is monitored through several complementary approaches, such as the immunological detection of GFP accumulation from GFP–Atg8 genomic fusion which is delivered to the vacuole to be degraded once autophagy is induced. The GFP moiety is very resistant to proteolysis compared to Atg8 which is rapidly degraded in the vacuole. Therefore, detection of free GFP processed from GFP–Atg8 is a very reliable tool to measure levels of complete autophagy through the autophagic flux [62], that is delivery and turnover of the cargo in the vacuole. Autophagic flux is the ratio of free GFP/GFP–Atg8+free GFP quantified upon Western blot detection by using anti-GFP antibody [23]. Another complementary approach is the microscopic observation of GFP accumulation in vacuoles. For all microscope panels, we have used a representative image of either log or one day samples in order to identify GFP–Atg8 localization. In general, both assays are sufficient as evidence of autophagy activity.

In some particular occasions we also use an alternative approach, consisting of measuring *pho8*Δ60 enzymatic activity to determine nonspecific autophagy, as described by Noda and Klionsky [63] and modified by Guedes et al. [54].

2.7. Glucose Determination

We followed the directions detailed in [64].

2.8. Statistical Analysis

We followed the same procedure as described in Montella et al. [23]. Error bars in the histograms represent the standard deviation (SD) calculated from three independent experiments. Significance of the data was determinate by p-values from a Student's unpaired t-test denoted as follows: * = $0.05 > p > 0.01$; ** = $0.01 > p > 0.001$; *** = $0.001 > p > 0.0001$; **** = $p > 0.0001$.

3. Results

3.1. Glucose, Amino Acids, Nitrogen and Iron Deprivation Determine the Induction of Bulk Autophagy during Diauxic Transition

In a previous paper we demonstrated that autophagy is required for normal life span extension [23]. We wanted to determine which pathways are involved in autophagy regulation in the process of ageing. To start our analysis, we took daily samples from log phase (day 0) till day 15, following a standard CLS analysis. We used SD medium to avoid the addition of excess amino acids and thus did not affect the metabolism of the yeast cells. In Figure 1A we can observe that there is a great induction of autophagy and autophagic flux when cells reach the diauxic shift upon one day of growth, which is maintained and gradually decreases until day 6. The induction of autophagic flux is related with the phosphorylation of Atg1 protein, hence, Atg1 receives the starvation signal in order to induce autophagy during the diauxic and posdiauxic shifts, and also with the enzymatic activity determined by the *pho8*Δ60 assay. Microscopic observation of the cultures confirmed the former results, since free GFP derived from GFP–Atg8 fusion protein was accumulated inside vacuoles which appear colored in green and surrounded by a red membrane stained by the fluorescent styryl dye FM4-64. In order to ascertain whether our results were compatible with bulk or selective autophagy we repeated this experiment in the mutants *atg7* (involved in general autophagy) and *atg11* (representing of selective autophagy) (Figures 1B and S1B). Our results demonstrate that free GFP liberated from GFP–Atg8 fusion protein and detected both in Western blot and in the fluorescence microscopy indicated bulk autophagy and was independent of any type of selective autophagy.

One day of culture in SD media is the transition between a fermentative to respiratory metabolism, the diauxic shift, a metabolic regulatory checkpoint determinant in the process of ageing. At this point, we could observe that glucose is nearly exhausted in the culture media (Figure 1C). Consequently, we performed refeeding experiments upon one day of culture and observed that only upon 6 h of glucose addition autophagy (determined by *pho8*Δ60 specific activity, identification of free GFP by Western blot, and in vivo fluorescence identification of vacuolar accumulation of GFP) significantly decreased, concluding that a severe decrease in glucose concentration provoked the induction of autophagy, and that there is an increase in autophagy (Figures 1D and S1C). We carried out the same strategy with other nutrients which could also be limiting: amino acids, nitrogen and iron. Upon refeeding of iron, nitrogen, and amino acids, we observed that one-day refeeding did not provoke changes in autophagy (Figures 1D and S1C), however upon two days of culture we observed a clear descent in GFP accumulation caused by amino acid replenishment (Figure 1E,D). Both nitrogen and iron replenishment provoked a descent in autophagy upon two days of culture.

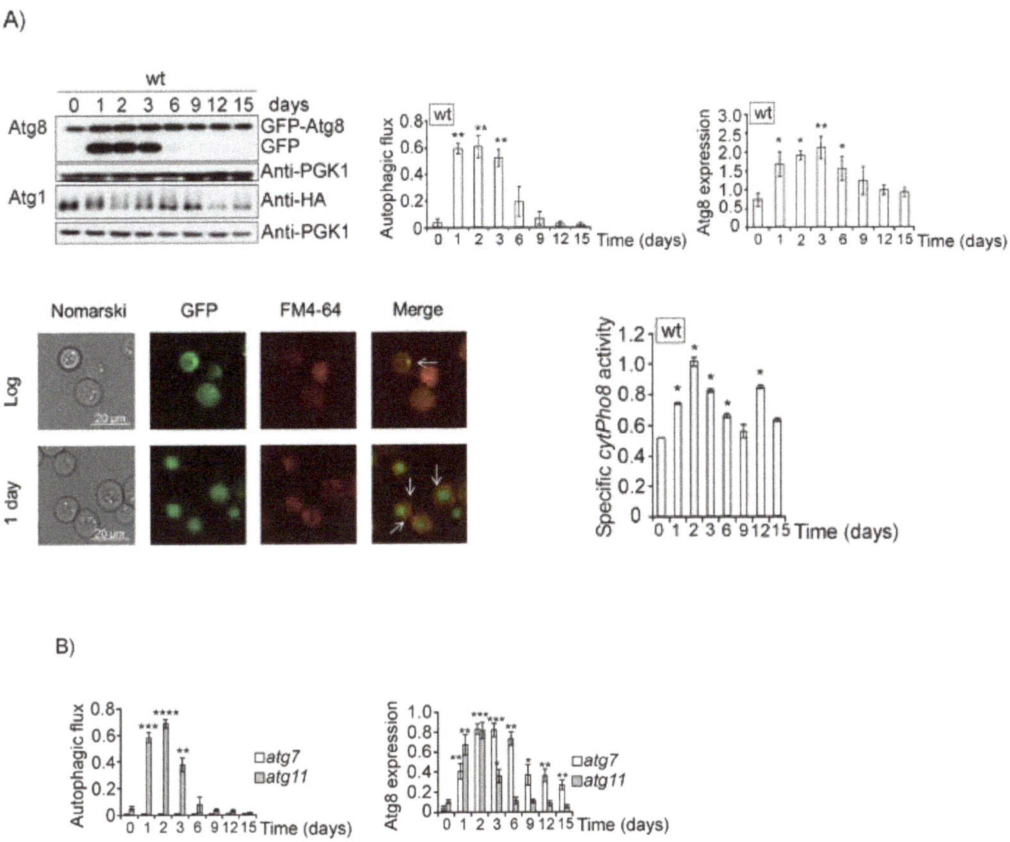

C)

Time	Glucose(%)	SD
SD medium	100.00	±0.56
Log	68.64	±1.90
8 hours	32.23	±1.45
1 day	6.44	±0.22
30 hours	5.02	±0.99
2 days	0.00	±0.00
3 days	0.00	±0.00
6 days	0.00	±0.00
9 days	0.00	±0.00
12 days	0.00	±0.00
15 days	0.00	±0.00

Figure 1. *Cont.*

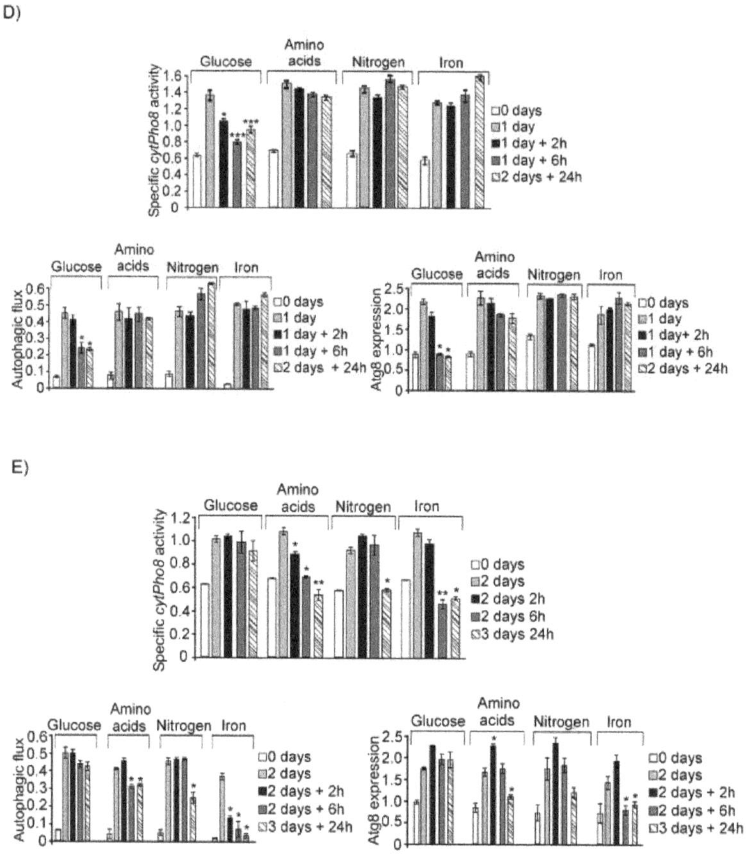

Figure 1. Sequential decreases of glucose and amino acids activates bulk autophagy during the diauxic shift in *Saccharomyces cerevisiae*. (**A**) wt cultures in which the fusion protein GFP–Atg8 or Atg1HA were integrated, were grown to log phase (OD_{600}: 0.6) in SD medium at 30 °C. Aliquots were collected at the indicated times for total protein extraction and Western blot analysis. GFP–Atg8 was monitored using an anti-GFP antibody. We used anti-PGK1 to detect Pgk1 as a loading control. Microscopic observation of GFP–Atg8 was carried out using a fluorescence microscope. GFP vacuolar accumulation was also determined upon the use of the fluorescent dye FM4-64. Autophagic flux was calculated as the ratio of free GFP and total GFP–Atg8 in the samples. Total GFP–Atg8 was determined as the addition of the form GFP–Atg8 and the band corresponding to free GFP, as a result of Atg8 vacuolar degradation, both detected by Western blot. Enzymatic autophagy activity was measured by using the alkaline phosphatase assay in the strain BY4741*pho8*Δ expressing a plasmid with the inactive Pho8 proenzyme targeted to the cytosol. Values of Atg1 proteins were determined upon Western blot analysis using anti-HA antibody. (**B**) *atg7* and *atg11* strains expressing the fusion protein GFP–Atg8 were grown in the same conditions as described in A. Autophagic flux and total Atg8 expression were determined as in (**A**). (**C**) Glucose content in the culture medium (%) was determined in wt cultures growing in SD media at 30 °C at the days indicated in the table for a total period of 15 days. Glucose in the sterile media, SD, at a final concentration of 2%, is the equivalent of 100% in the table. (**D**) wt cells bearing GFP–Atg8 in their genome were exponentially grown at OD_{600}:0.6 at 30 °C in SD media and a sample was collected for analysis. Upon one day of culture, 2% glucose, amino acids (60 mg/mL leucine, 20 mg/mL histidine, and 20 mg/mL tryptophan), 0.67% nitrogen or 10 mM iron, were respectively added to the cultures and samples were collected at 2, 6, and 24 h to perform *pho8*Δ60 enzymatic assays. Autophagic flux and GFP–Atg8 total expression were determined as previously detailed in A. Enzymatic autophagic activity was determined as in (**A**). (**E**) As in (**D**), but results correspond to two days of growth. Error bars in the histograms represent the standard deviation (SD) calculated from three independent experiments. Significance of the data was determinate by *p*-values from a Student's unpaired *t*-test denoted as follows: * = 0.05 > *p* > 0.01; ** = 0.01 > *p* > 0.001; *** = 0.001 > *p* > 0.0001; **** = *p* > 0.0001.

3.2. Both Mtl1 and Gcn2 Control Autophagy Induction during Diauxic Transition

In this context, we decided to explore the signaling pathways that could be involved in signaling autophagy when cells age.

We considered the possibility that amino acid and nitrogen depletion would provoke TORC1 inactivation at least partly. We analyzed TOR function by means of the readouts Rtg1, Sfp1, Msn2. Under TORC1 inactivation, Rtg1 and Msn2 were localized to the nucleus, whereas Sfp1 was located in the cytoplasm, confirming that Tor1 is not inactivated in our model (Figure 2A). In addition, we added rapamycin, a drug that specifically inactivates TORC1, to the previous cultures as a control, and we observed that Rtg1 and Msn2 localized to the nucleus and Sfp1 was located in the cytoplasm (Figure 2B). Moreover, rapamycin treatment caused a decrease in Atg13 phosphorylation and autophagy values calculated upon Pho8 assay did not increase, supporting the conclusion that TORC1 is not completely inactivated upon diauxic shift in the conditions of our study (Figure 2C).

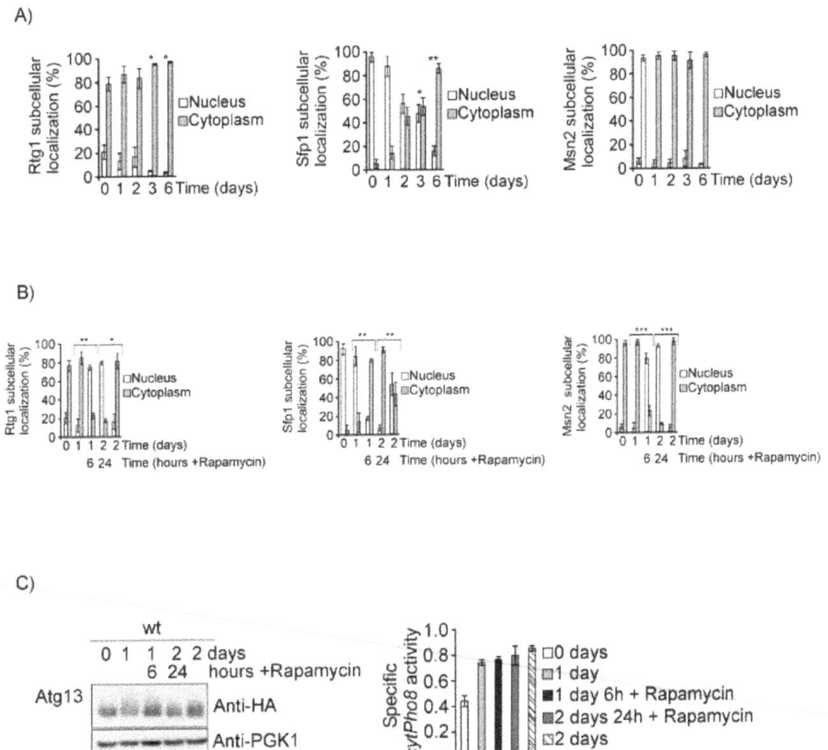

Figure 2. TORC1 is not inactivated during diauxic and posdiauxic shifts. (**A**) wt strains transformed with the plasmids Rtg1GFP, Sfp1GFP, or Msn2GFP respectively, were grown at 30 °C in SD media for the times indicated in the figures. Aliquots were collected for in vivo observation in the fluorescence microscope. Histograms represent the percentages of in vivo nuclear or cytoplasmic localization out of more than 1000 cells. (**B**) Cultures in A were treated with rapamycin (200 ng/mL) on day one of culture for 6 h and aliquots were collected for in vivo observation in the fluorescence microscope. Histograms are performed as in (**A**). (**C**) A wt strain expressing Atg13HA was exponentially grown at 30 °C in SD media. Rapamycin was added to the culture upon one day of growth at 200 ng/mL and samples were collected upon 6 and 24 h of exposure to the drug for total protein extraction, Western blot analysis, and identification of Atg13HA using anti-HA antibody. Autophagic enzymatic activity was determined through the alkaline phosphatase assay, as in (Figure 1A). Error bars in the histograms represent the standard deviation (SD) calculated from three independent experiments. Significance of the data was determinate by p-values from a Student's unpaired t-test denoted as follows: * = $0.05 > p > 0.01$; ** = $0.01 > p > 0.001$; *** = $0.001 > p > 0.0001$.

However, when *TOR1* was deleted we observed that autophagy (free GFP detected in the Western blot and fluorescence microscope accumulated in vacuoles) was extended to longer times (Figures 3A and S2A), concomitant with longer life extension as described in [54]. Taken together, we conclude that TORC1 is not inactivated in our system during the transition between fermentative and respiratory metabolism, therefore it is not the main pathway responsible for the bulk autophagy induction.

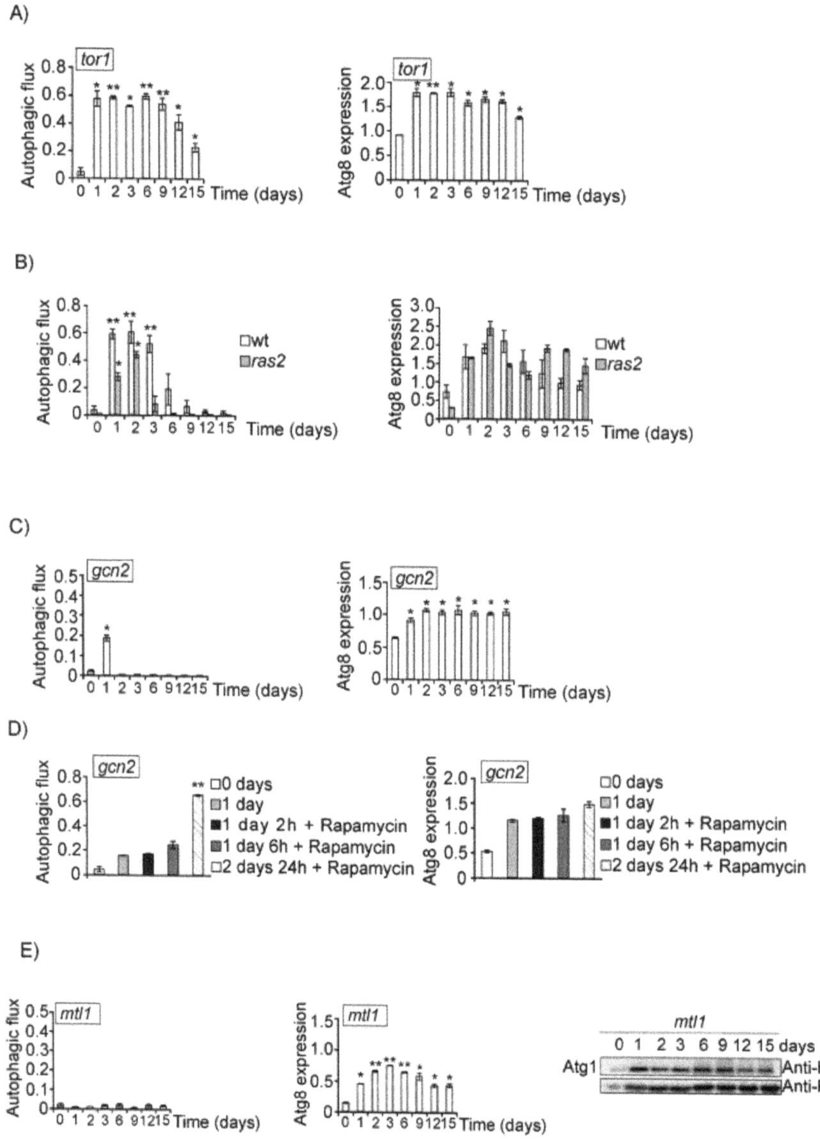

Figure 3. *Cont.*

F)

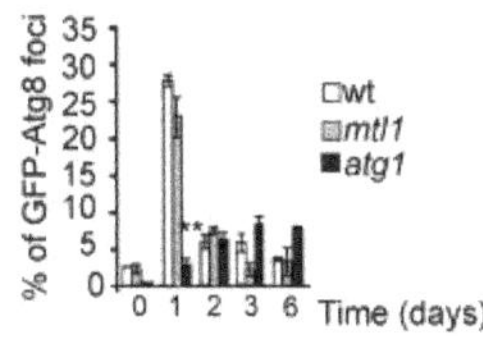

Figure 3. Mtl1 and Gcn2 control autophagy induction during glucose and amino acid starvation. Growth conditions, total Atg8 determination and autophagic flux determined in: (**A**) *tor1* mutant expressing GFP–Atg8; (**B**) wt strain and *ras2* mutant expressing GFP–Atg8; and (**C**) *gcn2* mutant expressing GFP–Atg8, performed as described in Figure 1B. (**D**) *gcn2* bearing GFP–Atg8 was exponentially grown at 30 °C in SD plus amino acids. Rapamycin (200 ng/mL) was added to the cultures upon 1 day of growth and samples were subsequently collected upon 2, 6 and 24 h of exposure to the drug. Aliquots were treated as in Figure 1B. (**E**) Growth conditions, total Atg8 and autophagic flux determinations in *mtl1* culture expressing the fusion protein GFP–Atg8 was performed as in Figure 1B. Identification of Atg1 protein in *mtl1* cultures transformed with the plasmid Atg1HA was performed upon Western blot analysis using the anti-HA antibody as in Figure 1A. (**F**) Percentage of Atg8 foci quantified in the experiments, described in Figures 1A and 3E, was calculated upon microscopic observation of more than 1000 cells. The axis label "% of GFP–Atg8 foci" refers to the percentage of cells with GFP–Atg8 foci. Error bars in the histograms represent the standard deviation (SD) calculated from three independent experiments. Significance of the data was determinate by p-values from a Student's unpaired t-test denoted as follows: * = $0.05 > p > 0.01$; ** = $0.01 > p > 0.001$.

We also explored a *ras2* mutant, since Ras2 is active in exponentially growing cells and becomes inactive as long as cells enter in respiratory metabolism and glucose becomes exhausted. *RAS2* deletion partially affected autophagy progression as compared to wt cultures (Figures 3B and S2B). Since other nutrients become depleted upon the diauxic shift, we took into consideration the Gcn2/eIF2alpha pathway which becomes activated under amino acid and other nutrient starvation [65]. The Gcn2 pathway is activated upon diauxic shifting, indicating the moment in which amino acid concentrations significantly decreased in the culture media. According to our results, and coincident with the above mentioned replenishment results, day 2 of growth should be the moment in which cells become starved for amino acids. We observed that upon the second day, autophagy disappears when Gcn2 is deleted, however the absence of Gcn2 the burst in autophagy observed upon 1 day of growth was not affected (Figures 3C and S2C). This result suggests that Gcn2 is required to induce macroautophagy upon two days of growth in SD medium, after the diauxic shift, probably due to a descent in the amino acid concentration one day after glucose starvation, suggesting that Gcn2 is not involved in autophagy signaling in response to carbon sources. Following this observation, we wanted to ascertain whether the regulatory function mediated by Gcn2 was dependent on TORC1. We treated *gcn2* mutant cultures on day 1 with rapamycin and observed induction of autophagy, suggesting that in the transition from fermentative to respiratory metabolism, TORC1 and *GCN2* are independent (Figures 3D and S2D). As expected, this conclusion is consistent with the previous observation that TORC1 function is not inactivated during the CLS experiment; as a consequence of that, this pathway is not relevant for autophagy induction during ageing in our experimental conditions.

Our results suggest that glucose is the principal nutrient in the culture medium we use, whose decrease causes autophagy induction during the diauxic shift (1 day of growth) therefore, we decided to analyze in more depth the role that Mtl1 could be playing in this process. Mtl1 is a cell wall receptor belonging to the CWI pathway, involved in glucose signaling during the diauxic shift and stationary phase [28,30]. Interestingly, in the absence of Mtl1, autophagy is undetectable by Western blot, with Atg1HA phosphorylation or in vivo GFP–Atg8 microscopic accumulation through all experiments, from day 1 to 15 (Figures 3E and S2E). In yeast, autophagy is initiated when the pre-autophagosome (PAS) structure is formed [66]. PAS can be detected in the fluorescence microscope as dotted accumulations of Atg proteins next to the vacuole. In *mtl1* diauxic cultures, PAS can be detected, as also observed in wt strain, and they are significantly higher than in the *atg1* mutant (Figure 3F), suggesting that Mtl1 does not block the initiation of the autophagy complex. This result suggests that Mtl1 is essential for receiving the signal of glucose concentration during the diauxic shift and to transmission of this signal to the autophagy machinery. We have also detected similar results when using alternative and fermentative carbon sources such as sucrose or fructose (Figure S2F).

3.3. Mtl1 CWI Cell Wall Receptor Signals Glucose Concentration to the Autophagy Machinery in a Manner Partly Dependent on Intracellular ATP Levels

In previous reports it has been demonstrated that several nutritional stressors (nitrogen, amino acids, iron …) cause the induction of autophagy. In order to analyze the specificity that Mtl1 could play in macroautophagy regulation, we used different nutrient concentrations: (glucose: 0.5%, 0.1%, 0.05% and 0%; amino acids: 0.1% and 0%; nitrogen: 0.06%, 0.01% and 0%; and iron: 0%). The decrease of each of the nutrients, glucose, amino acids, iron, or nitrogen concentrations induced the activation of macroautophagy in wt cells (Figures 4A and S3A) along with corresponding Atg1 phosphorylation (Figure 4D). In addition, the absence of Gcn2 precluded autophagy in a manner only dependent on amino acid availability (Figures 4B and S3B), whereas the absence of Mtl1 specifically abolished the glucose deprivation dependent autophagy (Figures 4C and S3C), supported by a lack of Atg1 phosphorylation (Figure 4D). We demonstrated that glucose concentrations below 0.5% caused a clear induction of autophagy specifically mediated by Mtl1, as in *mtl1* mutants autophagy was not induced.

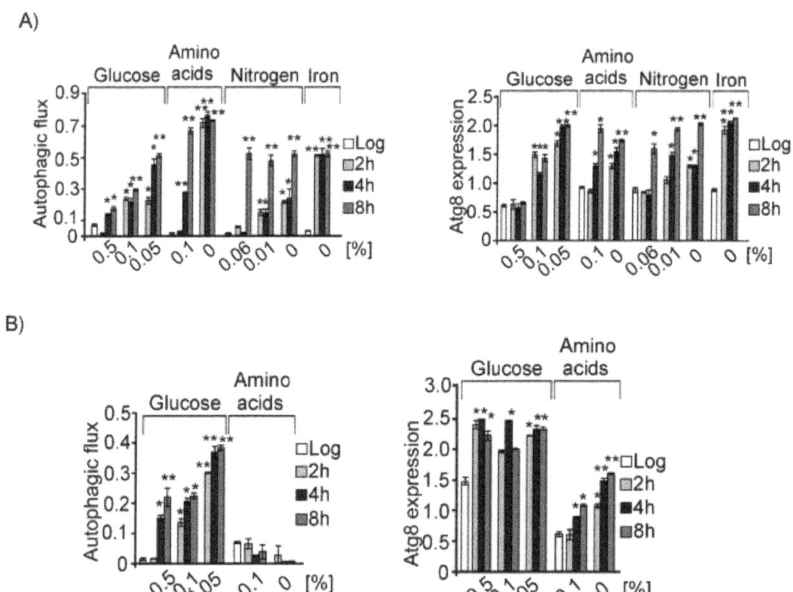

Figure 4. *Cont.*

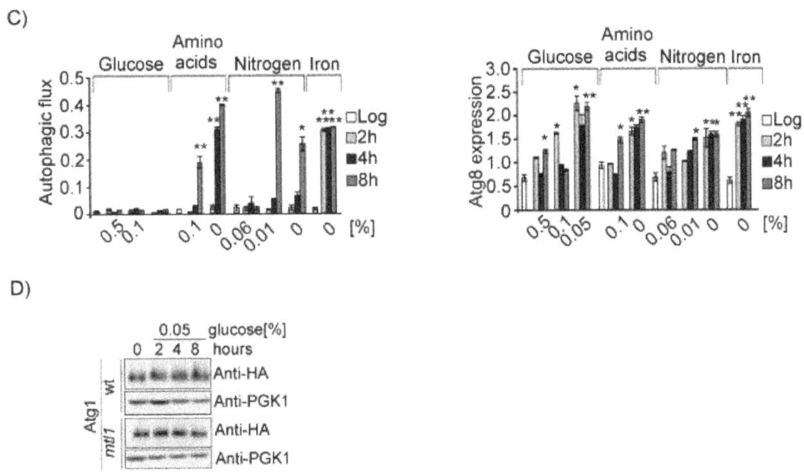

Figure 4. Mtl1 signals glucose limitation to autophagy machinery. (**A**) wt cells expressing GFP–Atg8 were exponentially grown in SD media. Aliquots were taken, washed, and transferred to different minimum media containing: 0.5, 0.1, or 0.05% glucose; 0.1 or 0% amino acids; 0.06, 0.01, or 0% nitrogen or medium without iron (0%). Autophagic flux and Atg8 expression were determined as in Figure 1B. The same experiments as in A were carried out in (**B**) *gcn2* mutant cultures expressing GFP–Atg8 and in a (**C**) *mtl1* strain expressing GFP–Atg8. (**D**) Atg1HA protein was identified by Western blot using anti-HA antibody, as in Figure 1A. Error bars in the histograms represent the standard deviation (SD) calculated from three independent experiments. Significance of the data was determinate by p-values from a Student's unpaired t-test denoted as follows: * = $0.05 > p > 0.01$; ** = $0.01 > p > 0.001$; *** = $0.001 > p > 0.0001$; **** = $p > 0.0001$.

However, placing cells at 0% glucose, we did not observe free GFP in either the Western blot or accumulated in vacuoles. Hence, bulk autophagy was not induced in wt or *mtl1* strains, as previously described by [67] (Figures 5A and S4A). These authors attributed this result to the sudden lack of ATP required for the autophagy machinery. We added ATP to wt and *mtl1* cultures completely depleted of glucose and observed that whereas in wt cultures autophagy induction was high, in the *mtl1* mutant the autophagy response was only partially restored (Figures 5B and S4B). Identical results were obtained during the diauxic shift in *mtl1* cultures when ATP was added to exponentially growing cells (Figures 5C and S4C). These results led us to the conclusion that the absence of *MTL1* provoked ATP starvation when glucose concentration drops below a threshold. In the former experiment, we can observe that autophagy induction occurs when glucose concentration drops from 2% to 0.5% in wt cultures (Figures 4A and S3A). However, in *mtl1* cultures, any decrease below 2% aborted autophagy (Figures 4C and S3C).

In order to ascertain the contribution of mitochondrial ATP to autophagy response during the diauxic shift and a reduction in glucose concentrations in exponentially growing cultures, we analyzed a *rho0* mutant lacking mitochondrial DNA. We observed that the absence of mitochondrial DNA did not preclude the induction of bulk autophagy upon one day of culture (Figures 5D and S4D) and upon glucose concentration reduction (Figures 5E and S5E). These results suggest that the functional role that Mtl1 plays in the autophagy response to glucose availability is not only linked to ATP accumulation.

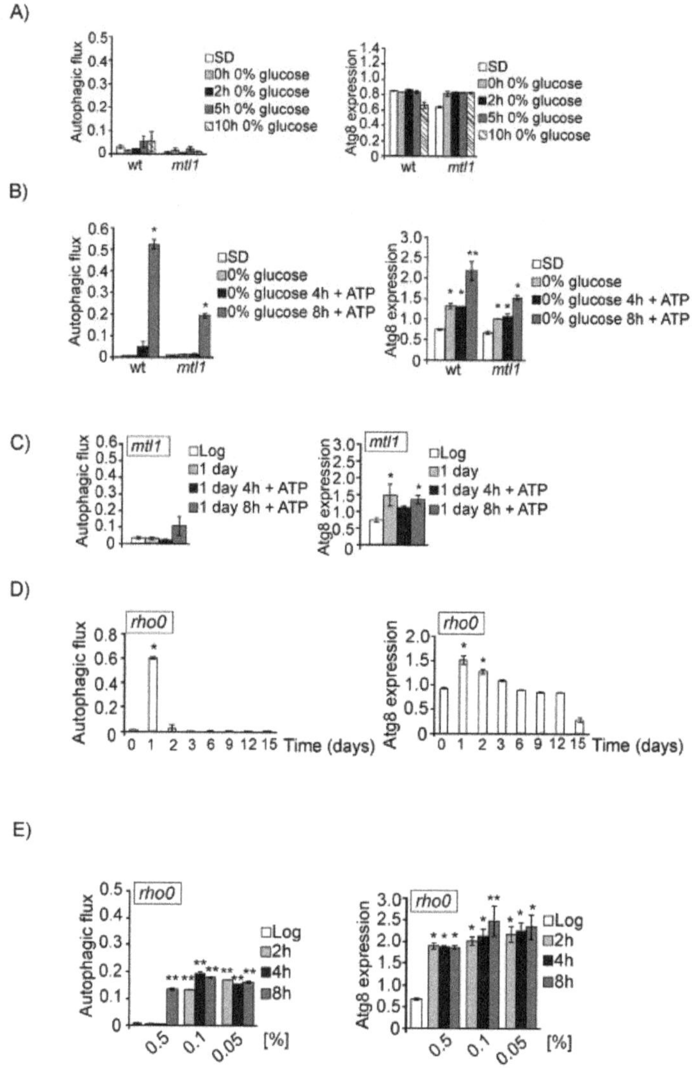

Figure 5. Mtl1 signals a decrease in glucose concentration to the autophagy machinery in a manner not fully dependent on ATP production by mitochondria. (**A**) wt and *mtl1* cells expressing GFP–Atg8 grown in SD media were transferred to minimum media devoid of glucose (0% glucose) to determine autophagic flux and Atg8 expression as in Figure 1B. (**B**) Upon transference to minimum media without glucose, ATP (at 200 mM final concentration) was added to the cultures described in (**A**) and samples were collected at the indicated times for determination of autophagy as in (**A**). (**C**) ATP (200 mM) was added to *mtl1* cultures growing in SD minimum medium for one day at the diauxic shift, and samples were collected at 4 and 8 h for autophagic flux and Atg8 expression determination, as in Figure 1B. (**D**) A *rho0* mutant expressing GFP–Atg8 was grown (as in Figure 1B) for autophagy determination. (**E**) *rho0* cells exponentially growing in SD were washed and transferred to several media containing different glucose concentrations to analyze autophagy, as in Figure 4A. Error bars in the histograms represent the standard deviation (SD) calculated from three independent experiments. Significance of the data was determinate by p-values from a Student's unpaired t-test denoted as follows: * = $0.05 > p > 0.01$; ** = $0.01 > p > 0.001$.

3.4. Both Ras2 or Sch9 Suppress mtl1 Deficiency in Bulk Autophagy Activation in All Metabolic Conditions That Imply Reduced Glucose Levels

We next tried to identify the pathway or pathways with which Mtl1 is connected to the signaling process converging on the autophagy machinery. In a previous study, we found that Mtl1 is negatively related to both Tor1 and Ras2 in response to oxidative stress and glucose starvation [53]. In addition, Mtl1 is also negatively related to Sch9, Slt2, and PKA during the diauxic shift or upon glucose depletion [30].

We demonstrated that Mtl1 mediates Bcy1 activating phosphorylation through TORC1 downregulation thus leading to PKA activation [30]. Therefore, and taking into consideration the hypothesis that *mtl1* mutants could cause impairment of the glucose signal through PKA, we also analyzed the overexpression of Bcy1, the PKA inhibitor, in both wt and *mtl1* strains. Overexpression of Bcy1 prolonged autophagy induction in wt cultures for longer times (Figures 6A and S5A) as compared to the wt empty strain (Figures 1A and S1A). However, Bcy1 overexpression did not restore autophagy in the *mtl1* mutant (Figures 3E, 6A, S2E and S5A). In previous papers, we observed clear impairment of Slt2 phosphorylation upon both oxidative stress and glucose deprivation in *mtl1* mutants [30]. Consistent with this information, we used a plasmid overexpressing the Pkc1 protein and a second plasmid bearing the *BCK1-20* allele which keeps Slt2 kinase constitutively activated, as Bck1 is the MAPKKK of the CWI pathway. Results depicted in Figure 3E, Figure 6A, Figures S2E and S5A demonstrate that the lack of bulk autophagy activation observed in the *mtl1* mutant, as a result of a decrease in glucose concentration during diauxic transition, is not caused by the lack of Slt2 kinase activity, since neither Pkc1 overexpression nor the *BCK1-20* allele restored the lack of autophagy induction in the *mtl1* mutant. We next decided to check Ras2 deletion in *mtl1*, since the GTPase is activated in the presence of glucose and is responsible for the synthesis of cAMP when glucose is the carbon source. Ras2 deletion in the *mtl1* mutant provoked the activation of bulk autophagy during the diauxic shift; in fact, the distribution levels of autophagy were equivalent between *ras2* and *mtl1ras2* (Figure 3B, Figure 6A, Figures S2B and S5A respectively) strains, suggesting that Mtl1 signals to Ras2 inactivation upon glucose starvation and diauxic transition signaling to activate bulk autophagy. Lastly, given the association between Mtl1 and Sch9 kinase [30], we also analyzed whether both proteins were also related to autophagy signaling. Deletion of *SCH9* restored autophagy in the *mtl1* mutant during the diauxic shift (Figures 6A and S5A). This result suggests a connection of Mtl1 with Sch9 towards autophagy. In order to corroborate the glucose specificity of these responses, exponentially grown cultures of *mtl1ras2* and *mtl1sch9* along with the corresponding controls were assayed for bulk autophagy response upon glucose starvation (Figures 6B and S5B). As expected, deletion of *RAS2* or *SCH9* suppressed the lack of autophagy induction in the absence of *MTL1*.

Snf1 is an AMPK family member which is highly conserved in eukaryotes. When glucose is exhausted at the beginning of the diauxic shift, Snf1 becomes activated to trigger a wide response of regulating activators and repressors to trigger respiratory metabolism (see review [21]). In order to detect a possible defect in Snf1 activation in the *mtl1* mutant, we analyzed Snf1 phosphorylation in Thr20 residues on the activation loop of the catalytic subunit in samples of wt, *mtl1, ras2, ras2mtl1, sch9* and *mtl1sch9*. In Figure 6C, it can be observed that Snf1 is correctly and similarly phosphorylated in all strains, concluding that *mtl1* defects in autophagy during the transition to stationary phase and upon glucose depletion are not a principal consequence of the lack of Snf1 activation as a response to glucose limitation.

In summary, our results suggest that either *RAS2* or *SCH9* deletion reverted the lack of autophagy in the *mtl1* mutant, suggesting that Mtl1 receives the signal of decreased glucose concentration and connects to both Ras2 and Sch9 inactivation, a mechanism that converges on macroautophagy induction. We have also observed similar results when the carbon source is either sucrose or fructose (Figure S2F).

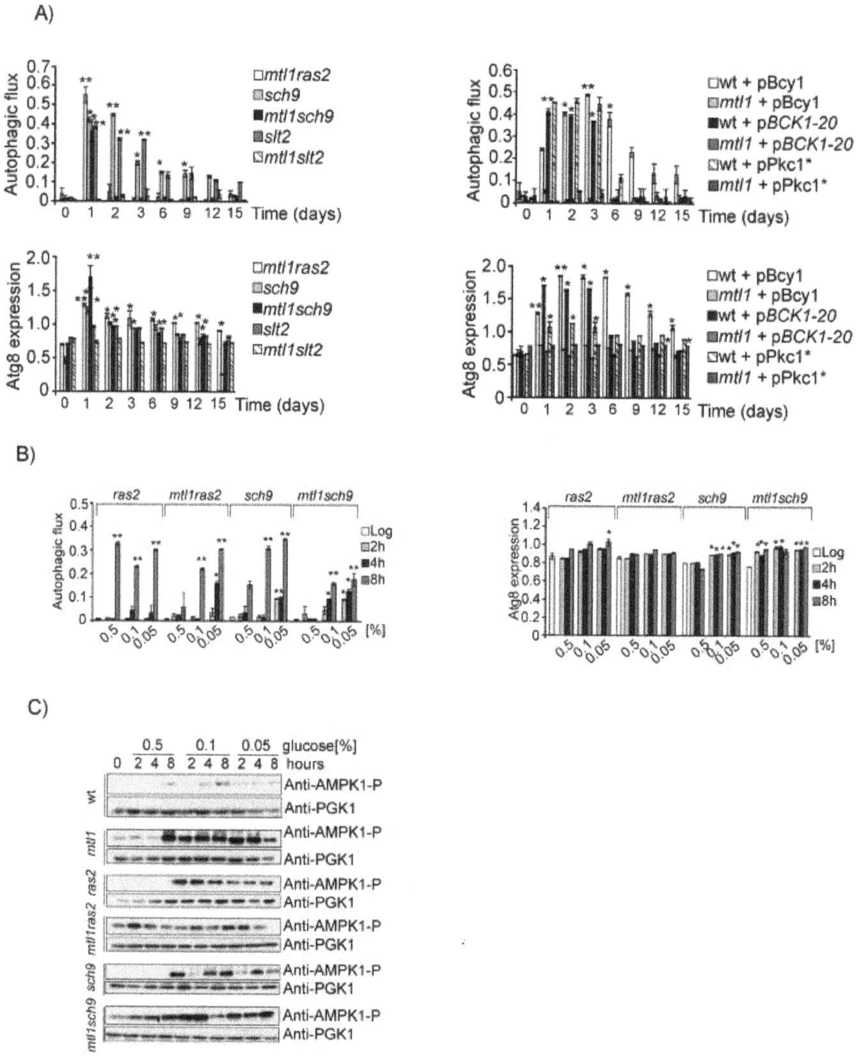

Figure 6. Both Ras2 and Sch9 suppress *mtl1* deficiency in autophagy signaling upon glucose concentration decreasing. (**A**) *mtl1ras2, sch9, mtl1sch9, slt2, mtl1slt2,* wt+pBcy1, *mtl1*+pBcy1, wt+p*BCK1-20*, *mtl1*+p*BCK1-20*, wt+pPkc1* and *mtl1*+pPkc1* strains expressing GFP–Atg8 were grown at 30 °C in SD media for 15 days. Samples were taken to determine autophagic flux and total Atg8 expression as described in Figure 1B. (**B**) Strains *ras2, mtl1ras2, sch9,* and *mtl1sch9* expressing GFP–Atg8 were exponentially grown in SD media to be subsequently transferred to minimum medium containing the indicated concentration of glucose. Samples were taken to determine autophagy as in Figure 4A. (**C**) wt samples (from Figure 4A) and *mtl1* samples (from Figure 4C) along with *mtl1, ras2, mtl1ras2, sch9* and *mtl1sch9* samples (from Figure 6B) were used for Western blot analysis and AMPK1 detection by using anti-AMPK1-P antibody. Error bars in the histograms represent the standard deviation (SD) calculated from three independent experiments. Significance of the data was determinate by p-values from a Student's unpaired t-test denoted as follows: * = $0.05 > p > 0.01$; ** = $0.01 > p > 0.001$.

3.5. Mtl1 Is Required for Mitochondrial Degradation Dependent on Atg33 and Independent of Atg32 during Chronological Ageing

We wondered whether Mtl1 involvement in autophagy regulation would be related to any carbon source, not only to glucose availability. To answer this question, we decided to analyze a non-fermentative carbon source, glycerol, that forces cells to directly enter into respiratory metabolism. Mtl1 was clearly not involved in the detection of glycerol concentration linked to autophagy activity, since in both wt and *mtl1* cells we detected similar levels and patterns of autophagy (Figures 7A and S6A). In a previous paper, we described that Mtl1 presented uncoupled respiration that provoked mitochondrial dysfunction and ROS accumulation [30]. In order to ascertain whether oxidative stress would be the cause of the autophagy problem, we added the antioxidant NAC to both wt and *mtl1* diauxic cultures (Figures 7B and S6B). In order to demonstrate that NAC was exerting its antioxidant function, samples were collected and stained with dihydroethidium (DHE) for in vivo visualization of cellular oxidation in the fluorescent microscope (Figure S6C). Our results indicate that oxidative stress is not the cause of autophagy impairment during diauxic shift in the *mtl1* mutant (Figures 7B and S6B,C).

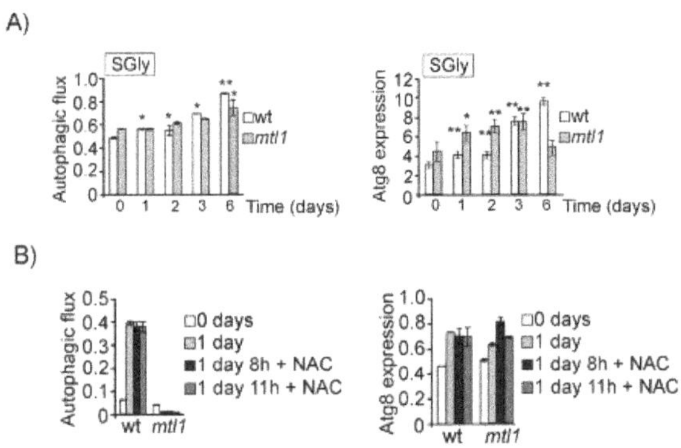

Figure 7. Mtl1 is not deficient in bulk autophagy in respiratory conditions. (**A**) wt and *mtl1* cultures were grown in minimum medium SGly (containing glycerol as unique carbon source) plus amino acids at 30 °C to stationary phase for 6 days. Samples were collected at the indicated times to identify macroautophagy as described in Figure 1B. (**B**) wt and *mtl1* cultures in SD medium growing to 1 day were treated with N-Acetyl cysteine (NAC) 5 mM for 8 and 11 h. Samples were collected for autophagy determinations as in (**A**). Error bars in the histograms represent the standard deviation (SD) calculated from three independent experiments. Significance of the data was determinate by p-values from a Student's unpaired t-test denoted as follows: * = $0.05 > p > 0.01$; ** = $0.01 > p > 0.001$.

There is an alternative possibility, that if the problem of *mtl1* is mitochondrial function, we would expect to detect severe deficiencies in mitophagy. For this purpose, we analyzed mitophagy in cells expressing a fusion of mitochondrial matrix protein to GFP, Idp1–GFP, transformed in either wt or *mtl1* strains [68]. We monitored the vacuole clipping of the fusion protein Idp1–GFP. The identification of free GFP with anti-GFP antibody in a Western blot would reveal the existence of mitophagy, since GFP is very resistant to degradation [69]. Mitophagy studies are usually carried out in respiratory carbon sources or alternatively in stationary cultures. When wt and *mtl1* cells were grown in SGly to the stationary phase, we observed mitophagy in both strains (Figures 8A and S7A) as opposed to *atg32* and *atg11* mutants, in which mitophagy was undetectable (Figures 8B and S7B). Atg32 is a mitochondrial outer protein required to initiate mitophagy as a selective type of autophagy (see review [70]). Atg11 is a critical protein for selective autophagy; it is essential in selective and non-selective autophagy processes (see review [71]). From the results shown in Figures 8B and S7B, we conclude that the *mtl1* mutant does not have any defects regarding mitophagy dependent on Atg32.

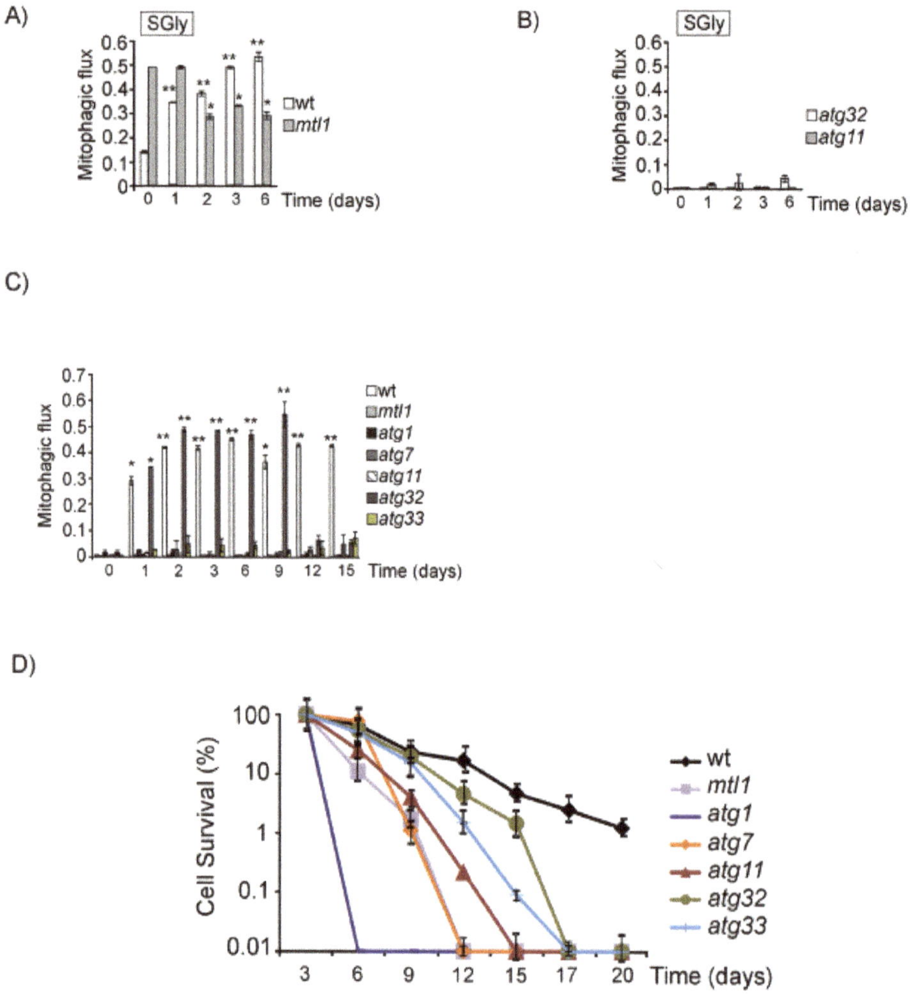

Strains	Maximum lifespan	SD	Average lifespan
wt	17.6	± 0.41	8.2
mtl1	9.9	± 0.21	4.5
atg1	5.2	± 0.63	1.6
atg7	6.8	± 0.14	2.3
atg11	10.2	± 0.70	5.3
atg32	12.5	± 0.53	7.2
atg33	11.4	± 0.17	6.0

Figure 8. *Cont.*

E)

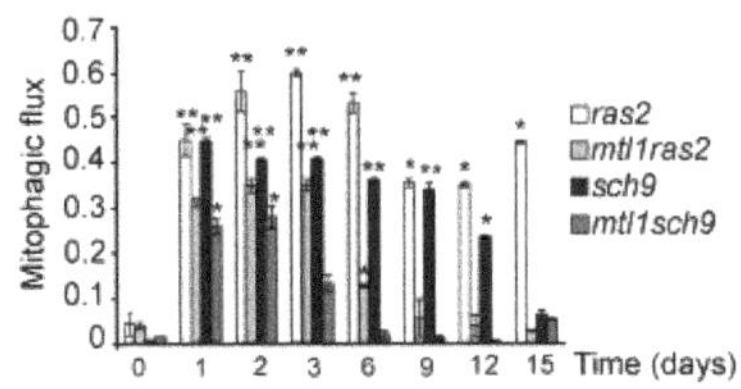

Figure 8. Mtl1 is needed for specific mitochondrial degradation during stationary phase. (**A**) wt and *mtl1* cultures transformed with the plasmid Idp1–GFP were grown in SGly media plus amino acids at 30 °C. Samples were taken at the indicated times to determine mitophagic flux. Mitophagic flux was calculated as the ratio between free GFP and total Idp1–GFP detected in the Western blot. (**B**) *atg32* and *atg11* cultures transformed with Idp1–GFP were grown in SGly medium plus amino acids at 30 °C. Samples were collected at the indicated times to determine mitophagic flux as in (**A**). (**C**) wt, *mtl1, atg1, atg7, atg11, atg32* and *atg33* strains bearing the plasmid Idp1–GFP were grown in SD media at 30 °C for 15 days in continuous shacking. Mitophagic flux was determined as in A. (**D**) Chronological life span curves for wt, *mtl1, atg1, atg7, atg11, atg32* and *atg33* strains cultured in SD media plus amino acids at 30 °C. Samples were taken at the indicated times to determine CLS, as described in the Materials and Methods. Numerical data regarding maximum life span (the day when cultures reach 10% survival) and average life span (the day at which 50% survival was recorded) for each strain is depicted. (**E**) *ras2, mtl1ras2, sch9* and *mtl1sch9* mutants, transformed with Idp1–GFP were treated as in (**C**) and mitophagic flux was determined as in (**A**). Error bars in the histograms represent the standard deviation (SD) calculated from three independent experiments. Significance of the data was determinate by *p*-values from a Student's unpaired *t*-test denoted as follows: * = $0.05 > p > 0.01$; ** = $0.01 > p > 0.001$.

We also checked mitophagy during the diauxic shift and stationary phase in cultures grown in SD with glucose as the only carbon source. We observed mitochondrial degradation as free GFP derived from Idp1–GFP accumulated in vacuoles in wt cultures (Figures 8C and S7C). However, this particular mitochondrial degradation was undetected in each of the *atg1, atg7* or *atg11* strains (Figures 8C and S7C). This particular mitophagy-like activity was not dependent on Atg32, since we observed similar results in both wt and *atg32* strains (Figures 8C and S7C). We discarded the possibility that our results reflected bulk autophagy, since *atg11* cultures did not demonstrate defects in bulk autophagy during the diauxic shift nor stationary phase (Figures 8C and S7C). It has been reported in yeast that Atg33 is a mitophagy mitochondrial outer membrane protein [68] required for the stationary phase. We observed a deficiency in mitophagy when we analyzed the *atg33* mutant (Figures 8C and S7C). More interesting was the finding that the *mtl1* mutant was as deficient as *atg11* and *atg33* in Idp1 mitophagy during the diauxic shift and stationary phase (Figures 8C and S7C). The three mutants turned out to have shorter chronological life spans than the corresponding wt (Figure 8D). Our results suggest that yeast cultures in SD demonstrate mitochondrial degradation in the vacuole upon the diauxic shift and during the stationary phase through a selective autophagy process independent of Atg32 but dependent on the Atg1, Atg7, Atg11 and Atg33 proteins. We also demonstrate that Mtl1 plays a relevant role in initiating this mechanism; one potential target would be Atg33 that will have to be investigated further.

We decided to check whether the absence of mitochondrial degradation during the chronological life span of *mtl1* was also alleviated by either *RAS2* or *SCH9* deletion, and obtained equivalent results to those described for bulk autophagy, inactivation of *RAS2* or *SCH9* restored mitophagy-like degradation during the diauxic shift to *mtl1* mutants (Figures 8C,E and S7C,D).

4. Discussion

Our results point to a situation by which gradual glucose depletion activates bulk autophagy, and for this response Mtl1 activity is essential. Unlike in wt cells, in the single mutant *mtl1* there was no a detectable response towards autophagy unless either *RAS2* or *SCH9* were deleted. Deletion of *RAS2* reverted the *mtl1* phenotype regarding bulk autophagy, which points to the importance of glucose availability and the switch from respiratory to fermentative metabolism, suggesting that during that transition the Ras2 pathway must be not active and Mtl1 is the connector between glucose and Ras2 activity. This is supported by the observation that these results also extended to other fermentative sugars (Figure S2F). Interestingly, the signal to autophagy in the models of glucose deprivation did not flow to TORC1, nor to the PKC1 pathway or PKA, but directly to the autophagy machinery to phosphorylate the Atg1 protein. Accordingly, some authors [2] already observed that TORC1 does not seem to play a principal function in glucose starvation.

According to former studies [67], the abrupt transition from 2% glucose to 0% glucose does not activate macroautophagy, because for this mechanism, ATP is essential and the mentioned transition causes cells to be suddenly exhausted for ATP. This is understandable, as cells are transferred from a culture containing high glucose concentrations to a culture without glucose nor other carbon source. However, during the transition from fermentative to respiratory metabolism, the decrease in glucose concentration occurs gradually. In wt cells, autophagy was activated already when the glucose concentration reached a value of 0.5% (27.77 mM), and reached maximum values when glucose levels decreased to 0.05% (0.13 mM), whereas in *mtl1* mutant autophagy was never induced. We hypothesized that this could have occurred because there might be a threshold for cells to sense glucose levels (or other alternative fermentable sugars), which would activate autophagy to obtain energy and nutrients probably linked to the induction of respiratory metabolism. Mtl1 could be the sensor for this threshold, and if unable to switch properly could consequently be unable to induce autophagy. In a previous study we observed that the *mtl1* mutant accumulates higher cAMP levels than wt cells. We believe that since the *mtl1* mutant has high cyclic AMP accumulation both in exponential or stationary cultures [28], cells would be depleted of ATP, with consequent Ras2 deletion potentially compensating for that depletion, thus avoiding the accumulation of cyclic AMP in *mtl1*. Nonetheless, this hypothesis was not sustained, since other nutrient stresses were capable of activating bulk autophagy in *mtl1* exponential cells (Figure 4C). It could be argued that during the diauxic shift, glucose reduction forces the switch from fermentative to respiratory metabolism and in these circumstances the main ATP source would be mitochondrial. Since in the absence of Mtl1 the ratio of cAMP/ATP would be higher than in wt cells, this would generate a signal of glucose starvation leading to the blockade of autophagy. However, our results demonstrate that the absence of mitochondrial DNA, and consequently the absence of ATP production through respiration metabolism (*rho0* mutant) did not preclude autophagy induction in glucose reduction conditions. Consequently, we also discarded this second hypothesis.

In humans, ATP increases autophagic flux in some cell lines but not in others [72]. Moreover, the addition of ATP to *mtl1* cultures only partly restored at a low degree of autophagy both in the diauxic transition and upon partial glucose depletion. This supports the model by which Mtl1 function couples glucose starvation to Ras activity and autophagy induction.

In agreement with the studies of [67], we observed that after full reduction of glucose, no nutritional stress could provoke bulk autophagy induction (rapamycin, amino acids, nitrogen, or iron).

In this study, we also present a nutritional model by which Gcn2 detects the signal of amino acid deprivation connecting to bulk autophagy in a manner independent of TORC1 activity. Hence, our data demonstrates that both Mtl1 and Gcn2 are key factors that induce bulk autophagy during the starvation process involving firstly glucose and secondly amino acid deprivation that occurs during the transition from fermentative to respiratory metabolism.

The observation that the *mtl1* mutant grows in the presence of glycerol as the only carbon source at similar rates to wt cells, indicates that mitochondrial function is sufficient to support the ATP requirements in this condition. Moreover, we also observed that Mtl1 did not participate in the induction of mitophagy in glycerol cultures dependent on Atg32 (Figure 8A,B).

We observed that either Mtl1 or Atg11 (involved in selective autophagy), are required to degrade mitochondria during transition to the stationary phase in synthetic media containing glucose as the only carbon source. This degradation process depends on the autophagy machinery, since in the absence of Atg7 or Atg1 it does not take place. Our data are consistent with a previous paper in which the authors showed that autophagy in response to carbon starvation requires Atg11 as a scaffold protein for the PAS [67]. Nevertheless, the mitophagy-like process that occurred in media containing glucose is independent of Atg32 but dependent on Atg33 (Figure 8C). This is not unexpected, since Atg33 was characterized as an autophagy protein specific to *Saccharomyces cerevisiae* whose role was linked to the induction of mitophagy during the stationary phase [68]. Whether or not the mitochondrial degradation that we observed during the stationary phase is a type of mitophagy dependent of Atg11 and Atg33 but independent of Atg32, should be further analyzed in future studies. Consequently, the availability of glucose as a carbon source has specific responses regarding autophagy in which Mtl1 is directly involved.

Snf1 is an AMPK orthologous to the mammalian AMP-kinase (reviewed in [73]) whose activity has been reported to be required in response to glucose starvation [21,74] to downregulate autophagy in the stationary phase [23], and is also required to induce autophagy in that context [67]. Snf1p is a catabolic regulator that is activated by an increase in the ADP/ATP ratio [75]. Nonetheless, in our studies Snf1, both in *mtl1* mutant or in wt cells, is highly activated during the diauxic shift when glucose levels are reduced and this activity is high during the stationary phase, therefore we cannot attribute lack of Snf1 activity to the defects observed in the *mtl1* mutant.

The conclusion to our data is that the transition from high to low concentrations of glucose triggers the connection between Mtl1 and autophagy and is not fully dependent on mitochondrial function and ATP accumulation. Another argument to support our hypothesis is that in the *mtl1* mutant, the presence of PAS is clearly detectable in the diauxic transition and in the presence of low glucose concentrations (Figure 3F), as opposed to that observed by previous authors [67] in the absence of ATP.

The linkage between Mtl1 and Ras2 has been previously described [28]. Our data suggest that Ras2 is the key regulator of bulk autophagy and the autophagy of mitochondria during the stationary phase. Whether or not this connection is related to mitochondria is at this moment unknown, since direct evidence for a regulation of mitochondria by Ras via cAMP–PKA is absent. Moreover, it is not unlikely that the Ras protein acts independently of adenylate cyclase and cAMP according to [76]. PKA is one of the effectors of Ras2 [77,78], and is also involved in autophagy regulation. However, concerning Mtl1 signaling in autophagy, PKA does not appear to be required as an intermediary molecule. Moreover, constitutive activation of the RAS-cAMP signaling pathway confers resistance to rapamycin [79–81]. This would explain why in an *mtl1* mutant, rapamycin does not provoke the induction of autophagy in the diauxic shift (not shown), taking into consideration the hypothesis that Mtl1 helps in the switch from fermentative to respiratory metabolism through the Ras/cAMP pathway.

Sch9 is a kinase effector of the TORC1 pathway [82]. In addition, Sch9 has been described as acting in a different pathway than PKA in glucose response. This is in agreement with our observations that Bcy1 overexpression or *SLT2* deletion did not suppress the lack of autophagy observed in the *mtl1* mutant. Our results are in line with the conclusion that TORC1, PKA, and Sch9 independently regulate autophagy during growth [17,83]. Deletion of Sch9 is sufficient to activate autophagy [17]. Here, we observed that the *sch9* mutant suppressed the lack of both bulk autophagy and mitophagy of *mtl1* during the diauxic shift and in conditions of low glucose, restricting the signal of glucose availability. Sch9 is also implicated in the selective autophagic degradation of ribosomes, mitochondria, and peroxisomes [84,85]. Although we still do not have a certain interpretation of the fact that

Sch9 regulates ribosomal gene transcription, and that ribosomal biogenesis is one of the mechanisms requiring more ATP consumption, one interpretation would be that diminishing the level of ribosomal biogenesis also diminishes ATP consumption, perhaps favoring the induction of autophagy. The molecular mechanisms underlying the involvement of cAMP in the induction of autophagy related to nutritional starvation conditions deserves future investigation.

Our findings suggest that the Mtl1 cell wall receptor of the CWI pathway is a glucose sensor required to activate both bulk autophagy and Atg33-Atg11 mitophagy in response to glucose concentration decreases. Activation occurs through either Ras2 or Sch9 inactivation converging on Atg1 phosphorylation.

Supplementary Materials: The following are available online at https://www.mdpi.com/article/10.3390/jof7110903/s1, Figure S1. Sequential descent of glucose and amino acids activates bulk autophagy during the diauxic shift in *Saccharomyces cerevisiae*. Figure S2. Mtl1 and Gcn2 control autophagy induction during glucose and amino acid starvation. Figure S3. Mtl1 signals glucose limitation to the autophagy machinery. Figure S4. Mtl1 signals decreases in glucose concentration to the autophagy machinery in a manner not fully dependent on ATP production by mitochondria. Figure S5. Both Ras2 and Sch9 suppress *mtl1* deficiency in autophagy signaling upon glucose concentration descent. Figure S6. Mtl1 is not deficient in bulk autophagy in respiratory conditions. Figure S7. Mtl1 is needed for specific mitochondrial degradation during stationary phase.

Author Contributions: Conceptualization, M.A.d.l.T.-R.; Methodology, M.A.d.l.T.-R., N.P.-C., S.M.-M.; Software, M.A.d.l.T.-R., N.P.-C., S.M.-M.; Validation M.A.d.l.T.-R., N.P.-C., S.M.-M.; Formal Analysis N.P.-C., S.M.-M., M.A.d.l.T.-R.; Investigation, M.A.d.l.T.-R., N.P.-C., S.M.-M.; Resources, M.A.d.l.T.-R.; Data Curation, S.M.-M., N.P.-C., M.A.d.l.T.-R.; Writing-Original Draft Preparation, M.A.d.l.T.-R.; Writing-Review & Editing, S.M.-M., N.P.-C., M.A.d.l.T.-R.; Visualization, M.A.d.l.T.-R., N.P.-C., S.M.-M.; Supervision, M.A.d.l.T.-R.; Project Administration, M.A.d.l.T.-R.; Funding Acquisition, M.A.d.l.T.-R. All authors have read and agreed to the published version of the manuscript.

Funding: This research was funded by Plan Nacional de I+D+I of the Spanish Ministry of Economy, Industry and Competitiveness (BIO2017-87828-C2-2-P).

Institutional Review Board Statement: Not applicable.

Informed Consent Statement: Not applicable.

Data Availability Statement: Not applicable.

Acknowledgments: We want to thank D. Abeliovitch (The Hebrew University of Jerusalem Cell Biology, Freiburg, Germany) for kindly sending us the plasmids pGFP–Atg8 and pIdp1–GFP. We want to acknowledge Inmaculada Montoliu and Roser Pané for their technical support. The research described in this publication was partly supported by the Plan Nacional de I+D+I of the Spanish Ministry of Economy, Industry and Competitiveness (BIO2017-87828-C2-2-P). Sandra Montella is funded by a fellowship from the Catalan Government (Spain).

Conflicts of Interest: The authors declare no conflict of interest.

References

1. de Virgilio, C. The essence of yeast quiescence. *FEMS Microbiol. Rev.* **2012**, *36*, 306–339. [CrossRef]
2. Hughes Hallett, J.E.; Luo, X.; Capaldi, A.P. State transitions in the TORC1 signaling pathway and information processing in *Saccharomyces cerevisiae*. *Genetics* **2014**, *198*, 773–786. [CrossRef] [PubMed]
3. Picazo, C.; Orozco, H.; Matallana, E.; Aranda, A. Interplay among Gcn5, Sch9 and mitochondria during chronological aging of wine yeast is dependent on growth conditions. *PLoS ONE* **2015**, *10*, e0117267. [CrossRef] [PubMed]
4. Santos, J.; Leão, C.; Sousa, M.J. Ammonium-dependent shortening of CLS in yeast cells starved for essential amino acids is determined by the specific amino acid deprived, through different signaling pathways. *Oxid. Med. Cell. Longev.* **2013**, *2013*, 161986. [CrossRef] [PubMed]
5. Liu, Z.; Wu, J.; Huang, D. New arahypins isolated from fungal-challenged peanut seeds and their glucose uptake-stimulatory activity in 3T3-L1 adipocytes. *Phytochem. Lett.* **2013**, *6*, 123–127. [CrossRef]
6. Fabrizio, P.; Pozza, F.; Pletcher, S.D.; Gendron, C.M.; Longo, V.D. Regulation of longevity and stress resistance by Sch9 in yeast. *Science* **2001**, *292*, 288–290. [CrossRef] [PubMed]
7. Longo, V.D. The Ras and Sch9 pathways regulate stress resistance and longevity. *Exp. Gerontol.* **2003**, *38*, 807–811. [CrossRef]

8. Kaeberlein, M.; Powers, R.W.; Steffen, K.K.; Westman, E.A.; Hu, D.; Dang, N.; Kerr, E.O.; Kirkland, K.T.; Fields, S.; Kennedy, B.K.; et al. Regulation of Yeast Replicative Life Span by TOR and Sch9 in Response to Nutrients. *Science* **2005**, *310*, 1193–1196. [CrossRef] [PubMed]
9. Yan, G.; Lai, Y.; Jiang, Y. The TOR Complex 1 Is a Direct Target of Rho1 GTPase. *Mol. Cell* **2012**, *45*, 743–753. [CrossRef] [PubMed]
10. Thomas, J.D.; Zhang, Y.J.; Wei, Y.H.; Cho, J.H.; Morris, L.E.; Wang, H.Y.; Zheng, X.S. Rab1A Is an mTORC1 Activator and a Colorectal Oncogene. *Cancer Cell* **2014**, *26*, 754–769. [CrossRef] [PubMed]
11. Yuan, W.; Guo, S.; Gao, J.; Zhong, M.; Yan, G.; Wu, W.; Chao, Y.; Jiang, Y. General control nonderepressible 2 (GCN2) kinase inhibits target of rapamycin complex 1 in response to amino acid starvation in *Saccharomyces cerevisiae*. *J. Biol. Chem.* **2017**, *292*, 2660–2669. [CrossRef]
12. Postnikoff, S.D.L.; Johnson, J.E.; Tyler, J.K. The integrated stress response in budding yeast lifespan extension. *Microb. Cell* **2017**, *4*, 368–375. [CrossRef] [PubMed]
13. Stephan, J.S.; Yeh, Y.-Y.; Ramachandran, V.; Deminoff, S.J.; Herman, P.K. The Tor and PKA signaling pathways independently target the Atg1/Atg13 protein kinase complex to control autophagy. *Proc. Natl. Acad. Sci. USA* **2009**, *106*, 17049–17054. [CrossRef] [PubMed]
14. Kamada, Y.; Funakoshi, T.; Shintani, T.; Nagano, K.; Ohsumi, M.; Ohsumi, Y. Tor-mediated induction of autophagy via an Apg1 protein kinase complex. *J. Cell Biol.* **2000**, *150*, 1507–1513. [CrossRef]
15. Kim, Y.C.; Guan, K. mTOR: A pharmacologic target for autophagy regulation. *J. Clin. Investig.* **2015**, *2*, 25–32. [CrossRef] [PubMed]
16. Budovskaya, Y.V.; Stephan, J.S.; Reggiori, F.; Klionsky, D.J.; Herman, P.K. The Ras/cAMP-dependent Protein Kinase Signaling Pathway Regulates an Early Step of the Autophagy Process in *Saccharomyces cerevisiae*. *J. Biol. Chem.* **2004**, *279*, 20663–20671. [CrossRef]
17. Yorimitsu, T.; Zaman, S.; Broach, J.R.; Klionsky, D.J. Protein Kinase A and Sch9 Cooperatively Regulate Induction of Autophagy in *Saccharomyces cerevisiae*. *Mol. Biol. Cell* **2007**, *18*, 4180–4189. [CrossRef] [PubMed]
18. Conrad, M.; Schothorst, J.; Kankipati, H.N.; Van Zeebroeck, G.; Rubio-Texeira, M.; Thevelein, J.M. Nutrient sensing and signaling in the yeast *Saccharomyces cerevisiae*. *FEMS Microbiol. Rev.* **2014**, *38*, 254–299. [CrossRef]
19. Cazzanelli, G.; Pereira, F.; Alves, S.; Francisco, R.; Azevedo, L.; Dias Carvalho, P.; Almeida, A.; Côte-Real, M.; Oliveira, M.J.; Lucas, C.; et al. The Yeast *Saccharomyces cerevisiae* as a Model for Understanding RAS Proteins and their Role in Human Tumorigenesis. *Cells* **2018**, *7*, 14. [CrossRef]
20. Celenza, J.L.; Carlson, M. A yeast gene that is essential for release from glucose repression encodes a protein kinase. *Science* **1986**, *233*, 1175–1180. [CrossRef] [PubMed]
21. Shashkova, S.; Welkenhuysen, N.; Hohmann, S. Molecular communication: Crosstalk between the Snf1 and other signaling pathways. *FEMS Yeast Res.* **2015**, *15*, fov026. [CrossRef]
22. Wang, Z.; Wilson, W.; Fujino, M.; Roach, P.J. Antagonistic controls of autophagy and glycogen accumulation by Snf1p, the yeast homolog of AMP-activated protein kinase, and the cyclin-dependent kinase Pho85p. *Mol. Cell. Biol.* **2001**, *21*, 5742–5752. [CrossRef] [PubMed]
23. Montella-Manuel, S.; Pujol-Carrion, N.; Mechoud, M.A.; de la Torre-Ruiz, M.A. Bulk autophagy induction and life extension is achieved when iron is the only limited nutrient in *Saccharomyces cerevisiae*. *Biochem. J.* **2021**, *478*, 811–837. [CrossRef]
24. Petkova, M.I.; Pujol-Carrion, N.; de la Torre-Ruiz, M.A. Mtl1 O-mannosylation mediated by both Pmt1 and Pmt2 is important for cell survival under oxidative conditions and TOR blockade. *Fungal Genet. Biol.* **2012**, *49*, 903–914. [CrossRef] [PubMed]
25. Kock, C.; Dufrêne, Y.F.; Heinisch, J.J. Up against the wall: Is yeast cell wall integrity ensured by mechanosensing in plasma membrane microdomains? *Appl. Environ. Microbiol.* **2015**, *81*, 806–811. [CrossRef] [PubMed]
26. Rajavel, M.; Philip, B.; Buehrer, B.M.; Errede, B.; Levin, D.E. Mid2 Is a Putative Sensor for Cell Integrity Signaling in. *Mol. Cell. Biol.* **1999**, *19*, 3969–3976. [CrossRef]
27. Vilella, F.; Herrero, E.; Torres, J.; De La Torre-Ruiz, M.A. Pkc1 and the upstream elements of the cell integrity pathway in *Saccharomyces cerevisiae*, Rom2 and Mtl1, are required for cellular responses to oxidative stress. *J. Biol. Chem.* **2005**, *280*, 9149–9159. [CrossRef]
28. Petkova, M.I.; Pujol-Carrion, N.; Arroyo, J.; García-Cantalejo, J.; De La Torre-Ruiz, M.A. Mtl1 is required to activate general stress response through TOR1 and RAS2 inhibition under conditions of glucose starvation and oxidative stress. *J. Biol. Chem.* **2010**, *285*, 19521–19531. [CrossRef] [PubMed]
29. Jin, C.; Parshin, A.V.; Daly, I.; Strich, R.; Cooper, K.F. The cell wall sensors Mtl1, Wsc1, and Mid2 are required for stress-induced nuclear to cytoplasmic translocation of cyclin C and programmed cell death in yeast. *Oxid. Med. Cell. Longev.* **2013**, *2013*, 320823. [CrossRef] [PubMed]
30. Sundaram, V.; Petkova, M.I.; Pujol-Carrion, N.; Boada, J.; de la Torre-Ruiz, M.A. Tor1, Sch9 and PKA downregulation in quiescence rely on Mtl1 to preserve mitochondrial integrity and cell survival. *Mol. Microbiol.* **2015**, *97*, 93–109. [CrossRef] [PubMed]
31. Levin, D.E. Regulation of cell wall biogenesis in *Saccharomyces cerevisiae*: The cell wall integrity signaling pathway. *Genetics* **2011**, *189*, 1145–1175. [CrossRef] [PubMed]
32. de la Torre-Ruiz, M.A.; Pujol, N.; Sundaram, V. Coping with oxidative stress. *Curr. Drug Targets* **2015**, *16*, 2–12. [CrossRef]
33. Staleva, L.; Hall, A.; Orlow, S.J. Oxidative Stress Activates FUS1 and RLM1 Transcription in the Yeast *Saccharomyces cerevisiae* in an Oxidant-dependent Manner. *Mol. Biol. Cell* **2004**, *15*, 5574–5582. [CrossRef]

34. Jendretzki, A.; Wittland, J.; Wilk, S.; Straede, A.; Heinisch, J.J. How do I begin? Sensing extracellular stress to maintain yeast cell wall integrity. *Eur. J. Cell Biol.* **2011**, *90*, 740–744. [CrossRef]
35. Yang, Z.; Klionsky, D.J. An overview of the molecular mechanism of autophagy. *Curr. Top. Microbiol. Immunol.* **2009**, *335*, 1–32. [CrossRef]
36. Kawamata, T.; Horie, T.; Matsunami, M.; Sasaki, M.; Ohsumi, Y. Zinc starvation induces autophagy in yeast. *J. Biol. Chem.* **2017**, *292*, 8520–8530. [CrossRef] [PubMed]
37. Yokota, H.; Gomi, K.; Shintani, T. Induction of autophagy by phosphate starvation in an Atg11-dependent manner in *Saccharomyces cerevisiae*. *Biochem. Biophys. Res. Commun.* **2017**, *483*, 522–527. [CrossRef]
38. Nakatogawa, H.; Suzuki, K.; Kamada, Y.; Ohsumi, Y. Dynamics and diversity in autophagy mechanisms: Lessons from yeast. *Nat. Rev. Mol. Cell Biol.* **2009**, *10*, 458–467. [CrossRef] [PubMed]
39. Yorimitsu, T.; Ke Wang, C.H.; Klionsky, D.J. Tap42-associated protein phosphatase type 2A negatively regulates induction of autophagy. *Autophagy* **2009**, *5*, 616–624. [CrossRef] [PubMed]
40. Fujioka, Y.; Suzuki, S.W.; Yamamoto, H.; Kondo-Kakuta, C.; Kimura, Y.; Hirano, H.; Akada, R.; Inagaki, F.; Ohsumi, Y.; Noda, N.N. Structural basis of starvation-induced assembly of the autophagy initiation complex. *Nat. Struct. Mol. Biol.* **2014**, *21*, 513–521. [CrossRef]
41. Yamamoto, H.; Fujioka, Y.; Suzuki, S.W.; Noshiro, D.; Suzuki, H.; Kondo-Kakuta, C.; Kimura, Y.; Hirano, H.; Ando, T.; Noda, N.N.; et al. The Intrinsically Disordered Protein Atg13 Mediates Supramolecular Assembly of Autophagy Initiation Complexes. *Dev. Cell* **2016**, *38*, 86–99. [CrossRef] [PubMed]
42. Ohsumi, Y. Historical landmarks of autophagy research. *Cell Res.* **2014**, *24*, 9–23. [CrossRef]
43. Feng, Y.; He, D.; Yao, Z.; Klionsky, D.J. The machinery of macroautophagy. *Cell Res.* **2014**, *24*, 24–41. [CrossRef]
44. Powers, R.W.; Kaeberlein, M.; Caldwell, S.D.; Kennedy, B.K.; Fields, S. Extension of chronological life span in yeast by decreased TOR pathway signaling. *Genes Dev.* **2006**, *20*, 174–184. [CrossRef]
45. Fabrizio, P.; Hoon, S.; Shamalnasab, M.; Galbani, A.; Wei, M.; Giaever, G.; Nislow, C.; Longo, V.D. Genome-wide screen in *Saccharomyces cerevisiae* identifies vacuolar protein sorting, autophagy, biosynthetic, and tRNA methylation genes involved in life span regulation. *PLoS Genet.* **2010**, *6*, e1001024. [CrossRef]
46. Gatica, D.; Lahiri, V.; Klionsky, D.J. Cargo Recognition and Degradation by Selective Autophagy. *Nat. Cell Biomol.* **2018**, *20*, 233–242. [CrossRef] [PubMed]
47. Mamaev, D.V.; Zvyagilskaya, R.A. Mitophagy in Yeast. *Biochemistry* **2019**, *84*, 225–232. [CrossRef] [PubMed]
48. Fukuda, T.; Kanki, T. Mechanisms and physiological roles of mitophagy in yeast. *Mol. Cells* **2018**, *41*, 35–44. [CrossRef] [PubMed]
49. Welter, E.; Montino, M.; Reinhold, R.; Schlotterhose, P.; Krick, R.; Dudek, J.; Rehling, P.; Thumm, M. Uth1 is a mitochondrial inner membrane protein dispensable for post-log-phase and rapamycin-induced mitophagy. *FEBS J.* **2013**, *280*, 4970–4982. [CrossRef]
50. Wach, A.; Brachat, A.; Pöhlmann, R.; Philippsen, P. New heterologous modules for classical or PCR-based gene disruptions in *Saccharomyces cerevisiae*. *Yeast* **1994**, *10*, 1793–1808. [CrossRef]
51. Pujol-Carrion, N.; Belli, G.; Herrero, E.; Nogues, A.; de la Torre-Ruiz, M.A. Glutaredoxins Grx3 and Grx4 regulate nuclear localisation of Aft1 and the oxidative stress response in *Saccharomyces cerevisiae*. *J. Cell Sci.* **2006**, *119*, 4554–4564. [CrossRef] [PubMed]
52. Gallego, C.; Eloi, G.; Colomina, N.; Enrique, H.; Aldea, M. The Cln3 cyclin is down-regulated by translational repression and degradation during the G1 arrest caused by nitrogen deprivation in budding yeast. *EMBO J.* **1997**, *16*, 7196–7206. [CrossRef] [PubMed]
53. Petkova, M.I.; Pujol-Carrion, N.; de la Torre-Ruiz, M.A. Signal flow between CWI/TOR and CWI/RAS in budding yeast under conditions of oxidative stress and glucose starvation. *Commun. Integr. Biol.* **2010**, *3*, 555–557. [CrossRef] [PubMed]
54. Guedes, A.; Ludovico, P.; Sampaio-Marques, B. Caloric restriction alleviates alpha-synuclein toxicity in aged yeast cells by controlling the opposite roles of Tor1 and Sir2 on autophagy. *Mech. Ageing Dev.* **2016**, *161*, 270–276. [CrossRef] [PubMed]
55. Ecker, N.; Mor, A.; Journo, D.; Abeliovich, H.; Ecker, N.; Mor, A. Deprivation is distinct from nitrogen starvation- induced macroautophagy Induction of autophagic flux by amino acid deprivation is distinct from nitrogen starvation-induced macroautophagy. *Autophagy* **2010**, *6*, 870–890. [CrossRef] [PubMed]
56. Görner, W.; Durchschlag, E.; Martinez-Pastor, M.T.; Estruch, F.; Ammerer, G.; Hamilton, B.; Ruis, H.; Schüller, C. Nuclear localization of the C2H2 zinc finger protein Msn2p is regulated by stress and protein kinase A activity. *Genes Dev.* **1998**, *12*, 586–597. [CrossRef]
57. Komeili, A.; Wedaman, K.P.; O'Shea, E.K.; Powers, T. Mechanism of metabolic control: Target of rapamycin signaling links nitrogen quality to the activity of the Rtg1 and Rtg3 transcription factors. *J. Cell Biol.* **2000**, *151*, 863–878. [CrossRef] [PubMed]
58. Hagai, A.; Zareic, M.; Kristoffer, T.G.R.; Richard, J.Y.; Joern, D. Involvement of mitochondrial dynamics in the segregation of mitochondrial matrix proteins during stationary phase mitophagy. *Physiol. Behav.* **2017**, *176*, 100–106. [CrossRef]
59. Kaiser, C.; Michaelis, S.; Mitchell, A. *Methods in Yeast Genetics*; Cold Spring Harbour Laboratory Press Ltd.: New York, NY, USA, 1994; pp. 107–121. ISBN 0-87969-451-3.
60. Pujol-Carrion, N.; Petkova, M.I.; Serrano, L.; de la Torre-Ruiz, M.A. The MAP kinase Slt2 is involved in vacuolar function and actin remodeling in *Saccharomyces cerevisiae* mutants affected by endogenous oxidative stress. *Appl. Environ. Microbiol.* **2013**, *79*, 6459–6471. [CrossRef] [PubMed]

61. Mechoud, M.A.; Pujol-Carrion, N.; Montella-Manuel, S.; De La Torre-Ruiz, M.A. Interactions of GMP with Human Glrx3 and with *Saccharomyces cerevisiae* Grx3 and Grx4 Converge in the Regulation of the Gcn2 Pathway. *Am. Soc. Microbiol.* **2020**, *86*, e00221-20. [CrossRef]
62. Shintani, T.; Klionsky, D.J. Cargo Proteins Facilitate the Formation of Transport Vesicles in the Cytoplasm to Vacuole Targeting Pathway. *J. Biol. Chem.* **2004**, *279*, 29889–29894. [CrossRef] [PubMed]
63. Noda, T.; Klionsky, D.J. The quantitative Pho8Delta60 assay of nonspecific autophagy. *Methods Enzymol.* **2008**, *6879*, 03203–03205. [CrossRef]
64. Hernandez-Lopez, M.; Prieto, J.; Randez-Gil, F. Osmotolerance and leavening ability in sweet dough. Comparative analysis between *Torulaspora delbrueckii* and *Saccharomyces cerevisiae* baker's yeast strains. *Antonie Leeuwenhoek* **2003**, *84*, 125–134. [CrossRef] [PubMed]
65. Dever, T.E.; Feng, L.; Wek, R.C.; Cigan, A.M.; Donahue, T.F.; Hinnebusch, A.G. Phosphorylation of initiation factor 2α by protein kinase GCN2 mediates gene-specific translational control of GCN4 in yeast. *Cell* **1992**, *68*, 585–596. [CrossRef]
66. Suzuki, K.; Kirisako, T.; Kamada, Y.; Mizushima, N.; Noda, T.; Ohsumi, Y. The pre-autophagosomal structure organized by concerted functions of APG genes is essential for autophagosome formation. *EMBO J.* **2001**, *20*, 5971–5981. [CrossRef] [PubMed]
67. Adachi, A.; Koizumi, M.; Ohsumi, Y. Autophagy induction under carbon starvation conditions is negatively regulated by carbon catabolite repression. *J. Biol. Chem.* **2017**, *292*, 19905–19918. [CrossRef]
68. Tomotake, K.; Daniel, J.K. Mitophagy in yeast occurs through a selective mechanism. *J. Biol. Chem.* **2008**, *283*, 32386–32393. [CrossRef]
69. Kolitsida, P.; Abeliovich, H. Methods for Studying Mitophagy in Yeast. *Methods Mol. Biol.* **2019**, *1880*, 669–678. [CrossRef] [PubMed]
70. Liu, Y.; Okamoto, K. Regulatory mechanisms of mitophagy in yeast. *Biochim. Biophys. Acta-Gen. Subj.* **2021**, *1865*, 129858. [CrossRef] [PubMed]
71. Zientara-Rytter, K.; Subramani, S. Mechanistic insights into the role of Atg11 in selecetive autophagy. *J. Mol. Biol.* **2020**, *432*, 104–122. [CrossRef]
72. Grisan, F.; Iannucci, L.F.; Surdo, N.C.; Gerbino, A.; Zanin, S.; Di Benedetto, G.; Pozzan, T.; Lefkimmiatis, K. PKA compartmentalization links cAMP signaling and autophagy. *Cell Death Differ.* **2021**, *28*, 2436–2449. [CrossRef] [PubMed]
73. Coccetti, P.; Nicastro, R.; Tripodi, F. Conventional and emerging roles of the energy sensor snf1/AMPK in *Saccharomyces cerevisiae. Microb. Cell* **2018**, *5*, 482–494. [CrossRef] [PubMed]
74. Hedbacker, K. SNF1/AMPK pathways in yeast. *Front. Biosci.* **2008**, *13*, 2408. [CrossRef] [PubMed]
75. Mayer, F.V.; Heath, R.; Underwood, E.; Sanders, M.J.; Carmena, D.; McCartney, R.R.; Leiper, F.C.; Xiao, B.; Jing, C.; Walker, P.A.; et al. ADP regulates SNF1, the *Saccharomyces cerevisiae* homolog of AMP-activated protein kinase. *Cell Metab.* **2011**, *14*, 707–714. [CrossRef]
76. Hlavatá, L.; Nyström, T. Ras Proteins Control Mitochondrial Biogenesis and Function in *Saccharomyces cerevisiae. Folia Microbiol.* **2003**, *48*, 725–730. [CrossRef] [PubMed]
77. Cebollero, E.; Reggiori, F. Regulation of autophagy in yeast *Saccharomyces cerevisiae. Biochim. Biophys. Acta Mol. Cell Res.* **2009**, *1793*, 1413–1421. [CrossRef] [PubMed]
78. Chen, Y.; Klionsky, D.J. The regulation of autophagy—Unanswered questions. *J. Cell Sci.* **2011**, *124*, 161–170. [CrossRef]
79. Pedruzzi, I.; Bürckert, N.; Egger, P.; De Virgilio, C. *Saccharomyces cerevisiae* Ras/cAMP pathway controls post-diauxic shift element-dependent transcription through the zinc finger protein Gis1. *EMBO J.* **2000**, *19*, 2569–2579. [CrossRef]
80. Pedruzzi, I.; Dubouloz, F.; Cameroni, E.; Wanke, V.; Roosen, J.; Winderickx, J.; De Virgilio, C. TOR and PKA Signaling Pathways Converge on the Protein Kinase Rim15 to Control Entry into G0. *Mol. Cell* **2003**, *12*, 1607–1613. [CrossRef]
81. Schmelzle, T.; Beck, T.; Martin, D.E.; Hall, M.N. Activation of the RAS/Cyclic AMP Pathway Suppresses a TOR Deficiency in Yeast. *Mol. Cell. Biol.* **2004**, *24*, 338–351. [CrossRef] [PubMed]
82. Urban, J.; Soulard, A.; Huber, A.; Lippman, S.; Mukhopadhyay, D.; Deloche, O.; Wanke, V.; Anrather, D.; Ammerer, G.; Riezman, H.; et al. Sch9 Is a Major Target of TORC1 in Saccharomyces cerevisiae. *Mol. Cell* **2007**, *26*, 663–674. [CrossRef] [PubMed]
83. Deprez, M.A.; Eskes, E.; Winderickx, J.; Wilms, T. The TORC1-Sch9 pathway as a crucial mediator of chronological lifespan in the yeast *Saccharomyces cerevisiae. FEMS Yeast Res.* **2018**, *18*, foy048. [CrossRef]
84. Lakhani, R.; Vogel, K.R.; Till, A.; Liu, J.; Burnett, S.F.; Gibson, K.M.; Subramani, S. Defects in GABA metabolism affect selective autophagy pathways and are alleviated by mTOR inhibition. *EMBO Mol. Med.* **2014**, *6*, 551–566. [CrossRef] [PubMed]
85. Waliullah, T.M.; Yeasmin, A.M.; Kaneko, A.; Koike, N.; Terasawa, M.; Totsuka, T.; Ushimaru, T. Rim15 and Sch9 kinases are involved in induction of autophagic degradation of ribosomes in budding yeast. *Biosci. Biotechnol. Biochem.* **2017**, *81*, 307–310. [CrossRef]

Journal of *Fungi*

MDPI

Article

Regulation of Pkc1 Hyper-Phosphorylation by Genotoxic Stress

Li Liu [1], Jiri Veis [2,3], Wolfgang Reiter [2,4], Edwin Motari [1,†], Catherine E. Costello [5], John C. Samuelson [1,6], Gustav Ammerer [2] and David E. Levin [1,6,*]

1 Department of Molecular and Cell Biology, Boston University Goldman School of Dental Medicine, Boston, MA 02118, USA; liul@bu.edu (L.L.); motari1@gmail.com (E.M.); jsamuels@bu.edu (J.C.S.)
2 Department of Biochemistry and Cell Biology, Max Perutz Labs, University of Vienna, 1030 Wien, Austria; jiri.veis@univie.ac.at (J.V.); reiterw6@univie.ac.at (W.R.); gustav.ammerer@univie.ac.at (G.A.)
3 Center for Medical Biochemistry, Max Perutz Laboratories, Medical University of Vienna, 1030 Wien, Austria
4 Mass Spectrometry Facility, Max Perutz Laboratories, University of Vienna, 1030 Wien, Austria
5 Department of Biochemistry, Center for Biomedical Mass Spectrometry, School of Medicine, Boston University, Boston, MA 02118, USA; cecmsms@bu.edu
6 Department of Microbiology, Boston University School of Medicine, Boston, MA 02118, USA
* Correspondence: delevin@bu.edu
† Current address: MilliporeSigma, Burlington, MA 01083, USA.

Abstract: The cell wall integrity (CWI) signaling pathway is best known for its roles in cell wall biogenesis. However, it is also thought to participate in the response to genotoxic stress. The stress-activated protein kinase Mpk1 (Slt2, is activated by DNA damaging agents through an intracellular mechanism that does not involve the activation of upstream components of the CWI pathway. Additional observations suggest that protein kinase C (Pkc1), the top kinase in the CWI signaling cascade, also has a role in the response to genotoxic stress that is independent of its recognized function in the activation of Mpk1. Pkc1 undergoes hyper-phosphorylation specifically in response to genotoxic stress; we have found that this requires the DNA damage checkpoint kinases Mec1 (Mitosis Entry Checkpoint) and Tel1 (TELomere maintenance), but not their effector kinases. We demonstrate that the casein kinase 1 (CK1) ortholog, Hrr25 (HO and Radiation Repair), previously implicated in the DNA damage transcriptional response, associates with Pkc1 under conditions of genotoxic stress. We also found that the induced association of Hrr25 with Pkc1 requires Mec1 and Tel1, and that Hrr25 catalytic activity is required for Pkc1-hyperphosphorylation, thereby delineating a pathway from the checkpoint kinases to Pkc1. We used SILAC mass spectrometry to identify three residues within Pkc1 the phosphorylation of which was stimulated by genotoxic stress. We mutated these residues as well as a collection of 13 phosphorylation sites within the regulatory domain of Pkc1 that fit the consensus for CK1 sites. Mutation of the 13 Pkc1 phosphorylation sites blocked hyper-phosphorylation and diminished *RNR3* (RiboNucleotide Reductase) basal expression and induction by genotoxic stress, suggesting that Pkc1 plays a role in the DNA damage transcriptional response.

Keywords: Hrr25; Mec1; Tel1; Pkc1; hydroxyurea; UV irradiation

Citation: Liu, L.; Veis, J.; Reiter, W.; Motari, E.; Costello, C.E.; Samuelson, J.C.; Ammerer, G.; Levin, D.E. Regulation of Pkc1 Hyper-Phosphorylation by Genotoxic Stress. *J. Fungi* **2021**, *7*, 874. https://doi.org/10.3390/jof7100874

Academic Editors: María Molina and Humberto Martín

Received: 11 September 2021
Accepted: 13 October 2021
Published: 17 October 2021

1. Introduction

The cell wall integrity (CWI) signaling pathway of the budding yeast *Saccharomyces cerevisiae* has been well characterized with regard to its regulation by cell wall stress [1–4]. This pathway regulates biosynthesis of cell wall polymers, organization of the actin cytoskeleton, exocytosis, and the protein kinase C 1 (Pkc1)-mediated stress-activated protein kinase (SAPK) cascade through activation of the small GTPase, Rho1. The SAPK cascade is a linear pathway comprised of Pkc1, a MEKK (Bck1), a pair of redundant MEKs (Mkk1/2) and a SAPK (Mpk1/Slt2). The activation of Mpk1, in response to cell wall stress or hyperactivation of upstream pathway components, drives transcription in support of cell wall biogenesis [5–9] through the SRF-like transcription factor Rlm1 [10,11] and the cell cycle transcription factor SBF [12–15]. Moreover, loss-of-function mutants in the SAPK cascade display cell lysis defects that are suppressed by external osmotic support [2], highlighting the central role of this signaling pathway in the maintenance of cell wall integrity.

Although the CWI pathway is best understood for its essential role in maintaining the structural integrity of the cell wall during growth, morphogenesis, and in response to cell wall stressors, there is evidence that this pathway is also important for survival of genomic stress. Defects in any component of the CWI pathway, from the cell surface sensors to Mpk1, cause hyper-sensitivity to a variety of DNA damaging agents [16–20]. The CWI pathway SAPK Mpk1 is phosphorylated by the DNA damage checkpoint kinases Mec1 and Tel1 in response to treatment with methylmethane sulfonate (MMS) [21] or caffeine [22]. The Mec1 and Tel1 protein kinases are orthologs of mammalian ATR and ATM, respectively [23], and have overlapping but distinct functions in the maintenance of yeast genome integrity. Tel1 signals the presence of double-strand breaks specifically [24], whereas Mec1 signals the presence of a variety of DNA damage types, including double-strand breaks [25]. Mec1 and Tel1 phosphorylate and activate the checkpoint effector kinases Chk1 and Rad53, which mediate cell cycle arrest and gene expression in support of DNA repair [26–29].

In addition to phosphorylation of Mpk1, Soriano–Carot et al. [30] detected a DNA damage-induced hyper-phosphorylation of Pkc1. They suggested the existence of a reciprocal regulatory circuit in which Pkc1 was required to activate the DNA damage checkpoint and the DNA damage checkpoint kinase Tel1 was required for the phosphorylation-induced Pkc1 band-shift in response to DNA damage. However, we found that the DNA damage checkpoint is activated normally in *pkc1Δ* mutants from various strain backgrounds [31]. Additionally, there is a bifurcation of pathway outputs at Pkc1, first revealed as a 100-fold increase in mitotic recombination frequency in pkc1 mutants that is not observed in mutants of pathway components below Pkc1 [32]. This is significant because mitotic recombination is the principal mode of double-stranded break repair of DNA in yeast [33]. Moreover, Pkc1 phosphorylates and activates CTP synthetase directly [34], revealing a role in nucleotide metabolism.

Finally, Mpk1 is activated by genotoxic stress through a pathway that does not require the activation of Pkc1 or the other protein kinases that function above Mpk1, but involves ubiquitination and degradation of Msg5, the protein phosphatase that normally maintains Mpk1 in a low activity state [31,35]. Intriguingly, Mpk1 activated in response to genotoxic stress does not drive cell wall stress transcription, suggesting that its activation in this context has a different function. These observations have led to the proposal that Pkc1 plays important roles in the response to genotoxic stress that are separate from its function in the activation of the CWI SAPK cascade [30], but these roles have not been elucidated. We hypothesize that the CWI pathway plays multiple unrecognized roles in the response to DNA damage—some driven by Mpk1, others by another pathway branch from Pkc1. In this study, we establish the pathway through which Pkc1 is hyper-phosphorylated in response to DNA damage and identify several sites within the Pkc1 regulatory domain whose phosphorylation is stimulated under conditions of genotoxic stress. Mutation of these sites impacts DNA damage-regulated gene expression.

2. Materials and Methods

2.1. Strains, Growth Conditions, and Transformations

The *S. cerevisiae* strains used in this study were derived from the EG123 background [36], the RDK2669 background (M. Smolka), the W303 background (J.C. Igual), or the Research Genetics background BY4742 (Research Genetics, Inc.; Huntsville, AL, USA) and are listed in Table 1.

Table 1. Yeast strains used in this study.

Strain	Relevant Genotype	Source or Reference
DL100	*MAT*a EG123 *ura3-52 leu2-3,112 trp1-1 his4 can1^r*	[36]
DL376	*MAT*a EG123 *ura3-52 leu2-3,112 trp1-1 his4 pkc1Δ::LEU2*	David Levin
DL1021	*MAT*a SEY6210 *leu2-3,112 ura3-52 his3-200 trp1-901 ede2-101 pkc1Δ::HIS3 suc2-9* (GPY1115)	Gerhard Paravicini
DL2772	*MATα* S288c (BY4742) *his3 leu2 lys2 ura3*	Research Genetics
DL3950	*MATα* MBS62 *sml1Δ::TRP1*	Marcus Smolka
DL3951	*MATα* MBS103 *sml1Δ::TRP1 tel1Δ::URA3*	Marcus Smolka
DL3952	*MATα* MBS104 *sml1Δ::TRP1 mec1Δ::KanMX*	Marcus Smolka
DL3953	*MATα* MBS72 *sml1Δ::TRP1 rad53Δ::HIS3*	Marcus Smolka
DL3954	*MATa/α MBS115 SML1/sml1Δ::TRP1 MEC1/mec1Δ::HIS3 TEL1/tel1Δ::URA3*	Marcus Smolka
DL4206	*MAT*a W303 *ade2 trp1 leu2 his3 ura3 can1*	Juan Carlos Igual
DL4277	*MATα* MBS *sml1Δ::TRP1 mec1Δ::HIS3 tel1Δ::URA3*	This study
DL4286	*MATα* BY4742 *chk1Δ::KanMX*	Research Genetics
DL4290	*MATa ura3-52 lys2-801 ade2-101 trp1-Δ63 his3-Δ200 leu2-Δ1 hrr25Δ::loxP-kanMX-loxP pGAL1-3HA-HRR25^{degron}* (KKY387)	[37]
DL4503	*MATα* MBS *sml1Δ::TRP1 mec1Δ::KanMX tel1Δ::URA3*	This study
DL4515	*MAT*a W303 *hrr25Δ::HPHMX4* (*pHRR25-HA*; p3484, *LEU2 2μ*)	This study
DL4527	*MAT*a W303 *hrr25Δ::HPHMX4* (*pHRR25-HA*; p3545, *HIS3 CEN*)	This study
DL4528	*MAT*a W303 *hrr25Δ::HPHMX4* (*phrr25-I82A-HA*; p3550, *HIS3 CEN*)	This study
DL4541	*MAT*a W303 *hrr25Δ::HPHMX4* (*pHRR25-GFP*; p3562, *HIS3 2μ*)	This study
DL4542	*MAT*a W303 *hrr25Δ::HPHMX4* (*phrr25-Δ404-GFP*; p3567, *HIS3 2μ*)	This study
DL4555	*MAT*a W303 *hrr25Δ::HPHMX4* (*phrr25-Δ404-HA*; p3546, *HIS3 CEN*)	This study
DL4556	*MAT*a W303 *hrr25Δ::HPHMX4* (*phrr25-3A-HA*; p3576, *HIS3 CEN*)	This study
JV826	*MAT*a BY4741 *PKC1-HTBeaq::NatMX*	This study

Yeast cell cultures were grown in YPD (1% Bacto yeast extract, 2% Bacto Peptone, 2% glucose) or minimal selective medium, SD (0.67% Yeast nitrogen base, 2% glucose) supplemented with the appropriate nutrients to select for plasmids, which are listed in Table 2. Yeast cells were transformed according to Geitz et al. [38]. Sorbitol (0.5 M) was used to prevent cell lysis for pkc1Δ strains. Cell wall stress was induced by treatment with calcofluor white (CFW, 40 μg/mL; Millipore Sigma, Burlington, MA), or by heat shock at 39 °C. Genotoxic stress was induced by treatment with hydroxyurea (HU; 250 mM, except where indicated otherwise; Fisher Scientific, Waltham, MA) or ultraviolet light (UV; 150 J/m^2 for Pkc1 band-shift, or 250 J/m^2 for viability assay). UV irradiation was carried out using an Analytik Jena UVP Crosslinker (Fisher Scientific). Cultures for viability tests were pelleted and resuspended in phosphate-buffered saline (PBS), dispersed on the surface of an empty petri dish for irradiation prior to dilution and plating. Cultures for Pkc1 band-shift were incubated in YPD for an additional 2 h for recovery after irradiation. To inhibit Hrr25 catalytic activity, 5 μM or 10 μM of PP1 analog IV, PP1 analog II (1NM PP1), or PP1 analog (Millipore Sigma) were used in agar plates. For Hrr25 inhibition in culture, PP1 analog IV (20 μM) was added to cultures at the time of addition of HU and incubated for 4 h.

Table 2. Plasmids used in this study.

Plasmid	Description	Source or Reference
p117	pRS313	[39]
p118	pRS314	[39]
p119	pRS315	[39]
p120	YEp351	[40]
p813	YEp351-*PKC1-HA*	David Levin
p1105	pRS425	[39]
p1202	pRS425-*GFP*	David Levin
p2062	pVDG7 *PKC1-GFP*	[41]
p2454	pRS413	[39]
p2947	p*RNR3-lacZ*	Stephen Elledge
p3064	pAG32-*RGC2*	[42]
p3149	pRS425-*3HA-ADH1*^{T}	[42]
p3357	pUG36-*HRR25-GFP*	Martha Cyert
p3358	pUG36-*GFP*	Martha Cyert
p3484	pRS425-*HRR25-HA*	This study
p3504	pRS313-*3HA-ADH1*^{T}	This study
p3517	YEp351-*pkc1-S577A-HA*	This study
p3521	YEp351-*pkc1-S577A, S626A-HA*	This study
p3522	YEp351-*pkc1-S577A, T626A, T753A-HA*	This study
p3523	YEp351-*pkc1-S577A, T626A, T753A, S804A-HA*	This study
p3525	pRS315-*HRR25-HA*	This study
p3538	pRS425-*hrr25-Δ404-HA*	This study
p3544	pRS423-*3HA-ADH1*^{T}	This study
p3545	pRS313-*HRR25-HA*	This study
p3546	pRS313-*hrr25-Δ404-HA*	This study
p3547	pRS423-*HRR25-HA*	This study
p3550	pRS313-*hrr25-I82A-HA*	This study
p3552	pRS423-*hrr25-I82A-HA*	This study
p3553	pRS423-*hrr25-I82G-HA*	This study
p3560	pRS423-*GFP*	This study
p3562	pRS423-*HRR25-GFP*	This study
p3567	pRS423-*hrr25-Δ404-GFP*	This study
p3570	pRS313-*hrr25-T453A-HA*	This study
p3572	pRS313-*hrr25-T453A, S405A-HA*	This study
p3574	YEp351-*pkc1-S577A, T626A, T753A, S761A, S804A-HA*	This study
p3576	pRS313-*hrr25-T453A, S405A, S438A-HA*	This study
p3597	YEp351-*pkc1-S577A, T626A, T753A, S761A, S772A, S804A-HA*	This study
p3603	YEp351-*pkc1-S577A, T626A, S686A, T753A, S761A, S772A, S804A-HA*	This study
p3604	YEp351-*pkc1-S577A, T626A, S686A, S750A, T753A, S761A, S772A, S804A-HA*	This study
p3605	YEp351-*pkc1-S152A, S577A, T626A, S686A, S750A, T753A, S761A, S772A, S804A-HA*	This study
p3606	YEp351-*pkc1-S152A, S577A, T626A, S657A, S686A, S750A, T753A, S761A, S772A, S804A-HA*	This study
p3608	YEp351-*pkc1-S152A, S577A, T626A, S657A, S686A, S750A, T753A, S761A, S772A, S781A, S804A-HA*	This study

Table 2. *Cont.*

Plasmid	Description	Source or Reference
p3610	YEp351-*pkc1-S75A, S152A, S577A, T626A, S657A, S686A, S750A, T753A, S761A, S772A, S781A, S804A-HA*	This study
p3612	YEp351-*pkc1-S75A, S152A, S577A, T626A, S657A, S686A, S750A, T753A, S761A, S772A, S781A, S797A, S804A-HA* (S/T13A)	This study
p3617	YEp351-*pkc1-S2A-HA*	This study
p3618	YEp351-*pkc1-S2A, S657A-HA*	This study
p3619	YEp351-*pkc1-S2A, S657A, S577A-HA* (S3A)	This study
p3623	pRS314-*PKC1-HA*	This study
p3624	pRS314-*pkc1-S2A, S657A, S577A-HA* (S3A)	This study
p3625	pRS314-*pkc1-S75A, S152A, S577A, T626A, S657A, S686A, S750A, T753A, S761A, S772A, S781A, S797A, S804A-HA* (S/T13A)	This study
pWR268	pFA6a-integrative *HTBeaq* tag, *NatMX*	[43]

2.2. Chromosomal Deletions and Strain Construction

A *sml1*Δ*::TRP1 mec1*Δ*::HIS3 tel1*Δ*::URA3* strain (DL4277) was generated as a meiotic segregant of DL3954 (*MAT***a**/α MBS115 *SML1/sml1*Δ*::TRP1 MEC1/mec1*Δ*::HIS3 TEL1/tel1*Δ*::URA3*). An *sml1*Δ*::TRP1 mec1*Δ*::KanMX tel1*Δ*::URA3* strain (DL4503) was created by homologous recombination of *mec1*Δ*::KanMX* at the *MEC1* locus in an *sml1*Δ*::TRP1 tel1*Δ*::URA3* strain (DL3951). The *mec1*Δ*::KanMX* allele was amplified by PCR from genomic DNA of yeast strain (DL3952) using Phusion high-fidelity DNA polymerase (Thermo Fisher Scientific, San Jose, CA). Transformants were selected for antibiotic G418 resistance and validated by genomic PCR across both integration junction sites. Endogenous tagging of Pkc1 with HTBeaq [43] was achieved by transformation of BY4741 with HB-tagging cassettes amplified from plasmid pWR268 [43], resulting in yeast strain JV826.

Plasmid-borne "gatekeeper" alleles of *HRR25* were tested for function initially in an *hrr25*Δ strain maintained with a plasmid-borne, *GAL1*-controlled *HRR25^{degron}* (DL4290) [37], which is viable only in galactose-containing medium. Later experiments with these mutant alleles were conducted in an *hrr25*Δ strain without the *HRR25^{degron}*. To create *hrr25*Δ*::HPHMX4* in W303 (DL4515), a plasmid bearing *HRR25-HA* (p3484; pRS425-*HRR25-HA*) was first introduced into DL4206 (*MAT***a** W303 *ade2 trp1 leu2 his3 ura3 can1*). The *HPHMX4* gene was amplified by PCR from plasmid p3064 (pAG32-*RGC2*) using primers with extensions homologous to the 5′ and 3′ ends of *HRR25* and used to transform DL4206 pRS425-*HRR25-HA* to hygromycin B resistance. Gene replacement was validated by PCR analysis across both integration junctions. The plasmid-borne *HRR25* in DL4515 was replaced with other alleles of *HRR25* carried on pRS313 or pRS423 using plasmid shuffle.

2.3. Plasmid Construction and Mutagenesis

The *HRR25* gene was epitope-tagged on its C-terminus with 3xHA and expressed from its natural promoter. The promoter region of *HRR25* (from position −979) and its entire coding sequence was amplified from genomic yeast DNA (DL2772) by high-fidelity PCR polymerase (Phusion) using primers designed with a HindIII site (upstream) and a NotI site (downstream), and subcloned into pRS425-*3HA-ADH1^T* (p3149) to generate pRS425-*HRR25-HA* (p3484). To create pRS425-*hrr25-*Δ*404-HA* (p3538), the promoter region of *HRR25* (from position −979) and its C-terminally truncated coding sequence was amplified from pRS425-*HRR25-HA* (p3484) by high-fidelity PCR polymerase (Phusion) using primers designed with a HindIII site (upstream) and a NotI site (downstream) and subcloned into pRS425-*3HA-ADH1^T* (p3149).

To express HA-tagged *HRR25* from *HIS3*-marked plasmids pRS313-*3HA-ADH1^T* (p3504) and pRS423-*3HA-ADH1^T* (p3544), first required subcloning of the *3HA-ADH1^T* sequence from pRS425-*3HA-ADH1^T* (p3149) into pRS313 or pRS423 [39] at SmaI and

SacI sites. Plasmid pRS313-*HRR25-HA* (p3545) was next created by subcloning a Sal1-Not1 fragment containing the promoter region of *HRR25* (from position −979) and its entire coding sequence from pRS425-*HRR25-HA* (p3484) into pRS313-*3HA-ADH1*^{T} (p3504). Plasmid pRS313-*hrr25-Δ404-HA* (p3546) was created similarly. Plasmid pRS423-*HRR25-HA* (p3547) was generated by subcloning the Sal1-Not1 fragment containing *HRR25* into pRS423-*3HA-ADH1*^{T} (p3544).

The *HRR25* gene was also epitope-tagged on its C-terminus with GFP and expressed from its natural promoter. A PCR fragment containing the GFP coding sequence was amplified from pRS425-*GFP* (p1202) with primers designed with Sma1 and Not1 sites (upstream) and a SacII site (downstream) and cloned into pRS423 to create GFP-tagging vector pRS423-*GFP* (p3560). An Apa1-Not1 fragment containing the promoter region of *HRR25* (from position −979) and its entire coding sequence digested from pRS423-*HRR25-HA* (p3547) was subcloned into pRS423-*GFP* (p3560) at the Apa1-Not1 sites to generate pRS423-*HRR25-GFP* (p3562). Plasmid pRS423-*hrr25-Δ404-GFP* (p3567) was similarly created from pRS313-*hrr25-Δ404-HA* (p3546).

Point mutations in *PKC1* were constructed initially in high copy plasmid YEp351-*PKC1-HA* (p813) for use in band-shift assays. However, for phenotypic analyses, we expressed mutant alleles of *PKC1-HA* from a centromeric plasmid. To create pRS314-*PKC1-HA*, *pkc1-3A-HA*, and *pkc1-S/T13A-HA*, *PKC1-HA* (from position -950) with its 3′ region (640 bp) were PCR amplified from YEp351-*PKC1-HA* (p813), YEp351-*pkc1-S3A-HA* (p3619) or YEp351-*pkc1-S/T13A-HA* (p3612), respectively, by PrimeSTAR, Max DNA Polymerase (Takara Bio USA, Mountainview, CA) using primers designed with a SalI site (upstream) and with a SacII site (downstream) and cloned into pRS314 (p118).

All point mutations in *HRR25* or *PKC1* were created using QuickChange or QuikChange Lightning Site-Directed Mutagenesis Kits (Agilent Technologies, Santa Clara) according to manufacturer's instructions. All DNA sequences derived from PCR mutagenesis were confirmed by sequence across the entire gene.

2.4. Protein Extraction

Protein extraction was carried out as described previously for co-immunoprecipitation [44], or using the rapid boiling method [45] for direct immunoblot experiments.

2.5. β-galactosidase Measurements

Measurements of β-galactosidase activity from *RNR3-lacZ* (p2947) expression experiments were conducted in triplicate and carried out as described in Zhao et al. [46].

2.6. Dephosphorylation Assay

Protein extracts were prepared from *sml1Δ* (DL3950) cells carrying PKC1-HA (p813), treated with 250 mM HU or untreated, by bead-beating in lysis buffer (10% glycerol, 20 mM Hepes pH 7.5, 150 mM NaCI, 0.5% triton X-100 and Mini protease inhibitor cocktail (Milipore Sigma), without phosphatase inhibitors, followed by centrifugation. The dephosphorylation reactions were performed using Lambda protein phosphatase (New England Biolabs, Ipswich, MA) according to manufacturer's protocol. Briefly, 10 µg of protein extract was incubated with 400 units of Lambda protein phosphatase in Lambda protein phosphatase buffer for 30 min at 30 °C. Phosphatase reactions were stopped by adding an equal volume of 2× SDS sample buffer and the resulting samples were boiled for 3 min prior to SDS-PAGE and immunoblot analysis.

2.7. Co-Immunoprecipitation

To detect association of Pkc1 with Hrr25 in wild-type cells, a plasmid expressing Pkc1-HA under the control of its own promoter (p813) was co-transformed with plasmids expressing either GFP-Hrr25 (p3357) or GFP alone (p3358) into DL100. Transformants were grown to mid-log phase in selective medium and starved for methionine for two hours to induce expression of GFP-Hrr25 or GFP, which were expressed under the control of the *MET25* promoter. Cultures were then treated with 250 mM HU for 4 h. To test the effect of *MEC1* and *TEL1* loss on the association of Pkc1 with Hrr25, a plasmid expressing Hrr25-GFP under the control of its own promoter (p3562) was co-transformed with p813 into strains DL3950 (*sml1Δ*) and DL4503 (*sml1Δ mec1Δ tel1Δ*). To test the association of Pkc1-HA with truncated Hrr25 (Hrr25-Δ404-GFP), plasmid p813 was transformed into strains bearing a chromosomal deletion of *HRR25* complemented by plasmids expressing either Hrr25-GFP (DL4541) or Hrr25-Δ404-GFP (DL4542). Transformants were grown to mid-log phase in selective medium and treated with 250 mM HU for 4 h. Protein extracts were made as described previously [44]. Extracts (100 μg of protein) were incubated with 10μL of GFP-trap agarose beads (Chromotek, Munich, Germany) at 4 °C for 2 h and the samples were washed with IP buffer (50 mM Tris-HCl, pH 7.5, 150 mM NaCl, 0.5% Triton) three times and boiled in SDS-PAGE buffer.

2.8. SDS-PAGE Electrophoresis and Immunoblot Analysis

Proteins were separated by SDS-PAGE (10% or 4–20% Mini-PROTEAN TGX gels, BioRad, Hercules, CA, USA) followed by immunoblot analysis using mouse monoclonal α-HA (BioLegend, San Diego, CA, USA) or α-GFP (Millipore Sigma) at a dilution of 1:10,000 and polyclonal rabbit α-Rad53 (Abcam) antibodies at a dilution of 1:2000 to detect Pkc1-HA or -GFP, Hrr25-HA or -GFP and Rad53, respectively. Secondary goat anti-mouse (Jackson ImmunoResearch, Westgrove, PA, USA) and donkey anti-rabbit (GE Healthcare, Chicago, IL, USA) antibodies were used at a dilution of 1:10,000. SDS-PAGE gels used to detect Pkc1 band-shifts were 10% and those used for Pkc1 association with Hrr25 were 4–20%.

2.9. Mass Spectrometric Analysis of Pkc1 Co-Immunoprecipitates

Pkc1-GFP was expressed from plasmid p2062 (gift of M. Cyert) in strain DL100 and isolated using GFP-trap agarose. Pkc1-GFP immunoprecipitates were run on SDS-PAGE gels prior to in-gel digestion of the contents of the entire lane with trypsin. After desalting, LC-MS/MS was performed using a nanoAcquity ultra-performance liquid chromatography (UPLC) capillary system (Waters Corp., Milford, MA, USA), coupled to an LTQ-Orbitrap hybrid mass spectrometer (Thermo Fisher Scientific) equipped with a TriVersa NanoMate ion source (Advion, Ithaca, NY, USA). Sample concentration and desalting were performed online using a nanoAcquity UPLC trapping column (180 μm by 20 mm; packed with 5-μm, 100-Å-pore-size Symmetry C_{18} material; Waters Corp.) at a flow rate of 15 μL/min for 1 min. Separation was accomplished on a nanoAcquity UPLC capillary column (100 μm by 100 mm; packed with 1.7-μm, 130-Å-pore-size bridged ethyl hybrid [BEH] C_{18} material; Waters Corp.). A linear gradient of A and B buffers (buffer A, 3% ACN–0.1% formic acid [FA]; buffer B, 97% ACN–0.1% FA) from 7% to 45% buffer B over 124 min was used at a flow rate of 0.5 μL/min to elute peptides into the mass spectrometer. Columns were washed and re-equilibrated between LC-MS/MS experiments. Electrospray ionization was carried out at 1.7 kV using the NanoMate, with the LTQ heated capillary set to 150 °C.

Mass spectra were acquired in the orbitrap mass analyzer in the positive-ion mode over the range of m/z 300 to 2000 at a resolution of 60,000 @ m/z 400. Mass accuracy after internal calibration was within 4 ppm. Simultaneously, tandem MS spectra were acquired using the LTQ for the five most abundant, multiply charged species in the mass spectrum with signal intensities of >8000 signal/noise levels. With MS/MS collision energies set at 35%, and helium used as the collision gas, MS/MS spectra were acquired over a range of m/z values dependent on the precursor ion. Dynamic exclusion was set such that MS/MS data for each species were acquired a maximum of twice. All spectra were recorded in profile mode for further processing and analysis. Xcalibur software was used for MS and MS/MS data analysis, while peptide and protein assignments were conducted using Mascot to search against the *ScerevisiaeOrfTrans* database.

2.10. Mass Spectrometric Analysis of Pkc1 Phospho-Sites

Histidine-biotin tandem affinity purifications of Pkc1-HTBeaq were based on methods described elsewhere [43], with modifications below. Stable isotope labeling using amino acids in cell culture (SILAC) [47] was achieved as described previously [43]. Cells expressing Pkc1 C-terminally tagged with a HTBeaq tag [43] were grown to mid-logarithmic phase ($OD_{600\ nm}$ = 2.0), treated with 200 mM HU for 4 h, harvested by filtration, and rapidly deep frozen in liquid N_2. In-solution digestion with trypsin and enrichment of phosphorylated peptides using TiO_2 was performed as described previously [43].

For these mass spectrometry analyses, a Q Exactive HF Orbitrap (Thermo Fisher Scientific) mass spectrometer was used with the following settings: Peptides were separated applying an increasing organic solvent (acetonitrile) gradient from 2.5% to 40% in 0.1% formic acid over 60 (YPD) or 120 (SILAC) minutes at a flow rate of 275 nl/min. MS1 resolution was set to 70,000 @ m/z 200, AGC 3×10^6. MS2 resolution was set to 17,500 @ m/z 200, AGC 1×10^5, 500 ms max. ion injection time (IIT). The mass spectrometer was configured to pick the twelve most abundant precursor ions for data-dependent MS2 scans, applying HCD for fragmentation with a normalized collision energy (NCE) of 27. Dynamic exclusion time was set to 30 s. MS raw files were processed using MaxQuant (Max Planck Institute of Biochemistry, Munich) [48] software version 1.5.2.8 using standard settings, except for the following modifications. Spectra were searched against the *Saccharomyces* Genome Database (http://www.yeastgenome.org/ accessed on 8 October 2015) containing 6717 entries (3 February 2011), including a list of 248 common laboratory contaminants as well as reversed versions of all sequences. The enzyme specificity was set to trypsin. A maximum of two missed cleavages was allowed. Phosphorylation of serine, threonine, and tyrosine residues, oxidation of methionine, and deamidation of asparagine were set as variable modifications. For stable isotope labeling using amino acids in cell culture (SILAC)–labeled samples, Lys6 and Arg6 were additionally selected. Carbamidomethylation of cysteine was searched as a fixed modification. A maximum of five modifications per peptide was allowed. The false discovery rate for peptide, protein, and site identification was set to 1%. All files were searched together. Minimum delta score for modified peptides was set to 6. The MS phospho-proteomics data have been deposited at the zenodo repository (https://zenodo.org/ accessed on 10 September 2021) and can be accessed via https://doi.org/10.5281/zenodo.5552273.

2.11. Notes on Reproducibility

All immunoblots, and co-IPs, were reproduced at least once in independent experiments with representative images shown.

3. Results and Discussion

Pkc1 undergoes a phosphorylation-induced band-shift in response to DNA damage [30]. We found that the Pkc1 band-shift was induced specifically in response to DNA damage by alkylating agent methylmethane sulfonate (MMS), by dNTP depletion by hydroxyurea (HU) treatment, or by UV irradiation, but not in response to cell wall stress treatments calcofluor white (CFW) or elevated growth temperature (Figure 1a). We confirmed that this band-shift results from hyper-phosphorylation by treatment with Lambda protein phosphatase (Figure 1b). We also found that it is dependent on the partially redundant DNA damage checkpoint kinases Mec1 and Tel1 (Figure 1c). This is in contrast to the findings of Soriano-Carot, et al. [30], who found that Tel1 was uniquely responsible for the Pkc1 band-shift. In any case, we found that both of the known effector kinase targets of the checkpoint kinases, Rad53 and Chk1 [49,50], were dispensable for the Pkc1 band-shift (Figure 1d). These results suggest that Mec1 and Tel1 either phosphorylate Pkc1 directly, or act on another protein kinase that phosphorylates Pkc1. Thus, the DNA damage checkpoint kinases, Mec1 and Tel1, regulate hyper-phosphorylation of Pkc1 in response to genotoxic stress.

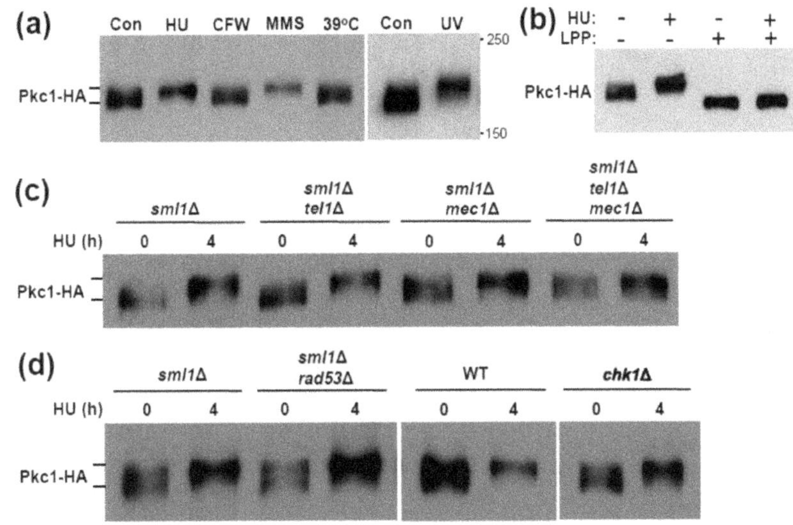

Figure 1. A phosphorylation band-shift in Pkc1 requires the *MEC1* and *TEL1* DNA damage checkpoint genes. (**a**) A Pkc1 band-shift is induced specifically by genotoxic stressors. Wild-type cells (DL3950; *sml1Δ*) expressing Pkc1-HA (from p813) were exposed to genotoxic stress (250 mM HU for 4 h, 0.04% MMS for 2 h, or 150 J/m^2 UV with a 2 h recovery period), cell wall stress (40 µg/mL CFW for 1 h, or heat shock at 39 °C for 1 h), or untreated (Con). Extracts were subjected to SDS-PAGE and immunoblot analysis for Pkc1-HA; (**b**) Phosphorylation is responsible for the HU-induced band-shift. Wild-type cells (DL3950) expressing Pkc1-HA were either treated with HU as above, or untreated. Extracts were treated with Lambda protein phosphatase (LPP), as described in Methods, prior to immunoblot analysis; (**c**,**d**) The HU-induced Pkc1 band-shift requires *MEC1* and *TEL1*, but not the checkpoint genes that they regulate (*RAD53* or *CHK1*). Cultures were treated with HU as above prior to immunoblot analysis for Pkc1-HA. Strains are DL3950 (*sml1Δ*), DL3951 (*sml1Δ tel1Δ*), DL3952 (*sml1Δ mec1Δ*), DL4277 (*sml1Δ mec1Δ tel1Δ*), DL3953 (*sml1Δ rad53Δ*), DL2772 (Res. Gen. WT), and DL4286 (Res. Gen. *chk1Δ*). The *sml1Δ* mutation is required to suppress the lethality of the *mec1Δ* and *rad53Δ* mutations.

3.1. Genotoxic Stress Induces Hrr25 Association with Pkc1

Next, we took a proteomics approach to identify protein kinases that associate with Pkc1 in response to HU-induced genotoxic stress. Cells expressing Pkc1-GFP were treated with 250 mM HU for 4 h, or were untreated, and Pkc1-GFP was immunoprecipitated from extracts and subjected to mass spectrometric (MS) analysis to identify co-precipitating proteins. Proteins that were found associated with Pkc1 only after HU treatment, or only without HU treatment are presented in Supplemental Tables S1 and S2, respectively. Although numerous non-specific associations were detected in both the untreated and treated samples, a minor signal from the DNA damage and repair kinase Hrr25 (HO and radiation repair) [51] was detected only in the HU-treated sample. *HRR25* encodes an isoform of casein kinase 1 (CK1) that has been implicated in the repair of DNA double strand breaks [51] and is required for the transcriptional induction of the ribonucleotide reductase genes *RNR2* and *RNR3* in response to DNA damage [52]. Many physical interactors have been identified for Pkc1, but Hrr25 is not among them (*Saccharomyces* Genome Database). Therefore, we validated the physical association between Pkc1 and Hrr25 by co-immunoprecipitation (co-IP). Not only did we detect a stable interaction between Pkc1-HA and Hrr25-GFP, but the interaction was reproducibly induced 5–10-fold in response to treatment with HU (Figure 2a). We detected no non-specific binding of Pkc1-HA to either GFP alone (Figure 2a), or to the GFP-trap beads (Supplemental Figure S1). Hrr25 was recruited similarly to Pkc1 in response to ultraviolet (UV) light exposure (Figure 2b), suggesting that this is part of the generalized response to genotoxic stress.

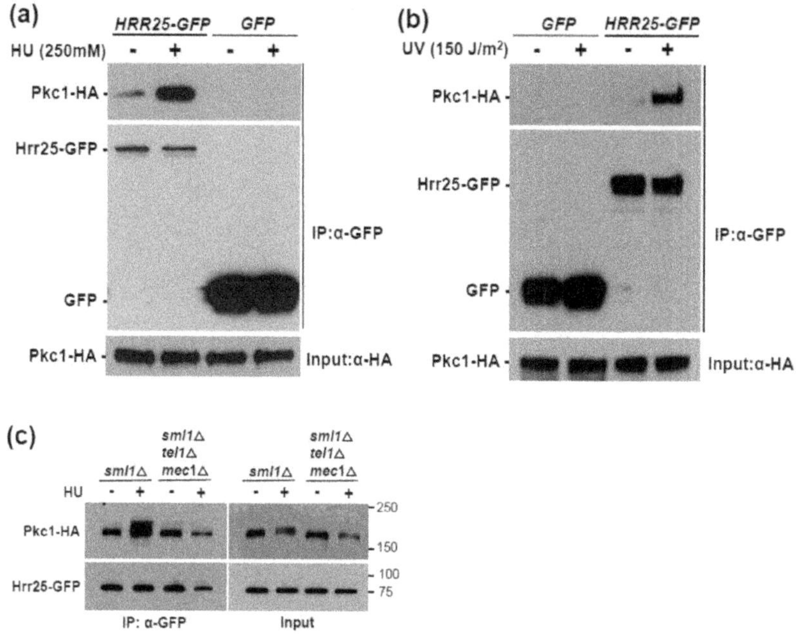

Figure 2. HU treatment induces association of Pkc1 with Hrr25. (**a**) Wild-type cells (DL100) co-expressing Pkc1-HA (from p813) and Hrr25-GFP (from p3357) or GFP (from p3358) were treated with 250 mM HU for 4 h. Hrr25-GFP or GFP was immunoprecipitated (IP) from extracts and samples were tested by immunoblot analysis for co-IP of Pkc1-HA. Input Pkc1-HA from extracts is shown at bottom; (**b**) UV treatment induces association of Pkc1 with Hrr25. Wild-type cells (DL100) co-expressing Pkc1-HA (from p813) and Hrr25-GFP (from p3357) or GFP (from p3358) were treated with UV light (150 J/m^2) and returned to culture for 2 h post-irradiation prior to extract preparation. Hrr25-GFP or GFP was immunoprecipitated (IP) from extracts and treated as above; (**c**) *MEC1* and *TEL1* are required for the HU-induced association of Hrr25 with Pkc1. Cultures co-expressing Pkc1-HA and Hrr25-GFP (from p3562) were treated with HU as above and processed for co-IP of Pkc1-HA with Hrr25-GFP. Strains were DL3950 (*sml1*Δ) and DL4277 (*sml1*Δ *tel1*Δ *mec1*Δ). Molecular mass markers (in kDa) are shown on the right.

We next tested the role of Mec1 and Tel1 in the induced association of Hrr25 with Pkc1. Here, we found that, not only are the DNA damage checkpoint kinases required for the Pkc1 phosphorylation band-shift, but they are also required for the induced association of Hrr25 with Pkc1 (Figure 2c), suggesting that Mec1 and Tel1 induce the Pkc1 phosphorylation band-shift indirectly by driving its association with Hrr25. Therefore, we sought to test this idea using mutants in *HRR25*. Because the *HRR25* gene is essential for viability, we constructed two analog-sensitive "gatekeeper" alleles [53] of *HRR25* (*hrr25-I82A* and *hrr25-I82G*). These were introduced on plasmids into an *hrr25*Δ strain that is maintained with a galactose-inducible form of *HRR25* and tested for their sensitivity to a collection of protein kinase inhibitory ATP-analogs in glucose-containing medium (YPD). We determined that the *hrr25-I82A* allele conferred optimum growth sensitivity to PP1 analog IV, which did not appreciably inhibit the growth of the strain expressing wild-type *HRR25* (Supplemental Figure S2). We next introduced the *hrr25-I82A* allele to an *hrr25*Δ strain by plasmid shuffle (See Materials and Methods, Section 2) to test the importance of this protein kinase in the HU-induced Pkc1 band-shift. The strain expressing the *hrr25-I82A* allele along with an epitope tagged form of Pkc1, was subjected to inhibition of Hrr25, together with HU treatment to induce genotoxic stress. Pkc1 failed to display an HU-induced band-shift in the *hrr25-I82A* strain in the presence of inhibitor, in contrast to the *HRR25* strain (Figure 3a). This suggests that Hrr25 catalytic activity is, indeed, required for the Pkc1-bandshift observed in response to genotoxic stress.

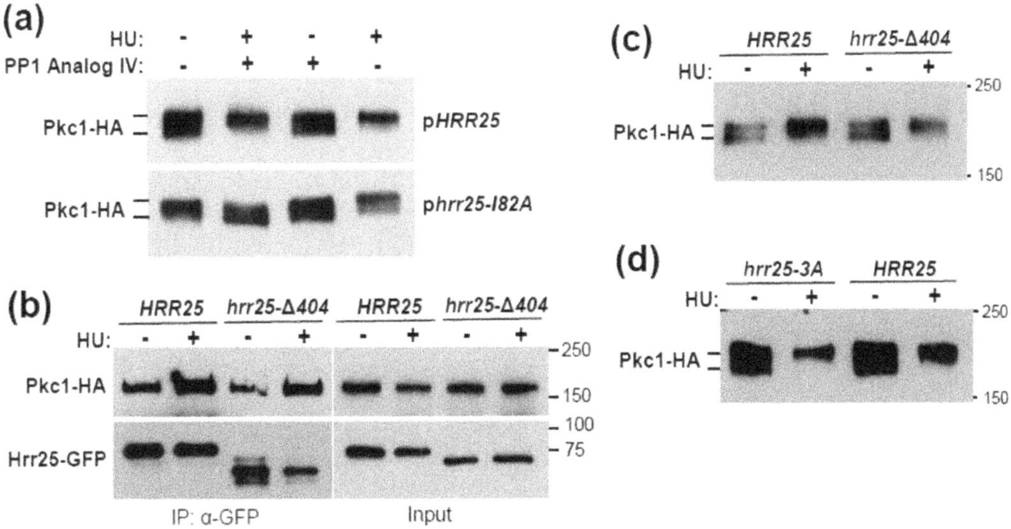

Figure 3. An analog-sensitive form of Hrr25 shows that its catalytic activity is required for the HU-induced Pkc1 band-shift. (**a**) An *hrr25*Δ strain complemented by plasmid-borne *HRR25* (DL4527) or *hrr25-182A* (DL4528; encoding an analog-sensitive form) and expressing Pkc1-HA (from p813) was treated simultaneously with HU (250 mM) and/or PP1 analog IV (20 μM) for 4 h. Extracts were processed for immunoblot analysis of Pkc1-HA; (**b**) A C-terminal truncation of Hrr25 lacking three potential Mec1/Tel1 phosphorylation sites associates normally with Pkc1. An *hrr25*Δ strain complemented by plasmid-borne *HRR25-GFP* (DL4541) or *hrr25-Δ404* (DL4542) and expressing Pkc1-HA (from p813) was treated with HU (250 mM for 4 h). Hrr25-GFP was immunoprecipitated from extracts and tested for co-IP of Pkc1-HA by immunoblot analysis. Molecular mass markers (in kDa) are on the right; (**c**) A C-terminal truncation of Hrr25 does not impact the HU-induced Pkc1 band-shift. An *hrr25*Δ strain complemented by plasmid-borne *HRR25* (DL4527) or *hrr25-Δ404* (DL4555) and expressing Pkc1-HA was treated with HU as above and processed for immunoblot analysis of Pkc1-HA (**d**) A mutant form of *HRR25* lacking three potential Mec1/Tel1 phosphorylation sites does not impact the HU-induced Pkc1 band-shift. An *hrr25*Δ strain complemented by plasmid-borne *HRR25* (DL4527) or *hrr25-3A* (DL4556) and expressing Pkc1-HA was treated with HU as above and processed for immunoblot analysis of Pkc1-HA.

Hrr25 possesses three potential Mec1/Tel1 sites (S/T-Q) [21], which all reside within the C-terminal tail of Hrr25 (residues S405, S438, and T453). Therefore, we generated a truncated version of Hrr25 that is missing the C-terminal domain from residues 405–494 (*hrr25-Δ404*), which removes all three potential Mec1/Tel1 sites. This allele complemented the lethality of the *hrr25*Δ mutation, therefore we tested it for HU-induced association with Pkc1 and for the HU-induced Pkc1 band-shift. The truncated form of Hrr25 was recruited normally to Pkc1 in response to HU treatment (Figure 3b) and retained its ability to drive the Pkc1 band-shift (Figure 3c). However, we considered the possibility that the Hrr25 C-terminal region might carry an auto-inhibitory domain, truncation of which could activate Hrr25 independently of Mec1/Tel1. To address this possibility, we mutated the three potential Mec1/Tel1 sites within this domain to Ala residues, yielding *hrr25-3A*, and tested the influence of this allele on the HU-induced Pkc1 band-shift. As with the truncated allele, the *hrr25-3A* mutant was able to mediate the HU-induced band-shift normally

(Figure 3d). Therefore, we conclude that Mec1 and Tel1 likely regulate HU-induced Hrr25 association with Pkc1 and the Pkc1 band-shift by means other than direct phosphorylation of Hrr25. Because neither Rad53, nor Chk1, the known effector kinases of Mec1 and Tel1, were required for the HU-induced Pkc1 band-shift, it remains unclear how the sensor kinases regulate the Hrr25 action on Pkc1. It is possible that Mec1 and Tel1 act on Pkc1, rather than on Hrr25, to regulate the association of these protein kinases. Nevertheless, our results strongly suggest that both Hrr25 and Pkc1 are indirect Mec1/Tel1 effectors.

3.2. Identification of Pkc1 Phospho-Sites in Response to HU Treatment

Hrr25 is an ortholog of mammalian casein kinase 1 delta (CK1δ) [54]. Two phosphorylation site motifs for this class of protein kinase have been described [55–58]. These protein kinases require either a priming phosphorylation at position −3 relative to the target S/T site (i.e., p-S/TXXS/T) or an acidic residue at position −3 relative to the target site (i.e., D/EXXS/T).

We conducted both a shotgun mass spectrometric (MS) analysis and a quantitative SILAC MS analysis of HU-induced phosphorylation sites on Pkc1 [43]. These analyses identified many phosphorylation sites on Pkc1, several of which had not been described previously (i.e., T570, T626, S666, T779, S781, and T785; Supplemental Figure S3). Among the phosphorylation sites identified in our SILAC MS experiments, only three appeared to be upregulated in response to HU treatment: S2, S577, and S657 (Supplemental Table S3). Of these, both S577 and S657 are potential Hrr25 phosphorylation sites because S577 has a priming phosphorylation site (p-S574) and S657 is positioned three residues beyond E654 (Figure 4a). Therefore, we started our mutational analysis with these three sites. The individual mutations did not display a detectable impact on the HU-induced Pkc1 band-shift (not shown), so we created a triple mutant. The *pkc1-3A* (*pkc1-S2A, S577A, S657A*) allele was able to complement the lethality of a *pkc1*Δ mutant for growth on rich medium in the absence of osmotic support. Therefore, we examined its behavior in response to genotoxic stress. The HU-induced Pkc1 band-shift was somewhat diminished in the *pkc1-3A* mutant (Figure 4b), revealing a cumulative effect of these phosphorylation sites. We next tested this mutant for its sensitivity to HU and UV treatment. However, this mutant did not display increased sensitivity to either treatment (Figure 4c). Therefore, we considered a larger collection of Pkc1 phosphorylation sites.

Among the phosphorylation sites identified in our analyses, together with those established previously [21,59–61], we identified 13 phosphorylation sites within Pkc1 that fit either of the two CK1 consensus sequences. Figure 4a shows a map of these sites, all of which reside within the N-terminal regulatory domain of Pkc1. However, most of these sites (11) are clustered in an area without recognizable regulatory features between residues 577 and 804. We mutated all of these phosphorylation sites to alanine residues in groups and in series. We were only able to detect an impact on the HU-induced Pkc1 band-shift once at least nine of these CK1 residues were mutated (data not shown). Blocking the HU-induced band-shift completely required mutation of all 13 sites in the *pkc1-S/T13A* mutant (Figure 4d). The UV-induced band-shift was also blocked in the *pkc1-S/T13A* mutant (Figure 4e). The DNA damage checkpoint was activated normally in the *pkc1-S/T13A* mutant, as judged by a strong Rad53 band-shift (Figure 4d,e). The *pkc1-S/T13A* allele also complemented the null mutant for growth in the absence of osmotic support. Therefore, we tested the sensitivity of this strain to genotoxic stress. However, like the *pkc1-3A* mutant, the *pkc1-S/T13A* mutant did not display enhanced sensitivity to either HU or UV treatment (Figure 4c), suggesting that the biological impact of Pkc1 hyper-phosphorylation during genotoxic stress is too subtle to detect by this viability assay.

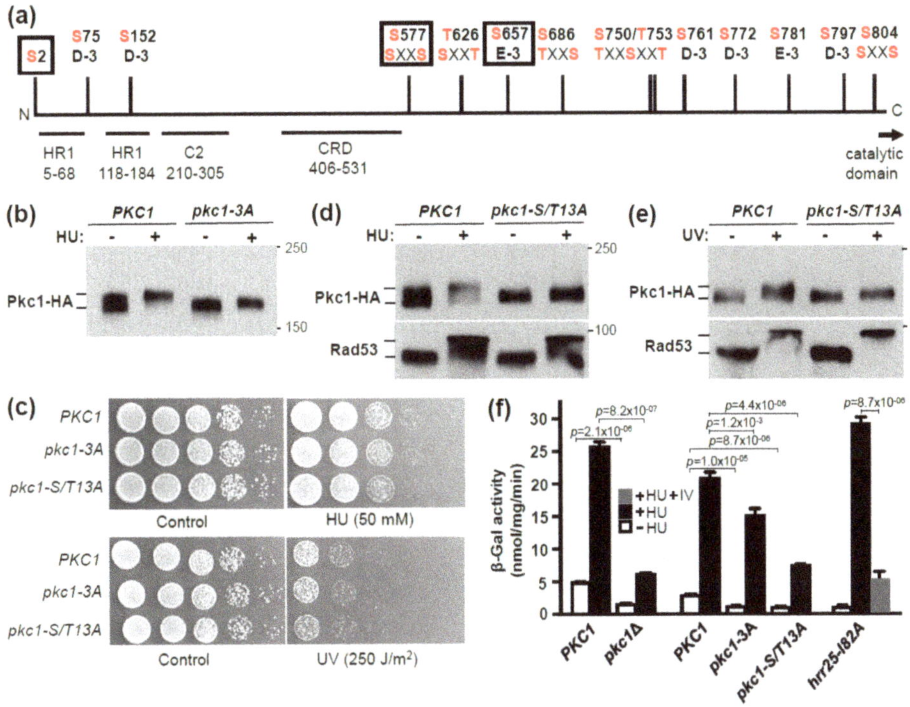

Figure 4. CK1 phosphorylation sites within the Pkc1 regulatory domain are responsible for the genotoxic stress-induced Pkc1 band-shift. (**a**) Phosphorylation sites within the Pkc1 regulatory domain mutated in this study. Phosphorylated residues are marked in red. The three residues mutated in the *pkc1-3A* allele (S2, S577, and S657) are marked by boxes and were identified by SILAC MS as increased in phosphorylation state in response to HU treatment. Two of these residues (S577 and S657) are within consensus CK1 phosphorylation sites, with either a priming phospho-Ser at position −3 (S577) or an acidic residue (Asp) at position −3 (S657). Other phosphorylated residues that reside within CK1 consensus sites are also indicated and were mutated in the *pkc1-S/T13A* allele. Known regulatory elements, Rho-binding domains (HR1), calcium/lipid-binding domain (C2), and Cys-rich domain (CRD) are also shown. The catalytic domain is C-terminal to the regulatory domain and starts at residue 824; (**b**) HU-induced phosphorylation band-shift of the Pkc1-3A mutant. Plasmids were *PKC1-HA* (p813) and *pkc1-3A-HA* (p3619) (**c**) The *pkc1-3A* and *pkc1-S/T13A* mutants do not show increased sensitivity to genotoxic stress. Serial 10-fold dilutions of cultures grown to mid-log phase in YPD were spotted onto plates (left to right) with or without HU. Cultures treated with UV were similarly diluted and spotted onto YPD plates. Plates were incubated at 25 °C for two days. Plasmids were *PKC1-HA* (p3623), *pkc1-3A* (p3624), and *pkc1-S/T13A* (p3625); (**d**) HU-induced phosphorylation band-shift of the Pkc1-S/T13A mutant and Rad53. Plasmids were *PKC1-HA* (p813) *pkc1-S/T13A* (p3612); (**e**) UV-induced phosphorylation band-shift of the Pkc1-S/T13A mutant and Rad53. Plasmids were *PKC1-HA* (p813) and *pkc1-S/T13A* (p3612). Strain DL1021 (*pkc1*Δ) was used for experiments shown in (**b**,**d**,**e**). Strain DL376 (*pkc1*Δ) was used for the experiment in panel (**c**); (**f**) HU-induced *RNR3-lacZ* expression is diminished in a *pkc1*Δ mutant, and in the *pkc1-3A* and *pkc1-S/T13A* mutants. Strain DL376 (*pkc1*Δ) was co-transformed with p*RNR3-lacZ* (p2947) and *PKC1-HA* (p3623), *pkc1-3A* (p3624), *pkc1-S/T13A* (p3625), or vector alone (p118). Cells were cultured in the presence of 0.5 M sorbitol for osmotic support (pair on left), or in the absence of sorbitol. Cultures were treated for 4 h with 250 mM HU and β-galactosidase activity was measured from extracts. The *hrr25-I82A* mutant (DL4528; right) was treated with HU plus or minus PP1 analog IV (20 μM) for 4 h. Each value is the mean and standard deviation from three independent cultures. Pair-wise *p*-values for HU-treated and untreated samples were calculated using student t-test and were all at least $p \leq 0.00001$, except the HU-treated *PKC1* and *pkc1-3A* pair, which was $p = 0.0012$. An additional p-value of $p < 0.00001$ was obtained for the *hrr25-I82A* mutant for HU-treated samples, with and without analog IV.

HRR25 has been implicated in the transcriptional response to DNA damage, most notably in the induction of the *RNR3* gene [52], but its role has not been clearly established. The cell cycle transcriptional regulatory factor SBF, comprised of Swi4 and Swi6, is similarly important for this transcriptional response and Hrr25 associates with and phosphorylates Swi6 in vitro [52]. However, *RNR3* induction in response to DNA damage was also shown to be largely under the control of the Mec1-Rad53-Dun1 pathway through the Crt1 transcriptional repressor, which is hyper-phosphorylated in response to genotoxic stress [26]. Therefore, we asked if the expression of an *RNR3-lacZ* reporter was influenced by *PKC1*. We found that both basal *RNR3-lacZ* expression and its induction by HU treatment were strongly diminished in a *pkc1Δ* mutant grown in the presence of osmotic support (Figure 4f). We also examined the effect of the *pkc1-3A* and *pkc1-S/T13A* alleles. *RNR3-lacZ* basal expression and induction were only modestly diminished in the *pkc1-3A* mutant, but were more strongly impaired in the *pkc1-S/T13A* mutant (Figure 4f). Although the basal and induced levels of *RNR3-lacZ* expression were reduced in each of these mutants, the relative induction was retained for all (approximately 6 to 8-fold). Finally, we tested the analog-sensitive *hrr25-I82A* mutant for induction of *RNR3-lacZ* expression and found that, as anticipated, inhibitor treatment strongly diminished HU induction (Figure 4f). These results suggest that Hrr25 may regulate the transcriptional response to DNA damage in part through phosphorylation of Pkc1 (Figure 5). This pathway would be independent of the Mec1-Rad53-Dun1 pathway, because Pkc1 hyper-phosphorylation in response to genotoxic stress does not require Rad53.

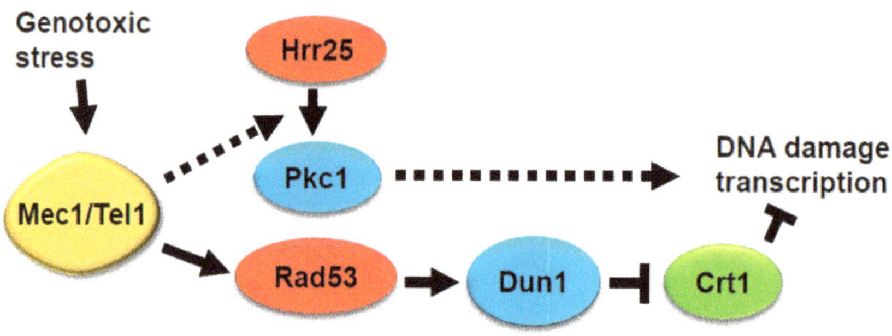

Figure 5. Proposed contribution of Hrr25 and Pkc1 to DNA damage-induced transcription. Pathway from Mec1 and Tel1 through Hrr25 and Pkc1 is added to the pathway established by Huang et al. [26]. Dashed arrows suggest indirect regulation.

4. Conclusions

We can draw several conclusions from this study. First, the checkpoint kinases Mec1 and Tel1 regulate hyper-phosphorylation of Pkc1 under conditions of genotoxic stress by inducing the association of CK1 homolog Hrr25 with Pkc1. This happens through a mechanism that does not require the phosphorylation of Hrr25 by Mec1 or Tel1 and suggests that Hrr25 and Pkc1 are indirect effectors of the checkpoint kinases. Second, a large collection of CK1 phosphorylation sites contribute to the genotoxic stress-induced Pkc1 band-shift. Finally, CK1 phosphorylation site mutants in Pkc1 are partially deficient in *RNR3-lacZ* basal and DNA damage-induced expression, suggesting that Pkc1 hyper-phosphorylation by Hrr25 contributes to this response.

Supplementary Materials: The following are available online at https://www.mdpi.com/article/10.3390/jof7100874/s1, Table S1: Pkc1 MS with HU only, Table S2: Pkc1 MS without HU only, Table S3: SILAC MS Pkc1 with and without HU, Figure S1: Binding of Hrr25 to Pkc1. Figure S2: Sensitivity of *hrr25* "gatekeeper" mutants to growth inhibition by inhibitory ATP analogs, Figure S3: Pkc1 phosphorylation sites identified in this study.

Author Contributions: D.E.L., L.L., W.R. and E.M. contributed to the design of the experimental approach and the interpretation of data. Conceptualization, D.E.L. and W.R.; Methodology, D.E.L., L.L., W.R., G.A. and E.M.; Software, J.V., W.R. and E.M.; Validation and Formal Analysis, D.E.L., L.L., W.R. and E.M.; Data Curation, L.L., J.V. and E.M.; Writing—Original Draft Preparation, D.E.L., L.L., W.R. and E.M.; Writing—Review & Editing, L.L., J.V., W.R., E.M., C.E.C., J.C.S., G.A. and D.E.L.; Supervision, Project Administration, and Funding Acquisition, D.E.L., G.A., C.E.C. and J.C.S. All authors have read and agreed to the published version of the manuscript.

Funding: This work was supported by grants from the NIH (R01 GM48533 to DEL; R01 GM129324 to JCS; and R24 GM134210 and S10 RR020946 to CEC). GA, JV, and WR were supported by the FWF Austrian Science Fund Special Research Program F34.

Data Availability Statement: The MS phospho-proteomics data have been deposited at the zenodo repository (https://zenodo.org/ accessed on 10 September 2021) and can be accessed via 10.5281/zenodo.5102666.

Acknowledgments: We thank David Hollenstein for general MS-related support, and Martha Cyert, Scott Emr, Juan Carlos Igual, Marcus Smolka, Stephen Elledge, and Gerhard Paravicini for yeast strains and plasmids.

Conflicts of Interest: The authors declare no conflict of interests. The funders had no role in the design of the study; in the collection, analyses, or interpretation of data; in the writing of the manuscript, or in the decision to publish the results.

References

1. Klis, F.M.; Mol, P.; Hellingwerf, K.; Brul, S. Dynamics of cell wall structure in *Saccharomyces cerevisiae*. *FEMS Microbiol. Rev.* **2002**, *26*, 239–256. [CrossRef]
2. Levin, D.E. Cell Wall Integrity Signaling in *Saccharomyces cerevisiae*. *Microbiol. Mol. Biol. Rev.* **2005**, *69*, 262–291. [CrossRef]
3. Levin, D.E. Regulation of cell wall biosynthesis in *Saccharomyces cerevisiae*: The cell wall integrity signaling pathway. *Genetics* **2011**, *189*, 1145–1175. [CrossRef]
4. Lesage, G.; Bussey, H. Cell Wall Assembly in *Saccharomyces cerevisiae*. *Microbiol. Mol. Biol. Rev.* **2006**, *70*, 317–343. [CrossRef] [PubMed]
5. Jung, U.S.; Levin, D. Genome-wide analysis of gene expression regulated by the yeast cell wall integrity signalling pathway. *Mol. Microbiol.* **1999**, *34*, 1049–1057. [CrossRef] [PubMed]
6. Roberts, C.J.; Nelson, B.; Marton, M.J.; Stoughton, R.; Meyer, M.R.; Bennett, H.A.; He, Y.D.; Dai, H.; Walker, W.L.; Hughes, T.R.; et al. Signaling and Circuitry of Multiple MAPK Pathways Revealed by a Matrix of Global Gene Expression Profiles. *Science* **2000**, *287*, 873–880. [CrossRef] [PubMed]
7. Jung, U.S.; Sobering, A.K.; Romeo, M.J.; Levin, D.E. Regulation of the yeast Rlm1 transcription factor by the Mpk1 cell wall integrity MAP kinase. *Mol. Microbiol.* **2002**, *46*, 781–789. [CrossRef] [PubMed]
8. Garcia, R.; Bermejo, C.; Grau, C.; Pérez, R.; Peña, J.M.R.; François, J.M.; Nombela, C.; Arroyo, J. The Global Transcriptional Response to Transient Cell Wall Damage in Saccharomyces cerevisiae and Its Regulation by the Cell Integrity Signaling Pathway. *J. Biol. Chem.* **2004**, *279*, 15183–15195. [CrossRef]
9. Boorsma, A.; De Nobel, H.; Ter Riet, B.; Bargmann, B.; Brul, S.; Hellingwerf, K.J.; Klis, F.M. Characterization of the transcriptional response to cell wall stress in *Saccharomyces cerevisiae*. *Yeast* **2004**, *21*, 413–427. [CrossRef] [PubMed]
10. Dodou, E.; Treisman, R. The Saccharomyces cerevisiae MADS-box transcription factor Rlm1 is a target for the Mpk1 mitogen-activated protein kinase pathway. *Mol. Cell. Biol.* **1997**, *17*, 1848–1859. [CrossRef]
11. Watanabe, Y.; Takaesu, G.; Hagiwara, M.; Irie, K.; Matsumoto, K. Characterization of a serum response factor-like protein in Saccharomyces cerevisiae, Rlm1, which has transcriptional activity regulated by the Mpk1 (Slt2) mitogen-activated protein kinase pathway. *Mol. Cell. Biol.* **1997**, *17*, 2615–2623. [CrossRef]
12. Madden, K.; Sheu, Y.-J.; Baetz, K.; Andrews, B.; Snyder, M. SBF Cell Cycle Regulator as a Target of the Yeast PKC-MAP Kinase Pathway. *Science* **1997**, *275*, 1781–1784. [CrossRef]
13. Baetz, K.; Moffat, J.; Haynes, J.; Chang, M.; Andrews, B. Transcriptional Coregulation by the Cell Integrity Mitogen-Activated Protein Kinase Slt2 and the Cell Cycle Regulator Swi4. *Mol. Cell. Biol.* **2001**, *21*, 6515–6528. [CrossRef] [PubMed]
14. Kim, K.Y.; Truman, A.W.; Levin, D.E. Yeast Mpk1 mitogen-activated protein kinase activates transcription through Swi4/Swi6 by a non-catalytic mechanism that requires upstream signal. *Mol. Cell. Biol.* **2008**, *28*, 2579–2589. [CrossRef]
15. Kim, K.-Y.; Levin, D.E. Mpk1 MAPK Association with the Paf1 Complex Blocks Sen1-Mediated Premature Transcription Termination. *Cell* **2011**, *144*, 745–756. [CrossRef] [PubMed]
16. LeDuc, A.; He, C.H.; Ramotar, D. Disruption of the Saccharomyces cerevisiae cell-wall pathway gene SLG1 causes hypersensitivity to the antitumor drug bleomycin. *Mol. Genet. Genom.* **2003**, *269*, 78–89. [CrossRef] [PubMed]
17. Queralt, E.; Igual, J.C. Functional Connection Between the Clb5 Cyclin, the Protein Kinase C Pathway and the Swi4 Transcription Factor in *Saccharomyces cerevisiae*. *Genetics* **2005**, *171*, 1485–1498. [CrossRef]

18. Zu, T.; Verna, J.; Ballester, R. Mutations in WSC genes for putative stress receptors result in sensitivity to multiple stress conditions and impairment of Rlm1-dependent gene expression in *Saccharomyces cerevisiae*. *Mol. Genet. Genom.* **2001**, *266*, 142–155. [CrossRef] [PubMed]
19. Soriano-Carot, M.; Bañó, M.C.; Igual, J.C. The yeast mitogen-activated protein kinase Slt2 is involved in the cellular response to genotoxic stress. *Cell Div.* **2012**, *7*, 1. [CrossRef]
20. Jiménez-Gutiérrez, E.; Alegría-Carrasco, E.; Sellers-Moya, A.; Molina, M.; Martín, H. Not just the wall: The other ways to turn the yeast CWI pathway on. *Int. Microbiol.* **2020**, *23*, 107–119. [CrossRef]
21. Albuquerque, C.P.; Smolka, M.B.; Payne, S.; Bafna, V.; Eng, J.; Zhou, H. A Multidimensional Chromatography Technology for In-depth Phosphoproteome Analysis. *Mol. Cell. Proteom.* **2008**, *7*, 1389–1396. [CrossRef]
22. Truman, A.; Kim, K.-Y.; Levin, D.E. Mechanism of Mpk1 Mitogen-Activated Protein Kinase Binding to the Swi4 Transcription Factor and Its Regulation by a Novel Caffeine-Induced Phosphorylation. *Mol. Cell. Biol.* **2009**, *29*, 6449–6461. [CrossRef]
23. Gobbini, E.; Cesena, D.; Galbiati, A.; Lockhart, A.; Longhese, M.P. Interplays between ATM/Tel1 and ATR/Mec1 in sensing and signaling DNA double-strand breaks. *DNA Repair* **2013**, *12*, 791–799. [CrossRef] [PubMed]
24. Nakada, D.; Shimomura, T.; Matsumoto, K.; Sugimoto, K. The ATM-related Tel1 protein of *Saccharomyces cerevisiae* controls a checkpoint response following phleomycin treatment. *Nucleic Acids Res.* **2003**, *31*, 1715–1724. [CrossRef] [PubMed]
25. Zou, L.; Elledge, S.J. Sensing DNA Damage Through ATRIP Recognition of RPA-ssDNA Complexes. *Science* **2003**, *300*, 1542–1548. [CrossRef] [PubMed]
26. Huang, M.; Zhou, Z.; Elledge, S.J. The DNA Replication and Damage Checkpoint Pathways Induce Transcription by Inhibition of the Crt1 Repressor. *Cell* **1998**, *94*, 595–605. [CrossRef]
27. Melo, J.; Toczyski, D. A unified view of the DNA-damage checkpoint. *Curr. Opin. Cell Biol.* **2002**, *14*, 237–245. [CrossRef]
28. Sidorova, J.M.; Breeden, L.L. Precocious G1/S transitions and genomic instability: The origin connection. *Mutat. Res.* **2003**, *532*, 5–19. [CrossRef]
29. Liang, F.; Wang, Y. DNA Damage Checkpoints Inhibit Mitotic Exit by Two Different Mechanisms. *Mol. Cell. Biol.* **2007**, *27*, 5067–5078. [CrossRef] [PubMed]
30. Soriano-Carot, M.; Quilis, I.; Bañó, C.; Igual, J.C. Protein kinase C controls activation of the DNA integrity checkpoint. *Nucleic Acids Res.* **2014**, *42*, 7084–7095. [CrossRef]
31. Liu, L.; Levin, D.E. Intracellular mechanism by which genotoxic stress activates yeast SAPK Mpk1. *Mol. Biol. Cell* **2018**, *29*, 2898–2909. [CrossRef] [PubMed]
32. Huang, K.N.; Symington, L.S. Mutation of the gene encoding protein kinase C 1 stimulates mitotic recombination in Saccharomyces cerevisiae. *Mol. Cell. Biol.* **1994**, *14*, 6039–6045. [CrossRef]
33. Li, X.; Heyer, W.-D. Homologous recombination in DNA repair and DNA damage tolerance. *Cell Res.* **2008**, *18*, 99–113. [CrossRef]
34. Yang, W.-L.; Bruno, M.E.C.; Carman, G.M. Regulation of Yeast CTP Synthetase Activity by Protein Kinase C. *J. Biol. Chem.* **1996**, *271*, 11113–11119. [CrossRef]
35. Lee, J.; Liu, L.; Levin, D.E. Stressing out or stressing in: Intracellular pathways for SAPK activation. *Curr. Genet.* **2019**, *65*, 417–421. [CrossRef]
36. Siliciano, P.G.; Tatchell, K. Transcription and regulatory signals at the mating type locus in yeast. *Cell* **1984**, *37*, 969–978. [CrossRef]
37. Kafadar, K.A.; Zhu, H.; Snyder, M.; Cyert, M.S. Negative regulation of calcineurin signaling by Hrr25p, a yeast homolog of casein kinase I. *Genes Dev.* **2003**, *17*, 2698–2708. [CrossRef]
38. Gietz, R.D.; Schiestl, R.H.; Willems, A.R.; Woods, R.A. Studies on the transformation of intact yeast cells by the LiAc/SS-DNA/PEG procedure. *Yeast* **1995**, *11*, 355–360. [CrossRef]
39. Sikorski, R.S.; Hieter, P. A system of shuttle vectors and yeast host strains designed for efficient manipulation of DNA in Saccharomyces cerevisiae. *Genetics* **1989**, *122*, 19–27. [CrossRef]
40. Hill, J.E.; Myers, A.; Koerner, T.J.; Tzagoloff, A. Yeast/*E. coli* shuttle vectors with multiple unique restriction sites. *Yeast* **1986**, *2*, 163–167. [CrossRef] [PubMed]
41. Denis, V.; Cyert, M.S. Molecular Analysis Reveals Localization of *Saccharomyces cerevisiae* Protein Kinase C to Sites of Polarized Growth and Pkc1p Targeting to the Nucleus and Mitotic Spindle. *Eukaryot. Cell* **2005**, *4*, 36–45. [CrossRef] [PubMed]
42. Lee, J.; Reiter, W.; Dohnal, I.; Gregori, C.; Beese-Sims, S.; Kuchler, K.; Ammerer, G.; Levin, D.E. MAPK Hog1 closes the S. cerevisiae glycerol channel Fps1 by phosphorylating and displacing its positive regulators. *Genes Dev.* **2013**, *27*, 2590–2601. [CrossRef] [PubMed]
43. Reiter, W.; Anrather, R.; Dohnal, I.; Pichler, P.; Veis, J.; Grøtli, M.; Posas, F.; Ammerer, G. Validation of regulated protein phosphorylation events in yeast by quantitative mass spectrometry analysis of purified proteins. *Proteomics* **2012**, *12*, 3030–3043. [CrossRef]
44. Kamada, Y.; Jung, U.S.; Piotrowski, J.; Levin, D. The protein kinase C-activated MAP kinase pathway of Saccharomyces cerevisiae mediates a novel aspect of the heat shock response. *Genes Dev.* **1995**, *9*, 1559–1571. [CrossRef] [PubMed]
45. Kushnirov, V.V. Rapid and reliable protein extraction from yeast. *Yeast* **2000**, *16*, 857–860. [CrossRef]
46. Zhao, C.; Jung, U.S.; Garrett-Engele, P.; Roe, T.; Cyert, M.S.; Levin, D.E. Temperature-Induced Expression of Yeast FKS2 Is under the Dual Control of Protein Kinase C and Calcineurin. *Mol. Cell. Biol.* **1998**, *18*, 1013–1022. [CrossRef]

47. Ong, S.-E.; Blagoev, B.; Kratchmarova, I.; Kristensen, D.B.; Steen, H.; Pandey, A.; Mann, M. Stable Isotope Labeling by Amino Acids in Cell Culture, SILAC, as a Simple and Accurate Approach to Expression Proteomics. *Mol. Cell. Proteom.* **2002**, *1*, 376–386. [CrossRef] [PubMed]
48. Cox, J.; Mann, M. MaxQuant enables high peptide identification rates, individualized p.p.b.-range mass accuracies and proteome-wide protein quantification. *Nat. Biotechnol.* **2008**, *26*, 1367–1372. [CrossRef]
49. Nyberg, K.A.; Michelson, R.J.; Putnam, C.W.; Weinert, T.A. Toward Maintaining the Genome: DNA Damage and Replication Checkpoints. *Annu. Rev. Genet.* **2002**, *36*, 617–656. [CrossRef]
50. Harper, J.W.; Elledge, S.J. The DNA Damage Response: Ten Years After. *Mol. Cell* **2007**, *28*, 739–745. [CrossRef]
51. Hoekstra, M.F.; Liskay, R.M.; Ou, A.C.; DeMaggio, A.J.; Burbee, D.G.; Heffron, F. HRR25, a putative protein kinase from budding yeast: Association with repair of damaged DNA. *Science* **1991**, *253*, 1031–1034. [CrossRef] [PubMed]
52. Ho, Y.; Mason, S.; Kobayashi, R.; Hoekstra, M.; Andrews, B. Role of the casein kinase I isoform, Hrr25, and the cell cycle-regulatory transcription factor, SBF, in the transcriptional response to DNA damage in Saccharomyces cerevisiae. *Proc. Natl. Acad. Sci. USA* **1997**, *94*, 581–586. [CrossRef] [PubMed]
53. Lopez, M.S.; Kliegman, J.I.; Shokat, K.M. The Logic and Design of Analog-Sensitive Kinases and Their Small Molecule Inhibitors. *Methods Enz.* **2014**, *548*, 189–213. [CrossRef]
54. Peng, Y.; Grassart, A.; Lu, R.; Wong, C.C.; Yates, J.; Barnes, G.; Drubin, D.G. Casein Kinase 1 Promotes Initiation of Clathrin-Mediated Endocytosis. *Dev. Cell* **2015**, *32*, 231–240. [CrossRef] [PubMed]
55. Flotow, H.; Roach, P.J. Synergistic phosphorylation of rabbit muscle glycogen synthase by cyclic AMP-dependent protein kinase and casein kinase I. Implications for hormonal regulation of glycogen synthase. *J. Biol. Chem.* **1989**, *264*, 9126–9128. [CrossRef]
56. Flotow, H.; Roach, P.J. Role of acidic residues as substrate determinants for casein kinase I. Role of acidic residues as substrate determinants for casein kinase I. *J. Biol. Chem.* **1991**, *266*, 3724–3727. [CrossRef]
57. Flotow, H.; Graves, P.; Wang, A.; Fiol, C.; Roeske, R.; Roach, P. Phosphate groups as substrate determinants for casein kinase I action. *J. Biol. Chem.* **1990**, *265*, 14264–14269. [CrossRef]
58. Pulgar, V.; Marin, O.; Meggio, F.; Allende, C.C.; Allende, J.E.; Pinna, L.A. Optimal sequences for non-phosphate-directed phoshorylation by protein kinase CK1 (casein kinase-1)—A re-evaluation. *Eur. J. Biochem.* **1999**, *260*, 520–526. [CrossRef]
59. Swaney, D.L.; Beltrao, P.; Starita, L.; Guo, A.; Rush, J.; Fields, S.; Krogan, N.J.; Villen, J. Global analysis of phosphorylation and ubiquitylation cross-talk in protein degradation. *Nat. Methods* **2013**, *10*, 676–682. [CrossRef]
60. MacGilvray, M.E.; Shishkova, E.; Place, M.; Wagner, E.R.; Coon, J.J.; Gasch, A.P. Phosphoproteome Response to Dithiothreitol Reveals Unique Versus Shared Features of *Saccharomyces cerevisiae* Stress Responses. *J. Proteome Res.* **2020**, *19*, 3405–3417. [CrossRef]
61. Lanz, M.C.; Yugandhar, K.; Gupta, S.; Sanford, E.J.; Faça, V.M.; Vega, S.; Joiner, A.M.N.; Fromme, J.C.; Yu, H.; Smolka, M.B. In-depth and 3-dimensional exploration of the budding yeast phosphoproteome. *EMBO Rep.* **2021**, *22*, e51121. [CrossRef] [PubMed]

Journal of *Fungi*

MDPI

Article

Clotrimazole-Induced Oxidative Stress Triggers Novel Yeast Pkc1-Independent Cell Wall Integrity MAPK Pathway Circuitry

Ángela Sellers-Moya, Marcos Nuévalos, María Molina * and Humberto Martín *

Departamento de Microbiología y Parasitología, Facultad de Farmacia, Instituto Ramón y Cajal de Investigaciones Sanitarias (IRYCIS), Universidad Complutense de Madrid, 28040 Madrid, Spain; angselle@ucm.es (Á.S.-M.); marnueva@ucm.es (M.N.)

* Correspondence: molmifa@ucm.es (M.M.); humberto@ucm.es (H.M.); Tel.: +34-91-3941888 (M.M. & H.M.)

Abstract: Azoles are one of the most widely used drugs to treat fungal infections. To further understand the fungal response to azoles, we analyzed the MAPK circuitry of the model yeast *Saccharomyces cerevisiae* that operates under treatment with these antifungals. Imidazoles, and particularly clotrimazole, trigger deeper changes in MAPK phosphorylation than triazoles, involving a reduction in signaling through the mating pathway and the activation of the MAPKs Hog1 and Slt2 from the High-Osmolarity Glycerol (HOG) and the Cell Wall Integrity (CWI) pathways, respectively. Clotrimazole treatment leads to actin aggregation, mitochondrial alteration, and oxidative stress, which is essential not only for the activation of both MAPKs, but also for the appearance of a low-mobility form of Slt2 caused by additional phosphorylation to that occurring at the conserved TEY activation motif. Clotrimazole-induced ROS production and Slt2 phosphorylation are linked to Tpk3-mediated PKA activity. Resistance to clotrimazole depends on HOG and CWI-pathway-mediated stress responses. However, Pkc1 and other proteins acting upstream in the pathway are not critical for the activation of the Slt2 MAPK module, suggesting a novel rewiring of signaling through the CWI pathway. We further show that the strong impact of azole treatment on MAPK signaling is conserved in other yeast species.

Keywords: yeast; cell wall integrity; MAPK; phosphorylation; azoles; clotrimazole

Citation: Sellers-Moya, Á.; Nuévalos, M.; Molina, M.; Martín, H. Clotrimazole-Induced Oxidative Stress Triggers Novel Yeast Pkc1-Independent Cell Wall Integrity MAPK Pathway Circuitry. *J. Fungi* **2021**, *7*, 647. https://doi.org/10.3390/jof7080647

Academic Editor: Robert A. Arkowitz

Received: 13 July 2021
Accepted: 5 August 2021
Published: 9 August 2021

1. Introduction

Signal transduction pathways mediated by mitogen-activated protein kinases (MAPKs) are essential for eukaryotic cells to respond to external stimuli, regulating vital cell processes such as proliferation, differentiation, survival, and apoptosis. As they play a central role in cellular homeostasis, compromised MAPK pathways may be involved in the development of serious diseases such as cancer or neurodegenerative disorders [1]. These pathways are composed of a module of three evolutionarily conserved kinases that are activated by sequential phosphorylation: a MAP kinase kinase kinase (MAPKKK), a MAP kinase kinase (MAPKK), and a MAP kinase (MAPK) [2]. For full activation, MAPKs need to be phosphorylated by MAPKKs on conserved T (threonine) and Y (tyrosine) within the activation domain [3]. Activated MAPKs then phosphorylate a diverse set of target proteins with different functions, including transcription factors, which are fundamental to the adaptive response.

In the yeast *Saccharomyces cerevisiae*, five MAPK routes govern mating, filamentous growth, osmolarity, spore wall assembly, and cell wall integrity [3]. The pheromone response pathway is crucial for yeast to initiate the process of mating differentiation [4]. Pheromones from the opposite mating partner are detected by G-protein-coupled receptors, transmitting the signal to the p21-activated kinase (PAK) Ste20 through G-protein activation. The MAPK module is composed of MAPKKK Ste11, MAPKK Ste7, and MAPK Fus3, which are attached to the scaffold protein Ste5. Kss1, the homolog MAPK of Fus3 responsible for filamentous and invasive growth, is also phosphorylated in response to pheromones but in a more transient manner than Fus3 [5]. The High-Osmolarity Glycerol (HOG) pathway is necessary for adaptation to hyperosmotic stress and is composed of two branches [6]. The first branch is named after the protein Sho1, a membrane-localized scaffold protein that interacts with the sensors Hkr1 and Msb2. The signal is transmitted to Ste20, which in turn activates MAPKKK Ste11. The second branch is named after the sensor histidine kinase Sln1, which signals through the phosphotransfer protein Ypd1 to the response regulator Ssk1, which activates the redundant MAPKKKs Ssk2 and Ssk22. Both branches converge in MAPKK Pbs2, leading to MAPK Hog1 activation [7].

The Cell Wall Integrity (CWI) pathway is necessary to maintain cell wall stability in response to cell surface stress, although it can also be activated by other stimuli not directly related to cell wall damage such as oxidative or genotoxic stress [8]. Perturbations in the cell wall or the plasma membrane are detected by a group of membrane-spanning sensors (the Wsc-type sensors Wsc1, Wsc2, and Wsc3, Mid2, and Mtl1) [9], which activate Rho1 GTPase through GDP/GTP exchange factors (GEFs) Rom2, Rom1, and Tus1 [10]. Activated Rho1 interacts with the protein kinase Pkc1, which in turn activates the MAPK module composed of MAPKKK Bck1, two redundant MAPKKs Mkk1 and Mkk2, and MAPK Slt2 [11]. Once activated, Slt2 can phosphorylate a diverse set of substrates, including the transcription factors Rlm1 and SBF complex (consisting of Swi4 and Swi6), which regulate the expression of genes implicated in cell-wall repair such as *MLP1* and *FKS1*, respectively [12]. In addition to Slt2 dual phosphorylation in its conserved TEY domain, other residues have been described to be phosphorylated under specific conditions. This is the case with caffeine treatment, which leads to DNA damage checkpoint kinases-mediated Slt2 phosphorylation at Ser423 and Ser428, affecting Slt2 interaction with Swi4 [13].

In recent years, the emergence of pathogenic fungi resistant to distinct antifungals has been alarming, as they have rapidly developed diverse resistance mechanisms against the different classes of drugs used in the clinic [14]. It is thus important to understand fungal cell responses to antifungal treatment, as this knowledge will aid the development of new therapeutic strategies to overcome this problem. Azoles are the most widely used class of antifungals. They are divided in two families, including imidazoles, which are commonly used for the treatment of superficial mycoses, and triazoles, which are preferably used in systemic mycosis due to their better pharmacokinetic profile, spectrum of activity, and safety [15]. Within imidazoles, clotrimazole is effective against skin, vulvovaginal, and oropharyngeal fungal infections and has become a drug of interest for the treatment of other diseases such as malaria and some cancers [16]. The main effect of azole antifungals is the inhibition of the ergosterol biosynthetic pathway, and thus alterations in the permeability and fluidity of the cell membrane [17]. Additionally, some of the azoles were reported to have a more complex mode of action, triggering other effects such as the induction of farnesol production [18] or the generation of reactive oxygen species (ROS) [19,20]. Regarding the effect of azoles on MAPK signaling, little is known. Activation of CWI, HOG, and Cek1 routes was observed in *Candida albicans* after exposure to the triazole fluconazole [21], and deletion of CWI pathway components encoding genes *PKC1*, *BCK1*, *SLT2*, and *SWI4* rendered cells of both *S. cerevisiae* and *C. albicans* more sensitive to fluconazole [22]. However, the effect of imidazoles on fungal MAPK pathways has not yet been explored.

The use of *S. cerevisiae* as a model system in antifungal research is currently considered opportune [23]. Here, we exploited the powerful genetics and biochemistry of this yeast to gain insight into the nature of the stress imposed by clotrimazole and the MAPK signaling circuitry that enables resistance to this azole.

2. Materials and Methods

2.1. Yeast Strains and Plasmids

The *Saccharomyces cerevisiae* strains used in this work were the wild-type BY4741 (*MAT*a *his3Δ1 leu2Δ0 met15Δ0 ura3Δ0*) and the isogenic *kanMX4* deletion mutants from Euroscarf (Frankfurt, Germany); the YMJ29 strain (BY4742 isogenic, *MATα his3Δ1 leu2Δ0* *lys2Δ0 ura3Δ0 mkk2Δ::kanMX4 mkk1Δ::SpHIS5*) [24]; the wild-type CML128 (*MAT*a *leu2-3 112 ura3-52 trp1 his4 can1r*) and its isogenic *pkc1::LEU2* mutant strain MML344 [25]; WT-RLM1Myc (BY4741 isogenic *RLM1-6MYC::HIS3*) and *slt2Δ*-RLM1Myc (BY4741 isogenic *slt2Δ::kanMX4 RLM1-6MYC::HIS3*) [26]; mutant *rlm1Δ* (BY4741 isogenic *rlm1Δ::TRP1*) [27]; and mutant *bcy1Δ* (BY4741 isogenic *bcy1Δ::KanMX4*) [28]. Mutant *pkc1Δ* (BY4741 isogenic *pkc1Δ::HIS3*) was kindly provided by Dr. Javier Arroyo. Strains YSTH1 (Y3656 isogenic, *MATα can1Δ::MFA1pr-HIS3-MFα1pr-LEU2 his3 leu2 ura3 met15 lys2 HO::natMX6::SSB1*) and YSTH2 (Y3656 isogenic, *HO::pMLP1-MKK1*^{S386P}*-tADH-natMX6::SSB1*) including IPAC [29] were a gift from Dr. Elena Jiménez-Gutiérrez. The WT-SWI6Myc strain (BY4741 isogenic *SWI6-6MYC::LEU2*) was obtained by amplifying the pRS305m plasmid containing a Myc_6 epitope [30] with primers 5′-CTGACGAAATGCAAGATTTTTAAAAAGCATG CTTCAGCTATGGAGCAAAAGCTCATTTCTG-3′ and 5′-CAAATAAAGTCATAAAAGTT AATGCAATGAAATCACATGCCCTTTTTAAGCAAGGATTTTC-3′ (regions homologous to the *SWI6* sequence are underlined) and integrating the PCR product into the *SWI6* locus. To obtain BY4741-T (BY4741 isogenic *trp1Δ::NatMX6*), the disruption cassette amplified with the primers 5′-GCACGTGAGTATACGTGATTAAGCACACAAAGGCAGCTTGG CGTACGCTGCAGGTCGAC-3′ and 5′-CAATACTTAAATAAATACTACTCAGTAATAAC CTATTTCTTAGCATCGATGAATTCGAGCTCG-3′ (regions homologous to *TRP1* are underlined) was integrated into the *TRP1* locus.

For the experiments with different yeast species, the wild-type strains used were *Saccharomyces cerevisiae* BY4741, *Pichia pastoris* GS115 (Invitrogen, Waltham, MA, USA), *Pichia anomala* 1026, *Kluyveromyces lactis* 1049, *Hansenula mrakii* K9, *Candida glabrata* ATCC 2001 CBS138, *Meyerozyma guilliermondii* TP11010 CNM-CL9533, *Candida albicans* 4482A, and *Candida tropicalis* 4458 CNM-CL9537. Strains were kindly provided by Ana Alastruey-Izquierdo from the Instituto de Salud Carlos III, Spain (CNM-strains) and by Elvira Marín (Departamento de Microbiología y Parasitología, Universidad Complutense de Madrid, Spain).

The plasmids used in this work were pRS316 [31], pRS316-*SLT2* [32], p2313 (YEp351-*SLT2-FLAG*) [33], pVD67 (pGRU2-*PKC1-GFP*) [34], pRS314 [31], pRS314-RLM1-HA, pRS314-*rlm1*-3m-HA, pRS314-*rlm1*-L324A/V326A-HA, pRS314-*rlm1*-10m-HA [27,35], p*MLP1-LacZ* [36], and YEplac112-Ilv6-mCherry [37].

2.2. Culture Conditions

Depending on the experimental approaches used, yeast cells were grown on YPD (2% glucose, 2% peptone, and 1% yeast extract) or SD medium (a 0.17% yeast nitrogen base, 0.5% ammonium sulfate, and 2% glucose) supplemented with the required amino acids, either broth or agar. For routine cultures, yeast cells were grown overnight at 24 or 30 °C on YPD or SD medium. The cultures were refreshed to an OD_{600} of 0.3 and grown on YPD for 2 h at the same temperature to allow cells to enter into the exponential phase. For further experiments, the cultures were either grown under the same conditions (the untreated samples) or supplemented with the corresponding compound. The compounds tested were clotrimazole, ketoconazole, itraconazole, fluconazole, chlorpromazine, N-acetylcysteine (Sigma-Aldrich, St. Louis, MO, USA), and Congo red (Merck, Darmstadt, Germany). Cells were collected at the indicated times and processed as mentioned in the following assays. For the experiments with different yeast species, YPD was supplemented with 1 M MES buffer.

2.3. Multi-Well-Plate Sensitivity Assay

Yeast cells from an overnight culture were diluted to a final OD_{600} of 0.005 and cultured at 30 °C in multi-well plates containing YPD with serial dilutions of the azoles clotrimazole, ketoconazole, itraconazole, and fluconazole at concentrations from 0.0976 to 100 μg/mL, as well as a control without compound. Growth was determined as OD_{595} after 24 h of static incubation using a microplate reader (Bio rad 680; Bio rad, Hercules, CA, USA).

2.4. Yeast Drop Dilution Growth Assays

Growth assays on solid media were performed by culturing cells in YPD to an OD_{600} of 0.5 and spotting samples (5 μL) of 10-fold dilutions of the cell suspensions onto the surface of YPD plates, followed by incubation at 30 °C for 72 h. YPD plates were supplemented with 1 μg/mL of clotrimazole, 30 μg/mL of Congo red, 20 mM of N-acetylcysteine, or 1 M of sorbitol (PanReac AppliChem ITW Reagents, Barcelona, Spain) when indicated.

2.5. Preparation of Yeast Extracts and Immunoblotting Analysis

The preparation of yeast extracts from routine cultures, fractionation by SDS-PAGE, and transfer to nitrocellulose membranes were described previously [38]. When indicated, staining of membranes with 0.5% Ponceau in 5% acetic acid was performed. Rabbit monoclonal anti-phospho-p44/42 (Erk1/2, Thr202/Tyr204, #4370; Cell Signaling; Danvers, MA, USA), rabbit monoclonal anti-phospho-p38 (Thr180/Tyr182, #9215; Cell Signaling), mouse monoclonal anti-Mpk1 (E-9; Santa Cruz Biotechnology, Dallas, TX, USA), rabbit polyclonal anti-Hog1 (y215; Santa Cruz), mouse monoclonal anti-Myc clone 4A6 (Millipore, Burlington, MA, USA), mouse monoclonal anti-HA (12CA5, Sigma-Aldrich,), and rabbit polyclonal anti-G6PDH (Sigma-Aldrich) antibodies were used to recognize dually phosphorylated Slt2, Kss1, and Fus3; dually phosphorylated Hog1; Slt2; Hog1; Myc-tagged proteins; HA-tagged proteins; and G6PDH, respectively. Detection of Glucose-6-phosphate dehydrogenase (G6PDH) encoded by the gene *ZWF1* was used as a loading control. The primary antibodies were detected using a fluorescence-conjugated secondary antibody with an Odyssey Infrared Imaging System (Li-Cor Biosciences; Lincoln, NE, USA).

2.6. Alkaline Phosphatase Assay

Collected cells were resuspended in cold lysis buffer (50 mM TrisHCl at pH 7.5, 150 mM NaCl, 50 mM NaF, 5 mM sodium pyrophosphate, 50 mM β-glycerol-phosphate, 0.1% NP 40, and 10% glycerol) supplemented with 1 mM phenylmethylsulfonyl fluoride (PMSF) and a protease inhibitors mixture (Thermo Scientific, Waltham, MA, USA). Cells were lysed using glass beads, and the protein concentration of the extracts was measured at 280 nm and normalized with lysis buffer. For immunoprecipitation, 200 μL of extracts were incubated with 40 μL of FLAG-Sepharose beads (Sigma-Aldrich) overnight at 4 °C. Beads were extensively washed with alkaline phosphatase buffer (50 mM Tris-HCl at pH 7.5, 100 mM NaCl, 10 mM $MgCl_2$, and 1 mM DTT, adjusted to a pH of 7.9) and finally resuspended with 400 μL of the same buffer. When indicated, 15 μL of alkaline phosphatase from calf intestine (ALP) or 40 μL of 100 mM sodium orthovanadate (Na_3VO_4), as a phosphatase inhibitor, to a final concentration of 10 mM were added. Samples were incubated at 37 °C for 1 h and centrifugated at 3000 rpm for 1 min. We added 2× SDS loading buffer. Proteins were boiled for 5 min and then analyzed by SDS-PAGE and immunoblotting.

2.7. Microscopy Techniques

For mCherry in vivo fluorescence microscopy, yeast transformants were cultured as usual and cells were collected by centrifugation of 1 mL of culture at 2500 rpm. For actin staining, cells were cultured as usual, fixed with *p*-formaldehyde, and stained with rhodamine–phalloidin as described previously [39]. Samples were prepared for visualization and the microscope used was an Eclipse TE2000U (Nikon, Tokyo, Japan) with the appropriate sets of filters. Digital images were acquired with an Orca C4742-95-12ER charge-coupled device camera (Hamamatsu, Hamamatsu city, Japan) and processed with HCImage software (Hamamatsu).

2.8. Quantitative RT-PCR Assays

RNA isolation and purification from yeast cells were conducted using the NucleoSpin RNA Mini kit for RNA purification (Macherey-Nagel, Düren, Germany), following the manufacturer's instructions. Real-time quantitative PCR (RT-qPCR) assays were performed as previously detailed [40]. For quantification, the abundance of each gene was determined relative to the standard transcript of *ACT1*, and the final data of relative gene expression were calculated following the $2^{-2\Delta\Delta Ct}$ method [41]. Primers were kindly provided by Dr. Javier Arroyo.

2.9. Flow Cytometry Evaluation of Oxidative Damage

For the detection of oxygen free radicals (ROS) and variations in the mitochondrial membrane potential, yeast cells were grown on YPD and cultured as usual, and 2.5 µg/mL of dihydroethidium (DHE) or 5 µg/mL of rhodamine 123, respectively, were added to 1 mL of each sample and incubated at 24 °C for 5 and 30 min. Then, samples were diluted 1:10 in PBS and analyzed on a FACScalibur flow cytometer (Becton Dickinson, Franklin Lakes, NJ, USA). Data were analyzed with FlowJo software (Becton Dickinson).

3. Results

3.1. Clotrimazole Activates the CWI and HOG Pathways and Attenuates the Mating Pathway

To test the changes in yeast MAPK circuitry imposed by antifungal azoles, we began by analyzing the sensitivity of the reference *S. cerevisiae* wild-type strain (BY4741) to four clinically relevant azoles: two imidazoles, clotrimazole (CLT) and ketoconazole (KTC); and two triazoles, itraconazole (ITC) and fluconazole (FLC). As observed in Figure 1A, CLT was the most potent azole, exhibiting inhibitory concentrations eight-fold lower than the rest of the azoles, with the sensitivity to FLC being the lowest.

We thus analyzed the impact of treating exponentially growing yeast cells with an inhibitory CLT concentration of 50 µg/mL and a comparable eight-fold-higher concentration of KTC, ITC, and FLC at two different time points on MAPK phosphorylation (Figure 1B). Treatment with both imidazoles, CLT and KTC, led to the activation of the MAPKs Hog1 and Slt2 of the HOG and CWI pathways by dual phosphorylation, respectively, after a short exposure of 30 min. Interestingly, at a longer exposure time of 240 min, in addition to the increase in Slt2 phosphorylation levels, a slower phospho-Slt2 migrating band was apparent after treatment with either CLT or KTC. The phosphorylated Slt2 form that ran with low mobility is hereafter called lowphospho-Slt2 (LP-Slt2), in contrast with the canonical phospho-Slt2 (P-Slt2). The stimulation of yeast cells with either of these two imidazole drugs induced a reduction in the basal phosphorylation levels of the MAPKs from the pheromone response pathway Fus3 and Kss1. After the longest exposure time, triazoles prompted some activation of Slt2, but neither the appearance of the LP-Slt2 form, nor the activation of Hog1, nor a reduction in Fus3 and Kss1 phosphorylation occurred. All these results indicate that imidazoles, but not triazoles, elicit specific and profound changes in yeast-MAPK-mediated signal transduction. Since CLT imposed the most severe changes in MAPK phosphorylation, we decided to focus on the effect of this imidazole on signaling in yeast cells.

We first identified the HOG pathway branch that senses and transmits the stimulus in response to CLT. In contrast to Sho1, removal of Ssk1 impaired signaling to Hog1 (Figure 1C), indicating that CLT is detected and transmitted through the branch of the HOG pathway in which the Sln1-Ypd1-Ssk1 phospho-relay system operates. We also explored the crosstalk between HOG and other MAPK pathways, as the HOG pathway regulates signaling through the CWI and mating pathways under some stimuli [29,42,43]. Lack of this MAPK caused only a slight reduction in Slt2 phosphorylation at 30 min without interfering with the appearance of the LP-Slt2 form, indicating that Slt2 activation does not depend on this crosstalk mechanism (Figure 1D). In contrast, CLT-imposed Fus3 and Kss1 dephosphorylation disappeared at 30 min and was partially impaired at 240 min in the absence of Hog1, suggesting that Hog1 modulates signaling through the mating pathway in response to this azole.

The importance of the essential components for signaling through the HOG and CWI MAPK pathways for resistance to CLT was also analyzed. As observed in Figure 1E, *hog1*Δ and *pbs2*Δ mutants exhibited a slight sensitivity to CLT, whereas *slt2*Δ and *bck1*Δ mutants displayed a significantly higher sensitivity to the antifungal drug. In contrast, only the CWI pathway kinases proved to be necessary for cells to grow under the cell-wall-altering agent Congo red (CR). These results confirm the relevance of these pathways for responding to this azole.

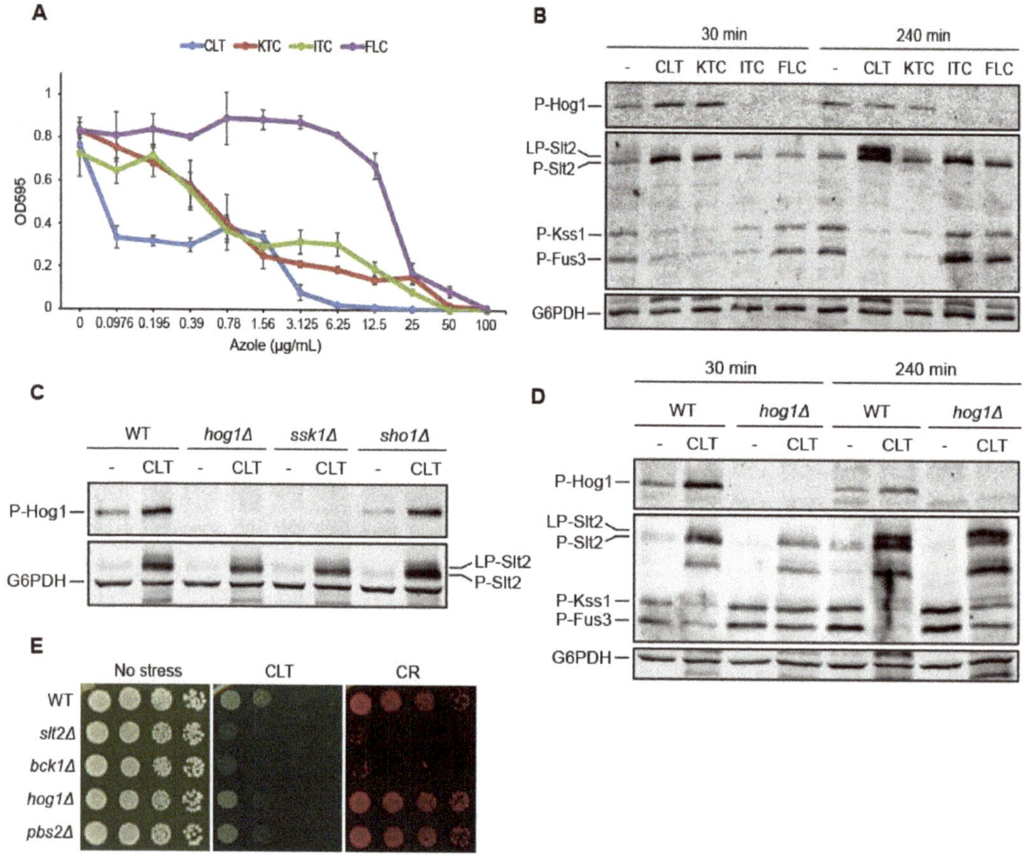

Figure 1. Impact of azole treatment on *S. cerevisiae* MAPK signaling. (**A**) Multi-well-plate sensitivity assay of the BY4741 strain for the indicated concentrations of clotrimazole (CLT), ketoconazole (KTC), itraconazole (ITC), and fluconazole (FLC). Cells were cultured at 30 °C for 24 h and the optical density at 595 nm (OD_{595}) was measured. Data are presented as the mean of three independent experiments. Error bars indicate standard deviation. (**B**) Western blotting analysis of extracts of BY4741 cells treated with 50 µg/mL of CLT; 400 µg/mL of KTC, ITC, and FLC; or without any azole (-) for the indicated times at 24 °C. Dually phosphorylated Hog1 was detected with anti-phospho-p38; dually phosphorylated Slt2, Kss1, and Fus3 were detected with anti-phospho-p44/42; and G6PDH, encoded by the gene *ZWF1* (as the loading control), with anti-G6PDH. (**C**,**D**) Western blotting analysis of cell extracts of BY4741 (WT) and the indicated isogenic mutant strains. Cells were cultured without (-) or with 50 µg/mL of CLT for 30 min (**C**) or 30 and 240 min (**D**) at 24 °C. Dually phosphorylated MAPKs and G6PDH were detected as described in (**B**). Representative blots from three independent experiments are shown. (**E**) Drop dilution growth assay of BY4741 (WT) and the indicated isogenic mutant strains. Ten-fold serial dilutions of cell suspensions were spotted onto YPD plates in the absence (no stress) or presence of 1 µg/mL of CLT or 30 µg/mL of Congo red (CR) and incubated at 30 °C for 72 h. A representative assay from three independent experiments is shown.

3.2. *Clotrimazole Induces a Phosphorylation-Dependent Mobility Shift in Slt2*

To further investigate the effect of CLT on MAPK signaling, we compared Slt2, Fus3, and Kss1 phosphorylation under CLT and CR treatment in an *slt2*Δ strain transformed with either a centromeric plasmid carrying the *SLT2* gene or the corresponding empty vector (Figure 2A). Whereas CLT led to the appearance of LP-Slt2 and stronger Slt2 phosphorylation than CR, neither the absence nor the presence of this MAPK had an impact on reducing the amount of phosphorylated Fus3 and Kss1 upon treatment with this azole. Interestingly, the use of anti-Slt2 antibodies showed that a significant proportion of the overall amount of Slt2 within the cell experienced the SDS-PAGE mobility shift.

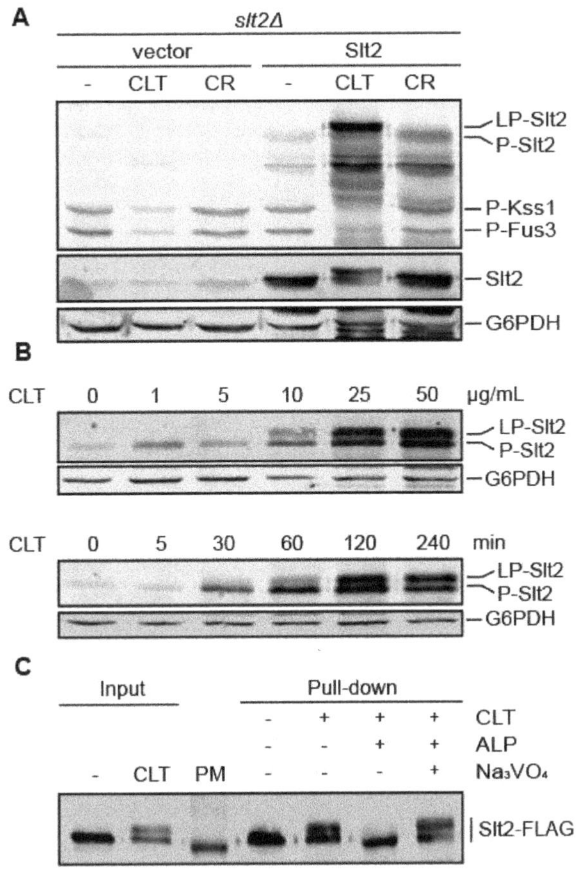

Figure 2. Characterization of Slt2 phosphorylation and LP-Slt2 emergence upon clotrimazole treatment. (**A**) Western blotting analysis of extracts of BY4741 *slt2*Δ cells transformed with either the empty vector pRS316 or pRS316-*SLT2* and cultured with 50 µg/mL of CLT, 30 µg/mL of CR, or without treatment (−) for 4 h at 24 °C. Dually phosphorylated Slt2, Kss1, and Fus3; Slt2 protein; and G6PDH (as a loading control) were detected with anti-phospho-p44/42, anti-Mpk1, and anti-G6PDH, respectively. (**B**) Western blotting analysis of extracts of BY4741 cells treated with increasing concentrations of CLT for 4 h (upper panel) and with 50 µg/mL of CLT for different times (lower panel) at 24 °C. Dually phosphorylated Slt2 and G6PDH were detected as described in (**A**). (**C**) Alkaline phosphatase assay of BY4741 *slt2*Δ cells transformed with p2313 plasmid (Slt2-FLAG) and cultured with (+) or without (−) 50 µg/mL of CLT for 4 h at 24 °C. Immunoprecipitated Slt2-FLAG with anti-FLAG antibodies (pull-down) was treated with (+) or without (−) alkaline phosphatase (ALP) from calf intestine in the absence (−) or presence (+) of sodium orthovanadate (Na_3VO_4), a phosphatase inhibitor. Cell extracts (input) were also analyzed. Slt2-FLAG was detected with anti-Mpk1 antibody. PM corresponds to the lane loaded with the molecular weight protein marker. Representative blots from three (**A**,**B**) and two (**C**) independent experiments are shown.

Having found that Slt2 shifted upward when cells were treated with imidazoles, we next sought to determine the dose-dependence and kinetics of this Slt2 modification upon CLT treatment. As shown in Figure 2B, Slt2 activation was dependent on CLT concentration and exposure time. Thus, treatment with CLT concentrations higher than 10 μg/mL and exposure times longer than 30 min resulted in a strong and progressive increase in both Slt2 phosphorylation and LP-Slt2 levels.

To gain insight into the origin of this LP-Slt2 emergence with CLT treatment, we assessed whether it can be promoted by phosphorylation events. Therefore, we analyzed the effect of phosphatase treatment on Slt2 electrophoretic mobility (Figure 2C). The removal of phosphates from Slt2 purified from CLT-treated cells led to the disappearance of the low-mobility form, suggesting that Slt2 can be phosphorylated in one or more sites additional to the conserved T and Y at the activation loop in response to CLT stimulation.

3.3. *Pkc1 Is Not Critical for Clotrimazole-Induced Signal Transduction through the CWI Pathway*

To characterize how the CWI pathway senses the stimulus and transmits the signal when cells are exposed to CLT, we studied the contribution of different components of this pathway. CR-treated cells were analyzed in parallel, since the proteins involved in signal transduction for this stimulus are well-known [44,45].

As observed for CR, CLT-induced Slt2 phosphorylation was abolished in the absence of both Mkk1 and Mkk2 or Bck1, indicating the essential role of protein kinases from the MAPK module in conveying signaling through the pathway (Figure 3A,B). As expected [46], Mkk1 played a central role in phosphorylating Slt2, but the low level of phosphorylated Slt2 observed in the *mkk1Δ* mutant was also accompanied by the appearance of a slow migrating band. This result reflects that this additional post-translational modification does not depend on the level of Slt2 activation.

Strikingly, and in contrast to what occurs with other CWI pathway stimuli such as CR, Pkc1 proved to be dispensable for CLT-induced Slt2 phosphorylation. As observed in Figure 3C, signaling through the CWI pathway in response to CLT was possible in the absence of Pkc1, leading to an increase in Slt2 phosphorylation, although P-Slt2 and LP-Slt2 levels were lower than in the wild-type strain. Therefore, when *pkc1Δ* cells were transformed with a plasmid bearing Pkc1-GFP, the phosphorylation levels of Slt2 increased in CLT-treated cells and CR-induced activation of Slt2 was restored (Figure 3D). Drop growth assays in the presence of sorbitol were also performed (Figure 3E). The *pkc1Δ* mutant, in contrast to the *slt2Δ* one, was sensitive to CLT even in the presence of an osmotic stabilizer, suggesting a role for Pkc1 in cell defense against this antifungal, in addition to its role in CWI pathway signal transmission. As this was the first time that Pkc1 did not seem to be essential for CWI pathway signaling, we wanted to confirm these results. We thus performed the same experiments mentioned above using a *pkc1Δ* mutant derived from another yeast strain with a different genetic background and observed similar results (Figure S1).

To further ascertain the specificity of these observations for CLT treatment and not for other stimuli of the pathway, we also analyzed the activation of Slt2 in the *pkc1Δ* mutant in response to caffeine, zymolyase, and tunicamycin. As shown in Figure 3F, none of these stimuli led to Slt2 activation in the absence of Pkc1, whereas CLT clearly prompted Slt2 phosphorylation to some extent. Altogether, these results show that, although Pkc1 seems to be important for the cell to respond to and survive CLT stress, an alternative route independent of Pkc1 is also capable of activating the MAPK module of the CWI pathway in response to this azole.

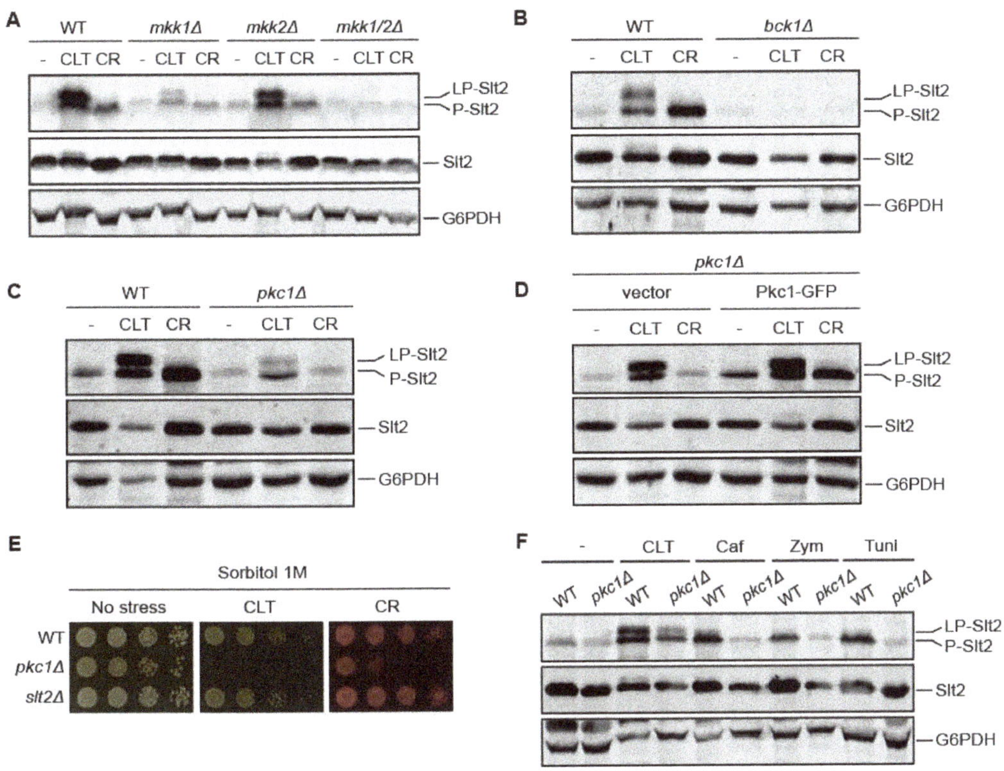

Figure 3. Involvement of protein kinases operating at the CWI pathway in clotrimazole-induced signaling to Slt2. (**A–C**) Western blotting analysis of extracts of BY4741 (WT) and the indicated isogenic mutant strains. Cells were cultured without stimulus (-), with 50 μg/mL of CLT, or with 30 μg/mL of CR for 4 h at 24 °C. Dually phosphorylated Slt2, Slt2 protein, and G6PDH (as a loading control) were detected with anti-phospho-p44/42, anti-Mpk1, and anti-G6PDH, respectively. (**D**) Western blotting analysis of *pkc1*Δ cells transformed with either the empty vector pRS316 or with pVD67 (Pkc1-GFP). Cells were cultured and proteins were detected as described in (**A–C**). (**E**) Sensitivity of BY4741 (WT) and the indicated mutant strains to CLT and CR per drop dilution growth assay. Ten-fold serial dilutions of cell suspensions were spotted onto YPD plates supplemented with 1 M sorbitol in the absence (no stress) or presence of 1 μg/mL of CLT or 30 μg/mL of CR, and incubated at 30 °C for 72 h. A representative assay from three independent experiments is shown. (**F**) Western blotting analysis of extracts of BY4741 (WT) and its isogenic *pkc1*Δ strain. Cells were cultured without stimulus (-), with 50 μg/mL of CLT, 8 mM of caffeine (Caf), 0.4 U/mL of zymolyase (Zym), or 5 μg/mL of tunicamycin (Tuni) for 2 h at 24 °C. Proteins were detected as described in (**A–C**). Representative blots from two independent experiments are shown.

3.4. The Mechanosensor Wsc1 Is Involved in CWI Signaling Induced by Clotrimazole

Since Pkc1 is not essential for CLT-induced Slt2 activation, we analyzed the importance of the CWI pathway components upstream of Pkc1 for signal transmission. Removal of any of the Rho1 GEFs, namely Rom1, Rom2, or Tus1, neither prevented Slt2 activation nor Slt2 band shift (Figure 4A). Notably, the effect of *ROM2* deletion on CR-induced Slt2 activation is in sharp contrast to that of CLT-induced Slt2 activation. Since Rom2 is likely the main GEF for Rho1 activation after cell-wall damage [45], this shows that CLT does not impact yeast cells as a typical cell-wall-stress stimulus.

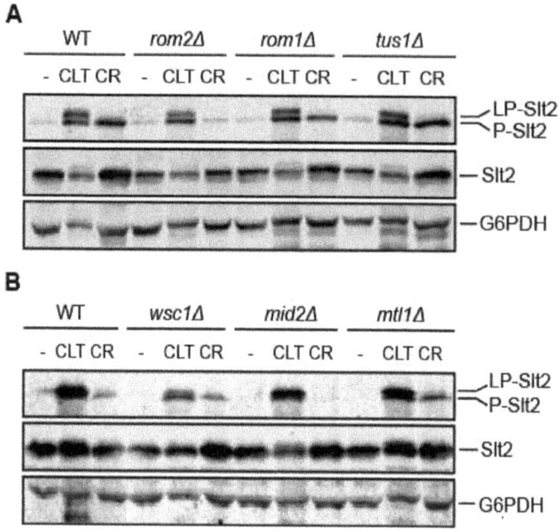

Figure 4. Involvement of components upstream of Pkc1 in clotrimazole-triggered signaling through the CWI pathway. (**A**,**B**) Western blotting analysis of extracts of BY4741 (WT) and the indicated isogenic mutant strains. Cells were cultured without stimulus (-), with 50 μg/mL of CLT, or with 30 μg/mL of CR for 4 h at 24 °C. Dually phosphorylated Slt2, Slt2 protein, and G6PDH (as a loading control) were detected with anti-phospho-p44/42, anti-Mpk1, and anti-G6PDH, respectively. Representative blots from three independent experiments are shown.

The analysis of cells lacking each of the three main cell-wall-stress sensors (Wsc1, Mid2, and Mtl1) revealed that unlike the observation for Mid2 and Mtl1, the lack of Wsc1 considerably reduced Slt2 activation (Figure 4B), but without completely compromising MAPK activation. These results indicate an active but partial role of this mechanosensor in the response to CLT.

3.5. The Transcriptional Rlm1-Driven CWI Pathway Response to Clotrimazole Is Weak

To explore the transcriptional response to CLT driven by the CWI pathway, we studied the involvement of the principal transcription factors activated by Slt2, namely Rlm1 and the SBF complex (composed of Swi4 and Swi6 proteins). First, we analyzed the sensitivity of the mutants *rlm1*Δ, *swi4*Δ, and *swi6*Δ to CLT (Figure 5A). As in the case of CR, cells deficient in Swi4 and Swi6 showed a high sensitivity to the azole, similar to that observed for *slt2*Δ cells, whereas the lack of Rlm1 did not seem to severely impact CLT sensitivity.

As shown in Figure 5B, both Rlm1 and Swi6 experienced Slt2-dependent phosphorylation after CLT exposure, which was detectable by a characteristic SDS-PAGE mobility shift [47,48]. Intriguingly, whereas the CLT-induced mobility shift in Swi6 followed the same pattern as that generated by CR, Rlm1 forms of slower mobility than those observed in the presence of CR appeared when cells were treated with this azole. This result suggests that under CLT treatment, Rlm1 is likely phosphorylated at sites additional to those targeted by other stimuli of the pathway. Moreover, this additional modification is dependent on the presence of Slt2 (Figure 5C). We thus used Rlm1 mutants in three and ten potential targets of Slt2-phosphorylation (Rlm1-3m-HA and Rlm1-10m-HA), which were previously described to be important for full Rlm1 activation [27], and analyzed their mobility shift after CLT treatment (Figure 5D). Both Rlm1 mutant versions migrated faster than the wild-type form, indicating that these phosphorylation sites are involved in a CLT-dependent mobility pattern. An Rlm1 version unable to interact with Slt2 (Rlm1-L3124AV326A-HA) [35] showed an even more reduced CLT-induced mobility shift. Note that the mobility shift was not completely abolished in either the Rlm1-10m-HA version or the Rlm1-L3124AV326A-HA version, again suggesting the presence of post-translational modifications in other residues apart from those already known for this transcription factor.

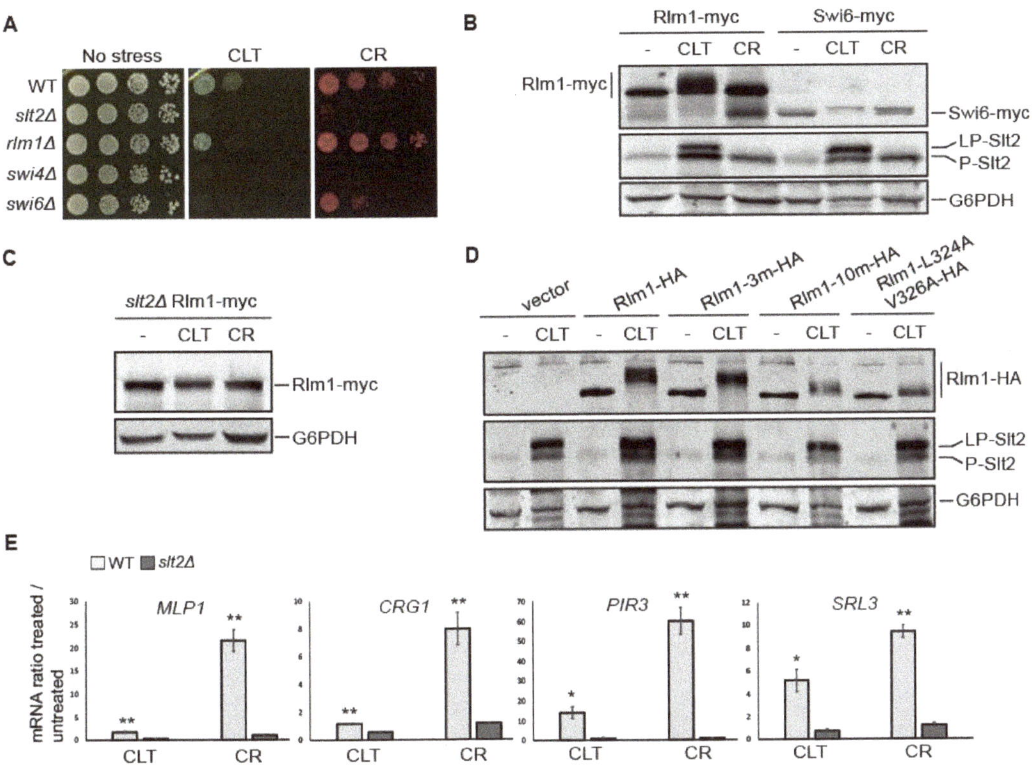

Figure 5. Analysis of the role of transcription factors Rlm1 and SBF in the clotrimazole response. (**A**) Drop dilution growth assay of BY4741 (WT) and the indicated isogenic mutant strains. Ten-fold serial dilutions of cell suspensions were spotted onto YPD plates in the absence (no stress) or presence of 1 μg/mL of CLT or 30 μg/mL of CR and incubated at 30 °C for 72 h. A representative assay from three independent experiments is shown. (**B**,**C**) Western blotting analysis of the extracts of WT-RLM1Myc, WT-SWI6Myc, and *slt2*Δ-RLM1Myc cells after treatment with 50 μg/mL of CLT, 30 μg/mL of CR, or without treatment (-) for 4 h at 24 °C. Rlm1-myc and Swi6-myc, dually phosphorylated Slt2, and G6PDH (as a loading control) were detected with anti-myc, anti-phospho-p44/42, and anti-G6PDH, respectively. (**D**) Western blotting analysis of extracts of the *rlm1*Δ mutant strain transformed with the pRS314 plasmid bearing the indicated versions of Rlm1-HA. Cells were cultured without stimulus (-) or with 50 μg/mL of CLT for 4 h at 24 °C. Rlm1-HA was detected with anti-HA antibody and dually phosphorylated Slt2 and G6PDH were detected as described in (**A**). Representative blots from two independent experiments are shown. (**E**) Transcriptional expression of *MLP1*, *CRG1*, *PIR3*, and *SRL3* was analyzed by RT-qPCR in BY4741 WT and isogenic *slt2*Δ strains after treatment with 50 μg/mL of CLT, 30 μg/mL of CR, or in basal conditions for 4 h at 24 °C. Values represent the expression ratio between treated and non-treated cells. Data correspond to the mean and standard deviation of three independent experiments. Student's t-test was calculated between the WT and *slt2*Δ strains. * p-value < 0.05; ** p-value < 0.01.

Surprisingly, when the expression of different Rlm1-dependent genes (*MLP1*, *CRG1*, *PIR3*, and *SRL3*) was analyzed by RT-qPCR (Figure 5E), the relative mRNA levels were notably lower in cells treated with CLT compared with those stimulated with CR, indicating that the observed additional phosphorylation of Rlm1 does not result in higher activity of this transcription factor. Nevertheless, while CLT-induced gene expression occurred at a low level, it showed a statistically significant reduction in the absence of Slt2, reflecting that this response is fully dependent on the CWI MAPK. To determine whether that induction was sufficient to generate a response of the pathway in cells exposed to this azole, we performed a sensitivity assay of cells bearing the *MLP1*-mediated positive feedback circuit called Integrity Pathway Activation Circuit (IPAC), which amplifies the signal and renders cells hypersensitive to CWI activators [29]. As confirmed in Figure S2, cells carrying the IPAC were considerably more sensitive to CLT than those without the circuit. This indicated that the minimal induction of the Rlm1-dependent *MLP1* promoter was sufficient to activate the circuitry and amplify the CLT-induced signal.

3.6. Oxidative Stress Is Necessary for Slt2 Activation by Clotrimazole

Since some azoles induce the accumulation of reactive oxygen species (ROS) in fungal cells [19,20], we studied whether CLT would induce oxidative stress and, if so, whether the produced ROS are responsible for CWI pathway activation and LP-Slt2 formation. To this end, we measured the mitochondrial membrane potential and ROS generation following CLT treatment by flow cytometry using the fluorescent dyes rhodamine 123 and dihydroethidium (DHE), respectively [49,50]. As a positive control, we treated cells with H_2O_2, a strong inducer of hydroxyl radical formation. After 4 h, we found that both CLT and H_2O_2 led to a significant increase in the fluorescence signal with both dyes (Figure 6A), confirming that CLT induces the formation of ROS and the alteration of mitochondrial activity. Furthermore, under CLT-treatment conditions, the mitochondrial morphology was clearly affected (Figure 6B). The branched tubular structure found in normal cells disappeared and deformed structures emerged, as revealed by analyzing the localization of the mitochondrial protein Ilv6 tagged with mCherry used as a mitochondrial marker.

To associate oxidative stress with CWI signaling following CLT stimulation, we used the antioxidant N-acetylcysteine (NAC). As shown in Figure 6C, the sensitivity of the *slt2*Δ mutant to CLT was suppressed by NAC addition to the medium, suggesting that oxidative stress may be responsible for CWI pathway activation by CLT. The sensitivity to CLT displayed by cells deficient in antioxidant systems such as superoxide dismutase (Sod1) or catalase (Ctt1) was also eliminated by NAC, further supporting the oxidative effect of CLT and the utility of NAC in preventing it. The growth of all these deletion mutants was slightly higher than that of the WT strain onto CLT+NAC plates (Figure 6C), suggesting that these mutant strains might have some additional compensatory mechanisms to promote growth under oxidative stress that keep operating when this stress disappears. Additionally, we found that simultaneously exposing exponentially growing yeast cells to NAC and CLT significantly reduced Slt2 activation and eliminated the characteristic CLT-induced Slt2 shift (Figure 6D). The reductions in P-Kss1 and P-Fus3 basal levels induced by CLT were also abolished by NAC.

As we have observed for CLT, chlorpromazine (CPZ) has been reported to alter the plasma membrane, induce ROS production, and activate the yeast CWI pathway [51,52]. Therefore, we analyzed whether CPZ has a similar impact on yeast MAPK signaling to that of CLT. As shown in Figure 6D, this agent also provoked the appearance of LP-Slt2 and the reductions in P-Kss1 and P-Fus3 levels. Moreover, both effects were prevented by the addition of NAC. Hog1 phosphorylation was also induced by CPZ, and NAC hindered both the CPZ- and CLT-induced Hog1 activation (Figure 6E). These results suggest that the alteration in the plasma membrane together with the oxidative damage generated by any of these two compounds trigger this characteristic MAPK signaling response, not observed for other stimuli.

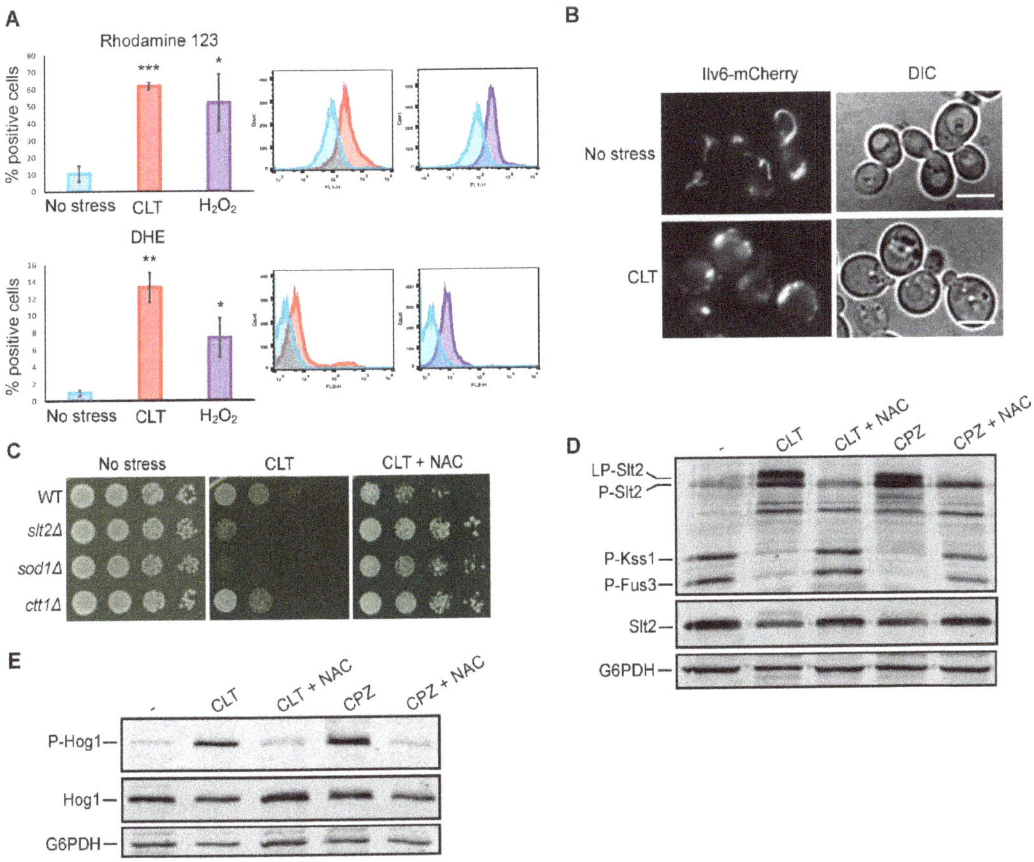

Figure 6. Effect of clotrimazole on mitochondrial morphology and oxidative stress: impact on MAPK phosphorylation. (**A**) Flow cytometry assay of the BY4741 strain after treatment with 50 µg/mL of CLT, 10 mM of H_2O_2, or basal conditions (no stress) for 4 h. Graphs represent the percentage of positive cells detected with flow cytometry that express a fluorescence signal for rhodamine 123 or dihydroethidium (DHE) fluorochromes. Data are presented as the mean of three independent experiments, and error bars indicate the standard deviation. Histograms from a representative experiment are shown. Student's t-test was calculated between treated (CLT or H_2O_2) and non-treated (no stress) cells. p-value of < 0.05 (*), < 0.01 (**), or < 0.001(***). (**B**) Mitochondrial morphology alteration upon CLT treatment. Fluorescence and DIC microscopy images of BY4741-T cells transformed with YEplac112-Ilv6-mCherry plasmid after treatment with 50 µg/mL of CLT or in basal conditions (no stress) for 4 h. Representative photographs from two independent experiments are shown. Scale bar = 5 µm. (**C**) Sensitivity of BY4741 (WT) and the indicated isogenic mutant strains to CLT per a drop dilution growth assay. Ten-fold serial dilutions of cell suspensions were spotted onto YPD plates in the absence (no stress) or presence of 1 µg/mL of CLT and incubated at 30 °C for 72 h. Plates were supplemented with 20 mM of N-acetylcysteine (NAC) when indicated. A representative assay from three independent experiments is shown. (**D**,**E**) Western blotting analysis of the extracts of BY4741 cells treated with 50 µg/mL of CLT or 250 µM of chlorpromazine (CPZ) in combination with 20 mM of NAC, when indicated. Cells were cultured for 4 h (**D**) or 30 min (**E**) at 24 °C. Dually phosphorylated Slt2, Kss1, and Fus3; Slt2 protein; dually phosphorylated Hog1; Hog1 protein; and G6PDH (as a loading control) were detected with anti-phospho-p44/42, anti-Mpk1, anti-phospho-p38, anti-Hog1, and anti-G6PDH, respectively. Representative blots from three (**D**) and two (**E**) independent experiments are shown.

3.7. Tpk3-Mediated PKA Activity Modulates Slt2 Phosphorylation in Response to Clotrimazole

A strong correlation exists between ROS levels and the state of the actin cytoskeleton in yeast [53]. Additionally, miconazole was shown to induce yeast actin aggregation, which precedes ROS accumulation [54]. CLT-treated cells also displayed aberrant actin aggregates, as revealed by the actin dye rhodamine–phalloidin (Figure 7A), suggesting that actin alteration can contribute to CWI pathway activation.

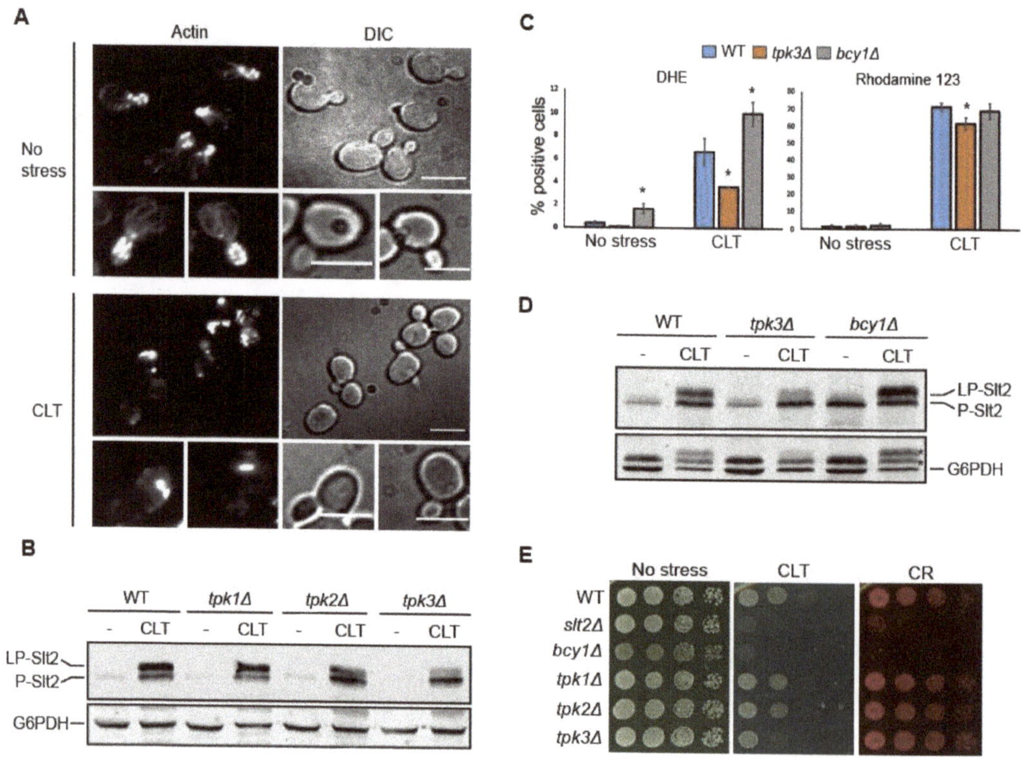

Figure 7. Involvement of the PKA kinase activity in clotrimazole-induced ROS production and Slt2 phosphorylation. (**A**) Rhodamine–phalloidin-based actin staining of BY4741 cells treated with 50 μg/mL of CLT or in basal conditions (no stress) for 4 h. Representative fluorescence and DIC photographs from three independent experiments are shown. Scale bar = 5 μm. (**B**) Western blotting analysis of the BY4741 (WT) strain and the indicated isogenic mutants. Cells were cultured without stimulus (-) or with 50 μg/mL of CLT for 4 h at 24 °C. Dually phosphorylated Slt2 and G6PDH (as a loading control) were detected with anti-phospho-p44/42 and anti-G6PDH, respectively. A representative blot from two independent experiments is shown. (**C**) Flow cytometry assay of the BY4741 (WT) strain and the isogenic *tpk3*Δ and *byc1*Δ mutants. Cells were either left untreated (no stress) or treated with 50 μg/mL of CLT for 4 h at 24 °C. Graphs represent the percentage of positive cells detected with flow cytometry that express a fluorescence signal for rhodamine 123 or dihydroethidium (DHE) fluorochromes. Data are presented as the mean of three independent experiments. Error bars indicate the standard deviation. Student's t-test was calculated between the WT and mutant strains. * p-value < 0.05. (**D**) Western blotting analysis of the same samples as in (**C**). Proteins were detected as in (**B**). Asterisks in the lower panel showing G6PDH levels indicate bands that correspond to P-Slt2 and LP-Slt2. (**E**) Sensitivity of BY4741 *slt2*Δ and the same strains as in (**B**,**D**) to CLT and CR per the drop dilution growth assay. Ten-fold serial dilutions of cell suspensions were spotted onto YPD plates in the absence (no stress) or presence of 1 μg/mL of CLT or 30 μg/mL of CR and incubated at 30 °C for 72 h. A representative assay from two independent experiments is shown.

Since actin aggregation was linked to Ras/PKA signaling [55], and the Ras/PKA pathway was related to the induction of mitochondrial ROS in response to distinct antifungal drug treatments [56], we next analyzed the involvement of this signaling pathway in Slt2 activation by CLT. First, we assessed the possible implication of each of the three PKA catalytic subunits, Tpk1, Tpk2, and Tpk3, on CWI signaling, as each of them can play specific functional roles [57,58]. As shown in Figure 7B, Tpk3 seemed to be partially involved in LP-Slt2 formation in response to CLT. Next, we measured the ROS levels and mitochondrial membrane potential of *tpk3Δ* cells in parallel to cells deficient in Bcy1, the negative regulatory subunit of the PKA kinase (Figure 7C). Compared with wild-type cells, the absence of Tpk3 led to a significant reduction in ROS levels, whereas the lack of Bcy1 resulted in even higher ROS generation after CLT exposure, as measured using DHE fluorescent dye. Nevertheless, differences in the mitochondrial membrane potential were not so remarkable, as monitored using rhodamine 123. When analyzing the same samples with Western blotting (Figure 7D), a reduction and a rise in LP-Slt2 levels were clearly observed in *tpk3Δ* and *bcy1Δ* mutants, respectively, confirming the relationship between the PKA pathway, oxidative stress, and the appearance of LP-Slt2 in the presence of CLT. Performing drop growth assays, we found that deletion of *TPK3* resulted in increased sensitivity to CLT but not to other stimuli of the pathway such as CR, which supports the specific implication of this PKA subunit in CLT-induced CWI pathway signaling (Figure 7E). Conversely, cells lacking Bcy1 were sensitive to either CLT or CR, which suggested that the lack of this protein can impair cell growth in any stress condition.

3.8. Conservation of Azole-Induced MAPK Signaling in Saccharomycotina

To investigate whether the MAPK-mediated response to azoles observed in *S. cerevisiae* is conserved in other yeast species, we compared the MAPK phosphorylation of distinct yeast species belonging to the Saccharomycotina subphylum (budding yeasts) in response to CLT and FLC. Wild-type strains of *Saccharomyces cerevisiae*, *Candida glabrata*, *Meyerozyma guilliermondii*, *Candida albicans*, *Candida tropicalis*, *Pichia anomala*, *Kluyveromyces lactis*, *Pichia pastoris*, and *Hansenula mrakii* species were grown in rich medium and treated with FLC or CLT, and dual phosphorylation of the corresponding Slt2 and Hog1 orthologues was analyzed at distinct time points by Western blotting using anti-phospho-p44/42 and anti-phospho-p38 antibodies, respectively, as these antibodies can recognize dually phosphorylated MAPKs at the activation loop from different organisms. As shown in Figure 8, the phosphorylated forms of Slt2 orthologues were detected with the anti-phospho-p44/42 antibody in all the yeast species, as the protein bands matched the predicted molecular weights of the corresponding MAPKs. In general, CLT induced a phosphorylation of the CWI pathway MAPK, which increased with time, with an activation peak at 30 min in most species. The exception was the phosphorylated MAPK from *M. guilliermondii*, whose levels decreased after CLT treatment. In the case of FLC, Slt2 orthologues were also phosphorylated, but the highest phosphorylation was generally reached at the longest time point of 240 min.

The anti-phospho-p38 antibody recognized all Hog1 orthologues excepting for *P. anomala*, whose extracts showed no signal against this antibody. As with Slt2 orthologues, CLT induced the highest level of HOG pathway phosphorylation at short time points (5 or 30 min), with the signal decreasing at the longest time point of 240 min. However, in the cases of *C. glabrata* and *M. guilliermondii*, MAPK phosphorylation levels decreased after CLT treatment. In contrast to CLT, the response to FLC through the HOG pathway was weak and heterogeneous between species. Despite a CLT-dependent Slt2 mobility shift seeming to be a feature of *S. cerevisiae* Slt2, activation of this MAPK upon azole treatment is conserved in other yeast species. Altogether, these results indicate that the CWI and HOG pathways respond to azoles in different budding yeasts. Since some of them are of clinical interest, the implication of these pathways in azole-resistance should be addressed.

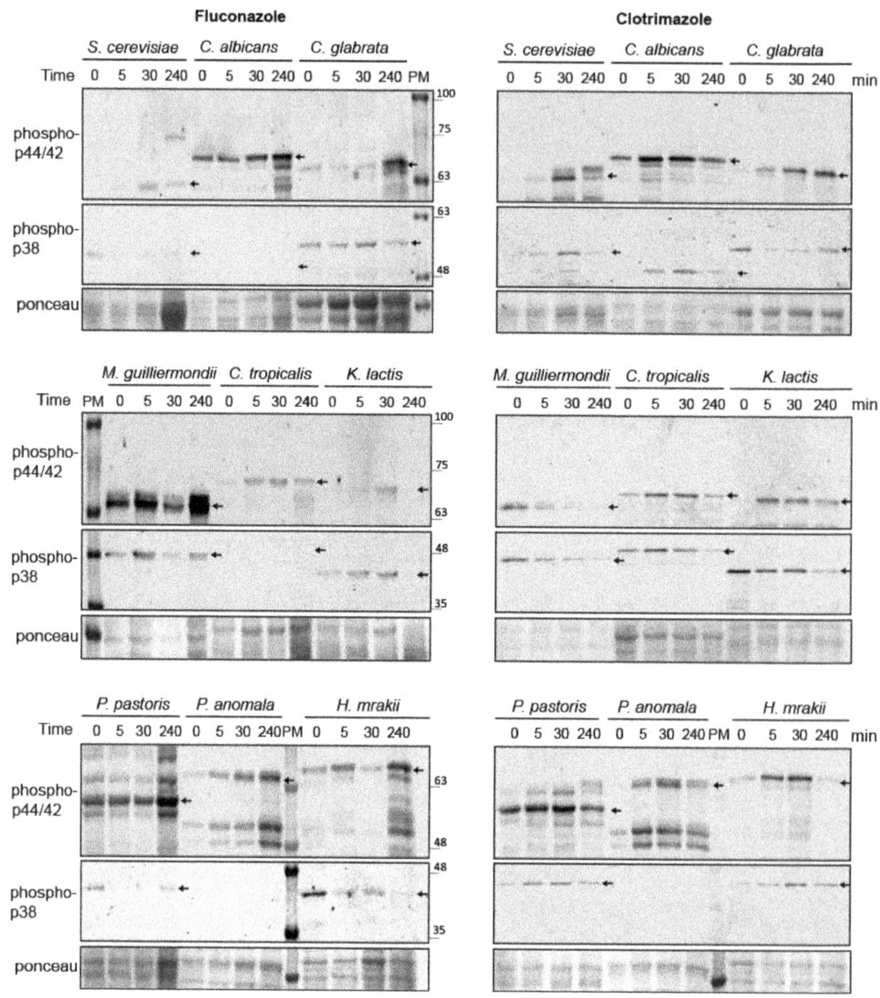

Figure 8. Activation of CWI and HOG pathways by azoles in other yeast species. Western blotting analysis of extracts of *Saccharomyces cerevisiae* BY4741, *Candida albicans* 4482A, *Candida glabrata* CBS138, *Meyerozyma guilliermondii* TP11010, *Candida tropicalis* 4458, *Kluyveromyces lactis* 1049, *Pichia pastoris* GS115, *Pichia anomala* 1026, and *Hansenula mrakii* K9 strains. Cells were cultured with 200 µg/mL of fluconazole (left panels) or with 50 µg/mL of clotrimazole (right panels) for the indicated times at 30 °C. Dually phosphorylated MAPK homologues of Slt2 and Hog1 were detected with anti-phospho-p44/42 and anti-phospho-p38, respectively, and the corresponding bands are indicated by black arrows in the blots. Ponceau staining was used for the loading control. PM corresponds to the lane loaded with the molecular weight protein marker. Representative blots from three independent experiments are shown.

4. Discussion

In many aspects, *S. cerevisiae* has guided our understanding of the eukaryotic cell for decades, and especially when studying fungal biology. Although it lacks pathogenicity, this yeast species is considered a useful model system for studying azole susceptibility, as it provides information on sterol metabolism and cellular signaling [23]. Surprisingly, although they are the most widely used antifungals, not much is known about the action of azoles on MAPK signaling in fungi. To gain more insight into the mode of action of azoles and, in particular, to discover how imidazole antifungals act on fungal MAPK pathways, we explored the effect of CLT on *S. cerevisiae* MAPK circuitries. Our results show that unlike triazoles, imidazoles trigger profound changes in MAPK signaling, affecting all four MAPKs operating in vegetative yeast cells. Whereas MAPKs Hog1 and Slt2 are activated, signaling through mating MAPKs Fus3 and Kss1 is significantly reduced in a Hog1-dependent manner following CLT treatment, suggesting that similar mechanisms known to downregulate the mating pathway by hyperosmotic stress [43] are likely operating under CLT treatment. However, the actual crosstalk mechanisms that regulate mutual modulation between these two pathways are still poorly understood. Among them, Hog1-dependent Ste50 phosphorylation was found to negatively regulate signaling toward Kss1 and Fus3 [43,59]. Since CLT activates the Sln1 branch of the HOG pathway and Ste50 operates both in the alternative Sho1 branch and in the mating pathway, this Ste50 negative feedback phosphorylation may be a mechanism prioritizing the most necessary among pathways sharing components against a condition that threatens survival. The sensitivity of mutants to CLT with disrupted signaling through the HOG or CWI pathways revealed the importance of prioritization of these pathways specialized in responding to stress conditions.

Knowledge is limited of the mode of activation of the HOG pathway by other stressors distinct from hyper-osmotic shock. As mentioned above, CLT activates the HOG pathway through stimulation of the Sln1 branch of the pathway. Other stimuli use this branch for activating Hog1, such as citric acid [60] or acetic acid [61]. Interestingly, sphingolipid or ergosterol depletion also triggers this pathway [62], and the alteration in the membrane fluidity during hypoxia, probably through ergosterol deficiency, also leads to Hog1 activation through the Sln1 branch [63]. Therefore, the depletion of ergosterol generated by CLT may account for the observed Hog1 activation. However, the treatment of yeast cells with the triazoles FLC and ITC did not result in Hog1 phosphorylation, suggesting that other biological effects are also necessary for triggering the HOG pathway. Here, we showed that oxidative stress underlies CLT-induced Hog1 phosphorylation. Accordingly, FLC was reported to lack significant ROS-inducing capacity in yeast [64], and compounds that generate oxidative stress, such as hydrogen peroxide [65], also signal to Hog1 through the Sln1 branch.

The activation of Hog1 is important for the subsequent activation of the CWI pathway in response to some stresses. This is the case with SDS [29] and zymolyase [42] treatments. Zymolyase, a mixture of cell-wall-digesting enzymes, promotes Slt2 activation partly because of Hog1-driven glycerol accumulation [66]. However, genome-wide expression studies found no induction of genes coding the prototypical glycerol-producing proteins *GPD1* and *GPD2* after treatment with antifungals targeting ergosterol biosynthesis, including clotrimazole [67]. Consistently, we found that Slt2 activation in response to CLT does not rely on previous Hog1 activation, suggesting independence and limited crosstalk between the HOG and CWI pathways under treatment with imidazoles. The most remarkable traits in imidazole-induced MAPK signaling are related to the CWI pathway. Treatment of yeast cells with either clotrimazole or ketoconazole promoted Slt2 MAPK phosphorylation not only at the TEY activation loop but also at additional sites, producing an Slt2 form of low electrophoretic mobility (LP-Slt2). Although phosphorylation sites other than T190 and Y192 were reported [13], no changes in mobility such as that induced by CLT have been observed to date for Slt2. The strong Slt2-dependent phosphorylation of Rlm1 and Swi6 suggests that LP-Slt2 is fully catalytically active, at least on Rlm1 and Swi6. Therefore, the extra phosphorylation on Slt2 does not seem to negatively impact the activity of MAPK Slt2. Although the effects on the kinase activity and functionality merit further research, this additional post-translational modification is likely to impose an additional layer of regulation on Slt2.

Another striking feature of CLT-induced signaling is Pkc1-independent Slt2 activation. To the best of our knowledge, this is the first stimulus of the CWI pathway for which Pkc1 is not fully required for Slt2 activation, indicating that a novel signaling circuitry is operating under this stress. Under this condition, none of the Rho1 GEFs are necessary for signaling through the route, which also suggests the formation of alternative protein complexes. Nevertheless, the mechanosensor Wsc1 seems to be at least partially implicated in CLT sensing and transduction of the signal. We unsuccessfully attempted to find an alternative protein kinase distinct to Pkc1 that is able to activate the Slt2 MAPK module. Therefore, although the most presumable possibility is a surrogate kinase operating in this non-canonical Pkc1-independent signaling, other phosphorylation-independent mechanisms that can enhance signaling through the MAPK module cannot be ruled out. This is another example of the extraordinary plasticity of MAPK pathways, as recently observed with RAF-independent ERK phosphorylation in mammalian cells [68]. Although dispensable for Slt2 activation, we showed that Pkc1 is essential for cell viability under CLT stress. This finding is in line with the increased efficacy of antifungal drugs targeting the cell membrane, such as fluconazole, following the pharmacological inhibition of PKC [22]. These authors suggested that PKC1 enables tolerance to drugs that affect the cell membrane via the MAPK cascade, because the lack of MAPK module components results in sensitivity to these compounds [22]. However, in contrast to the absence of Slt2, we showed that the lack of Pkc1 promotes a non-osmotically remediable CLT sensitivity, indicating additional functions to the activation of the MAPK cascade for Pkc1 in this condition. Several targets, in addition to the Slt2 MAPK cascade, were already found for Pkc1, as well as roles for this kinase independent of CWI signaling [69] that can account for this phenotype.

We suggest that imidazole-induced CWI pathway rewiring occurs due to a combination of membrane alteration and intense oxidant activity. Different evidence supports this model. First, both FLC and CLT are antifungal agents that target Erg11, affecting the ergosterol biosynthesis pathway, as readily observed by their similar chemical–genetic profiles [70]. However, whereas FLC induces only a limited amount of ROS [64], here, we showed that CLT, which promotes LP-Slt2 formation, generates a high level of ROS in yeast, similar to the strong oxidant hydrogen peroxide (H_2O_2). Second, the antioxidant N-acetylcysteine eliminates the Slt2 band shift and significantly reduces Slt2 phosphorylation in response to CLT, indicating the importance of oxidative stress in CWI pathway activation in response to this imidazole. Notably, oxidative stress generated by other compounds, such as hydrogen peroxide or diamide, leads to Slt2 phosphorylation, but does not trigger the phosphorylation pattern observed for CLT (our unpublished results, [71]). Third, CPZ, one of the most-used antipsychotic and neuroleptic drugs, damages cell membranes and affects transport activity. CPZ also induces the unfolded protein response and inhibits protein synthesis [72]. Interestingly, CPZ also induces oxidative stress with high ROS generation not only in mammalian [73] but also in yeast cells [51], in which it also promotes an Slt2 phosphorylation pattern similar to CLT, suggesting the involvement of the same MAPK signaling circuitry. Fourth, the activity of the PKA pathway was linked to mitochondrial damage and ROS generation via Tpk3, which is a master regulator of mitochondrial activity. Cells lacking this catalytic subunit of the PKA show decreased respiratory activity and mitochondrial content [58]. Actin aggregation induces Ras/cAMP/PKA signaling, leading to the production of ROS from dysfunctional mitochondria [53,55]. Under this condition, the deletion of *TPK3* is sufficient to prevent the production of ROS [55]. Here, we showed that Slt2 phosphorylation correlates with PKA activation in response to CLT, with the PKA catalytic subunit Tpk3 being important for LP-Slt2 formation. Together, it is thus likely that the actin aggregation promoted by CLT can contribute to PKA-mediated mitochondrial alteration and ROS production, leading to CWI pathway activation. In this regard, distinct components of the CWI pathway, including Bck1, Mkk1, and Slt2, were found to localize to mitochondria under oxidative stress [74]. Moreover, Slt2 is required for mitophagy [75]. Therefore, it is tempting to speculate that the CWI pathway may sense oxidative stress at the mitochondria, and thus be involved in their mitophagy. In this line, it was observed that fungicidal drugs induce the switch from fermentative growth to ROS-producing mitochondrial respiration [56], and mitophagy is usually induced by prolonged respiration [76].

In addition to *S. cerevisiae*, the sub-phylum Saccharomycotina contains over 1000 other known species of budding yeasts, with more than 300 with a known genome, including the opportunistic pathogens *Candida albicans* and *Candida glabrata* [77]. Conservation occurs in many aspects between *S. cerevisiae* and pathogen yeasts, including some canonical drug resistance mechanisms [78]. Our studies with different budding yeasts of this sub-phylum highlight the evolutionary conservation of the CWI and HOG pathways as stress-responsive pathways following azole treatment. Defining the differences and similarities between *S. cerevisiae* and pathogenic fungi underpinning the response to azoles will enable the rational design of combination therapies to tackle resistance to these antifungal drugs.

Supplementary Materials: The following are available online at https://www.mdpi.com/article/10.3390/jof7080647/s1, Figure S1: Analysis of the role of Pkc1 in clotrimazole signaling through the CWI pathway in the CML128 background. Figure S2: Effect of clotrimazole on cells bearing the Integrity Pathway Activation Circuit (IPAC).

Author Contributions: Conceptualization, Á.S.-M., M.M. and H.M.; methodology, Á.S.-M. and M.N.; validation, Á.S.-M., M.N., M.M. and H.M.; formal analysis, Á.S.-M., M.N., M.M. and H.M.; investigation, Á.S.-M., M.N., M.M. and H.M.; resources, M.M. and H.M.; data curation, Á.S.-M.; writing—original draft preparation, Á.S.-M., M.M. and H.M.; writing—review and editing, Á.S.-M., M.M. and H.M.; visualization, Á.S.-M.; supervision, M.M. and H.M.; project administration, M.M. and H.M.; funding acquisition, M.M. and H.M. All authors have read and agreed to the published version of the manuscript.

Funding: This research was funded by Ministerio de Ciencia e Innovación (Spain), grant number PID2019-105342GB-I00; Comunidad de Madrid and European Structural and Investment Funds, grant number S2017/BMD-3691 (InGEMICS-CM); and Á.S.-M. was recipient of an FPI contract from Ministerio de Ciencia, Innovación y Universidades (Spain).

Institutional Review Board Statement: Not applicable.

Informed Consent Statement: Not applicable.

Data Availability Statement: The data presented in this study are available on request from the corresponding authors.

Acknowledgments: We thank Esmeralda Alonso-Rodríguez for her initial work on this subject. We thank Javier Arroyo, María Angeles de la Torre, Elena Jiménez-Gutiérrez, Ana Alastruey-Izquierdo, and Elvira Marín for providing some of the materials used in this study. We also thank Lucía Sastre for technical support, and other colleagues from Unit 3 of the Departamento de Microbiología y Parasitología at UCM for their useful comments and discussions throughout our work. We acknowledge the Servicio de Genómica y Proteómica (UCM, Madrid, Spain) for conducting the DNA sequencing and RT-qPCR assays.

Conflicts of Interest: The authors declare no conflict of interest.

References

1. Kim, E.K.; Choi, E.J. Compromised MAPK signaling in human diseases: An update. *Arch. Toxicol.* **2015**, *89*, 867–882. [CrossRef] [PubMed]
2. Cargnello, M.; Roux, P.P. Activation and Function of the MAPKs and Their Substrates, the MAPK-Activated Protein Kinases. *Microbiol. Mol. Biol. Rev.* **2011**, *75*, 50–83. [CrossRef] [PubMed]
3. Chen, R.E.; Thorner, J. Function and Regulation in MAPK Signaling Pathways. *Biochim. Biophys. Acta Mol. Cell Res.* **2007**, *1773*, 1311–1340. [CrossRef]
4. Li, Y.; Roberts, J.; Aghdam, Z.A.; Hao, N. Mitogen-activated protein kinase (MAPK) dynamics determine cell fate in the yeast mating response. *J. Biol. Chem.* **2017**, *292*, 20354–20361. [CrossRef]
5. Sabbagh, W.; Flatauer, L.J.; Bardwell, A.J.; Bardwell, L. Specificity of MAP kinase signaling in yeast differentiation involves transient versus sustained MAPK activation. *Mol. Cell* **2001**, *8*, 683–691. [CrossRef]
6. Saito, H.; Posas, F. Response to hyperosmotic stress. *Genetics* **2012**, *192*, 289–318. [CrossRef] [PubMed]
7. Hohmann, S. An integrated view on a eukaryotic osmoregulation system. *Curr. Genet.* **2015**, *61*, 373–382. [CrossRef]
8. Jiménez-Gutiérrez, E.; Alegría-Carrasco, E.; Sellers-Moya, Á.; Molina, M.; Martín, H. Not just the wall: The other ways to turn the yeast CWI pathway on. *Int. Microbiol.* **2020**, *23*, 107–119. [CrossRef]

9. Jendretzki, A.; Wittland, J.; Wilk, S.; Straede, A.; Heinisch, J.J. How do I begin? Sensing extracellular stress to maintain yeast cell wall integrity. *Eur. J. Cell Biol.* **2011**, *90*, 740–744. [CrossRef] [PubMed]
10. Krause, S.A.; Cundell, M.J.; Poon, P.P.; McGhie, J.; Johnston, G.C.; Price, C.; Gray, J.V. Functional specialisation of yeast Rho1 GTP exchange factors. *J. Cell Sci.* **2012**, *125*, 2721–2731. [CrossRef]
11. González-Rubio, G.; Sellers-Moya, Á.; Martín, H.; Molina, M. A walk-through MAPK structure and functionality with the 30-year-old yeast MAPK Slt2. *Int. Microbiol.* **2021**, 1–13. [CrossRef]
12. Sanz, A.B.; García, R.; Rodríguez-Peña, J.M.; Arroyo, J. The CWI pathway: Regulation of the transcriptional adaptive response to cell wall stress in yeast. *J. Fungi* **2018**, *4*, 1. [CrossRef]
13. Truman, A.W.; Kim, K.-Y.; Levin, D.E. Mechanism of Mpk1 Mitogen-Activated Protein Kinase Binding to the Swi4 Transcription Factor and Its Regulation by a Novel Caffeine-Induced Phosphorylation. *Mol. Cell. Biol.* **2009**, *29*, 6449–6461. [CrossRef] [PubMed]
14. Fisher, M.C.; Hawkins, N.J.; Sanglard, D.; Gurr, S.J. Worldwide emergence of resistance to antifungal drugs challenges human health and food security. *Science* **2018**, *360*, 739–742. [CrossRef]
15. Allen, D.; Wilson, D.; Drew, R.; Perfect, J. Azole antifungals: 35 years of invasive fungal infection management. *Expert Rev. Anti-Infect. Ther.* **2015**, *13*, 787–798. [CrossRef] [PubMed]
16. Crowley, P.D.; Gallagher, H.C. Clotrimazole as a pharmaceutical: Past, present and future. *J. Appl. Microbiol.* **2014**, *117*, 611–617. [CrossRef]
17. Odds, F.C.; Brown, A.J.P.; Gow, N.A.R. Antifungal agents: Mechanisms of action. *Trends Microbiol.* **2003**, *11*, 272–279. [CrossRef]
18. Hornby, J.M.; Nickerson, K.W. Enhanced production of farnesol by *Candida albicans* treated with four azoles. *Antimicrob. Agents Chemother.* **2004**, *48*, 2305–2307. [CrossRef]
19. Kobayashi, D.; Kondo, K.; Uehara, N.; Otokozawa, S.; Tsuji, N.; Yagihashi, A.; Watanabe, N. Endogenous reactive oxygen species is an important mediator of miconazole antifungal effect. *Antimicrob. Agents Chemother.* **2002**, *46*, 3113–3117. [CrossRef]
20. François, I.E.J.A.; Cammue, B.P.A.; Borgers, M.; Ausma, J.; Dispersyn, G.D.; Thevissen, K. Azoles: Mode of antifungal action and resistance development. Effect of miconazole on endogenous reactive oxygen species production in *Candida albicans*. *Anti-Infect. Agents Med. Chem.* **2006**, *5*, 3–13. [CrossRef]
21. Navarro-García, F.; Eisman, B.; Fiuza, S.M.; Nombela, C.; Pla, J. The MAP kinase Mkc1p is activated under different stress conditions in *Candida albicans*. *Microbiology* **2005**, *151*, 2737–2749. [CrossRef]
22. Lafayette, S.L.; Collins, C.; Zaas, A.K.; Schell, W.A.; Betancourt-Quiroz, M.; Leslie Gunatilaka, A.A.; Perfect, J.R.; Cowen, L.E. PKC signaling regulates drug resistance of the fungal pathogen *Candida albicans* via circuitry comprised of mkc1, calcineurin, and hsp90. *PLoS Pathog.* **2010**, *6*, 79–80. [CrossRef]
23. Demuyser, L.; Van Dijck, P. Can *Saccharomyces cerevisiae* keep up as a model system in fungal azole susceptibility research? *Drug Resist. Update* **2019**, *42*, 22–34. [CrossRef]
24. Jimenez-Sanchez, M.; Cid, V.J.; Molina, M. Retrophosphorylation of Mkk1 and Mkk2 MAPKKs by the Slt2 MAPK in the yeast cell integrity pathway. *J. Biol. Chem.* **2007**, *282*, 31174–31185. [CrossRef] [PubMed]
25. De la Torre-Ruiz, M.A.; Torres, J.; Arin, J.; Herrero, E. Sit4 Is Required for Proper Modulation of the Biological Functions Mediated by Pkc1 and the Cell Integrity Pathway in *Saccharomyces cerevisiae*. *J. Biol. Chem.* **2002**, *277*, 33468–33476. [CrossRef] [PubMed]
26. García, R.; Sanz, A.B.; Rodríguez-Peña, J.M.; Nombela, C.; Arroyo, J. Rlm1 mediates positive autoregulatory transcriptional feedback that is essential for Slt2-dependent gene expression. *J. Cell Sci.* **2016**, *129*, 1649–1660. [CrossRef]
27. Sanz, A.B.; García, R.; Rodríguez-Peña, J.M.; Nombela, C.; Arroyo, J. Slt2 MAPK association with chromatin is required for transcriptional activation of Rlm1 dependent genes upon cell wall stress. *Biochim. Biophys. Acta Gene Regul. Mech.* **2018**, *1861*, 1029–1039. [CrossRef] [PubMed]
28. García, R.; Bravo, E.; Diez-Muñiz, S.; Nombela, C.; Rodríguez-Peña, J.M.; Arroyo, J. A novel connection between the Cell Wall Integrity and the PKA pathways regulates cell wall stress response in yeast. *Sci. Rep.* **2017**, *7*, 5703. [CrossRef]
29. Jiménez-Gutiérrez, E.; Alegría-Carrasco, E.; Alonso-Rodríguez, E.; Fernández-Acero, T.; Molina, M.; Martín, H. Rewiring the yeast cell wall integrity (CWI) pathway through a synthetic positive feedback circuit unveils a novel role for the MAPKKK Ssk2 in CWI pathway activation. *FEBS J.* **2020**, *287*, 4881–4901. [CrossRef]
30. Flández, M.; Cosano, I.C.; Nombela, C.; Martín, H.; Molina, M. Reciprocal Regulation between Slt2 MAPK and Isoforms of Msg5 Dual-specificity Protein Phosphatase Modulates the Yeast Cell Integrity Pathway. *J. Biol. Chem.* **2004**, *279*, 11027–11034. [CrossRef]
31. Sikorski, R.S.; Hieter, P. A system of shuttle vectors and yeast host strains designed for efficient manipulation of DNA in *Saccharomyces cerevisiae*. *Genetics* **1989**, *122*, 19–27. [CrossRef] [PubMed]
32. Alonso-Rodríguez, E.; Fernández-Piñar, P.; Sacristán-Reviriego, A.; Molina, M.; Martín, H. An analog-sensitive version of the protein kinase Slt2 allows identification of novel targets of the yeast cell wall integrity pathway. *J. Biol. Chem.* **2016**, *291*, 5461–5472. [CrossRef]
33. Kim, K.Y.; Cosano, I.C.; Levin, D.E.; Molina, M.; Martín, H. Dissecting the transcriptional activation function of the cell wall integrity MAP kinase. *Yeast* **2007**, *24*, 335–342. [CrossRef]
34. Denis, V.; Cyert, M.S. Molecular analysis reveals localization of *Saccharomyces cerevisiae* protein kinase C to sites of polarized growth and Pkc1p targeting to the nucleus and mitotic spindle. *Eukaryot. Cell* **2005**, *4*, 36–45. [CrossRef] [PubMed]
35. Jung, U.S.; Sobering, A.K.; Romeo, M.J.; Levin, D.E. Regulation of the yeast Rlm1 transcription factor by the Mpk1 cell wall integrity MAP kinase. *Mol. Microbiol.* **2002**, *46*, 781–789. [CrossRef] [PubMed]

36. García, R.; Rodríguez-Peña, J.M.; Bermejo, C.; Nombela, C.; Arroyo, J. The high osmotic response and cell wall integrity pathways cooperate to regulate transcriptional responses to zymolyase-induced cell wall stress in *Saccharomyces cerevisiae*. *J. Biol. Chem.* **2009**, *284*, 10901–10911. [CrossRef] [PubMed]
37. Fernández-Acero, T.; Bertalmio, E.; Luna, S.; Mingo, J.; Bravo-Plaza, I.; Rodríguez-Escudero, I.; Molina, M.; Pulido, R.; Cid, V.J. Expression of Human PTEN-L in a Yeast Heterologous Model Unveils Specific N-Terminal Motifs Controlling PTEN-L Subcellular Localization and Function. *Cells* **2019**, *8*, 1512. [CrossRef]
38. Martín, H.; Rodríguez-Pachón, J.M.; Ruiz, C.; Nombela, C.; Molina, M. Regulatory mechanisms for modulation of signaling through the cell integrity Slt2-mediated pathway in *Saccharomyces cerevisiae*. *J. Biol. Chem.* **2000**, *275*, 1511–1519. [CrossRef]
39. Jiménez, J.; Cid, V.J.; Cenamor, R.; Yuste, M.; Molero, G.; Nombela, C.; Sánchez, M. Morphogenesis beyond Cytokinetic Arrest in *Saccharomyces cerevisiae*. *J Cell Biol.* **1998**, *143*, 1617–1634. [CrossRef]
40. García, R.; Bermejo, C.; Grau, C.; Pérez, R.; Rodríguez-Peña, J.M.; Francois, J.; Nombela, C.; Arroyo, J. The Global Transcriptional Response to Transient Cell Wall Damage in *Saccharomyces cerevisiae* and Its Regulation by the Cell Integrity Signaling Pathway. *J. Biol. Chem.* **2004**, *279*, 15183–15195. [CrossRef]
41. Livak, K.J.; Schmittgen, T.D. Analysis of relative gene expression data using real-time quantitative PCR and the 2-ΔΔCT method. *Methods* **2001**, *25*, 402–408. [CrossRef]
42. Bermejo, C.; Rodríguez, E.; García, R.; Rodríguez-Peña, J.M.; Rodríguez De La Concepción, M.L.; Rivas, C.; Arias, P.; Nombela, C.; Posas, F.; Arroyo, J. The sequential activation of the yeast HOG and SLT2 pathways is required for cell survival to cell wall stress. *Mol. Biol. Cell* **2008**, *19*, 1113–1124. [CrossRef] [PubMed]
43. Nagiec, M.J.; Dohlman, H.G. Checkpoints in a yeast differentiation pathway coordinate signaling during hyperosmotic stress. *PLoS Genet.* **2012**, *8*, e1002437. [CrossRef]
44. Bermejo, C.; García, R.; Straede, A.; Rodríguez-Peña, J.M.; Nombela, C.; Heinisch, J.J.; Arroyo, J. Characterization of sensor-specific stress response by transcriptional profiling of wsc1 and mid2 deletion strains and chimeric sensors in *Saccharomyces cerevisiae*. *Omics J. Integr. Biol.* **2010**, *14*, 679–688. [CrossRef]
45. Levin, D.E. Regulation of cell wall biogenesis in *Saccharomyces cerevisiae*: The cell wall integrity signaling pathway. *Genetics* **2011**, *189*, 1145–1175. [CrossRef]
46. Tatjer, L.; Sacristán-Reviriego, A.; Casado, C.; González, A.; Rodríguez-Porrata, B.; Palacios, L.; Canadell, D.; Serra-Cardona, A.; Martín, H.; Molina, M. Wide-Ranging Effects of the Yeast Ptc1 Protein Phosphatase Acting Through the MAPK Kinase Mkk1. *Genetics* **2016**, *202*, 141–156. [CrossRef] [PubMed]
47. Marín, M.J.; Flández, M.; Bermejo, C.; Arroyo, J.; Martín, H.; Molina, M. Different modulation of the outputs of yeast MAPK-mediated pathways by distinct stimuli and isoforms of the dual-specificity phosphatase Msg5. *Mol. Genet. Genom.* **2009**, *281*, 345–359. [CrossRef]
48. Kim, K.-Y.; Truman, A.W.; Caesar, S.; Schlenstedt, G.; Levin, D.E. Yeast Mpk1 cell wall integrity mitogen-activated protein kinase regulates nucleocytoplasmic shuttling of the Swi6 transcriptional regulator. *Mol. Biol. Cell* **2010**, *21*, 1609–1619. [CrossRef] [PubMed]
49. Büttner, S.; Eisenberg, T.; Carmona-Gutierrez, D.; Ruli, D.; Knauer, H.; Ruckenstuhl, C.; Sigrist, C.; Wissing, S.; Kollroser, M.; Fröhlich, K.U.; et al. Endonuclease G Regulates Budding Yeast Life and Death. *Mol. Cell.* **2007**, *25*, 233–246. [CrossRef]
50. Kwolek-Mirek, M.; Zadrag-Tecza, R. Comparison of methods used for assessing the viability and vitality of yeast cells. *FEMS Yeast Res.* **2014**, *14*, 1068–1079. [CrossRef]
51. Caldara, M.; Graziano, S.; Gullì, M.; Cadonici, S.; Marmiroli, N. Off-target effects of neuroleptics and antidepressants on *Saccharomyces cerevisiae*. *Toxicol. Sci.* **2017**, *156*, 538–548. [CrossRef] [PubMed]
52. Kamada, Y.; Jung, U.S.; Piotrowski, J.; Levin, D.E. The protein kinase C-activated MAP kinase pathway of *Saccharomyces cerevisiae* mediates a novel aspect of the heat shock response. *Genes Dev.* **1995**, *9*, 1559–1571. [CrossRef] [PubMed]
53. Gourlay, C.W.; Carpp, L.N.; Timpson, P.; Winder, S.J.; Ayscough, K.R. A role for the actin cytoskeleton in cell death and aging in yeast. *J. Cell Biol.* **2004**, *164*, 803–809. [CrossRef] [PubMed]
54. Thevissen, K.; Ayscough, K.R.; Aerts, A.M.; Du, W.; De Brucker, K.; Meert, E.M.K.; Ausma, J.; Borgers, M.; Cammue, B.P.A.; François, I.E.J.A. Miconazole induces changes in actin cytoskeleton prior to reactive oxygen species induction in yeast. *J. Biol. Chem.* **2007**, *282*, 21592–21597. [CrossRef]
55. Gourlay, C.W.; Ayscough, K.R. Actin-Induced Hyperactivation of the Ras Signaling Pathway Leads to Apoptosis in *Saccharomyces cerevisiae*. *Mol. Cell. Biol.* **2006**, *26*, 6487–6501. [CrossRef]
56. Belenky, P.; Camacho, D.; Collins, J.J. Report Fungicidal Drugs Induce a Common Oxidative-Damage Cellular Death Pathway. *Cell Rep.* **2012**, *3*, 350–358. [CrossRef]
57. Robertson, L.S.; Causton, H.C.; Young, R.A.; Fink, G.R. The yeast A kinases differentially regulate iron uptake and respiratory function. *Proc. Natl. Acad. Sci. USA* **2000**, *97*, 5984–5988. [CrossRef]
58. Chevtzoff, C.; Vallortigara, J.; Avéret, N.; Rigoulet, M.; Devin, A. The yeast cAMP protein kinase Tpk3p is involved in the regulation of mitochondrial enzymatic content during growth. *Biochim. Biophys. Acta Bioenerg.* **2005**, *1706*, 117–125. [CrossRef]
59. Hao, N.; Zeng, Y.; Elston, T.C.; Dohlman, H.G. Control of MAPK specificity by feedback phosphorylation of shared adaptor protein Ste50. *J. Biol. Chem.* **2008**, *283*, 33798–33802. [CrossRef]

60. Lawrence, C.L.; Botting, C.H.; Antrobus, R.; Coote, P.J. Evidence of a New Role for the High-Osmolarity Glycerol Mitogen-Activated Protein Kinase Pathway in Yeast: Regulating Adaptation to Citric Acid Stress. *Mol. Cell. Biol.* **2004**, *24*, 3307–3323. [CrossRef] [PubMed]
61. Mollapour, M.; Piper, P.W. Hog1p mitogen-activated protein kinase determines acetic acid resistance in *Saccharomyces cerevisiae*. *FEMS Yeast Res.* **2006**, *6*, 1274–1280. [CrossRef]
62. Tanigawa, M.; Kihara, A.; Terashima, M.; Takahara, T.; Maeda, T. Sphingolipids Regulate the Yeast High-Osmolarity Glycerol Response Pathway. *Mol. Cell. Biol.* **2012**, *32*, 2861–2870. [CrossRef]
63. Hickman, M.J.; Spatt, D.; Winston, F. The Hog1 mitogen-activated protein kinase mediates a hypoxic response in *Saccharomyces cerevisiae*. *Genetics* **2011**, *188*, 325–338. [CrossRef] [PubMed]
64. Román, E.; Prieto, D.; Martin, R.; Correia, I.; Mesa Arango, A.C.; Alonso-Monge, R.; Zaragoza, O.; Pla, J. Role of catalase overproduction in drug resistance and virulence in *Candida albicans*. *Future Microbiol.* **2016**, *11*, 1279–1297. [CrossRef] [PubMed]
65. Lee, Y.M.; Kim, E.; An, J.; Lee, Y.; Choi, E.; Choi, W.; Moon, E.; Kim, W. Dissection of the HOG pathway activated by hydrogen peroxide in *Saccharomyces cerevisiae*. *Environ. Microbiol.* **2017**, *19*, 584–597. [CrossRef]
66. Laz, E.V.; Lee, J.; Levin, D.E. Crosstalk between *Saccharomyces cerevisiae* SAPKs Hog1 and Mpk1 is mediated by glycerol accumulation. *Fungal Biol.* **2020**, *124*, 361–367. [CrossRef]
67. Bammert, G.F.; Fostel, J.M. Genome-wide expression patterns in *Saccharomyces cerevisiae*: Comparison of drug treatments and genetic alterations affecting biosynthesis of ergosterol. *Antimicrob. Agents Chemother.* **2000**, *44*, 1255–1265. [CrossRef]
68. Scheffler, L.; Feicht, S.; Babushku, T.; Kuhn, L.B.; Ehrenberg, S.; Frankenberger, S.; Lehmann, F.M.; Hobeika, E.; Jungnickel, B.; Baccarini, M.; et al. ERK phosphorylation is RAF independent in naïve and activated B cells but RAF dependent in plasma cell differentiation. *Sci. Signal.* **2021**, *14*, eabc1648. [CrossRef] [PubMed]
69. Heinisch, J.J.; Rodicio, R. Protein kinase C in fungi-more than just cell wall integrity. *FEMS Microbiol. Rev.* **2018**, *42*, 22–39. [CrossRef]
70. Parsons, A.B.; Lopez, A.; Givoni, I.E.; Williams, D.E.; Gray, C.A.; Porter, J.; Chua, G.; Sopko, R.; Brost, R.L.; Ho, C.H.; et al. Exploring the Mode-of-Action of Bioactive Compounds by Chemical-Genetic Profiling in Yeast. *Cell* **2006**, *126*, 611–625. [CrossRef] [PubMed]
71. Vilella, F.; Herrero, E.; Torres, J.; De La Torre-Ruiz, M.A. Pkc1 and the upstream elements of the cell integrity pathway in *Saccharomyces cerevisiae*, Rom2 and Mtl1, are required for cellular responses to oxidative stress. *J. Biol. Chem.* **2005**, *280*, 9149–9159. [CrossRef] [PubMed]
72. De Filippi, L.; Fournier, M.; Cameroni, E.; Linder, P.; De Virgilio, C.; Foti, M.; Deloche, O. Membrane stress is coupled to a rapid translational control of gene expression in chlorpromazine-treated cells. *Curr. Genet.* **2007**, *52*, 171–185. [CrossRef] [PubMed]
73. Elmorsy, E.; Al-Ghafari, A.; Aggour, A.M.; Khan, R.; Amer, S. The role of oxidative stress in antipsychotics induced ovarian toxicity. *Toxicol. In Vitro* **2017**, *44*, 190–195. [CrossRef]
74. Fairn, G.D.; MacDonald, K.; McMaster, C.R. A chemogenomic screen in *Saccharomyces cerevisiae* uncovers a primary role for the mitochondria in farnesol toxicity and its regulation by the Pkc1 pathway. *J. Biol. Chem.* **2007**, *282*, 4868–4874. [CrossRef] [PubMed]
75. Mao, K.; Wang, K.; Zhao, M.; Xu, T.; Klionsky, D.J. Two MAPK-signaling pathways are required for mitophagy in *Saccharomyces cerevisiae*. *J. Cell Biol.* **2011**, *193*, 755–767. [CrossRef]
76. Okamoto, K.; Kondo-Okamoto, N.; Ohsumi, Y. Article Mediates Degradation of Mitochondria via Selective Autophagy. *Dev. Cell* **2009**, *17*, 87–97. [CrossRef] [PubMed]
77. Shen, X.; Opulente, D.A.; Kominek, J.; Kurtzman, C.P.; Hittinger, C.T.; Rokas, A.; Shen, X.; Opulente, D.A.; Kominek, J.; Zhou, X.; et al. Tempo and Mode of Genome Evolution in the Budding Yeast Subphylum. *Cell* **2018**, *175*, 1533–1545. [CrossRef]
78. Shapiro, R.S.; Robbins, N.; Cowen, L.E. Regulatory Circuitry Governing Fungal Development, Drug Resistance, and Disease. *Microbiol. Mol. Biol. Rev.* **2011**, *75*, 213–267. [CrossRef]

Journal of *Fungi*

MDPI

Article

Slt2 Is Required to Activate ER-Stress-Protective Mechanisms through TORC1 Inhibition and Hexosamine Pathway Activation

Isabel E. Sánchez-Adriá [1], Gemma Sanmartín [1], Jose A. Prieto [1], Francisco Estruch [2] and Francisca Randez-Gil [1,*]

[1] Department of Biotechnology, Instituto de Agroquímica y Tecnología de los Alimentos, Consejo Superior de Investigaciones Científicas, Avda. Agustín Escardino 7, 46980 Paterna, Valencia, Spain; i.s.adria@iata.csic.es (I.E.S.-A.); gemma.sanmartin@iata.csic.es (G.S.); prieto@iata.csic.es (J.A.P.)

[2] Departament of Biochemistry and Molecular Biology, Universitat de València, Dr. Moliner 50, 46100 Burjassot, Valencia, Spain; Francisco.Estruch@uv.es

* Correspondence: randez@iata.csic.es

Abstract: Slt2, the MAPK of the cell wall integrity (CWI) pathway, connects different signaling pathways and performs different functions in the protective response of *S. cerevisiae* to stress. Previous work has evidenced the relation of the CWI pathway and the unfolded protein response (UPR), a transcriptional program activated upon endoplasmic reticulum (ER) stress. However, the mechanisms of crosstalk between these pathways and the targets regulated by Slt2 under ER stress remain unclear. Here, we demonstrated that ectopic expression of *GFA1*, the gene encoding the first enzyme in the synthesis of UDP-GlcNAc by the hexosamine biosynthetic pathway (HBP) or supplementation of the growth medium with glucosamine (GlcN), increases the tolerance of *slt2* mutant cells to different ER-stress inducers. Remarkably, GlcN also alleviates the sensitivity phenotype of cells lacking *IRE1* or *HAC1*, the main actors in controlling the UPR. The exogenous addition of GlcN reduced the abundance of glycosylated proteins and triggered autophagy. We also found that TORC1, the central stress and growth controller, is inhibited by tunicamycin exposure in cells of the wild-type strain but not in those lacking Slt2. Consistent with this, the tunicamycin-induced activation of autophagy and the increased synthesis of ATP in response to ER stress were absent by knock-out of *SLT2*. Altogether, our data placed Slt2 as an essential actor of the ER stress response by regulating the HBP activity and the TORC1-dependent signaling.

Keywords: *Saccharomyces cerevisiae*; CWI pathway; UPR; glucosamine; tunicamycin; N-glycosylation; autophagy

Citation: Sánchez-Adriá, I.E.; Sanmartín, G.; Prieto, J.A.; Estruch, F.; Randez-Gil, F. Slt2 Is Required to Activate ER-Stress-Protective Mechanisms through TORC1 Inhibition and Hexosamine Pathway Activation. *J. Fungi* **2022**, *8*, 92. https://doi.org/10.3390/jof8020092

Academic Editors: María Molina and Humberto Martín

Received: 14 December 2021
Accepted: 15 January 2022
Published: 18 January 2022

1. Introduction

The fungal cell wall is an external rigid structure that gives shape and integrity to the cell [1]. Its composition, made mainly of polysaccharides and glycoproteins, is continuously remodeled to allow the growth and the morphological changes required during the cell cycle [2–4]. Substantial changes in composition and thickness also occur in response to environmental physical stress, such as osmotic and heat stresses, in order to avoid cell membrane rupture and lysis [5]. In consonance with this, the biosynthetic pathways involved in its formation are strictly regulated in response to different signaling pathways [3,6].

In *Saccharomyces cerevisiae*, the cell wall integrity (CWI) pathway is the key pathway in controlling cell wall dynamics [7]. Signals are initiated at the plasma membrane (PM) through the cell-surface sensors Wsc1-3, Mid2, and Mtl1 [8,9], and transmitted to the downstream MAPK Slt2, which activates the transcriptional response [3]. In addition, the CWI pathway is activated not only in response to cell wall damage but also by compounds or conditions not apparently related to the cell wall. Rapamycin, alkaline pH, cadmium, and genotoxic or oxidative stresses, among others, are some of the diverse stimuli described for this pathway [7]. Furthermore, Slt2 has been reported to be involved in the regulation of different targets and cellular responses [10], i.e., mitophagy and pexophagy [11] or endoplasmic reticulum (ER) inheritance [12]. All of this evidences the role of the CWI pathway and Slt2 as central players in the protective response of *S. cerevisiae* to stress. Nevertheless, we are far from having a complete view of how the CWI pathway and Slt2 connect different signaling pathways and perform functions other than those directly related to the cell wall.

Previous work has evidenced the relation of the CWI pathway and the unfolded protein response (UPR), a transcriptional program activated upon ER stress [13]. When environmental conditions or chemical agents such as tunicamycin or β-mercaptoethanol increase the load of unfolded proteins, an ER-resident sensor Ire1 (inositol-requiring protein-1) triggers the nuclear import of Hac1, a transcription factor that upregulates the transcription of genes such as ER chaperones and folding enzymes [14,15]. Nevertheless, the activity of the Ire1-Hac1 system appears to account only for a part of the ER stress response [16]. ER stress activates the CWI pathway signaling [17], which finally results in the phosphorylation of Slt2 [18]. Consistent with this, the *slt2* mutant displays ER stress hypersensitivity, and constitutive activation of the CWI pathway provides increased ER stress resistance [18]. On the other hand, *ire1* mutants as well as strains expressing misfolded proteins have defects in cell wall integrity and cell wall stress activates the UPR in a process dependent on both Ire1 and Slt2 [19]. However, activation of Slt2 by ER stress is independent of Ire1/Hac1, and the ER stress sensitivity phenotype of the *slt2* mutant is not reversed by the presence of sorbitol, a cell-wall-stabilizing agent. Finally, an *slt2 ire1* double mutant shows additive sensitivity to ER stress compared with the parental mutants [18]. Hence, Slt2 is well-known in the ER-stress-protective response in *S. cerevisiae*, but the mechanisms of crosstalk between the CWI pathway and the UPR, and the targets regulated by Slt2 under ER stress remains unclear.

Recently, the hexosamine biosynthetic pathway (HBP), a highly conserved metabolic route from bacteria to humans [20], has emerged as one of the key sensors for cellular nutrition because the synthesis of its final product, UDP-N-acetylglucosamine (UDP-GlcNAc), is critically dependent on intermediates from a number of metabolic branches, including glucose, amino acids, fatty acids, and nucleotide [21]. The first committed and rate-limiting step of the HBP is mediated by glutamine:fructose-6-phosphate amidotransferase, encoded by the yeast *GFA1* gene [22], which converts fructose-6-phosphate and glutamine into glucosamine-6-phosphate (GlcN-6-P) and glutamate. Through four enzymatic steps, the HBP provides UDP-GlcNAc, an essential amino sugar donor for glycosylation of proteins and lipids, and for the biosynthesis of chitin [23]. Transcription of *GFA1* is upregulated by CWI pathway signaling [24], which results in higher levels of chitin [25]. Quite remarkably, X-box binding protein 1 (Xbp1s), the human homolog of yeast *HAC1*, is a direct transcriptional activator of the HBP [26]. Thus, the HBP appears to connect the cell wall and ER stress signaling by regulating the UDP-GlcNAc supply. Nevertheless, the physiological relevance of this role in the phenotype of CWI-pathway and UPR mutants remains unclear.

In addition to transcriptional regulation, Gfa1 activity is regulated by UDP-GlcNAc feedback inhibition [20] and post-translational modification, two mechanisms that in human GFAT-1 appear to be coordinated [27,28]. Recently, the yeast kinase Isr1 has also been found to negatively regulate the HBP by phosphorylating Gfa1 [29]. Overexpression of *ISR1* is lethal, which is rescued by co-overexpression of *GFA1* or exogenous glucosamine, while *isr1* mutant cells display tunicamycin resistance, implying increased protein glycosylation by enhanced UDP-GlcNAc availability [29]. Consistent with this, genetic mutagenesis screening in *Caenorhabditis elegans* has identified gain-of-function mutations in GFAT-1 that suppress tunicamycin-induced ER stress [30]. Likewise, increased synthesis of UDP-GlcNAc by exogenous supplementation of HBP intermediates provided similar results [30]. Evidence of a link between HBP metabolites and cellular protein quality control, leading to improved protein homeostasis, has been also reported [30]. However, neither N-glycosylation nor UPR signaling appeared to be affected by increased UDP-GlcNAc [30]. Hence, the regulation of the flux through the HBP is key in allowing cells to face proteotoxicity, although the exact molecular mechanisms that operate under this condition need to be clarified.

Here, we have studied the ER-stress-sensitive phenotype of cells lacking Slt2 and its relationship with the activity of the HBP and the role of the MAPK in controlling the activity of Gfa1. Furthermore, the direct and indirect effects of exogenous glucosamine in promoting ER-stress resistance have also been analyzed. Overall, our work highlights the importance of Slt2 in regulating the HBP and TORC1 activity upon ER stress, which determines the load of ER-incoming proteins and the bioenergetics of the protective response.

2. Materials and Methods

2.1. Strains and Plasmids

The *S. cerevisiae* strains, oligonucleotides, and plasmids used in this study are listed in the Supplementary Materials (Tables S1–S3). The *CHS3* deletion in the *slt2* mutant strain (Table S1) was carried out by PCR-based gene replacement using the hphMX4 module in the pAG32 plasmid (Table S3) as a template and synthetic oligonucleotides (Table S2). Detection of the correct gene disruption and tagging was performed by diagnostic PCR [31], using a set of oligonucleotides (Table S2), designed to bind outside of the replaced gene sequence and within the marker module.

2.2. Media, Culture Conditions, and Stress Sensitivity Tests

Previously described standard methods were followed for media preparation [32]. Yeast cells were cultured at 30 °C in YPD (1% yeast extract, 2% peptone, and 2% glucose), SCD (0.67% yeast nitrogen base without amino acids (ForMedium, Hunstanton, UK), plus 2% glucose), MPD (0.17% yeast nitrogen base without amino acids and ammonium sulphate (ForMedium), 0.1% L-proline, plus 2% glucose), or SCD-Ino (0.69% yeast nitrogen base without amino acids and inositol (ForMedium), plus 2% glucose). Yeast transformants carrying the geneticin (kanMX4), nourseothricin (natMX4), or hygromycin B (hphMX4) resistant module were selected on YPD agar plates containing 200 mg/L of G-418 (Sigma; St. Louis, MI, USA), 50 mg/L of nourseothricin (clonNAT; WERNER Bioagents, Jena-Cospeda, Germany), or 300 mg/L of hygromycin B (Sigma), respectively [33,34]. *Escherichia coli* DH5α host strain was grown in Luria-Bertani (LB) medium (1% peptone, 0.5% yeast extract and 0.5% NaCl) supplemented with ampicillin (50 mg/L). All amino acids, sugars, and antibiotics were filter sterilized and added to the autoclaved medium. Solid media contained 2% agar. Yeast cells were transformed by the lithium acetate method [35].

For plate phenotype experiments, cultures were diluted to OD_{600} = 0.8 and 10-fold serial dilutions spotted (2 μL) onto YPD- or MPD-agar solid media, lacking or containing glucosamine (GlcN; Sigma; cat# G4875), calcofluor white (CFW; Sigma; cat# F3543), dithiothreitol (DTT; Serva Electrophoresis GmbH, Heidelberg, Germany; cat# 20710), or tunicamycin (Tn; Enzo Life Sciences, Farmingdale, NY, USA; cat# BML-CC104) as indicated. SDS (0.003% final concentration) was added to the culture medium when proteasome inhibitors, MG132 (Selleckchem; Houston, TX, USA; cat# S2619), bortezomib (Selleckchem; cat# S1013), and delanzomib (Selleckchem; cat# S1157) were tested. Stock solutions of tunicamycin (25 mg/mL, DMSO), cycloheximide (100 mg/mL, DMSO), CFW (10 mg/mL, water), GlcN (100 mg/mL, water), and proteasome inhibitors (100 mM, DMSO) were prepared, sampled in small volumes, and stored at −20 °C until use. For each experiment, a fresh sample was thawed and diluted to the working concentration. Unless otherwise indicated, colony growth was inspected after 2–4 days of incubation at 30 °C.

2.3. Microscopy and Chitin Staining

To visualize the amount of chitin, exponentially growing cells (OD_{600} = 0.5) were fixed with formaldehyde, washed with PBS, and treated with 0.06% diethanolamine as formerly described [36]. Samples (50 μL) were then incubated overnight in the dark at 4 °C with 5 μL of 1 mg/mL CFW. Finally, cells were washed five times with PBS, resuspended in immunofluorescence mounting solution, and stored at 4 °C until their visualization under a Zeiss 510 Meta Confocal microscope with a 63 × Plan-Apochromat 1.4 NA Oil

DIC objective lens (Zeiss, Oberkochen, Germany). Image processing was conducted with ImageJ (http://rsb.info.nih.gov/ij/, accessed on 7 October 2021.

2.4. qRT-PCR

Total RNA and cDNA were prepared and quantitative RT-PCR (qPCR) experiments were carried out as described previously [37]. Briefly, qPCR was performed in a DNA Engine Peltier Thermal Cycler (Bio-Rad, Hercules, CA, USA) using the SYBR Premix Ex Taq Tli RNaseH Plus Green with ROX (Takara, Shiga, Japan) and specific oligonucleotides (Table S2). Relative quantification of gene expression was determined using the comparative Ct (threshold cycle number) method analysis [38]. Fold change values were calculated as the $2^{(-\Delta dCt)}$, where dCt = Ct[Target] − Ct[Housekeeping], and ΔdCt = (ΔExperimental condition) − (ΔControl). Samples were run in triplicate and normalized to *ACT1* mRNA as a housekeeping gene. Each graph is representative of at least three independent experiments.

2.5. Galactosidase Assay

SCD-Ura-grown overnight seed cultures were refreshed at OD_{600} = 0.1 in YPD with or without the addition of 11.5 mM GlcN and cultivated at 30 °C. When OD_{600} reached 0.5, aliquots (15 units) were withdrawn for their analysis (control), and cultures were exposed to 2 μg/mL of tunicamycin for 90 min. Cells were centrifuged, washed with Z buffer (60 mM Na_2HPO_4, 40 mM NaH_2PO_4, 10 mM KCl, 1 mM $MgSO_4$), and protein extracts were prepared and processed for galactosidase activity as previously described [39]. One galactosidase unit is defined as the amount of enzyme that is able to convert 1 nmol of the substrate o-NPG per min under the assay conditions. The given values represent the mean ± SD of three independent experiments, each conducted in triplicate.

2.6. Preparation of Protein Extracts and Western Blot Analysis

Proteins were extracted, separated, and analyzed by SDS-PAGE and Western blot as previously described [40]. GFP-Atg8 and Gfa1-TAP were visualized by using a monoclonal anti-GFP antibody (1:3000; Roche Diagnostics, Indianapolis, IN, USA; cat# 11814460001) and soluble peroxidase-anti-peroxidase (α-PAP) antibody (1:1000; Sigma; cat# P1291), respectively. Rabbit anti-phospho Rps6 (1:10,000; kindly provided by T. Moustafa) was used to check the activity of TORC1. Total CPY and Gas1 were probed with rabbit polyclonal anti-CPY (1:10,000; antibodies-online, Aachen, Germany; cat# ABIN607698) and anti-Gas1 (1:10,000; a gift from H. Riezman). Hac1 and Kar2 were detected with a mouse monoclonal XBP1-antibody (1:1000; Santa Cruz Biotechnology, Santa Cruz, CA, USA; cat# sc-8015) and rabbit monoclonal anti-Kar2 (1:3000; Santa Cruz Biotechnology; cat# 33630), respectively. N-glycosylated proteins were visualized with horseradish peroxidase-conjugated concanavalin A (ConA-HRP; 1:10,000; Sigma; cat# L6397). Mouse monoclonal phosphoglycerate kinase 1 (Pgk1; 1:3000; ThermoFisher, Waltham, MA, USA; cat# 459250) and rabbit glucose-6-phosphate dehydrogenase (G6Pdh) antibody (1:3000; Cell Signaling, Danvers, MA, USA; cat# 8866) were used as loading control. The secondary antibodies used were HRP-conjugated goat anti-rabbit (1:2000; Cell Signaling; cat# 7074) or rabbit anti-mouse (1:5000; Dako, Carpinteria, CA, USA; cat# P0260). Blots were carried out and images were captured as described elsewhere [41].

2.7. Cycloheximide Treatment

Pulse analysis of Gas1 degradation in wild-type and *slt2* mutant cells grown in YPD-lacking or containing GlcN (OD_{600} = 1.0) was carried out by adding cycloheximide (CHX) at a concentration of 100 μg/mL. Aliquots were immediately withdrawn (control), and cultures were shaken at 30 °C for an additional 60, 120, or 180 min. Samples at each time point were centrifuged, washed, and processed as described above for Western blot analysis of Gas1.

2.8. ATP Assay

Overnight-grown YPD seed cultures of the BY4741 wild-type and *slt2* mutant strain were refreshed at OD_{600} = 0.1 in the same medium lacking or containing GlcN and cultivated at 30 °C until OD_{600} = 0.3. Aliquots were withdrawn for their immediate analysis (control), and cultures were split 1:2 and incubated in the presence or absence of 2 μg/mL tunicamycin. At different times during growth, 100 μL samples were analyzed for ATP levels using the CellTiter-Glo® Luminescent Assay following the manufacturer's instructions (Promega, Madison, WI, USA). The ATP level in the cell suspensions was calculated after correcting for the reagent background using the signal produced by an ATP standard as reference. Values provided are expressed as nmol of ATP per OD_{600} and represent the mean (± SD) of triplicate assays. ATP kinetics for each strain and condition was repeated at least three times.

2.9. Statistical Analysis

Sample averages were compared using a Student's *t*-test with Excel software (Microsoft, Redmond, WA, USA). Different letters represent significant differences at a $p < 0.05$ probability level.

3. Results

3.1. Activation of the HBP Provides Tunicamycin Tolerance and Rescues the ER-Stress-Sensitivity Phenotype of the slt2 Mutant

We first examined the growth of the *slt2* strain in the presence of tunicamycin, a natural inhibitor of Alg7, which induces unfolded protein stress. *ALG7* encodes the first enzyme in the N-linked glycosylation pathway [42]. As expected from previous reports [18], deletion of *SLT2* in the BY4741 wild-type strain resulted in strong sensitivity to the drug (Figure 1A). Tunicamycin induced the activation of a UPRE::*lacZ* reporter [43] both in wild-type and *slt2* mutant cells, although the activation levels were lower in the latter (Figure 1B). Similar behavior was observed when analyzing the induction of Hac1 and Kar2 by tunicamycin in wild-type and *slt2* mutant cells (Figure S1A). Kar2, a UPR-dependent ER chaperone protein [16], and Hac1, the UPR transcription factor [44,45], are well-known readouts of the UPR signaling. Finally, the overexpression of a functional mature form of *HAC1* [46] did not provide any growth advantage to *slt2* mutant cells in the presence of tunicamycin (Figure S1B).

Then, we analyzed the implication of the HBP in the phenotype of the *slt2* strain. As mentioned, transcription of *GFA1*, the gene encoding the first enzyme in the synthesis of UDP-GlcNAc by the HBP [22], has been reported to be regulated by several stress conditions, including cell-wall and ER stress [24]. As shown in Figure 1C, the induction of *GFA1*, both by tunicamycin and by calcofluor white (CWF), a fluorochrome that binds chitin [47], was mainly dependent on the presence of Slt2. Consistent with this, the abundance of the Gfa1 protein was lower in cells of the *slt2* mutant exposed to either CFW or tunicamycin (Figure 1D). This suggested that impaired HBP flux by Gfa1 downregulation might cause ER stress sensitivity accounting for the phenotype of the *slt2* strain. Indeed, exogenous addition of glucosamine, GlcN (Figure 1A), or expression of *GFA1* from a multicopy plasmid [48] rescued the ER-stress-sensitivity phenotype of the *slt2* mutant (Figure S1C). Moreover, GlcN supplementation stimulated the growth of wild-type cells in the presence of tunicamycin, indicating that the aminosugar effects are not restricted to the CWI MAPK mutant (Figure S2). Interestingly, we also note that the addition of GlcN reduced the tunicamycin-induced UPR response and the transcriptional activation of *GFA1* in both wild-type and *slt2* cells (Figure 1B,C). The aminosugar GlcN is taken up by glucose transporters and phosphorylated by *S. cerevisiae* hexokinase [49], thus increasing the synthesis of UDP-GlcNAc and bypassing the need for Gfa1 activity [50]. We conclude that the role of the CWI pathway in ER stress is largely dependent on the HBP activity.

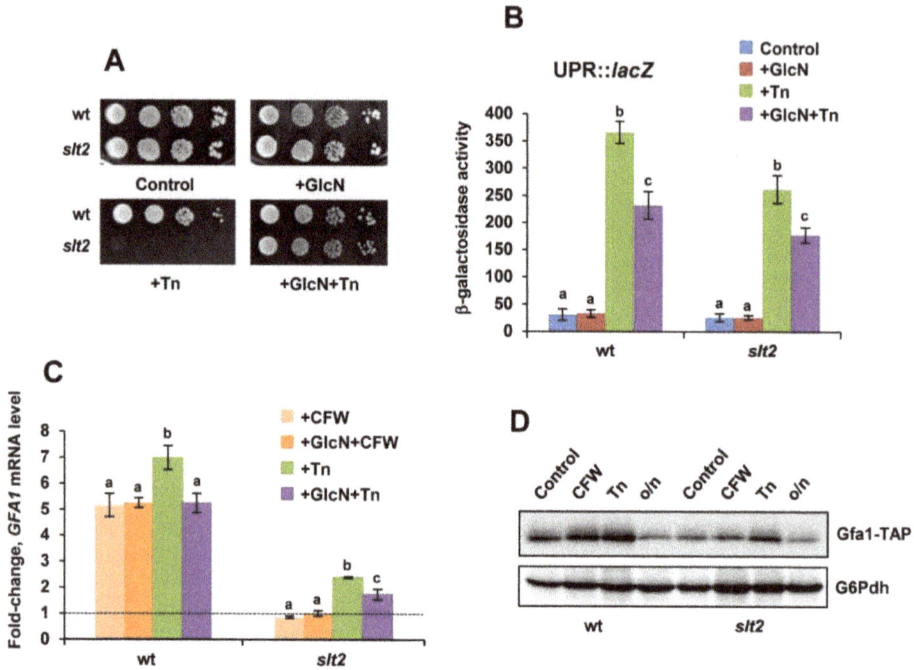

Figure 1. Activation of the HBP reduces the UPR response and provides ER-stress tolerance. (**A**) Serial dilutions (1–10^{-3}) of YPD-grown cultures (OD_{600} ~ 0.8) of the BY4741 wild-type (wt) and its isogenic *slt2* mutant were spotted (2 μL) onto YPD plates that lacked or contained glucosamine, GlcN (11.5 mM) and/or tunicamycin, Tn (0.5 μg/mL) and were incubated at 30 °C for 2–4 days. (**B**) The activity of a UPRE::*lacZ* reporter was assayed in YPD-grown cells (OD_{600} ~ 0.5) of the indicated strains exposed to 11.5 mM GlcN and/or 2 μg/mL tunicamycin (Tn) for 90 min. Aliquots of yeast cultures were harvested and cells were assayed for β-galactosidase activity. Data represent the mean value ± SD of three independent experiments. The activity values with different letters were significantly different at $p < 0.05$. (**C**) YPD-grown cells of the indicated strains (OD_{600} ~ 0.5) were treated with 40 μg/mL calcofluor white (CFW) and/or 2 μg/mL tunicamycin (Tn) for 90 min and samples from untreated (control) and treated cultures were processed for qPCR analysis of *GFA1* mRNA. Expression differences between untreated and treated samples for the wt and *slt2* strain are shown as fold-change. Data represent the mean (±SD) of at least three independent experiments. Statistically significant ($p < 0.05$) differences are denoted with different letters. (**D**) Protein extracts from Gfa1-TAP-tagged cells of the wild-type (wt) and *slt2* mutant strain were obtained by NaOH-treatment and analyzed by regular SDS-PAGE and Western blot by using soluble peroxidase-anti-peroxidase (α-PAP) antibody as described in the Materials and Methods section. YPD cultures were grown until mid-log phase (OD_{600} ~ 0.5) at 30 °C (control) and treated with 40 μg/mL CFW or 2 μg/mL Tn for 90 min. Untreated cultures grown overnight (o/n) were also tested. The level of glucose-6-phosphate dehydrogenase (G6Pdh) was used as a loading control for crude extracts. A representative experiment is shown.

3.2. The HBP Links Different Signaling Pathways in the ER Stress Response

In addition to the CWI and the UPR, other signaling pathways, among them, the osmosensing high osmotic glycerol (HOG) pathway [51], have been identified as playing a role in the protective response to ER stress [18,39]. Interestingly, the HOG and CWI pathways are positively coordinated to regulate many stress responses [6]. As shown in Figure 2A, deletion of *HOG1* caused a strong sensitivity to tunicamycin as reported [39]. However, the presence of exogenous GlcN reversed the tunicamycin sensitivity of the *hog1* strain (Figure 2A). The result led us to examine the phenotype of cells lacking *IRE1* or *HAC1*, the main actors in controlling the transcriptional and post-translational response to ER stress [15]. As expected, *ire1* and *hac1* mutant cells showed a strong sensitivity to the presence of 0.4 μg tunicamycin/mL, a phenotype that was not alleviated by the addition of GlcN at doses (5.25 mM) that provide some protection to *slt2* (Figure 2B). However, at higher GlcN concentrations (11.5 mM), *ire1* and *hac1* mutants grew as a wild-type strain (Figure 2B).

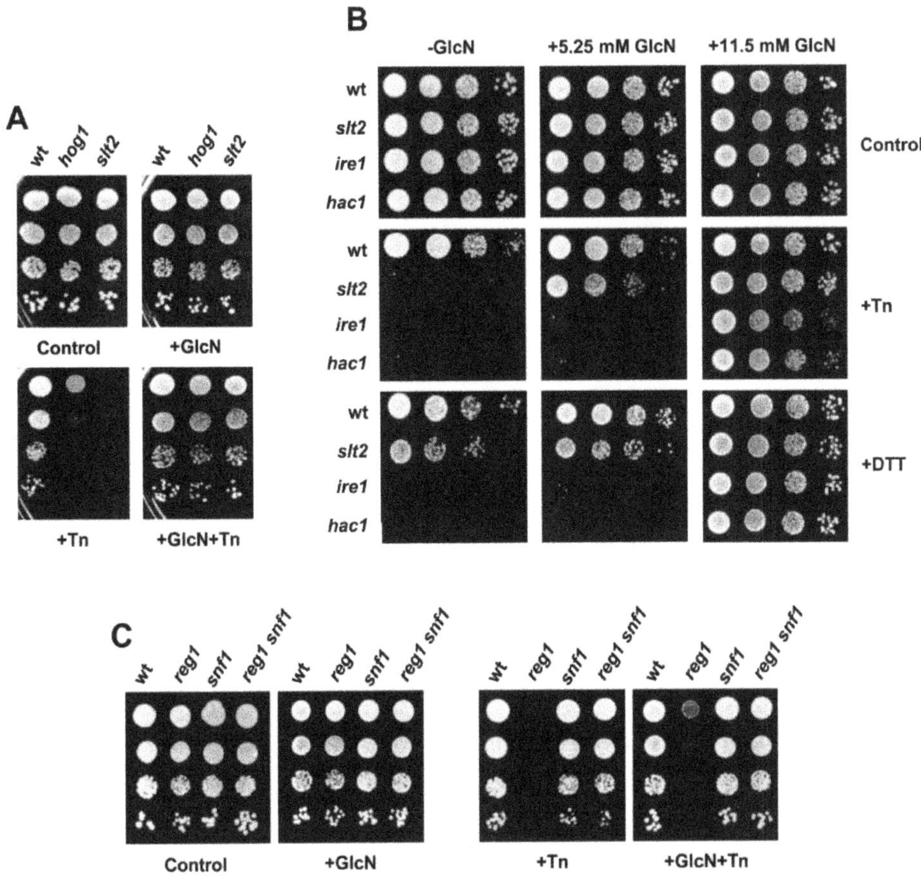

Figure 2. Glucosamine supplementation confers tolerance to several ER stress inducers and to mutants in different signaling pathways. (**A**) Drop test of BY4741 wild-type (wt), *hog1*, and *slt2* mutant strains. Cultures were diluted (1–10^{-3}) and spotted (2 μL) onto YPD plates lacking or containing 11.5 mM glucosamine (GlcN) and/or 0.5 μg/mL tunicamycin (Tn). (**B**) Cultures of the indicated strains, BY4741 wild type (wt), *slt2*, *ire1*, and *hac1*, were assayed for growth on YPD plates lacking or containing GlcN at the indicated concentrations and/or 0.5 μg/mL Tn or 17 mM dithiothreitol (DTT). (**C**) The ER-stress tolerance of the BY4741 wild-type (wt), *reg1*, *snf1*, and *reg1 snf1* strains was inspected on YPD plates lacking or containing glucosamine (GlcN) and/or tunicamycin (Tn). Drug concentrations and cultures processing conditions were as in panel (**A**). In all cases, a representative experiment is shown.

We speculated that GlcN alleviates the tunicamycin sensitivity of different yeast mutants just by increasing the synthesis of UDP-GlcNAc. Tunicamycin is structurally related to UDP-GlcNAc, and thus both could compete by binding to the active site of Alg7. However, tunicamycin inhibition has been reported to be noncompetitive in nature [52]. Consistent with this, the exogenous addition of GlcN had similar effects when dithiothreitol

(DTT) was used as an ER-stress inducer (Figure 2B). The reducing agent DTT disrupts protein folding by preventing disulfide bond formation. Furthermore, GlcN was effective in partially overcoming the requirement of inositol of CWI pathway mutants lacking Bck1, the MAPKK of the CWI pathway [3], or Slt2 (Figure S3). Depletion of the membrane lipid component inositol triggers the UPR [53], likely by adversely affecting the integrity of glycosylphosphatidylinositol (GPI)-anchored proteins [53] and impairment of Ca^{2+} fluxes, thereby contributing to protein misfolding [54]. However, GlcN was unable to rescue the inositol auxotrophy of *PHO85* mutations (data not shown), a well-known regulator of sphingolipid biosynthesis [37].

Then, we examined the effects of GlcN in the nutrient-sensing Snf1-mediated catabolite repression pathway [55]. The Snf1 kinase, the *S. cerevisiae* ortholog of AMP-activated protein kinase, AMPK [56], has been reported to be involved in the regulation of the UPR [57]. Indeed, cells lacking Reg1, a regulatory subunit of the Glc7 protein phosphatase, which causes the inappropriate activation of Snf1 [56] displayed hypersensitivity to tunicamycin ([57]; Figure 2C). On the contrary, *snf1* and *reg1 snf1* mutant cells exhibited only a weak growth defect, if any (Figure 2C). We also found that GlcN provided improved tunicamycin tolerance to *snf1* and *reg1 snf1*, but this effect was scarce in cells devoted to a functional Reg1 protein (Figure 2C). Overall, the results suggest that the HBP mediates a specific protective response.

3.3. The Synthesis of Chitin Does Not Confer Protection against ER Stress

The synthesis of cell wall components, particularly chitin, is a common response to stress conditions that threaten cell integrity [3]. In particular, chitin levels have been reported to increase when GlcN is added to the culture medium of yeast cells [50]. In addition, genes involved in chitin biosynthesis have been identified among those that were upregulated in tunicamycin-resistant mutants isolated by adaptive aneuploidy [58]. Consequently, we analyzed whether chitin synthesis, as measured by CFW-fluorescence microscopy, could be in part responsible for the positive effects on ER-stress sensitivity of increasing levels of intermediates of the HBP. Single *slt2*, *chs3*, and double *slt2 chs3* mutants were examined for chitin levels and tunicamycin sensitivity in the presence or absence of GlcN (Figure 3). As expected, GlcN increased the synthesis of chitin in all the strains analyzed, except in the single *chs3* mutant (Figure 3A). To our surprise, cells of the *slt2 chs3* strain exhibited enhanced chitin fluorescence as compared with the single *chs3* strain, suggesting that other chitin synthase genes were induced in the absence of Slt2. Chs3, the major chitin synthase, is responsible for more than 90% of the chitin in *S. cerevisiae* [59], but additional enzymes, Chs1 and Chs2 exist [1]. A lack of Slt2 reduced chitin levels in the lateral cell wall, but deposits still remain visible at the bud-neck region (Figure 3A), a change that was also apparent in tunicamycin-treated cells. Changes in trafficking, abundance, and subcellular location of chitin synthase enzymes could explain these results [1,50].

Elevated or reduced levels of chitin did not appear to have a great impact on the tunicamycin sensitivity of yeast cells. Although increased chitin synthesis by GlcN addition (Figure 3A) correlated with improved tolerance to tunicamycin-induced ER stress in the *slt2* mutant, the single *chs3* displayed higher tolerance than the wild type in the absence of GlcN (Figure 3B). Neither the improved synthesis of chitin in the double *slt2 chs3* mutant increased its tolerance to the drug as compared with the single *chs3* (Figure 3B). We also note that the addition of GlcN to the culture medium, either lacking or containing tunicamycin, caused a strong growth defect in *slt2 chs3* cells (data not shown). Although we have no obvious explanation for this result, it seems that the addition of GlcN may lead to energy imbalances, as UDP-GlcNAc recycling could be impaired in the context of some cell-wall mutants. We conclude that chitin synthesis is not a major determinant of ER-stress tolerance.

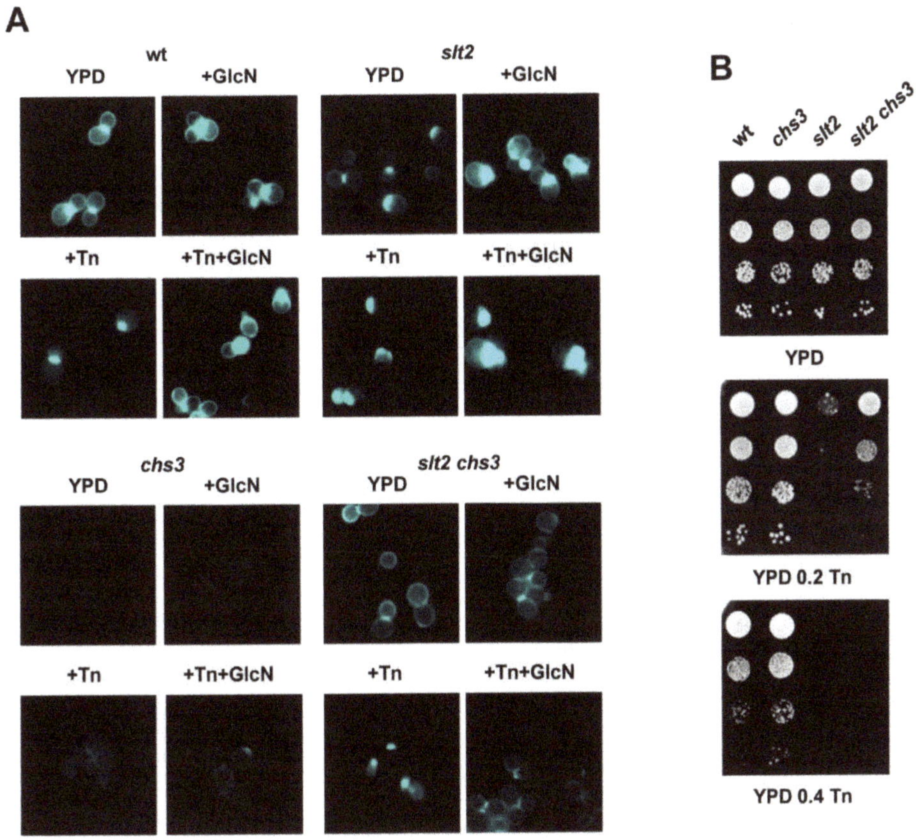

Figure 3. Increased synthesis of chitin in response to tunicamycin exposure is not essential for ER stress tolerance. (**A**) Exponentially growing cells of the BY4741 wild-type (wt), *slt2*, *chs3*, and *slt2 chs3* mutant strains were treated with 11.5 mM glucosamine (GlcN) and/or 2 µg/mL tunicamycin (Tn) for 90 min and processed for chitin staining with calcofluor white (CFW) and microscopy visualization as described in the Materials and Methods section. (**B**) The same strains were checked for growth in YPD medium lacking or containing tunicamycin at the indicated concentrations (µg/mL). Cultures were diluted and spotted as indicated in Figure 1. In all cases, representative experiments are shown.

3.4. The Exogenous Addition of GlcN Reduces the Abundance of Glycosylated Proteins

Previous work by Denzel and coworkers [30] demonstrated that increased synthesis of N-glycan precursors in the HBP improves ER protein homeostasis and extends lifespan in *C. elegans*, phenotypes that were ascribed to improved protein homeostasis, although the molecular mechanisms involved were not clarified. One possibility is that increased UDP-GlcNAc could modulate in some way N-glycosylation, the main target of tunicamycin. However, a global change in steady-state protein glycosylation in gfat-1 gain-of-function mutants of *C. elegans* has not been observed [30]. We, therefore, were interested to examine this aspect in yeast cells. Protein glycosylation as measured by N-glycan labeling with concanavalin A was recorded by Western blot in samples from wild-type and *slt2* mutant cells cultivated in the presence of GlcN and/or tunicamycin (Figure 4A). As expected, tunicamycin exposure caused a bulk reduction in N-glycosylated proteins in both wild-type and *slt2* strains. Unexpectedly, increased HBP flux by GlcN addition had a similar effect, although the reduction seemed to be less intense (Figure 4A).

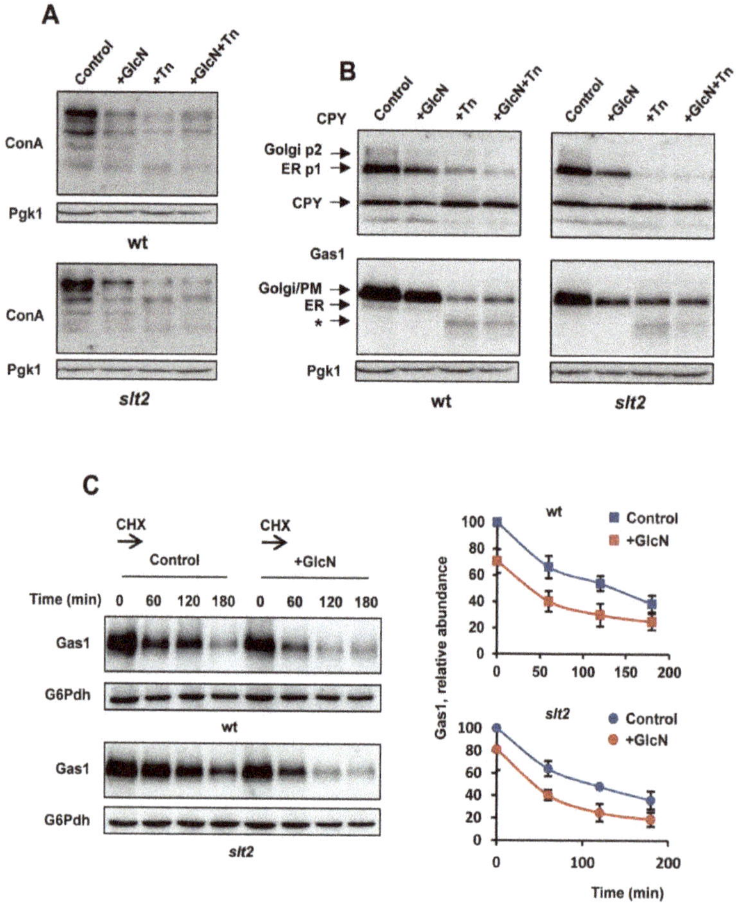

Figure 4. Increased HBP flux by glucosamine supplementation reduces the abundance of N-glycosylated proteins. (**A**) Protein extracts from the indicated strains, BY4741 wild type (wt) and *slt2*, were separated by SDS-PAGE and analyzed by Western blot for concanavalin A staining (ConA). Aliquots from YPD-grown cells (OD_{600} ~ 0.5) were withdrawn (control) and cultures were shaken at 30 °C for an additional 180 min in the presence of 11.5 mM glucosamine (GlcN) and/or 2 µg/mL tunicamycin (Tn). Samples at each time point were centrifuged, washed, and processed as described in Section 2. The level of phosphoglycerate kinase (Pgk1) was used as a loading control for crude extracts. A representative experiment is shown. (**B**) The indicated strains were cultivated under the same conditions and protein extracts were processed by SDS-PAGE and Western blot analysis of CPY (upper panel) and Gas1 (lower panel). The arrows show the ER-localized "p1" form of proCPY (67 KDa), the Golgy-localized "p2" form of CPY (69 KDa), and the vacuolar 61 kDa active, mature form of the enzyme. Likewise, arrows in lower panel show 105 KDa ER-form ("p1") and the 125 KDa ("p2") mature forms of Gas1, respectively. Bands labeled with (*) corresponded with degraded forms of Gas1. (**C**) Pulse analysis of Gas1 degradation in wild-type and *slt2* mutant cells grown in YPD-lacking or containing 11.5 mM GlcN (OD_{600} ~ 1.0) was carried out by adding cycloheximide (CHX) at a concentration of 100 µg/mL. Aliquots at the indicated times were withdrawn and protein extracts were processed by SDS-PAGE and Western blot as in panel (**B**). The level of glucose-6-phosphate dehydrogenase (G6Pdh) was used as a loading control for crude extracts. The graph shows the abundance of Gas1 at each time point relative to that of the control for GlcN-treated and -untreated samples of each strain analyzed. Data are the mean (±SD) of three independent biological replicates.

Then, we checked the protein abundance and electrophoretic profile of two ER-client proteins that are often used as model secretory glycoproteins, CPY, the yeast vacuolar carboxypeptidase Y [60] and Gas1, a β-1,3-glucanosyltransferase that localizes to the cell surface via glycosylphosphatidylinositol (GPI) anchor [61,62]. As can be seen, the fraction of ER- and vacuole-localized pro-CPY forms was sharply reduced in tunicamycin-treated cells of wild-type and *slt2* cells (Figure 4B). Likewise, the abundance of Gas1 decreased in cells exposed to the drug, with the appearance of degraded forms of higher electrophoretic mobility. More importantly, GlcN exposure caused again similar effects, although the combined exposure to both tunicamycin and GlcN did not appear to result in a further reduction of protein abundance (Figure 4B). Consistent with this, a gradual loss of Gas1 abundance was observed when control cells were exposed to a pulse of GlcN (Figure S4). We also noted that decreased protein abundance by tunicamycin or GlcN exposure did not appear to affect glycolytic enzymes, such as phosphoglycerate kinase (Pgk1, Figure 4B), glucose-6-phosphate dehydrogenase (G6Pdh, Figure S4), or hexokinase PII, Hxk2 (data not shown), used as loading controls.

Finally, we wonder whether the GlcN-induced loss of Gas1 abundance reflects enhanced protein degradation. To examine this, cycloheximide (CHX) was added to wild-type and *slt2* YPD cultures lacking (control) or containing GlcN. Aliquots of cells were collected immediately and at specific time points, and protein samples were analyzed by Western blot for Gas1 relative abundance. As shown in Figure 4C, the apparent rate of CHX-induced disappearance of Gas1 was similar in GlcN-treated and -untreated cultures of either wild-type or *slt2*.

3.5. *TORC1 Remains Active in Tunicamycin-Exposed Cells of the Slt2 Mutant Strain*

TORC1 integrates multiple signaling pathways and plays a key role as a central stress and growth controller [63]. TORC1 activity promotes the cellular translation capacity and restricts the abundance of the proteolytic machinery [64,65], and thus, regulation of TORC1 is crucial to ensure protein homeostasis under stress conditions. Quite remarkably, tunicamycin has been reported to inhibit TORC1 signaling [66,67], and to increase—via Slt2—proteasome abundance [68]. Therefore, we first examined whether increased levels of HBP intermediates regulate TORC1 signaling. As expected, tunicamycin quickly inhibited the TORC1 activity, as measured by phosphorylation of the 40S ribosomal protein S6 (Rps6), in cells of the wild-type strain, an effect that was not mainly affected by the simultaneous addition of GlcN (Figure 5). The phosphorylation of Rps6 is a well-established readout of TORC1-dependent signaling [69,70]. To our surprise, TORC1 activity was insensitive to tunicamycin in the *slt2* mutant, both in the presence or absence of GlcN (Figure 5). We conclude that Slt2 is an essential effector of TORC1 activity in tunicamycin-exposed cells.

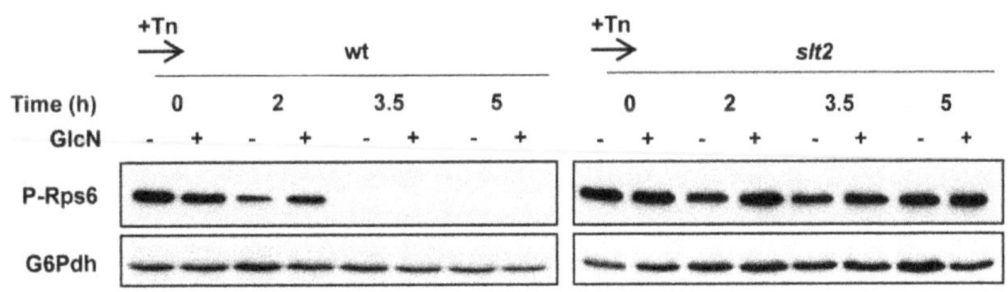

Figure 5. TORC1 activity is not inhibited by tunicamycin in absence of Slt2. Analysis of TORC1 activity as measured by phosphorylation of the 40S ribosomal protein S6. Protein extracts obtained from YPD-grown cultures (OD_{600} ~ 0.5) were treated with 2 μg/mL tunicamycin (Tn) for the indicated times in the presence (+) or absence (−) of 11.5 mM glucosamine (GlcN), separated by SDS-PAGE and analyzed by Western blot for phospho-Rps6 (P-Rps6). The level of glucose-6-phosphate dehydrogenase (G6Pdh) was used as a loading control for crude extracts. A representative experiment is shown.

3.6. *The Proteasome Homeostasis Is Not Critical to Tunicamycin Survival*

Downstream TORC1, Slt2 has been reported to control Adc17 and proteasome abundance [68]. Adc17, a stress-inducible RAC (regulatory particle assembly chaperone), is crucial for proteasome assembly and to maintain proteasome levels [71]. Thus, the lack of induction of Adc17 in *slt2* mutant cells has been claimed to be the main determinant of the tunicamycin-sensitive phenotype of the MAPK mutant [68]. To confirm this idea, we first tested the phenotype of cells lacking Adc17. Previous work has reported that the *adc17* mutant is tunicamycin sensitive, although the phenotype was weak and visible by drop test only under extremely high (5 μg/mL) tunicamycin concentrations [68]. As shown in Figure S5, *adc17* mutant cells grew as well as the wild type in the presence of 0.5 μg/mL of tunicamycin, a drug dose that fully inhibits the growth of *slt2*. Likewise, no apparent effect on tunicamycin sensitivity was observed by knockdown of different RACs genes such as *NAS6*, *HSM3*, *RPN4*, and *RPN14* (Figure S5).

Then, we analyzed tunicamycin sensitivity in the presence of proteasome inhibitors. Previous work by Denzel et al. [30] reported that gfat-1 gain-of-function mutants of *C. elegans* display enhanced proteasome activity. Thus, we reasoned that impaired proteasome activity would cause increased tunicamycin sensitivity. We first tested the effect of MG132, short peptide aldehydes that block active sites of the proteasome [72]. The use of proteasome inhibitors in wild-type *S. cerevisiae* cells is hampered by the impermeability of the cell wall or membrane [73], an issue that can be overcome by the use of a synthetic medium containing L-proline and SDS [74]. As it is shown, proteasome inhibition by MG132 did not result in increased toxicity of tunicamycin in either of the strains analyzed, wild type, *slt2*, *ire1*, or *hac1* (Figure 6A). On the contrary, the inhibitor caused a slight improvement in growth at low doses of tunicamycin, a subtle effect that could be explained by the activation of compensatory mechanisms. Evidence indicates that proteasome inhibition or impairment activates autophagy [75]. Consistent with this, similar results (Figure 6B) were obtained by using bortezomib (also named PS-341), a reversible inhibitor of the proteasome containing a peptide-like backbone and boronate group, or its structurally related inhibitor delanzomib [76].

3.7. *Knock-Out of SLT2 and Glucosamine Treatment Has Distinct Effects on the Tunicamycin-Induced Autophagic Response*

Beyond proteasome regulation, TORC1 suppresses autophagy [64], a self-degradation mechanism that improves proteostasis through the clearance of aggregated proteins [77]. Autophagy induction is triggered by TORC1 inhibition in response to either nutrient starvation or stress conditions, including ER stress [78]. Hence, we tested the effect of HBP activation on autophagy induction, as measured by the GFP-Atg8 processing assay [79]. Wild-type and *slt2* cells were transformed with a plasmid encoding GFP-Atg8 [80] and the abundance of GFP-Atg8 and free GFP after 3 and 6 h in the presence or absence of GlcN and/or tunicamycin was analyzed by Western blot. The ubiquitin-like protein Atg8 is one of the major Atg proteins that is involved in autophagosome expansion [81], and accordingly, *ATG8* is up-regulated following the induction of autophagy at the transcriptional and translational level [79,82]. Finally, the appearance of free GFP monitors the autophagic flux, as Atg8 is rapidly degraded in the vacuole but GFP is not [83]. As expected, we observed increased levels of GFP-Atg8 in tunicamycin-exposed wild-type cells, an effect that was more pronounced after 6 h of treatment (Figure 7). We also noted that the level of GFP-Atg8 was insensitive to tunicamycin treatment in the *slt2* mutant strain (Figure 7), in good correspondence with the absence of TORC1 inhibition in this strain. Interestingly, glucosamine treatment also increased slightly the abundance of GFP-Atg8 in both wild-type and *slt2* cells after 6 h of the onset of the experiment (Figure 7B), suggesting that enhanced activity of the HBP stimulates the autophagy, a result previously reported in *C. elegans* [30].

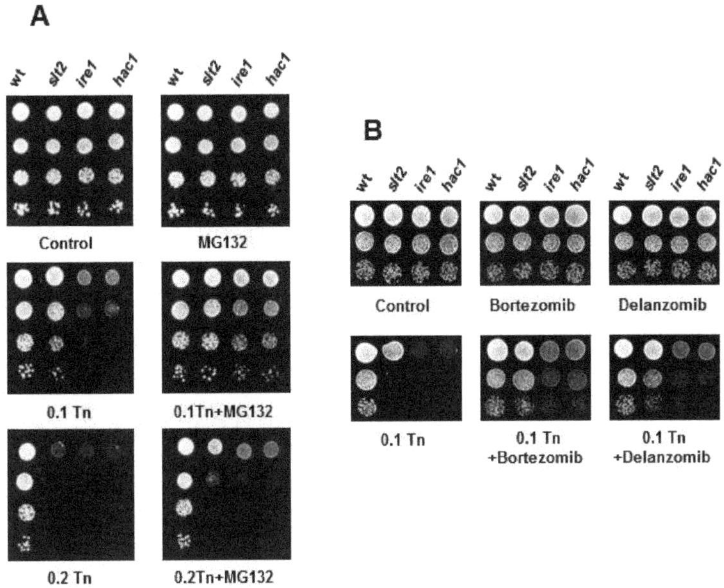

Figure 6. Effects of proteasome inhibitors on tunicamycin tolerance. (**A**) Cultures of the indicated strains, BY4741 wild-type (wt) *slt2*, *ire1*, and *hac1* were assayed for growth on MPD-SDS (0.003%) plates lacking (control) or containing tunicamycin (Tn) at the indicated concentration (μg/mL) and/or 75 μM MG132. (**B**) The same strains that in (**A**) were tested for growth in the presence of tunicamycin (Tn) at the indicated concentration (μg/mL) and/or 50 μM of bortezomib or delanzomib. A representative experiment is shown.

With regard to the autophagic flux, a band corresponding with free GFP was only observed after 6 h of tunicamycin exposure of wild-type cells (Figure 7B). Likewise, the presence of GlcN appeared to increase the proteolysis of GFP-Atg8 expressed in the wild-type strain, but the effect was weak. Finally, the tunicamycin treatment did not cause the proteolysis of GFP-Atg8 in the *slt2* mutant (Figure 7B). We conclude that Slt2 is required to trigger autophagy in response to ER stress. The activity of the HBP also appears to play a role in stimulating this protective mechanism.

3.8. The Bioenergetics Response of Yeast Cells to ER Stress Depends on a Functional Slt2 MAPK

Previous work indicated that ER-to-mitochondria Ca^{2+} transfer increases during the early phase of tunicamycin exposure to stimulate mitochondrial bioenergetics [84]. As a result, ATP levels, oxygen consumption, and reductive power increase in order to face the energy demand for protein folding and clearance of protein aggregates under ER stress [85]. Evidence also suggests that TORC1 is a central signaling effector of this response as its inhibition by rapamycin mimics the bioenergetics effects of tunicamycin [86]. Therefore, we decided to assess ATP levels in tunicamycin-treated cells of the wild-type and *slt2* strain in the presence or absence of GlcN. As shown in Figure 8, the level of ATP at the onset of the experiment (0 h) was slightly higher in GlcN-containing YPD-grown cells of both, wild-type and *slt2* mutant strain, a result that could be explained in light of the reduced abundance of glycoproteins in these cells (Figure 4). Protein translation is one of the energetically most expensive processes [87]. As expected, the level of ATP increased in wild-type cells after 4 and 6 h of tunicamycin treatment, but not in cells lacking the MAPK Slt2 (Figure 8). We also observed no significant differences by the combined exposure to both tunicamycin and GlcN, indicating that the aminosugar does not interfere, at least at the doses used, with the ATP overproduction upon ER stress.

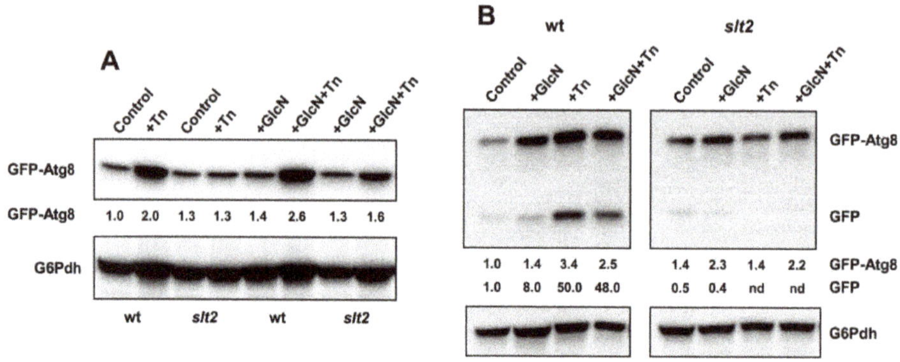

Figure 7. HBP activation and *SLT2* knock-out causes distinct responses to autophagy. (**A**) Overnight SCD-Leu grown cultures of pRS415-GFP-ATG8 transformants of the BY4741 wild-type (wt) and *slt2* strains were refreshed in YPD (OD_{600} = 0.1) lacking or containing 11.5 mM glucosamine (GlcN) and grown at 30 °C until OD_{600} ~ 0.3. Aliquots were withdrawn for their immediate analysis (Control), and cultures were split in two and incubated at 30 °C in the presence (Tn; GlcN+Tn) or absence (control; GlcN) of 2 µg/mL tunicamycin (Tn) for 3 h. Protein extracts were prepared as described in Section 2, separated by SDS-PAGE, and analyzed by Western blot for GFP-Atg8 and free GFP using anti-GFP antibody. The image shows only a part of the gel where GFP-Atg8 was localized as free GFP was hardly detected. (**B**) The same strains were assayed as above except that the tunicamycin (Tn) treatment was extended for 6 h. In all cases, the bands corresponding with GFP-Atg8 or free GFP (GFP) are indicated. The level of glucose-6-phosphate dehydrogenase (G6Pdh) was used as a loading control for crude extracts. The values at the bottom of the images represent the GFP-Atg8 and free GFP abundance relative to that of the wild-type strain under control conditions that was set at 1.0. Representative experiments are shown.

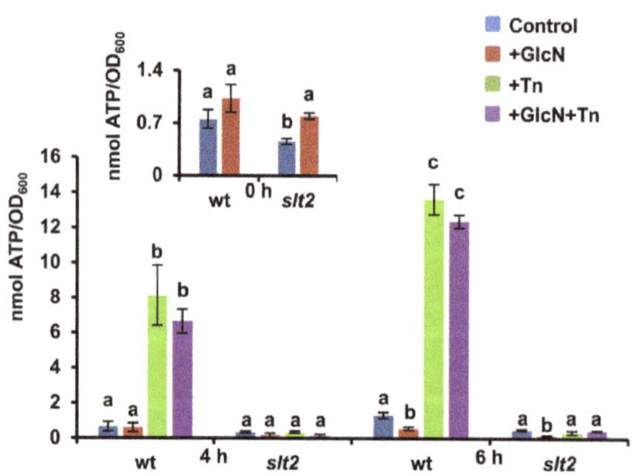

Figure 8. The tunicamycin-induced increase in ATP levels is absent in *slt2* mutant cells. YPD-grown overnight seed cultures of the BY4741 wild-type and *slt2* mutant strain were refreshed at OD_{600} = 0.1 in the same medium lacking or containing 11.5 mM glucosamine (GlcN) and grown at 30 °C until OD_{600} = 0.3. Aliquots were withdrawn for their immediate analysis (time 0), and cultures were split in two and incubated at 30 °C in the presence (Tn; GlcN+Tn) or absence (Control; GlcN) of 2 µg/mL tunicamycin (Tn) for 4 and 6 h. Data are expressed as nmol of ATP per OD_{600} and represent the mean (±SD) of triplicate assays. ATP kinetics for each strain and condition was repeated at least three times. Different letters represent significant differences at $p < 0.05$ probability level for each strain and growth condition compared with the corresponding control.

4. Discussion

The idea that the role of Slt2 in the ER-stress protective response was connected to its function as a transcriptional activator of *GFA1* ([24]; this work), came from the finding that increased biosynthesis of UDP-GlcNAc in *C. elegans* improves protein homeostasis [30]. Slt2, the MAPK of the CWI pathway, has been reported to play an important role in ER-stress tolerance, although the exact mechanism is unclear as *slt2* mutant cells show only a weak defect in the activation of the UPR ([18]; this work), the transcriptional program addressed to mitigate the accumulation of unfolded proteins [13–15]. Here, we demonstrated that ectopic expression of *GFA1* or supplementation of the growth medium with the aminosugar GlcN confers increased tolerance to different ER-stress inducers and rescues the ER-stress growth defect of *slt2* mutant cells. GlcN is converted by the action of hexokinase to GlcN-6-phosphate [49], which increases the level of HBP intermediates and relieves the need for Gfa1 [50]. Quite remarkably, we also found that activation of the HBP by GlcN addition alleviated the tunicamycin-sensitivity phenotype of cells lacking Hog1, the MAPK of the osmolarity HOG pathway [51], Ire1 or Hac1. Like Slt2, strains lacking Hog1 display sensitivity to tunicamycin with no or minor effects on the UPR-mediated regulation [39]. On the contrary, Ire1 and Hac1 are essential effectors of the UPR and the lack of any of them strongly impairs the transcriptional upregulation of hundreds of genes in response to ER stress [16,44,45]. However, all of them have in common a role in the transcriptional activation of *GFA1* [18,26]. Altogether, our results stress the importance of the hexosamine pathway in the ER-stress protective response in *S. cerevisiae* and the role of Gfa1 as a central effector, whose activity is coordinately controlled by a number of conserved signaling pathways.

Regulating protein degradation is an integral part of the UPR program to relieve ER stress [15,88]. Consistent with this, enhanced proteolysis involving ER-associated degradation (ERAD), proteasomal activity, and autophagy was observed in *C. elegans* in response to increased flux through the HBP [30]. Similarly, HBP activation was found to reduce aggregated polyQ and toxicity in tissue cultures [89]. Although the mechanism involved in these effects was not clarified, it was suggested that enhanced HBP would generate a weak ER stress that, in turn, would increase autophagy through eIF2α phosphorylation [89], the α subunit of the eukaryotic initiation factor-2 [90]. Previous work reported that glucosamine treatment can lead to ER stress [91] and eIF2α phosphorylation [92], whereby reducing polyQ [93] and SDS-insoluble Huntingtin aggregates [94] in animal models.

Unlike this view, we found evidence that is against the idea that GlcN causes ER stress and that increased autophagic activity is on the basis of the role of the HBP in the ER stress response. (1) The addition of GlcN to the culture medium reduced the tunicamycin-induced UPR response and the transcriptional activation of *GFA1* and *KAR2*, the gene encoding the best-known ER chaperone, which is indispensable when facing ER stress [95]. Furthermore, overexpression of *GFA1* or GlcN supplementation leads to a general growth improvement in yeast cells exposed to ER stress. We conclude that in *S. cerevisiae* the aminosugar provides protection against ER stress, instead of being an ER-stress inducer; (2) autophagy induction by GlcN supplementation was weak and the increase in autophagic flux was hardly visible. Neither deletion of *ATG* genes essential for bulk autophagy, such as *ATG1*, *ATG8*, *ATG12*, *ATG13*, or *ATG33*, resulted in increased sensitivity to tunicamycin (Figure S6). Moreover, we did not find changes by GlcN supplementation in the degradation kinetics of Gas1 induced by cycloheximide treatment, suggesting that the aminosugar does not stimulate protein degradation. (3) We observed a global downregulation in the abundance of concanavalin A-labeled N-glycosylated proteins in cells treated with GlcN, an effect that was not observed in previous studies in *C. elegans* [30,89], and that was confirmed by Western blot analysis of CPY and Gas1, two well-known RE-client proteins [60–62]; and (4) no differences in the abundance of several glycolytic proteins were observed when GlcN-treated and -untreated cells were compared, suggesting that the downregulation of N-glycosylated proteins abundance was specific and not due to the activation of global proteolysis mechanisms.

The finding that activated HBP by GlcN supplementation causes a decrease in the abundance of N-glycosylated ER-client proteins was in some way surprising as enhanced flux through the HBP increases the content of UDP-GlcNAc that serves as a precursor for N-glycosylation. Nevertheless, different evidence suggests that this observation could be the result of energetic adaptations. Effective glycosylation and folding of proteins require both biosynthetic precursors and ATP. In mammalian cells, numerous surface proteins and growth factors are N-glycosylated, and the extent of this modification is feedback regulated by glucose availability [96,97], which ensures that cells do not engage in anabolic metabolism when nutrients are limiting [98]. An energetic checkpoint that only allows effective receptor glycosylation and folding when ATP is in excess has also been identified [99]. However, the HBP is a non-energy-generating pathway that consumes glucose [100]. Indeed, AMPK, the major energy-sensing effector, the human homolog of yeast Snf1, has been reported to inhibit by phosphorylation GFAT1 [101] in order to reduce the HBP flux when ATP becomes scarce. Based on this, the energy consumed by the HBP activity in cells overexpressing *GFA1* or by GlcN supplementation could be much higher than in optimal growth conditions. Indeed, GlcN treatment rapidly and transiently lowers ATP levels [100] as the aminosugar acts as a glucose analog that is phosphorylated by hexokinase [49]. In addition, GlcN causes transcriptional reprogramming [102] and represses the respiration rate (QO_2), even more rapidly than glucose [103], which contributes to reducing energy supply. Consistent with all of this, GlcN has been reported to increase the life span of *C. elegans* and aging mice by mimicking a low-carbohydrate diet [104] and it has been proposed as a promising candidate for pharmacological caloric restriction mimetics [105]. Thus, the results of our study showing that the ATP balance in GlcN-treated and -untreated control cells is rather similar suggests that compensatory mechanisms operate to reduce the demand of ATP when the HBP flux is overloaded, which is consistent with a downregulation of N-glycosylated proteins after GlcN addition to the culture medium. How increased HBP activity causes a reduction in ER-client proteins remains unknown. Emerging evidence indicates that ERAD, the principal quality-control mechanism, not only mediates the elimination of structurally abnormal proteins in the ER but also contributes to the regulation of native proteins [106]. More work is required to clarify the GlcN effects on N-glycosylation and if these are functionally linked to ERAD or other quality control of protein folding mechanisms.

ER stress induced by tunicamycin exposure inhibited TORC1 signaling and activated the autophagy and mitochondrial bioenergetics in wild-type cells of *S. cerevisiae*. Indeed, TORC1 inhibition has been reported to be involved [86] in stimulating mitochondrial uptake of Ca^{2+} release by tunicamycin exposure, thereby increasing respiration and ATP production [85], in order to face the enhanced energy requirements under ER stress conditions [84]. A prominent role of TORC1 inhibition in enhancing overall protein degradation by the ubiquitin–proteasome system and autophagy has also been widely established [64–66,68]. In particular, inhibition of TORC1 by ER stress was found to induce Adc17 and to increase proteasome abundance in yeast [68]. In our work, the lack of Adc17 and other proteasome subunits did not have noticeable effects on tunicamycin growth. Neither the absence of important *ATG* genes essential for bulk autophagy had apparent consequences on ER-stress sensitivity, at least at the doses of tunicamycin tested in our study. Likewise, autophagy mutants do not appreciably compromise cell survival or genome integrity in genotoxic stress conditions [65]. Nevertheless, these results should be taken with caution as impairment of a protein degradation mechanism may lead to increased activity of alternative systems. Indeed, recent data indicate the presence of connections and reciprocal regulation mechanisms between autophagy and the ubiquitin–proteasome system [107].

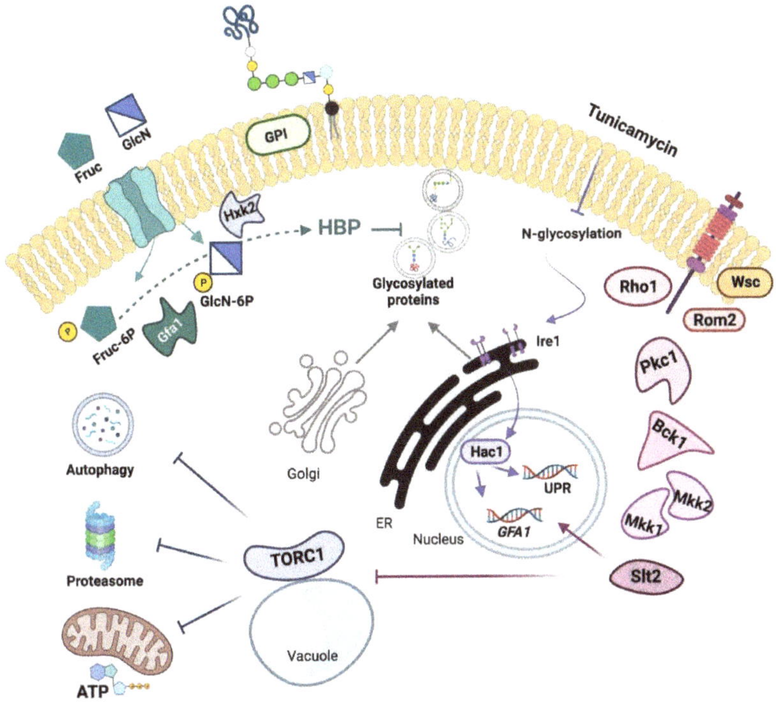

Figure 9. Schematic representation of the ER-stress signaling network and its interaction with Slt2 and the hexosamine pathway. The CWI signaling pathway from the anchored Wsc sensors to the final MAPK Slt2 (see [3,5–10] as representative reviews) and the metabolic steps from fructose (Fruc) or glucosamine (GlcN) to glucosamine-6-phosphate (GlcN-6P) catalyzed by Gfa1 [22] and Hxk2 [49,50], respectively, are shown. Gfa1 catalyzes the first committed and rate-limiting step of the HBP, which provides UDP-N-acetylglucosamine for, among others, glycosylation and GPI-anchoring of proteins [20,21]. Tunicamycin exposure induces ER stress by inhibiting the N-glycosylation of proteins [42], which triggers a protective response, the UPR, a transcription program mediated by Ire1 and Hac1, that upregulates the transcription of hundreds of genes [13–16], among them *GFA1* [24]. Expression of *GFA1* also depends on Slt2 [26], which link the CWI and the UPR in providing increased flux through the HBP under ER-stress conditions, a response that reduces the abundance of ER-client proteins. ER-stress also inhibits TORC1 [66,67], a stress and growth controller [63], which downregulates protein homeostasis mediated by autophagy [64,78] and proteasome [68] activities and the bioenergetics response [84,86], which provides the energy required for protein folding and clearance of protein aggregates under ER stress [85]. Remarkably, Slt2 is required to both *GFA1* expression and TORC1 inhibition in response to ER-stress, which accounts for the strong growth defect of cells devoted of Slt2 in media containing ER-stress-inducers, a phenotype that can be relieved by *GFA1* overexpresion or GlcN supplementation. Arrows and bars denote positive and negative interactions, respectively. For additional details, see the text.

Remarkably, we found that autophagy or enhanced production of ATP was absent in cells lacking Slt2. The most likely explanation is the impaired inhibition of TORC1 associated with the *SLT2* mutation under ER stress. Recently, the role of Slt2 in regulating rapamycin-induced autophagy and TORC1 inactivation has been ruled out [65]. Neither TORC1 inactivation after DNA damage was found to be dependent on Slt2, although autophagy induction was partially reduced by the loss of MAPK [65]. Hence, TORC1 inactivation under ER stress appears to differ mechanistically with respect to other stressful conditions in the requirement of Slt2. Interestingly, previous work has placed

Slt2 downstream of TORC1 inhibition in response to caffeine [108] and rapamycin treatment [36], which would mediate the PKA regulation by TORC1 [109]. Rather, our work places Slt2 upstream of TORC1 signaling, although a direct interaction between them seems unlikely. Crosstalk between CWI, PKA, Snf1, and TORC1 signaling has been extensively documented [110] and these are obviously potential effectors in the ER-stress signaling network. Nonetheless, as we graphically summarize in Figure 9, our study highlights the importance of Slt2 in controlling the adaptive response to ER stress by HBP-dependent and -independent mechanisms. Indeed, Slt2 is required both to induce *GFA1* transcription and to inhibit TORC1 in response to ER-stress, which accounts for the strong defect of *slt2* mutant cells under these conditions. This phenotype can be alleviated by hyperactivation of the HBP via *GFA1* overexpression or GlcN supplementation, which reduces the load of ER-incoming proteins and the ER stress.

Supplementary Materials: The following supporting information can be downloaded at: https://www.mdpi.com/article/10.3390/jof8020092/s1, Figure S1: Overexpression of *GFA1* provides enhanced ER stress tolerance; Table S1: *Saccharomyces cerevisiae* strains used in this study; Figure S2: Glucosamine supplementation stimulates the growth of *Saccharomyces cerevisiae* in the presence of the ER-stress inducer tunicamycin; Table S2: Oligonucleotides used in this study; Figure S3: Glucosamine addition partially overcomes the inositol requirement of CWI pathway mutants; Table S3: Plasmids used in this study; Figure S4: Glucosamine reduces Gas1 abundance; Figure S5: Knock-out of RACs does not cause increased ER-stress sensitivity; Figure S6: *ATG* genes essential for bulk autophagy are dispensable for ER stress tolerance.

Author Contributions: Conceptualization, F.R.-G.; formal analysis, I.E.S.-A. and G.S.; funding acquisition, F.E. and F.R.-G.; investigation, I.E.S.-A., G.S. and F.R.-G.; methodology, I.E.S.-A. and G.S.; supervision, J.A.P. and F.E.; writing—original draft, J.A.P. and F.R.-G. All authors have read and agreed to the published version of the manuscript.

Funding: This research has been supported by the Grants BIO2015-71059-R and PID2020-115623RB-100 funded by the Spanish Ministry of Science and Innovation (MICIN)/AEI/10.13039/501100011033, and by "ERDF A way of making Europe".

Institutional Review Board Statement: Not applicable.

Informed Consent Statement: Not applicable.

Data Availability Statement: Not applicable.

Acknowledgments: We would like to thank Matthias Rose (Frankfurt University), Cesar Roncero (Salamanca University), Daniel Klionsky (Michigan University), Tarek Moustafa (Medical University of Graz), Howard Riezman (University of Geneva) and Kazutoshi Mori (Kyoto University) for providing plasmids, antibodies and yeast strains. The schematic representation displayed in Figure 9 was created by using the software from BioRender.com.

Conflicts of Interest: The authors declare that the research was conducted in the absence of any commercial or financial relationships that could be construed as a potential conflict of interest.

References

1. Cid, V.J.; Durán, A.; del Rey, F.; Snyder, M.P.; Nombela, C.; Sánchez, M. Molecular basis of cell integrity and morphogenesis in *Saccharomyces cerevisiae*. *Microbiol. Rev.* **1995**, *59*, 345–386. [CrossRef]
2. Klis, F.M.; Boorsma, A.; De Groot, P.W. Cell wall construction in *Saccharomyces cerevisiae*. *Yeast* **2006**, *23*, 185–202. [CrossRef]
3. Levin, D.E. Regulation of cell wall biogenesis in *Saccharomyces cerevisiae*: The cell wall integrity signaling pathway. *Genetics* **2011**, *189*, 1145–1175. [CrossRef]
4. Orlean, P. Architecture and biosynthesis of the *Saccharomyces cerevisiae* cell wall. *Genetics* **2012**, *192*, 775–818. [CrossRef]
5. Fuchs, B.B.; Mylonakis, E. Our paths might cross: The role of the fungal cell wall integrity pathway in stress response and cross talk with other stress response pathways. *Eukaryot. Cell.* **2009**, *8*, 1616–1625. [CrossRef]
6. Rodríguez-Peña, J.M.; García, R.; Nombela, C.; Arroyo, J. The high-osmolarity glycerol (HOG) and cell wall integrity (CWI) signalling pathways interplay: A yeast dialogue between MAPK routes. *Yeast* **2010**, *27*, 495–502. [CrossRef]
7. Jiménez-Gutiérrez, E.; Alegría-Carrasco, E.; Sellers-Moya, Á.; Molina, M.; Martín, H. Not just the wall: The other ways to turn the yeast CWI pathway on. *Int. Microbiol.* **2020**, *23*, 107–119. [CrossRef]

8. Rodicio, R.; Heinisch, J.J. Together we are strong—cell wall integrity sensors in yeasts. *Yeast* **2010**, *27*, 531–540. [CrossRef] [PubMed]
9. Jendretzki, A.; Wittland, J.; Wilk, S.; Straede, A.; Heinisch, J.J. How do I begin? Sensing extracellular stress to maintain yeast cell wall integrity. *Eur. J. Cell Biol.* **2011**, *90*, 740–744. [CrossRef] [PubMed]
10. González-Rubio, G.; Sellers-Moya, Á.; Martín, H.; Molina, M. A walk-through MAPK structure and functionality with the 30-year-old yeast MAPK Slt2. *Int. Microbiol.* **2021**, *24*, 521–543. [CrossRef] [PubMed]
11. Mao, K.; Klionsky, D.J. MAPKs regulate mitophagy in *Saccharomyces cerevisiae*. *Autophagy* **2011**, *7*, 1564–1565. [CrossRef] [PubMed]
12. Du, Y.; Walker, L.; Novick, P.; Ferro-Novick, S. Ptc1p regulates cortical ER inheritance via Slt2p. *EMBO J.* **2006**, *25*, 4413–4422. [CrossRef]
13. Mori, K. The unfolded protein response: The dawn of a new field. *Proc. Jpn. Acad. Ser. B Phys. Biol. Sci.* **2015**, *91*, 469–480. [CrossRef]
14. Walter, P.; Ron, D. The unfolded protein response: From stress pathway to homeostatic regulation. *Science* **2011**, *334*, 1081–1086. [CrossRef]
15. Read, A.; Schröder, M. The unfolded protein response: An overview. *Biology* **2021**, *10*, 384. [CrossRef]
16. Travers, K.J.; Patil, C.K.; Wodicka, L.; Lockhart, D.J.; Weissman, J.S.; Walter, P. Functional and genomic analyses reveal an essential coordination between the unfolded protein response and ER-associated degradation. *Cell* **2000**, *101*, 249–258. [CrossRef]
17. Bonilla, M.; Cunningham, K.W. Mitogen-activated protein kinase stimulation of Ca(2+) signaling is required for survival of endoplasmic reticulum stress in yeast. *Mol. Biol. Cell.* **2003**, *14*, 4296–4305. [CrossRef] [PubMed]
18. Chen, Y.; Feldman, D.E.; Deng, C.; Brown, J.A.; De Giacomo, A.F.; Gaw, A.F.; Shi, G.; Le, Q.T.; Brown, J.M.; Koong, A.C. Identification of mitogen-activated protein kinase signaling pathways that confer resistance to endoplasmic reticulum stress in *Saccharomyces cerevisiae*. *Mol. Cancer Res.* **2005**, *3*, 669–677. [CrossRef] [PubMed]
19. Scrimale, T.; Didone, L.; de Mesy Bentley, K.L.; Krysan, D.J. The unfolded protein response is induced by the cell wall integrity mitogen-activated protein kinase signaling cascade and is required for cell wall integrity in *Saccharomyces cerevisiae*. *Mol. Biol. Cell* **2009**, *20*, 164–175. [CrossRef]
20. Milewski, S. Glucosamine-6-phosphate synthase–the multi-facets enzyme. *Biochim. Biophys. Acta* **2002**, *1597*, 173–192. [CrossRef]
21. Denzel, M.S.; Antebi, A. Hexosamine pathway and (ER) protein quality control. *Curr. Opin. Cell Biol.* **2015**, *33*, 14–18. [CrossRef]
22. Watzele, G.; Tanner, W. Cloning of the glutamine:fructose-6-phosphate amidotransferase gene from yeast. Pheromonal regulation of its transcription. *J. Biol. Chem.* **1989**, *264*, 8753–8758. [CrossRef]
23. Namboori, S.C.; Graham, D.E. Acetamido sugar biosynthesis in the Euryarchaea. *J. Bacteriol.* **2008**, *190*, 2987–2996. [CrossRef] [PubMed]
24. Zheng, J.; Khalil, M.; Cannon, J.F. Glc7p protein phosphatase inhibits expression of glutamine-fructose-6-phosphate transaminase from *GFA1*. *J. Biol. Chem.* **2000**, *275*, 18070–18078. [CrossRef]
25. Lagorce, A.; Le Berre-Anton, V.; Aguilar-Uscanga, B.; Martin-Yken, H.; Dagkessamanskaia, A.; François, J. Involvement of *GFA1*, which encodes glutamine-fructose-6-phosphate amidotransferase, in the activation of the chitin synthesis pathway in response to cell-wall defects in *Saccharomyces cerevisiae*. *Eur. J. Biochem.* **2002**, *269*, 1697–1707. [CrossRef]
26. Wang, Z.V.; Deng, Y.; Gao, N.; Pedrozo, Z.; Li, D.L.; Morales, C.R.; Criollo, A.; Luo, X.; Tan, W.; Jiang, N.; et al. Spliced X-box binding protein 1 couples the unfolded protein response to hexosamine biosynthetic pathway. *Cell* **2014**, *156*, 1179–1192. [CrossRef]
27. Ruegenberg, S.; Horn, M.; Pichlo, C.; Allmeroth, K.; Baumann, U.; Denzel, M.S. Loss of GFAT-1 feedback regulation activates the hexosamine pathway that modulates protein homeostasis. *Nat. Commun.* **2020**, *11*, 687. [CrossRef] [PubMed]
28. Ruegenberg, S.; Mayr, F.A.M.C.; Atanassov, I.; Baumann, U.; Denzel, M.S. Protein kinase A controls the hexosamine pathway by tuning the feedback inhibition of GFAT-1. *Nat. Commun.* **2021**, *12*, 2176. [CrossRef]
29. Alme, E.B.; Stevenson, E.; Krogan, N.J.; Swaney, D.L.; Toczyski, D.P. The kinase Isr1 negatively regulates hexosamine biosynthesis in *S. cerevisiae*. *PLoS Genet.* **2020**, *16*, e1008840. [CrossRef] [PubMed]
30. Denzel, M.S.; Storm, N.J.; Gutschmidt, A.; Baddi, R.; Hinze, Y.; Jarosch, E.; Sommer, T.; Hoppe, T.; Antebi, A. Hexosamine pathway metabolites enhance protein quality control and prolong life. *Cell* **2014**, *156*, 1167–1178. [CrossRef]
31. Huxley, C.; Green, E.D.; Dunham, I. Rapid assessment of *S. cerevisiae* mating type by PCR. *Trends Genet.* **1990**, *6*, 236.
32. Guthrie, C.; Fink, G.R. Guide to yeast genetics and molecular biology. *Methods Enzymol.* **1991**, *194*, 21–37.
33. Wach, A.; Brachat, A.; Pöhlmann, R.; Philippsen, P. New heterologous modules for classical or PCR-based gene disruptions in *Saccharomyces cerevisiae*. *Yeast* **1994**, *10*, 1793–1808. [CrossRef] [PubMed]
34. Goldstein, A.L.; McCusker, J.H. Three new dominant drug resistance cassettes for gene disruption in *Saccharomyces cerevisiae*. *Yeast* **1999**, *15*, 1541–1553. [CrossRef]
35. Ito, H.; Fukuda, Y.; Murata, K.; Kimura, A. Transformation of intact yeast cells treated with alkali cations. *J. Bacteriol.* **1983**, *153*, 163–168. [CrossRef]
36. Torres, J.; Di Como, C.J.; Herrero, E.; De La Torre-Ruiz, M.A. Regulation of the cell integrity pathway by rapamycin-sensitive TOR function in budding yeast. *J. Biol. Chem.* **2002**, *277*, 43495–43504. [CrossRef]
37. Prieto, J.A.; Estruch, F.; Córcoles-Sáez, I.; Del Poeta, M.; Rieger, R.; Stenzel, I.; Randez-Gil, F. Pho85 and PI(4,5)P_2 regulate different lipid metabolic pathways in response to cold. *Biochim. Biophys. Acta Mol. Cell. Biol. Lipids.* **2020**, *1865*, 158557. [CrossRef] [PubMed]

38. Schmittgen, T.D.; Livak, K.J. Analyzing real-time PCR data by the comparative C(T) method. *Nat. Protoc.* **2008**, *3*, 1101–1108. [CrossRef]
39. Torres-Quiroz, F.; García-Marqués, S.; Coria, R.; Randez-Gil, F.; Prieto, J.A. The activity of yeast Hog1 MAPK is required during endoplasmic reticulum stress induced by tunicamycin exposure. *J. Biol. Chem.* **2010**, *285*, 20088–20096. [CrossRef]
40. Hernández-López, M.J.; García-Marqués, S.; Randez-Gil, F.; Prieto, J.A. Multicopy suppression screening of *Saccharomyces cerevisiae* Identifies the ubiquitination machinery as a main target for improving growth at low temperatures. *Appl. Environ. Microbiol.* **2011**, *77*, 7517–7525. [CrossRef]
41. Córcoles-Sáez, I.; Hernández, M.L.; Martínez-Rivas, J.M.; Prieto, J.A.; Randez-Gil, F. Characterization of the *S. cerevisiae inp51* mutant links phosphatidylinositol 4,5-bisphosphate levels with lipid content, membrane fluidity and cold growth. *Biochim. Biophys. Acta.* **2016**, *1861*, 213–226. [CrossRef] [PubMed]
42. Back, S.H.; Schröder, M.; Lee, K.; Zhang, K.; Kaufman, R.J. ER stress signaling by regulated splicing: IRE1/HAC1/XBP1. *Methods* **2005**, *35*, 395–416. [CrossRef] [PubMed]
43. Mori, K.; Kawahara, T.; Yoshida, H.; Yanagi, H.; Yura, T. Signalling from endoplasmic reticulum to nucleus: Transcription factor with a basic-leucine zipper motif is required for the unfolded protein-response pathway. *Genes Cells* **1996**, *1*, 803–817. [CrossRef] [PubMed]
44. Cox, J.S.; Walter, P. A novel mechanism for regulating activity of a transcription factor that controls the unfolded protein response. *Cell* **1996**, *87*, 391–404. [CrossRef]
45. Sidrauski, C.; Walter, P. The transmembrane kinase Ire1p is a site-specific endonuclease that initiates mRNA splicing in the unfolded protein response. *Cell* **1997**, *90*, 1031–1039. [CrossRef]
46. Valkonen, M.; Penttilä, M.; Saloheimo, M. Effects of inactivation and constitutive expression of the unfolded- protein response pathway on protein production in the yeast *Saccharomyces cerevisiae*. *Appl. Environ. Microbiol.* **2003**, *69*, 2065–2072. [CrossRef]
47. Roncero, C.; Durán, A. Effect of Calcofluor white and Congo red on fungal cell wall morphogenesis: In vivo activation of chitin polymerization. *J. Bacteriol.* **1985**, *163*, 1180–1185. [CrossRef]
48. Gomez, A.; Perez, J.; Reyes, A.; Duran, A.; Roncero, C. Slt2 and Rim101 contribute independently to the correct assembly of the chitin ring at the budding yeast neck in *Saccharomyces cerevisiae*. *Eukaryot. Cell* **2009**, *8*, 1449–1459. [CrossRef]
49. Ballou, C.E.; Maitra, S.K.; Walker, J.W.; Whelan, W.L. Developmental defects associated with glucosamine auxotrophy in *Saccharomyces cerevisiae*. *Proc. Natl. Acad. Sci. USA* **1977**, *74*, 4351–4355. [CrossRef]
50. Bulik, D.A.; Olczak, M.; Lucero, H.A.; Osmond, B.C.; Robbins, P.W.; Specht, C.A. Chitin synthesis in *Saccharomyces cerevisiae* in response to supplementation of growth medium with glucosamine and cell wall stress. *Eukaryot. Cell* **2003**, *2*, 886–900. [CrossRef]
51. Saito, H.; Posas, F. Response to hyperosmotic stress. *Genetics* **2012**, *192*, 289–318. [CrossRef] [PubMed]
52. Heifetz, A.; Keenan, R.W.; Elbein, A.D. Mechanism of action of tunicamycin on the UDP-GlcNAc:dolichyl-phosphate Glc-NAc-1-phosphate transferase. *Biochemistry* **1979**, *18*, 2186–2192. [CrossRef]
53. Promlek, T.; Ishiwata-Kimata, Y.; Shido, M.; Sakuramoto, M.; Kohno, K.; Kimata, Y. Membrane aberrancy and unfolded proteins activate the endoplasmic reticulum stress sensor Ire1 in different ways. *Mol. Biol. Cell* **2011**, *22*, 3520–3532. [CrossRef] [PubMed]
54. Fu, S.; Yang, L.; Li, P.; Hofmann, O.; Dicker, L.; Hide, W.; Lin, X.; Watkins, S.M.; Ivanov, A.R.; Hotamisligil, G.S. Aberrant lipid metabolism disrupts calcium homeostasis causing liver endoplasmic reticulum stress in obesity. *Nature* **2011**, *473*, 528–531. [CrossRef]
55. Conrad, M.; Schothorst, J.; Kankipati, H.N.; Van Zeebroeck, G.; Rubio-Texeira, M.; Thevelein, J.M. Nutrient sensing and signaling in the yeast *Saccharomyces cerevisiae*. *FEMS Microbiol. Rev.* **2014**, *38*, 254–299. [CrossRef] [PubMed]
56. Hedbacker, K.; Carlson, M. SNF1/AMPK pathways in yeast. *Front. Biosci.* **2008**, *13*, 2408–2420. [CrossRef]
57. Ferrer-Dalmau, J.; Randez-Gil, F.; Marquina, M.; Prieto, J.A.; Casamayor, A. Protein kinase Snf1 is involved in the proper regulation of the unfolded protein response in *Saccharomyces cerevisiae*. *Biochem. J.* **2015**, *468*, 33–47. [CrossRef]
58. Beaupere, C.; Dinatto, L.; Wasko, B.M.; Chen, R.B.; VanValkenburg, L.; Kiflezghi, M.G.; Lee, M.B.; Promislow, D.E.L.; Dang, W.; Kaeberlein, M.; et al. Genetic screen identifies adaptive aneuploidy as a key mediator of ER stress resistance in yeast. *Proc. Natl. Acad. Sci. USA* **2018**, *115*, 9586–9591. [CrossRef]
59. Bulawa, C.E. *CSD2*, *CSD3*, and *CSD4*, genes required for chitin synthesis in *Saccharomyces cerevisiae*: The *CSD2* gene product is related to chitin synthases and to developmentally regulated proteins in *Rhizobium* species and *Xenopus laevis*. *Mol. Cell. Biol.* **1992**, *12*, 1764–1776.
60. Hasilik, A.; Tanner, W. Biosynthesis of the vacuolar yeast glycoprotein carboxypeptidase Y. Conversion of precursor into the enzyme. *Eur. J. Biochem.* **1978**, *85*, 599–608. [CrossRef]
61. Conzelmann, A.; Riezman, H.; Desponds, C.; Bron, C. A major 125-kd membrane glycoprotein of *Saccharomyces cerevisiae* is attached to the lipid bilayer through an inositol-containing phospholipid. *EMBO J.* **1988**, *7*, 2233–2240. [CrossRef]
62. Nuoffer, C.; Jenö, P.; Conzelmann, A.; Riezman, H. Determinants for glycophospholipid anchoring of the *Saccharomyces cerevisiae* GAS1 protein to the plasma membrane. *Mol. Cell. Biol.* **1991**, *11*, 27–37.
63. Albert, V.; Hall, M.N. mTOR signaling in cellular and organismal energetics. *Curr. Opin. Cell Biol.* **2015**, *33*, 55–66. [CrossRef]
64. Su, K.H.; Dai, C. mTORC1 senses stresses: Coupling stress to proteostasis. *Bioessays* **2017**, *39*, 1600268. [CrossRef]
65. Ueda, S.; Ozaki, R.; Kaneko, A.; Akizuki, R.; Katsuta, H.; Miura, A.; Matsuura, A.; Ushimaru, T. TORC1, Tel1/Mec1, and Mpk1 regulate autophagy induction after DNA damage in budding yeast. *Cell. Signal.* **2019**, *62*, 109344. [CrossRef]

66. Lempiäinen, H.; Uotila, A.; Urban, J.; Dohnal, I.; Ammerer, G.; Loewith, R.; Shore, D. Sfp1 interaction with TORC1 and Mrs6 reveals feedback regulation on TOR signaling. *Mol. Cell* **2009**, *33*, 704–716. [CrossRef]
67. Ahmed, K.; Carter, D.E.; Lajoie, P. Hyperactive TORC1 sensitizes yeast cells to endoplasmic reticulum stress by compromising cell wall integrity. *FEBS Lett.* **2019**, *593*, 1957–1973. [CrossRef]
68. Rousseau, A.; Bertolotti, A. An evolutionarily conserved pathway controls proteasome homeostasis. *Nature* **2016**, *536*, 184–189. [CrossRef]
69. González, A.; Shimobayashi, M.; Eisenberg, T.; Merle, D.A.; Pendl, T.; Hall, M.N.; Moustafa, T. TORC1 promotes phosphorylation of ribosomal protein S6 via the AGC kinase Ypk3 in *Saccharomyces cerevisiae*. *PLoS ONE* **2015**, *10*, e0120250. [CrossRef]
70. Yerlikaya, S.; Meusburger, M.; Kumari, R.; Huber, A.; Anrather, D.; Costanzo, M.; Boone, C.; Ammerer, G.; Baranov, P.V.; Loewith, R. TORC1 and TORC2 work together to regulate ribosomal protein S6 phosphorylation in *Saccharomyces cerevisiae*. *Mol. Biol. Cell* **2016**, *27*, 397–409. [CrossRef]
71. Hanssum, A.; Zhong, Z.; Rousseau, A.; Krzyzosiak, A.; Sigurdardottir, A.; Bertolotti, A. An inducible chaperone adapts proteasome assembly to stress. *Mol. Cell* **2014**, *55*, 566–577. [CrossRef]
72. Gaczynska, M.; Osmulski, P.A. Small-molecule inhibitors of proteasome activity. *Methods Mol. Biol.* **2005**, *301*, 3–22.
73. Lee, D.H.; Goldberg, A.L. Selective inhibitors of the proteasome-dependent and vacuolar pathways of protein degradation in *Saccharomyces cerevisiae*. *J. Biol. Chem.* **1996**, *271*, 27280–27284. [CrossRef]
74. Pannunzio, V.G.; Burgos, H.I.; Alonso, M.; Mattoon, J.R.; Ramos, E.H.; Stella, C.A. A simple chemical method for rendering wild-type yeast permeable to brefeldin A that does not require the presence of an *erg6* mutation. *J. Biomed. Biotechnol.* **2004**, *3*, 150–155. [CrossRef]
75. Zhu, K.; Dunner, K., Jr.; McConkey, D.J. Proteasome inhibitors activate autophagy as a cytoprotective response in human prostate cancer cells. *Oncogene* **2010**, *29*, 451–462. [CrossRef]
76. Fricker, L.D. Proteasome Inhibitor Drugs. *Annu. Rev. Pharmacol. Toxicol.* **2020**, *60*, 457–476. [CrossRef]
77. Menzies, F.M.; Fleming, A.; Rubinsztein, D.C. Compromised autophagy and neurodegenerative diseases. *Nat. Rev. Neurosci.* **2015**, *16*, 345–357. [CrossRef]
78. Yorimitsu, T.; Nair, U.; Yang, Z.; Klionsky, D.J. Endoplasmic reticulum stress triggers autophagy. *J. Biol. Chem.* **2006**, *281*, 30299–30304. [CrossRef]
79. Kirisako, T.; Baba, M.; Ishihara, N.; Miyazawa, K.; Ohsumi, M.; Yoshimori, T.; Noda, T.; Ohsumi, Y. Formation process of autophagosome is traced with Apg8/Aut7p in yeast. *J. Cell. Biol.* **1999**, *147*, 435–446. [CrossRef]
80. Suzuki, K.; Kirisako, T.; Kamada, Y.; Mizushima, N.; Noda, T.; Ohsumi, Y. The pre-autophagosomal structure organized by concerted functions of APG genes is essential for autophagosome formation. *EMBO J.* **2001**, *20*, 5971–5981. [CrossRef]
81. Xie, Z.; Nair, U.; Klionsky, D.J. Atg8 controls phagophore expansion during autophagosome formation. *Mol. Biol. Cell* **2008**, *19*, 3290–3298. [CrossRef]
82. Huang, W.P.; Scott, S.V.; Kim, J.; Klionsky, D.J. The itinerary of a vesicle component, Aut7p/Cvt5p, terminates in the yeast vacuole via the autophagy/Cvt pathways. *J. Biol. Chem.* **2000**, *275*, 5845–5851. [CrossRef]
83. Cheong, H.; Klionsky, D.J. Biochemical methods to monitor autophagy-related processes in yeast. *Methods Enzymol.* **2008**, *451*, 1–26.
84. Bravo-Sagua, R.; Parra, V.; López-Crisosto, C.; Díaz, P.; Quest, A.F.; Lavandero, S. Calcium transport and signaling in mitochondria. *Compr. Physiol.* **2017**, *7*, 623–634.
85. Bravo, R.; Vicencio, J.M.; Parra, V.; Troncoso, R.; Munoz, J.P.; Bui, M.; Quiroga, C.; Rodriguez, A.E.; Verdejo, H.E.; Ferreira, J.; et al. Increased ER-mitochondrial coupling promotes mitochondrial respiration and bioenergetics during early phases of ER stress. *J. Cell Sci.* **2011**, *124*, 2143–2152. [CrossRef]
86. Bravo-Sagua, R.; López-Crisosto, C.; Parra, V.; Rodriguez-Peña, M.; Rothermel, B.A.; Quest, A.F.; Lavandero, S. mTORC1 inhibitor rapamycin and ER stressor tunicamycin induce differential patterns of ER-mitochondria coupling. *Sci. Rep.* **2016**, *6*, 36394. [CrossRef]
87. Hu, X.P.; Dourado, H.; Schubert, P.; Lercher, M.J. The protein translation machinery is expressed for maximal efficiency in *Escherichia coli*. *Nat. Commun.* **2020**, *11*, 5260. [CrossRef]
88. McCaffrey, K.; Braakman, I. Protein quality control at the endoplasmic reticulum. *Essays Biochem.* **2016**, *60*, 227–235.
89. Horn, M.; Denzel, S.I.; Srinivasan, B.; Allmeroth, K.; Schiffer, I.; Karthikaisamy, V.; Miethe, S.; Breuer, P.; Antebi, A.; Denzel, M.S. Hexosamine pathway activation improves protein homeostasis through the integrated stress response. *iScience* **2020**, *23*, 100887. [CrossRef]
90. Komar, A.A.; Merrick, W.C. A retrospective on eIF2A-and not the alpha subunit of eIF2. *Int. J. Mol. Sci.* **2020**, *21*, 2054. [CrossRef]
91. Lombardi, A.; Ulianich, L.; Treglia, A.S.; Nigro, C.; Parrillo, L.; Lofrumento, D.D.; Nicolardi, G.; Garbi, C.; Beguinot, F.; Miele, C.; et al. Increased hexosamine biosynthetic pathway flux dedifferentiates INS-1E cells and murine islets by an extracellular signal-regulated kinase (ERK)1/2-mediated signal transmission pathway. *Diabetologia* **2012**, *55*, 141–153. [CrossRef]
92. Kline, C.L.; Schrufer, T.L.; Jefferson, L.S.; Kimball, S.R. Glucosamine-induced phosphorylation of the alpha-subunit of eukaryotic initiation factor 2 is mediated by the protein kinase R-like endoplasmic-reticulum associated kinase. *Int. J. Biochem. Cell. Biol.* **2006**, *38*, 1004–1014. [CrossRef]

93. Kouroku, Y.; Fujita, E.; Tanida, I.; Ueno, T.; Isoai, A.; Kumagai, H.; Ogawa, S.; Kaufman, R.J.; Kominami, E.; Momoi, T. ER stress (PERK/eIF2alpha phosphorylation) mediates the polyglutamine-induced LC3 conversion, an essential step for autophagy formation. *Cell Death. Differ.* **2007**, *14*, 230–239. [CrossRef]
94. Krzyzosiak, A.; Sigurdardottir, A.; Luh, L.; Carrara, M.; Das, I.; Schneider, K.; Bertolotti, A. Target-based discovery of an inhibitor of the regulatory phosphatase PPP1R15B. *Cell* **2018**, *174*, 1216–1228. [CrossRef]
95. Hsu, C.L.; Prasad, R.; Blackman, C.; Ng, D.T. Endoplasmic reticulum stress regulation of the Kar2p/BiP chaperone alleviates proteotoxicity via dual degradation pathways. *Mol. Biol. Cell* **2012**, *23*, 630–641. [CrossRef]
96. Lau, K.S.; Partridge, E.A.; Grigorian, A.; Silvescu, C.I.; Reinhold, V.N.; Demetriou, M.; Dennis, J.W. Complex N-glycan number and degree of branching cooperate to regulate cell proliferation and differentiation. *Cell* **2007**, *129*, 123–134. [CrossRef]
97. Wellen, K.E.; Lu, C.; Mancuso, A.; Lemons, J.M.; Ryczko, M.; Dennis, J.W.; Rabinowitz, J.D.; Coller, H.A.; Thompson, C.B. The hexosamine biosynthetic pathway couples growth factor-induced glutamine uptake to glucose metabolism. *Genes Dev.* **2010**, *24*, 2784–2799. [CrossRef]
98. Metallo, C.M.; Vander Heiden, M.G. Metabolism strikes back: Metabolic flux regulates cell signaling. *Genes Dev.* **2010**, *24*, 2717–2722. [CrossRef]
99. Fang, M.; Shen, Z.; Huang, S.; Zhao, L.; Chen, S.; Mak, T.W.; Wang, X. The ER UDPase ENTPD5 promotes protein N-glycosylation, the Warburg effect, and proliferation in the PTEN pathway. *Cell* **2010**, *143*, 711–724. [CrossRef]
100. Marshall, S.; Bacote, V.; Traxinger, R.R. Discovery of a metabolic pathway mediating glucose-induced desensitization of the glucose transport system. Role of hexosamine biosynthesis in the induction of insulin resistance. *J. Biol. Chem.* **1991**, *266*, 4706–4712. [CrossRef]
101. Eguchi, S.; Oshiro, N.; Miyamoto, T.; Yoshino, K.; Okamoto, S.; Ono, T.; Kikkawa, U.; Yonezawa, K. AMP-activated protein kinase phosphorylates glutamine: Fructose-6-phosphate amidotransferase 1 at Ser243 to modulate its enzymatic activity. *Genes Cells* **2009**, *14*, 179–189. [CrossRef]
102. Gancedo, J.M. The early steps of glucose signalling in yeast. *FEMS Microbiol. Rev.* **2008**, *32*, 673–704. [CrossRef]
103. Furst, A.; Michels, C.A. An evaluation of D-glucosamine as a gratuitous catabolite repressor of *Saccharomyces carlsbergensis*. *Mol. Gen. Genet.* **1977**, *155*, 309–314. [CrossRef]
104. Weimer, S.; Priebs, J.; Kuhlow, D.; Groth, M.; Priebe, S.; Mansfeld, J.; Merry, T.L.; Dubuis, S.; Laube, B.; Pfeiffer, A.F.; et al. D-Glucosamine supplementation extends life span of nematodes and of ageing mice. *Nat. Commun.* **2014**, *5*, 3563. [CrossRef]
105. Hofer, S.J.; Davinelli, S.; Bergmann, M.; Scapagnini, G.; Madeo, F. Caloric restriction mimetics in nutrition and clinical trials. *Front. Nutr.* **2021**, *8*, 717343. [CrossRef]
106. Qi, L.; Tsai, B.; Arvan, P. New insights into the physiological role of endoplasmic reticulum-associated degradation. *Trends Cell. Biol.* **2017**, *27*, 430–440. [CrossRef]
107. Kocaturk, N.M.; Gozuacik, D. Crosstalk between mammalian autophagy and the ubiquitin-proteasome system. *Front. Cell Dev. Biol.* **2018**, *6*, 128. [CrossRef]
108. Kuranda, K.; Leberre, V.; Sokol, S.; Palamarczyk, G.; François, J. Investigating the caffeine effects in the yeast *Saccharomyces cerevisiae* brings new insights into the connection between TOR, PKC and Ras/cAMP signalling pathways. *Mol. Microbiol.* **2006**, *61*, 1147–1166. [CrossRef]
109. Soulard, A.; Cremonesi, A.; Moes, S.; Schütz, F.; Jenö, P.; Hall, M.N. The rapamycin-sensitive phosphoproteome reveals that TOR controls protein kinase A toward some but not all substrates. *Mol. Biol. Cell* **2010**, *21*, 3475–3486. [CrossRef]
110. Shashkova, S.; Welkenhuysen, N.; Hohmann, S. Molecular communication: Crosstalk between the Snf1 and other signaling pathways. *FEMS Yeast Res.* **2015**, *15*, fov026. [CrossRef]

Journal of *Fungi*

MDPI

Article

Systematic Identification of Essential Genes Required for Yeast Cell Wall Integrity: Involvement of the RSC Remodelling Complex

Ana Belén Sanz, Sonia Díez-Muñiz, Jennifer Moya †, Yuliya Petryk, César Nombela, José M. Rodríguez-Peña * and Javier Arroyo

Departamento de Microbiología y Parasitología, Facultad de Farmacia, Universidad Complutense de Madrid, Instituto Ramón y Cajal de Investigaciones Sanitarias (IRYCIS), 28040 Madrid, Spain; absanzsa@ucm.es (A.B.S.); soniadiez@farm.ucm.es (S.D.-M.); jmoyavaquero@gmail.com (J.M.); yupetryk@ucm.es (Y.P.); cnombela@ucm.es (C.N.); jarroyo@ucm.es (J.A.)

* Correspondence: josemanu@ucm.es

† Current address: Centro Nacional de Biotecnología, CSIC, 28049 Madrid, Spain.

Abstract: Conditions altering the yeast cell wall lead to the activation of an adaptive transcriptional response mainly governed by the cell wall integrity (CWI) mitogen-activated protein kinase (MAPK) pathway. Two high-throughput screenings were developed using the yTHC collection of yeast conditional mutant strains to systematically identify essential genes related to cell wall integrity, and those required for the transcriptional program elicited by cell wall stress. Depleted expression of 52 essential genes resulted in hypersensitivity to the dye Calcofluor white, with chromatin organization, Golgi vesicle transport, rRNA processing, and protein glycosylation processes, as the most highly representative functional groups. Via a flow cytometry-based quantitative assay using a CWI reporter plasmid, 97 strains exhibiting reduced gene-reporter expression levels upon stress were uncovered, highlighting genes associated with RNA metabolism, transcription/translation, protein degradation, and chromatin organization. This screening also led to the discovery of 41 strains displaying a basal increase in CWI-associated gene expression, including mainly putative cell wall-related genes. Interestingly, several members of the RSC chromatin remodelling complex were uncovered in both screenings. Notably, Rsc9 was necessary to regulate the gene expression of CWI-related genes both under stress and non-stress conditions, suggesting distinct requirements of the RSC complex for remodelling particular genes.

Keywords: cell wall; stress response; screening; transcription; yeast; essential genes

Citation: Sanz, A.B.; Díez-Muñiz, S.; Moya, J.; Petryk, Y.; Nombela, C.; Rodríguez-Peña, J.M.; Arroyo, J. Systematic Identification of Essential Genes Required for Yeast Cell Wall Integrity: Involvement of the RSC Remodelling Complex. *J. Fungi* **2022**, *8*, 718. https://doi.org/10.3390/jof8070718

Academic Editor: Aaron Neiman

Received: 29 April 2022
Accepted: 6 July 2022
Published: 8 July 2022

1. Introduction

The fungal cell wall is an essential structure responsible for cell morphology; it contributes to preserving osmostability, protects cells against environmental stresses, and delineates cellular immunogenicity. In the eukaryotic model *Saccharomyces cerevisiae*, the cell wall consists of three major components: an inner layer of glucans (β-1,3 and β-1,6-glucan); a small amount of chitin (polymer of N-acetylglucosamine) and an outer layer of mannoproteins (such as agglutinins and flocculins involved in cell adhesion; and others related to enzymatic or structural functions). All these components must be correctly assembled to build a functional structure (for review, see [1,2]). Despite its apparent rigidity, the cell wall is very dynamic, so its composition and organization vary during growth and development. Additionally, when yeast cells face conditions where the cell wall integrity is compromised an adaptive response is triggered to maintain cell viability. This response includes changes in the synthesis and crosslinking of different polymers, and increased expression levels of genes functionally related to cell wall maintenance [3,4]. Cell wall stress adaptive responses in yeast are mainly regulated by the cell wall integrity (CWI)

MAPK pathway, which includes a conserved MAPK module [5,6]. This module is activated through a cascade of phosphorylation events, which ultimately lead to the phosphorylation of the MAPK of the pathway, Slt2/Mpk1. In turn, phosphorylated Slt2 activates the transcription factor Rlm1, the main factor responsible for the transcriptional reprogramming elicited under these conditions [7–9]. CWI signalling is triggered by a variety of cell wall stressors (such as cell wall-interfering compounds) [4] or mutations in genes affecting proteins involved in cell wall homeostasis. Yeast mutants deleted for functionally relevant cell wall-related genes display constitutive Slt2 hyperactivation [10,11]. It was recently claimed that nucleosome remodelling at the promoters of Rlm1-dependent genes through cooperation between the ATP-dependent chromatin-remodelling SWI/SNF and acetyltransferase SAGA complexes is required for the binding of phosphorylated Rlm1, to target promoters and assembly of the transcription initiation machinery under cell wall stress [9,12]. In fact, mutants in different SWI and SAGA subunits display cell wall-related phenotypes. Additionally, various works also reported the association of cell wall integrity with mutants in some subunits of the RSC (Remodel the Structure of Chromatin) complex [13–16], which belongs to the SWI/SNF family of ATP-dependent chromatin remodellers [17]. However, information is lacking regarding their possible connection with CWI-dependent genes.

Largely due to its potential as a selective target for antifungal drugs, in addition to other biotechnological purposes, a great effort has been made to characterize genes involved in biogenesis, regulation, and maintenance of the yeast cell wall. This objective was fuelled by the availability of the complete collection of haploid yeast mutant strains individually deleted in each of the approximately 4800 non-essential genes identified in this organism [18]. To date, several genome-wide phenotypic or chemical-genomic profiling screens have been conducted using this yeast knock-out collection to identify cell wall-related mutants. Hypersensitivity to K1 and K2 killer toxins that form lethal pores in the plasma membrane [19,20], Congo red (chitin-binding dye), caspofungin (β-1,3-glucan synthase inhibitor), zymolyase (enzymatic cocktail including β-1,3-glucanase and chitinase activities) [21,22], freeze-thaw or heat stress [23,24] are some examples of the screenings developed to date. These experiments have permitted the identification of a broad functional catalogue of genes potentially related to maintaining cellular integrity under cell wall stress conditions. Moreover, from their comparative analysis, it has been possible to infer damage-specific signalling mechanisms and uncover genes required to withstand both specific and general cell wall injuries.

Following a complementary approach, we performed large-scale screenings using the whole collection of haploid deletion strains in all non-essential genes of *S. cerevisiae* transformed with a reporter system based on transcriptional fusion of the *MLP1/KDX1* (*YKL161C*) promoter to the coding sequence of the *nat1* gene encoding nourseothricin N-acetyltransferase from *Streptomyces noursei* [25]. Firstly, we used this collection of mutant strains to identify genes whose absence produces constitutive activation of the CWI pathway under non-stress growth conditions [11]. Globally, this strategy uncovered mutants related to signalling and potential effectors required to maintain cell viability upon cell wall stress. Additionally, we used the same collection in a second screening to identify elements involved in the adaptive transcriptional response elicited upon cell wall stress and required for proper gene expression under stress conditions affecting cellular integrity [9]. This approach allowed the identification of several protein complexes related to the regulation of gene expression upon zymolyase-mediated cell wall damage conditions, including the SWI/SNF ATP-dependent chromatin-remodelling complex.

All these functional studies did not include the approximately one thousand yeast genes essential for viability under standard laboratory growth conditions, which account for nearly 19% of the genes identified in the *S. cerevisiae* genome [18,26]. Considering that around 20% of yeast genes may show a direct or indirect functional relationship with cell wall homeostasis [27], and the fact that this structure is essential for cell survival, this functional group would be expected to be significantly represented within the group of essential genes.

In practice, functional genomics applied to essential genes is scarce due to the difficulty of handling these yeast strains. In this context, several strategies have been developed to facilitate these studies, particularly temperature-sensitive mutations or conditionally degradable fusion proteins containing a degron sequence [28]. These approaches show, as the main drawbacks, that repression conditions can affect cell physiology (heat shock) and sequence alteration of the ORFs under study. As an alternative, the yeast Tet-promoters Hughes collection (yTHC) [29] is available. This collection, which has been successfully used in several large-scale screens [28] including those related to MAPK signalling [30], encompasses approximately 75% of essential yeast genes. It is based on replacing the native promoter region of essential genes with the tetracycline (tet)-regulatable promoter to create mutant conditional alleles. In this way, repression of expression is controlled by adding doxycycline to the growth medium, without affecting yeast physiology.

Using the yTHC collection, we developed two different types of screenings to identify cell wall-related genes. One aimed to identify essential genes required to cope with cell wall damage conditions imposed by treatment with the chitin-binding dye Calcofluor white (CW). The second screening was designed to identify elements required for the adaptive transcriptional response induced by cell wall stress. Using both strategies, we uncovered yeast genes related to cell wall maintenance, including those associated with the biogenesis of this structure and the regulatory elements required for this process. We have delved deeper into the study of the participation of the RSC chromatin remodelling complex in regulating the expression of CWI pathway-dependent genes. All this information increases our basic knowledge of the gene catalogue required to withstand cell wall damage. In addition, it will facilitate the establishment of the mode of action of antifungal drugs that target products encoded by essential genes.

2. Materials and Methods

2.1. Yeast Strains and Media

All the experiments in this work were performed using the *Saccharomyces cerevisiae* Tet-promoters collection (yTHC) from the Hughes Laboratory, University of Toronto (Canada), provided by Open Biosystems (Dharmacon Inc., GE Healthcare, Hatfield, UK). It includes 821 essential yeast genes whose expression is regulated by doxycycline, since in each strain the endogenous promoter has been replaced with the $TetO_7$ promoter. This collection was created using the strain R1158 (BY4741 background) *URA3::CMV-t*TA *MATa his3-1 leu2-0 met15-0*.

Yeast strains were grown on YPD medium (2% glucose, 2% peptone, 1% yeast extract). For selection of yeast transformants bearing the pBS1 plasmid, SD-His medium was used (0.17% yeast nitrogen base, 0.5% ammonium sulphate, 2% glucose, supplemented with the required amino acids). When required, doxycycline (Calbiochem, Darmstadt, Germany) was added to the culture medium at a final concentration of 10 μg/mL.

2.2. Calcofluor White Sensitivity Screening

The complete yTHC collection, including the wild-type strain, was grown in 96-well microtiter plates in 200 μL of YDP medium for 24–48 h at 30 °C. Afterwards, 5 μL of these cultures (optical density ~0.15 at 595 nm) were inoculated in a new set of four 96-well microtiter plates containing 145 μL of YPD, YPD plus 10 μg/mL of doxycycline (DOX), YPD plus 40 μg/mL of Calcofluor white (CW, Fluorescent brightener 28, Sigma-Aldrich, Merck, Darmstadt, Germany) and YPD+DOX+CW. After 48 h of static incubation at 30 °C, the level of growth was determined by measuring the absorbance at 595 nm in each well using a microplate reader (Model 680, Bio-Rad, Hercules, CA, USA). The level of sensitivity of each mutant was calculated as follows. First, we obtained the absorbance ratios after growing in YPD+DOX+CW vs. YPD+CW and YPD+DOX vs. YPD (as an indicator of non-CW mediated effects). Finally, both ratios were divided (presence vs. absence of CW) to obtain a numerical value corresponding to the level of sensitivity of each mutant allele to CW when gene expression is blocked in the presence of doxycycline. After an initial global screening, a mutant strain was considered hypersensitive to the drug when this final ratio was ≤0.2. To focus on mutants hypersensitive to CW when gene expression is turned off by doxycycline, those mutants largely affected in growth in the presence of the drug in the absence of doxycycline were not further considered. Finally, those mutants selected from the first screening (CW hypersensitive) were retested in two additional independent sensitivity assays as described above. Eventually, those mutants showing mean ratios ≤ 0.2 were selected as positive hits of the screening.

2.3. *Screening for Altered MLP1-GFP Expression Using Flow Cytometry*

Initially, the full collection of Tet-promoter mutant strains was individually transformed following the lithium acetate protocol with the pBS1 plasmid using SC-His plates as selection medium. This plasmid is a variant of the *MLP1pro-MLP1-GFP* plasmid [25] in which the *Sma*I-digested *HIS3* ORF (1.7 Kb) from plasmid p34H-*HIS3* [31] was introduced into the *Stu*I site within the *URA3* gene, thus changing the selection marker. To identify mutants showing alterations in *MLP1* expression upon cell wall stress, one transformant from each mutant strain was grown in YPD medium supplemented with 10 μg/mL of doxycycline at 24 °C overnight and then divided into two parts. One part continued growing under the same conditions (non-treated culture), whereas the other was supplemented with CW (10 μg/mL). Cells were harvested after 3 h of incubation, washed twice with PBS, and finally resuspended in PBS supplemented with 0.4 μg/mL propidium iodide (PI) (Sigma Aldrich, St. Louis, MO, USA) and analysed using a FACSCan flow cytometer (Becton Dickinson, Bergen County, NJ, USA) equipped with an argon-ion laser emitting at 488 nm by acquiring green fluorescence emission through a 530/30 bandpass (BP) filter. For each sample, the mean fluorescence intensities (MFIs) and the percentage of GFP-positive cells within the PI-negative population (dead cells are stained with PI) were obtained using the BD CellQuest 3.3 software (Becton Dickinson) on Mac® OS 9.2.2. Following this, the MFI ± CW ratio was calculated for each mutant strain (R1) and the wild-type strain (R2). Finally, those mutants displaying a quotient R1/R2 ≤ 0.25 and/or a percentage of GFP-positive cells in the presence of CW lower than 15% (mean value for the wild-type strain is around 75%) were initially selected as positive hits from the screening. To confirm these data, these mutants were retested, as described above, using two additional independent transformants. Eventually, those mutants showing a mean MFI R1/R2 from the three replicates ≤0.25 and/or lower than 15% of GFP-positive cells (as mean value) were considered positive hits.

2.4. *Western Blotting Assays*

The procedures used for immunoblotting analyses, including cell collection and lysis, collection of proteins, fractionation by SDS–PAGE, and transfer to nitrocellulose membranes were performed as previously described [32] using the Odyssey Infrared Imaging System (LI-COR). The detection of phosphorylated Slt2 was accomplished using the anti-phospho-p44/p42 MAPK monoclonal antibody (thr202/tyr204; Cat. No. 4370; Cell Signaling Technology, Beverly, MA, USA). To monitor protein loading, Glucose-6-phosphate dehydrogenase levels were determined using an anti-G6PDH polyclonal antibody (Cat. No. A9521; Sigma-Aldrich Corp, St. Louis, MO, USA). The secondary antibodies used were IRDye 800CW goat anti-rabbit (Cat. No. 926-32211) and IRDye 680LT goat anti-mouse (Cat. No. 926-68020), both from LI-COR Biosciences (Lincoln, NE, USA). All antibodies were used at the dilutions recommended by the manufacturers. Protein band quantification was carried out by densitometric analysis using the software Image Studio Lite (LI-COR Biosciences). Phospho-Slt2 protein levels were normalized against the loading control.

2.5. *Quantitative RT-PCR Assays*

RNA isolation and RT-qPCR assays were performed as previously described [8]. For quantification, the abundance of each transcript was determined using the amount of the standard transcript *ACT1* for input cDNA normalization, and final data on relative gene expression between the conditions tested were calculated following the $2^{-\Delta\Delta Ct}$ method [33]. Oligonucleotide sequences are available upon request.

3. Results and Discussion

3.1. High-Throughput Screening for Essential Genes Required to Cope with Cell Wall Damage

We designed a screening using the yTHC collection of yeast conditional mutant strains containing 821 of the 1105 total reported essential genes to obtain the complete repertoire of essential genes related to cell wall maintenance. The rationale of this screening was that those strains with a defective cell wall would display a phenotype of hypersensitivity to agents that disturb this essential fungal structure [27]. To identify strains hypersensitive to the cell wall-interfering compound CW, under restrictive conditions (repression of gene expression due to the presence of doxycycline in the culture medium), we first set up the optimal growth conditions for the yeast collection in a 96 multi-well plate format using sublethal concentrations of CW. Eventually, the assay was performed at static incubation at 30 °C, in the presence or not of 10 µg/mL of doxycycline and 40 µg/mL CW. The effect on cell growth was quantified by measuring the optical density at 595 nm after 48 h (see Figure 1). As explained in detail in Materials and Methods, the mutant strains were considered hypersensitive to CW when their growth was reduced by this agent in the presence of doxycycline in the culture medium in comparison with the same conditions but in the absence of CW. Following this approach, for each conditional allele we obtained a growth ratio (equivalent to sensitivity ratio) from 0 (maximum sensitivity due to down-regulation of gene expression) to ~1 (absence of effect on growth, as observed for the wild-type strain), indicative of the relative hypersensitivity level to CW (Figure 1 and Table S1). The mutant strains that exhibited higher sensitivity levels (arbitrary cut-off growth ratio ≤ 0.2) were selected as preliminary positive hits from this first round of screening. Next, this group of mutants underwent two additional screening rounds under the same experimental conditions to finally select those with a reproducible CW hypersensitivity phenotype (see Material and Methods for details). Finally, 52 mutant strains were considered positive hits of the screening (Tables 1 and S2). The CW hypersensitivity of some selected mutants was validated by serial dilution assays in the presence or absence of doxycycline (Figure S1).

To determine the functional categories of the genes selected in the screening, we manually grouped the gene set using the functional annotation deposited in the *Saccharomyces Genome Database* (SGD) tool "GO Slim Mapper: process". For the 52 CW hypersensitive mutants, the more represented groups corresponded to chromatin organization (8), Golgi vesicle transport (8), rRNA processing (8), and protein glycosylation/glycosylphosphatidylinositol (GPI) anchor biosynthesis (7) (Table 1). In other cell wall-related screenings carried out on the complete collection of viable haploid yeast deletion mutants, similar functional groups were uncovered [22,34]. Moreover, these screenings including treatments with agents interfering with cell-wall construction via different mechanisms of action, such as Congo red, zymolyase, or caspofungin, uncovered a high percentage of specific mutants for each drug [22]. We can speculate that the situation may be similar in the case of mutants in essential genes. In any case, additional work will be necessary, using alternative cell-wall interfering compounds to confirm this hypothesis.

It is reasonable to speculate that some of the mutants identified in our screening could be ascribed to the so-called "Type I" of essential genes, which are those postulated to execute "core" functions indispensable to the organism [35], like those included in the group of rRNA processing. The hypersensitivity to cell wall interfering agents, in this case, could be somehow associated with pleiotropic effects on cell wall architecture. However, other genes seem more directly related to cell wall biogenesis or maintenance, such as those involved in protein glycosylation or GPI biosynthesis, which are required for the correct function of cell wall proteins or those related to vesicle transport required for the homeostasis of this structure (see next sections). In addition, recent studies have shown that the MAPK CWI signalling pathway regulates the expression of genes necessary for survival in the presence of agents that damage the cell wall in collaboration with the chromatin remodelling complexes SWI/SNF [9] and SAGA [12]. In fact, we found components of SWI/SNF (*ARP7*) and SAGA (*TAF5* and *TAF10*), and, remarkably, we also identified many members of the RSC chromatin remodelling complex (*ARP7, RSC6, RSC8, RSC58,* and *STH1*). This result implies that the RSC complex plays an important role under cell wall stress conditions. In this respect, previous studies have reported that some *rsc* mutants show cell wall-related phenotypes [13–16].

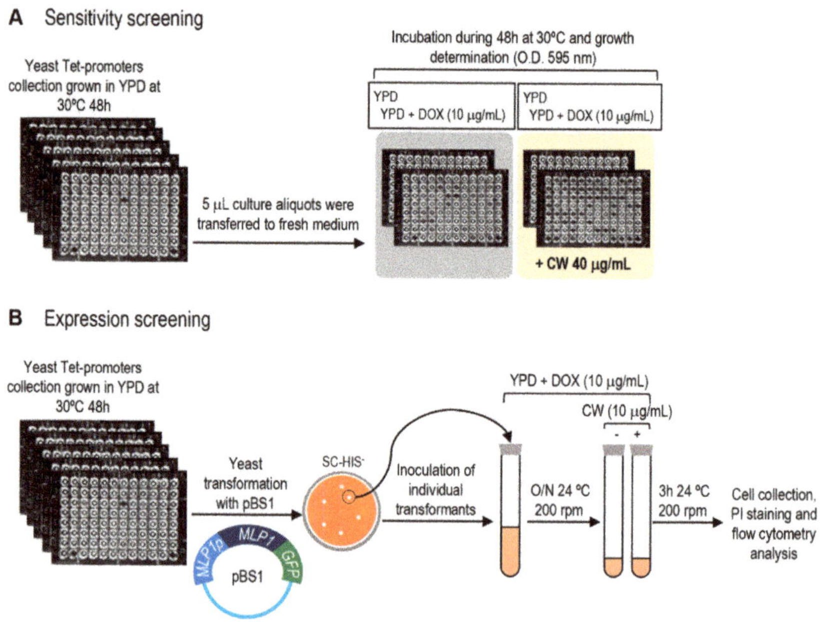

Figure 1. Overview of the screening strategies designed to identify cell wall-related mutants in essential genes. (**A**) Hypersensitive phenotype to Calcofluor white was studied in a 96-well microtiter plate format. (**B**) Detection of altered expression levels of the CWI-reporter gene *KDX1/MLP1*. The complete yeast Tet-promoter collection was transformed with the plasmid pBS1 (*MLP1p-MLP1-GFP*) and individual transformants were tested for Mlp1-GFP expression in the presence or absence of Calcofluor white (CW) by flow cytometry. DOX: exposure to doxycycline at the indicated concentration. PI: Propidium Iodide.

Table 1. List of genes selected from the screening for Calcofluor white (CW) hypersensitivity, ordered according to their Gene Ontology annotations. CW sensitivity indicates the level of growth affectation in the presence of CW, from 0.0 (maximum effect) to 1.0 (absence of effect).

Gene	CW Sensitivity	Biological Process	Gene	CW Sensitivity	Biological Process
CDC20	0.10	Cell cycle	*CFT1*	0.14	mRNA processing
CKS1	0.09	Cell cycle	*DCP2*	0.14	mRNA processing
GUK1	0.16	Cell wall	*PRP31*	0.16	mRNA processing
ARP7	0.19	Chromatin organization	*NUP145*	0.18	Nuclear transport
RSC58	0.20	Chromatin organization	*PRE5*	0.18	Protein degradation
RSC6	0.08	Chromatin organization	*PRE6*	0.18	Protein degradation
RSC8	0.20	Chromatin organization	*RPT6*	0.17	Protein degradation
SPT16	0.20	Chromatin organization	*OST2*	0.07	Protein glycosylation

Table 1. *Cont.*

Gene	CW Sensitivity	Biological Process	Gene	CW Sensitivity	Biological Process
STH1	0.12	Chromatin organization	*RFT1*	0.10	Protein glycosylation
TAF10	0.15	Chromatin organization	*ROT1*	0.11	Protein glycosylation
TAF5	0.20	Chromatin organization	*SWP1*	0.18	Protein glycosylation
ARC40	0.18	Cytoskeleton organization	*VRG4*	0.06	Protein glycosylation
PFY1	0.06	Cytoskeleton organization	*SRP68*	0.19	Protein targeting
MCM7	0.18	DNA replication	*SRP72*	0.13	Protein targeting
SLD5	0.17	DNA replication	*YPP1*	0.17	Protein targeting
BET3	0.13	Golgi vesicle transport	*BRX1*	0.15	rRNA processing
DOP1	0.15	Golgi vesicle transport	*DBP9*	0.20	rRNA processing
SEC14	0.14	Golgi vesicle transport	*EBP2*	0.17	rRNA processing
SEC18	0.12	Golgi vesicle transport	*FAP7*	0.04	rRNA processing
SEC2	0.11	Golgi vesicle transport	*FCF1*	0.17	rRNA processing
SFT1	0.09	Golgi vesicle transport	*SPB1*	0.19	rRNA processing
TIP20	0.14	Golgi vesicle transport	*NOP2*	0.15	rRNA processing
YIP1	0.16	Golgi vesicle transport	*NOP15*	0.19	rRNA processing
GAB1	0.12	GPI biosynthesis	*MOT1*	0.18	Transcription
MCD4	0.18	GPI biosynthesis	*RPL18A*	0.17	Translation
PHS1	0.10	Lipid metabolism	*TIF35*	0.14	Translation

3.2. Identification of Essential Genes Required for the Transcriptional Adaptive Cell Wall Stress Response

In yeast, cell wall damage triggers a compensatory mechanism principally based on the transcriptional induction of a group of effector genes, finely regulated via the CWI pathway governed by the MAPK Slt2 [4]. Therefore, identifying essential genes involved in this process is of great interest to complete the catalogue of genes associated with this adaptive cellular response. To this end, complementary to the phenotypic study, we designed and developed a flow cytometry screening to discover yeast mutants for essential genes affected in the transcriptional induction caused by treatment with CW (Figure 1). The yTHC collection was individually transformed with a plasmid incorporating the fusion of the *MLP1/KDX1* gene coding sequence (encoding a pseudokinase paralog of Slt2) to that of the GFP protein, expressed under the control of the native *MLP1* promoter [25]. *MLP1* shows low basal gene expression levels but is highly expressed under cell wall stress, making this gene of particular usefulness for transcriptional studies. Moreover, it is controlled via the CWI pathway, and therefore its up-regulation is largely dependent on Slt2 and the transcription factor Rlm1 [8,25,36,37]. The wild-type strain of this collection (R1158 background) transformed with the plasmid containing the *MLP1-GFP* fusion presented, under non-stress conditions, 3.09 $\pm$ 1.92% of positive cells for GFP fluorescence with a mean fluorescence intensity (MFI) of 22.68 $\pm$ 3.5. Upon induction conditions (3 h exposure to CW), these values rose sharply to 73.69 $\pm$ 7.0% of GFP positive cells and MFI values around 154.50 $\pm$ 39.20. These data indicate that the *MLP1-GFP* construction offers optimum fluorescence levels to identify mutant strains affected in the expression of this CWI-reporter under basal and stress conditions. As an example, GFP fluorescence images of a wild-type and a mutant strain in which *MLP1* induction is blocked are shown (Figure S2).

As described in Materials and Methods and illustrated in Figure 1, the complete collection of yeast mutants expressing Mlp1-GFP was grown in a rich medium including 10 μg/mL of doxycycline and then exposed, or not, to CW for 3 h. Next, GFP fluorescence levels, as the readout of Mlp1-GFP amounts for each mutant tested, were quantified by flow cytometry. The wild-type strain was always included as a control of *MLP1* induction in each experiment. Thus, the effect of the down-regulation of the expression of each conditional allele on Mlp1-GFP levels after CW treatment was determined, calculating first the ratio of MFIs between treatment and control conditions (absence of CW) for the wild-type and the mutant strains. Finally, the ratio obtained for each mutant strain was divided by that calculated for the wild-type strain. Flow cytometry data analyses uncovered mutants with changes in global MFIs and mutants in which the number of cells expressing detectable fluorescence signals was affected. Representative examples of the different flow cytometry patterns obtained are shown in Figure 2A. From these observations, in a first round of screening, we preliminarily selected those mutants showing at least a 75% decrease in Mlp1-GFP levels with respect to those of the wild-type strain (fluorescence ratio mutant/fluorescence ratio WT ≤ 0.25) and/or showing a percentage of GFP positive cells lower than 15% (Table S3). Next, two additional experiments were conducted using different yeast transformants of this group of mutants to finally define the positive hits of the screening (see Material and Methods for details and Table S4).

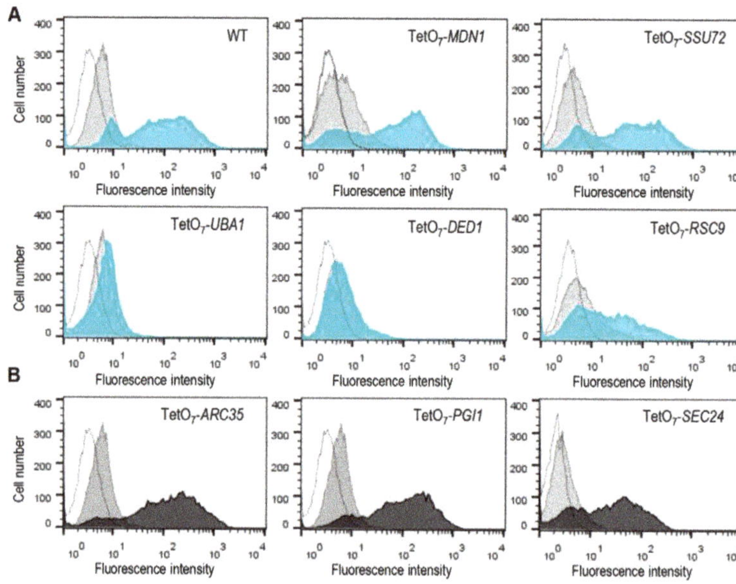

Figure 2. Identification of mutants from the yTHC collection that exhibit altered Mlp1-GFP levels by flow cytometry analysis. (**A**) Representative experiments of mutants in which Mlp1-GFP fluorescence levels were found to be affected (lower panels), or not affected (upper panels), compared with those of the wild-type strain. For each strain, histograms show the distribution of GFP fluorescence under non-stress conditions (grey-filled) and after treatment with Calcofluor white (10 µg/mL for 3 h) (blue-filled) in the presence of doxycycline (10 µg/mL). (**B**) Examples of mutants displaying increased GFP signal under non-stress conditions (absence of CW) in the presence of doxycycline are shown. For each conditional allele, histograms show basal GFP fluorescence corresponding to wild-type (grey-filled) and mutant (black-filled) strains handled in parallel. Control cells that do not express GFP are shown in each histogram overlay (unfilled peaks).

We sought functional categories within the group of 97 mutants selected (Figure 3A and Table S4). The most represented group (17%) corresponded to that associated with RNA metabolism, including RNA processing and degradation (exosome component). The identification of this group of genes, in conjunction with those associated with transcription, ribosome biogenesis, translation, and protein folding, could be explained as a consequence of alterations in the cellular machinery required for the correct expression and translation of the gene reporter Mlp1 under cell wall stress. However, it cannot be ruled out that CWI pathway signalling or regulatory defects could also be present in some mutants. Other groups of interest corresponded to those related to lipid metabolism, protein degradation, nucleocytoplasmic RNA transport, or chromatin remodelling. Within the first group, we found genes involved in ergosterol biosynthesis (*ERG1* and *ERG26*) or sphingolipid metabolism (*LCB2* and *TSC13*). Their identification is in accordance with the fact that sphingolipids are a structurally diverse class of lipids implicated in many cell signalling functions and interconnected with the ergosterol pathway [38]. The protein degradation group includes several subunits of the 26S proteasome (*PRE5*, *PRE6*, *RPN11*, and *RPT2*) and some ubiquitin-related enzymes (*UBA1*, *CDC34*, and *HRT1*). Interestingly, proteasome up-regulation has been proposed under different stress conditions, such as heat shock or oxidative stress [39,40]. Moreover, a control of proteasome abundance mediated by the MAPK Slt2 has also been reported [41]. Since proteasome-mediated degradation regulates the turnover of numerous cellular proteins involved in the majority of cellular processes [42], it is difficult, in these mutants, to discern between the existence of direct effects on CWI pathway regulation, mainly due to ubiquitination, or pleiotropic effects that indirectly alter the adaptive transcriptional response. Interestingly, proteasome inhibition leads to a slight increase in CW sensitivity (our unpublished results). Additional work will be required to clarify the molecular mechanisms involved in the CWI pathway-proteosome interaction. Still, the identification of only certain elements of this large protease complex allows us to hypothesize that they are particularly important under cell wall stress conditions.

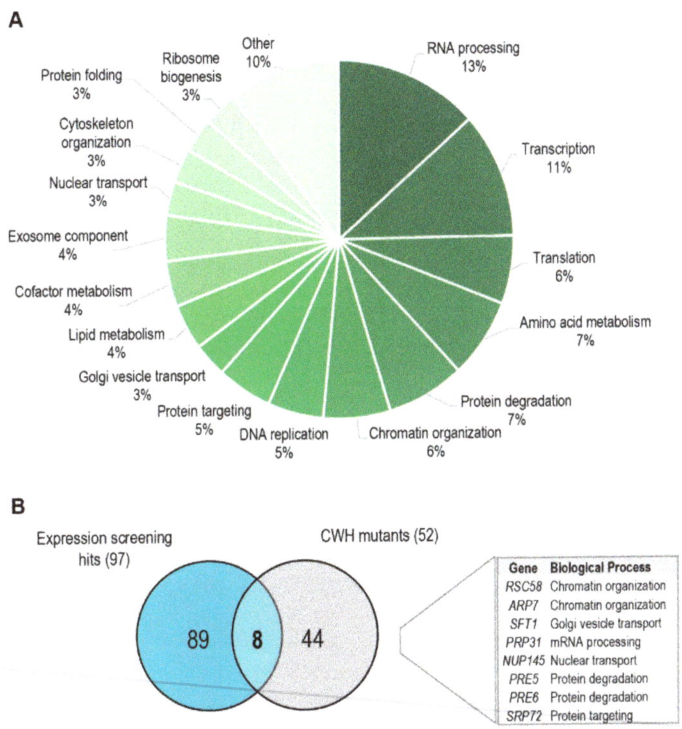

Figure 3. Functional classification of genes identified in the expression screening. (**A**) Relative distribution of functional categories based on Gene Ontology (biological process) from the 97 positive hits. (**B**) The common group of essential gene mutants uncovered in both the phenotypic and expression screenings are shown. CWH: Calcofluor white hypersensitive.

Other mutants of interest are those related to nucleocytoplasmic RNA transport (*NUP145* or *MEX67*). Considering that, under heat shock stress, a selective nuclear export of heat-shock mRNAs and retention of regular transcripts mediated by Mex67 have recently been reported [43], similar or alternative mechanisms may take place under other stress conditions. Finally, it is important to highlight the presence of genes functionally related to chromatin remodelling, including essential subunits of the SWI/SNF (*ARP7*) and SAGA (*TAF9*) chromatin-modifying complexes, which were previously associated with the regulation of the expression of CWI-dependent genes upon cell wall stress [9,12]. Interestingly, within this group of chromatin-associated mutants, we also found essential elements of the RSC remodelling complex (*ARP7*, *RSC58*, and *RSC9*). In agreement, different ATP-dependent chromatin remodelling complexes work together to regulate the expression of stress-responsive genes [44,45].

Comparative analyses of the genes identified in both the phenotypic and transcriptional screenings under the imposed cut-off thresholds revealed a reduced number of mutants, including two subunits of the RSC complex and two proteasome members (Figure 3B). This low overlap agrees with previous studies with the complete collection of viable deletion mutants after zymolyase treatment. In these assays, 154 mutants exhibited hypersensitivity to zymolyase [22], and 159 mutants showed a reduction in the *MLP1*-reporter expression under zymolyase treatment [9], but only 35 mutants were found common in both studies. In agreement, most of the genes included in the cell wall damage compensatory mechanism work cooperatively to support cell wall integrity, but they are not essential individually to maintain cell viability under stress situations, except for those involved in signalling or transcriptional regulation, where the overall response is impaired. The elements identified of the RSC complex would belong to this category.

3.3. Impact of the RSC Complex on the Expression of CWI Pathway-Dependent Genes

We uncovered several elements of the RSC chromatin remodelling complexes showing CW hypersensitivity and/or alteration of Mlp1-GFP levels under cell wall stress conditions (Tables S2 and S4). The RSC complex consists of 17 subunits (both essential and dispensable for cell viability) that can work in smaller subcomplexes, hindering its functional characterization [46]. This complex is mainly involved in transcription regulation, remodelling nucleosomes in promoter and transcribed coding sequences [47,48]. It also plays a role in chromosomal transactions such as DNA replication and repair and chromosome segregation [47,49]. In agreement with the functional complexity of RSC, previous large-scale transcriptional studies carried out under basal growth conditions of several yeast RSC subunit mutants (*rsc3*, *rsc30*, *rsc4* or *rsc14*) have demonstrated a small overlap in the genes regulated by the different subunits [13,50,51]. Additionally, in these studies, some RSC subunits modulated the expression of a reduced group of cell wall-related genes.

We aimed to further investigate the relationship between possible alterations in the cell wall of some essential *rsc* mutants identified in our screening and modulation of the adaptive transcriptional response mediated by the CWI pathway when cell wall integrity is compromised. Firstly, mRNA levels of the CWI reporter gene *MLP1* were quantified via RT-qPCR in *rsc9* and *rsc58* mutants. These subunits belong to the substrate recruitment module (SRM) of the RSC complex [46]. In the case of Rsc9, this element has been involved both in the repression and activation of specific stress-regulated genes (i.e., hydrogen peroxide and rapamycin [52] or osmostress [53]). The transcriptional induction of *MLP1* in the presence of CW was significantly affected in these mutants under shut-off gene conditions (Figure 4A), validating the experimental approach developed in our screening and suggesting that the RSC complex regulates transcription upon cell wall stress. In the absence of stress, both subunits negatively regulate *MLP1* transcription, as inferred from the increase in the amount of *MLP1* transcripts in both mutants, particularly in the *rsc58* strain (Figure 4B). Next, we investigated the impact of *rsc* mutations on the expression of additional CWI-dependent genes. To this end, the mRNA levels of 15 other genes induced by cell wall stress [8,36,37,54,55] were analysed under the conditions indicated above, focusing on the *rsc9* mutant (Figure 4C). Interestingly, we observed two types of genes regarding the effect caused by the absence of Rsc9. One group, including *YPL088W*, *PRM10*, and *YLR042C*, showed a behaviour similar to *MLP1* in terms of the impact on their expression levels after treatment with CW compared with the up-regulation observed in the wild-type strain; and a second, more extensive group, in which no significant alteration in transcript levels was observed. Quantification of the transcript levels of these genes under non-stress conditions in the *rsc9* mutant relative to those of the wild-type strain showed a derepression of genes *YLR194C*, *SRL3*, *BAG7*, *CRG1*, *SED1*, *FKS2*, *NCA3* and *NQM1* in the absence of *RSC9* (Figure 4D). This basal regulation was not manifested in the case of the *VMA8* gene, which encodes a subunit of vacuolar ATPase non-transcriptionally regulated by cell wall stress. Therefore, our results indicate that the RSC complex regulates the gene expression of certain CWI-related genes both under stress and non-stress conditions. As described above, positive or negative transcriptional modulation of specific genes under different stress conditions, probably a consequence of their interaction with other regulators (repressors or activators), has been previously described for Rsc9 [52] and other RSC subunits [13,51]. Under basal (non-stress) conditions, RSC could also be involved in stabilizing the nucleosomes present at some CWI-dependent genes. Similarly, some UPR genes show high basal level expression when RSC is disrupted, indicating the presence of inherently altered chromatin in their absence [45]. Upon cell wall stress conditions, Slt2 phosphorylates and activates the transcription factor Rlm1 to recruit both elements to the promoters of CWI-responsive genes in complex with SWI/SNF [9]. SWI/SNF activity is necessary to evict nucleosomes positioned in this region and permit pre-initiation complex (PIC) assembly in cooperation with the SAGA complex, which mediates histone acetylation for nucleosome reorganization [6,12]. The results described here suggest that the RSC complex could cooperate with SWI/SNF and SAGA complexes for the chromatin remodelling necessary for the transcriptional activation of CWI-dependent genes upon stress. There seems to be distinct requirements of the RSC complex for remodelling particular genes in a stress-dependent manner. Thus, the influence of the RSC complex via its subunit Rsc9 on the expression levels of genes regulated by the CWI pathway shows a highly gene-specific regulatory pattern, probably a consequence of each particular promoter architecture, chromatin status, and interaction with the transcriptional machinery.

3.4. Identification of Essential Genes Associated with the Constitutive Activation of the Cell Wall Integrity Pathway

One of the objectives of this work was to uncover essential genes that could be directly or indirectly related to the presence of alterations in the fungal cell wall. On many occasions, they were revealed by the constitutive activation of the CWI route without the need to apply any type of external stimulus. Presumably, this type of mutant should be included among those that exhibited an increment in Mlp1-GFP levels in the absence of CW when the essential gene expression is blocked (presence of doxycycline). We arbitrarily ascribed to this group those mutants exhibiting a relevant level of MFI (higher than three-fold) with respect to the value obtained for the wild-type strain (Tables 2 and S5). Representative flow cytometry profiles of these mutants are shown in Figure 2B. Eventually, 41 yeast strains met these criteria. After functional analyses, as described above, we found that approximately 68% belonged to four functional groups closely related to cell wall homeostasis: Protein glycosylation, Golgi vesicle transport, Glycosylphosphatidylinositol (GPI)-anchor assembly, and Cytoskeleton organization. GPI-anchored proteins (GPI-APs) are luminal secretory cargos that are attached by a post-translational glycolipid modification, the GPI anchor, to the plasma membrane and/or cell wall in yeast, consisting of a process conserved among eukaryotes [56]. A number of the GPI proteins in yeast serve enzymatic functions required for the biosynthesis and continuous shape adaptations of the cell wall, some seem to be structural elements of the cell wall, and others mediate cell adhesion [57]. The secretory pathway plays a key role in cell wall construction and remodelling, which explains the identification of this type of mutant in our screening. In fact, some of them, Sec24 and Sar1 belonging to the coat protein complex II (COPII), are involved in the specialized transport of GPI-APs to the ER [56]. The relationship between protein glycosylation and cell wall integrity is well established since most cell wall proteins are glycoproteins that have passed through the secretory pathway in transit to the cell wall. Thus, the isolation of this type of mutant is not unexpected given the importance of glycosylation for the proper conformation, localization, and function of cell wall proteins [58]. Regarding actin cytoskeleton-related genes, potential cell wall alterations could be a consequence of the requirement of this structure for correct spatiotemporal vesicular trafficking and cell wall assembly [59]. In accordance with the putative existence of alterations in the cell wall of mutants displaying incremented basal levels of Mlp1-GFP, about 17% were also selected in the screening for CW hypersensitivity (Table 2). In agreement with this idea, this value was significantly higher (50%) when a less strict cut-off point (0.4) was considered for CW hypersensitivity (Table 2).

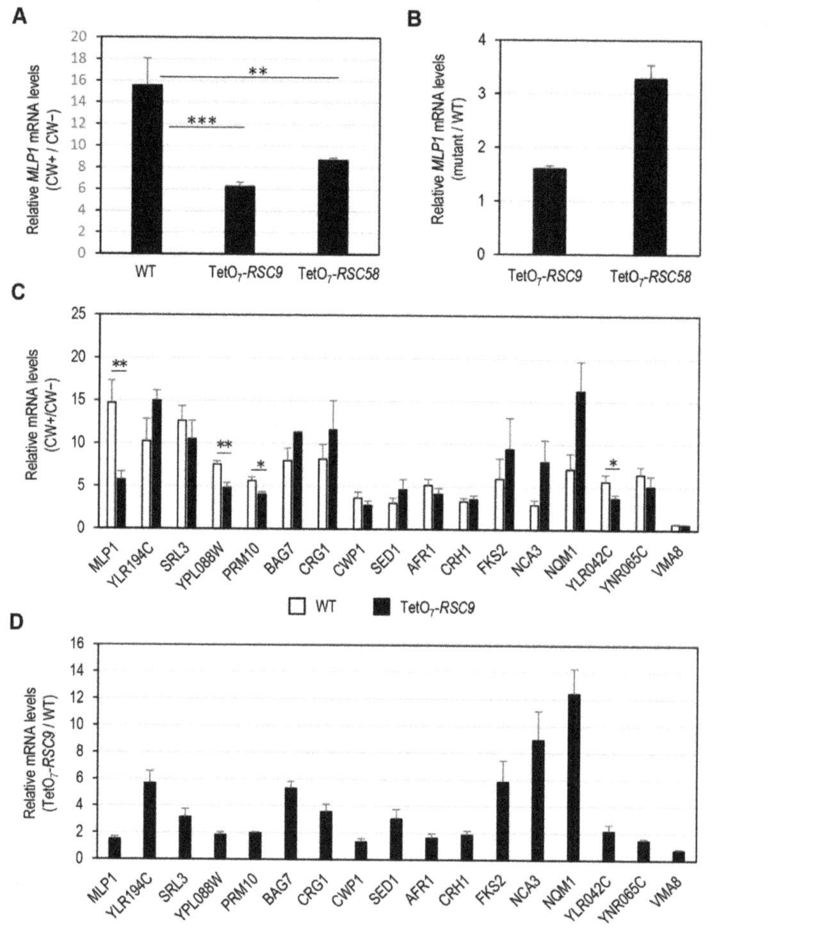

Figure 4. Gene expression analysis of CWI-responsive genes in the wild-type strain and mutants in essential subunits of the RSC complex. (**A**) *MLP1* mRNA levels of the wild-type (WT), $TetO_7$-*RSC9*, and $TetO_7$-*RSC58* strains treated with doxycycline (DOX, 10 µg/mL) and exposed or not to CW (10 µg/mL for 3 h) were analysed by RT-qPCR. Values represent the ratio between CW-treated and nontreated cells grown in the presence of doxycycline. (**B**) Basal mRNA levels of *MLP1* in $TetO_7$-*RSC9* and $TetO_7$-*RSC58* strains in the presence of DOX. Values represent the ratio between mutant and WT cells. (**C**) Analysis of mRNA levels of several selected CWI-responsive genes in the WT and $TetO_7$-*RSC9* strains as indicated in A. A CWI-independent gene, *VMA8*, was included as a control. (**D**) Impact of depletion of Rsc9 on the basal expression levels of the indicated CWI-related genes, represented as indicated in panel B. Data correspond to the mean and standard deviation of at least three independent experiments. Statistical significance was determined using a two-tailed, unpaired, Student's t-test, comparing it with data from the WT strain (* $p \leq 0.05$, ** $p \leq 0.01$, *** $p \leq 0.001$).

Table 2. List of genes corresponding to those mutants showing increased Mlp1-GFP levels under non-stress conditions. Mlp1-GFP ratio refers to the MFI (mean fluorescence intensity) value for each mutant respect to the wild-type. CW sensitivity denotes the sensitivity level to CW obtained from the phenotypic screen previously described. Those mutants with ratio ≤ 0.2 are highlighted in red and those with ratio ≤ 0.4 in orange. N.M.: non-measurable. References including previously described cell wall associations are listed.

Gene	Mlp1-GFP Ratio	CW Sensitivity	Biological Process	Cell Wall Association
PGI1	5.23	0.22	Carbohydrate metabolism	
CDC12	5.29	0.30	Cell cycle	
MET30	4.84	0.93	Cell cycle	
MTW1	4.65	N.M.	Chromosome segregation	
SPC34	7.09	1.41	Chromosome segregation	
ARP2	7.63	1.48	Cytoskeleton organization	
COF1	7.17	0.73	Cytoskeleton organization	[60]
ARC35	6.29	0.37	Cytoskeleton organization	
LAS17	3.57	0.87	Cytoskeleton organization	[61,62]
TRS20	3.01	0.38	Golgi vesicle transport	[63]
SEC5	3.95	0.44	Golgi vesicle transport	
SEC3	5.07	0.24	Golgi vesicle transport	[64]
SEC4	4.94	0.27	Golgi vesicle transport	[65]
YPT1	3.78	0.41	Golgi vesicle transport	[66]
SEC15	6.93	N.M.	Golgi vesicle transport	
SEC24	3.64	0.32	Golgi vesicle transport	
SEC10	9.46	0.50	Golgi vesicle transport	
SAR1	3.27	N.M.	Golgi vesicle transport	
GPI17	3.44	0.21	GPI biosynthesis	
GPI16	6.27	0.51	GPI biosynthesis	
GAB1	6.47	0.12	GPI biosynthesis	
GPI12	3.03	0.39	GPI biosynthesis	
PGA1	12.99	N.M.	GPI biosynthesis	
RER2	3.28	0.27	Lipid metabolism	[67]
ERO1	4.62	N.M.	Protein folding	
ROT1	7.97	0.11	Protein folding	[68,69]
RFT1	4.39	0.10	Protein glycosylation	
ALG14	5.72	0.23	Protein glycosylation	
WBP1	5.95	0.42	Protein glycosylation	
PMI40	3.04	0.91	Protein glycosylation	
SEC53	3.14	0.44	Protein glycosylation	
VRG4	2.96	0.06	Protein glycosylation	
SWP1	3.36	0.18	Protein glycosylation	
ALG11	3.07	0.21	Protein glycosylation	[70]
OST2	5.74	0.07	Protein glycosylation	
DPM1	3.34	0.46	Protein glycosylation	[67]
SEC63	4.22	0.82	Protein targeting	
CDC42	9.02	2.19	Signaling	[71,72]
RPA190	3.60	0.33	Transcription from RNA pol I	
MOT1	3.82	0.18	Transcription from RNA pol II	
YNL171C	5.45	1.23	Unknown	

To further investigate whether the increase in Mlp1 expression in these mutants could be associated with the activation of the CWI pathway, we analysed the levels of phospho-Slt2 (typical readout of the level of activation of this pathway) in two mutants from each of the four functional groups described above. To this end, we obtained total protein extracts from these strains grown in the presence of doxycycline and treated or not with CW. Next, these extracts were analysed by Western blotting using an antibody that recognizes the dually phosphorylated (active) form of Slt2. As shown in Figure 5A, in the mutant strains tested, the levels of phospho-Slt2 under basal growth conditions were clearly higher than those observed in the wild-type strain. These results support the notion that the increase in Mlp1 levels detected in these mutants, when the conditional allele is off, is a consequence of the activation of the CWI pathway. Interestingly, as shown in Figure 5A, in the presence of CW, in some mutants the activation level (phospho-Slt2) does not increase beyond basal levels, while in others the ability to over-activate the pathway seems to be maintained. As expected, this effect was not observed in other mutants not affecting Mlp1-GFP levels under non-stress conditions (Figure 5B). Future gene-focused studies will be required to decipher whether the constitutive activation of the CWI pathway is due to signalling via the cell surface CWI pathway associated sensors or is mediated by downstream elements of the pathway, which eventually induces Slt2 phosphorylation.

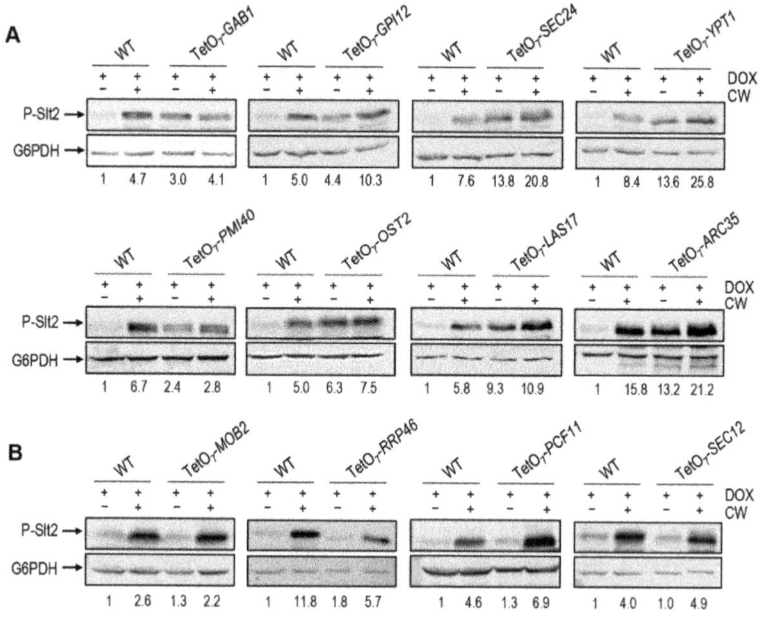

Figure 5. Detection of increased levels of Slt2 phosphorylation (P-Slt2) in the absence of cell wall stress. (**A**) Example of mutants displaying increased P-Slt2 basal levels. (**B**) Examples of mutants without P-Slt2 alteration. Cells exposed to doxycycline 10 μg/mL (DOX) treated or not with Calcofluor white 10 μg/mL for 3 h (CW) of the indicated yeast strains were taken and processed. Western blots detecting P-Slt2 with G6PDH as loading control are shown. Numbers correspond to the P-Slt2 fold-change obtained from densitometric quantification of the P-Slt2 bands normalized against G6PDH bands, using the values of the wild-type strain (WT) grown in the absence of CW as a reference (fold-change set to 1.0).

Globally, it is noteworthy that the percentage of essential genes directly or indirectly related to cell wall integrity revealed in this work is similar to that predicted from pioneering cell wall phenotypic studies using viable knock-out mutant strains [27]; this supports the importance of maintaining the integrity of this fungal structure. The results described herein improve our knowledge regarding the putative connection between essential genes and cell wall integrity. This will greatly facilitate further in-depth studies directed towards investigating at which cellular level (signalling, transcriptional or post-transcriptional) a specific essential gene impacts on the adaptive response to withstand cell wall injuries. Moreover, the results of this work will be useful for the development of targeted studies on specific genes, particularly in the case of the search and design of antifungal agents directed against the cell wall.

Supplementary Materials: The following supporting information can be downloaded at: https://www.mdpi.com/article/10.3390/jof8070718/s1, Table S1: Calcofluor white hypersensitivity screening dataset; Table S2: Positive hits from the Calcofluor white hypersensitivity screening; Table S3: Mlp1-GFP expression screening dataset; Table S4: Positive hits from the Mlp1-GFP expression screening; Table S5: Mutant strains showing increased basal levels of Mlp1-GFP; Figure S1: CW susceptibility assays of some selected hits from the CW sensitivity screening. Figure S2: Fluorescence microscopy images of strains expressing Mlp1-GFP.

Author Contributions: Conceptualization, J.M.R.-P., A.B.S. and J.A.; methodology, S.D.-M., A.B.S. and J.M.R.-P.; formal analysis, A.B.S. and J.M.R.-P., investigation, S.D.-M., J.M., Y.P. and A.B.S.; writing–original draft preparation, J.M.R.-P. and A.B.S.; Writing-Review & Editing, J.A. and C.N.; visualization, A.B.S. Supervision, J.M.R.-P., C.N. and J.A.; project administration, J.A. Funding Acquisition, J.A. and C.N. All authors have read and agreed to the published version of the manuscript.

Funding: This work was supported by the grants BIO2016-79289-P and PID2019-105223GB-I00 (Ministerio de Ciencia e Innovación, MICINN, Spain) and S2017/BMD3691-InGEMICS (Comunidad de Madrid and European Structural and Investment Funds) to J.A and the Special Chair on Fermented Beverages and Health to C.N., J.M. and Y.P. were the recipient of the research contracts (Laboratory technician) PEJ15/BIO/TL-0538 and PEJ-2018-TL/BMD-11487 from Consejería de Educación e Investigación de la Comunidad de Madrid y del Fondo Social Europeo, respectively.

Institutional Review Board Statement: Not applicable.

Informed Consent Statement: Not applicable.

Data Availability Statement: The data presented in this study are available on request from the corresponding author.

Acknowledgments: All members of our research group (UCM-920640: Yeast Functional Genomics) at the Department of Microbiology and Parasitology (UCM, Madrid, Spain) are acknowledged for their support. Thanks are due to all members of the Genomics Unit (UCM, Spain) for their help with RT-qPCR and Carlos Pérez Martín and Amalia Vázquez Gutierrez of the Flow Cytometry and Fluorescence Microscopy Unit (UCM, Spain) for expert help in flow cytometry.

Conflicts of Interest: The authors declare no conflict of interest.

References

1. Orlean, P. Architecture and biosynthesis of the *Saccharomyces cerevisiae* cell wall. *Genetics* **2012**, *192*, 775–818. [CrossRef] [PubMed]
2. Cabib, E.; Arroyo, J. How carbohydrates sculpt cells: Chemical control of morphogenesis in the yeast cell wall. *Nat. Rev. Microbiol.* **2013**, *11*, 648–655. [CrossRef] [PubMed]
3. Popolo, L.; Gualtieri, T.; Ragni, E. The yeast cell-wall salvage pathway. *Med. Mycol.* **2001**, *39* (Suppl. 1), 111–121. [CrossRef]
4. Sanz, A.B.; García, R.; Rodríguez-Peña, J.M.; Arroyo, J. The CWI Pathway: Regulation of the Transcriptional Adaptive Response to Cell Wall Stress in Yeast. *J. Fungi* **2017**, *4*, 1. [CrossRef] [PubMed]
5. Levin, D.E. Regulation of cell wall biogenesis in *Saccharomyces cerevisiae*: The cell wall integrity signaling pathway. *Genetics* **2011**, *189*, 1145–1175. [CrossRef] [PubMed]
6. Sanz, A.B.; García, R.; Pavon-Verges, M.; Rodriguez-Pena, J.M.; Arroyo, J. Control of Gene Expression via the Yeast CWI Pathway. *Int. J. Mol. Sci.* **2022**, *23*, 1791. [CrossRef] [PubMed]
7. Jung, U.S.; Levin, D.E. Genome-wide analysis of gene expression regulated by the yeast cell wall integrity signalling pathway. *Mol. Microbiol.* **1999**, *34*, 1049–1057. [CrossRef]
8. García, R.; Bermejo, C.; Grau, C.; Pérez, R.; Rodríguez-Peña, J.M.; Francois, J.; Nombela, C.; Arroyo, J. The global transcriptional response to transient cell wall damage in *Saccharomyces cerevisiae* and its regulation by the cell integrity signaling pathway. *J. Biol. Chem.* **2004**, *279*, 15183–15195. [CrossRef]
9. Sanz, A.B.; García, R.; Rodríguez-Peña, J.M.; Díez-Muñiz, S.; Nombela, C.; Peterson, C.L.; Arroyo, J. Chromatin remodeling by the SWI/SNF complex is essential for transcription mediated by the yeast cell wall integrity MAPK pathway. *Mol. Biol. Cell* **2012**, *23*, 2805–2817. [CrossRef]
10. Lagorce, A.; Hauser, N.C.; Labourdette, D.; Rodríguez, C.; Martin-Yken, H.; Arroyo, J.; Hoheisel, J.D.; Francois, J. Genome-wide analysis of the response to cell wall mutations in the yeast *Saccharomyces cerevisiae*. *J. Biol. Chem.* **2003**, *278*, 20345–20357. [CrossRef]
11. Arias, P.; Díez-Muñiz, S.; García, R.; Nombela, C.; Rodríguez-Peña, J.M.; Arroyo, J. Genome-wide survey of yeast mutations leading to activation of the yeast cell integrity MAPK pathway: Novel insights into diverse MAPK outcomes. *BMC Genom.* **2011**, *12*, 390. [CrossRef] [PubMed]
12. Sanz, A.B.; García, R.; Rodríguez-Peña, J.M.; Nombela, C.; Arroyo, J. Cooperation between SAGA and SWI/SNF complexes is required for efficient transcriptional responses regulated by the yeast MAPK Slt2. *Nucleic Acids Res.* **2016**, *44*, 7159–7172. [CrossRef] [PubMed]
13. Angus-Hill, M.L.; Schlichter, A.; Roberts, D.; Erdjument-Bromage, H.; Tempst, P.; Cairns, B.R. A Rsc3/Rsc30 zinc cluster dimer reveals novel roles for the chromatin remodeler RSC in gene expression and cell cycle control. *Mol. Cell* **2001**, *7*, 741–751. [CrossRef]
14. Chai, B.; Hsu, J.M.; Du, J.; Laurent, B.C. Yeast RSC function is required for organization of the cellular cytoskeleton via an alternative PKC1 pathway. *Genetics* **2002**, *161*, 575–584. [CrossRef]

15. Wilson, B.; Erdjument-Bromage, H.; Tempst, P.; Cairns, B.R. The RSC chromatin remodeling complex bears an essential fungal-specific protein module with broad functional roles. *Genetics* **2006**, *172*, 795–809. [CrossRef]
16. Wang, S.L.; Cheng, M.Y. The defects in cell wall integrity and G2-M transition of the htl1 mutant are interconnected. *Yeast* **2012**, *29*, 45–57. [CrossRef]
17. Rando, O.J.; Winston, F. Chromatin and transcription in yeast. *Genetics* **2012**, *190*, 351–387. [CrossRef]
18. Giaever, G.; Chu, A.M.; Ni, L.; Connelly, C.; Riles, L.; Veronneau, S.; Dow, S.; Lucau-Danila, A.; Anderson, K.; Andre, B.; et al. Functional profiling of the *Saccharomyces cerevisiae* genome. *Nature* **2002**, *418*, 387–391. [CrossRef]
19. Page, N.; Gerard-Vincent, M.; Menard, P.; Beaulieu, M.; Azuma, M.; Dijkgraaf, G.J.; Li, H.; Marcoux, J.; Nguyen, T.; Dowse, T.; et al. A *Saccharomyces cerevisiae* genome-wide mutant screen for altered sensitivity to K1 killer toxin. *Genetics* **2003**, *163*, 875–894. [CrossRef]
20. Serviene, E.; Luksa, J.; Orentaite, I.; Lafontaine, D.L.; Urbonavicius, J. Screening the budding yeast genome reveals unique factors affecting K2 toxin susceptibility. *PLoS ONE* **2012**, *7*, e50779. [CrossRef]
21. Rodríguez-Peña, J.M.; Díez-Muñiz, S.; Bermejo, C.; Nombela, C.; Arroyo, J. Activation of the yeast cell wall integrity MAPK pathway by zymolyase depends on protease and glucanase activities and requires the mucin-like protein Hkr1 but not Msb2. *FEBS Lett.* **2013**, *587*, 3675–3680. [CrossRef] [PubMed]
22. García, R.; Botet, J.; Rodríguez-Peña, J.M.; Bermejo, C.; Ribas, J.C.; Revuelta, J.L.; Nombela, C.; Arroyo, J. Genomic profiling of fungal cell wall-interfering compounds: Identification of a common gene signature. *BMC Genom.* **2015**, *16*, 683. [CrossRef] [PubMed]
23. Ando, A.; Nakamura, T.; Murata, Y.; Takagi, H.; Shima, J. Identification and classification of genes required for tolerance to freeze-thaw stress revealed by genome-wide screening of *Saccharomyces cerevisiae* deletion strains. *FEMS Yeast Res.* **2007**, *7*, 244–253. [CrossRef] [PubMed]
24. Auesukaree, C.; Damnernsawad, A.; Kruatrachue, M.; Pokethitiyook, P.; Boonchird, C.; Kaneko, Y.; Harashima, S. Genome-wide identification of genes involved in tolerance to various environmental stresses in *Saccharomyces cerevisiae*. *J. Appl. Genet.* **2009**, *50*, 301–310. [CrossRef] [PubMed]
25. Rodríguez-Peña, J.M.; Díez-Muñiz, S.; Nombela, C.; Arroyo, J. A yeast strain biosensor to detect cell wall-perturbing agents. *J. Biotechnol.* **2008**, *133*, 311–317. [CrossRef]
26. Giaever, G.; Nislow, C. The yeast deletion collection: A decade of functional genomics. *Genetics* **2014**, *197*, 451–465. [CrossRef]
27. De Groot, P.W.; Ruiz, C.; Vázquez de Aldana, C.R.; Duenas, E.; Cid, V.J.; Del Rey, F.; Rodríguez-Peña, J.M.; Pérez, P.; Andel, A.; Caubín, J.; et al. A genomic approach for the identification and clasification of genes involved in cell wall formation and its regulation in *Saccharomyces cerevisiae*. *Comp. Funct. Genom.* **2001**, *2*, 124–142. [CrossRef]
28. Norman, K.L.; Kumar, A. Mutant power: Using mutant allele collections for yeast functional genomics. *Brief Funct. Genom.* **2016**, *15*, 75–84. [CrossRef]
29. Mnaimneh, S.; Davierwala, A.P.; Haynes, J.; Moffat, J.; Peng, W.T.; Zhang, W.; Yang, X.; Pootoolal, J.; Chua, G.; Lopez, A.; et al. Exploration of essential gene functions via titratable promoter alleles. *Cell* **2004**, *118*, 31–44. [CrossRef]
30. Cappell, S.D.; Baker, R.; Skowyra, D.; Dohlman, H.G. Systematic analysis of essential genes reveals important regulators of G protein signaling. *Mol. Cell* **2010**, *38*, 746–757. [CrossRef]
31. Tatjer, L.; Sacristan-Reviriego, A.; Casado, C.; Gonzalez, A.; Rodriguez-Porrata, B.; Palacios, L.; Canadell, D.; Serra-Cardona, A.; Martin, H.; Molina, M.; et al. Wide-Ranging Effects of the Yeast Ptc1 Protein Phosphatase Acting Through the MAPK Kinase Mkk1. *Genetics* **2016**, *202*, 141–156. [CrossRef] [PubMed]
32. Bermejo, C.; Rodríguez, E.; García, R.; Rodríguez-Peña, J.M.; Rodríguez de la Concepcion, M.L.; Rivas, C.; Arias, P.; Nombela, C.; Posas, F.; Arroyo, J. The sequential activation of the yeast HOG and SLT2 pathways is required for cell survival to cell wall stress. *Mol. Biol. Cell.* **2008**, *19*, 1113–1124. [CrossRef]
33. Livak, K.J.; Schmittgen, T.D. Analysis of relative gene expression data using real-time quantitative PCR and the 2(-Delta Delta C(T)) Method. *Methods* **2001**, *25*, 402–408. [CrossRef] [PubMed]
34. Zhao, F.; Li, J.; Lin, K.; Chen, H.; Lin, Y.; Zheng, S.; Liang, S.; Han, S. Genome-wide screening of *Saccharomyces cerevisiae* deletion mutants reveals cellular processes required for tolerance to the cell wall antagonist calcofluor white. *Biochem. Biophys. Res. Commun.* **2019**, *518*, 1–6. [CrossRef] [PubMed]
35. Chen, P.; Wang, D.; Chen, H.; Zhou, Z.; He, X. The nonessentiality of essential genes in yeast provides therapeutic insights into a human disease. *Genome Res.* **2016**, *26*, 1355–1362. [CrossRef] [PubMed]
36. García, R.; Rodríguez-Peña, J.M.; Bermejo, C.; Nombela, C.; Arroyo, J. The high osmotic response and cell wall integrity pathways cooperate to regulate transcriptional responses to zymolyase-induced cell wall stress in *Saccharomyces cerevisiae*. *J. Biol. Chem.* **2009**, *284*, 10901–10911. [CrossRef]
37. García, R.; Bravo, E.; Díez-Muñiz, S.; Nombela, C.; Rodríguez-Peña, J.M.; Arroyo, J. A novel connection between the Cell Wall Integrity and the PKA pathways regulates cell wall stress response in yeast. *Sci. Rep.* **2017**, *7*, 5703. [CrossRef]
38. Montefusco, D.J.; Matmati, N.; Hannun, Y.A. The yeast sphingolipid signaling landscape. *Chem. Phys. Lipids* **2014**, *177*, 26–40. [CrossRef]
39. Hahn, J.S.; Neef, D.W.; Thiele, D.J. A stress regulatory network for co-ordinated activation of proteasome expression mediated by yeast heat shock transcription factor. *Mol. Microbiol.* **2006**, *60*, 240–251. [CrossRef]

40. Ma, M.; Liu, Z.L. Comparative transcriptome profiling analyses during the lag phase uncover *YAP1*, *PDR1*, *PDR3*, *RPN4*, and *HSF1* as key regulatory genes in genomic adaptation to the lignocellulose derived inhibitor HMF for *Saccharomyces cerevisiae*. *BMC Genom.* **2010**, *11*, 660. [CrossRef]
41. Rousseau, A.; Bertolotti, A. An evolutionarily conserved pathway controls proteasome homeostasis. *Nature* **2016**, *536*, 184–189. [CrossRef] [PubMed]
42. Livneh, I.; Cohen-Kaplan, V.; Cohen-Rosenzweig, C.; Avni, N.; Ciechanover, A. The life cycle of the 26S proteasome: From birth, through regulation and function, and onto its death. *Cell Res.* **2016**, *26*, 869–885. [CrossRef]
43. Zander, G.; Hackmann, A.; Bender, L.; Becker, D.; Lingner, T.; Salinas, G.; Krebber, H. mRNA quality control is bypassed for immediate export of stress-responsive transcripts. *Nature* **2016**, *540*, 593–596. [CrossRef] [PubMed]
44. Erkina, T.Y.; Zou, Y.; Freeling, S.; Vorobyev, V.I.; Erkine, A.M. Functional interplay between chromatin remodeling complexes RSC, SWI/SNF and ISWI in regulation of yeast heat shock genes. *Nucleic Acids Res.* **2010**, *38*, 1441–1449. [CrossRef]
45. Sahu, R.K.; Singh, S.; Tomar, R.S. The ATP-dependent SWI/SNF and RSC chromatin remodelers cooperatively induce unfolded protein response genes during endoplasmic reticulum stress. *Biochim. Biophys. Acta Gene Regul. Mech.* **2021**, *1864*, 194748. [CrossRef] [PubMed]
46. Ye, Y.; Wu, H.; Chen, K.; Clapier, C.R.; Verma, N.; Zhang, W.; Deng, H.; Cairns, B.R.; Gao, N.; Chen, Z. Structure of the RSC complex bound to the nucleosome. *Science* **2019**, *366*, 838–843. [CrossRef]
47. Lorch, Y.; Kornberg, R.D. Chromatin-remodeling for transcription. *Q. Rev. Biophys.* **2017**, *50*, e5. [CrossRef]
48. Biernat, E.; Kinney, J.; Dunlap, K.; Rizza, C.; Govind, C.K. The RSC complex remodels nucleosomes in transcribed coding sequences and promotes transcription in *Saccharomyces cerevisiae*. *Genetics* **2021**, *217*, iyab021. [CrossRef]
49. Sing, T.L.; Hung, M.P.; Ohnuki, S.; Suzuki, G.; San Luis, B.J.; McClain, M.; Unruh, J.R.; Yu, Z.; Ou, J.; Marshall-Sheppard, J.; et al. The budding yeast RSC complex maintains ploidy by promoting spindle pole body insertion. *J. Cell Biol.* **2018**, *217*, 2445–2462. [CrossRef]
50. Kasten, M.; Szerlong, H.; Erdjument-Bromage, H.; Tempst, P.; Werner, M.; Cairns, B.R. Tandem bromodomains in the chromatin remodeler RSC recognize acetylated histone H3 Lys14. *EMBO J.* **2004**, *23*, 1348–1359. [CrossRef]
51. Conde, R.; Cueva, R.; Larriba, G. Rsc14-controlled expression of *MNN6*, *MNN4* and *MNN1* regulates mannosylphosphorylation of *Saccharomyces cerevisiae* cell wall mannoproteins. *FEMS Yeast Res.* **2007**, *7*, 1248–1255. [CrossRef] [PubMed]
52. Damelin, M.; Simon, I.; Moy, T.I.; Wilson, B.; Komili, S.; Tempst, P.; Roth, F.P.; Young, R.A.; Cairns, B.R.; Silver, P.A. The genome-wide localization of Rsc9, a component of the RSC chromatin-remodeling complex, changes in response to stress. *Mol. Cell* **2002**, *9*, 563–573. [CrossRef]
53. Mas, G.; de Nadal, E.; Dechant, R.; Rodríguez de la Concepcion, M.L.; Logie, C.; Jimeno-Gonzalez, S.; Chavez, S.; Ammerer, G.; Posas, F. Recruitment of a chromatin remodelling complex by the Hog1 MAP kinase to stress genes. *EMBO J.* **2009**, *28*, 326–336. [CrossRef] [PubMed]
54. Garcia, R.; Itto-Nakama, K.; Rodriguez-Pena, J.M.; Chen, X.; Sanz, A.B.; de Lorenzo, A.; Pavon-Verges, M.; Kubo, K.; Ohnuki, S.; Nombela, C.; et al. Poacic acid, a beta-1,3-glucan-binding antifungal agent, inhibits cell-wall remodeling and activates transcriptional responses regulated by the cell-wall integrity and high-osmolarity glycerol pathways in yeast. *FASEB J.* **2021**, *35*, e21778. [CrossRef] [PubMed]
55. Rodríguez-Peña, J.M.; Pérez-Diaz, R.M.; Alvarez, S.; Bermejo, C.; García, R.; Santiago, C.; Nombela, C.; Arroyo, J. The 'yeast cell wall chip'—A tool to analyse the regulation of cell wall biogenesis in *Saccharomyces cerevisiae*. *Microbiology* **2005**, *151*, 2241–2249. [CrossRef]
56. Muñiz, M.; Riezman, H. Trafficking of glycosylphosphatidylinositol anchored proteins from the endoplasmic reticulum to the cell surface. *J. Lipid Res.* **2016**, *57*, 352–360. [CrossRef]
57. Pittet, M.; Conzelmann, A. Biosynthesis and function of GPI proteins in the yeast *Saccharomyces cerevisiae*. *Biochim. Biophys. Acta* **2007**, *1771*, 405–420. [CrossRef]
58. Bowman, S.M.; Free, S.J. The structure and synthesis of the fungal cell wall. *Bioessays* **2006**, *28*, 799–808. [CrossRef]
59. Mishra, M.; Huang, J.; Balasubramanian, M.K. The yeast actin cytoskeleton. *FEMS Microbiol. Rev.* **2014**, *38*, 213–227. [CrossRef]
60. Curwin, A.J.; von Blume, J.; Malhotra, V. Cofilin-mediated sorting and export of specific cargo from the Golgi apparatus in yeast. *Mol. Biol. Cell* **2012**, *23*, 2327–2338. [CrossRef]
61. Li, H.; Page, N.; Bussey, H. Actin patch assembly proteins Las17p and Sla1p restrict cell wall growth to daughter cells and interact with cis-Golgi protein Kre6p. *Yeast* **2002**, *19*, 1097–1112. [CrossRef] [PubMed]
62. Sukegawa, Y.; Negishi, T.; Kikuchi, Y.; Ishii, K.; Imanari, M.; Ghanegolmohammadi, F.; Nogami, S.; Ohya, Y. Genetic dissection of the signaling pathway required for the cell wall integrity checkpoint. *J. Cell Sci.* **2018**, *131*, jcs219063. [CrossRef] [PubMed]
63. Mahfouz, H.; Ragnini-Wilson, A.; Venditti, R.; De Matteis, M.A.; Wilson, C. Mutational analysis of the yeast TRAPP subunit Trs20p identifies roles in endocytic recycling and sporulation. *PLoS ONE* **2012**, *7*, e41408. [CrossRef] [PubMed]
64. Guo, Q.; Duan, Y.; Meng, N.; Liu, Y.; Luo, G. The N-terminus of Sec3 is required for cell wall integrity in yeast. *Biochimie* **2020**, *177*, 30–39. [CrossRef]
65. Suda, Y.; Tachikawa, H.; Inoue, I.; Kurita, T.; Saito, C.; Kurokawa, K.; Nakano, A.; Irie, K. Activation of Rab GTPase Sec4 by its GEF Sec2 is required for prospore membrane formation during sporulation in yeast *Saccharomyces cerevisiae*. *FEMS Yeast Res.* **2018**, *18*, fox095. [CrossRef]

66. Bialek-Wyrzykowska, U.; Bauer, B.E.; Wagner, W.; Kohlwein, S.D.; Schweyen, R.J.; Ragnini, A. Low levels of Ypt protein prenylation cause vesicle polarization defects and thermosensitive growth that can be suppressed by genes involved in cell wall maintenance. *Mol. Microbiol.* **2000**, *35*, 1295–1311. [CrossRef]
67. Orlowski, J.; Machula, K.; Janik, A.; Zdebska, E.; Palamarczyk, G. Dissecting the role of dolichol in cell wall assembly in the yeast mutants impaired in early glycosylation reactions. *Yeast* **2007**, *24*, 239–252. [CrossRef]
68. Bickle, M.; Delley, P.A.; Schmidt, A.; Hall, M.N. Cell wall integrity modulates RHO1 activity via the exchange factor ROM2. *EMBO J.* **1998**, *17*, 2235–2245. [CrossRef]
69. Machi, K.; Azuma, M.; Igarashi, K.; Matsumoto, T.; Fukuda, H.; Kondo, A.; Ooshima, H. Rot1p of *Saccharomyces cerevisiae* is a putative membrane protein required for normal levels of the cell wall 1,6-beta-glucan. *Microbiology* **2004**, *150*, 3163–3173. [CrossRef]
70. Absmanner, B.; Schmeiser, V.; Kampf, M.; Lehle, L. Biochemical characterization, membrane association and identification of amino acids essential for the function of Alg11 from *Saccharomyces cerevisiae*, an alpha1,2-mannosyltransferase catalysing two sequential glycosylation steps in the formation of the lipid-linked core oligosaccharide. *Biochem. J.* **2010**, *426*, 205–217. [CrossRef]
71. Rodríguez-Peña, J.M.; Rodríguez, C.; Alvarez, A.; Nombela, C.; Arroyo, J. Mechanisms for targeting of the *Saccharomyces cerevisiae* GPI-anchored cell wall protein Crh2p to polarised growth sites. *J. Cell Sci.* **2002**, *115*, 2549–2558. [CrossRef] [PubMed]
72. Rodríguez-Pachon, J.M.; Martin, H.; North, G.; Rotger, R.; Nombela, C.; Molina, M. A novel connection between the yeast Cdc42 GTPase and the Slt2-mediated cell integrity pathway identified through the effect of secreted *Salmonella* GTPase modulators. *J. Biol. Chem.* **2002**, *277*, 27094–27102. [CrossRef] [PubMed]

Journal of *Fungi*

MDPI

Article

Defining Functions of Mannoproteins in *Saccharomyces cerevisiae* by High-Dimensional Morphological Phenotyping

Farzan Ghanegolmohammadi [1,2], Hiroki Okada [3], Yaxuan Liu [1], Kaori Itto-Nakama [1], Shinsuke Ohnuki [1], Anna Savchenko [1,4], Erfei Bi [3], Satoshi Yoshida [5] and Yoshikazu Ohya [1,*]

1 Department of Integrated Biosciences, Graduate School of Frontier Sciences, The University of Tokyo, Chiba 277-8562, Japan; 2344785075@edu.k.u-tokyo.ac.jp or farzang@mit.edu (F.G.); liuyaxuan_200@outlook.com (Y.L.); kaori.nakama-itto@edu.k.u-tokyo.ac.jp (K.I.-N.); ohnuki@edu.k.u-tokyo.ac.jp (S.O.); a.savchenko@maastrichtuniversity.nl (A.S.)

2 Department of Biological Engineering, Massachusetts Institute of Technology, Cambridge, MA 02139, USA

3 Department of Cell and Developmental Biology, Perelman School of Medicine, University of Pennsylvania, Philadelphia, PA 19104, USA; okad@pennmedicine.upenn.edu (H.O.); ebi@pennmedicine.upenn.edu (E.B.)

4 Cardiovascular Research Institute Maastricht, Maastricht University Medical Center, ER 6229 Maastricht, The Netherlands

5 School of International Liberal Studies, Nishi-Waseda Campus, Waseda University, Tokyo 169-8050, Japan; satosh@waseda.jp

* Correspondence: ohya@edu.k.u-tokyo.ac.jp

Abstract: Mannoproteins are non-filamentous glycoproteins localized to the outermost layer of the yeast cell wall. The physiological roles of these structural components have not been completely elucidated due to the limited availability of appropriate tools. As the perturbation of mannoproteins may affect cell morphology, we investigated mannoprotein mutants in *Saccharomyces cerevisiae* via high-dimensional morphological phenotyping. The mannoprotein mutants were morphologically classified into seven groups using clustering analysis with Gaussian mixture modeling. The pleiotropic phenotypes of cluster I mutant cells (*ccw12*Δ) indicated that *CCW12* plays major roles in cell wall organization. Cluster II (*ccw14*Δ, *flo11*Δ, *srl1*Δ, and *tir3*Δ) mutants exhibited altered mother cell size and shape. Mutants of cluster III and IV exhibited no or very small morphological defects. Cluster V (*dse2*Δ, *egt2*Δ, and *sun4*Δ) consisted of endoglucanase mutants with cell separation defects due to incomplete septum digestion. The cluster VI mutant cells (*ecm33*Δ) exhibited perturbation of apical bud growth. Cluster VII mutant cells (*sag1*Δ) exhibited differences in cell size and actin organization. Biochemical assays further confirmed the observed morphological defects. Further investigations based on various omics data indicated that morphological phenotyping is a complementary tool that can help with gaining a deeper understanding of the functions of mannoproteins.

Keywords: mannoprotein; cell wall; budding yeast; morphology; CalMorph

Citation: Ghanegolmohammadi, F.; Okada, H.; Liu, Y.; Itto-Nakama, K.; Ohnuki, S.; Savchenko, A.; Bi, E.; Yoshida, S.; Ohya, Y. Defining Functions of Mannoproteins in *Saccharomyces cerevisiae* by High-Dimensional Morphological Phenotyping. *J. Fungi* **2021**, *7*, 769. https://doi.org/10.3390/jof7090769

Academic Editors: María Molina and Humberto Martín

Received: 28 July 2021
Accepted: 14 September 2021
Published: 17 September 2021

1. Introduction

The cell wall is a rigid structure that plays essential roles in establishing cell morphology and dictating the oval shape of budding yeast, *Saccharomyces cerevisiae*, and it also confers robustness on the cell by stabilizing internal osmotic conditions and serving as a site for cell wall enzymes to exert their effects [1–3]. Electron microscopic analysis has revealed that the yeast cell wall is a highly organized composite consisting of internal interconnected filamentous polysaccharides (1,3-β-glucan, 1,6-β-glucan, and chitin) and external non-filamentous glycoproteins (mannoproteins), which form a firm extracellular matrix similar to reinforced concrete [4,5]. Whereas 1,3-β-glucan is the major filamentous cell wall component [2,6] mainly responsible for dictating the yeast cell shape, non-filamentous mannoproteins, of which 36 members have been identified to date, have also been suggested to play fundamental roles in the cell wall [3].

Previous studies have indicated that individual deletions of genes encoding mannoproteins may result in subtle growth defects [7–10]. This can be partly explained by gene duplication, as 26 of 36 mannoprotein genes are duplicated [11]. Another reason is that cell wall defects caused by the lack of mannoproteins affect cell morphology rather than growth phenotypes [12]. Thus, the morphological phenotyping of mannoprotein mutants would provide more information on their functions, highlighting the importance of morphology as another metric with which to study the genes involved in cell wall assembly. In general, mannoproteins play a collective role in maintaining the cell wall structure [13], but differences in the localization, structure, and probably also the function of mannoproteins in the cell wall [3,8] suggest that the perturbation of individual genes may result in different morphological phenotypes [12]. Each mannoprotein is likely to have a distinct role in the cell wall, but the details have not been elucidated due to limited quantitative morphological analysis of mannoprotein mutants.

This study was performed to determine a responsibility assignment matrix (hereafter we call it responsibility matrix) through morphological clustering analysis of mannoproteins in relation to their molecular functions. For this purpose, high-dimensional morphological phenotyping was performed after extracting the morphological features of each mannoprotein mutant with the image processing program CalMorph [14]. Analyses of morphological abnormalities based on a powerful parametric approach revealed specific morphological phenotypes that will help with uncovering the responsibility matrix of mannoproteins in the yeast cell wall.

2. Materials and Methods

2.1. Strains and Growth Conditions

Of 36 mannoprotein gene-deletion mutants of the budding yeast, *S. cerevisiae* [3], we studied 32 mutants (Table S1) that were straightforward to observe as single cells under a microscope. These 32 mutants were isogenic derivatives of BY4741 (*MATa his3 leu2 met15 ura3*) and were purchased from EUROSCARF (Oberursel, Germany). Cells with mutations in the other four mannoprotein genes (*aga1*Δ, *flo5*Δ, *flo9*Δ, and *dan4*Δ) were not studied due to heavy cell aggregation. Mutant and wild-type (WT) strains were cultured under optimal growth conditions at 25 °C in nutrient-rich yeast extract peptone dextrose (YPD) medium containing 1% (*w*/*v*) Bacto yeast extract (BD Biosciences, San Jose, CA, USA), 2% (*w*/*v*) polypeptone (Wako Chemicals, Richmond, VA, USA), and 2% (*w*/*v*) dextrose, as described previously [14]. A WT diploid strain (BY4743) and the homozygous gene deletion mutants in the BY4743 background used for Western blotting were purchased from EUROSCARF (Oberursel, Germany).

2.2. Fluorescence Staining, Microscopy, and Image Processing

To minimize variation due to inconsistencies in data acquisition, we followed a precise protocol for the preparation, fixation, and fluorescence staining of yeast cells in the early *log* phase of growth (<5.0 × 10^6 cells; 5 biological replicates), as described previously [15–18]. Briefly, yeast cells were fixed for 30 min in growth medium supplemented with formaldehyde (final concentration, 3.7%) and potassium phosphate buffer (100 mM, pH 6.5) at 25 °C. Yeast cells were then collected via centrifugation at room temperature and further incubated in potassium phosphate buffer containing 4% formaldehyde for 45 min. The fixed cells were subsequently prepared for fluorescence microscopy. First, actin staining was performed by treating the cells overnight with 15 U/mL rhodamine-phalloidin (Invitrogen, Carlsbad, CA, USA) and 1% Triton-X in phosphate-buffered saline (PBS). Second, cell wall mannoproteins were stained by treating cells for 10 min with 1 mg/mL fluorescein isothiocyanate-conjugated concanavalin A (Sigma-Aldrich, St. Louis, MO, USA) in P buffer (10 mM sodium phosphate and 150 mM NaCl, pH 7.2). Finally, after washing twice with P buffer, the yeast cells were mixed with mounting buffer (1 mg/mL *p*-phenylenediamine, 25 mM NaOH, 10% PBS, and 90% glycerol) containing 20 mg/mL 4′,6-diamidino-2-phenylindole (Sigma-Aldrich) to stain DNA.

Images of triple-stained cells were captured using an Axio Imager microscope equipped with a 6100 ECplan-Neofluar lens (Carl Zeiss, Oberkochen, Germany), a CoolSNAP HQ cooled charged coupled device (CCD) camera (Roper Scientific Photometrics, Tucson, AZ, USA), and AxioVision software ver. 4.5 (Carl Zeiss). The obtained images were quantified using CalMorph with regard to 501 morphological parameters related to the cell-cycle phase, actin cytoskeleton, cell wall, and nuclear DNA. The descriptions for each trait have been reported previously [14], and the CalMorph user manual is available at http://www.yeast.ib.k.u-tokyo.ac.jp/CalMorph/download.php?path=CalMorph-manual.pdf, accessed on 21 September 2019. Only those experiments containing at least 200 cells, detected by CalMorph, were considered for statistical analysis.

2.3. Data Analysis

All statistical analyses were performed using R (http://www.r-project.org, accessed on 21 September 2019). To assess the effects of genetic perturbation on the morphology of the mutants, we compared the cell morphological traits of the mutants with the corresponding WT distribution (i.e., null distribution) for each trait using an ANOVA model based on a generalized linear model (GLM). The GLM is an extension of the normal linear model in which predictors are linear but link functions are nonlinear to cope with violations of some standard assumptions of linear models [19]. These properties allow the analysis to cover probability distributions other than the Gaussian distribution. CalMorph generated 501 morphological parameters with which we established models based on the probability distributions for 490 unimodal parameters using the UNImodal MOrphological data pipeline (UNIMO; unpublished). Briefly, we first categorized CalMorph parameters into the following four data types: non-negative parameters, ratios, coefficients of variation (CVs; further converted to noise values, see below), and proportions. Then, we showed that these parameters could be explained well by 10 unimodal distributions to accommodate the statistical model used in the GLM: gamma, inverse gamma, and Weibull distributions for non-negative parameters; beta and logit-normal distributions for ratios; Gaussian, logistic, and reverse Gumbel distributions for noise parameters; and binomial and beta-binomial distributions for proportions. The best fit unimodal probability distribution for each parameter was eventually determined using the Akaike information criterion (AIC). CVs ($\frac{Population\ standard\ deviation\ (\sigma)}{Population\ mean\ (\mu)}$) are nonlinearly dependent on mean values [9]. We used LOESS (locally estimated scatterplot smoothing) regression with a smooth span (f) to uncouple this concomitant dependency. AIC values were used to choose the best-fitting model among various smooth spans ($0.10 \leq f \leq 0.99$). Finally, noise parameters were calculated as the residuals, i.e., observed value minus predicted value.

To estimate Z-values, once maximum-likelihood estimation converged, we transformed each morphological parameter to a Z-value via the Wald test (one-sample two-sided test) using the summary.gamlss R function [20]. The false discovery rate (FDR), the rate of type I error associated with rejecting the null hypothesis due to multiple comparisons, was estimated based on 2000 permutations.

2.4. Dimensionality Reduction and Clustering

To extract the most effective parameters, we performed principal component analysis (PCA), the most commonly used method for reducing dimensionality [21,22], on the obtained Z-values using the prcomp function (stats package). We then calculated the cumulative contribution ratio (CCR) to describe variation in the data. Based on the result, we used the first five principal components (CCR = 81.34%) for clustering analysis (Figure S1A).

Mixture model clustering is a probability-based approach in which we assume the dataset is best described as a mixture of probability models. In Gaussian mixture modeling (GMM), the most commonly used model-based clustering method [23], Gaussian distributions are fitted to the dataset. Gaussian distributions are randomly initialized and their parameters optimized iteratively to achieve a better fit. The expectation maximization algorithm estimates all parameters to assign members into *c* clusters. We employed the mclust package [23] to determine the underlying Gaussian mixture distributions (Figure S1B,C).

2.5. Kinetics of Cluster V Mannoproteins (Dse2 and Egt2)

2.5.1. Yeast Media and Culture Conditions

Standard culture media and genetic techniques were used [24]. Yeast strains were grown routinely at 25 °C in synthetic complete (SC) minimal medium lacking specific amino acid(s) and/or uracil or YPD. Neutralized SC medium (pH 7.0) was used for live-cell imaging of green fluorescent protein (GFP) molecules exposed to the extracellular environment to prevent quenching of the GFP signal caused by the acidity of the standard SC medium.

2.5.2. Constructions of Strains

New strains were constructed either by integrating a plasmid carrying a modified gene at a genomic locus or by transferring a deletion or tagged allele of a gene from a plasmid or from one strain to another via PCR amplification and yeast transformation; see footnotes in Table S2 [25–27].

2.5.3. Primers and Plasmids

All PCR primers and plasmids used in this study are listed in Table S3. All PCR primers were purchased from Integrated DNA Technologies (Coralville, IA, USA). All new constructs were validated via sequencing performed at the DNA Sequencing Facility, University of Pennsylvania. The plasmids pFA6a-GFPEnvy-KanMX6, pFA6a-link-GFPEnvy-KanMX6, and pRS316-ENVY-FKS1(1-789) were described previously [28]. The plasmids bWL715 (pHIS3p:mRuby2-Tub1+3′UTR::HPH [29]) and pFA6a-URA3-KanMX6 [30] were generous gifts from Wei-Lih Lee (Dartmouth College) and John Pringle (Stanford University), respectively.

The following plasmids were generated for this study. To generate pFA6a-link-GFPEnvy-CaURA3, a ~0.7-kb *Pac*I-*Asc*I fragment containing *GFP*^{Envy} from pFA6a-link-GFPEnvy-SpHis5 [27] was subcloned to replace the ~0.7-kb *Pac*I-*Asc*I region of pFA6a-link-yomApple-CaURA3 (#44879; Addgene, Watertown, MA, USA). To generate proHIS3-ymScarlet-I-TUB1-tTUB1-HPH (integrative, *hphMX*, expresses Tub1 N-terminally tagged with ymScarlet-I under the control of the *HIS3* promoter), two DNA fragments carrying either the ~0.7-kb ymScarlet-I insert or a ~6.5-kb plasmid backbone were amplified via PCR using the plasmid YIp128-proACT1-lifeact-ymScarlet-I-tADH1 (lab stock, integrative, *LEU2*, expresses Lifeact C-terminally tagged with ymScarlet-I under the control of the *ACT1* promoter) as the template DNA and the primers P1409 and P1412, or using the plasmid bWL715 as the template DNA and the primers P1410 and P1411, respectively. The resultant PCR products were then assembled using a Quick-Fusion cloning kit (Bimake, Houston, TX, USA). To generate pRS305-ENVY-FKS1(1-789), a ~4.2-kb DNA fragment carrying the partial open reading frame (ORF) of *GFP*^{Envy}*-FKS1* (from ~1 kb of the *FKS1* promoter region, *GFP*^{Envy}, and the *FKS1* ORF until residue 789 followed by a new stop codon) was amplified via PCR using pRS316-ENVY-FKS1(1-789) as the template DNA and the primers P222 and P512. The resultant PCR product was then subcloned into *Apa*I- and *Sac*I-digested pRS305 (integrative, *LEU2*) using a Quick-Fusion cloning kit.

2.5.4. Imaging and Data Analysis

Time-lapse microscopy was conducted as described previously with slight modifications [31]. Cells were cultured to an exponential phase at 25 °C in SC medium, briefly sonicated at 15% power for 5 s to declump the cells (model Q55; Qsonica, Newtown, CT, USA), concentrated via centrifugation, and spotted onto concanavalin A-coated glass-bottom dishes. After a sufficient amount of cells had adhered to the bottom of each dish (> 50% cell cover in a microscopic field), the SC medium was replaced with neutralized SC liquid medium, and the dishes were then incubated at room temperature (23 °C) for

15 min to allow the cells to acclimatize. Images were acquired at room temperature with a spinning-disk confocal microscope (Eclipse Ti2-U; Nikon, Tokyo, Japan) with a 100× /1.49NA oil objective (CFI Apo TIRF 100×; Nikon) combined with a confocal scanner unit (CSU-X1; Yokogawa, Tokyo, Japan). An EMCCD camera (Evolve 512 Delta; Photometrics, Tucson, AZ, USA) was used for image capturing. Solid-state lasers for excitation (488 nm for GFP, and 561 nm for red fluorescent protein) were housed in a laser merge module (ILE-400; Spectral Applied Research, Richmond Hill, ON, Canada). The imaging system was controlled using MetaMorph (version 7.10.4.431, Molecular Devices, San Jose, CA, USA). Images were taken every 2 min with 11 z-stacks with a step size of 0.8 μm. Sum or maximum intensity projections were calculated using NIH ImageJ (1.51 h) [32]. To quantify fluorescence intensities, the integrated density at a division site was calculated from the sum intensity projection of an image stack by subtracting the fluorescence intensity in the background area from the total intensity in an ImageJ-drawn polygon covering the division site.

2.6. Biochemistry

Whole-cell protein extracts were prepared as described previously [33]. Briefly, cells were pelleted, treated with NaOH (0.1 N), and incubated on ice (5 min). Then, cells were pelleted, resuspended in SDS sample buffer including 62.5 mM Tris-HCl (pH 6.8), glycerol (10%), SDS (2%), β-mercaptoethanol (2%), and bromophenol blue (0.005%), boiled for 5 min, and pelleted. Afterward, the supernatants were loaded in a mini-gel (4–15%; Bio-Rad, CA, USA), and Western blotting was performed with rabbit anti–phospho-p42/44 MAPK (T202/Y204) antibody (Cell Signaling Technology, Danvers, MA, USA) and rabbit anti-yeast Rho1 antibody (Abmart, Berkeley Heights, NJ, USA). HRP-conjugated secondary antibodies were obtained from Millipore, and proteins were detected with an enhanced chemiluminescence system (ECL plus; Amersham, Darmstadt, Germany).

2.7. Similarity of Mannoprotein Mutants and Drug-Treated Wild-Type Cells in Morphology

Morphological profiles of ccw12Δ (I), ccw14Δ (II), cwp2Δ (IV), sun4Δ (V), ecm33Δ (VI), and sag1Δ (VII) were compared with WT cells treated with unicamycin, echinocandin B, nikkomycin Z, and hydroxyurea. Morphological data of the drug-treated cells were obtained from [34]. To investigate the profile similarity, first, CalMorph values were transformed to Z-values (Wald test) using the UNIMO pipeline (490 parameters). Then, the obtained Z-values of the WT replicates were exposed to PCA. Finally, Z-values of the mutants/drug-treated cells were projected onto PC axes of the WT. Pearson correlation coefficient (r) was calculated between each pair using first 94 PC scores (CCR = 99%).

2.8. Mannoprotein Analysis Based on Omics Studies

2.8.1. Estimation of Fitness

To estimate the fitness of 32 mannoprotein mutants, a previously reported dataset containing the logarithmic strain growth rate coefficients of gene-deletion mutants grown on basal medium (LSC_{basal}) was employed [10]. *p*-values were calculated to determine whether the fitness of each strain was significantly lower than that of the WT based on one tail of the estimated probability distribution, as described previously [12], using the pnorm function (stats package), and the results were corrected for family-wise error using the qvalue function in the qvalue package [35].

2.8.2. Analysis of Protein Abundance and Protein–Protein Interactions

To determine the abundance of 32 mannoproteins at the protein level, mean values from 21 datasets were used as reported previously [36]. Protein–protein interaction (PPI) data were obtained from the BioGRID database [37]. We examined physical interactions (between interactors A and B) for each mutant of *S. cerevisiae* S288C (Taxonomy ID: 559292). Two types of PPI networks were considered: PPIs among the 32 mannoproteins and PPIs between each of the 32 mannoproteins and the whole proteome (i.e., the protein–protein interactome profile). Networks were visualized using Cytoscape 3.8.2 [38].

2.8.3. Genetic Interaction Analysis

Genetic interaction (GI) data were collected as reported previously [39]. Significant interactions based on both queries and array analysis were considered for further analysis ($p < 0.05$). Two types of GI networks were considered: GIs among the 32 mannoprotein genes and GIs between each of the 32 mannoprotein genes and the whole genome (i.e., the genetic interactome profile). Finally, networks were visualized using Cytoscape 3.8.2 [38].

2.8.4. Chemical-Genetic Profile Analysis

The chemical-genetic profiles of the 32 mannoproteins were obtained through text mining of the *Saccharomyces* Genome Database (SGD).

3. Results

3.1. Effects of Genetic Perturbations on Cell Morphology

We analyzed the morphology of mutants with deletions of individual genes encoding 32 mannoproteins using the image processing program CalMorph. To perform morphological phenotyping, it is necessary to consider the diversity in yeast morphological measurements. We applied different probability distribution models to accurately estimate the true value of each morphological parameter [17]. The use of 490 unimodal morphological parameters enabled a powerful approach, revealing biological information that may be masked with commonly used imaging methods. We found that cell morphology was remarkably altered: of 490 parameters, perturbations were detected in 136 parameters, consisting of 16, 77, and 43 parameters related to actin, cell, and nuclear DNA morphology, respectively (Wald test, FDR = 0.05; Table S4). This observation implies profound effects of mannoproteins on cell morphology, suggesting that mannoproteins may play roles in dictating cell shape and the progression of the cell cycle.

To understand the morphological alterations more holistically, we reduced the number of dimensions of the morphological space to five via PCA of the Z-values of 136 significantly changed parameters; the first five principal components accounted for 81.34% of the variation (Figure S1A and Table S5). We then used GMM, one of the most commonly used model-based clustering methods for normally distributed data, to cluster the mannoprotein mutants (Figure S1B). The posterior probabilities associated with the data were evaluated in our GMM analysis to validate our clustering results (Figure S1C). Using GMM, we successfully clustered 32 mannoprotein mutants into seven groups (Figure 1). The *ccw12*Δ mutant, the single member of cluster I, was the mutant with the most abundant covalently linked cell wall protein. Members of cluster II (*ccw14*Δ, *flo11*Δ, *srl1*Δ, and *tir3*Δ) were mutants of serine-rich mannoproteins. Cluster III (nine mutants) and IV (13 mutants) accounted for more than half of the mannoprotein mutants, with their members exhibiting no or very small effects on cell morphology. Members of cluster V (*dse2*Δ, *egt2*Δ, and *sun4*Δ) were endoglucanase mutants. The *ecm33*Δ mutant in cluster VI had a mutation in a glycosylphosphatidylinositol (GPI)-anchored protein thought to be involved in bud morphogenesis. In the single member of cluster VII, the *sag1*Δ mutant, no morphological defects in vegetative growth had been reported previously.

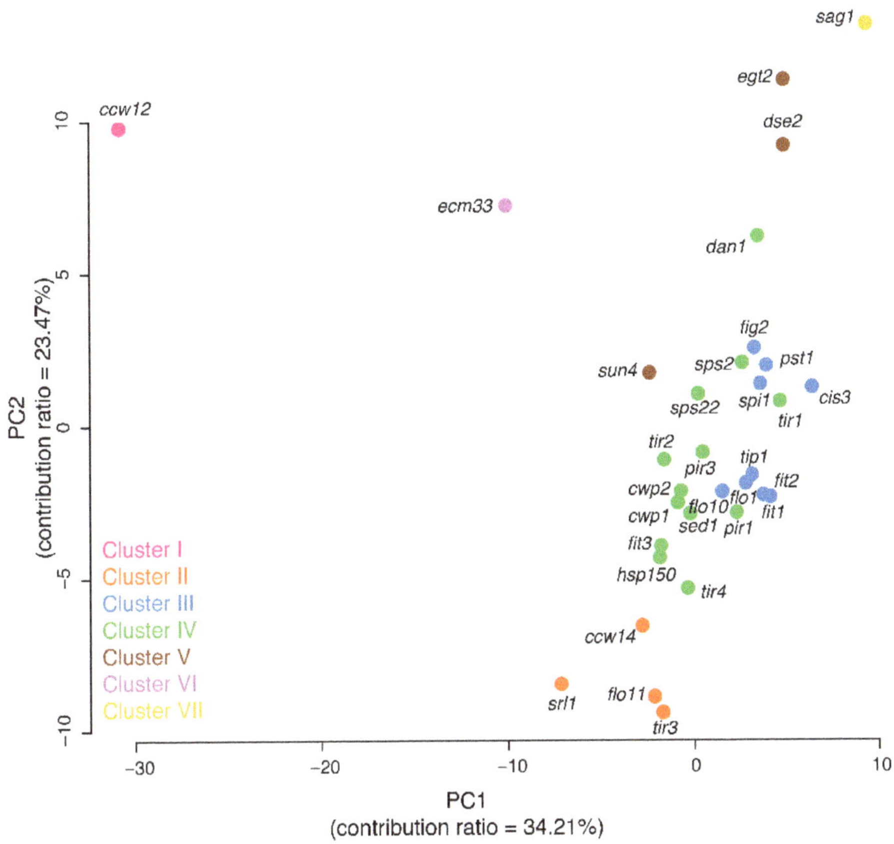

Figure 1. Two-dimensional principal component (PC) analysis score plot (biplot) illustrating clustering of the 32 mannoproteins. Each circle represents a single mannoprotein deletion mutant. The mixture likelihood values at individual points, based on the first five PC scores (CCR = 81.34%; Figure S1A) and a seven-component EEI model (Figure S1B), revealed data trends, including seven clusters. Mutants are color-coded.

3.2. Phenotype of the Cluster I Mutant (*ccw12*Δ)

Among all mannoprotein mutants, *ccw12*Δ cells exhibited the greatest morphological alterations with 81 significantly changed parameters (Wald test, FDR = 0.05; Table S6). The *ccw12*Δ cells were larger in size at the S/G2 (C11-1_A1B and C101_A1B) and M phases (C11-1_C) and had a rounder cell shape (C115_A, C115_A1B, and C115_C) and wider neck at both the S/G2 (C109_A1B) and M phases (C109_C) (Figure S2). In addition, the *ccw12*Δ mutation affected bud morphogenesis, resulting in a rounded bud shape (C114_A1B and C114_C) and a disturbed budding direction (C106_A1B and C106_C) (Figure S3). Further phenotypic analysis using chitin staining revealed a significantly elevated population of cells exhibiting abnormal chitin staining ($p < 0.05$, *t*-test, Figure 2A,B), demonstrating that the loss of *CCW12* function had a detrimental impact on cell wall organization and assembly.

Cell wall damage is accompanied by the activation of the cell wall integrity (CWI) pathway and the phosphorylation of Slt2 MAPK [40–42]. We found that the *ccw12*Δ mutant exhibited a marked increase in Slt2 phosphorylation, indicating that the cell wall was damaged in this mutant (Figures 3 and S4).

Taken together, and given that Ccw12 is important for CWI, these observations indicate that this gene deletion causes pleiotropic defects in cell growth and morphology, possibly because of a severe loss of mannoprotein structures and functions.

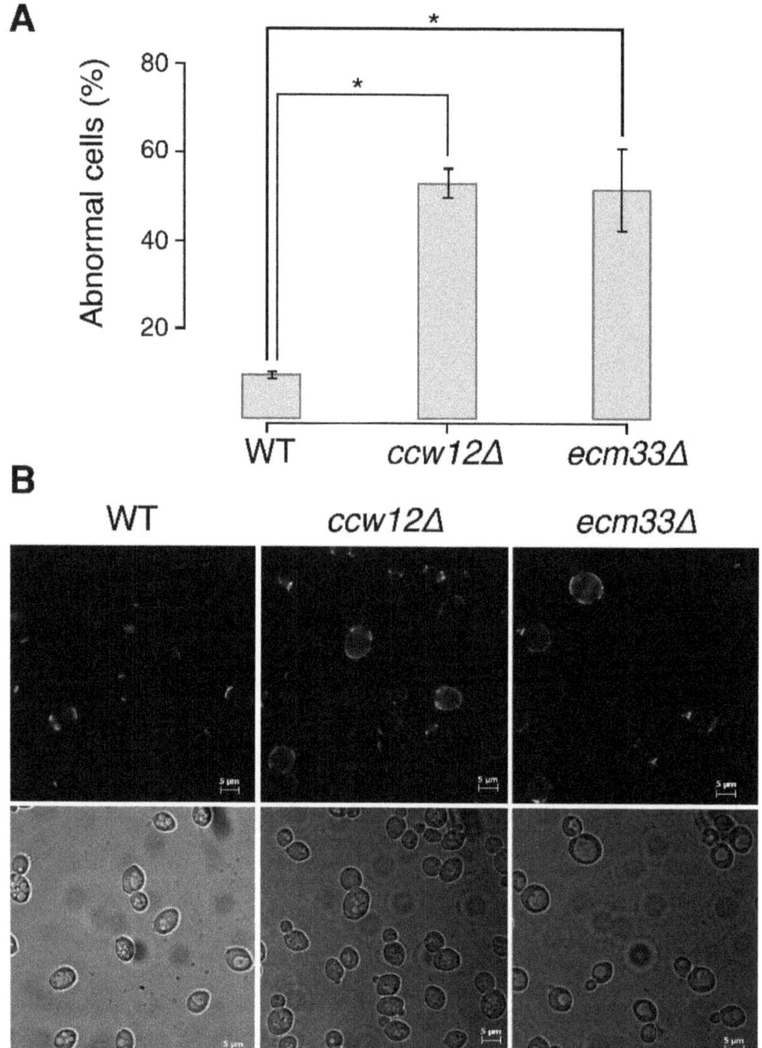

Figure 2. Abnormalities in *ccw12*Δ (cluster I) and *ecm33*Δ (cluster VI) cells. (**A**). Wild-type (WT) and mutant cells were grown in yeast extract peptone dextrose medium at 25 °C with shaking at 200 rpm until *log* phase. Cells (2.0×10^6 cells) were suspended in 1 mL of phosphate-buffered saline (PBS) and mixed well with 5 µL of 5 mg/mL wheat germ agglutinin in PBS to stain chitin. After incubation at room temperature (30 min), the stained cells were washed three times and observed under a fluorescence microscope with a 4′,6-diamidino-2-phenylindole filter. The bar plot shows the percentages of abnormal *ccw12*Δ (cluster I) and *ecm33*Δ (cluster VI) cells in comparison with WT cells. Error bars indicate standard deviations. * $p < 0.05$ (t test). (**B**). Examples of chitin staining in WT, *ccw12*Δ (cluster I), and *ecm33*Δ (cluster VI) cells.

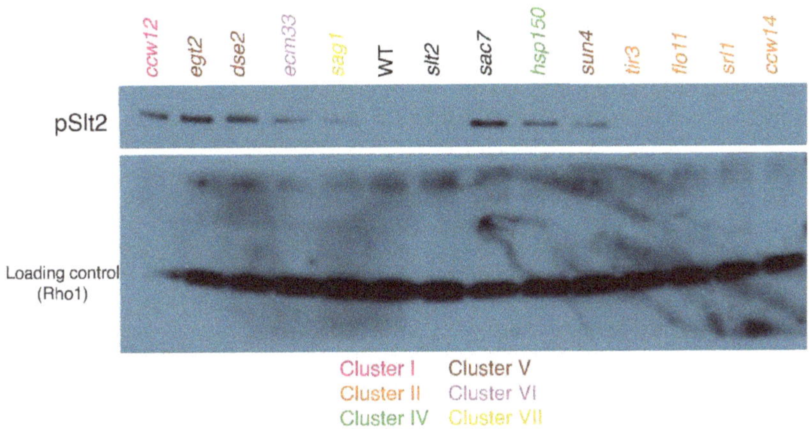

Figure 3. Western blotting of phosphorylated Slt2 (pSlt2, upper panel) and loading control Rho1 (lower panel). BY4743 (WT); *ccw12Δ/ccw12Δ* (cluster I); *ccw14Δ/ccw14Δ*, *flo11Δ/flo11Δ*, *srl1Δ/srl1Δ*, and *tir3Δ/tir3Δ* (cluster II); *hsp150Δ/hsp150Δ* (cluster IV); *dse2Δ/dse2Δ*, *egt2Δ/egt2Δ*, and *sun4Δ/sun4Δ* (cluster V); *ecm33Δ/ecm33Δ* (cluster VI); and *sag1Δ/sag1Δ* (cluster VII) cells were examined for the presence of phosphorylated Slt2. Rabbit antibody against phospho-p42/44 MAPK (T202/Y204) and rabbit antibody against yeast Rho1 were used to detect the phosphorylated Slt2 and Rho1, respectively. *slt2Δ/slt2Δ* and *sac7Δ/sac7Δ* were used as negative and positive controls, respectively, for phosphorylated Slt2. Mutants are color-coded according to Gaussian mixture model clustering of morphological data (see Figure 1).

3.3. Phenotype of Cluster II Mutants (ccw14Δ, srl1Δ, flo11Δ, and tir3Δ)

Cluster II mutants tended to produce larger mother cells at the M phase. The most noticeable morphological mutant in this cluster was *srl1Δ*, which had a significantly larger mother cell size (C11-1_C), mother cell outline length (C12-1_C), and long axis (C103_C; Wald test, FDR = 0.05; Table S6). Both the mother cell outline length (C12-1_C) and long axis length (C103_C) of all cluster II mutants were larger than those in the other clusters, and nearly equivalent to those of the cluster I mutant (*ccw12Δ*) (Figure S5). Therefore, we considered that the cluster II mutants exhibited perturbations in the mother cell size and shape at the M phase. There was no obvious increase in Slt2 phosphorylation, suggesting little cell wall damage in the cluster II mutants (Figures 3 and S4).

3.4. Phenotype of Cluster V Mutants (dse2Δ, egt2Δ, and sun4Δ)

Among the cluster V mutants, *egt2Δ* exhibited the greatest morphological changes, with significant differences in 31 parameters (Wald test, FDR = 0.05, Table S6). Morphological analysis of the cluster V mutants revealed common morphological features, such as the accumulation of cells at the M phase (D202 and D213) with actin patches localized at the bud neck (A109 and A118) (Figure S6). As actin patches are localized to the bud neck in cytokinesis, the morphological features of the cluster V mutants are suggestive of defects in cell separation. Consistent with this, cluster V genes (*DSE2*, *EGT2*, and *SUN4*) all encode cell wall mannoproteins similar to glucanase. It should be noted that the mutants exhibited no significant changes in bud cell size (C11-2_C and C12-2_C) or nuclear size (D14-2_C and D17-2_C) (Figure S7A), suggesting no defects in cell division but defects in physical attachment between mother and daughter cells. Mother cells frequently started the next budding cycle while still attached to old daughter cells (Figure S7B). The phosphorylation of Slt2 was increased in all cluster V mutants, suggesting cell wall damage (Figures 3 and S4).

Glucanases are localized at the site of division in cytokinesis. To understand the precise timing of the function of glucanases in cell separation, we tagged cluster V gene products with GFP and performed quantitative time-lapse imaging to obtain information on real-time protein abundance at the division site (Figure 4) [28]. The accumulation peaks of both GFP-Egt2 and Dse2-GFP occurred after those of two secondary septum (SS)-forming enzymes, GFP-Fks1 and Chs3-GFP, suggesting that cluster V genes likely function after SS formation. Dse2-GFP exhibited accumulation kinetics remarkably similar to those of Cts1-GFP ($r = 0.98$, Table S7), a chitinase required for the degradation of the primary septum (PS) during cell separation. These observations imply that Dse2 may function in the same process as Cts1. Interestingly, the peak of GFP-Egt2 at the division site occurred between the peaks of the SS-forming enzymes and the peak of the PS-degrading enzyme, suggesting that Egt2 may be involved in cell wall remodeling or maturation, which is required for cell separation. Taken together, these results further support the involvement of cluster V genes in cell separation and explain the major cluster V mutant phenotype of mother cells with unseparated old daughter cells.

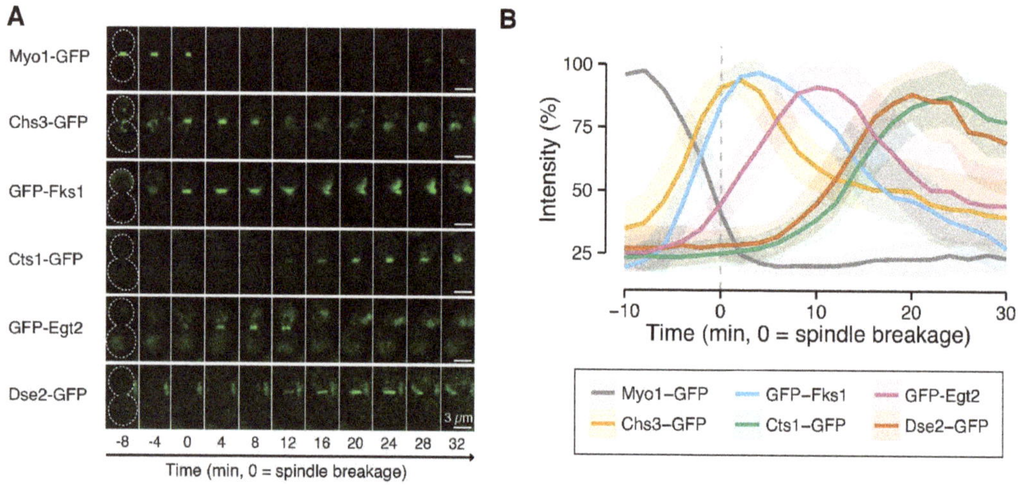

Figure 4. Kinetics of Dse2 and Egt2 (cluster V) proteins and proteins involved in cytokinesis and cell separation. (**A**). Images of green fluorescent protein (GFP)-tagged cluster V proteins and proteins involved in cytokinesis and cell separation. Montages of cells were created from frames selected from time-lapse series consisting of images taken at 2-min intervals. The white dotted line represents the cell outline. The following strains were used: YEF10861 (*MYO1-GFP mScarlet-TUB1*), YEF10856 (*CHS3-GFP mScarlet-TUB1*), YEF10857 (*GFP-FKS1 mScarlet-TUB1*), YEF10862 (*CTS1-GFP mScarlet-TUB1*), YEF10879 (*GFP-EGT2 mScarlet-TUB1*), and YEF10858 (*DSE2-GFP mScarlet-TUB1*). (**B**). Kinetics of the GFP-tagged proteins indicated in (**A**). The vertical dashed line shows timing of spindle breakage. Bold lines and associated shaded bands represent mean and SD values, respectively. $n > 23$ for each strain.

3.5. Phenotype of the Cluster VI Mutant (ecm33Δ)

The *ecm33*Δ cells exhibited significant differences in 22 morphological parameters (Wald test, FDR = 0.05; Table S6) and were characterized by round mother cells (C115_A1B and C115_C), an altered neck position (C105_A1B and C105_C), and altered bud direction (C106_A1B and C106_C) (Figure S8). In addition, a reduced region of actin at the neck during the M phase (A9_C) and a lower proportion of cells exhibiting an isotropic pattern of actin (A117) were observed, suggesting that the defects in this mutant manifest before isotropic bud growth (Figure S8A). Consistent with these observations, the bud/mother cell size ratio (C118_C) (Figure S9) and ratio of cells with a large bud within budded cells (C125_C; Table S6) were both significantly decreased in *ecm33*Δ. We observed a uniform distribution of chitin on the *ecm33Δ* cell surface (Figure 2B). The phosphorylation of Slt2 was increased in *ecm33Δ*, suggesting cell wall damage in the cluster VI mutant (Figures 3 and S4). These findings suggest possible roles of *ECM33* in bud growth and cell wall assembly.

3.6. Phenotype of the Cluster VII Mutant (sag1Δ)

The *sag1Δ* cells exhibited significant differences in 28 morphological parameters (Wald test, FDR = 0.05). The sag1Δ mutation caused a smaller cell size at the G1 phase (C11-1_A, related to C103_A, C104_A, and C12-1_A) (Figure S10A and Table S6). Accordingly, the nucleus was also smaller at the G1 phase in *sag1Δ* cells (D102_A, D14-1_A, and D179_A) (Table S6). We observed the same trend (smaller bud size) at the M phase (C11-2_C, related to C107_C, C108_C, C12-2_C, C102_C, and C101_C) (Figure S10B and Table S6). Moreover, delocalized actin patches were observed frequently in *sag1Δ* cells (A111 and A112) (Figures S10C and S11), suggesting the perturbation of actin polarization and polarized bud growth. The size of the actin region in *sag1Δ* was more heterogeneous at the S/G2 phase (ACV7-1_A1B). We observed increased phosphorylation of Slt2 in *sag1Δ*, suggesting cell wall damage in the cluster VII mutant (Figure 3 and Figure S4). Although *SAG1* is thought to play an important role only in the mating aggregation process [43,44], this is the first study revealing its effects on cell morphology during the vegetative growth phase.

3.7. Mannoprotein Gene Duplication

Many mannoprotein genes have been generated by gene duplication (Table S1 and Figure S12). Therefore, the effects of gene duplication were examined by measuring its impact on the morphological phenotype of each mutant (Table S8). More than 80% of the mutants with duplicated genes belonged to clusters III and IV and exhibited no obvious changes in their morphological phenotypes. The remaining *ccw12Δ* (I), *ecm33Δ* (VI), and *tir3Δ* (II) mutants exhibited changes in the morphological phenotype, but no obvious changes were observed in the deletion mutations of their counterparts. This is probably because gene duplication can result in functional bias. On the other hand, among strains with deleted genes unrelated to gene duplication, a significantly lower percentage of the mutants exhibited no obvious changes in the morphological phenotype (Table S8). Taken together, the duplication of mannoprotein genes resulted in a reduction in their functional effects, which made it difficult to examine the morphological phenotype of these gene-deletion strains.

3.8. Comparisons of Morphology and Fitness among Mannoprotein Mutants

Associations between the comprehensive morphological phenotypes of the 32 mannoprotein mutants and the fitness of these mutants were assessed. Our morphological analysis including 490 morphological parameters revealed 12 mannoprotein mutants with significant abnormalities in at least one morphological parameter (Wald test, FDR = 0.05; Table S6). On the other hand, the fitness analysis of the gene-deleted strains revealed only one mutant (*ccw12Δ*) with a significantly decreased growth rate in normal medium (Wald test, FDR = 0.05) (Figure 5A). The *ccw12Δ* mutant exhibited the greatest changes in its morphological phenotype. More differences were found in the morphological phenotype among the mutants than in fitness aspects, probably because of the high sensitivity of morphological phenotyping [12]. The morphological phenotype was also considered to be more greatly affected by the disruption of cell wall proteins.

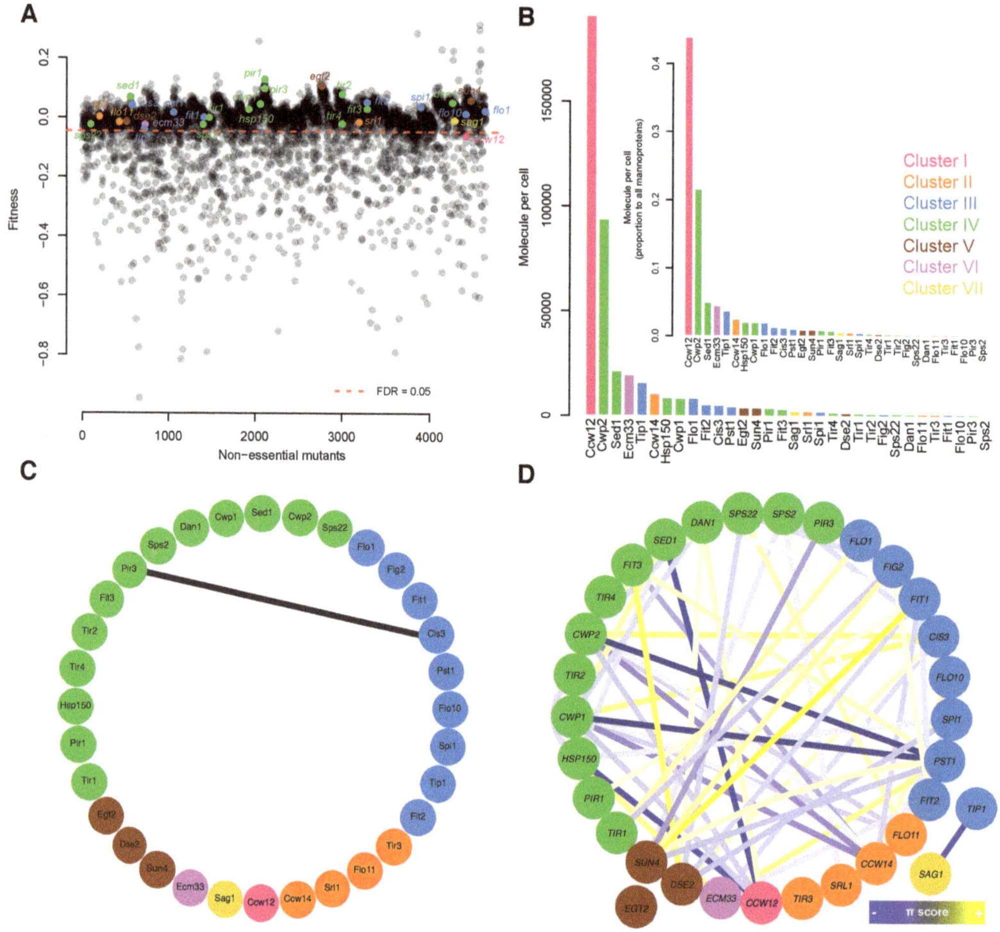

Figure 5. Mannoprotein analyses based on omics studies. (**A**). Scatter plot representing fitness-related defects. The dashed red line indicates a false discovery rate (FDR) of 5%. Data are from Warringer et al. [10]. Fitness data for *ccw14Δ* were not available in the dataset. (**B**). Bar plot showing the average cell wall mannoprotein abundances. Inset: A subset of the data. Data were obtained from [36]. (**C**). Protein–protein interactions among mannoprotein proteins are shown. Black line shows physical interaction. Data are from Oughtred et al. [37]. (**D**). Genetic interactions (GIs) among mannoprotein genes presented as blue (negative GI) or yellow (positive GI) lines ($p < 0.05$). Data are from Costanzo et al. [39]. *EGT2* did not have any significant GIs. In all sections, mutants are color-coded according to Gaussian mixture model clustering of morphological data (see Figure 1).

3.9. Comparisons of Mannoprotein Mutants and Glycosylation-Defective Cells in Morphology

The remarkable differences in morphological phenotype found for *ccw12Δ* can be explained in terms of protein expression levels (Figure 5B). Yeast cells contain approximately 190,000 Ccw12 protein molecules per cell, accounting for more than 40% of all mannoproteins. The second most highly expressed mannoprotein is Cwp2, with approximately 93,000 molecules expressed per cell. As no morphological abnormalities were detected in *cwp2Δ*, the expression level of a mannoprotein originally expressed at a high level would more markedly affect the morphology.

In order to know which cell wall metabolic pathways are relevant to mannoprotein function, we compare the morphology of *ccw12Δ* with those of the cells treated with cell wall agents. For this purpose, we used tunicamycin, echinocandin B, nikkomycin Z, and hydroxyurea, which affect protein glycosylation, 1,3-β-glucan synthesis, chitin synthesis, and DNA replication, respectively (Figure S13). We found that *ccw12*Δ is the most similar to the tunicamycin-treated cells ($r = 0.813$), implying a close relationship between protein glycosylation and mannoprotein function. *ccw12*Δ was also similar to the echinocandin B-treated cells ($r = 0.723$), but not similar to the cells treated with nikkomycin Z ($r = 0.347$) or hydroxyurea ($r = 0.315$). Taken together, these observations indicate that defects in *CCW12* resulted in serious damage to yeast cells, similar to defects in protein glycosylation and 1,3-β-glucan, which is the main filamentous component of the yeast cell wall.

3.10. Comparison of Morphological Clustering Results with Those from Analyses of Other Omics Data

We compared our clustering data with other omics data on interactions. A survey of comprehensive data on PPIs identified only one unidirectional interaction between Pir3 and Cis3 (Figure 5C). However, neither *pir3*Δ nor *cis3*Δ exhibited detectable changes in the morphological phenotype in the present study. Therefore, we could not infer the biological significance of the interaction between these two proteins based on morphological phenotyping. In addition, studying the PPI profile at the proteomic level did not reveal any similar patterns of PPI frequency among members of the same cluster (Figure S14A,B and Table S9). There were no associations between interactome profiles in each cluster either (Figure S14C).

With regard to GIs, we identified 26 positive and 42 negative relationships among the 32 mannoprotein genes. As with the PPI network, GIs among the 32 mannoprotein genes could not be directly linked to molecular functionality (Figure 5D). However, the lack of detectable morphological and fitness defects in many of the individual mannoprotein mutants may be explained by negative GIs comprising more than half (~61.7%) of all GIs. The lack of defects may be due to the existence of parallel pathways with the same or similar biological functions, such as the preservation of the cell wall structure. There were no noticeable GI patterns based on frequency of an interactome profile (Figure S15A,B and Table S10) or correlations between the members of a given cluster (Figure S15C).

Perturbations upon exposure to 106 different chemical compounds were tested in mannoprotein mutants, and the data are summarized in the SGD (Table S11). The chemical-genetic profiles of the mannoproteins were then visualized as a scatter plot in two-dimensional space representing the deletion mutants and chemicals (Figure S16). The comparison of the frequency of each mutant revealed that the chemical response phenotypes of *ccw12*Δ and *ecm33*Δ have been frequently tested. Of the seven clusters, only members of cluster V (*dse2*Δ, *egt2*Δ, and *sun4*Δ) exhibited similar fitness defects with (S)-lactic acid (5.1% *w*/*v*) and miconazole (1000 μg/mL). Otherwise, the results of chemical-genetic profiling did not appear to be linked to molecular functionality.

Morphological phenotyping of the mannoprotein mutants clearly accentuated unique aspects of the functional network that cannot be identified using other omics technologies. Thus, morphological phenotyping, as a complementary tool, provides deeper knowledge on cell wall organization, remodeling, and protein function. We succeeded in clustering 32 mannoproteins into seven groups based on their morphology and elucidated their specific functions in the cell (Figure 6).

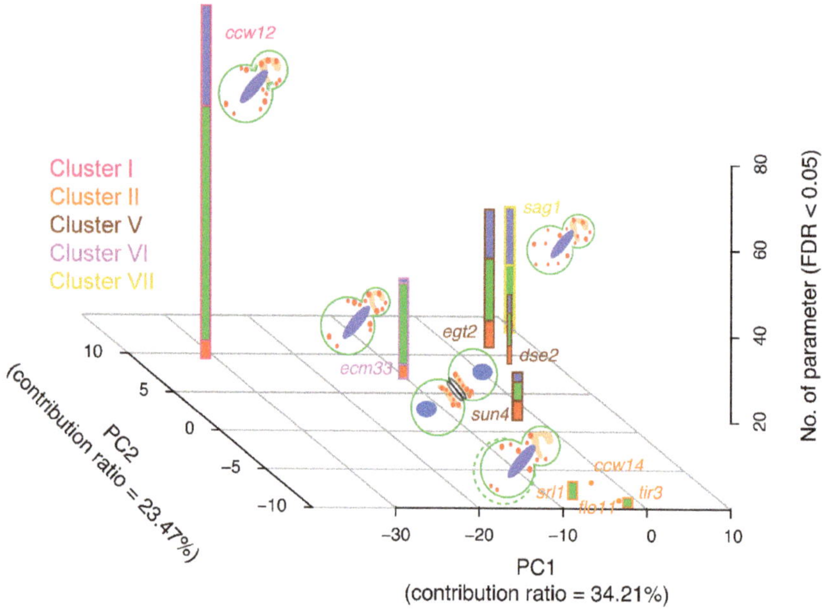

Figure 6. Schematic representation of the mannoprotein responsibility matrix. Stacked bars show the numbers of disturbed CalMorph parameters (Wald test, FDR = 0.05) related to actin, the cell wall, and the nucleus (illustrated in red, green, and blue, respectively) in mutants of clusters I, II, V, VI, and VII. Prominent implications of morphological defects caused by each mutation are illustrated in a small budding yeast cell where actin, the cell wall, and the nucleus are shown in red, green, and blue, respectively. Green dashed circle represents larger mother cell. Mutants are color-coded according to Gaussian mixture model clustering of morphological data (see Figure 1).

4. Discussion

In this study, we used high-dimensional morphological phenotyping to gain a system-level understanding of 32 cell wall mannoproteins in *S. cerevisiae*. We found 12 mannoprotein mutants with significant abnormalities in at least one morphological parameter. Nearly 30% of the 490 unimodal morphological parameters examined were affected in the mannoprotein mutants, implying distinct roles of mannoproteins in cell morphology. Multivariate analysis revealed seven groups of mutants categorized according to the effects of the mutation on their functions. The results indicate that high-dimensional morphological phenotyping of mannoprotein mutants is an effective approach for determining the responsibility matrix of yeast mannoproteins, which is difficult to obtain with other omics technologies.

4.1. Ccw12 Is a Major Cell Wall Stabilizer

The highly pleiotropic morphological defects of *ccw12*Δ cells, including the wide neck, a typical phenotype of cell wall mutants [45], and altered cell shape for both mother and daughter compartments [46], clearly indicated the important role of Ccw12 as a major structural component of the cell wall [47]. This small (133 amino acid residues) and highly glycosylated GPI-anchored protein has been previously shown to impact the maintenance of newly synthesized areas of the cell wall [13] and cell fitness [10]. In addition, *ccw12*Δ has been reported to affect 473 genes acting in various cellular pathways, including 32 genes directly involved in the construction and remodeling of the cell wall [47]. Here, we confirmed that *ccw12*Δ cells exhibited the most significant morphological defects with differences found for 81 parameters. Ccw12 is localized at the presumptive budding site, around the bud, and at the septum [47], which explains the defect in the neck width of *ccw12*Δ. An abnormal round cell morphology was also reported previously for *ccw12*Δ cells [13]. The defect in *CCW12* impacted another component of the cell wall because staining using wheat germ agglutinin (WGA) revealed the abnormal localization of chitin. Whereas chitin is located at the budding site in the WT strain, a uniform distribution of chitin on the cell surface was observed in *ccw12*Δ cells. Taken together, these results further confirmed that Ccw12 plays a major role in the maintenance of a rigid cell shape and the stabilization of the cell wall structure.

4.2. Cluster V Member Genes Encode Endoglucanases

After cytokinesis, mother and daughter cells undergo cell separation, which requires enzymatic digestion of the cell wall [48,49]. Dse2 is a well-known hydrolytic enzyme (glucanase) that functions exclusively in efficient cell separation from the daughter cell side [49,50]. Other cluster V member genes (*EGT2* and *SUN4*) have also been reported to encode glycosidases, and our results clearly showed that genetic perturbation prevented efficient daughter cell separation in all cluster V mutants. Consistent with the mutant phenotypes, co-localization studies have revealed that Dse2, Egt2, and Sun4 form a complex at the birth scar [51]. Cluster V mutants exhibited no defects in cell-cycle progression or daughter cell growth in the next cell cycle, indicating lesser effects of these genes in cell proliferation. Due to redundancy arising from intertwining pathways and the proteins involved, it was not clear how precisely diverse cell wall digestion systems are integrated to achieve effective cell separation; for example, *SUN4* genetically interacts negatively with some septin construction genes, including *CDC11* and *CDC12*, making its role in cell separation complex. Interestingly, our kinetic analysis revealed the temporal order among glucanases/chitinases. Dse2, Egt2, and Cts1 were deposited at the division site after septum synthesis was completed. However, Egt2 preceded Dse2 and Cts1. Therefore, Egt2 may be involved in cell wall maturation and making the wall architecture conducive for cell separation, whereas Dse2 and Cts1 are septum-hydrolyzing enzymes that arrive at the division site during the last step of cell separation. This finding suggests that glucanase- and chitinase-mediated cell separation is accomplished in a stepwise process. Consistent with the above observations, the expression of cell-separation genes is also regulated in a strict temporal order [52], as observed in our kinetic analysis. Early enzymes, such as Egt2, may function to remodel the cell wall or septum structure to facilitate the delivery of Cts1 to the PS [49].

4.3. ECM33 Plays a Role in Bud Growth

The molecular function of Ecm33 has not been fully elucidated. Previous studies suggested that it may play roles in determining cell shape [7], cell wall biogenesis [53,54], and apical growth [55]. Consistent with those previous reports, we confirmed that the roundness of mother cells (C115) and bud site selection (C106) were perturbed at both the S/G2 and M phases in *ecm33*Δ. *ECM33* also has strong negative GIs with mannosyltransferase genes including *MNN11*, *ANP1*, and *HOC1*, which can explain the role of *ECM33* in cell wall assembly.

The smaller proportion of *ecm33*Δ cells exhibiting an isotropic pattern of actin suggests that *ECM33* functions before isotropic bud growth. However, apical bud growth seemed normal because the long and short axis lengths of the buds as well as their ratio were not significantly altered in *ecm33*Δ cells. Therefore, one possibility is that the apical and isotropic bud growth switch is delayed in the mutant. It has also been reported that *ECM33* deletion triggers the activation of the CWI pathway through the phosphorylation of Slt2 [54]. Although the CWI pathway is involved in cell-cycle checkpoints, such as the cell wall integrity checkpoint and cell morphological checkpoint, it is unlikely that any cell-cycle checkpoints were activated because cell-cycle progression appeared to be normal in *ecm33*Δ cells. However, further studies are needed to determine how Ecm33 impacts cell-cycle progression.

4.4. SAG1 Deletion Perturbs Actin Distribution during Vegetative Growth

SAG1 (*AGα1*) encodes a cell-adhesion molecule called α-agglutinin in *MATα* cells [42,55], but the function of this molecule during the vegetative growth of *MATa* cells has yet to be identified. In this study, we examined the morphological phenotype of *sag1Δ MATa* cells. The results showed that the *sag1Δ* mutation affects the mother and bud cell sizes at the G1 and M stages of the cell cycle, respectively, in *MATa* cells. It also perturbed actin polarization and polarized bud growth. As Sag1 binds directly to Aga1, it would be interesting to investigate the phenotype of *aga1Δ*. However, it was difficult to analyze *aga1Δ* because the mutant cells were not suitable for morphological phenotyping due to their propensity to aggregate. The construction of weak alleles of *AGA1* will be necessary to examine the morphological phenotype and investigate its relationship with Sag1.

5. Conclusions

This study provided a comprehensive analysis of morphological phenotypes of yeast mannoprotein mutants. The morphology of each cluster of mutants could be explained by the molecular functions of the mannoproteins. The cluster I gene (*CCW12*) encodes a mannoprotein that accounts for 40% of the total mannoproteins in a cell, plays a major structural role, and contributes the most to cell morphogenesis. The cluster II genes (*CCW14, FLO11, SRL1*, and *TIR3*) do not play structural roles but have similar effects on cell size and cell shape. The cluster V genes (*DSE2, EGT2*, and *SUN4*) encode glucosidases, which are required for cell separation. The cluster VI gene (*ECM33*) is required for bud growth and cell wall assembly. Finally, the cluster VII gene (*SAG1*) is required for cell aggregation and is important for determining cell size and actin organization. Cluster III and cluster IV genes do not play major roles in cell morphogenesis. The results presented here increase our understanding of the mechanistic and functional roles of glycoproteins in cell morphogenesis. Morphology-based analysis seems to be a practical means of relating morphological defects to underlying molecular mechanisms, indicating the sensitivity of our approach for determining the responsibility matrix of mannoproteins regarding maintaining the cell wall structure.

Supplementary Materials: The following are available online at https://www.mdpi.com/article/10.3390/jof7090769/s1, Figure S1. Multivariate analysis of mannoprotein morphological data, Figure S2. Specific morphological features of ccw12Δ cells (cluster I), Figure S3. Specific morphological features of ccw12Δ buds (cluster I), Figure S4. Complete gel of western blotting of phosphorylated Slt2, Figure S5. Specific morphological features of mother cells of all mutants in cluster II, Figure S6. Specific morphological features shared among members of cluster V, Figure S7. Morphological defects in cluster V do not affect the cell cycle, Figure S8. Specific morphological features of ecm33Δ (cluster VI), Figure S9. Morphological parameters related to cell size in ecm33Δ (cluster VI) versus other mannoprotein mutants, Figure S10. Specific morphological features of sag1Δ (cluster VII), Figure S11. Specific actin-related morphological features of sag1Δ (cluster VII) in other mutants, Figure S12. Gene homology among mannoproteins, Figure S13. Morphological similarity of mannoproteins and drug-treated wild-type cells, Figure S14. Protein–protein interaction (PPIs) network, Figure S15. Genetic interaction network, Figure S16. Chemical-genetic profile of mannoproteins, Table S1. List of 36 cell wall mannoprotein strains, Table S2. List of strains to study kinetics of mannoproteins (Dse2 and Egt2; cluster V) and proteins involved in the cytokinesis and cell separation, Table S3. Oligonucleotides (A) and plasmids (B) used in this study, Table S4. Parameters showed statistically significant difference at least in one mutant compared with the null distribution (Wald test, FDR = 0.05), Table S5. Significant loadings of first five PC spaces (used for GMM clustering) after Bonferroni correction (t-test, $p < 0.05$), Table S6. Significant parameters of each mannoprotein mutant (Wald test, FDR = 0.05), Table S7. Pearson correlation coefficient (r)1 between Dse2-GFP and other proteins, Table S8. Duplicated and non-duplicated genes of mannoproteins considering their functional responsibility and diversification, Table S9. Protein-Protein interactions of 32 mannoproteins. Data obtained from Oughtred et al. (2019), Table S10. Genetic interactions of 32 mannoproteins. Data obtained from Costanzo et al. (2016), Table S11. Chemical-genetic profile of 32 mannoprotein mutants. Data obtained from the *Saccharomyces* Genome Database (SGD).

Author Contributions: Conceptualization, F.G. and Y.O.; Experiments, Y.L., K.I.-N., H.O. and S.Y.; Methodology, F.G. and S.O.; Formal Analysis, F.G. and S.O.; Data Curation, A.S., E.B. and S.Y.; Writing—Draft Preparation, F.G., H.O. and Y.O.; Writing—Review and Editing, Y.O., F.G., E.B. and S.Y.; Visualization, F.G.; Supervision, E.B. and Y.O.; Project Administration, Y.O. All authors have read and agreed to the published version of the manuscript.

Funding: This work was supported by Grants-in-Aid for Scientific Research from the Ministry of Education, Culture, Sports, Science and Technology, Japan to Y.O. (19H03205) and a MEXT scholarship to F.G. (160693) as well as a National Institutes of Health grant to E.B. (GM115420).

Institutional Review Board Statement: Not applicable.

Informed Consent Statement: Not applicable.

Data Availability Statement: Any additional data will be available upon request to the corresponding author.

Acknowledgments: We thank Kuninori Suzuki and other members of the Laboratory of Signal Transduction for their participation in helpful discussions.

Conflicts of Interest: The authors declare no conflict of interest.

References

1. Klis, F.M.; Boorsma, A.; De Groot, P.W.J. Cell Wall Construction in *Saccharomyces cerevisiae*. *Yeast* **2006**, *23*, 185–202. [CrossRef] [PubMed]
2. Lesage, G.; Bussey, H. Cell Wall Assembly in *Saccharomyces cerevisiae*. *Am. Soc. Microbiol.* **2006**, *70*, 317–343. [CrossRef] [PubMed]
3. Orlean, P. Architecture and Biosynthesis of the *Saccharomyces cerevisiae* Cell Wall. *Genetics* **2012**, *192*, 775–818. [CrossRef] [PubMed]
4. Douglas, C.M.; Foor, F.; Marrinan, J.A.; Morin, N.; Nielsen, J.B.; Dahl, A.M.; Mazur, P.; Baginsky, W.; Li, W.; El-Sherbeini, M. The *Saccharomyces cerevisiae FKS1* (*ETG1*) Gene Encodes an Integral Membrane Protein Which Is a Subunit of 1, 3-Beta-D-Glucan Synthase. *Proc. Natl. Acad. Sci. USA* **1994**, *91*, 12907–12911. [CrossRef]
5. Inoue, S.B.; Takewakt, N.; Takasuka, T.; Mio, T.; Adachi, M.; Fujii, Y.; Miyamoto, C.; Arisawa, M.; Furuichi, Y.; Watanabe, T. Characterization and Gene Cloning of 1, 3-β-d-Glucan Synthase from *Saccharomyces cerevisiae*. *Eur. J. Biochem.* **1995**, *231*, 845–854. [CrossRef]
6. Levin, D.E. Regulation of Cell Wall Biogenesis in *Saccharomyces cerevisiae*: The Cell Wall Integrity Signaling Pathway. *Genetics* **2011**, *189*, 1145–1175. [CrossRef] [PubMed]
7. De Groot, P.W.J.; Ruiz, C.; de Aldana, C.R.; Duevnas, E.; Cid, V.J.; Del Rey, F.; Rodriquez-Pena, J.M.; Pérez, P.; Andel, A.; Caubin, J.; et al. A Genomic Approach for the Identification and Classification of Genes Involved in Cell Wall Formation and Its Regulation in *Saccharomyces cerevisiae*. *Comp. Funct. Genom.* **2001**, *2*, 124–142. [CrossRef] [PubMed]
8. Gonzalez, M.; Goddard, N.; Hicks, C.; Ovalle, R.; Rauceo, J.M.; Jue, C.K.; Lipke, P.N. A Screen for Deficiencies in GPI-Anchorage of Wall Glycoproteins in Yeast. *Yeast* **2010**, *27*, 583–596. [CrossRef]
9. Levy, S.F.; Siegal, M.L. Network Hubs Buffer Environmental Variation in *Saccharomyces cerevisiae*. *PLoS Biol.* **2008**, *6*, e264. [CrossRef]
10. Warringer, J.; Ericson, E.; Fernandez, L.; Nerman, O.; Blomberg, A. High-Resolution Yeast Phenomics Resolves Different Physiological Features in the Saline Response. *Proc. Natl. Acad. Sci. USA* **2003**, *100*, 15724–15729. [CrossRef]
11. Diss, G.; Gagnon-Arsenault, I.; Dion-Coté, A.M.; Vignaud, H.; Ascencio, D.I.; Berger, C.M.; Landry, C.R. Gene Duplication Can Impart Fragility, Not Robustness, in the Yeast Protein Interaction Network. *Science* **2017**, *355*, 630–634. [CrossRef]
12. Suzuki, G.; Wang, Y.; Kubo, K.; Hirata, E.; Ohnuki, S.; Ohya, Y. Global Study of Holistic Morphological Effectors in the Budding Yeast *Saccharomyces cerevisiae*. *BMC Genom.* **2018**, *19*, 1–14. [CrossRef] [PubMed]
13. Ragni, E.; Sipiczki, M.; Strahl, S. Characterization of Ccw12p, a Major Key Player in Cell Wall Stability of *Saccharomyces cerevisiae*. *Yeast* **2007**, *24*, 309–319. [CrossRef] [PubMed]
14. Ohya, Y.; Sese, J.; Yukawa, M.; Sano, F.; Nakatani, Y.; Saito, T.L.; Saka, A.; Fukuda, T.; Ishihara, S.; Oka, S.; et al. High-Dimensional and Large-Scale Phenotyping of Yeast Mutants. *Proc. Natl. Acad. Sci. USA* **2005**, *102*, 19015–19020. [CrossRef] [PubMed]
15. Ghanegolmohammadi, F.; Yoshida, M.; Ohnuki, S.; Sukegawa, Y.; Okada, H.; Obara, K.; Kihara, A.; Suzuki, K.; Kojima, T.; Yachie, N.; et al. Systematic Analysis of Ca^{2+} Homeostasis in *Saccharomyces cerevisiae* Based on Chemical-Genetic Interaction Profiles. *Mol. Biol. Cell* **2017**, *28*, 3415–3427. [CrossRef]
16. Ohnuki, S.; Kashima, M.; Yamada, T.; Ghanegolmohammadi, F.; Zhou, Y.; Goshima, T.; Maruyama, J.-I.; Kitamoto, K.; Hirata, D.; Akao, T.; et al. Genome Editing to Generate Nonfoam-Forming Sake Yeast Strains. *Biosci. Biotechnol. Biochem.* **2019**, *83*, 1583–1593. [CrossRef] [PubMed]
17. Ohnuki, S.; Ohya, Y. High-Dimensional Single-Cell Phenotyping Reveals Extensive Haploinsufficiency. *PLoS Biol.* **2018**, *16*, e2005130. [CrossRef]

18. Chadani, T.; Ohnuki, S.; Isogai, A.; Goshima, T.; Kashima, M.; Ghanegolmohammadi, F.; Nishi, T.; Hirata, D.; Watanabe, D.; Kitamoto, K. Genome Editing to Generate Sake Yeast Strains with Eight Mutations That Confer Excellent Brewing Characteristics. *Cells* **2021**, *10*, 1299. [CrossRef]
19. Nelder, J.A.; Wedderburn, R.W.M. Generalized Linear Models. *J. R. Stat. Soc. Ser. A* **1972**, *135*, 370–384. [CrossRef]
20. Rigby, R.A.; Stasinopoulos, M.D.; Heller, G.Z.; De Bastiani, F. *Distributions for Modeling Location, Scale, and Shape: Using GAMLSS in R*; CRC Press: Boca Raton, FL, USA, 2019.
21. Caicedo, J.C.; Cooper, S.; Heigwer, F.; Warchal, S.; Qiu, P.; Molnar, C.; Vasilevich, A.S.; Barry, J.D.; Bansal, H.S.; Kraus, O.; et al. Data-Analysis Strategies for Image-Based Cell Profiling. *Nat. Methods* **2017**, *14*, 849–863. [CrossRef]
22. Ghanegolmohammadi, F.; Ohnuki, S.; Ohya, Y. Single-Cell Phenomics in Budding Yeast: Technologies and Applications. In *Single-Cell Omics*; Academic Press: Cambridge, MA, USA, 2019; pp. 355–379.
23. Scrucca, L.; Fop, M.; Murphy, T.B.; Raftery, A.E. McLust 5: Clustering, Classification and Density Estimation Using Gaussian Finite Mixture Models. *R J.* **2016**, *8*, 289. [CrossRef] [PubMed]
24. Fink, G.R.; Guthrie, C. *Guide to Yeast Genetics and Molecular Biology*; Academic Press: Boca Raton, FL, USA, 1991.
25. Longtine, M.S.; Mckenzie, A., III; Demarini, D.J.; Shah, N.G.; Wach, A.; Brachat, A.; Philippsen, P.; Pringle, J.R. Additional Modules for Versatile and Economical PCR-Based Gene Deletion and Modification in *Saccharomyces cerevisiae*. *Yeast* **1998**, *14*, 953–961. [CrossRef]
26. Lee, S.; Lim, W.A.; Thorn, K.S. Improved Blue, Green, and Red Fluorescent Protein Tagging Vectors for *S. cerevisiae*. *PLoS ONE* **2013**, *8*, e67902. [CrossRef] [PubMed]
27. Slubowski, C.J.; Funk, A.D.; Roesner, J.M.; Paulissen, S.M.; Huang, L.S. Plasmids for C-Terminal Tagging in *Saccharomyces cerevisiae* That Contain Improved GFP Proteins, Envy and Ivy. *Yeast* **2015**, *32*, 379–387. [CrossRef]
28. Okada, H.; MacTaggart, B.; Ohya, Y.; Bi, E. The Kinetic Landscape and Interplay of Protein Networks in Cytokinesis. *Iscience* **2021**, *24*, 101917. [CrossRef]
29. Markus, S.M.; Omer, S.; Baranowski, K.; Lee, W.L. Improved Plasmids for Fluorescent Protein Tagging of Microtubules in *Saccharomyces cerevisiae*. *Traffic* **2015**, *16*, 773–786. [CrossRef] [PubMed]
30. Onishi, M.; Ko, N.; Nishihama, R.; Pringle, J.R. Distinct Roles of Rho1, Cdc42, and Cyk3 in Septum Formation and Abscission During Yeast Cytokinesis. *J. Cell Biol.* **2013**, *202*, 311–329. [CrossRef] [PubMed]
31. Okada, H.; MacTaggart, B.; Bi, E. Analysis of Local Protein Accumulation Kinetics by Live-Cell Imaging in Yeast Systems. *STAR Protoc.* **2021**, *2*, 100733. [CrossRef]
32. Schneider, C.A.; Rasband, W.S.; Eliceiri, K.W. NIH Image to ImageJ: 25 Years of Image Analysis. *Nat. Methods* **2012**, *9*, 671–675. [CrossRef]
33. Jonasson, E.M.; Rossio, V.; Hatakeyama, R.; Abe, M.; Ohya, Y.; Yoshida, S. Zds1/Zds2–PP2ACdc55 Complex Specifies Signaling Output from Rho1 GTPase. *J. Cell Biol.* **2016**, *212*, 51–61. [CrossRef]
34. Okada, H.; Ohnuki, S.; Roncero, C.; Konopka, J.B.; Ohya, Y. Distinct Roles of Cell Wall Biogenesis in Yeast Morphogenesis As Revealed by Multivariate Analysis of High-Dimensional Morphometric Data. *Mol. Biol. Cell* **2014**, *25*, 222–233. [CrossRef]
35. Storey, J.D.; Bass, A.J.; Dabney, A.; Robinson, D. Qvalue: Q-Value Estimation for False Discovery Rate Control. R Package Version 2.24.0. 2021. Available online: http://github.com/jdstorey/qvalue (accessed on 13 September 2021).
36. Ho, B.; Baryshnikova, A.; Brown, G.W. Unification of Protein Abundance Datasets Yields a Quantitative *Saccharomyces cerevisiae* Proteome. *Cell Syst.* **2018**, *6*, 192–205. [CrossRef] [PubMed]
37. Oughtred, R.; Stark, C.; Breitkreutz, B.-J.; Rust, J.; Boucher, L.; Chang, C.; Kolas, N.; O'Donnell, L.; Leung, G.; McAdam, R.; et al. The BioGRID Interaction Database: 2019 Update. *Nucleic Acids Res.* **2019**, *47*, 529–541. [CrossRef]
38. Shannon, P.; Markiel, A.; Ozier, O.; Baliga, N.S.; Wang, J.T.; Ramage, D.; Amin, N.; Schwikowski, B.; Ideker, T. Cytoscape: A Software Environment for Integrated Models of Biomolecular Interaction Networks. *Genome Res.* **2003**, *13*, 2498–2504. [CrossRef]
39. Costanzo, M.; VanderSluis, B.; Koch, E.N.; Baryshnikova, A.; Pons, C.; Tan, G.; Wang, W.; Usaj, M.; Hanchard, J.; Lee, S.D.; et al. A Global Genetic Interaction Network Maps a Wiring Diagram of Cellular Function. *Science* **2016**, *353*, aa1420. [CrossRef] [PubMed]
40. González-Rubio, G.; Fernández-Acero, T.; Martín, H.; Molina, M. Mitogen-Activated Protein Kinase Phosphatases (MKPs) in Fungal Signaling: Conservation, Function, and Regulation. *Int. J. Mol. Sci.* **2019**, *20*, 1709. [CrossRef] [PubMed]
41. Jiménez-Gutiérrez, E.; Alegría-Carrasco, E.; Alonso-Rodríguez, E.; Fernández-Acero, T.; Molina, M.; Martín, H. Rewiring the Yeast Cell Wall Integrity (CWI) Pathway through a Synthetic Positive Feedback Circuit Unveils a Novel Role for the MAPKKK Ssk2 in CWI Pathway Activation. *FEBS J.* **2020**, *287*, 4881–4901. [CrossRef]
42. González-Rubio, G.; Sellers-Moya, Á.; Martín, H.; Molina, M. Differential Role of Threonine and Tyrosine Phosphorylation in the Activation and Activity of the Yeast MAPK Slt2. *Int. J. Mol. Sci.* **2021**, *22*, 1110. [CrossRef]
43. Doi, S.; Tanabe, K.; Watanabe, M.; Yamaguchi, M.; Yoshimura, M. An α-Specific Gene, SAG1 Is Required for Sexual Agglutination in *Saccharomyces cerevisiae*. *Curr. Genet.* **1989**, *15*, 393–398. [CrossRef]
44. Huang, G.; Dougherty, S.D.; Erdman, S.E. Conserved WCPL and CX4C Domains Mediate Several Mating Adhesin Interactions in *Saccharomyces cerevisiae*. *Genetics* **2009**, *182*, 173–189. [CrossRef] [PubMed]
45. Kubo, K.; Okada, H.; Shimamoto, T.; Kimori, Y.; Mizunuma, M.; Bi, E.; Ohnuki, S.; Ohya, Y. Implications of Maintenance of Mother–Bud Neck Size in Diverse Vital Processes of *Saccharomyces cerevisiae*. *Curr. Genet.* **2019**, *65*, 253–267. [CrossRef]

46. Ohnuki, S.; Kobayashi, T.; Ogawa, H.; Kozone, I.; Ueda, J.-Y.; Takagi, M.; Shin, K.-Y.; Hirata, D.; Nogami, S.; Ohya, Y. Analysis of the Biological Activity of a Novel 24-Membered Macrolide JBIR-19 in *Saccharomyces cerevisiae* by the Morphological Imaging Program CalMorph. *FEMS Yeast Res.* **2012**, *12*, 293–304. [CrossRef]
47. Ragni, E.; Piberger, H.; Neupert, C.; Garcia-Cantalejo, J.; Popolo, L.; Arroyo, J.; Aebi, M.; Strahl, S. The Genetic Interaction Network of *CCW12*, a *Saccharomyces cerevisiae* Gene Required for Cell Wall Integrity during Budding and Formation of Mating Projections. *BMC Genom.* **2011**, *12*, 1–18. [CrossRef]
48. Juanes, M.A.; Piatti, S. The Final Cut: Cell Polarity Meets Cytokinesis at the Bud Neck in *S. cerevisiae*. *Cell. Mol. Life Sci.* **2016**, *73*, 3115–3136. [CrossRef] [PubMed]
49. Weiss, E.L. Mitotic Exit and Separation of Mother and Daughter Cells. *Genetics* **2012**, *192*, 1165–1202. [CrossRef] [PubMed]
50. Colman-Lerner, A.; Chin, T.E.; Brent, R. Yeast Cbk1 and Mob2 Activate Daughter-Specific Genetic Programs to Induce Asymmetric Cell Fates. *Cell* **2001**, *107*, 739–750. [CrossRef]
51. Kuznetsov, E.; Váchová, L.; Palková, Z. Cellular Localization of Sun4p and Its Interaction with Proteins in the Yeast Birth Scar. *Cell Cycle* **2016**, *15*, 1898–1907. [CrossRef] [PubMed]
52. Guo, X.; Bernard, A.; Orlando, D.A.; Haase, S.B.; Hartemink, A.J. Branching Process Deconvolution Algorithm Reveals a Detailed Cell-Cycle Transcription Program. *Proc. Natl. Acad. Sci. USA* **2013**, *110*, 968–977. [CrossRef]
53. Lussier, M.; White, A.-M.; Sheraton, J.; di Paolo, T.; Treadwell, J.; Southard, S.B.; Horenstein, C.I.; Chen-Weiner, J.; Ram, A.F.J.; Kapteyn, J.C.; et al. Large Scale Identification of Genes Involved in Cell Surface Biosynthesis and Architecture in *Saccharomyces cerevisiae*. *Genetics* **1997**, *147*, 435–450. [CrossRef]
54. Pardo, M.; Monteoliva, L.; Vazquez, P.; Martinez, R.; Molero, G.; Nombela, C.; Gil, C. PST1 and ECM33 Encode Two Yeast Cell Surface GPI Proteins Important for Cell Wall Integrity. *Microbiology* **2004**, *150*, 4157–4170. [CrossRef]
55. Bidlingmaier, S.; Snyder, M. Large-Scale Identification of Genes Important for Apical Growth in *Saccharomyces cerevisiae* by Directed Allele Replacement Technology (DART) Screening. *Funct. Integr. Genom.* **2002**, *1*, 345–356. [CrossRef] [PubMed]

Journal of *Fungi*

MDPI

Article

The Role of Dimorphism Regulating Histidine Kinase (Drk1) in the Pathogenic Fungus *Paracoccidioides brasiliensis* Cell Wall

Marina Valente Navarro [1], Yasmin Nascimento de Barros [2], Wilson Dias Segura [2], Alison Felipe Alencar Chaves [3], Grasielle Pereira Jannuzzi [4], Karen Spadari Ferreira [2], Patrícia Xander [2] and Wagner Luiz Batista [1,2,*]

1 Department of Microbiology, Immunology and Parasitology, Federal University of São Paulo, São Paulo 04023-062, Brazil; marinavnavarro@hotmail.com
2 Department of Pharmaceutical Sciences, Federal University of São Paulo, Diadema 09913-030, Brazil; barros_yasmin@hotmail.com (Y.N.d.B.); segurawd@gmail.com (W.D.S.); karen.spdari@unifesp.br (K.S.F.); patricia.xander@unifesp.br (P.X.)
3 Center of Toxins, Immune-Response and Cell Signaling—CeTICS, Instituto Butantan, São Paulo 05503-900, Brazil; felipealison@gmail.com
4 Department of Clinical and Toxicological Analyses, University of São Paulo, São Paulo 05508-000, Brazil; grasi_jannuzzi@hotmail.com
* Correspondence: batista@unifesp.br; Tel.: +55-11-3319-3594; Fax: +55-11-3319-3300

Abstract: Dimorphic fungi of the *Paracoccidioides* genus are the causative agents of paracoccidioidomycosis (PCM), an endemic disease in Latin America with a high incidence in Brazil. This pathogen presents as infective mycelium at 25 °C in the soil, reverting to its pathogenic form when inhaled by the mammalian host (37 °C). Among these dimorphic fungal species, dimorphism regulating histidine kinase (Drk1) plays an essential role in the morphological transition. These kinases are present in bacteria and fungi but absent in mammalian cells and are important virulence and cellular survival regulators. Hence, the purpose of this study was to investigate the role of PbDrk1 in the cell wall modulation of *P. brasiliensis*. We observed that PbDrk1 participates in fungal resistance to different cell wall-disturbing agents by reducing viability after treatment with iDrk1. To verify the role of *PbDRK1* in cell wall morphogenesis, qPCR results showed that samples previously exposed to iDrk1 presented higher expression levels of several genes related to cell wall modulation. One of them was *FKS1*, a β-glucan synthase that showed a 3.6-fold increase. Furthermore, confocal microscopy analysis and flow cytometry showed higher β-glucan exposure on the cell surface of *P. brasiliensis* after incubation with iDrk1. Accordingly, through phagocytosis assays, a significantly higher phagocytic index was observed in yeasts treated with iDrk1 than the control group, demonstrating the role of PbDrk1 in cell wall modulation, which then becomes a relevant target to be investigated. In parallel, the immune response profile showed increased levels of proinflammatory cytokines. Finally, our data strongly suggest that PbDrk1 modulates cell wall component expression, among which we can identify β-glucan. Understanding this signalling pathway may be of great value for identifying targets of antifungal molecular activity since HKs are not present in mammals.

Keywords: histidine kinase; dimorphism; *Paracoccidioides*; paracoccidioidomycosis; cell wall

Citation: Navarro, M.V.; de Barros, Y.N.; Segura, W.D.; Chaves, A.F.A.; Jannuzzi, G.P.; Ferreira, K.S.; Xander, P.; Batista, W.L. The Role of Dimorphism Regulating Histidine Kinase (Drk1) in the Pathogenic Fungus *Paracoccidioides brasiliensis* Cell Wall. *J. Fungi* **2021**, *7*, 1014. https://doi.org/10.3390/jof7121014

Academic Editors: María Molina and Humberto Martín

Received: 30 September 2021
Accepted: 24 November 2021
Published: 26 November 2021

1. Introduction

Paracoccidioidomycosis (PCM) is a systemic granulomatous human disease endemic in Latin America. It is caused by *Paracoccidioides* spp., a thermally-dimorphic fungus that presents as an infective mycelium in the environment, and it switches to a pathogenic yeast form in the mammalian host [1,2]. Its clinical manifestations occur in two distinct forms, acute or subacute and chronic [3], affecting mainly the lungs, but it is capable of spreading to other tissues [4,5]. Primary infection usually occurs by inhaling propagules (conidia) produced during the mycelial form [6]. Once inhaled, fungal propagules will be recognized by cells of the innate immune system [7]. The recognition of fungal cell wall components begins with pathogen-associated molecular patterns (PAMPs) through pathogen recognition receptors (PRRs). These receptors include Toll-like receptors (TLRs), mannose receptors, complement pathway molecules and lectin family receptors (CLRs), such as dectin-1 [8]. The interaction of these molecules with fungal yeasts leads to activation of the innate immune response, consequently activating mediators involved in eliminating these pathogens and controlling the adaptive immune response [9].

However, fungi have mechanisms to prevent their elimination by the host's immune system [10]. Several fungi that engage in morphological transitions are of great medical importance, such as *Talaromyces marneffei* (*Penicillium marneffei*), *Blastomyces dermatitidis*, *Coccidioides immitis*, *Histoplasma capsulatum*, *Sporothrix schenckii*, and *Paracoccidioides* spp. [11]. The transition in *Paracoccidioides* spp. and other dimorphic fungi is essential for the establishment of the disease [12]. This alters not only the cell morphology but also the composition of the cell wall elements. In *Paracoccidioides* spp. mycelium, there is a prevalence of β-1,3-glucan and β-1,6-glucan, and in the yeast form, there is a prevalence of α-1,3-glucan and chitin [13]. This ensures fungal survival in the host environment since the content of α-1,3-glucan is correlated with the degree of fungal virulence [14]. In addition, masking the presence of β-1,3-glucan molecules, a highly immunogenic structure, is recognized by the dectin-1 receptor of phagocytic cells [15]. The morphological switch is believed to be an additional evasion strategy against phagocytic cells and mechanisms for recognition of the cell wall components [15]. In this context, the fungal cell wall plays an important role in immunological recognition.

In *Paracoccidioides* spp., different genes are expressed according to the phase (yeast or mycelium) [16]. The mechanism of this transition has been unclear. However, genes related to the control of the mycelium-yeast transition (M-Y) have recently been identified in *B. dermatitidis* and *H. capsulatum*, including dimorphism-regulating histidine kinase (*DRK1*). *DRK1* is mainly expressed in the yeast phase [11] of *B. dermatitidis* [17], *S. schenckii* [18], and *T. marneffei* [19] and, more recently, it was characterized in *P. brasiliensis* [20].

Histidine kinases (HKs) were discovered in the 1980s in *Escherichia coli* [21], and were believed to be present only in bacteria. In the 1990s, they were also discovered in plants, fungi, archaea, cyanobacteria and amoebas [22]. In fungi, the functions attributed to HKs have not been explored very well [23]. HKs are classified based on phylogenetic analyses; in fungi, there are 16 groups, determined by the C- and N-terminal regions and the domains present in each group [22,24,25]. This is a signal transduction mechanism that contains a conserved kinase domain and a conserved regulatory domain. After an extracellular stimulus, the HK domain is autophosphorylated on a histidine residue, followed by a phosphate group transfer to the regulatory domain in an aspartate residue, which catalyses a downstream reaction of the effector domain that leads to downstream signalling [26].

Among the characterized pathways in fungi, we can mention the response to osmotic stress [25], oxidative protection against phagocytic cells [27] and regulation of the dimorphism in pathogenic fungi [17,18,20,28]. In *P. brasiliensis*, *DRK1* is known to be a group III histidine kinase essential to the dimorphic transition process [20].

Since HKs are known to regulate morphological switches in *P. brasiliensis*, this work aimed to characterize the PbDrk1 protein, which is involved in the transition from mycelium to yeast. This investigation is of great interest since HKs use a phosphorylation mechanism where the amino acid phosphoryl-receiving groups are aspartate and histidine residues, unlike the serine, threonine and tyrosine residues that are prevalent in mammals. Thus, knowing that these molecular sensors are absent in humans, it is extremely important to study the components that are part of these activation pathways, as they may represent potential molecular targets in the development of new antifungal agents.

2. Materials and Methods

2.1. Fungal Isolates and Growth Conditions

The *P. brasiliensis* isolate Pb18 was grown in yeast peptone dextrose modified medium (mYPD) (0.5% yeast extract, 1% peptone, and 0.5% glucose, pH 6.7) for 4 to 5 days at 37 °C and shaking at 150 rpm. For mycelium growth, viable yeast cells were cultivated in mYPD at 25 °C for 7 days at 150 rpm. Viability was assessed by Trypan blue 0.4% counting on Neubauer's chamber, using the formula: cell viability(%) = viable cells / total cells × 100. All chemicals were purchased from Sigma-Aldrich (St. Louis, MO, USA) unless otherwise mentioned.

2.2. Histidine Kinase Inhibitor Susceptibility

About 1×10^6 yeast were incubated with different concentrations (100, 50, 25, 12.5, and 6.25 μg/mL) of Fludioxonil (Thermo Scientific, Waltham, MA, USA), a specific inhibitor of class III histidine kinase (iDrk1). The inhibitor was solubilized in DMSO (dimethylsulfoxide). Yeasts were incubated for 24 h under constant agitation of 180 rpm at 37 °C. Each yeast culture was diluted (10, 50, 100, 500, and 1000 times) in YPDmod broth, and 10 μL of each suspension was plated to YPDmod agar medium. Plates were photographed after 7 days of growth at 37 °C. This assay was performed in biological triplicate.

2.3. Dimorphic Transition Assay

Yeast cells of Pb18 were grown in mYPD agar (pH 6.5) at 37 °C for 4 to 5 days and inoculated in mYPD broth medium (pH 6.5). Then yeasts were incubated at 25 °C for five to six days to reverse yeast to mycelium entirely. After the complete transition, yeasts were centrifuged at 3000× *g* and washed with PBS buffer (pH 7.2). The mycelium was then seeded in 6-well plates and 20 μM iDrk1 (Fludioxonil) was added. The samples were monitored every 24 h under an optical microscope (Zeiss) at 100× magnification. Every 24 h, the culture medium was supplemented with 20 μM fludioxonil (because of inhibitor photodegradation). With each addition, a new solution was prepared to guarantee its activity. This assay was performed in a biological duplicate.

2.4. Cell Wall Disturbing Agents Spot Test

The sensitivity of Pb18 to cell wall disruptors was investigated using the spot assay. About 1×10^6 yeasts were incubated with iDrk1 (25 μg/mL) for 24 h at 37 °C at 150 rpm. Each yeast culture was diluted (10, 50, 100, and 500 times) in YPDmod broth, and 10 μL of each suspension was applied to mYPD agar medium supplemented with different cell wall disrupting agents, such as: Congo Red (Congo Red) (2.5 μM), Calcofluor White (1 μg/mL) and sodium chloride (150 mM). The plates were incubated for seven days at 37 °C, and then photographed. This assay was performed in biological triplicate.

2.5. RNA Extraction and Real-Time Quantitative PCR Analysis

Pb18 yeasts were grown for four to five days in YPDmod pH 6.5 medium at 37 °C and 150 rpm, counted and the volume was adjusted to 30 mL at a concentration of 1×10^6 cells/mL. Then, yeasts were incubated with 25 μg/mL of iDrk1 for 24 h at 37 °C. After this incubation period, RNA extraction was performed. Samples were centrifuged at 3000× *g* for 10 min at 4 °C and washed 3 times with PBS (pH 7.2). Then, in 15 mL tubes, approximately 500 μL of glass beads (425–600 μm—Sigma-Aldrich, San Louis, MO, USA) and 1.5 mL of TRizol® (Invitrogen, Waltham, MA, USA) were added to the sample. The tubes were vigorously vortexed for 6 cycles of 1 min, alternating with 1 min on ice. RNA extraction was performed as previously described [29]. Quantification was performed using spectrophotometry (NanoDrop 2000/2000c, Thermo Fisher Scientific, Waltham, MA, USA). For complementary DNA (cDNA) synthesis, 500 ng of RNA was initially submitted to the DNase I enzyme (Thermo Fisher Scientific, Waltham, MA, USA) and then to ProtoScript First Strand cDNA Synthesis kit (New England BioLabs, Ipswich, MA, USA), according to the manufacturer's instructions. To assess gene expression by real-time quantitative PCR, the reaction was performed with SYBR® Green Master Mix (Applied Biosystems, Foster City, CA, USA) according to the manufacturer's instructions. The endogenous expression genes for ribosomal protein 60S L34 (L34r) and 18S were used as normalizing controls. For each gene of interest and normalizer gene, a negative reaction control was also added. The samples were prepared in triplicate to a 96-well plate (0.2 mL MicroAmp™ Optical 96-Well Reaction Plate—Applied Biosystems) compatible with the equipment used, and the plate was sealed with an optical adhesive (MicroAmp™ Optical Adhesive Film—Thermo Fisher Scientific, Waltham, MA, USA). The equipment used was the ABI StepOne Plus Real-Time PCR System (Applied Biosystems) with the following conditions: 10 min at 95 °C, followed by 40 cycles of 15 s at 95 °C and 1 min at 60 °C. The dissociation curve included an additional cycle of 15 s at 95 °C, 20 s at 60 °C, and 15 s at 95 °C. The curves of oligonucleotides efficiency were evaluated from a cDNA obtained previously and serially diluted (100, 10, 1, and 0.1 ng/μL). The Ct values of each dilution point were determined and used to make the standard curve and finally calculate the primer efficiency ($E = 10^{(-1/slope)-1} \times 100$). The relative expression was determined based on the $2^{-\Delta\Delta Ct}$ method [30]. The sequences used for each gene are listed in Table 1.

Table 1. Oligonucleotides used for real-time quantitative PCR analysis.

Gene	Sequence (5′–3′)	Gene ID
L34	Foward: AAAGGAACCGCACCAAAATG Reverse: AGACCTGGGAGTATTCACGG	PADG_04402
18S	Foward: CGGAGAGAGGGAGCCTGAGAA Reverse: GGGATTGGGTAATTTGCGC	ADG_12090
FKS1	Forward: GTTCCCATCACCGATCCTATTT Reverse: GAAGGAGAGCAAGAAGACGATAC	PADG_11846
KRE6	Foward: TTCCGACGAGTTCAACAAAGA Reverse: CTGCGTCACTCCATACCAAATA	PADG_07170
PHR2	Foward: ACTGAGGACAAACACCATCAG Reverse: ACAGATCTGCAACGACGTAAA	PADG_04918
GEL3	Foward: CGTTGTCAGCGGAGGTATCGTC Reverse: AGGGCAGGTTCGGAGTTCAGTG	PADG_04918
AGN1	Foward: AAATGCGGCACGGAGGAGA Reverse: AAGGGTGGTATCAAGTGCCGAGT	PADG_03169
CHT3	Foward: GCGAGGAATTGGGTGATAGAA Reverse: AGGGTTGACGCTATCAGAAATAA	PADG_08156
CHS2	Foward: CCCGAACCTACTGCACTTTATC Reverse: TGCCCTTACCCGCTTTAATC	PADG_08636
CHS3	Foward: CGCTATGGTTAAGGATCCCGAGA Reverse: GCATCCAGGCAAGCAAGTAACA	O94191_PARB
CHS4	Foward: ACCGGATGAGGCCACTATTACAGA Reverse: GTCTGCAATCGCTGCTCAACG	PADG_07911
CHS5	Foward: AGAGTATCAAGGCTGAGCTGGAACG Reverse: CGGAAAGGACGGCTTCGGTT	A9XTF9_PARBR

2.6. Quantification of Cell Wall Components

About 1×10^6 yeast cells were grown and incubated for 24 h with 25 μg/mL of iDrk1 in mYPD at 37 °C under stirring at 180 rpm. Then, yeasts were collected, homogenized in blocking solution (0.5% BSA, 5% rabbit serum, 5 mM EDTA, 2 mM NaN_3 in PBS, pH 7.2) and incubated for 30 min at room temperature. Then, yeasts were incubated with 1 μg/mL of the β-glucan binding probe for 1 h on ice. This probe corresponds to the human Dectin-1 receptor fused to the FC portion of mouse IgG1 (Sino Biological, Beijing, China). Samples were incubated with Alexa-488-conjugated anti-mouse IgG secondary antibody (Molecular Probes, Eugene, OR, USA) at a 1:200 ratio for 45 min on ice. Between each incubation, step cells were washed three times with wash buffer (0.5% BSA, 5 mM EDTA, 2 mM NaN_3 in PBS, pH 7.2). Determination of chitin oligomers was performed with the WGA (wheat germ agglutinin) marker conjugated to FITC (Sigma-Aldrich, San Louis, MO,

USA), at a concentration of 25 μg/mL in 1 mL of PBS (pH 7.2). Mannan determination was performed with FITC-conjugated Concanavalin A (Sigma-Aldrich, San Louis, MO, USA), at a concentration of 25 μg/mL in 500 μL of PBS (pH 7.2). Both samples were incubated for 1 h protected from light under agitation at 800 rpm. Next, samples were centrifuged at 3000× g for 5 min at 4 °C and washed three times with 1 mL of PBS (pH 7.2). Finally, all samples were homogenized in 500 μL of PBS (pH 7.2) and analyzed by flow cytometry (BD FACSCaliburTM, Becton Dickinson) using the FL-1 detection channel. Each sample was prepared in experimental triplicate. A total of 10,000 events were counted and quantification graphs were generated from the median fluorescence intensity (MFI). The data obtained were analyzed using FlowJo software version 10.6.2 (FlowJo, LLC, FlowJo™, Ashland, OR, USA) [31].

2.7. Confocal Microscopy

To evaluate the exposion of β-glucan molecules through fluorescence confocal microscopy, cells were prepared as described on the previous section and then centrifuged and homogenized in PBS:glycerol (3:1). The slides were prepared with 30 μL of the sample, sealed and analyzed under a confocal microscope (SP8 Lightning, Leica Microsystems). Fluorescence intensity was quantified using ImageJ analysis software (version 1.53i) through corrected total cell fluorescence (CTCF).

2.8. PKA Activity and cAMP Quantification

PKA activity was performed using the PKA Colorimetric Activity Kit (Thermo Fisher Scientific, Waltham, MA, USA), according to the manufacturer's specifications. For this purpose, 1×10^7 Pb18 cells were incubated with or without 25 μg/mL of iDrk1 or 1 mM H_2O_2 for 30 min in YPDmod at 37 °C and 180 rpm. Yeasts were collected and washed with PBS (pH 7.2), centrifuged for 3000× g for 10 min and homogenized in Tris-based lysis buffer contained in the kit. Lysis was assessed by adding glass beads and vigorous vortexing (5 cycles of 1 min interspersed with 1 min of incubation on ice). Samples were then centrifuged and the supernatant was recovered. For PKA activity assay, 1 mg of protein was used. Intracellular levels of cyclic AMP (cAMP) were quantified by the Cyclic AMP ELISA Kit (Cayman Chemical, Ann Harbor, MI, USA). Pb18 yeasts were prepared as described above. Cellular lysis was performed according to the manufacturer's specifications, using 10 mM HCl and glass beads. For cAMP quantification, it was used about 0.5 μg of protein. The total protein extracts quantification was performed by Bradford assay [32]. Absorbance was read at 405 nm (BioTek—Synergy HT) and data were plotted in triplicate from the standard curves absorbance values.

2.9. Glycogen Accumulation

A 10 mL suspension of 5×10^6 Pb18 yeasts/mL was incubated with or without 25 μg/mL of iDrk1 for 24 or 48 h, followed by centrifugation at 2500 rpm for 3 min (the supernatant was discarded). The pellet was homogenized in 1 mL of iodine solution (0.2% iodine and 0.4% potassium iodide) and incubated for 3 min at room temperature. The samples were again centrifuged at 2500 rpm for 3 min, the supernatant was discarded and the pellet homogenized in 30 μL of the iodine solution. The samples were plated to a 96-well plate, 200 μL of PBS (pH 7.2) was added to each well and the samples were photographed [33].

2.10. Phagocytosis Assay

In vitro phagocytosis was performed with the J774 macrophage cell line. About 2.5×10^5 viable cells were plated in a 24-well plate containing RPMI (Gibco, Gaithersburg, MD, USA) supplemented with 10% FBS. A 15 mm diameter circular sterilized coverslip was added to each well. After adhesion, macrophages were primed with 100 ng/mL of LPS 30 min before interaction with *P. brasiliensis*. J774 cell line is activated by LPS showing changes in morphology, such as cytoplasm expansion, contributing to a better performance in phagocytosis assays [34,35]. The interaction was carried out in a 2:1 ratio (yeast:macrophages) and incubated for 24 h at 37 °C and 5% CO_2. Previously, yeasts were incubated in the presence or absence of iDrk1 25 μg/mL for 24 h at 37 °C. After the interaction, each well was washed with sterile PBS (pH 7.2), coverslips were stained with hematology dyes (Newprov, Paraná, Brazil) and the supernatant was recovered for cytokine assay. The phagocytic index was determined from the protocol established by Popi et al. [36]. After 48 h, wells intended to quantify colony-forming units (CFU) were washed with 1 mL of ice-cold sterile ultrapure water and vigorously homogenized. The supernatant was plated in BHI medium supplemented with 10% FBS, and incubated for seven days at 37 °C. After growth, colony-forming units were counted. This experiment was carried out in biological triplicate.

The phagocytosis assay was also assessed by flow cytometry. Samples were prepared as described above. Prior to interaction, yeasts were labeled with CFSE (CellTrace™ CFSE Cell Proliferation—Thermo Fisher). For each 1×10^7 Pb18 yeasts cells, 3 μL of CFSE (3 μg/μL) were added, with final volume of 500 μL of PBS (pH 7.2) and 0.5% BSA. Yeasts were incubated for 10 min at 37 °C, then centrifuged at 3000× *g* for 5 min and washed three times with PBS (pH 7.2). After 24 h, the samples were washed and the cells were detached from the wells using a cell scraper and homogenized in 500 μL of PBS (pH 7.2). Data acquisition was immediately performed in a flow cytometer (BD FACSCaliburTM, Becton Dickinson) using the FL-1 detection channel. After the first acquisition, 8 μL of Trypan Blue 0.4% was added to each sample to quench the signal of yeasts that could be adhered to the cell surface instead of internalized and again, the data was acquired. A total of 10,000 events were counted for each sample and the data were analyzed using the FlowJo software version 10.6.2 (FlowJo, LLC, FlowJo™, Ashland, OR, USA) [37].

2.11. Cytokines Determination

Supernatants from the phagocytosis assay were collected. The cytokines TNFα and IL12p70 were measured using the DuoSet ELISA kits (R&D Systems, Mineápolis, MN, EUA). The assay was performed in 96-well EIA/RIA plates according to the manufacturer's specifications with modifications. First, plates were coated with 50 μL of capture antibody and incubated at room temperature for 16 h. Wells were blocked with 200 μL of diluent solution (1% BSA in PBS pH 7.2) and incubated at room temperature for 1 h. Then, 50 μL of culture supernatant were added, in triplicate, and 50 μL of the cytokine standard to perform the standard curve. Plates were then incubated for 2 h at room temperature. Next, 50 μL of Streptavidin-HRP solution (provided in the kit) were added to each well and the plate was incubated for 20 min protected from direct light. Finally, Tetramethylbenzidine (TBM) substrate was added and the plates were incubated for 20 min at room temperature, protected from light. At the end of the incubation, 50 μL of stop solution (2N H_2SO_4) was added. Between each incubation step, the wells were washed three times with 200 μL of wash buffer (0.05% Tween-20 in PBS pH 7.2). Absorbance was read at 430 nm (BioTek—Synergy HT).

2.12. Statistical Analysis

The data contained in this work were validated with the reproducibility of at least three independent experiments. For comparison analysis, a Student's t-test and significance analysis were performed, as were the one-way variance (ANOVA), followed by Tukey's test. Differences were considered significant when $p < 0.05$.

3. Results

3.1. Susceptibility of P. brasiliensis to Drk1 Pharmacological Inhibitors

Studies analysing the role of Drk1 in fungi [17,20] demonstrated that the use of specific inhibitors of group III histidine kinases (iprodione or fludioxonil) are efficient in promoting biological responses. Fludioxonil is a product derived from pyrrolnitrine, a compound isolated from *Pseudomonas pyrrocinia* [38]. The use of this inhibitor is already well established in studies with pathogenic fungi [17,39–42]. Initially, a susceptibility test was assessed with the Pb18 isolate. For this purpose, the yeasts were incubated for 24 h with different concentrations (100 to 6.25 µg/mL) of iDrk1 (fludioxonil) and then inoculated in mYPD medium. After 7 days of incubation, it was observed that there was no reduction in fungal viability at the concentrations tested (Figure 1A). Thus, based on the literature data [17], which demonstrated that a concentration of 25 µg/mL was able to inhibit the activity of Drk1 in *Blastomyces dermatitidis*, this concentration was established for the following experiments (Figure 1A).

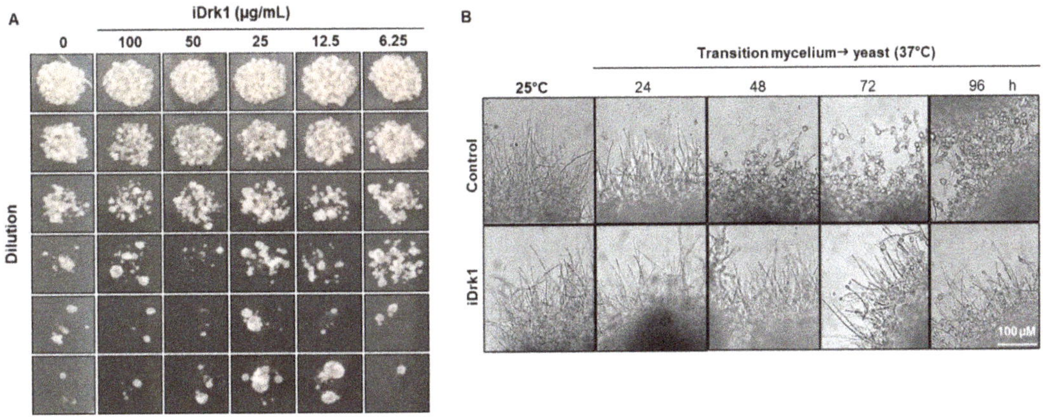

Figure 1. (**A**) Susceptibility of *P. brasiliensis* yeast cells to Drk1 pathway inhibitor (iDrk). A total of 1×10^6 cells/mL were incubated with different concentrations of iDrk1 and incubated for 24 h. Then, the yeasts cells were diluted, plated in solid mYPD medium and incubated in for 7 days at 37 °C. (**B**) Dimorphic transition assay of *P. brasiliensis* in the presence or not of iDrk1. The fungus was cultivated as mycelial, at 25 °C and 150 rpm. Subsequently, it was plated in 6-well plates and incubated at 37 °C in mYPD medium supplemented or not with 20 µM iDrk1. Cultures were monitored every 24 h and observed under an optical microscope to evaluate the dimorphic transition.

3.2. Role of PbDrk1 in P. brasiliensis Cell Wall Maintenance

The transition from mycelium to yeast triggers the cell wall morphogenesis machinery, which involves synthesizing several cell wall sugars and proteins critical to survival during infection and evasion of the immune system [43]. In *Penicillium marneffei* pathogenic fungus, the role of *DRK1* is essential for stress adaptation, hyphal morphogenesis, and cell wall integrity [44]. Previously, we showed *PbDRK1* transcription to be phase-specific for the yeast form and demonstrated that PbDrk1 participates in dimorphic switching in *P. brasiliensis* when iprodione (another Drk1 inhibitor) is used [20]. To confirm the action of fludioxonil (iDrk1) in the *P. brasiliensis* yeast-mycelium switch, a dimorphic transition assay was performed. Initially, the fungus was cultivated in the mycelial form at 25 °C and then it was incubated at 37 °C in the presence or absence of iDrk1. The dimorphic transition was then followed every 24 h under an optical microscope. After 24 h at 37 °C, it was possible to observe the formation of yeasts in the distal portion of the hyphae in the control group, and at 96 h, there was a predominance of yeasts. In the group treated with iDrk1, at 96 h, there was a predominance of hyphae. Thus, the addition of fludioxonil also impairs the fungus's ability to make the complete transition from mycelium to yeast (Figure 1B).

In addition to the already known mechanism involving the dimorphic transition [17,18,20,45], it was observed that the strain deleted for Drk1 presents with sensitivity for conidia germination [44]. Based on this evidence, Pb18 yeasts were treated with iDrk1 and inoculated in YPDmod medium containing cell wall stressing agents such as Congo red (CR), Calcofluor White (CFW), and sodium chloride (Figure 2A). CR and CFW dyes are classically used in studies involving the synthesis and organization of fungal cell wall [46]. Both molecules have two groups of sulfonic acids that exert antifungal activity [47]. The action of CFW and CR occurs through binding to nascent chitin chains, preventing the access of enzymes that promote the binding of chitin with β-1-3-glucan and β-1-6-glucan chains. As a result, the cell wall becomes weakened, which can compromise its viability [47]. Sodium chloride promotes osmotic stress and changes the structure of the cell wall [48]. After seven days of incubation, it was observed that, under all conditions, previous exposure to iDrk1 induced a fairly dramatic viability reduction, especially under osmotic stress (Figure 2A). Previous data from our group [20] showed an increase in the number of *PbDRK1* transcripts when the fungus is subjected to osmotic stress. The data show that the inhibition of Drk1 results in an increased sensitivity to cell wall stressors and to osmotic stress.

3.3. Modulation of Cell Wall Gene Expression in P. brasiliensis

Previous data demonstrated that *PbDRK1* is mainly expressed in the yeast phase [20]. However, the associated pathways remain poorly understood. Therefore, to investigate the possible targets regulated by *PbDRK1*, Pb18 yeasts were incubated in the absence or presence of iDrk1 for 24 h. Then, the RNA extraction protocol was applied, followed by a cDNA synthesis reaction. Finally, RT-qPCR analysis of several genes involved in cell wall synthesis was performed: *CHT3*, *CHS2*, *CHS3*, *CHS4*, and *CHS5* (genes involved in the synthesis and maintenance of chitins) and *FKS1*, *KRE6*, *PHR2*, *GEL3*, and *AGN1* (genes involved in glucan synthesis and maintenance).

As shown in Figure 2B, genes encoding chitin synthase enzymes *CHT3*, *CHS2* and *CHS3* exhibited a 3.7-, 5.8- and 2.0-fold increase, respectively, in samples treated with iDrk1. However, there were no significant changes in the *CHS4* and *CHS5* genes (Figure 2B). On the other hand, the *CHS4* and *CHS5* genes were more highly expressed in the mycelial phase of Pb18 [49]. Thus, these data suggest that iDrk1 may modulate the expression of some chitin synthesis genes.

Significant increases in the transcript levels of several genes related to β-glucan synthesis and cell wall integrity have been observed (Figure 2C). A 4-fold increase in the *FKS1* gene, which encodes a β-(1,3)-glucan synthase, was observed. The *KRE6* gene showed a 2.7-fold increase in expression levels when compared with the no-treatment control. This gene is important in the β-(1,6)-glucan synthesis process and for molecules that are part of the β-(1,3)-glucan net composition [13,50]. The *PHR2* gene showed a 5.5-fold increase in expression levels (Figure 2C). This gene is involved in the maintenance of the cell wall and fungal virulence [51]. Finally, the expression of the *GEL3* gene exhibited a fourfold increase compared to the control without treatment. This gene is part of the process of elongation of the β-(1,3)-glucan chains and cell wall integrity [52,53]. On the other hand, the *AGN1* gene, which is involved in the maintenance and synthesis of α-(1,3)-glucan and highly expressed in the yeast phase [54], showed no difference in transcript levels (Figure 2C). These findings support the hypothesis that PbDrk1 could regulate negatively genes involved with the cell wall synthesis.

3.4. Modulation of the Cell Wall Components

As previously demonstrated, the inhibition of PbDrk1 was responsible for modulating the expression of genes involved in the synthesis of important cell wall components, such as *FKS1*. Associated with this, we observed a greater sensitivity of the fungus to cell wall disturbances when treated with iDrk1. Therefore, we decided to evaluate the levels of β-glucan, chitin and mannan in the Pb18 cell wall after treatment with iDrk1. For this purpose, β-glucan labelling was performed using the Dectin-1-Fc probe. Next, chitin levels were determined using the WGA (wheat germ agglutinin) marker conjugated with FITC [55]. This lectin molecule has an excellent affinity for N-acetyl-β-D-glucosaminyl residues and N-acetyl-β-D-glucosamine oligomers. Finally, to verify the mannan levels, concanavalin A conjugated with FITC was used [31]. This molecule has a high affinity for terminal residues of α-D-mannose and α-D-glucose.

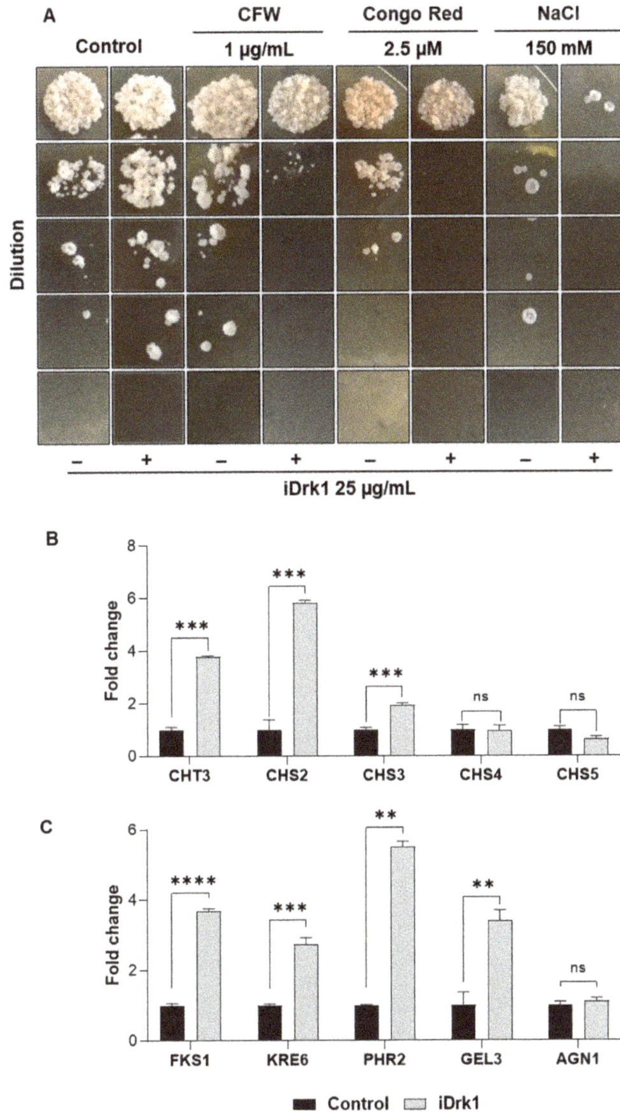

Figure 2. (**A**) Growth of *P. brasiliensis* in the presence of cell wall disrupting agents. A total of 1×10^6 cells/mL were subjected to 25 µg/mL of iDrk1 and incubated for 24 h. Next, yeast cells were diluted, plated in solid mYPD medium containing different agents that disturb the cell wall, such as Calcofluor White (CFW), Congo Red and NaCl. Finally, cells were incubated in for seven days at 37 °C. Expression of cell wall morphogenesis related genes in *P. brasiliensis*. Pb18 cells were subjected to 25 µg/mL of iDrk1 for 24 h and then total RNA extraction was performed. The related genes (**B**) synthesis of chitin, such as *CHT3*, *CHS2*, *CHS3*, *CHS4* and *CHS5* and (**C**) synthesis of glucans such as *FKS*, *KRE6*, *PHR2*, *GEL3* and *AGN1* were analyzed, where p-value ≤ 0.01 (**), p-value ≤ 0.001 (***) and p-value ≤ 0.0001 (****).

Yeasts were incubated for 24 h in the presence or absence of iDrk1. Then, labelling was performed for each fluorescent marker described above. Fluorescence analyses were obtained through flow cytometry, where 10,000 events were obtained for each sample.

Quantification was performed using the median fluorescence intensity (MFI). Figure 3A,B show a significant increase in β-glucan and chitin levels after 24 h of treatment with iDrk compared to the untreated control. On the other hand, no change in mannan levels was observed. In addition to the flow cytometry measurements, β-glucan localization was also evaluated by confocal microscopy. Yeast labelling was performed as described in the previous section using the Dectin-1-Fc probe and a secondary antibody anti-mouse IgG conjugated to Alexa 488. This assay made it possible to observe an increase in labelling in the fungus previously incubated with iDrk1 compared to the control. Most of the marking was observed on the yeast surface (Figure 3C). Fluorescence quantification (Figure 3D) was obtained from the corrected total fluorescence values (CTCF). These results indicate that PbDrk inhibition increases the β-glucan levels in *P. brasiliensis* yeast cells.

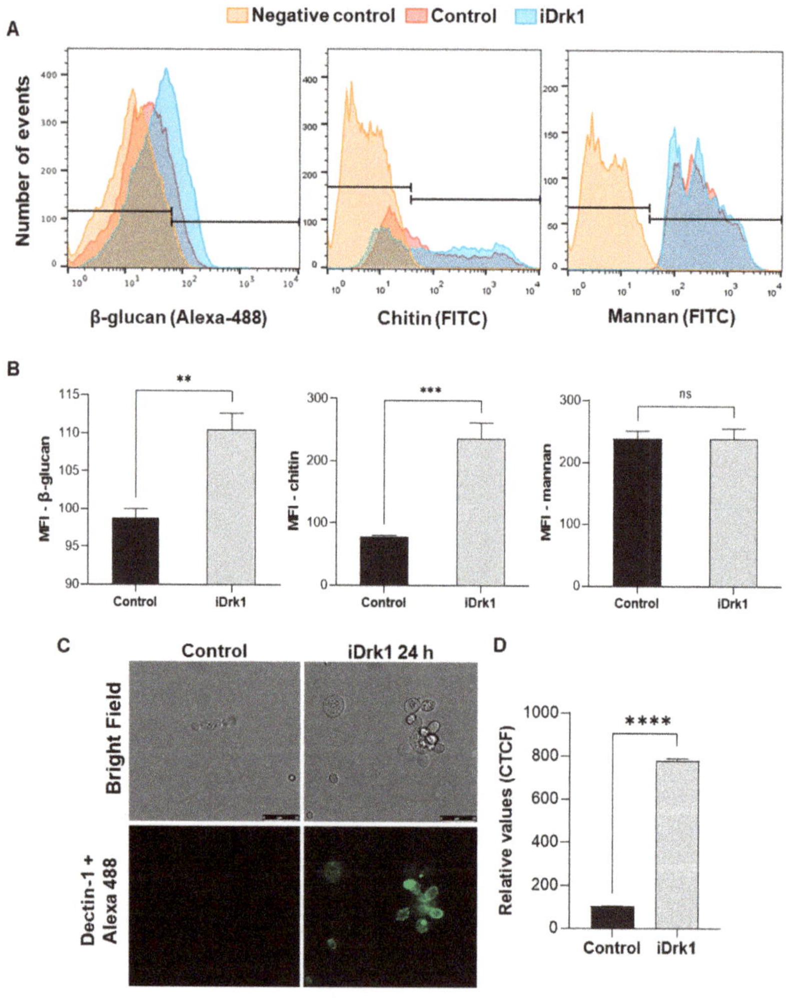

Figure 3. Quantification of *P. brasiliensis* cell wall components after incubation with iDrk1. Pb 18 cells were subjected to 25 µg/mL of iDrk1 for 24 h and then labeled with fc-Dectin-1 and Alexa 488 for dosage of β-glucan, WSA conjugated with

FITC for dosage of chitin oligomers and Concanavalin A conjugated with FITC for mannan dosage. (**A**) Histograms from each cell wall component by flow cytometry, where negative control is represented as yeast that were not fluorescence labelled. (**B**) Quantification through the median fluorescence intensity (MFI) where p-value $\leq$ 0.004 (**). Evaluation of β-glucan exposure in *P. brasiliensis* cell wall after incubation with iDrk1. Cells of Pb18 were subjected to 25 μg/mL of iDrk1 for 24 h and then labeled with fc-Dectin-1 and Alexa 488. (**C**) Confocal microscopy analysis. (**D**) Corrected quantification of total fluorescence (CTCF) from confocal microscopy analysis where p-value $\leq$ 0.01 (**), p-value $\leq$ 0.001 (***) and p-value $\leq$ 0.0001 (****).

3.5. Inhibition of PbDrk1 Induces Increased Phagocytosis of P. brasiliensis and Alters Cytokine Production by Macrophages

Fungal cell wall β-glucan is also an important pathogen-associated molecular pattern (PAMP). To avoid the innate immune response, many fungal pathogens depend on the synthesis of the cell wall α-glucan, which functions as a stealth molecule to mask the β-glucans itself or links other masking structures to the cell wall [56]. As demonstrated in the previous results, when we inhibited the PbDrk1 pathway, the Pb18 yeast underwent changes in their cell wall composition. Under normal conditions, the main components of the outermost layer of the Pb18 cell wall are α-glucan molecules [13,57]. Thus, we determined whether the increase in β-glucan would make the fungus more susceptible to recognition by cells of the immune system. To answer this question, a phagocytosis assay was performed with J774 murine macrophages. After 24 h of interaction, the supernatant was collected for subsequent cytokine dosage. In this assay, it was possible to observe that fungi treated with iDrk1 had a higher phagocytic index than the untreated control (Figure 4A). To confirm these observations, a phagocytosis assay was performed using flow cytometry [37]. The sample treatment was performed as described above, but before the cell-yeast interaction, the Pb18 yeasts were labelled with the intracellular dye CFSE. This result is similar to that observed in the previous experiment. In Figure 4B, we can see the greater signal intensity in the sample where the fungus was previously treated with iDrk1. Furthermore, we observed a reduced number of colony-forming units (CFUs) in the group treated with iDrk1 (Figure 4C). This dataset suggests that PbDrk1 inhibition of *P. brasiliensis* may have contributed to an increase in phagocytosis and in the susceptibility of fungal cells to macrophage elimination.

The quantification of the proinflammatory cytokines TNFα and IL-12p70 was evaluated by ELISA. Before the phagocytosis assay, J774 cells were primed with LPS. As a control, the supernatant from the macrophages activated only with LPS was analysed. In Figure 4D, we can see a significant increase in TNFα levels in the supernatant of cell cultures incubated with *P. brasiliensis* yeasts previously exposed to iDrk1. The same result was observed for the cytokine IL-12p70 (Figure 4E).

3.6. Regulation of cAMP-PKA and Glycogen Accumulation in P. brasiliensis

Perception of the environment is fundamental for fungal survival in the host. The cyclic AMP-dependent protein kinase A pathway (cAMP-PKA) is highly conserved and is involved in several biological processes, both in human pathogenic and phytopathogenic fungi [58]. In addition, it also contributes to gene expression regulation and cell wall remodeling [59]. Through stimuli external to the cell, the enzyme adenylate cyclase converts ATP to adenosine 3′,5′-cyclic monophosphate (cAMP), an important secondary messenger that binds to the catalytic subunit of protein kinase A (PKA). This generates a conformational change that releases PKA catalytic subunits and activates transcription factors and other signalling pathways involved in cell wall integrity, stress response and virulence [58,60,61]. Thus, Pb18 yeasts treated or not with iDrk1 for 24 h were lysed, and the protein extract was used to quantify the PKA activity and cAMP dosage. As shown in Figure 5A, it was possible to verify a reduction in PKA activity after exposure to iDrk1. As a positive control for the reaction, *P. brasiliensis* was subjected to oxidative stress with 300 mM hydrogen peroxide (H_2O_2) for 30 min. This condition is known to generate an increase in PKA activity [62]. It was also possible to observe no significant difference in the dosage of cAMP levels between the treated and control samples (Figure 5B).

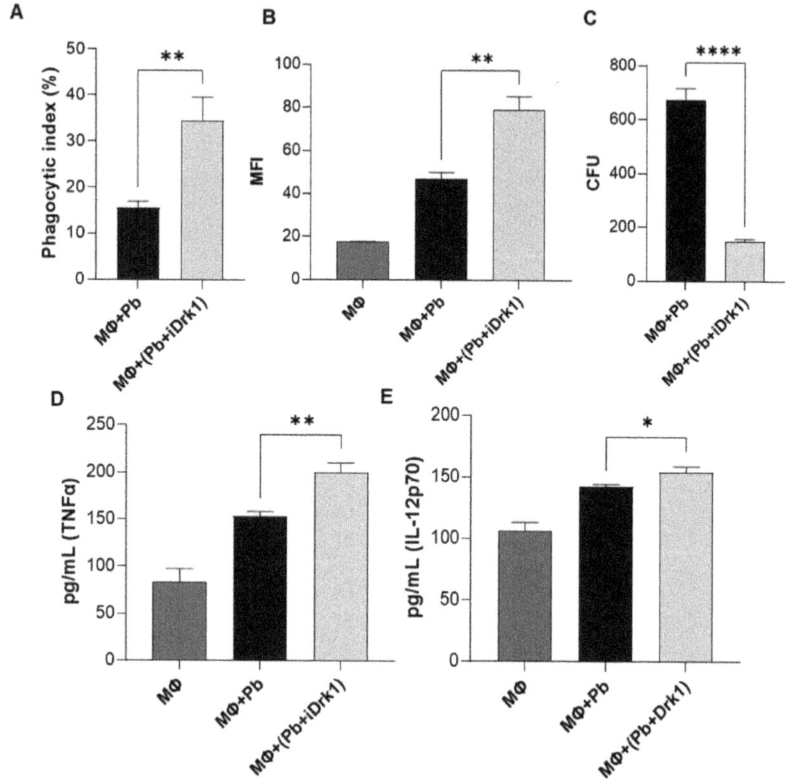

Figure 4. Phagocytosis assay. 2.5 × 10^5 J774 cells (MΦ) were plated in RPMI medium supplemented with 10% FBS. Then the cells were primed for 30 min with LPS 100 ng/mL. In parallel, *P. brasiliensis* yeasts were subjected to 25 μg/mL of iDrk1 for 24 h. The interaction was carried out in a 2:1 ratio (yeast:macrophages) for 24 h at 37 °C and 5% CO_2. After this period, the cells were stained and visualized under an optical microscope to count the internalized yeasts. (**A**) Phagocytic index was calculated, where p-value ≤ 0.005 (**). (**B**) Phagocytosis assay performed by labeling *P. brasiliensis* yeasts with CFSE prior to interaction with macrophages. Fluorescence quantification was performed by flow cytometry and relative quantification was obtained through the median intensity (MFI) where p-value ≤ 0.005 (**). (**C**) CFU determination from yeasts recovered from macrophages, where p-value ≤ 0.0001 (****). Cytokine assay from J774 cell supernatant after interaction with *P. brasiliensis* submitted or not with iDrk1. Macrophages were plated in RPMI medium supplemented with 10% FCS. Then, cells were primed for 30 min with LPS 100 ng/mL. In parallel, *P. brasiliensis* yeasts were subjected to 25 μg/mL of iDrk1 for 24 h. The interaction was carried out in a 2:1 ratio (yeast:macrophages) for 24 h at 37 °C and 5% CO_2. After this period, the supernatant was collected and the cytokines (**D**) TNFα, (**E**) IL-12p70 were measured, where p-value ≤ 0.05 (*), p-value ≤ 0.01 (**).

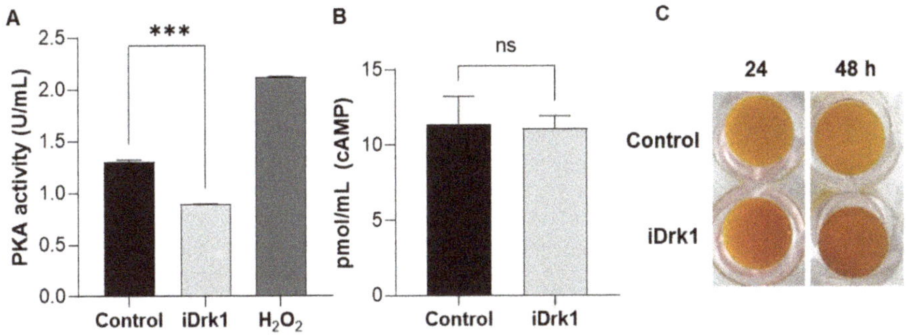

Figure 5. PKA activity and cAMP quantification in *P. brasiliensis* after incubation with iDrk1 (**A**) Quantification of PKA activity and (**B**) dosage of *P. brasiliensis* cAMP after 24 h incubation with 25 μg/mL of iDrk1, where p-value ≤ 0.001 (***). (**C**) After 24 and 48 h of yeast incubation in the presence of the inhibitor, the cells were stained with a 0.2% iodine and 0.4% potassium iodide solution to assess glycogen accumulation.

The cell stress response is orchestrated by connecting several pathways that converge to promote cell survival. As previously described, the cell wall integrity pathway also activates responses that bind to cAMP-PKA. Activation of PKA regulates cellular functions related to glycogen metabolism [63], and it antagonizes its intracellular accumulation [64]. Thus, Pb18 yeasts incubated in the presence or absence of iDrk1 for 24 and 48 h were stained with iodine solution and photographed (Figure 5C). The greater the accumulation of intracellular glycogen, the darker the sample becomes [33]. In Figure 5C, there was an accumulation of intracellular glycogen in relation to the control. Combined with the gene expression analysis and quantification of the cell wall components, this result suggests that iDrk1 may modulate the cell wall component synthesis, generating a cellular response that leads to a decrease in PKA activity. This decreased activity may lead to intracellular glycogen accumulation [65,66].

4. Discussion

Currently, the number of antifungal substances available for the treatment of phytopathogens is approximately nine times greater than those available for the treatment of mycoses in mammals [67]. In this scenario, it is important to emphasize the need for studies that unravel the mechanisms of fungal pathogenicity. With a better understanding of these pathways and the discovery of new targets, it will be possible to develop new drugs with antifungal potential. Thus, this study aimed to understand better the role of a histidine kinase (PbDrk1), an important regulator in the dimorphic switch and morphogenesis of the cell wall of *P. brasiliensis*.

We evaluated pharmacological inhibitors as a strategy to elucidate the role of PbDrk1. The inhibitors iprodione and fludioxonil (both group III histidine kinase inhibitors) are substances widely used in agriculture to combat phytopathogens [68]. Molecular biology and cell signalling studies have shown that these molecules act specifically on group III histidine kinases through interactions with HAMP domains, which are present only in this group of kinases [39,69]. In *Saccharomyces cerevisiae*, group III histidine kinases are absent. When an orthologous gene to histidine kinase group III of *Neurospora crassa* is introduced, *S. cerevisiae* becomes sensitive to pyrrolnitrine [68,70]. Other pathogenic fungi with group III histidine kinases, such as *B. dermatitidis* and *Candida albicans*, were submitted to this inhibitor and presented various sensitivities [17]. For Pb18 yeast cells, fludioxonil was used at 100 μg/mL, showing no reduced cell viability. Thus, the concentration established in the other tests at 25 μg/mL was selected since it is an intermediate concentration and was already established in *B. dermatitidis* [17].

The fungal cell wall plays a fundamental role in the host-parasite interaction since its composition can influence the immune response [13]. In *Paracoccidioides* spp., during the dimorphic transition, the cell wall morphogenesis machinery is activated to remodel its components. In *H. capsulatum* and *B. dermatitidis*, it was shown that *DRK1* acts as a regulator of dimorphism and virulence. A study carried out in *B. dermatitidis* silenced the *DRK1* gene and demonstrated a blockade of the dimorphic transition from mycelium to yeast at 37 °C. Furthermore, this silencing impaired the expression of *BAD1* (a virulence gene activated during transition) [11]. In *H. capsulatum*, Drk1 regulates genes specifically expressed in the yeast phase, such as *CBP1*, *YPS-3* and *AGS1* [11]. In *C. albicans*, the deletion of *NIK1*, a group III histidine kinase, makes yeast incapable of transitioning from yeast to hyphae, consequently making it less virulent [71]. In *Penicillium marneffei*, the deletion of the *DRKA* and *SLNA* genes was also essential for the dimorphic transition during macrophage infection and conidial germination, respectively [44].

The ability of *Paracoccidioides* spp. to cause PCM depends on its dimorphic transition and establishment in the host, either by the resistance and evasion of the immune response machinery or by its ability to be a facultative intracellular parasite [8]. The use of iDrk1 (fludioxonil) in *P. brasiliensis* cultivated in the form of mycelium prevented it from performing the dimorphic transition efficiently, confirming previous findings [20]. A study carried out in *Penicillium marneffei* showed the importance of *DRK1* in cell wall morphogenesis. Strains deleted for this gene were not able to grow in a medium supplemented with cell wall stressors (Congo red) and osmotic stress agents (NaCl and sorbitol) [44]. In that same study, *DRK1* mutant hyphal growth and conidial germination were affected, and transmission electron microscopy images showed significant thickening of the cell wall [44]. These data support the results obtained in Pb18. In this work, yeasts submitted to iDrk1 and later inoculated in culture medium supplemented with cell wall stressors had severely impaired growth.

In *Paracoccidioides* spp. the cell wall chitin content represents a significant fraction of the cell dry weight [49]. The *CHT3*, *CHS2*, and *CHS3* genes showed significant increases compared with the control that was not treated with iDrk1. In *C. albicans*, it was observed that overexpression of *CHT3* induces greater sensitivity to cell wall stressors, and the *CHS2* gene acts as an important regulator [72,73]. In *P. brasiliensis*, it is known that the *CHS3* gene is mainly expressed in the yeast phase [49] and is related to cell growth [74]. These results suggest that PbDrk1 inactivity may modulate the expression of some genes involved in chitin synthesis, leading to cell wall instability.

In addition to chitin, one of the main elements that comprise the cell wall of *Paracoccidioides* spp. are glucan molecules. As already mentioned, the predominance of β-glucan during the mycelial phase is reverted to α-glucan in the yeast phase [2]. This mechanism is seen as one strategy to evade the immune system. The level of α-glucan in the cell wall can be related to the degree of virulence [14]. Among the genes analysed, *AGN1* was characterized as phase-specific in *P. brasiliensis* yeasts, regulating the synthesis of α-(1,3)-glucan [54]. Our data showed that PbDrk1 does not participate in the modulation of this gene. On the other hand, genes involved in the synthesis and maintenance of β-(1,3)-glucan and β-(1,6)-glucan chains, such as *FKS1*, *KRE6*, *PHR2*, and *GEL3*, showed significant increases after exposure to iDrk1. These results indicate that PbDrk1 regulates the expression of cell wall synthesis genes directly related to the dimorphic transition. The expression data were confirmed by evaluating the levels of β-glucan and chitin in Pb18 treated with a Drk1 inhibitor. The glucan modulation transition is essential for masking β-glucan molecules, as the host's immune system directly recognizes these molecules by phagocytic cells.

In the host, the fungus is initially recognized by PRRs and PAMPs [75]. Among the receptors that may be involved, we highlight the Dectin-1 receptor that specifically binds to β-glucan [76]. Chitin molecules are also recognized by the immune system via TLR-2, inducing the production of cytokines and the recruitment of phagocytic cells by the recognition by Dectin-1 [77]. The outermost region of the *P. brasiliensis* cell wall comprises a thick layer of mannan [2]. This layer is believed to protect fungal cells from immunological recognition, hiding the main immunogenic molecules [78]. Confocal microscopy assays indicated an increase in β-glucan labelling. These images showed that yeasts treated with iDrk1 showed greater fluorescence intensity when labelled with the Dectin-1 receptor. Thus, the increase in β-glucan and increased expression and synthesis of chitin in cells treated with iDrk1 indicate that this pathway may be related to virulence and fungal cell wall architecture.

In parallel, the cAMP-PKA pathway also plays a role in cell wall remodeling. In *Cryptococcus neoformans*, PKA is involved in the phosphorylation of components that, when translocated to the nucleus, regulate the expression of genes involved in cell wall integrity [59]. In fungi, this pathway is related to cell growth, differentiation, stress response, pathogenicity, and cell wall integrity, among others [61]. In *C. neoformans*, it was shown that the deletion of several genes involved in cell wall integrity led to impairment of the cAMP-PKA pathway [79]. In *C. albicans*, this pathway plays a central regulator of its morphological transition and, consequently, pathogenicity [80]. In *P. lutzii*, it was reported that the specific inhibition of PKA impedes dimorphic transition [81]. Recently, Garcia et al.; (2017) showed a new insight into the signalling pathways involved in the regulation of cell wall integrity. Alterations of the β-1,3-glucan network in the fungal cell wall induced the activation of the CWI pathway and in parallel inhibited the PKA signalling [33]. Thus, when submitting Pb18 yeasts to iDrk1, we observed a reduction in PKA activity, supporting literature data on the PKA influence on cell wall gene expression and glycogen accumulation [33,59] and a possible correlation with the PbDrk1 pathway.

On the other hand, there was no significant difference in cAMP levels in the presence or absence of iDrk1. Cyclic AMP is a secondary messenger whose intracellular levels are regulated by adenylate cyclase and the phosphodiesterase enzyme balance [65]. Both are regulated by other signalling pathways and not exclusively via PKA [82]. Thus, it is possible to infer that PbDrk1 modulates PKA activity but does not correlate with cAMP levels. In this case, the maintenance of cAMP levels can be regulated by other pathways, such as calcineurin, MAP kinases, and G protein subunits [80,83]. Finally, we can infer that PbDrk1 modulates intracellular levels of intracellular glycogen via PKA. In *Aspergillus fumigatus*, deletion of the PKA catalytic subunit led to an increase in intracellular glycogen levels [65]. Furthermore, in *S. cerevisiae*, it was found that glycogen is present in two fractions, one of them in the cell cytosol in its soluble form and the other associated with cell wall components, covalently linked specifically to β-(1,3)-glucan and β-(1,6)-glucan [84]. This evidence complements the observed results of increased β-glucan exposure on the surface of *P. brasiliensis* after exposure to iDrk1. Together, the results obtained thus far point to the participation of PbDrk1 in the *P. brasiliensis* cell wall modulation.

During the immune response in a fungal infection, the action of pro- and anti-inflammatory cytokines is essential to determine disease progression and/or pathogen clearance [85]. Host resistance to infection by *Paracoccidioides* spp. is associated with a Th1 response, which induces macrophage activation and actively controls fungal dissemination [86]. The individual's susceptibility to the disease is associated with a Th2 response [87]. It is known that the participation of the Dectin-1 receptor together with TLR in the recognition pathway of *Paracoccidioides* spp. triggers the production of several proinflammatory cytokines [88], such as TNFα. This molecule is produced by cells involved in the immune system, including activated macrophages and regulatory T cells, acting as central mediators in inflammation and the regulation of the immune response [89]. It is possible to observe the increased production of this cytokine by macrophages that phagocytosed yeasts exposed to iDrk1, indicating a more exacerbated proinflammatory response when compared to the control. Furthermore, a significant increase in the expression of the cytokine IL12p70 was also observed, produced by monocytes, macrophages, and dendritic cells, and it is an important element in the activation of the Th1 response [90].

5. Conclusions

The data obtained during this work will contribute to a better understanding of a regulatory pathway that has not yet been explored in this model. The PbDrk1 protein has already been shown to be a key element in the dimorphic transition pathway [20]. It is now possible to state that it is a likely virulence factor in regulating cell wall genes. The activity of PbDrk1 negatively modulates the synthesis of molecules such as chitin and β-glucans, contributing to its masking and favouring the pathogenicity of *P. brasiliensis*. Histidine kinases are proteins absent in mammals, and the inhibition of their activity makes fungal cells susceptible to immune system cells. In other fungal models, the loss of this gene represents a decrease in virulence. Finally, the study of PbDrk1 and associated pathways would enable a different approach to the development of new antifungal drugs.

Author Contributions: Conceptualization, W.L.B. and M.V.N.; methodology, M.V.N., Y.N.d.B., W.D.S., A.F.A.C., K.S.F., G.P.J. and P.X.; software, M.V.N., Y.N.d.B., W.D.S.; validation, M.V.N. and Y.N.d.B.; formal analysis, M.V.N.; writing—review and editing, W.L.B. and M.V.N.; supervision, W.L.B.; project administration, W.L.B. and M.V.N.; funding acquisition, W.L.B. All authors have read and agreed to the published version of the manuscript.

Funding: This research was funded by Fundação de Amparo à Pesquisa do Estado de São Paulo (FAPESP), grant number 2017/04592-0 and Conselho Nacional de Desenvolvimento Científico e Tecnológico (CNPq) grant number 141726/2017-2; 311008/2020-8 and Coordenação de Aperfeiçoamento de Pessoal de Nível Superior (CAPES).

Institutional Review Board Statement: Not applicable.

Informed Consent Statement: Not applicable.

Acknowledgments: We are thankful to Sandro Rogério de Almeida (Universidade de São Paulo—USP, Brazil) for technical assistance and material donation to ELISA and phagocytosis experiments. We are thankful for the support of Núcleo de Microscopia Confocal Unifesp Diadema for the confocal microscopy experiments.

Conflicts of Interest: The authors declare no conflict of interest.

References

1. Shikanai-Yasuda, M.A.; Mendes, R.P.; Colombo, A.L.; de Queiroz Telles, F.; Kono, A.; Paniago, A.M.M.; Nathan, A.; do Valle, A.C.F.; Bagagli, E.; Benard, G.; et al. II Consenso Brasileiro em Paracoccidioidomicose-2017. *Epidemiol. Serviços Saúde* **2018**, *27*, e0500001. [CrossRef] [PubMed]
2. Chaves, A.F.A.; Navarro, M.V.; de Barros, Y.N.; Silva, R.S.; Xander, P.; Batista, W.L. Updates in *Paracoccidioides* biology and genetic advances in fungus manipulation. *J. Fungi* **2021**, *7*, 116. [CrossRef] [PubMed]
3. Delboni Nunes, N.E.; Schmidt, E.B.; Massaroni Peçanha, M.A.; Zanotti, R.L.; Gonçalves Ferreira, C.U.; Lamas de Araújo, M.; Potratz, F.F.; Peçanha, P.M.; Batista Ferreira, M.E.; Delmaestro, D.; et al. Paracoccidioidomycosis: Epidemiological and Clinical Aspects in 546 Cases Studied in the State of Espírito Santo, Brazil. *Am. J. Trop. Med. Hyg.* **2017**, *97*, 836–844. [CrossRef]
4. de Almeida, S.M. Central nervous system paracoccidioidomycosis: An overview. *Braz. J. Infect. Dis.* **2005**, *9*, 126–133. [CrossRef]
5. de Almeida, F.A.; Neves, F.F.; Mora, D.J.; Dos Reis, T.A.; Sotini, D.M.; Ribeiro, B.D.M.; Andrade-Silva, L.E.; Nascentes, G.N.; Ferreira-Paim, K.; Silva-Vergara, M.L. Paracoccidioidomycosis in Brazilian Patients With and Without Human Immunodeficiency Virus Infection. *Am. J. Trop. Med. Hyg.* **2017**, *96*, 368–372. [CrossRef]
6. Goughenour, K.D.; Rappleye, C.A. Antifungal therapeutics for dimorphic fungal pathogens. *Virulence* **2016**, *8*, 211–221. [CrossRef]
7. Santos, L.A.; Grisolia, J.C.; Burger, E.; de Araujo Paula, F.B.; Dias, A.L.T.; Malaquias, L.C.C. Virulence factors of *Paracoccidioides brasiliensis* as therapeutic targets: A review. *Antonie van Leeuwenhoek* **2020**, *5*, 593–604. [CrossRef]
8. Giusiano, G. The Trojan Horse Model in *Paracoccidioides*: A Fantastic Pathway to Survive Infecting Human Cells. *Front. Cell. Infect. Microbiol.* **2020**, *10*, 605679. [CrossRef]
9. Calich, V.L.G.; Mamoni, R.L.; Loures, F.V. Regulatory T cells in paracoccidioidomycosis. *Virulence* **2019**, *10*, 810–821. [CrossRef]
10. Boyce, K.J.; Andrianopoulos, A. Fungal dimorphism: The switch from hyphae to yeast is a specialized morphogenetic adaptation allowing colonization of a host. *FEMS Microbiol. Rev.* **2015**, *39*, 797–811. [CrossRef]
11. Nemecek, J.C.; Wüthrich, M.; Klein, B.S. Global Control of Dimorphism and Virulence in Fungi. *Science* **2006**, *312*, 583–588. [CrossRef]
12. Campos, C.B.L.; Di Benedette, J.P.T.; Morais, F.V.; Ovalle, R.; Nobrega, M.P. Evidence for the role of calcineurin in morphogenesis and calcium homeostasis during mycelium-to-yeast dimorphism of *Paracoccidioides brasiliensis*. *Eukaryot. Cell* **2008**, *7*, 1856–1864. [CrossRef]
13. Puccia, R.; Vallejo, M.C.; Matsuo, A.L.; Longo, L.V.G. The *Paracoccidioides* Cell Wall: Past and Present Layers Toward Understanding Interaction with the Host. *Front. Microbiol.* **2011**, *2*, 257. [CrossRef]
14. Hogan, L.H.; Klein, B.S. Altered expression of surface alpha-1,3-glucan in genetically related strains of *Blastomyces dermatitidis* that differ in virulence. *Infect. Immun.* **1994**, *62*, 3543–3546. [CrossRef]

15. Klein, B.S.; Tebbets, B. Dimorphism and virulence in fungi. *Curr. Opin. Microbiol.* **2007**, *10*, 314–319. [CrossRef]
16. Felipe, M.S.S.; Andrade, R.V.; Arraes, F.B.M.M.; Nicola, A.M.; Maranhão, A.Q.; Torres, F.A.G.G.; Silva-Pereira, I.; Poças-Fonseca, M.J.; Campos, E.G.; Moraes, L.M.P.P.; et al. Transcriptional profiles of the human pathogenic fungus *Paracoccidioides brasiliensis* in mycelium and yeast cells. *J. Biol. Chem.* **2005**, *280*, 24706–24714. [CrossRef]
17. Lawry, S.M.; Tebbets, B.; Kean, I.; Stewart, D.; Hetelle, J.; Klein, B.S. Fludioxonil Induces Drk1, a Fungal Group III Hybrid Histidine Kinase, to Dephosphorylate its Downstream Target, Ypd1. *Antimicrob. Agents Chemother.* **2016**, *61*, e01414-16. [CrossRef]
18. Hou, B.; Zhang, Z.; Zheng, F.; Liu, X. Molecular cloning, characterization and differential expression of DRK1 in *Sporothrix schenckii*. *Int. J. Mol. Med.* **2013**, *31*, 99–104. [CrossRef]
19. Boyce, K.J.; Cao, C.; Andrianopoulos, A. Two-Component Signaling Regulates Osmotic Stress Adaptation via SskA and the High-Osmolarity Glycerol MAPK Pathway in the Human Pathogen *Talaromyces marneffei*. *mSphere* **2016**, *1*, e00086-15. [CrossRef]
20. Chaves, A.F.A.; Navarro, M.V.; Castilho, D.G.; Calado, J.C.P.; Conceição, P.M.; Batista, W.L. A conserved dimorphism-regulating histidine kinase controls the dimorphic switching in *Paracoccidioides brasiliensis*. *FEMS Yeast Res.* **2016**, *16*, fow047. [CrossRef]
21. Mizuno, T.; Wurtzel, E.T.; Inouye, M. Osmoregulation of gene expression. II. DNA sequence of the envZ gene of the ompB operon of *Escherichia coli* and characterization of its gene product. *J. Biol. Chem.* **1982**, *257*, 13692–13698. [CrossRef]
22. Hérivaux, A.; So, Y.-S.; Gastebois, A.; Latgé, J.-P.; Bouchara, J.-P.; Bahn, Y.-S.; Papon, N. Major Sensing Proteins in Pathogenic Fungi: The Hybrid Histidine Kinase Family. *PLoS Pathog.* **2016**, *12*, e1005683. [CrossRef]
23. Wuichet, K.; Cantwell, B.J.; Zhulin, I.B. Evolution and phyletic distribution of two-component signal transduction systems. *Curr. Opin. Microbiol.* **2010**, *13*, 219–225. [CrossRef]
24. Catlett, N.L.; Yoder, O.C.; Turgeon, B.G. Whole-Genome Analysis of Two-Component Signal Transduction Genes in Fungal Pathogens. *Eukaryot. Cell* **2003**, *2*, 1151–1161. [CrossRef]
25. Defosse, T.A.; Sharma, A.; Mondal, A.K.; de Bernonville, T.D.; Latgé, J.P.; Calderone, R.; Giglioli-Guivarc'h, N.; Courdavault, V.; Clastre, M.; Papon, N. Hybrid histidine kinases in pathogenic fungi. *Mol. Microbiol.* **2015**, *95*, 914–924. [CrossRef]
26. Shor, E.; Chauhan, N. A Case for Two-Component Signaling Systems As Antifungal Drug Targets. *PLoS Pathog.* **2015**, *11*, e1004632. [CrossRef]
27. Chauhan, N.; Latge, J.-P.; Calderone, R. Signalling and oxidant adaptation in *Candida albicans* and *Aspergillus fumigatus*. *Nat. Rev. Microbiol.* **2006**, *4*, 435–444. [CrossRef]
28. Holbrook, E.D.; Rappleye, C.A. *Histoplasma capsulatum* pathogenesis: Making a lifestyle switch. *Curr. Opin. Microbiol.* **2008**, *11*, 318–324. [CrossRef]
29. Castilho, D.G.; Navarro, M.V.; Chaves, A.F.A.A.; Xander, P.; Batista, W.L. Recovery of the *Paracoccidioides brasiliensis* virulence after animal passage promotes changes in the antioxidant repertoire of the fungus. *FEMS Yeast Res.* **2018**, *18*, foy007. [CrossRef]
30. Livak, K.J.; Schmittgen, T.D. Analysis of Relative Gene Expression Data Using Real-Time Quantitative PCR and the $2-\Delta\Delta CT$ Method. *Methods* **2001**, *25*, 402–408. [CrossRef]
31. Nogueira, M.; Istel, F.; Jenull, S.; Walker, L.; Gow, N.; Lion, T. Quantitative Analysis of *Candida* Cell Wall Components by Flow Cytometry with Triple-Fluorescence Staining. *J. Microbiol. Mod. Tech.* **2017**, *2*, 101. [CrossRef]
32. Bradford, M. A Rapid and Sensitive Method for the Quantitation of Microgram Quantities of Protein Utilizing the Principle of Protein-Dye Binding. *Anal. Biochem.* **1976**, *72*, 248–254. [CrossRef]
33. García, R.; Bravo, E.; Diez-Muñiz, S.; Nombela, C.; Rodríguez-Peña, J.M.; Arroyo, J. A novel connection between the Cell Wall Integrity and the PKA pathways regulates cell wall stress response in yeast. *Sci. Rep.* **2017**, *7*, 5703. [CrossRef] [PubMed]
34. Crosby, L.; Casey, W.; Morgan, K.; Ni, H.; Yoon, L.; Easton, M.; Misukonis, M.; Burleson, G.; Ghosh, D.K. Murine J774 macrophages recognize LPS/IFN-g, non-CpG DNA or two-CpG DNA-containing sequences as immunologically distinct. *Nitric Oxide Biol. Chem.* **2010**, *22*, 242–257. [CrossRef]
35. Romera, L.M.D.; Kaihami, G.H.; Jannuzzi, G.P.; de Almeida, J.R.F.; de Almeida, S.R. The Critical Role of Notch1–TLR 4 Signaling in the Inflammatory and Fungicidal Activity of Macrophages Against *Paracoccidioides brasiliensis* Strain Pb18. *Mycopathologia* **2017**, *182*, 797–807. [CrossRef]
36. Popi, A.F.; Daniel Lopes, J.; Mariano, M. GP43 from *Paracoccidioides brasiliensis* inhibits macrophage functions. An evasion mechanism of the fungus. *Cell. Immunol.* **2002**, *218*, 87–94. [CrossRef]
37. Nuutila, J.; Lilius, E.M. Flow cytometric quantitative determination of ingestion by phagocytes needs the distinguishing of overlapping populations of binding and ingesting cells. *Cytom. Part A* **2005**, *65*, 93–102. [CrossRef]
38. Arima, K.; Imanaka, H.; Kousaka, M.; Fukuta, A.; Tamura, G. Pyrrolnitrin, a New Antibiotic Substance, Produced by *Pseudomonas*. *Agric. Biol. Chem.* **1964**, *28*, 575–576. [CrossRef]
39. Brandhorst, T.T.; Kean, I.R.L.; Lawry, S.M.; Wiesner, D.L.; Klein, B.S. Phenylpyrrole fungicides act on triosephosphate isomerase to induce methylglyoxal stress and alter hybrid histidine kinase activity. *Sci. Rep.* **2019**, *9*, 5047. [CrossRef]
40. Cui, S.; Hassan, R.Y.A.; Heintz-Buschart, A.; Bilitewski, U. Regulation of *Candida albicans* interaction with macrophages through the activation of HOG pathway by genistein. *Molecules* **2016**, *21*, 162. [CrossRef]
41. Kamei, M.; Yamashita, K.; Takahashi, M.; Fukumori, F.; Ichiishi, A.; Fujimura, M. Deletion and expression analysis of beta-(1,3)-glucanosyltransferase genes in *Neurospora crassa*. *Fungal Genet. Biol.* **2013**, *52*, 65–72. [CrossRef]
42. McCormick, A.; Jacobsen, I.D.; Broniszewska, M.; Beck, J.; Heesemann, J.; Ebel, F. The two-component sensor kinase TcsC and its role in stress resistance of the human-pathogenic mold *Aspergillus fumigatus*. *PLoS ONE* **2012**, *7*, e38262. [CrossRef]
43. Yoshimi, A.; Miyazawa, K.; Abe, K. Function and Biosynthesis of Cell Wall α-1,3-Glucan in Fungi. *J. Fungi* **2017**, *3*, 63. [CrossRef]

44. Boyce, K.J.; Schreider, L.; Kirszenblat, L.; Andrianopoulos, A. The two-component histidine kinases DrkA and SlnA are required for in vivo growth in the human pathogen *Penicillium marneffei*. *Mol. Microbiol.* **2011**, *82*, 1164–1184. [CrossRef]
45. Zhang, Z.; Hou, B.; Xin, Y.; Liu, X. Protein Profiling of the Dimorphic Pathogenic Fungus, *Sporothrix schenckii*. *Mycopathologia* **2012**, *173*, 1–11. [CrossRef]
46. Roncero, C.; Durán, A. Effect of Calcofluor white and Congo red on fungal cell wall morphogenesis: In vivo activation of chitin polymerization. *J. Bacteriol.* **1985**, *163*, 1180–1185. [CrossRef]
47. Ram, A.F.J.; Klis, F.M. Identification of fungal cell wall mutants using susceptibility assays based on Calcofluor white and Congo red. *Nat. Protoc.* **2006**, *1*, 2253–2256. [CrossRef]
48. Ene, I.V.; Walker, L.A.; Schiavone, M.; Lee, K.K.; Martin-Yken, H.; Dague, E.; Gow, N.A.R.; Munro, C.A.; Brown, A.J.P. Cell wall remodeling enzymes modulate fungal cell wall elasticity and osmotic stress resistance. *MBio* **2015**, *6*, e00986-15. [CrossRef]
49. Niño-Vega, G.A.; Munro, C.A.; San-Blas, G.; Gooday, G.W.; Gow, N.A. Differential expression of chitin synthase genes during temperature-induced dimorphic transitions in *Paracoccidioides brasiliensis*. *Med. Mycol.* **2000**, *38*, 31–39. [CrossRef]
50. Zambuzzi-Carvalho, P.; Tomazett, P.; Santos, S.; Ferri, P.; Borges, C.; Martins, W.; de Almeida Soares, C.; Pereira, M. Transcriptional profile of *Paracoccidioides* induced by oenothein B, a potential antifungal agent from the Brazilian Cerrado plant Eugenia uniflora. *BMC Microbiol.* **2013**, *13*, 227. [CrossRef]
51. Degani, G.; Popolo, L. The Glucan-Remodeling Enzyme Phr1p and the Chitin Synthase Chs1p Cooperate to Maintain Proper Nuclear Segregation and Cell Integrity in *Candida albicans*. *Front. Cell. Infect. Microbiol.* **2019**, *9*, 400. [CrossRef]
52. Bolouri Moghaddam, M.-R.; Vilcinskas, A.; Rahnamaeian, M. The insect-derived antimicrobial peptide metchnikowin targets *Fusarium graminearum* β(1,3)glucanosyltransferase Gel1, which is required for the maintenance of cell wall integrity. *Biol. Chem.* **2017**, *398*, 491–498. [CrossRef]
53. da Silva Castro, N.; Maia, Z.A.; Pereira, M.; de Almeida Soares, C.M. Screening for glycosylphosphatidylinositol-anchored proteins in the *Paracoccidioides brasiliensis* transcriptome. *Genet. Mol. Res.* **2005**, *4*, 326–345.
54. Villalobos-Duno, H.; San-Blas, G.; Paulinkevicius, M.; Sánchez-Martín, Y.; Nino-Vega, G. Biochemical Characterization of *Paracoccidioides brasiliensis* α-1,3-Glucanase Agn1p, and Its Functionality by Heterologous Expression in *Schizosaccharomyces pombe*. *PLoS ONE* **2013**, *8*, e66853. [CrossRef]
55. Gonçales, R.A.; Ricci-Azevedo, R.; Vieira, V.C.S.; Fernandes, F.F.; de Oliveira Thomaz, S.M.; Carvalho, A.; Vendruscolo, P.E.; Cunha, C.; Roque-Barreira, M.C.; Rodrigues, F. Paracoccin overexpression in *Paracoccidioides brasiliensis* enhances fungal virulence by remodeling chitin properties of the cell wall. *J. Infect. Dis.* **2020**, *224*, 164–174. [CrossRef]
56. Wagener, J.; Striegler, K.; Wagener, N. α- and β-1,3-glucan synthesis and remodeling. *Curr. Top. Microbiol. Immunol.* **2020**, *425*, 53–82. [CrossRef]
57. Kanetsuna, F.; Carbonell, L.M.; Azuma, I.; Yamamura, Y. Biochemical Studies on the Thermal Dimorphism of *Paracoccidioides brasiliensis*. *J. Bacteriol.* **1972**, *110*, 208–218. [CrossRef]
58. Lin, C.J.; Wu, C.Y.; Yu, S.J.; Chen, Y.L. Protein kinase A governs growth and virulence in *Candida tropicalis*. *Virulence* **2018**, *9*, 331–347. [CrossRef]
59. Caza, M.; Kronstad, J.W. The cAMP/Protein Kinase a Pathway Regulates Virulence and Adaptation to Host Conditions in *Cryptococcus neoformans*. *Front. Cell. Infect. Microbiol.* **2019**, *9*, 212. [CrossRef] [PubMed]
60. Bravo Ruiz, G.; Lorenz, A. What do we know about the biology of the emerging fungal pathogen of humans *Candida auris*? *Microbiol. Res.* **2021**, *242*, 126621. [CrossRef] [PubMed]
61. Zhu, W.; Zhou, M.; Xiong, Z.; Peng, F.; Wei, W. The cAMP-PKA Signaling Pathway Regulates Pathogenicity, Hyphal Growth, Appressorial Formation, Conidiation, and Stress Tolerance in *Colletotrichum higginsianum*. *Front. Microbiol.* **2017**, *8*, 1416. [CrossRef] [PubMed]
62. Liu, Y.; Yang, F.; Li, S.; Dai, J.; Deng, H. Glutaredoxin Deletion Shortens Chronological Life Span in *Saccharomyces cerevisiae* via ROS-Mediated Ras/PKA Activation. *J. Proteome Res.* **2018**, *17*, 2318–2327. [CrossRef] [PubMed]
63. Freitas, F.Z.; de Paula, R.M.; Barbosa, L.C.B.; Terenzi, H.F.; Bertolini, M.C. cAMP signaling pathway controls glycogen metabolism in *Neurospora crassa* by regulating the glycogen synthase gene expression and phosphorylation. *Fungal Genet. Biol.* **2010**, *47*, 43–52. [CrossRef] [PubMed]
64. Smith, A.; Ward, M.P.; Garrett, S. Yeast PKA represses Msn2p/Msn4p-dependent gene expression to regulate growth, stress response and glycogen accumulation. *EMBO J.* **1998**, *17*, 3556–3564. [CrossRef]
65. de Assis, L.J.; Manfiolli, A.; Mattos, E.; Fabri, J.H.T.M.; Malavazi, I.; Jacobsen, I.D.; Brock, M.; Cramer, R.A.; Thammahong, A.; Hagiwara, D.; et al. Protein Kinase A and High-Osmolarity Glycerol Response Pathways Cooperatively Control Cell Wall Carbohydrate Mobilization in *Aspergillus fumigatus*. *MBio* **2018**, *9*, e01952-18. [CrossRef]
66. Schmelzle, T.; Beck, T.; Martin, D.E.; Hall, M.N. Activation of the RAS/cyclic AMP pathway suppresses a TOR deficiency in yeast. *Mol. Cell. Biol.* **2004**, *24*, 338–351. [CrossRef]
67. Fisher, M.C.; Hawkins, N.J.; Sanglard, D.; Gurr, S.J. Worldwide emergence of resistance to antifungal drugs challenges human health and food security. *Science* **2018**, *360*, 739–742. [CrossRef]
68. Kilani, J.; Fillinger, S. Phenylpyrroles: 30 years, two molecules and (nearly) no resistance. *Front. Microbiol.* **2016**, *7*, 2014. [CrossRef]
69. Schumacher, M.M.; Enderlin, C.S.; Selitrennikoff, C.P. The osmotic-1 locus of *Neurospora crassa* encodes a putative histidine kinase similar to osmosensors of bacteria and yeast. *Curr. Microbiol.* **1997**, *34*, 340–347. [CrossRef]

70. Motoyama, T.; Kadokura, K.; Ohira, T.; Ichiishi, A.; Fujimura, M.; Yamaguchi, I.; Kudo, T. A two-component histidine kinase of the rice blast fungus is involved in osmotic stress response and fungicide action. *Fungal Genet. Biol.* **2005**, *42*, 200–212. [CrossRef]
71. Yamada-Okabe, T.; Mio, T.; Ono, N.; Kashima, Y.; Matsui, M.; Arisawa, M.; Yamada-Okabe, H. Roles of three histidine kinase genes in hyphal development and virulence of the pathogenic fungus *Candida albicans*. *J. Bacteriol.* **1999**, *181*, 7243–7247. [CrossRef]
72. Dünkler, A.; Wendland, J.J.; Dünkler, A.; Wendland, J.J. *Candida albicans* Rho-type GTPase-encoding genes required for polarized cell growth and cell separation. *Eukaryot. Cell* **2007**, *6*, 844–854. [CrossRef]
73. Preechasuth, K.; Anderson, J.C.; Peck, S.C.; Brown, A.J.P.; Gow, N.A.R.; Lenardon, M.D. Cell wall protection by the *Candida albicans* class I chitin synthases. *Fungal Genet. Biol.* **2015**, *82*, 264–276. [CrossRef]
74. Barreto, L.; Sorais, F.; Salazar, V.; San-Blas, G.; Niño-Vega, G.A. Expression of *Paracoccidioides brasiliensis* CHS3 in a *Saccharomyces cerevisiae* CHS3 null mutant demonstrates its functionality as a chitin synthase gene. *Yeast* **2010**, *27*, 293–300. [CrossRef]
75. De Quaglia e Silva, J.C.; Della Coletta, A.M.; Gardizani, T.P.; Romagnoli, G.G.; Kaneno, R.; Dias-Melicio, L.A. Involvement of the Dectin-1 Receptor upon the Effector Mechanisms of Human Phagocytic Cells against *Paracoccidioides brasiliensis*. *J. Immunol. Res.* **2019**, *2019*, 1529189. [CrossRef]
76. Romagnolo, A.G.; de Quaglia e Silva, J.C.; Della Coletta, A.M.; Gardizani, T.P.; Martins, A.T.L.; Romagnoli, G.G.; Kaneno, R.; de Campos Soares, A.M.V.; De Faveri, J.; Dias-Melicio, L.A. Role of Dectin-1 receptor on cytokine production by human monocytes challenged with *Paracoccidioides brasiliensis*. *Mycoses* **2018**, *61*, 222–230. [CrossRef]
77. Elieh Ali Komi, D.; Sharma, L.; Dela Cruz, C.S. Chitin and Its Effects on Inflammatory and Immune Responses. *Clin. Rev. Allergy Immunol.* **2018**, *54*, 213–223. [CrossRef]
78. Yadav, B.; Mora-Montes, H.M.; Wagener, J.; Cunningham, I.; West, L.; Haynes, K.; Brown, A.J.P.; Gow, N.A.R. Differences in fungal immune recognition by monocytes and macrophages: N-mannan can be a shield or activator of immune recognition. *Cell Surf.* **2020**, *6*, 100042. [CrossRef]
79. Donlin, M.J.; Upadhya, R.; Gerik, K.J.; Lam, W.; VanArendonk, L.G.; Specht, C.A.; Sharma, N.K.; Lodge, J.K. Cross talk between the cell wall integrity and cyclic AMP/protein kinase A pathways in *Cryptococcus neoformans*. *MBio* **2014**, *5*, e01573-14. [CrossRef]
80. Cao, C.; Wu, M.; Bing, J.; Tao, L.; Ding, X.; Liu, X.; Huang, G. Global regulatory roles of the cAMP/PKA pathway revealed by phenotypic, transcriptomic and phosphoproteomic analyses in a null mutant of the PKA catalytic subunit in *Candida albicans*. *Mol. Microbiol.* **2017**, *105*, 46–64. [CrossRef]
81. Sestari, S.J.; Brito, W.A.; Neves, B.J.; Soares, C.M.A.; Salem-Izacc, S.M. Inhibition of protein kinase A affects *Paracoccidioides* lutzii dimorphism. *Int. J. Biol. Macromol.* **2018**, *113*, 1214–1220. [CrossRef] [PubMed]
82. Sassone-Corsi, P. The cyclic AMP pathway. *Cold Spring Harb. Perspect. Biol.* **2012**, *4*, a011148. [CrossRef] [PubMed]
83. Koschinski, A.; Zaccolo, M. Activation of PKA in cell requires higher concentration of cAMP than in vitro: Implications for compartmentalization of cAMP signalling. *Sci. Rep.* **2017**, *7*, 14090. [CrossRef] [PubMed]
84. Arvindekar, A.U.; Patil, N.B. Glycogen: A covalently linked component of the cell wall in *Saccharomyces cerevisiae*. *Yeast* **2002**, *19*, 131–139. [CrossRef] [PubMed]
85. Kurokawa, C.S.; Araujo, J.P.; Soares, A.M.V.C.; Sugizaki, M.F.; Peraçoli, M.T.S. Pro- and anti-inflammatory cytokines produced by human monocytes challenged in vitro with *Paracoccidioides brasiliensis*. *Microbiol. Immunol.* **2007**, *51*, 421–428. [CrossRef] [PubMed]
86. Tristão, F.S.M.; Rocha, F.A.; Carlos, D.; Ketelut-Carneiro, N.; Souza, C.O.S.; Milanezi, C.M.; Silva, J.S. Th17-Inducing Cytokines IL-6 and IL-23 Are Crucial for Granuloma Formation during Experimental Paracoccidioidomycosis. *Front. Immunol.* **2017**, *8*, 949. [CrossRef] [PubMed]
87. Fortes, M.R.P.; Miot, H.A.; Kurokawa, C.S.; Marques, M.E.A.; Marques, S.A. Immunology of paracoccidioidomycosis. *An. Bras. Dermatol.* **2011**, *86*, 516–525. [CrossRef] [PubMed]
88. Burger, E. Paracoccidioidomycosis protective immunity. *J. Fungi* **2021**, *7*, 137. [CrossRef]
89. Davis, M.R.; Thompson, G.R.; Patterson, T.F. Fungal Infections Potentiated by Biologics. *Infect. Dis. Clin. N. Am.* **2020**, *34*, 389–411. [CrossRef]
90. Thompson, A.; Orr, S.J. Emerging IL-12 family cytokines in the fight against fungal infections. *Cytokine* **2018**, *111*, 398–407. [CrossRef]

Journal of *Fungi*

MDPI

Article

Specific Functional Features of the Cell Integrity MAP Kinase Pathway in the Dimorphic Fission Yeast *Schizosaccharomyces japonicus*

Elisa Gómez-Gil, Alejandro Franco, Beatriz Vázquez-Marín, Francisco Prieto-Ruiz, Armando Pérez-Díaz, Jero Vicente-Soler, Marisa Madrid, Teresa Soto * and José Cansado *

Yeast Physiology Group, Department of Genetics and Microbiology, Campus de Excelencia Internacional de Ambito Regional (CEIR) Campus Mare Nostrum, Universidad de Murcia, 30071 Murcia, Spain; elisa.gomez2@um.es (E.G.-G.); afranco@um.es (A.F.); beatriz.vazquez@um.es (B.V.-M.); francisco.prieto@um.es (F.P.-R.); armandojesus.perez@um.es (A.P.-D.); jerovic@um.es (J.V.-S.); marisa@um.es (M.M.)

* Correspondence: teresaso@um.es (T.S.); jcansado@um.es (J.C.)

Abstract: Mitogen activated protein kinase (MAPK) signaling pathways execute essential functions in eukaryotic organisms by transducing extracellular stimuli into adaptive cellular responses. In the fission yeast model *Schizosaccharomyces pombe* the cell integrity pathway (CIP) and its core effector, MAPK Pmk1, play a key role during regulation of cell integrity, cytokinesis, and ionic homeostasis. *Schizosaccharomyces japonicus*, another fission yeast species, shows remarkable differences with respect to *S. pombe*, including a robust yeast to hyphae dimorphism in response to environmental changes. We show that the CIP MAPK module architecture and its upstream regulators, PKC orthologs Pck1 and Pck2, are conserved in both fission yeast species. However, some of *S. pombe*'s CIP-related functions, such as cytokinetic control and response to glucose availability, are regulated differently in *S. japonicus*. Moreover, Pck1 and Pck2 antagonistically regulate *S. japonicus* hyphal differentiation through fine-tuning of Pmk1 activity. Chimeric MAPK-swapping experiments revealed that *S. japonicus* Pmk1 is fully functional in *S. pombe*, whereas *S. pombe* Pmk1 shows a limited ability to execute CIP functions and promote *S. japonicus* mycelial development. Our findings also suggest that a modified N-lobe domain secondary structure within *S. japonicus* Pmk1 has a major influence on the CIP signaling features of this evolutionarily diverged fission yeast.

Keywords: fission yeast; MAPK; cell integrity pathway; *S. japonicus*; *S. pombe*; protein kinase C; Pmk1; dimorphism; hyphae

Citation: Gómez-Gil, E.; Franco, A.; Vázquez-Marín, B.; Prieto-Ruiz, F.; Pérez-Díaz, A.; Vicente-Soler, J.; Madrid, M.; Soto, T.; Cansado, J. Specific Functional Features of the Cell Integrity MAP Kinase Pathway in the Dimorphic Fission Yeast *Schizosaccharomyces japonicus*. *J. Fungi* **2021**, 7, 482. https://doi.org/10.3390/jof7060482

Academic Editors: María Molina and Humberto Martín

Received: 18 May 2021
Accepted: 11 June 2021
Published: 14 June 2021

1. Introduction

The evolutionarily conserved MAPK signaling pathways orchestrate multiple and fundamental cellular functions in eukaryotic organisms that assure cell proliferation and survival during unperturbed conditions and in response to environmental stimuli. The intensely studied fission yeast model *Schizosaccharomyces pombe* has three MAPK cascades, known as the stress-activated protein kinase (SAPK) pathway, the cell integrity pathway (CIP), and the pheromone sexual differentiation pathway [1]. Specifically, the CIP regulates a wide range of processes in *S. pombe*, including cell wall composition and integrity, vacuole fusion, cytokinesis, and ionic homeostasis [1–4]. The core effector of the CIP is MAPK Pmk1, an ERK1/2 ortholog that is the downstream target of a MAPK module composed by MAPKKK Mkh1 and MAPKK Pek1 [1,5]. *S. pombe* Pmk1 becomes activated in a cell-cycle-dependent manner during cytokinesis [6], during nutritional changes such as glucose fasting [3], or in response to environmental insults including heat, cell wall damage, osmotic-saline, and oxidative stresses [5,6]. The architecture of the CIP upstream of the Mkh1–Pek1–Pmk1 MAPK module includes two redundant protein kinase C (PKC) orthologs named Pck1 and Pck2 [6–11]. Pck1 and Pck2 share overlapping roles in cell viability and partially complement each other, and their simultaneous deletion elicits a synthetic lethal phenotype [1]. Pck2 operates downstream of the Rho GTPases Rho1 and Rho2, and is the major positive regulator of CIP activity during unperturbed growth and stress, whereas a Rho1–Pck1 branch contributes scarcely to Pmk1 activation during vegetative growth and in response to cell wall stress [6–12]. A complex cross-regulatory relationship between the CIP and TOR signaling that involves the AKT (Gad8)-dependent translation of both Pck1 and Pck2, allows activation of the CIP during growth and stress [13]. However, Rho1 and Rho2 can also synergistically regulate the biosynthesis of the two major *S. pombe* cell wall polymers α- and β-D-glucan through Pck1 and Pck2 and independently of CIP signaling [1]. The CIP and SAPK pathways functionally interact In Vivo, and share common targets including the transcription factor Atf1 and the RNA binding protein Rnc1 [14].

S. japonicus, another fission yeast species, is becoming an emerging model organism to explore developmental and physiological evolutionary changes within the genus *Schizosaccharomyces* thanks to its distinctive features with respect to *S. pombe*. Indeed, *S. japonicus* cells display unusual membrane lipid composition [15], and divide through semi-open mitosis, as opposed to the typical closed mitosis of *S. pombe* cells [16]. In addition, *S. japonicus* utilizes glucose exclusively via fermentation and is essentially respiration defective, whereas *S. pombe* is a respiro-fermentative organism [17]. Most importantly, and unlike the remaining species of the genus, *S. japonicus* yeast cells have the ability to undergo a vigorous and reversible dimorphic switch to hyphae in response to environmental stimuli. Known inducers of hyphal differentiation in this organism include changes in the composition of the growth medium, and DNA damage triggered with camptothecin (CPT), a specific topoisomerase I inhibitor [18–21].

The *S. japonicus* proteome carries gene orthologs encoding the main components of the three MAP kinase pathways present in *S. pombe* [22]. We have recently described that the stress-responsive functions of the SAPK pathway in *S. pombe* are largely conserved in *S. japonicus*, and include regulation of G2/M progression during the cell cycle, cellular adaptation to multiple stress conditions, and the positive control of chronological lifespan and sexual differentiation [23]. Notably, the SAPK pathway acts in concert with a quorum sensing mechanism mediated by aromatic alcohols that represses *S. japonicus* yeast to hypha differentiation in response to increased population density [23]. In contrary, *S. japonicus* mutants lacking the putative CIP MAP kinase module components Mkh1, Pek1, and MAPK Pmk1 display defective mycelia formation in solid medium [22], suggesting that this signaling pathway may play instead a positive role to modulate yeast to hypha dimorphic transition. However, the precise biological functions of the *S. japonicus* CIP during growth, stress, and mycelial differentiation, together with the possible role of PKC orthologs Pck1 and Pck2 as upstream regulators of this signaling cascade, are currently unknown. In this work we show that, despite the existence of some degree of functional conservation, certain biological features of CIP signaling have evolved separately in *S. japonicus* and *S. pombe*, and that Pck1 and Pck2 modulate Pmk1 MAPK activity during *S. japonicus* dimorphic switch in an opposite fashion. Finally, structural changes at the N-lobe of *S. japonicus* Pmk1 likely played a significant role in modulating CIP signaling to cope with the specific developmental requirements of this fission yeast species.

2. Materials and Methods

2.1. Strains, Growth Conditions, and Reagents

References and genotypes of *S. pombe* and *S. japonicus* strains employed in this work are listed in Table S1. They were grown with shaking at 30 °C in yeast extract plus supplements (YES) or Edinburgh minimal medium (EMM2) with 2% glucose, and supplemented with adenine, leucine, histidine, or uracil (100 mg/L, Sigma-Aldrich, St. Louis, MO, USA) [24]. Mutant strains with single/multiple gene deletions, or expressing genomic fusions, were obtained by transformation, or after tetrad or random spore dissection and analysis of appropriate crosses in sporulation medium (SPA) [25]. In stress treatments, log-phase cultures (OD_{600} = 0.5; ~10^6 cells/mL) were supplemented with either KCl (osmotic-saline stress; Sigma-Aldrich), caspofungin (cell wall stress; Sigma-Aldrich), or hydrogen peroxide (oxidative stress; Sigma-Aldrich). In glucose starvation experiments, cells grown in YES medium with 7% glucose were recovered by filtration, and resuspended in the same medium lacking glucose and osmotically equilibrated with 3% glycerol. In yeast to hyphae induction experiments with *S. japonicus*, log-phase cultures grown in YES medium with 6% glucose were treated with camptothecin (CPT; Sigma-Aldrich) to a final concentration of 0.2 μM. YEMA [18] and RGE [26] media supplemented with 2% agar were used to quantify *S. japonicus* mycelial growth. *S. japonicus* transformation by electroporation was performed exactly as described [27].

2.2. Gene Disruption and Gene Fusion

Sequences of *S. japonicus* genes were obtained from the annotated database at *Ensembl-Fungi* (http://fungi.ensembl.org/Schizosaccharomyces_japonicus/Info/Index?db=core (accessed on 14 June 2021)). The *S. japonicus pck1+*, *pck2+*, and *pmk1+* null mutants were obtained by entire deletion of the corresponding coding sequences by PCR-mediated strategy using plasmids pFK14 (*S. japonicus ura4+* gene cloned into pGEMT-easy vector; [28]) or pFA6a-natMX6 [29] as templates, and their replacement with either *ura4+* or *natMX6* cassettes flanked by long 5′ and 3′UTRs of respective genes following a PCR approach [30]. Oligonucleotides employed to obtain each one of the transformation cassettes are shown in Table S2.

To construct the integrative plasmid pFA6a-$Pmk1^{Sj}$:GFP-*natMX6*, the *$pmk1^{+Sj}$* ORF plus 5′ regulatory sequences were amplified by PCR using *S. japonicus* genomic DNA as a template, the 5′-oligonucleotide Pmk1Sj:GFP-*Sal*I-5′, which hybridizes −309 to −288 bp upstream of the *$pmk1^{+Sj}$* start codon and contains a *Sal*I site, and the 3′-oligonucleotide Pmk1Sj:GFP-*Pac*I-3′, which hybridizes at the 3′ end of *$pmk1^{+Sj}$* ORF before the stop codon, and contains a *Pac*I site. The obtained DNA fragment was cloned into plasmid pFA6a-GFP:*natMX6* [29] digested with *Sal*I and *Pac*I. Plasmid pFA6a-$Pmk1^{Sj}$:GFP-*natMX6* was then digested at the unique *Swa*I site within *$pmk1^{+Sj}$* ORF, and transformed into strain NIG2028. *Nat^R* transformants were obtained, and the correct integration and expression of the fusion was verified by DNA sequencing and Western blot analysis.

2.3. Construction of *S. pombe* and *S. japonicus* Strains Expressing Wild-Type and Chimeric Pmk1–HA Fusions

To obtain the integrative plasmid pSL-$Pmk1^{Sj}$:HA, the *$pmk1^{+Sj}$* ORF plus 5′ regulatory sequences were amplified by PCR using *S. japonicus* genomic DNA as a template, the 5′-oligonucleotide PromPmk1Jp-*Not*I-5′, which hybridizes at positions −498 to −475 bp upstream of the *$pmk1^+$* start codon and contains a *Not*I site, and the 3′-oligonucleotide Pmk1Jp:HA-*BamH*I-3′, which hybridizes at the 3′ end of *$pmk1^{+Sj}$* ORF and incorporates an additional 33-nucleotide sequence encoding one HA epitope (sequence -GYPYDVPDYAG), and contains a *BamH*I site. The obtained PCR fragment was digested with *Not*I and *BamH*I and cloned into plasmid pSL, which is a derivative of plasmid pS0550 (kindly provided by S. Oliferenko) that incorporates the *$nmt1^+$* 3′ regulatory sequences. To obtain integrative plasmid pSL-$Pmk1^{Sj}$(ins 82-105)–HA, we followed a modular PCR-based approach employing *S. japonicus* genomic DNA and different nucleotide pairs. A first DNA fragment was amplified with the oligonucleotide Pmk1Jp-EXON-5′, which hybridizes 620 to 691 bp downstream of *$pmk1^{+Sp}$* ORF and incorporates a 71-nucleotide sequence encoding the 24 aa motif present in $Pmk1^{Sp}$ N-lobe, and 3′-oligonucleotide Pmk1Jp:HA-*BamH*I-3′. A second DNA fragment was amplified with 5′-oligonucleotide PromPmk1Jp-*Not*I-5′ and 3′-oligonucleotide Pmk1Jp-EXON-3′, which hybridizes 424 to 448 bp downstream of *$pmk1^{+Sj}$* ORF plus and incorporates a reverse complement 71-nucleotide sequence from $Pmk1^{Sp}$ ORF described above. Both DNA fragments were gel purified and used as templates in a second-round PCR in the presence of PromPmk1Jp-*Not*I-5′ and Pmk1Jp:HA-*BamH*I-3′. The resulting PCR fragment was digested with *Not*I/*BamH*I and cloned into plasmid pSL. To obtain integrative plasmid pSL-Pmk1^{Sp}:HA6H, the *pmk1^{+Sp}* ORF plus regulatory sequences were amplified by PCR using genomic DNA from *S. pombe* strain MM2 as template [5], and 5′-oligonucleotide PromPmk1Sp-*Xba*I-5′, which hybridizes at positions 330 to 313 upstream of the *pmk1^{+Sp}* start codon and contains a *Xba*I site, and the 3′-oligonucleotide Pmk1Sp–HA-3′, which hybridizes at the 3′ end of *pmk1^{+Sp}* ORF and incorporates a 48-nucleotide sequence encoding one HA epitope plus six consecutive histidine residues, and a *Sma*I site. The DNA fragment was digested with *Xba*I and *Sma*I and cloned into plasmid pSL. For plasmid pSL-Pmk1^{Sp} (Δ82-105):HA, the *pmk1^{+Sp}* ORF plus regulatory sequences were amplified in by PCR using *S. pombe* genomic DNA as template, the 5′-oligonucleotide PromPmk1Sp-*Xba*I-5′, and 3′-oligonucleotide Pmk1Sp-ΔExon-3′, which hybridizes −326 to −346 bp upstream and 774 to 792 bp downstream of *pmk1^{+Sp}* ORF. A second DNA fragment was amplified with the 5′-oligonucleotide Pmk1Sp-ΔExon-5′, which is complementary and the reverse of Pmk1Sp-ΔExon-3′, and 3′-oligonucleotide Pmk1Sp–HA-3′. Both fragments were gel purified and used in a second-round PCR as templates in the presence of the external oligonucleotides PromPmk1Sp-*Xba*I-5′ and Pmk1Sp–HA-3′. The resulting PCR fragment was digested with *Xba*I and *Sma*I and cloned into plasmid pSL. All of the above plasmids were digested at the unique *Afe*I site within *ura4^{+Sj}*, and the linear fragments were transformed into the *S. japonicus pmk1$^+$::NatR* strain EGJ83. *ura4$^+$* transformants were obtained, and the correct integration and expression of the HA fusions was verified by sequencing and Western blot analysis.

To express the chimeric constructs described above in *S. pombe*, the respective Pmk1–HA fused cassettes plus 5′ regulatory sequences were gel purified from plasmids pSL-Pmk1^{Sj}:HA, pSL-Pmk1^{Sj}(ins 82-105)–HA, pSL-Pmk1^{Sp}:HA, and pSL-Pmk1^{Sp}(Δ82-105):HA after digestion with *Xba*I and *Sma*I, and cloned into *S. pombe* integrative plasmid pJK148. The resulting plasmid constructs were digested at the unique *Nru*I site within *Leu1^{+Sp}*, and the linear fragments were transformed into the *S. pombe pmk1$^+$::NatR* strain MM1904. *Leu1$^+$* transformants were obtained, and the correct integration and expression of the fusions was verified by sequencing and Western blot analysis.

2.4. Detection and Quantification of Activated and Total Pmk1 Levels

Extracts from *S. pombe* and *S. japonicus* cells collected during unperturbed growth, or after stress or CPT treatments were prepared under native conditions employing chilled acid-washed glass beads and lysis buffer (10% glycerol, 50 mM Tris/HCl pH 7.5, 15 mM Imidazole, 150 mM NaCl, 0.1% Nonidet NP-40, plus specific protease and phosphatase inhibitor, Sigma-Aldrich), as described [31]. Dually phosphorylated Pmk1 was detected employing a rabbit polyclonal anti-phospho-p44/42antibody (9191L; Cell Signaling). Total Pmk1 levels were detected with either mouse monoclonal anti-HA (Pmk1–HA fusions; clone 12CA5; Roche Molecular Biochemicals, Basel, Switzerland), or anti-GFP antibody (Pmk1–GFP fusions; Roche Molecular Biochemicals). Rabbit polyclonal anti-Cdk1/Cdc2 antibody (PSTAIR; Millipore) was used for loading control. Immunoreactive bands were revealed with anti-mouse or anti-rabbit peroxidase-conjugated secondary antibodies (both from Sigma-Aldrich), and the ECL system (GE-Healthcare, Uppsala, Sweden). Densitometric quantification of Western blot signals from 16-bit .jpg digital images of blots was performed using ImageJ [32]. Relative units for Pmk1 activation were estimated by determining the signal ratio of the anti-phospho-p44/42 blot (activated Pmk1) with respect to the anti-HA or anti-GFP blot (total Pmk1) at each time point. Unless otherwise stated, results shown correspond to experiments performed as biological duplicates or triplicates. Mean relative units ± SD and/or representative results are shown. *p*-values were analyzed by unpaired Student's t-test.

2.5. Co-immunoprecipitation

Whole-cell extracts (5 mg total protein) were prepared in lysis buffer (50 mM Tris/HCl pH 7.5, 0.5% sodium deoxicholate, 150 mM NaCl, 1% NP-40 and protease inhibitor (Sigma-Aldrich), and incubated with anti-HA monoclonal antibody (12CA5; Roche Molecular

Biochemicals) and protein A sepharose beads for 4 h at 4 °C. The beads were washed two times with lysis buffer, two times with washing buffer 2 (50 mM Tris/HCl pH 7.5, 0.05% sodium deoxicholate, 500 mM NaCl, 0.1% NP-40), one time with washing buffer 3 (50 mM Tris/HCl pH 7.5, 0.05% sodium deoxicholate, 0.1% NP-40), and resuspended in sample buffer. Proteins were separated by SDS-PAGE, transferred to nitrocellulose filters (GE Healthcare), and hybridized with either mouse monoclonal anti-HA antibody (Pmk1–HA fusions; clone 12CA5; Roche Molecular Biochemicals), or anti-GFP antibody (Pek1–GFP fusions; Roche Molecular Biochemicals), followed by immunodetection with anti-mouse peroxidase-conjugated secondary antibody (Sigma-Aldrich), and the ECL system (GE Healthcare).

2.6. Plate Assay of Stress Sensitivity for Growth

S. pombe and *S. japonicus* wild-type and mutant strains were grown in YES liquid medium to OD_{600} = 0.5, and appropriate decimal dilutions were spotted per duplicate on YES solid medium or in the same medium supplemented with varying concentrations of caspofungin (Sigma-Aldrich), and magnesium chloride plus the calcineurin inhibitor FK506 (Tracholimus; Sigma-Aldrich). Plates were incubated at 30 °C for 3 days and then photographed. All the assays were repeated at least three times with similar results. Representative experiments are shown in the corresponding figures.

2.7. Quantification of Mycelial Growth during Nutritional Stress

Approximately 2.10^6 cells from log-phase cultures (OD_{600}= 0.5) of wild-type and mutant strains growing in YES medium were spotted on YEMA or RGE plates, incubated at 30 °C for 7 days, and then photographed and saved as 16-bit .jpg digital images. The area of mycelial expansion was outlined for each strain ($n \geq 6$) by freehand, and measured with ImageJ using the *analyze tool* [32].

2.8. Microscopy Observation

Fluorescence images in Figure 1C were acquired using a Leica Thunder imager by either fluorescence or DIC microscopy as single medial plane images from a set of 20 stacks (0.3 µm each), and processed through ImageJ [32]. Fluorescence images in Figure 1A, Figure 2D,E and Figure 4E were obtained with a Leica DM4000B microscope and processed using IM500 Image Manager software. Calcofluor white was employed for cell wall/septum staining as described [24]. To quantify the increase in cell length during hyphal induction experiments with CPT, samples were taken at the indicated times and fixed immediately with formaldehyde [33]. After staining with calcofluor white the length of mononuclear late G2 cells was measured. Three biological replicates ($n \geq 400$) were scored for each strain genotype.

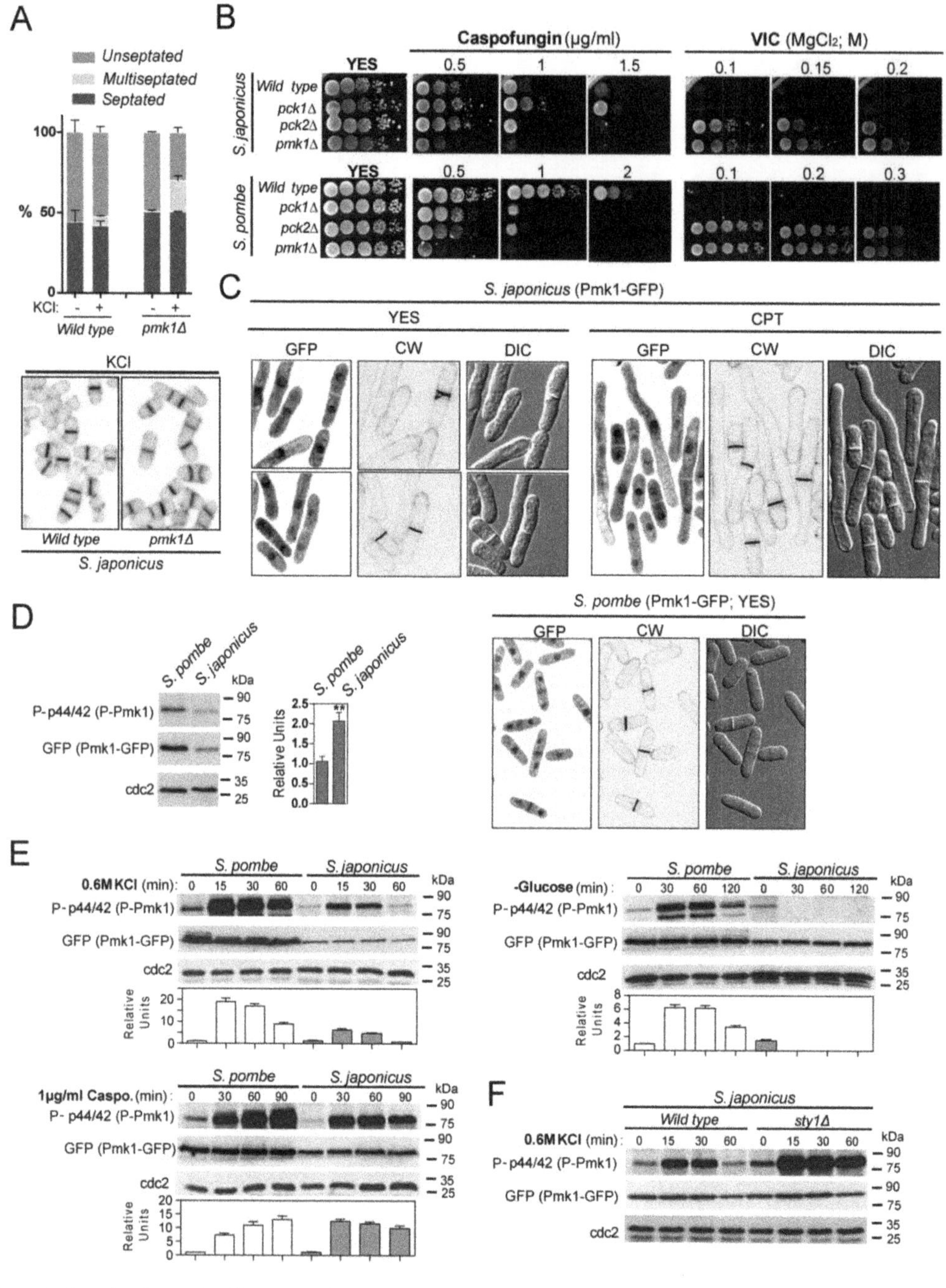

Figure 1. Functional roles, localization, and activation pattern of the CIP MAPK Pmk1 in *S. japonicus.* (**A**) *Upper*: Exponentially growing *S. japonicus* wild-type and *pmk1*Δ strains were incubated for 10 h in YES medium plus 0.6M KCl, and the percentage of unseptated, septated, and multiseptated cells (represented as mean ± SD from biological triplicates; number

of total cells ≥ 200) was determined by fluorescence microscopy after calcofluor white staining. *Lower*: Representative images are shown; (**B**) Decimal dilutions of the indicated strains growing in YES medium were spotted onto YES plates supplemented with caspofungin (0.5–2 μg/mL) (upper panels), and 0.5 μg/mL FK506 plus 0.1, 0.15, or 0.2 M $MgCl_2$ (VIC assay; lower panels), and incubated for 3 days at 30 °C before being photographed. Representative experiments are shown; (**C**) *S. japonicus* and *S. pombe* wild-type cells expressing genomic Pmk1–GFP fusions were grown in YES medium to mid-log phase or treated for 6 h with 0.2 μM CPT (*S. japonicus*), and observed both by fluorescence and DIC microscopy after calcofluor white staining. Images were acquired as single medial plane images and are inverted for fluorescence; (**D**) *S. pombe* and *S. japonicus* wild-type strains expressing genomic Pmk1–GFP fusions were grown in YES medium to mid-log phase, and activated/total Pmk1 were detected with anti-phospho-p44/42 and anti-GFP antibodies, respectively. Anti-Cdc2 was used as a loading control. Relative units as mean ± SD (biological triplicates) for Pmk1 phosphorylation (anti-phospho-p44/42 blot) were determined with respect to the internal control (anti-GFP blot). **, $p < 0.005$; as calculated by unpaired Student's *t*-test; (**E**) *S. pombe* and *S. japonicus* wild-type strains expressing genomic Pmk1–GFP fusions were grown in YES medium to mid-log phase, and treated with 0.6 M KCl, 1 μg/mL caspofungin, or recovered by filtration and resuspended in osmotically equilibrated medium lacking glucose for the indicated times. Activated/total Pmk1 was detected with anti-phospho-p44/42 and anti-GFP antibodies, respectively. Anti-Cdc2 was used as a loading control. Relative Pmk1 activity units as mean ± SD (biological duplicates) were determined as described in (**C**); (**F**) *S. japonicus* wild-type and *sty1Δ* strains expressing a genomic Pmk1–GFP fusion were grown in YES medium to mid-log phase, and treated with 0.6 M KCl for the indicated times. Activated/total Pmk1 were detected with anti-phospho-p44/42 and anti-GFP antibodies, respectively. Relative Pmk1 activity units as mean ± SD (biological duplicates) were determined as above.

3. Results

3.1. Environmental Control of S. japonicus Cell Integrity MAPK Pathway by the PKC Orthologs Pck1 and Pck2

To explore the evolutionary conservation of the cell integrity pathway within the fission yeast clade, we obtained a *S. japonicus* strain deleted in MAPK Pmk1 by homologous recombination (see Materials and Methods). *S. pombe pmk1Δ* cells display a multiseptated phenotype when cultured in medium supplemented with sorbitol, calcofluor, KCl, or caffeine [34–36]. While the percentage of septated and multiseptated cells was very similar in *S. japonicus* wild-type and *pmk1Δ* cells growing in the presence of sorbitol, calcofluor and caffeine, multiseptation increased significantly in *pmk1Δ* cells treated with KCl (Figure 1A), suggesting that Pmk1 might play a less meaningful role during control of septation in this organism. As compared to the wild-type strain, *S. japonicus pmk1Δ* cells were growth sensitive to the β-glucan synthase inhibitor caspofungin but to a lesser extent than the *S. pombe pmk1Δ* mutant (Figure 1B), indicating that Pmk1 likely regulates cell wall synthesis and/or integrity in both fission yeast species. Calcineurin and Pmk1 play antagonistic roles during control of calcium homeostasis in *S. pombe*, and lack of Pmk1 activity allows cell growth in the presence of $MgCl_2$ when calcineurin activity is blocked with the specific inhibitor FK506. This phenotype, known as "VIC" (viable in the presence of immunosuppressant and chloride ion) [37], was also displayed by *S. japonicus pmk1Δ* cells (Figure 1B), revealing that a functional crosstalk between calcineurin and Pmk1 signaling is conserved within the *Schizosaccharomyces* genus.

Pmk1 MAPKs in *S. pombe* and *S. japonicus* share a high amino-acid sequence identity, particularly at the N- and C-lobes (~85% identity within a 336 amino-acid sequence), and decreases gradually towards the C-tail (Figure S2). The glycine-ATP-phosphate-binding loop, the invariant lysine and glutamic acid residues required for full kinase activity, the ERK-type TEY activation loop, and the common docking (CD) domain required for MAPK binding to upstream activators and downstream targets, are conserved in *S. japonicus* Pmk1 with respect to the corresponding *S. pombe* counterpart (Figure S2). However, a striking feature of *S. japonicus* Pmk1 secondary structure is the absence at its N-lobe of a 24 aa motif (sequence –ITCIYDLDIINPYNFNEVYIYEEL-), which encompasses the full subdomain IV and part of subdomain V (hinge region), and includes the putative gatekeeper residue E104 (Figure S2). A global BlastP search of yeast and fungal proteomes indicated that this unusual N-lobe conformation is highly specific of *S. japonicus* Pmk1, as it was found in only three hypothetical cell integrity MAPKs from ascomycetes belonging to order *Chaetothyriales*

(Gen Pep: DV737_g786 and DV736_g4926), and the species *Ophiocordyceps camponoti* (Gen Pep: CP532_1836) (Figure S3).

S. pombe Pmk1 displays a nucleo-cytoplasmic localization during unperturbed growth and is targeted to the septum during cytokinesis [5]. We found that, in contrast to *S. pombe*, in *S. japonicus* a genomic Pmk1–GFP fusion shows nucleo–cytoplasmic location but it is not targeted to the septum, as confirmed after observation of septum staining with calcofluor white and DIC microscopy of yeast cells growing exponentially in rich medium, or in hyphae produced after treatment with CPT (Figure 1C). Incubation with an anti-phospho-p44/42 ERK antibody, which detects dually phosphorylated Pmk1 in *S. pombe* [5], revealed a single phosphorylation signal with the predicted molecular weight of Pmk1 (~47 Da) in extracts from *S. japonicus* wild-type cells, but not in those from the *pmk1Δ* mutant (Figure S2). We note that the relative basal Pmk1 phosphorylation level was significantly higher (2–2.5 times) in vegetatively growing *S. japonicus* versus *S. pombe* cells expressing genomic Pmk1–GFP fusions (Figure 1D). *S. pombe* Pmk1 becomes activated in response to several stimuli including saline stress, glucose starvation, and cell wall damage induced with caspofungin [3,5,7,12]. Pmk1 also became activated in *S. japonicus* cells treated with salt (KCl) or caspofungin, but to different magnitudes and dynamics than in *S. pombe* (Figure 1E). Remarkably, Pmk1 activity decreased quickly during glucose starvation (Figure 1E), suggesting that the CIP is not operative in *S. japonicus* under this specific stimulus. In *S. pombe*, transcriptional induction of tyrosine and serine/threonine phosphatases by the SAPK pathway through the Sty1–Atf1 branch promotes the dephosphorylation of activated Pmk1 during vegetative growth and in response to stress, thus revealing a cross-inhibitory mechanism between the two MAPK cascades [4]. Indeed, both basal and the induced phosphorylation of Pmk1 under saline stress increased significantly in a *S. japonicus sty1Δ* mutant [23], as compared to wild-type cells (Figure 1F), suggesting that the interplay between the SAPK and CIP pathways is conserved in both *Schizosaccharomyces* species.

The two non-essential PKC orthologs, Pck1 and Pck2 [1,38], are core upstream activators of the CIP during growth and stress in *S. pombe* [6,7]. *S. japonicus* proteome also carries two putative orthologs for Pck1 (SJAG_01787) and Pck2 (SJAG_00610), which share ~52 and ~66% amino-acids sequence identity with the respective *S. pombe* counterparts. Both kinases have the conserved putative HR1 rho-binding repeat domains that in *S. pombe* mediate their binding to Rho GTPases Rho1 and Rho2 [39], plus the C2 (Ca^{2+}-binding), Ps (pseudosubstrate), and C1 (diacylglycerol-binding) motifs present in *S. pombe* Pck1 and Pck2 (Figures S4 and S5) [39]. The canonical phosphorylated residues at the kinase domain within the AL (activation loop), TM (Turn motif), and HM (hydrophobic motif), are also conserved (Figures S4 and S5) [39]. *S. japonicus* single mutants lacking either Pck1 or Pck2 were viable; however, we were unable to obtain a double-mutant strain lacking both kinases, suggesting that they partially complement each other and share essential functional roles for cell viability as in *S. pombe* [40,41]. As compared to wild-type cells, Pmk1 basal activity was strongly reduced (>80%) in *S. japonicus pck2Δ* cells during vegetative growth, and increased approximately by two-fold in the *pck1Δ* mutant (Figure 2A). Accordingly, in *pck2Δ* cells reduced Pmk1 activity was accompanied by a less severe VIC phenotype than that of *pmk1Δ* cells (Figure 1B). Thus, *S. japonicus* Pck2 is a major activator of the Pmk1 MAPK module during unperturbed growth. *S. pombe* Pmk1 activation during osmotic–saline stress with KCl is fully dependent on Pck2 function, whereas Pck1 does not play a meaningful role during this response [6]. This signaling mechanism is also conserved within the *S. japonicus* CIP, since Pck2 deletion completely abolished Pmk1 activation in the presence of KCl (Figure 2B). Despite showing very low MAPK levels, the growth sensitivity to caspofungin of *S. japonicus pck2Δ* cells was negligible as compared to that of *pmk1Δ* cells (Figure 1B). Notably, and in contrast to the *S. pombe pck1Δ* mutant, *S. japonicus* cells lacking Pck1 were growth resistant to this compound (Figure 1B). In *S. pombe*, Pck2 is a major factor responsible for Pmk1 activation during cell wall damage induced with caspofungin, while Pck1 likely plays a secondary role [7]. Instead, caspofungin-induced Pmk1 activation in

S. japonicus was strictly Pck2-dependent, as Pck1 absence did not reduce MAPK activation and/or dynamics (Figure 2B). Together, the above findings suggest that respective positive and negative modulation of CIP activity by Pck2 and Pck1 has an opposite impact on *S. japonicus* cell wall integrity, and that MAPK activation in response to cell wall damage is transduced to the MAPK module exclusively through Pck2.

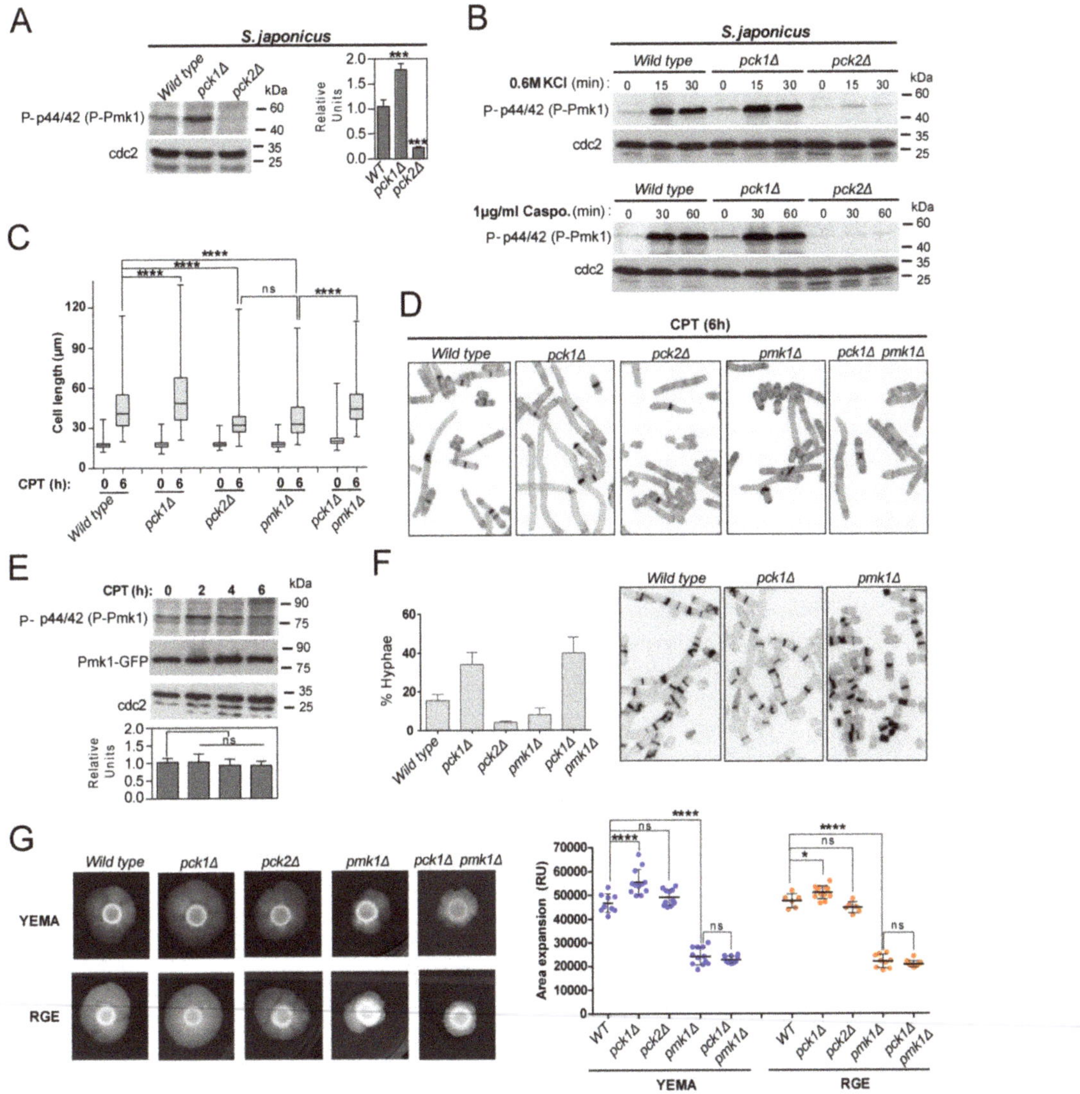

Figure 2. Antagonistic control of Pmk1 activity by PKC orthologs Pck1 and Pck2 modulates *S. japonicus* hyphal differentiation. (**A**) *S. japonicus* strains of the indicated genotypes were grown in YES medium to mid-log phase, and activated Pmk1 were detected with anti-phospho-p44/42, whereas anti-Cdc2 was used as a loading control. Relative units as mean ± SD (biological triplicates) for Pmk1 phosphorylation (anti-phospho-p44/42 blot) were determined with respect to the internal

control (anti-Cdc2 blot). ***, $p < 0.001$; as calculated by unpaired Student's t-test; (**B**) *S. japonicus* strains of the indicated genotypes were grown in YES medium to mid-log phase, treated with either 0.6 M KCl (upper panels) or 1 μg/mL caspofungin (lower panels) for the indicated times, and activated Pmk1 was detected with anti-phospho-p44/42, whereas anti-Cdc2 was used as a loading control. Results from representative experiments are shown; (**C**) Exponentially growing *S. japonicus* cells of the indicated genotypes were inoculated at an initial cell density of 10^6 cells/mL in YES medium with 6% glucose and incubated for 0 and 6 h with 0.2 μM CPT. Cell length at each time point is represented as box and whisker plots. Data obtained after quantification of one experiment performed per triplicate ($n \geq 400$ cells/strain) is shown. ****, $p < 0.0001$; ns, not significant, as calculated by one-way ANOVA; (**D**) Representative images of strains analyzed in (**C**) were obtained by fluorescence microscopy after calcofluor white staining; (**E**) Exponentially growing *S. japonicus* wild-type cells (YES medium with 6% glucose) expressing a Pmk1–GFP fusion were treated with 0.2 μM CPT for the indicated times. Activated/total Pmk1 were detected with anti-phospho-p44/42 and anti-GFP antibodies, respectively. Anti-Cdc2 was used as a loading control. Relative units as mean ± SD (biological duplicates) for Pmk1 phosphorylation (anti-phospho-p44/42 blot) were determined with respect to the internal control (anti-GFP blot). ns, not significant, as calculated by unpaired Student's t-test; (**F**) *Left*: *S. japonicus* strains of the indicated genotypes were grown in YES medium plus 0.2 μM CPT for 12 h, and the percentage of hyphae were quantified. Percentages are expressed as mean ± SD and correspond to biological duplicates ($n \geq 200$ cells/sample). *Right*: Representative images of strains were obtained by fluorescence microscopy after calcofluor white staining; (**G**) *Left*: cells from log-phase cultures of the indicated strains growing in YES medium (~2.10^6 cells in each case) were spotted on YEMA and RGE plates, incubated at 30 °C for 7 days, and then photographed. *Right*: the total area of mycelial expansion (expressed as relative units, RU) was measured for each strain genotype ($n \geq 6$) and is represented as scatter plot. *, $p < 0.05$; ****, $p < 0.0001$; ns, not significant, as calculated by one-way ANOVA.

3.2. *Fine-Tuning of Pmk1 Activity by Pck1 and Pck2 Modulates Antagonistically S. japonicus Hyphal Differentiation*

As compared to wild-type cells, *S. japonicus* null mutants lacking the putative CIP MAP kinase module components Mkh1 (MAPKKK), Pek1 (MAPKK), and Pmk1 (MAPK), show defective mycelia formation when growing in a malt-extract-based solid medium (YEMA) [22], suggesting that the CIP plays a positive role to modulate yeast to hypha dimorphic transition. However, the possible role of the upstream activators of the MAP kinase module Pck1 and Pck2, and the functional relevance of this control, have not been addressed. We note that the average cell length after 6 h of incubation in rich medium supplemented with 0.2 μM CPT, which induces a quick yeast to hyphae transition, was significantly higher in *S. japonicus pck1Δ* cells than in wild-type cells (81.5 + 15.4 μm in *pck1Δ* cells versus 55.1 + 24.9 μm in wild-type cells; Figure 2C,D). On the contrary, CPT-induced differentiation was markedly impaired to a similar extent in both *pck2Δ* and *pmk1Δ* mutants (34.2 + 12.6 and 37.6 + 16.2 μm in *pck2Δ* and *pmk1Δ* cells, respectively; Figure 2C,D). Total and activated basal Pmk1 levels remained unchanged during the early stages of hyphal initiation (Figure 2E). Interestingly, the enhanced hyphal differentiation of *pck1Δ* cells was suppressed in the absence of Pmk1 function only partially (47.2 + 16.2 μm in *pck1Δ pmk1Δ* double mutant; Figure 2C,D). In line with the above findings, the *pck1Δ* mutant produced a higher ratio of filaments and/or hypha than the wild-type strain after 12 h of incubation with CPT, whereas this percentage became significantly reduced in *pck2Δ* and *pmk1Δ* mutants, but not in a *pck1Δ pmk1Δ* strain (Figure 2F).

The mycelial area of expansion of the *pmk1Δ* mutant, but not that of the *pck2Δ* strain, decreased significantly with respect to the wild type when incubated for six days in YEMA or RGE plates (Figure 2G). In contrary, mycelium production was enhanced in *pck1Δ* cells growing in both inducing media, and this phenotype was fully suppressed by simultaneous deletion of Pmk1 (Figure 2G). Taken together, the above findings suggest that Pmk1 plays a major positive role to control both the early and late phases of hyphal differentiation in *S. japonicus*, whereas the positive impact of Pck2 signaling seems restricted to the initial differentiation step. Conversely, Pck1 negatively modulates this process by a mechanism that fully depends on the reduction of Pmk1 activity during the late stages of mycelium production (plate assays), but not at the early stages of the dimorphic switch (CPT treatment).

3.3. Species-Specific N-lobe Domain Architecture Accounts for Differential Control of Pmk1 Function in S. pombe and S. japonicus during Growth, Stress, and Hyphal Differentiation

The unusual secondary structure of *S. japonicus* Pmk1 at its N-lobe lacking most parts of subdomains IV and V (Figure S2), together with the results obtained so far, prompted us to thoroughly explore the possible impact of MAPK conformation on the evolutionary conservation of Pmk1 functions in the two fission yeast species. To this end, we constructed *S. pombe* and *S. japonicus pmk1Δ* mutants expressing distinct genomic Pmk1–HA fused chimeras under the control of their respective wild-type promoters (named as Pmk1^{Sp} and Pmk1^{Sj} thereafter) (Figure 3A). In contrast to the wild-type Pmk^{Sp}–HA fusion, expression in *S. pombe pmk1Δ* cells of a MAPK version lacking the conserved N-lobe 24 aa motif (Pmk1^{Sp} (Δ82-105)–HA; Figure 3A), did not show detectable phosphorylation during unperturbed growth (Figure 3B), suggesting that this motif is essential for kinase activation and/or activity. Accordingly, Pmk1^{Sp} (Δ82-105)–HA failed to suppress both the VIC and multiseptated phenotypes of *S. pombe pmk1Δ* cells (Figure 3C,D), and was not activated in response to saline stress (Figure 3E). Indeed, the relative basal activity of a *S. japonicus* Pmk1 fusion expressed in *S. pombe* (Pmk1^{Sj}–HA; Figure 3A) was significantly higher than that of the endogenous MAPK (Figure 3B). Pmk1^{Sj}–HA suppressed the VIC and multiseptated phenotypes of *S. pombe pmk1Δ* cells to a similar extent as Pmk1^{Sp}–HA (Figure 3C,D), but displayed a slightly lower activation threshold in response to an osmotic-saline stress (Figure 3E). Moreover, insertion of the N-lobe 24 aa motif within *S. japonicus* Pmk1 (Pmk1^{Sj} (ins82-105)–HA), resulted in a total lack of MAPK activity and functions (Figure 3B–E). Taking into account these findings, we hypothesized that the differences in biological activity among the different Pmk1 chimeras might result from changes in their binding affinity to the upstream activating MAPKK Pek1 [1,5]. Indeed, co-IP experiments revealed that the In Vivo association of Pmk1^{Sj}–HA to a genomic Pek1–GFP fusion was enhanced as compared to that of Pmk1^{Sp}–HA, and absent in cells expressing either Pmk1^{Sp}(Δ82-105)–HA or Pmk1^{Sj}-ins(82-105)–HA fusions (Figure 3F). Together, these observations indicate that Pmk1^{Sj} is fully functional in *S. pombe*, and that absence of the subdomain IV–V motif within the ATP-binding pocket at the N-terminal lobe is required for MAPKK binding and functional activation of the MAPK. They also suggest that the enhanced MAPKK binding and altered activity and activation pattern of Pmk1^{Sj} with regard to the *S. pombe* isoform (i.e. increased basal activity and reduced activation in response to stress), may rely upon structural constraints determined by its atypical N-terminal lobe.

Insertion of the *S. pombe* 24 aa motif within *S. japonicus* Pmk1 (Pmk1^{Sj} (ins82-105)–HA), resulted in a clear reduction in basal MAPK activity as compared to the wild-type Pmk1^{Sj}–HA fusion when expressed in a *S. japonicus pmk1Δ* background (Figure 4A). In contrary, MAPK activation during osmotic–saline stress remained unaffected (Figure 4B). However, Pmk1^{Sj}(ins82-105)–HA did not abolish the VIC phenotype of *S. japonicus pmk1Δ* cells as the Pmk1^{Sj}–HA version did (Figure 4C). Interestingly, a genomic Pmk1^{Sp}–HA chimera expressed in *S. japonicus* showed a relative basal MAPK activity comparable to that of the endogenous Pmk1^{Sj}–HA construct (Figure 4A), and became activated to a similar threshold in response to KCl stress (Figure 4D). Nonetheless, Pmk1^{Sp} was not fully functional in *S. japonicus*, because it suppressed the VIC phenotype of the *pmk1Δ* mutant only partially (Figure 4B). Finally, deletion of the Pmk1^{Sp} N-lobe motif (Pmk1^{Sp} (Δ82-105)–HA) resulted in a biologically inactive and non-functional MAPK chimera (Figure 4A–C).

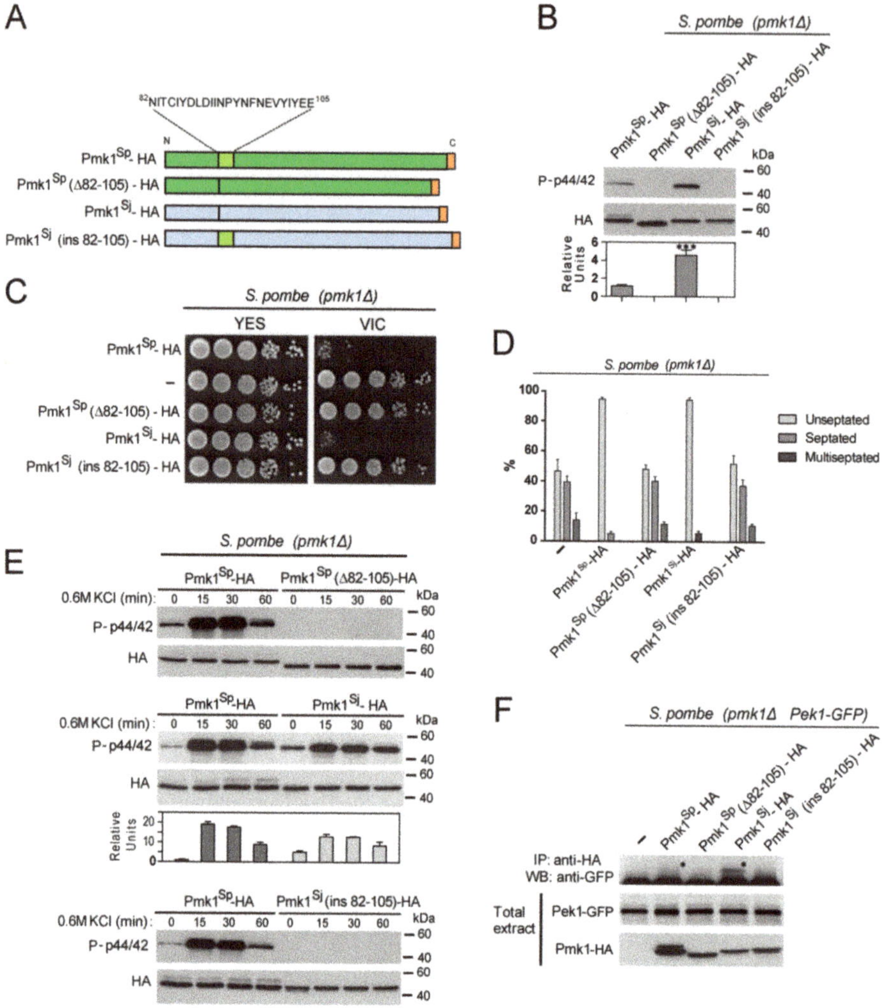

Figure 3. *S japonicus* Pmk1 is fully functional in *S. pombe*. (**A**) *S. pombe* and *S. japonicus* Pmk1–HA chimeric constructs. Pmk1^{Sp}–HA: *S. pombe* wild-type Pmk1 construct showing the N-lobe subdomain IV–V conserved 24 aa motif (light green) is shown. Pmk1^{Sp} (Δ82-105)–HA: *S. pombe* Pmk1 version lacking the conserved N-lobe conserved motif. Pmk1^{Sj}–HA: *S. japonicus* wild-type Pmk1 construct that lacks the conserved N-lobe conserved motif. Pmk1^{Sj} (ins82-105)–HA: *S. japonicus* Pmk1 version with an insertion of the *S. pombe* Pmk1 N-lobe 24 aa motif (light green); (**B**) *S. pombe pmk1*Δ cells expressing the Pmk1–HA chimeric constructs described in (**A**) were grown in YES medium to mid-log phase, and activated/total Pmk1 were detected with anti-phospho-p44/42 and anti-HA antibodies, respectively. Relative units as mean ± SD (biological triplicates) for Pmk1 phosphorylation (anti-phospho-p44/42 blot), were determined with respect to the anti-HA blot (total Pmk1). ***, $p < 0.001$; as calculated by unpaired Student's *t*-test; (**C**) Decimal dilutions of *S. pombe pmk1*Δ cells (-) and those expressing the indicated Pmk1–HA chimeric constructs, were spotted onto YES plates supplemented with 0.5 µg/mL FK506 plus 0.2 M $MgCl_2$ (VIC assay), and incubated for 3 days at 28 °C before being photographed. A representative experiment is shown; (**D**) *S. pombe pmk1*Δ cells (-) and those expressing the indicated Pmk1–HA chimeric constructs were grown for 24 h in YES medium plus 1M sorbitol, and the percentage of septated/multiseptated cells (represented as mean ± SD from biological triplicates; number of cells ≥ 200) was determined by fluorescence microscopy after calcofluor white staining;

(**E**) *S. pombe* strains expressing the indicated Pmk1–HA chimeric constructs were grown in YES medium to mid-log phase and treated with 0.6 M KCl for the indicated times. Activated/total Pmk1 were detected with anti-phospho-p44/42 and anti-HA antibodies, respectively. Relative units as mean ± SD (biological duplicates) for Pmk1 phosphorylation (anti-phospho-p44/42 blot) were determined with respect to the anti-HA blot (total Pmk1). Either representative results or those obtained from biological triplicates are shown; (**F**) Co-immunoprecipitation of Pmk1–HA chimeric constructs and a Pek1–GFP (MAPKK) genomic fusion from yeast extracts obtained from vegetatively growing cultures from the indicated strains. Results from a representative experiment are shown. IP, immunoprecipitation; WB, Western blot; *, specific Pek1-GFP fusion.

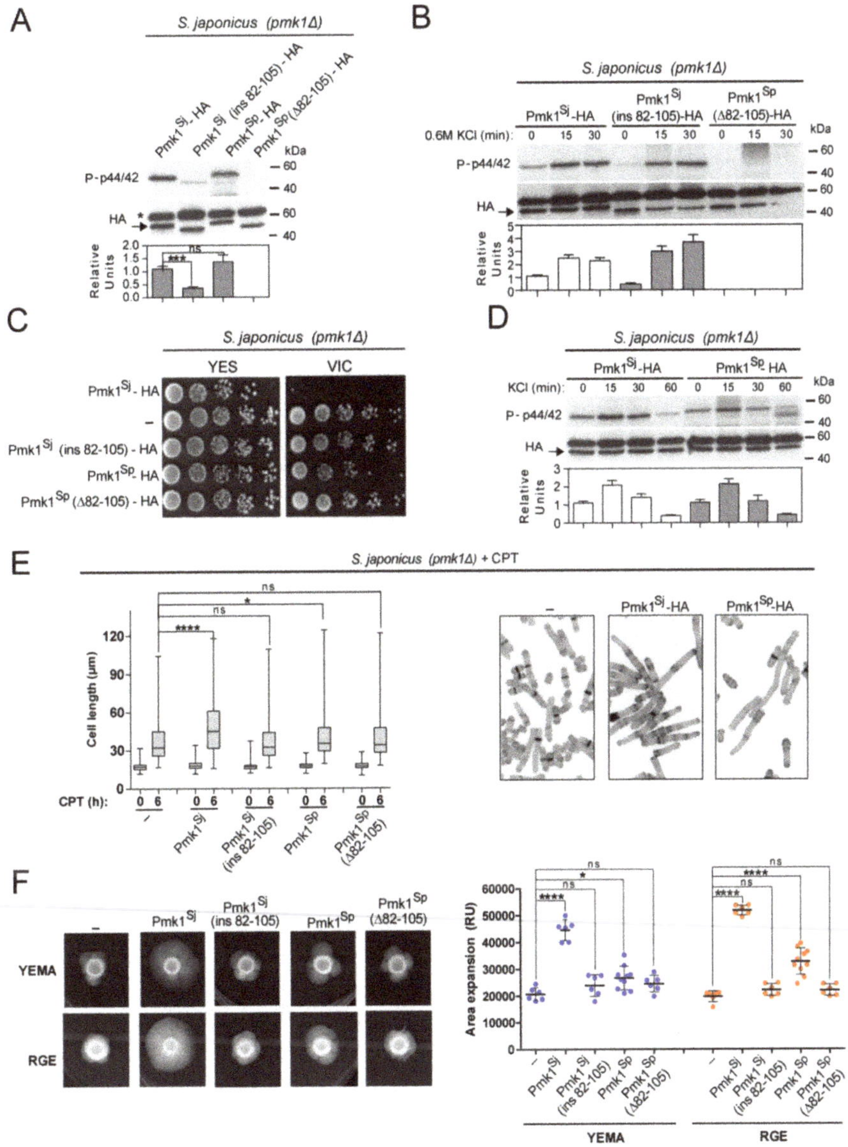

Figure 4. *S. pombe* Pmk1 shows little ability to execute CIP functions and to promote mycelial development in *S. japonicus*.

(**A**) *S. japonicus pmk1Δ* cells expressing the Pmk1–HA chimeric constructs described in Figure 3A were grown in YES medium to mid-log phase, and activated/total Pmk1 were detected with anti-phospho-p44/42 and anti-HA antibodies, respectively. Relative units as mean ± SD (biological triplicates) for Pmk1 phosphorylation (anti-phospho-p44/42 blot) were determined with respect to the anti-HA blot (total Pmk1). ***, $p < 0.001$; ns, not significant, as calculated by unpaired Student's *t*-test; (**B**) *S. japonicus pmk1Δ* strains expressing the Pmk1–HA chimeric constructs were grown in YES medium to mid-log phase and treated with 0.6 M KCl for the indicated times. Activated/total Pmk1 were detected with anti-phospho-p44/42 and anti-HA antibodies, respectively. Relative units as mean ± SD (biological duplicates) for Pmk1 phosphorylation (anti-phospho-p44/42 blot) were determined with respect to the anti-HA blot (total Pmk1); (**C**) Decimal dilutions of *S. japonicus pmk1Δ* cells (-), and those expressing Pmk1–HA chimeric constructs, were spotted onto YES plates supplemented with 0.5 µg/mL FK506 plus 0.2 M $MgCl_2$ (VIC), and incubated for 3 days at 30 °C before being photographed. A representative experiment is shown; (**D**) *S. japonicus pmk1Δ* strains expressing the Pmk1–HA chimeric constructs were grown in YES medium to mid-log phase and treated with 0.6 M KCl for the indicated times. Activated/total Pmk1 were detected with anti-phospho-p44/42 and anti-HA antibodies, respectively. Relative units as mean ± SD (biological duplicates) for Pmk1 phosphorylation (anti-phospho-p44/42 blot) were determined with respect to the anti-HA blot (total Pmk1); (**E**) *Left*: Exponentially growing *S. japonicus pmk1Δ* cells (-), and those expressing the indicated Pmk1–HA chimeric constructs were inoculated at an initial cell density of 10^6 cells/mL in YES medium with 6% glucose and incubated for 0 and 6 h with 0.2 µM CPT. Cell length at each time point is represented as box and whisker plots. Data obtained after quantification of one experiment performed per triplicate ($n \geq 400$ cells/strain) are shown. *, $p < 0.05$; ****, $p < 0.0001$; ns, not significant, as calculated by one-way ANOVA. *Right*: representative images were obtained by fluorescence microscopy after calcofluor white staining; (**F**) *Left*: cells from log-phase cultures of *S. japonicus pmk1Δ* cells (-), and those expressing Pmk1–HA chimeric constructs growing in YES medium (~2.10^6 total cells), were spotted on YEMA and RGE plates, incubated at 30 °C for 7 days, and then photographed. *Right*: the total area of mycelial expansion (expressed as relative units) was measured for each strain genotype ($n \geq 6$) and is represented as a scatter plot. *, $p < 0.05$; ****, $p < 0.0001$; ns, not significant, as calculated by one-way ANOVA.

As compared to endogenous Pmk1^{Sj}–HA, expression of Pmk1^{Sp}–HA showed a limited ability to promote filamentation in *S. japonicus pmk1Δ* cells during the initial steps of yeast to hypha differentiation induced with CPT (Figure 4E), and also at later stages, as confirmed by comparatively scoring the mycelial area of expansion of strains expressing both constructs growing on YEMA and RGE plates (Figure 4F). As might be expected, neither Pmk1^{Sj} (ins82-105)–HA nor Pmk1^{Sp} (Δ82-105)–HA fusions were able to alleviate the myceliation defect of the *pmk1Δ* mutant (Figure 4F). Therefore, these observations suggest that in *S. japonicus*, evolutionary divergence of MAPK Pmk1 secondary structure at the N-lobe has had a significant impact on its biological functions including the positive modulation of dimorphism.

4. Discussion

In this work, we have investigated the signaling architecture and biological functions of the cell integrity MAPK signaling pathway (CIP) in the dimorphic fission yeast *S. japonicus*. Comparative phenotypic assessment of *S. japonicus* and *S. pombe pmk1Δ* mutants revealed that control of ionic homeostasis and cell wall integrity is a conserved regulatory feature of the CIP in both fission yeast species. In contrast to *S. pombe*, Pmk1 did not localize to the septum in *S. japonicus*, and *pmk1Δ* cells only displayed cytokinetic defects in response to specific environmental cues (i.e. with KCl), suggesting that the CIP may not play a similar role during control of septation in both fission yeasts [1,34]. Pmk1-dependent regulation of cell separation in *S. pombe* operates in a manner mostly dependent on its localization at the division area; however, transcriptional control by nuclear Pmk1 has also been proposed to play some role during cytokinetic control [34]. Moreover, it has been recently shown that CIP/Pmk1 signaling participates in a checkpoint that relies on the SIN (septation initiation network), which ensures that *S. pombe* cytokinesis terminates successfully [35]. This regulatory mechanism could not have evolved in *S. japonicus*, since both fission species show marked differences in the timely coordination of mitosis and cytokinesis, with *S. japonicus* delaying the assembly of the CAR until late anaphase when chromosome segregation is complete [42]. Hence, it might be possible that Pmk1-dependent regulation of cytokinesis in *S. japonicus* is exerted at a transcriptional level.

S. japonicus Pmk1 becomes activated as with its *S. pombe* MAPK counterpart in response to environmental cues including saline and cell wall stress, and is negatively regulated by the SAPK pathway, as initially described in *S. pombe* [4]. However, unlike *S. pombe* Pmk1, which becomes strongly activated upon glucose withdrawal [3], Pmk1 activity undergoes a sudden decrease in *S. japonicus* cells after carbon source removal from the growth medium. Pmk1 activation in the absence of glucose is required in *S. pombe* for adaptive growth from fermentative to respiratory metabolism, and the use of alternative carbon sources such as glycerol [3]. Lack of Pmk1 activation in *S. japonicus* might thus be explained from an evolutionary perspective, since this fission yeast species is respiration defective, and utilizes glucose exclusively via fermentation [17].

The role of PKC orthologs Pck1 and Pck2 as key upstream regulators of the CIP pathway is maintained in *S. japonicus* as originally shown in *S. pombe* [1,6,7,39], although with several differences. As in *S. pombe*, *S. japonicus* Pck2 is the major activator of Pmk1 during unperturbed growth, whereas Pck1 deletion prompts increased Pmk1 phosphorylation. In *S. pombe*, this response has been shown to arise from a cell wall stress that is transduced to the CIP by Rho GTPases Rho1 and Rho2 [7], and whose existence in *S. japonicus* remains unknown. However, Pmk1 activation during cell wall damage, which in *S. pombe* relies on both Pck1 and Pck2, is in *S. japonicus* exclusively Pck2 dependent. These findings, together with the observation that *pck1Δ* cells are growth resistant to caspofungin, suggest that opposite control of CIP/Pmk1 activity by Pck1 and Pck2 has a meaningful impact on *S. japonicus* cell wall structure and integrity.

S. japonicus yeast cells readily differentiate to hyphae in response to different environmental cues [18–21]. This led us to thoroughly study the putative role of CIP signaling during regulation of the dimorphic transition in this organism. As proposed earlier [22], Pmk1 MAPK is an essential and positive modulator of both the early and late stages of *S. japonicus* yeast to hyphae differentiation. This role is similar to that played by cell integrity MAPKs during hyphal or pseudohyphal differentiation in other yeast species, including *Candida albicans* [43,44], *Cryptococcus neoformans* [45], and *Saccharomyces cerevisiae* [46]. However, our data also suggest that Pck1 and Pck2 may regulate CIP signaling during *S. japonicus* dimorphic switch in an intricate fashion. Deletion of Pck2, which strongly reduces Pmk1 basal activity, had a notoriously negative impact during the early steps of yeast to hyphae differentiation, and was very similar to that imposed by Pmk1 deletion. This suggests that the cellular requirement for basal CIP/Pmk1 signaling is maximal during the early stages of *S. japonicus* hyphal differentiation. On the contrary, Pck2 absence did not affect late mycelia expansion, indicating that only a minimal CIP activity threshold is necessary at the later stages of mycelia formation. On the other hand, Pck1 deletion, which increases basal Pmk1 activity, resulted both in increased hyphal elongation and myceliation during the whole differentiation process. Intriguingly, Pck1 negative regulation of *S. japonicus* yeast to hyphae switch is totally dependent on Pmk1 downregulation during late mycelial expansion, but appears to operate independently of the MAPK at the early stages of yeast differentiation. Altogether, these findings point to the existence of an antagonistic and biologically relevant control of *S. japonicus* hyphal differentiation by Pck1 and Pck2, which operates by fine-tuning of Pmk1 activity during the different stages of mycelia formation, but may also function independently of CIP signaling.

MAPKs adopt a typical kinase fold including N- and C-terminal lobes that mediate, respectively, ATP and effector/substrate binding, and are connected via a conserved hinge region at subdomain V [47]. Activation loop phosphorylation by the upstream MAPKKs results in MAPKs rotation of the N- and C-terminal lobes through this hinge region to facilitate phosphoryl transfer, and also triggers global changes in their conformational exchange dynamics [47]. Therefore, the atypical secondary structure of the N-lobe in *S. japonicus* Pmk1, which lacks a highly conserved motif corresponding to subdomain IV and part of the hinge region including the putative gatekeeper residue, raised the question of to what extent has evolutionary divergence modeled the biological functions of this MAPK? By performing kinase swapping experiments, we found that $Pmk1^{Sj}$ is totally functional when expressed in *S. pombe*, as evidenced by its ability to become fully activated during growth and stress, and to suppress the altered ionic homeostasis and defective cytokinesis of *S. pombe pmk1Δ* cells. On the contrary, the ability of $Pmk1^{Sp}$ to relieve the VIC phenotype and the hyphal differentiation defect of *S. japonicus pmk1Δ* cells was minimal. This behavior does not result from limited binding and activation of $Pmk1^{Sp}$ by the MAPKK, since its basal activity and activation pattern in response to stress were very similar to those shown by $Pmk1^{Sj}$. Thus, with its atypical structure, $Pmk1^{Sj}$ seems more efficient for downstream substrate and/or effector docking than $Pmk1^{Sp}$. Significantly, $Pmk1^{Sj}$ conformational change, elicited by insertion of the highly conserved $Pmk1^{Sp}$ N-lobe motif, resulted in a total lack of kinase function when expressed in *S. pombe* due to abrogated binding and activation by Pek1 MAPKK, but also in *S. japonicus*, despite becoming activated during growth and stress to a certain threshold. These results suggest that evolutionary divergence within the *Schizosaccharomyces* clade in the architecture of the Pmk1 N-terminal lobe has exerted a profound impact on MAPK binding/docking to both upstream (MAPKK) and downstream substrates/effectors. In support of this prediction, bioinformatics and structural studies with mammalian MAPKs have shown that co-evolution between the subdomains IV–V, the D-domain docking site, and the C-tail, play a major role for tight and allosteric regulation of MAPK activity [48].

In conclusion, while some of the functional roles of the CIP earlier described in *S. pombe* are conserved in *S. japonicus*, others have evolved separately in both fission yeast species. This includes the apparent inability of *S. japonicus* Pmk1 to become activated in response to glucose deprivation and the differential regulation of cytokinesis. Our findings also reveal an antagonistic role of PKC orthologs Pck1 and Pck2 in controlling MAPK basal activity during the regulation of yeast to hyphae dimorphic switch, and suggest that conformational changes in the N-lobe of *S. japonicus* Pmk1 have modeled CIP signaling to fulfill the specific developmental requirements of this fission yeast species.

Supplementary Materials: The following are available online at https://www.mdpi.com/article/10.3390/jof7060482/s1; Figure S1: Septation phenotypes of *S. japonicus pmk1Δ* cells; Figure S2: Conserved structural features of Pmk1 MAPKs in *S. pombe* and *S. japonicus*; Figure S3: Conserved sequences surrounding the N-lobe subdomains IV and V in several fungal CIP MAPKs; Figure S4: Conserved structural features of Pck1 orthologs in *S. pombe* and *S. japonicus*; Figure S5: Conserved structural features of Pck2 orthologs in *S. pombe* and *S. japonicus*; Table S1: *S. japonicus* and *S. pombe* strains used in this work; Table S2: Oligonucleotides used in this work.

Author Contributions: Conceptualization, T.S.; Data curation, E.G.-G., A.F., J.V.-S., and J.C.; Formal analysis, E.G.-G., A.F., F.P.-R., A.P.-D., M.M., T.S., and J.C.; Funding acquisition, B.V.-M., T.S., and J.C.; Investigation, E.G.-G., B.V.-M., F.P.-R., A.P.-D., J.V.-S., M.M., T.S., and J.C.; Methodology, E.G.-G., A.F., B.V.-M., A.P.-D., J.V.-S., M.M., and T.S.; Project administration, F.P.-R., J.V.-S., T.S., and J.C.; Resources, A.F. and J.C.; Supervision, T.S. and J.C.; Validation, E.G.-G., M.M., T.S., and J.C.; Visualization, E.G.-G.; Writing—original draft, J.C.; Writing—review and editing, E.G.-G., A.F., and T.S. All authors have read and agreed to the published version of the manuscript.

Funding: This research was funded by the Agencia Estatal de Investigación and Ministerio de Ciencia, Innovación y Universidades, Spain, grant number BFU2017-82423-P, Fundación Séneca de la Región de Murcia, Spain, grant number 20856/PI/18, and European Regional Development Fund (ERDF) co-funding from the European Union.

Data Availability Statement: The data presented in this study are available on request from the corresponding authors.

Acknowledgments: We thank Snezhana Oliferenko and Hironori Niki for *S. japonicus* yeast strains and plasmids. Elisa Gómez-Gil and Francisco Prieto-Ruiz are, respectively, Formación de Personal Investigador (FPI) and Formación de Profesorado Universitario (FPU) PhD fellows from the Universidad de Murcia and Ministerio de Educación y Formación Profesional, Spain.

Conflicts of Interest: The authors declare no conflict of interest. The funders had no role in the design of the study; in the collection, analyses, or interpretation of data; in the writing of the manuscript, or in the decision to publish the results.

References

1. Perez, P.; Cansado, J. Cell Integrity Signaling and Response to Stress in Fission Yeast. *Curr. Protein Pept. Sci.* **2010**, *11*, 680–692. [CrossRef] [PubMed]
2. Kabeche, R.; Madrid, M.; Cansado, J.; Moseley, J.B. Eisosomes Regulate Phosphatidylinositol 4,5-Bisphosphate (PI(4,5)P2) Cortical Clusters and Mitogen-activated Protein (MAP) Kinase Signaling upon Osmotic Stress. *J. Biol. Chem.* **2015**, *290*, 25960–25973. [CrossRef] [PubMed]
3. Madrid, M.; Fernandez-Zapata, J.; Sanchez-Mir, L.; Soto, T.; Franco, A.; Vicente-Soler, J.; Gacto, M.; Cansado, J. Role of the fission yeast cell integrity MAPK pathway in response to glucose limitation. *BMC Microbiol.* **2013**, *13*. [CrossRef] [PubMed]
4. Madrid, M.; Nunez, A.; Soto, T.; Vicente-Soler, J.; Gacto, M.; Cansado, J. Stress-activated protein kinase-mediated down-regulation of the cell integrity pathway mitogen-activated protein kinase Pmk1p by protein phosphatases. *Mol. Biol. Cell* **2007**, *18*, 4405–4419. [CrossRef]
5. Madrid, M.; Soto, T.; Khong, H.; Franco, A.; Vicente, J.; Perez, P.; Gacto, M.; Cansado, J. Stress-induced response, localization, and regulation of the Pmk1 cell integrity pathway in Schizosaccharomyces pombe. *J. Biol. Chem.* **2006**, *281*, 2033–2043. [CrossRef]
6. Barba, G.; Soto, T.; Madrid, M.; Nunez, A.; Vicente, J.; Gacto, M.; Cansado, J.; Grp, Y.P.; Grp, Y.P. Activation of the cell integrity pathway is channelled through diverse signalling elements in fission yeast. *Cell. Signal.* **2008**, *20*, 748–757. [CrossRef]
7. Sanchez-Mir, L.; Soto, T.; Franco, A.; Madrid, M.; Viana, R.; Vicente, J.; Gacto, M.; Perez, P.; Cansado, J. Rho1 GTPase and PKC Ortholog Pck1 Are Upstream Activators of the Cell Integrity MAPK Pathway in Fission Yeast. *PLoS ONE* **2014**, *9*. [CrossRef]
8. Nunez, A.; Franco, A.; Madrid, M.; Soto, T.; Vicente, J.; Gacto, M.; Cansado, J. Role for RACK1 Orthologue Cpc2 in the Modulation of Stress Response in Fission Yeast. *Mol. Biol. Cell* **2009**, *20*, 3996–4009. [CrossRef]
9. Soto, T.; Villar-Tajadura, M.; Madrid, M.; Vicente, J.; Gacto, M.; Perez, P.; Cansado, J. Rga4 Modulates the Activity of the Fission Yeast Cell Integrity MAPK Pathway by Acting as a Rho2 GTPase-activating Protein. *J. Biol. Chem.* **2010**, *285*, 11516–11525. [CrossRef]
10. Villar-Tajadura, M.; Coll, P.; Madrid, M.; Cansado, J.; Santos, B.; Perez, P. Rga2 is a Rho2 GAP that regulates morphogenesis and cell integrity in S-pombe. *Mol. Microbiol.* **2008**, *70*, 867–881. [CrossRef]
11. Sanchez-Mir, L.; Franco, A.; Martin-Garcia, R.; Madrid, M.; Vicente-Soler, J.; Soto, T.; Gacto, M.; Perez, P.; Cansado, J. Rho2 palmitoylation is required for plasma membrane localization and proper signaling to the fission yeast cell integrity MAPK pathway. *Mol. Cell. Biol.* **2014**. [CrossRef]
12. Garcia, P.; Tajadura, V.; Sanchez, Y. The Rho1p exchange factor Rgf1p signals upstream from the Pmk1 mitogen-activated protein kinase pathway in fission yeast. *Mol. Biol. Cell* **2009**, *20*, 721–731. [CrossRef]
13. Madrid, M.; Vazquez-Marin, B.; Franco, A.; Soto, T.; Vicente-Soler, J.; Gacto, M.; Cansado, J. Multiple crosstalk between TOR and the cell integrity MAPK signaling pathway in fission yeast. *Sci. Rep.* **2016**, *6*, 37515. [CrossRef]
14. Prieto-Ruiz, F.; Vicente-Soler, J.; Franco, A.; Gomez-Gil, E.; Sanchez-Marinas, M.; Vazquez-Marin, B.; Aligue, R.; Madrid, M.; Moreno, S.; Soto, T.; et al. RNA-Binding Protein Rnc1 Regulates Cell Length at Division and Acute Stress Response in Fission Yeast through Negative Feedback Modulation of the Stress-Activated Mitogen-Activated Protein Kinase Pathway. *mBio* **2020**, *11*. [CrossRef]
15. Makarova, M.; Peter, M.; Balogh, G.; Glatz, A.; MacRae, J.I.; Lopez Mora, N.; Booth, P.; Makeyev, E.; Vigh, L.; Oliferenko, S. Delineating the Rules for Structural Adaptation of Membrane-Associated Proteins to Evolutionary Changes in Membrane Lipidome. *Curr. Biol.* **2020**, *30*, 367–380.e368. [CrossRef]
16. Gu, Y.; Yam, C.; Oliferenko, S. Divergence of mitotic strategies in fission yeasts. *Nucleus* **2012**, *3*, 220–225. [CrossRef]
17. Kaino, T.; Tonoko, K.; Mochizuki, S.; Takashima, Y.; Kawamukai, M. Schizosaccharomyces japonicus has low levels of CoQ10 synthesis, respiration deficiency, and efficient ethanol production. *Biosci. Biotechnol. Biochem.* **2018**, *82*, 1031–1042. [CrossRef]
18. Sipiczki, M.; Takeo, K.; Yamaguchi, M.; Yoshida, S.; Miklos, I. Environmentally controlled dimorphic cycle in a fission yeast. *Microbiology* **1998**, *144 Pt 5*, 1319–1330. [CrossRef]
19. Papp, L.; Sipiczki, M.; Holb, I.J.; Miklos, I. Optimal conditions for mycelial growth of Schizosaccharomyces japonicus cells in liquid medium: It enables the molecular investigation of dimorphism. *Yeast* **2014**, *31*, 475–482. [CrossRef]
20. Aoki, K.; Furuya, K.; Niki, H. Schizosaccharomyces japonicus: A Distinct Dimorphic Yeast among the Fission Yeasts. *Cold Spring Harb. Protoc.* **2017**, *2017*, 082651. [CrossRef]
21. Niki, H. Induction of Hyphal Growth in Schizosaccharomyces japonicus. *Cold Spring Harb. Protoc.* **2017**, *2017*, 091868. [CrossRef]
22. Nozaki, S.; Furuya, K.; Niki, H. The Ras1-Cdc42 pathway is involved in hyphal development of Schizosaccharomyces japonicus. *FEMS Yeast Res.* **2018**, *18*. [CrossRef]
23. Gomez-Gil, E.; Franco, A.; Madrid, M.; Vazquez-Marin, B.; Gacto, M.; Fernandez-Breis, J.; Vicente-Soler, J.; Soto, T.; Cansado, J. Quorum sensing and stress-activated MAPK signaling repress yeast to hypha transition in the fission yeast Schizosaccharomyces japonicus. *PLoS Genet.* **2019**, *15*, e1008192. [CrossRef]
24. Moreno, S.; Klar, A.; Nurse, P. Molecular Genetic-Analysis Of Fission Yeast Schizosaccharomyces pombe. *Methods Enzymol.* **1991**, *194*, 795–823.

25. Petersen, J.; Russell, P. Growth and the Environment of Schizosaccharomyces pombe. *Cold Spring Harb. Protoc.* **2016**, *2016*, 079764. [CrossRef]
26. Kinnaer, C.; Dudin, O.; Martin, S.G. Yeast-to-hypha transition of Schizosaccharomyces japonicus in response to environmental stimuli. *Mol. Biol. Cell* **2019**, *30*, 975–991. [CrossRef]
27. Aoki, K.; Niki, H. Transformation of Schizosaccharomyces japonicus. *Cold Spring Harb. Protoc.* **2017**, *2017*, 091850. [CrossRef]
28. Furuya, K.; Niki, H. Isolation of heterothallic haploid and auxotrophic mutants of Schizosaccharomyces japonicus. *Yeast* **2009**, *26*, 221–233. [CrossRef]
29. Hentges, P.; Van Driessche, B.; Tafforeau, L.; Vandenhaute, J.; Carr, A.M. Three novel antibiotic marker cassettes for gene disruption and marker switching in Schizosaccharomyces pombe. *Yeast* **2005**, *22*, 1013–1019. [CrossRef]
30. Krawchuk, M.D.; Wahls, W.P. High-efficiency gene targeting in Schizosaccharomyces pombe using a modular, PCR-based approach with long tracts of flanking homology. *Yeast* **1999**, *15*, 1419–1427. [CrossRef]
31. Madrid, M.; Jimenez, R.; Sanchez-Mir, L.; Soto, T.; Franco, A.; Vicente-Soler, J.; Gacto, M.; Perez, P.; Cansado, J. Multiple layers of regulation influence cell integrity control by the PKC ortholog Pck2 in fission yeast. *J. Cell. Sci.* **2015**, *128*, 266–280. [CrossRef] [PubMed]
32. Schneider, C.; Rasband, W.; Eliceiri, K. NIH Image to ImageJ: 25 years of image analysis. *Nat. Methods* **2012**, *9*, 671–675. [CrossRef] [PubMed]
33. Hagan, I.M. Immunofluorescence Microscopy of Schizosaccharomyces pombe Using Chemical Fixation. *Cold Spring Harb. Protoc.* **2016**, *2016*. [CrossRef] [PubMed]
34. Sanchez-Mir, L.; Franco, A.; Madrid, M.; Vicente-Soler, J.; Antonia Villar-Tajadura, M.; Soto, T.; Perez, P.; Gacto, M.; Cansado, J. Biological Significance of Nuclear Localization of Mitogen-activated Protein Kinase Pmk1 in Fission Yeast. *J. Biol. Chem.* **2012**, *287*, 26038–26051. [CrossRef]
35. Edreira, T.; Celador, R.; Manjón, E.; Sánchez, Y. A novel checkpoint pathway controls actomyosin ring constriction trigger in fission yeast. *eLife* **2020**, *9*. [CrossRef]
36. Zaitsevskaya-Carter, T.; Cooper, J.A. Spm1, a stress-activated MAP kinase that regulates morphogenesis in S.pombe. *EMBO J.* **1997**, *16*, 1318–1331. [CrossRef]
37. Sugiura, R.; Toda, T.; Dhut, S.; Shuntoh, H.; Kuno, T. The MAPK kinase Pek1 acts as a phosphorylation-dependent molecular switch. *Nature* **1999**, *399*, 479–483. [CrossRef]
38. Toda, T.; Dhut, S.; SupertiFurga, G.; Gotoh, Y.; Nishida, E.; Sugiura, R.; Kuno, T. The fission yeast pmk1(+) gene encodes a novel mitogen-activated protein kinase homolog which regulates cell integrity and functions coordinately with the protein kinase C pathway. *Mol. Cell. Biol.* **1996**, *16*, 6752–6764. [CrossRef]
39. Madrid, M.; Vazquez-Marin, B.; Soto, T.; Franco, A.; Gomez-Gil, E.; Vicente-Soler, J.; Gacto, M.; Perez, P.; Cansado, J. Differential functional regulation of PKC orthologs in fission yeast. *J. Biol. Chem.* **2017**. [CrossRef]
40. Arellano, M.; Valdivieso, M.; Calonge, T.; Coll, P.; Duran, A.; Perez, P. Schizosaccharomyces pombe protein kinase C homologues, pck1p and pck2p, are targets of rho1p and rho2p and differentially regulate cell integrity. *J. Cell Sci.* **1999**, *112*, 3569–3578. [CrossRef]
41. Calonge, T.; Nakano, K.; Arellano, M.; Arai, R.; Katayama, S.; Toda, T.; Mabuchi, I.; Perez, P. Schizosaccharomyces pombe Rho2p GTPase regulates cell wall alpha-glucan biosynthesis through the protein kinase Pck2p. *Mol. Biol. Cell* **2000**, *11*, 4393–4401. [CrossRef]
42. Gu, Y.; Oliferenko, S. Comparative biology of cell division in the fission yeast clade. *Curr. Opin. Microbiol.* **2015**, *28*, 18–25. [CrossRef]
43. Kumamoto, C.A. A contact-activated kinase signals Candida albicans invasive growth and biofilm development. *Proc. Natl. Acad. Sci. USA* **2005**, *102*, 5576–5581. [CrossRef]
44. Monge, R.A.; Román, E.; Nombela, C.; Pla, J. The MAP kinase signal transduction network in Candida albicans. *Microbiology* **2006**, *152*, 905–912. [CrossRef]
45. Kraus, P.R.; Fox, D.S.; Cox, G.M.; Heitman, J. The Cryptococcus neoformans MAP kinase Mpk1 regulates cell integrity in response to antifungal drugs and loss of calcineurin function. *Mol. Microbiol.* **2003**, *48*, 1377–1387. [CrossRef]
46. Birkaya, B.; Maddi, A.; Joshi, J.; Free, S.J.; Cullen, P.J. Role of the cell wall integrity and filamentous growth mitogen-activated protein kinase pathways in cell wall remodeling during filamentous growth. *Eukaryot. Cell* **2009**, *8*, 1118–1133. [CrossRef]
47. Kumar, G.S.; Clarkson, M.W.; Kunze, M.B.A.; Granata, D.; Wand, A.J.; Lindorff-Larsen, K.; Page, R.; Peti, W. Dynamic activation and regulation of the mitogen-activated protein kinase p38. *Proc. Natl. Acad. Sci. USA* **2018**, *115*, 4655–4660. [CrossRef]
48. Nguyen, T.; Ruan, Z.; Oruganty, K.; Kannan, N. Co-conserved MAPK features couple D-domain docking groove to distal allosteric sites via the C-terminal flanking tail. *PLoS ONE* **2015**, *10*, e0119636. [CrossRef]

MDPI
St. Alban-Anlage 66
4052 Basel
Switzerland
Tel. +41 61 683 77 34
Fax +41 61 302 89 18
www.mdpi.com

Journal of Fungi Editorial Office
E-mail: jof@mdpi.com
www.mdpi.com/journal/jof

www.ingramcontent.com/pod-product-compliance
Lightning Source LLC
Chambersburg PA
CBHW061551220926
51850CB00005B/307

* 9 7 8 3 0 3 6 5 6 9 7 3 4 *